Lecture Notes in Physics

For information about Vols. 1–110, please contact your bookseller or Springer-Verlag.

Vol. 111: H.-J. Schmidt, Axiomatic Characterization of Physical Geometry. V, 163 pages. 1979.

Vol. 112: Imaging Processes and Coherence in Physics. Proceedings, 1979. Edited by M. Schlenker et al. XIX, 577 pages. 1980.

Vol. 113: Recent Advances in the Quantum Theory of Polymers. Proceedings 1979. Edited by J.-M. André et al. V, 306 pages. 1980.

Vol. 114: Stellar Turbulence. Proceedings, 1979. Edited by D. F. Gray and J. L. Linsky. IX, 308 pages. 1980.

Vol. 115: Modern Trends in the Theory of Condensed Matter. Proceedings, 1979. Edited by A. Pekalski and J. A. Przystawa. IX, 597 pages. 1980.

Vol. 116: Mathematical Problems in Theoretical Physics. Proceedings, 1979. Edited by K. Osterwalder. VIII, 412 pages. 1980.

Vol. 117: Deep-Inelastic and Fusion Reactions with Heavy Ions. Proceedings, 1979. Edited by W. von Oertzen. XIII, 394 pages. 1980.

Vol. 118: Quantum Chromodynamics. Proceedings, 1979. Edited by J. L. Alonso and R. Tarrach. IX, 424 pages. 1980.

Vol. 119: Nuclear Spectroscopy. Proceedings, 1979. Edited by G. F. Bertsch and D. Kurath. VII, 250 pages. 1980.

Vol. 120: Nonlinear Evolution Equations and Dynamical Systems. Proceedings, 1979. Edited by M. Boiti, F. Pempinelli and G. Soliani. VI, 368 pages. 1980.

Vol. 121: F. W. Wiegel, Fluid Flow Through Porous Macromolecular Systems. V, 102 pages. 1980.

Vol. 122: New Developments in Semiconductor Physics. Proceedings, 1979. Edited by F. Beleznay et al. V, 276 pages. 1980.

Vol. 123: D. H. Mayer, The Ruelle-Araki Transfer Operator in Classical Statistical Mechanics. VIII, 154 pages. 1980.

Vol. 124: Gravitational Radiation, Collapsed Objects and Exact Solutions. Proceedings, 1979. Edited by C. Edwards. VI, 487 pages. 1980.

Vol. 125: Nonradial and Nonlinear Stellar Pulsation. Proceedings, 1980. Edited by H. A. Hill and W. A. Dziembowski. VIII, 497 pages. 1980.

Vol. 126: Complex Analysis, Microlocal Calculus and Relativistic Quantum Theory. Proceedings, 1979. Edited by D. Iagolnitzer. VIII, 502 pages. 1980.

Vol. 127: E. Sanchez-Palencia, Non-Homogeneous Media and Vibration Theory. IX, 398 pages. 1980.

Vol. 128: Neutron Spin Echo. Proceedings, 1979. Edited by F. Mezei. VI, 253 pages. 1980.

Vol. 129: Geometrical and Topological Methods in Gauge Theories. Proceedings, 1979. Edited by J. Harnad and S. Shnider. VIII, 155 pages. 1980.

Vol. 130: Mathematical Methods and Applications of Scattering Theory. Proceedings, 1979. Edited by J. A. DeSanto, A. W. Sáenz and W. W. Zachary. XIII, 331 pages. 1980.

Vol. 131: H. C. Fogedby, Theoretical Aspects of Mainly Low Dimensional Magnetic Systems. XI, 163 pages. 1980.

Vol. 132: Systems Far from Equilibrium. Proceedings, 1980. Edited by L. Garrido. XV, 403 pages. 1980.

Vol. 133: Narrow Gap Semiconductors Physics and Applications. Proceedings, 1979. Edited by W. Zawadzki. X, 572 pages. 1980.

Vol. 134: $\gamma\gamma$ Collisions. Proceedings, 1980. Edited by G. Cochard and P. Kessler. XIII, 400 pages. 1980.

Vol. 135: Group Theoretical Methods in Physics. Proceedings, 1980. Edited by K. B. Wolf. XXVI, 629 pages. 1980.

Vol. 136: The Role of Coherent Structures in Modelling Turbulence and Mixing. Proceedings 1980. Edited by J. Jimenez. XIII, 393 pages. 1981.

Vol. 137: From Collective States to Quarks in Nuclei. Edited by H. Arenhövel and A. M. Saruis. VII, 414 pages. 1981.

Vol. 138: The Many-Body Problem. Proceedings 1980. Edited by R. Guardiola and J. Ros. V, 374 pages. 1981.

Vol. 139: H. D. Doebner, Differential Geometric Methods in Mathematical Physics. Proceedings 1981. VII, 329 pages. 1981.

Vol. 140: P. Kramer, M. Saraceno, Geometry of the Time-Dependent Variational Principle in Quantum Mechanics. IV, 98 pages. 1981.

Vol. 141: Seventh International Conference on Numerical Methods in Fluid Dynamics. Proceedings. Edited by W. C. Reynolds and R. W. MacCormack. VIII, 485 pages. 1981.

Vol. 142: Recent Progress in Many-Body Theories. Proceedings. Edited by J. G. Zabolitzky, M. de Llano, M. Fortes and J. W. Clark. VIII, 479 pages. 1981.

Vol. 143: Present Status and Aims of Quantum Electrodynamics. Proceedings, 1980. Edited by G. Gräff, E. Klempt and G. Werth. VI, 302 pages. 1981.

Vol. 144: Topics in Nuclear Physics I. A Comprehensive Review of Recent Developments. Edited by T.T.S. Kuo and S.S.M. Wong. XX, 567 pages. 1981.

Vol. 145: Topics in Nuclear Physics II. A Comprehensive Review of Recent Developments. Proceedings 1980/81. Edited by T. T. S. Kuo and S. S. M. Wong. VIII, 571-1.082 pages. 1981.

Vol. 146: B. J. West, On the Simpler Aspects of Nonlinear Fluctuating. Deep Gravity Waves. VI, 341 pages. 1981.

Vol. 147: J. Messer, Temperature Dependent Thomas-Fermi Theory. IX, 131 pages. 1981.

Vol. 148: Advances in Fluid Mechanics. Proceedings, 1980. Edited by E. Krause. VII, 361 pages. 1981.

Vol. 149: Disordered Systems and Localization. Proceedings, 1981. Edited by C. Castellani, C. Castro, and L. Peliti. XII, 308 pages. 1981.

Vol. 150: N. Straumann, Allgemeine Relativitätstheorie und relativistische Astrophysik. VII, 418 Seiten. 1981.

Vol. 151: Integrable Quantum Field Theory. Proceedings, 1981. Edited by J. Hietarinta and C. Montonen. V, 251 pages. 1982.

Lecture Notes in Physics

189

Nonlinear Phenomena

Proceedings of the CIFMO School and Workshop
held at Oaxtepec, México
November 29 – December 17, 1982

Edited by K. B. Wolf

Springer-Verlag
Berlin Heidelberg GmbH 1983

Editor

Kurt Bernardo Wolf
Instituto de Investigaciones
en Matemáticas Aplicadas y en Sistemas (IIMAS)
Universidad Nacional Autónoma de México, México D. F.

ISBN 978-3-540-12730-7 ISBN 978-3-540-38721-3 (eBook)
DOI 10.1007/978-3-540-38721-3

Originally published by Springer-Verlag Berlin Heidelberg New York in 1983

2153/3140-543210

The School and Workshop on Non-Linear Phenomena

The School and Workshop on Non-Linear Phenomena was planned as an activity sponsored by the Centro Internacional de Física y Matemáticas Orientadas (CIFMO). The Organizing Committee invited a number of distinguished research workers engaged in the subject to give lectures on their approach to the field from various directions of mathematical treatment, leading to various areas of application. These were directions which were achieving a certain maturity and coherence, and areas of importance both theoretical and technological. Included among the former we have the Inverse Scattering and Direct Linearizing Transforms, Riemann–Hilbert theory, symplectic geometry and canonical transformations, special functions, separation of variables, stochasticity, and group theory; among the latter, classical and quantum integrable systems, analysis of Painlevé equation systems, Yang–Mills fields, the Einstein and Einstein–Maxwell equations of general relativity, chaotic behaviour, and soliton solutions in oceanography and in other more general systems.

The preliminary program of the School was circulated widely, and based upon the response, the Workshop sessions were set up. These were mostly reports of work in progress or recently completed. The free and friendly athmosphere which prevailed during the event was conducive to strong interaction and constructive critique among the participants. Perhaps a good indicator of the climate is the number of cross-references in the contributions which follow. The School lecture notes and Workshop presentations constitute the two parts of this volume.

The Organizing Committee:
MARK J. ABLOWITZ, CHARLES P. BOYER, JERZY PLEBAŃSKI,
PAVEL WINTERNITZ, and KURT BERNARDO WOLF

Sponsors and Contributors

The School and Workshop on Non-Linear Phenomena was made possible with the financial support of the **Universidad Nacional Autónoma de México (UNAM)**, through its **Dirección General de Asuntos del Personal Académico**, and through the Departamento de Física Matemática of the **Instituto de Investigaciones en Matemáticas Aplicadas y en Sistemas (IIMAS)**; a special grant from the **Fondo de Fomento Educativo** of the **Banco de Cédulas Hipotecarias (BCH)** is also gratefully acknowledged.

The funding of the event was requested with the personal interest of **Dr. Tomás Garza Hernández**, Director of IIMAS up to April 1982, and strongly supported by **Dr. Alejandro Velasco Levy**, present Director, with whose unwavering help we overcame the deteriorating national economic frame. Thanks are also due to **Dr. Jaime Constantiner**, former director of BCH and member board of the Foundation, who, up to last year's bank nationalization, had been one of the staunchest patron of the sciences in Mexico.

Mexican institutions which provided their indirect but firm support to the import of the event, contributing through registration fees or in kind, are the **Universidad Autónoma Metropolitana** and the **Facultad de Ciencias, UNAM**. We thank the **Dirección General de Actividades Socioculturales** who granted, as a special courtesy, a concert of the Orquesta de Cámara de la Ciudad de México, which offered a programme of contemporary Mexican music with works of Carlos Chávez, Salvador Contreras, Blas Galindo, and Pablo Moncayo, directed by Maestro Manuel Bernal. Music, as Maestro Bernal emphasized, no less than Science, are manifestations of culture which do not yet know of national boundaries.

Centro Internacional de Física y Matemáticas Orientadas

(CIFMO)

On February 21–23, 1980, the Hacienda Vista Hermosa, in Jojutla, Morelos, was host to an informal gathering of scientists and functionaries of various Mexican and international bodies. The meeting was convened in order to discuss a proposal for the establishment in Mexico of a scientific center along the lines of the International Centre for Theoretical Physics in Trieste. Professor Abdus Salam had personally proposed such a mechanism to spread scientific endeavors beyond the borders of the developed countries. This idea had been championed by **Dr. Edmundo de Alba**, himself a scientist and then Director General de Investigación Científica y Superación Académica of the Secretaría de Educación Pública (SEP),[1] and **Prof. Jorge Flores Valdés**, then Director of the Instituto de Física, UNAM. Present at the Jojutla meeting were **Dr. A. Malek**, Regional Representative for UNESCO in Latin America, **Prof. Luciano Bertocchi**, Adjunct Director of ICTP (Trieste, Italy), **Prof. B. M. Udgaonkar**, Director of the Tata Institute of Fundamental Research (Bombay, India), **Prof. Henri Hogbe-Nlend** Director of the International Centre for Pure and Applied Mathematics (Nice, France), **Dr. Agustín Ayala Castañares**, Coordinador de la Investigación Científica, UNAM, **Fís. Javier Garzón**, and **Lic. Alfredo Ramírez Araiza**, from CONACyT,[2] and **Lic. Juan Antonio Mateos** from SEP. Representatives of various Mexican scientific institutes were: **Prof. Jorge Flores**, Director of the Instituto de Física, **Prof. Tomás Garza**, Director of IIMAS, **Prof. Daniel Reséndiz**, Director of the Instituto de Ingeniería, UNAM, **Prof. Eliezer Braun**, **Prof. Feliciano Sánchez**, and **Dr. Edmundo de Alba**. Scientists present were Profs. **Shlomo P. Neuman**, University of Arizona, **Sergio Mascarenhas**, Universidade de São Paulo, **Ismael Herrera**, **Marcos Moshinsky**, and **Kurt Bernardo Wolf**, from UNAM.

The final proposal document stated that the Center to be set up, under the name of **Centro Internacional de Física y Matemáticas Orientadas (CIFMO)** in the city of Cuernavaca, Morelos, would have as its main objectives the identification of important social and economic problems in which the knowledge of physics and mathematics be relevant; to contribute to the training of cadres, specialists of high quality which may be able to apply this knowledge in a direct and effective way in Mexico and other developing countries, and to strenghten the Mexican scientific institutions through active interaction with their counterparts throughout the world. This document was signed in Cuernavaca, Morelos, on March 23^{d}, 1981, by **Prof. Octavio Rivera Serrano** Rector of UNAM, **Dr. Edmundo Flores**, then Director of CONACyT, and **Lic. Fernando Solana**, then Minister of Education (SEP). The honor witness was **Lic. José López Portillo**, then President of the Republic.[3]

The Director of CIFMO would be **Prof. Arcadio Poveda**, who was leaving the direction of the Instituto de Astronomía at UNAM, and **Dr. Enrique Daltabuit**, Academic Secretary.

---

[1] Ministry for Public Education.

[2] Consejo Nacional de Ciencia y Tecnología, the Mexican Council for Science and Technology.

[3] Gaceta UNAM, N° 23, of March 26th, 1981.

As it so often happens with development plans, however, the final budgetary approval was withheld for reasons which became clear by the end of the year. By then, CIFMO had organized one Workshop on Applications of Amorphous Silicium to Solar Cells, and one on Fiber Optics and their Application to Informatics. The School and Workshop on Nonlinear Phenomena had consolidated its program, when word came of a developing economic crisis in the country, fueled by oil pricing, public incompetence, and inconceivable corruption. Indeed, within a year, the national currency stood at one-sixth of its former international value, professors' real salaries were cut by 40%, and any plans for expeditious establishment of new scientific institutions had to be shelved. We may still boldly dream, nevertheless, that in a not too-distant future, the recommendations of the Jojutla meeting will yet provide the blueprint for an international science center in Mexico. The gentle climate of the Valley of Morelos and the disposition for scientific endeavor will surely outlast the deepest of economic crises.

—THE EDITOR

This Volume

The preparation of this Proceedings volume was done entirely at IIMAS, using the computer facilities of the Departamento de Computación. The text-processing language is Donald Knuth's TEX, supplemented by the Fácil TEX macro package developed by **Dr. Max Díaz**, IIMAS' TEX Wizard. Although certainly neither unique nor new, this *is* the first TEX book produced in Mexico.[4] The typography was done by **Miguel Navarro Saad**, **Alberto Hernández**, and **Pablo Castañeda** (as well as the editor). I would like to thank them for their enthusiasm and perseverance in spite of the scanty remuneration we were able to offer. Invaluable technical help from **Arturo Olvera Chávez** is also gratefully acknowledged. Perhaps —this is our hope— enough local expertise has been developed in the TEX publication of scientific literature so that we can help establish a credible precedent for these endeavours in Mexico.

[4]This volume partially covers the commitment on the part of IIMAS stated in Project IVT/EE/NAL/81/1250 "Tipografía Científica Automatizada" presented to the Consejo Nacional de Ciencia y Tecnología, CONACyT, in January 1981 (aproval communicated on August 23^{d}, 1982, frozen due to lack of funds since).

PARTICIPANTS OF THE SCHOOL AND WORKSHOP IN OAXTEPEC

From left to right, back row: José S. Florio, Guillermo Ramírez, Jerzy P. Plebański, Charles P. Boyer, Mark J. Ablowitz, Athanasios Fokas, Frederick J. Ernst, Ling-Lie Chau, Paul Winternitz, Rosalia Santoleri, Radha Balakrishnan, Bernadette Dorizzi, Kurt Bernardo Wolf, Basile Grammaticos, Harvey Segur, Eduardo Piña, *and* José Luis Rius. *Front row:* Lidia Jiménez, Daniel David, Jaime Granados Samaniego, Héctor García Luna, Luis Guerrero, Francisco Javier Chinea, Ziemowit Popowicz, Arturo Olvera, Ernesto Lacomba, Julio Herrera, *and* Enrique Sámano Tirado. *(December 14, 1982.)*

School and Workshop Participants:

MARK J. ABLOWITZ
Department of Mathematics
and Computer Science
Clarkson College of Technology
Potsdam NY 13676, USA

PEDRO ARMENDÁRIZ
Departamento de Matemáticas
Universidad Autónoma Metropolitana
Iztapalapa, 09340 México D.F. MEXICO

RADHA BALAKRISHNAN
Department of Theoretical Physics
University of Madras
Madras–600 025, INDIA

CHARLES P. BOYER
Instituto de Investigaciones
en Matemáticas Aplicadas y en Sistemas
Universidad Nacional Autónoma de México
México 01000 DF, MEXICO

FRANCESCO CALOGERO
Istituto di Fisica
Università di Roma "La Sapienza"
Roma 00185, ITALIA

LING-LIE CHAU
High Energy Theory Group
Brookhaven National Laboratory
Upton, Long Island, NY 11973, USA

FRANCISCO JAVIER CHINEA
Departamento de Métodos Matemáticos,
Facultad de Ciencias
Universidad Complutense de Madrid
Madrid 3, ESPAÑA

DANIEL DAVID
Centre de Recherches
en Mathématiques Appliquées
Université de Montréal
Montréal, Québec H3C 3J7, CANADA

MANUEL DE LLANO
Instituto de Física
Universidad Nacional Autónoma de México
01000 México D.F. MEXICO

BERNADETTE DORIZZI
Centre National
d'Etudes des Tèlècommunications
38–40 Av. G. Leclerc,
Issy-le-Moulineaux 92131, FRANCE

FREDERICK J. ERNST
Physics Department
Illinois Institute of Technology
IIT Center, Chicago, Illinois 60616, USA

DANIEL FINLEY
Department of Physics
The University of New Mexico
Albuquerque NM 87131, USA

MITCHELL J. FEIGENBAUM
Physics Department
Cornell University
Ithaca NY 14853, USA

JOSÉ S. FLORIO
División de Estudios de Posgrado
Facultad de Ingeniería
Universidad Nacional Autónoma de México
01000 México D. F. MEXICO

Athanasios Fokas
Department of Mathematics
and Computer Science
Clarkson College of Technology
Potsdam NY 13676, USA

Basile Grammaticos
Centre National
d'Etudes des Tèlècommunications
38–40 Av. G. Leclerc,
Issy-le-Moulineaux 92131, FRANCE

Jaime Alejandro Granados Samaniego
Departamento de Física
Universidad Autónoma Metropolitana
Azcapotzalco, 02200 México D.F. MEXICO

Luis E. Guerrero
Departamento de Física
Facultad de Ciencias
Universidad Central de Venezuela
Caracas 1071, VENEZUELA

Brosl Hasslacher
MS285
Los Alamos National Laboratory,
Los Alamos NM 87545, USA

Julio Herrera
Centro de Estudios Nucleares
Universidad Nacional Autónoma de México
01000 México D. F. MEXICO

Darryl D. Holm
DHL001-P
Los Alamos National Laboratory,
Los Alamos NM 87545, USA

Lidia Jiménez
Departamento de Física
Universidad Autónoma Metropolitana
Iztapalapa, 09340 México D.F. MEXICO

Ernesto Lacomba
Departamento de Matemáticas
Universidad Autónoma Metropolitana
Iztapalapa, 09340 México D.F. MEXICO

Héctor Luna García
Departamento de Física
Universidad Autónoma Metropolitana
Azcapotzalco, 02200 México D.F. MEXICO

Willard Miller, Jr.
School of Mathematics
University of Minnesota
Minneapolis, Minn. 55455, USA

Marcos Moshinsky
Instituto de Física
Universidad Nacional Autónoma de México,
and El Colegio Nacional
México 01000 DF, MEXICO

Julio Roberto Murillo Torres
Facultad de Ciencias
Universidad Nacional Autónoma de México
01000 México D. F. MEXICO

Arturo Olvera Chávez
Instituto de Investigaciones
en Matemáticas Aplicadas y en Sistemas
Universidad Nacional Autónoma de México
01000 México D. F. MEXICO

Eduardo Piña
Departamento de Física
Universidad Autónoma Metropolitana
Iztapalapa, 09340 México D.F. MEXICO

Jerzy Plebański
Centro de Investigación y Estudios Avanzados,
Instituto Politécnico Nacional,
and Instituto de Investigaciones
en Matemáticas Aplicadas y en Sistemas
Universidad Nacional Autónoma de México
México 01000, DF, MEXICO

Ziemowit Popowicz
Institute of Theoretical Physics
University of Wrocław
Cybulskiego 36, 50–205 Wrocław, POLAND

Guillermo Ramírez
Departamento de Física
Universidad Autónoma Metropolitana
Iztapalapa, 09340 México D.F. MEXICO

PEDRO RIPA
Centro de Investigación y Estudios Superiores de Ensenada
Apdo. Postal 2732, Ensenada B.C.N., MEXICO

JOSÉ LUIS RIUS ALONSO
Instituto de Física
Universidad Nacional Autónoma de México
01000 México D.F. MEXICO

ENRIQUE SÁMANO TIRADO
Departamento de Física
Universidad Autónoma Metropolitana
Azcapotzalco, 02200 México D.F. MEXICO

ROSALIA SANTOLERI
Istituto di Fisica
Università di Roma "La Sapienza"
Roma 00185, ITALIA

HARVEY SEGUR
Aeronautical Research Assocs. of Princeton, Inc.
50 Washington Road, P.O.Box 2229,
Princeton NJ 08540, USA

STANLY STEINBERG
Department of Mathematics and Statistics
The University of New Mexico
Albuquerque NM 87131, USA

RODOLFO SUÁREZ CORTÉS
Departamento de Matemáticas
Universidad Autónoma Metropolitana
Azcapotzalco, 02200 México D.F. MEXICO

PAUL WINTERNITZ
Centre de Recherches
en Mathématiques Appliquées
Université de Montréal
Montréal, Québec H3C 3J7, CANADA

KURT BERNARDO WOLF
Instituto de Investigaciones
en Matemáticas Aplicadas y en Sistemas
Universidad Nacional Autónoma de México
01000 México D. F. MEXICO

WORKSHOP

* Presentor.

Contents

SCHOOL

---

* Presentor.

School

Invited Speakers:

Mark J. Ablowitz

Charles P. Boyer

Francesco Calogero

Ling–Lie Chau

Frederick J. Ernst

Mitchell J. Feigenbaum

Athanasios S. Fokas

Willard Miller, Jr.

Marcos Moshinsky

Jerzy Plebański

Harvey Segur

Pavel Winternitz

Comments on the Inverse Scattering Transform and Related Nonlinear Evolution Equations

Mark J. Ablowitz*

and

Athanassios S. Fokas

Department of Mathematics and Computer Science
Clarkson College of Technology
Potsdam, New York, USA

Contents:

* Presentor.

§1 Introduction

In these lectures we review some of the recent work that we have been involved with. The overall point of view has been towards developing an effective procedure, which is commonly referred to as the *Inverse Scattering Transform* (IST), to solve a class of nonlinear evolution equations. All of these equations have a related feature; namely they may be obtained via the compatibility of two associated linear operators. These equations also have many other common features, *e.g.*, special solutions often referred to as solitons, infinite number of symmetries and conserved quantities, reduction to ordinary differential equations (ODE's) of Painlevé type, Bäcklund transformations, and relations to Riemann-Hilbert boundary value problems (RHBVP's) and the so-called $\bar{\partial}$ (DBAR) problems. Mathematically speaking, we have seen that these studies have involved a wide range of fields beyond real and complex analysis: algebraic geometry, differential geometry, group theory, algebra, etc. Similarly, these equations have had a number of applications in physics: water waves, internal waves, nonlinear optics, relativity, quantum field theory, and many others. In fact, at this school we will hear about a number of such applications.

Since the monograph of Ablowitz and Segur [1] details the history and many of the important ideas in this field (through 1979 or so), we recommend that the interested reader consult this reference for suitable background information.

There are various types of equations which we are interested in. A few of these are listed below in order of dimensionality:

1. Ordinary Differential Equations *(1+0)*:

$$u'' - xu - 2u^3 = \alpha = \text{ constant,} \tag{I.1}$$

(Painlevé II: PII)

$$\ddot{u}_j = 8 \sum_{\substack{j,k=1 \\ j\neq k}}^{n} \frac{1}{(u_j - u_k)^3}, \tag{I.2}$$

(many body problems).

2. Partial Differential Equations (PDE's) in one Spatial Dimension *(1+1)*:

$$u_t + 6uu_x + u_{xxx} = 0, \tag{II.1}$$

(Korteweg–de Vries: KdV)

$$iu_t + u_{xx} + 2\sigma_1 u^2 u^* = 0, \qquad \sigma_1 = \pm 1, \tag{II.2}$$

where u^* is the complex conjugate of u,

(Nonlinear Schrödinger: NLS)

$$u_{xt} = \sin u, \tag{II.3}$$

(sine–Gordon)

$$u_{it} + c_i u_{ix} = \gamma_i u_j^* u_k^*, \tag{II.4}$$

where c_i and γ_i are constants, and $i, j, k = 1, 2, 3$ permuted; there is also an n-component version (see [1]).

(Three Wave)

$$u_{tt} - u_{xx} + (u^2)_{xx} + \sigma_1 u_{xxxx} = 0, \qquad \sigma_1 = \pm 1, \tag{II.5}$$

(Boussinesq Equation).

3. Differential-Difference, Partial Difference in one Spatial Dimension *(1+1, with discretizations)*:

$$u_{ntt} = e^{-(u_n - u_{n-1})} - e^{-(u_{n+1} - u_n)}, \qquad (III.1)$$

(Toda lattice)

$$iu_{nt} = (u_{n+1} + u_{n-1} - 2u_n) + \sigma_1 u_n u_n^*(u_{n+1} + u_{n-1}), \qquad \sigma_1 = \pm 1, \qquad (III.2)$$

(Discrete NLS).

Corresponding to every exactly solvable PDE in (*II*) there is a class of solvable differential-difference equations; similarly for every solvable differential-difference equation there is a class of solvable partial difference equations.[1] The sense in which we mean these statements have to do with how well the linearized version of the nonlinear equation is approximated by its discrete analogue. So for example the linearized version of (*II*.2) is

$$\mathcal{L}u = iu_t + u_{xx}.$$

The linearized differential-difference version, accurate to $O(\Delta x^2)$ in the continuum limit, is

$$\mathcal{L}_{\Delta x} = iu_{nt} + \frac{1}{\Delta x^2}(\delta^2 u_n),$$

where

$$\delta^2 u_n = u_{n+1} + u_{n-1} - 2u_n$$

is a second central difference. The linearized partial difference version of the equations, accurate to Δx^2 and Δt^2 for the PDE or to Δt^2 for its differential-difference analogue, is

$$\mathcal{L}_{\Delta x, \Delta t} = i\frac{\Delta^m u_n^m}{\Delta t} + \frac{1}{2\Delta x^2}(\delta^2 u_n^m + \delta^2 u_n^{m+1}),$$

where $\Delta^m u_n^m = u_n^{m+1} - u_n^m$, and $\delta^2 u_n^m$ is the second central difference evaluated at time step $m\Delta t$ and space step $n\Delta x$. Naturally, when approximating a given nonlinear equation [*e.g.*, (*II*.2)] one must add, to the linear discretizations, appropriate nonlinear terms. Procedures to do this as well as the development of accurate numerical schemes are outlined in [2].[2]

4. Singular Integro-Differential Equations *(1 1/2+1)*:

$$u_t + \frac{1}{\delta}u_x + 2uu_x + (Tu)_{xx} = 0, \qquad (IV.1)$$

where

$$Tu = \frac{1}{2\delta}\int_{-\infty}^{\infty} dy \coth\left(\frac{\pi}{2\delta}\right)(\xi - x)\, u(\xi), \qquad \delta = \text{constant}$$

(Intermediate Long Wave Equation: ILW).

As $\delta \to 0$ the ILW equation goes to the KdV equation:

$$u_t + 2uu_x + \tfrac{1}{3}\delta\, u_{xxx} = 0$$

[1] The discretized versions can be made to relax to their continuous counterparts as the mesh is refined.

[2] Some information is also given in [1].

[*i.e.*, (*II*.1)], whereas as $\delta \to \infty$ we have

$$u_t + 2uu_x + (Hu)_{xx} = 0 \tag{IV.2}$$

(The Benjamin–Ono equation: B–O), where Hu is the Hilbert Transform:

$$Hu = \frac{1}{\pi} \int_{-\infty}^{\infty} d\xi \, \frac{u(\xi)}{\xi - x}.$$

We shall see that the scattering theory associated with (*IV*.2) presents a certain novelty which in a sense bridges the gap between the one-dimensional IST and two-dimensional IST. Moreover, as with discrete problems, it can be expected that there be singular-integral versions for every solvable continuous nonlinear evolution equation.

5. Multidimensional PDE's *(2+1)*:

$$(u_t + 6uu_x + u_{xxx})_x = -3\sigma_1^2 u_{yy}, \qquad \sigma_1^2 = -1 \text{ (KPI)}; \quad \sigma_1^2 = 1 \text{ (KPII)}, \tag{V.1}$$

(Kadomtsev–Petviashvili equation: KP)

$$\begin{aligned} iu_t + (-\sigma_1 u_{xx} + u_{yy}) &= 2(\sigma_1\sigma_2 u^2 u^* + \phi u), \\ \sigma_1 \phi_{xx} + \phi_{yy} &= -2\sigma_2 (uu^*)_{xx}, \end{aligned} \qquad \sigma_1 = \pm 1, \quad \sigma_2 = \pm 1 \tag{V.2}$$

(Davey–Stewartson equation: D–S)

$$u_{it} + c_{ix} u_{ix} + c_{iy} u_{iy} = \gamma_i u_j^* u_k^* \tag{V.3}$$

where c_{ix}, c_{iy}, and γ_i are constants and $i, j, k = 1, 2, 3$ permuted; again, there is an n-component version

(Three wave equation).

There is also a natural extension of (*V*.3) to three dimensions. Although other nonlinear wave equations exist in more than two spatial, or three total dimensions, with apparently special properties —most notably the self-dual Yang-Mills equation—, nevertheless the IST for these equations, treated as initial value problems, essentially remains open. The new results that we will briefly discuss here have to do with the scattering theory —direct and inverse— associated with some of these equations.

§2 Methodology of the Inverse Scattering Transform

From a conceptual point of view the steps associated with the implementation of IST are now quite clear. Generally speaking, the class of solvable equations for which IST can be initiated results from the compatibility of two operators, which we shall call L and M:

$$Lv = \lambda v, \tag{1a}$$

$$v_t = Mv. \tag{1b}$$

The associated nonlinear evolution equation via compatibility is given by:

$$L_t + [L, M] = 0, \qquad [L, M] = LM - ML, \tag{1c}$$

if and only if $\lambda_t = 0$. Equations (1*a*–*c*) constitute Lax's famous result[3] Some pertinent remarks should be made about (1*a*–*c*).

---

[3]The interested reader may see [1] for details.

i. L and M may be matrix valued operators of arbitrary order.

ii. More complicated scattering problems than (1a) have been considered in the literature. Specifically by this we mean that the dependence on the scattering parameter λ can be complicated;

iii. The operator L may be differential [cases **1**, **2**, and **5**], integro-differential or otherwise viewed as a RHBVP in physical space as in **4**, discrete as in **3**, or even a purely linear algebraic system as in (I.2). We will not discuss problems such as **iii** in this presentation.

A review of some of the work associated with *(I.2)* appears in [1, §3.5]. Suffice it to say that the reduction to action angle variables is ascertained by direct computation involving the eigenvalues of suitable matrices. Hence, in a sense the scattering-inverse problem of many body systems such as (I.2) is elementary, and one would not need the more powerful methods associated with the other equations.

iv. When dealing with multidimensional PDE's such as (V.1–3) the auxiliary scattering parameter λ no longer has the importance that it had for one spatial dimension. Rather, here a parameter is inserted via a suitable boundary condition at infinity. In this lecture we will simply survey the main steps associated with the multidimensional analysis. Fokas in his lectures (see also [3]) will detail rather closely the essential ideas relevant to two spatial dimensions.

At this juncture it is appropriate to outline the essential conceptual steps associated with the IST:

i. Investigate the analytic properties of suitable scattering functions associated with the operator L in (1a). We seek eigenfunctions with specific normalizations, *i.e.*, the identity when the relevant scattering parameter is infinite. To effect this, the eigenfunction typically must be scaled by a suitable exponential factor.

ii. We use these analytic properties to construct an *inverse problem*, *i.e.*, inverse scattering, for the eigenfunctions. The inverse problem depends on certain scattering data. The scattering data is in essence a functional involving specific eigenfunctions of L. For the standard one-spatial dimension scattering problems the inverse problem may be viewed as a RHBVP *in scattering space*. For the B–O equation (IV.2) the RHBVP is nonlocal as opposed to equations (2) and (3). When considering the two-spatial dimensional problems a new concept emerges. In certain cases the required eigenfunctions are bounded everywhere but nowhere analytic. In this situation it is important to consider a generalization of the RHBVP, namely the DBAR problem. We shall see that the RHBVP may be viewed as a limit of the DBAR problem. The DBAR problem concept was first introduced by Beals and Coifman [4(iii)] in connection with the direct and inverse scattering of certain first order one-dimensional matrix systems. However, in these problems this concept is not crucial since the result has always reduced to a RHBVP. On the other hand, we find the DBAR approach is necessary when dealing with equations (V.1), $\sigma_1^2 = +1$ (KPII), and (V.2), $\sigma_1 = -1$. The other equations listed in **5** all have as their inverse problem a nonlocal RHBVP (as is the situation for B–O).

iii. The operator M, or the nonlinear evolution equation itself, can be used to fix the time evolution of the scattering data for times greater than *zero*.

iv. The solution to the nonlinear evolution equation, for all time, is obtained from the inverse problem for the eigenfunctions. The potential is obtained either directly via (1a) or through certain integrals over the scattering data and eigenfunctions (suitable integrals have been obtained in all concrete cases).

It should be remarked that the Gelfand–Levitan–Marchenko (GLM) approach has been abandoned via this formulation, though in the well known cases it is readily obtained via a sequence of straightforward transformations. Our point of view is that the GLM equation is a consequence of either the Riemann–Hilbert or DBAR problem and is therefore not as fundamental. Moreover we note that we are primarily concerned with the solution to a well posed initial value problem in infinite space, and presently it is necessary for the potentials to have sufficient decay at infinity in order to carry through the analysis. The RHBVP or the DBAR problem is a consequence of this assumption and *not* the starting point. It is this way that the difference between the approach of Zakharov and Shabat [5] is manifest. In their analysis concrete formulae in terms of initial data are not obtained.

Earlier work in IST falls essentially into two categories:

a. Generating a class of nonlinear evolution equations associated with a given scattering operator L.

b. Analyzing and solving the direct and inverse scattering of the associated operator L.

In this way the steps i–iv discussed above have been carried out. However, the analysis had been largely confined to second order and, in a few cases, third order operators both continuous and discrete (second order). The recent results to be mentioned here are those of:

α. Beals and Coifman [4] and Beals [6] on arbitrary n^{th} order systems and n^{th} order scalar operators, respectively.

β. Kodama, Ablowitz, and Satsuma [7] on the operators associated with (IV.1) (ILW).

γ. Fokas and Ablowitz [8] on the operator associated with (IV.2) (B–O).

δ. Fokas and Ablowitz [9] and Ablowitz, Bar Yaacov, and Fokas [10] on the operators associated with KPI and KPII, respectively. In this regard we mention the significant contribution of Manakov [11] which preceded the work of [9] and [10].

ϵ. Fokas and Ablowitz [12] on the first order systems associated with (V.2, 3). In this regard we note the work of Kaup [13] related to (V.3), though the analysis we develop for these equations is quite different from that of [13].

§3 Scattering Operators

Here we only list the operators associated with many of the above mentioned equations (the enumeration should be clear from the context), *i.e.*, *(II.1–L)* is the scattering operator associated with *(II.1)*, etc.

$$v_{xx} + uv = \lambda v, \qquad (II.1\text{–}L)$$

$$\begin{cases} v_{1x} = -i\varsigma v_1 + qv_2, & (II.2\text{–}L) \\ v_{2x} = i\varsigma v_2 + rv_1, & \text{also } (II.3\text{–}L) \end{cases}$$

Though (I.1) (PII) is related to the above 2×2 scattering problem (II.2–L) or, through a transformation, to (II.1–L), the analysis presents certain pecularities that are different from the others to be mentioned here. We will return to make a brief comment about this equation. We write

$$\mathbf{v}_x = i\varsigma \mathbf{J}\mathbf{v} + \mathbf{q}\mathbf{v}, \qquad (II.4\text{–}L)$$

where $\mathbf{v}$ is an $n \times l$ vector, $\mathbf{J}$ and $\mathbf{q}$ are $n \times n$ matrices, with $\mathbf{J} = \operatorname{diag}(\lambda_1, \ldots, \lambda_n)$ and $q_{ii} = 0$. The 2×2 version of (II.4–L) is (II.2–L). The three-wave problem has $n = 3$.

$$v_{xxx} + qv_x + rv = \lambda v. \qquad (II.5\text{–}L)$$

$$a_n v_{n+1} = a_{n-1} v_{n-1} + b_n v_n = \lambda v_n, \qquad (III.1\text{–}L)$$

$$\begin{cases} v_{1,n+1} = z v_{1,n} + Q_n v_{2,n}, \\ v_{2,n+1} = \dfrac{1}{z} v_{2,n} + R_n v_{1,n}, \end{cases} \qquad (III.2\text{–}L)$$

$$i W_x^+ + (\varsigma + 1/2\delta)(W^+ - W^-) = -u W^+; \qquad W^\pm = W(x \mp i\delta). \qquad (IV.1\text{–}L)$$

Alternatively, $W^\pm$ may be interpreted as the boundary values of functions analytic in strips (+: $0 < \operatorname{Im} x < 2\delta$, –: $-2\delta < \operatorname{Im} x < 0$).

$$i W_x^+ + \lambda(W^+ - W^-) = -u W^+; \qquad (V.2\text{–}L)$$

$W^\pm$ are the boundary values of functions analytic in the +: upper half x-plane and –: lower half x-plane.

$$\sigma v_y + v_{xx} + uv = 0, \qquad \sigma = i \text{ (KPI)}, \quad \sigma = -1 \text{ (KPII)}, \qquad (V.1\text{–}L)$$

$$\mathbf{v}_x = \mathbf{J}\mathbf{v}_y + \mathbf{q}\mathbf{v}, \qquad (V.2\text{–}L) \text{ or } (V.3\text{–}L)$$

$$\mathbf{J} = \operatorname{diag}(J_1, \ldots, J_n), \qquad q_{ii} = 0.$$

The D–S equation has $n = 2$ and the three-wave problem has $n = 3$. In the above formulae the potentials u, q, r, $\mathbf{q}$, a_n, b_n, Q_n, and R_n are functions of position and time[4]

§4 The Korteweg–de Vries equation

The basic point of view may be easily understood by first reviewing the situation for KdV. After this it will suffice to give the relevant RHBVP associated with many of the other equations. The DBAR problem associated with the KPII equation will be given, as well as some of the pertinent ideas associated with $(II.4\text{–}L)$. Details can be found in the references.

Associated with the operator $(II.1\text{–}L)$ are two complete sets of eigenfunctions which are bounded for all values of x, and having appropriate analytic extensions. They are defined by the equation and the boundary conditions, *i.e.*, the four eigenfunctions having the following asymptotic behavior, letting $\lambda = -k^2$ in $(II.1\text{–}L)$:

when $x \to -\infty$

$$\phi(x,k) \sim e^{-ikx}, \qquad (2a)$$
$$\overline{\phi}(x,k) \sim e^{ikx}, \qquad (2b)$$

and when $x \to \infty$

$$\psi(x,k) \sim e^{ikx}, \qquad (2c)$$
$$\overline{\psi}(x,k) \sim e^{-ikx}. \qquad (2d)$$

It is convenient to introduce the function $W(x,k) = v(x,k)e^{ikx}$, and therefore we have a modified set of eigenfunctions:

[4]Time acts only parametrically in these equations.

when $x \to -\infty$

$$M(x,k) \sim 1, \tag{3a}$$
$$\overline{M}(x,k) \sim e^{2ikx}, \tag{3b}$$

and when $x \to \infty$

$$N(x,k) \sim e^{2ikx}, \tag{3c}$$
$$\overline{N}(x,k) \sim 1. \tag{3d}$$

We will need only three of them. Completeness of any two gives:

$$\frac{M(x,k)}{a(k)} = \overline{N}(x,k) + r(k)\,N(x,k). \tag{4a}$$

$1/a(k)$ and $r(k)$ are called the ***transmission/reflection*** coefficients respectively. It may be easily shown that for u decaying fast enough as $|x| \to \infty$,[5] $M(x,k)$ and $a(k)$ are analytic and tend to unity for $\operatorname{Im} k > 0$, and $\overline{N}(x,k)$ is analytic and tends to unity for $\operatorname{Im} k < 0$. These statements may be ascertained from their integral equation formulation, *e.g.*, for $\overline{N}(x,k)$:

$$\overline{N}_{xx} - 2ik\overline{N}_x + u\overline{N} = 0, \tag{5a}$$
$$\overline{N}(x,k) = 1 + \int_{-\infty}^{\infty} d\xi\, G_-(x-\xi;k)\,u(\xi)\,\overline{N}(\xi,k), \tag{5b}$$
$$G_{-xx} - 2ikG_{-x} = -\delta(x-\xi) \Rightarrow G_-(x-\xi;k) = \frac{1}{2\pi}\int_{C_A} dp\,\frac{e^{ip\lambda(x-\xi)}}{p(p-2k)}, \tag{5c}$$

where C_A is a contour $P = P_R + i\epsilon$, with $-\infty < P_R < \infty$. Hence:

$$\overline{N}(x,k) = 1 - \int_x^{\infty} d\xi\,\frac{1-e^{2ik(x-\xi)}}{2ik}\,u(\xi)\,\overline{N}(\xi,k). \tag{5d}$$

$G_-(x-\xi,k)$ is analytic in the lower half k-plane, hence (5d), being a Volterra equation, implies that $\overline{N}(x,k)$ is also analytic for $\operatorname{Im} k < 0$. We also note the following discrete symmetry condition:

$$N(x,k) = e^{2ikx}\overline{N}(x,-k). \tag{4b}$$

We often refer to this type of condition as the analytic connection. Since $\overline{N}(x,k)$ is analytic for $\operatorname{Im} k > 0$, we have that $\overline{N}(x,-k)$ is analytic for $\operatorname{Im} k < 0$. With these analytic properties (4a,b) may be viewed as a RHBVP. There are numerous references and studies associated with such problems (see for example references in [14]). When a RHBVP is only a scalar equation, the solutions (when they exist) are expressible in terms of quadratures. On the other hand, when a RHBVP is a system of order grater than two then, generally speaking, the solutions are characterized via Fredholm integral equations. Equations (4a,b) are, in essence, a second order system for the functions $M(x,k)$, $\overline{M}(x,-k)$, $N(x,k)$, and $\overline{N}(x,-k)$. Though in this particular case a Fredholm integral equation is readily obtainable by taking a ***minus projection*** of (4a,b). Assuming for convenience $a(k) \neq 0$ for $\operatorname{Im} k > 0$, we operate on (4a) by

[5] We will not dwell on the particularities of just what function class u must be in, but [1], [4], [6] and their corresponding references will give the reader all necessary information.

$(2\pi i)^{-1}\int d\xi\,[\xi-(k-i\epsilon)]^{-1}$ after substracting unity from both sides, and find:

$$\overline{N}(x,k)=1+\frac{1}{2\pi i}\int_{-\infty}^{\infty}d\xi\,\frac{r(\xi)\,N(x,\xi)}{\xi-(k-i\epsilon)}. \tag{6a}$$

Using (4b) and taking $k\mapsto -k$:

$$N(x,k)=e^{2ikx}\left(1-\frac{1}{2\pi i}\int_{L}d\xi\,\frac{r(\xi)\,N(x,\xi)}{\xi+k}\right), \tag{6a}$$

where $L:-\infty+i\epsilon<\xi<\infty+i\epsilon$. It is straightforward to incorporate the contributions from the locations $a(k_j)=0$, $j=1,\ldots,N$. If u decays very rapidly, the only modification necessary is to deform L to pass over all the poles of $r(\xi)$. The potential u can either be obtained from equation (II.1–L) or via the relation:

$$u=-\frac{\partial}{\partial x}\frac{1}{\pi}\int_{L}d\xi\,r(\xi)\,N(x,\xi). \tag{6c}$$

[This relation is obtained by considering the asymptotic values of (6b) and the analogue of (5) for N as $k\to\infty$.]

Nevertheless it should be stressed that the RHBVP (4a, b) contains all the essential information in the problem. The remaining questions can be readily answered. Hereafter, we shall only give the RHBVP associated with a given scattering problem when it suffices to pin down the IST, *i.e.*, the DBAR problem is not essential.

We first make some remarks before moving on to other equations:

i. The time evolution of the scattering data may be obtained by analyzing the asymptotic behavior of the associated time evolution operator [u in (1b)]. For KdV this is:

$$v_t=(u_x+4ik^3)v+(4k^2-2u)v_x. \tag{7a}$$

Calling $v(x,k)=\phi(x,k)=a(k)\overline{\psi}(k,u)+b(k)\psi(x,k)$, where $r(k)=b(k)/a(k)$ and substitution into (7a), as $x\to\infty$, yields:

$$a_t=0,\qquad b_t=8ik^3b\quad\text{or}\quad r_t=8ik^3r. \tag{7b}$$

Similarly the residues of r in the upper half plane satisfy the same equation as does r. (This gives the time evolution of the so-called ***norming constants***.)

ii. Pure soliton solutions are obtained by assuming a special form for $r(k,t)$:

$$r(\kappa,t)=\begin{cases}0, & k\ \text{real},\\ \displaystyle\sum_{l=1}^{n}\frac{c_{l,0}e^{8\kappa_l^3t}}{k-i\kappa_l}, & \operatorname{Im}k>0.\end{cases} \tag{7c}$$

Then (6b, c) yields a linear algebraic system

$$N_j+\sum_{l=1}^{n}\frac{\exp(-2\kappa_lx+8\kappa_l^3t)\,c_{l,0}}{\kappa_l+\kappa_j}N_l=e^{-2\kappa_jx}, \tag{7d}$$

$$u=-2\frac{\partial}{\partial x}\sum_{l=1}^{n}\exp(8\kappa_l^3t)\,c_{l,0}N_l, \tag{7e}$$

where $N_j := N(x, k = ik_j)$. $(7d, e)$ are formulae which give pure soliton solutions. For $n = 1$:

$$N_1 = \frac{\exp(-2\kappa_1 x)}{1 + (c_{1,0}/2\kappa_1)\exp(-2\kappa_1 x + 8\kappa_1^3 t)}, \tag{7f}$$

$$u = 2\kappa_1^2 \operatorname{sech}^2 \kappa_1(x - 4\kappa_1^2 t - x_0), \qquad \exp(-2\kappa_1 x_0) := \frac{c_{1,0}}{2\kappa_1}. \tag{7g}$$

iii. The Gelfand–Levitan–Marchenko equation is obtained from $(6b)$ by assuming

a. triangularization:

$$N(x, k) = e^{2ikx}\left[1 + \int_x^\infty dk\, K(x, s) e^{ik(s-x)}\right],$$

substitution into $(6b)$, and

b. operating with

$$\frac{1}{2\pi}\int_{-\infty}^{\infty} dk\, e^{ik(x-y)}, \qquad y > x;$$

then we find:

$$K(x, y) + F(x + y) + \int_x^\infty ds\, K(x, s) F(s + y) = 0, \qquad y > x, \tag{8a}$$

where

$$F(x) = \frac{1}{2\pi}\int_{\mathcal{L}_0}^{\infty} dk\, r(k) e^{ik(x+y)}. \tag{8b}$$

Here, $\mathcal{L}_0$ is a contour passing above all the poles of $r(k)$. From $(6c)$ we have:

$$u = 2\frac{\partial}{\partial x} K(x, x). \tag{8c}$$

This clearly shows that the GLM equation is a direct consequence of the RHBVP. Nevertheless, we do not claim that there are no advantages to the GLM equation but only that, in some sense, it is not as fundamental as the RHBVP.

iv. There exists an integral equation motivated by $(6b)$ which provides a very general class of solutions to $(II.1)$; namely

$$\phi(x, t; k) + ie^{i(kx+k^3t)}\int_L d\lambda(l)\frac{\phi(x, t; l)}{l + k} = e^{i(kx+k^3t)}, \tag{9a}$$

where L and $d\lambda(k)$ are arbitrary contours and measure respectively, and u is given by

$$u(x, t) = -\frac{\partial}{\partial x}\int_L d\lambda(k)\,\phi(x, t; k). \tag{9b}$$

Equations $(9a, b)$ relax to $(6b, c)$ by taking $d\lambda(k) = r_0(k)\,dk/4\pi$, $\phi(x, k) = N(x, k/2)e^{ik^3t}$, and $L = \mathcal{L}$. The proof that $(9a, b)$ satisfy $(II.1)$ is given in [15] and only relies on the fact that $(9a)$ has no nontrivial homogeneous solutions.

A special class of solutions to (9a) gives a full three-parameter family of solutions to the self-similar ODE related to (*II*.1), *i.e.*, if

$$u(x,t) = V(z)(3t)^{-2/3}, \qquad z = x(3t)^{-1/3}, \tag{10a}$$

then V satisfies

$$V''' + 6VV' - (2V + zV') = 0. \tag{10b}$$

Moreover there is a one to one transformation between solutions of (10b) and those of PII or (*I*.1), *i.e.*,

$$V'' - zV - 2V^3 - \alpha = 0, \tag{10c}$$

and the transformation $U \longleftrightarrow V$ (see [16])

$$V = \frac{U' + \alpha}{2U - z}, \tag{10d}$$

$$U = -(V' + V^2). \tag{10e}$$

The three-parameter family of solutions to (10b, c) is obtained via solutions to the singular integral equation:

$$\phi(z;t) + \frac{1}{i\pi}\exp\left[i(tz + \tfrac{1}{3}t^3)\right]\int_{L*} d\tau\, \frac{\phi(z;\tau)}{\tau + t} = \exp\left[i(tz + \tfrac{1}{3}t^3)\right], \tag{10f}$$

$$U(z) = \frac{1}{\pi}\frac{\partial}{\partial z}\int_{L*} d\tau\, \phi(z;\tau), \tag{10g}$$

where $\int_{L*} = \sum_{j=1}^{3} \rho_j \int_{L_j}$, L_j being any three linearly independent contours associated with the linearized problem (three rays are sufficient).

The question of relating this approach with the initial value problem of (*I*.1) is presently under serious scrutiny. We shall report on this in a future communication [19]. Suffice it to say that both (10f) as well as the initial value problem can be formulated as a discontinuous RHBVP in sectors for a certain range of parameter space.

§5 Analytic problems associated with the $n \times n$ scattering problems

The RHBVP associated with (*II*.2–*L*) is formulated as follows. Define two complete sets of eigenfunctions to (*II*.2–*L*) with the boundary conditions

$$\begin{array}{ll} x \to -\infty: & x \to \infty: \\ \phi \sim \begin{pmatrix}1\\0\end{pmatrix} e^{-i\varsigma x}, & \psi \sim \begin{pmatrix}0\\1\end{pmatrix} e^{i\varsigma x}, \\ \bar{\phi} \sim \begin{pmatrix}0\\-1\end{pmatrix} e^{i\varsigma x}, & \bar{\psi} \sim \begin{pmatrix}1\\0\end{pmatrix} e^{-i\varsigma x}, \end{array} \tag{11a}$$

then completeness gives:

$$\begin{aligned}\phi(x,\varsigma) &= a(\varsigma)\overline{\psi}(x,\varsigma) + b(\varsigma)\psi(x,\varsigma),\\ \overline{\phi}(x,\varsigma) &= -\overline{a}(\varsigma)\psi(x,\varsigma) + \overline{b}(\varsigma)\overline{\psi}(x,\varsigma).\end{aligned} \tag{11b}$$

We define:

$$\begin{aligned} M &= \phi e^{i\varsigma x}, & N &= \psi e^{-i\varsigma x},\\ \overline{M} &= \overline{\phi} e^{-i\varsigma x}, & \overline{N} &= \overline{\psi} e^{i\varsigma x}.\end{aligned} \tag{11c}$$

Then, as $x \to \infty$:

$$M \sim \begin{pmatrix}1\\0\end{pmatrix}, \quad \overline{M} \sim \begin{pmatrix}0\\-1\end{pmatrix}, \qquad a \to 1, \quad \overline{a} \to 1.$$

The relevant integral equation formulation yields:

$$\begin{aligned} &M,\ N,\ a \text{ are analytic for } \operatorname{Im}\varsigma > 0,\\ &\overline{M},\ \overline{N},\ \overline{a} \text{ are analytic for } \operatorname{Im}\varsigma < 0.\end{aligned}$$

Then (11b) implies

$$\begin{aligned} \left(\frac{M}{a}\right) &= \overline{N} + \rho e^{2i\varsigma x} N,\\ \left(\frac{\overline{M}}{\overline{a}}\right) &= -N + \overline{\rho} e^{-2i\varsigma x}\overline{N},\end{aligned} \tag{11d}$$

where $\rho = b/a$ and $\overline{\rho} = \overline{b}/\overline{a}$. Equation (11d) is the RHBVP for the sectionally meromorphic functions M/a, $\overline{M}/\overline{a}$, N, and $\overline{N}$. The potentials can be obtained either from (*II*.2–*L*) or from readily obtained integral formulae. The reader should also note the analysis associated with (*II*.4–*L*) below.

In the case of (*II*.4–*L*) we shall derive the relevant RHBVP by introducing the concept of the DBAR problem [4]. For convenience assume $J = \mathrm{diag}(\lambda_1, \ldots, \lambda_n)$, where $\lambda_1 < \lambda_2 < \cdots < \lambda_n$ and λ_i are all real. Otherwise the analysis is a bit more complicated and the resulting RHBVP is in multisectors as oppossed to only two (upper/lower) half planes. Introduce the transformation, extending (*II*.4–*L*) to a matrix formulation,

$$\mathbf{v}(x,\varsigma) = \mathbf{m}(x,\varsigma)\exp(ix\varsigma\mathbf{J}), \tag{12a}$$

whereupon (*II*.4–*L*) is given by:

$$\frac{d\mathbf{m}}{dx} = i\varsigma[\mathbf{J},\mathbf{m}] + \mathbf{q}\mathbf{m}. \tag{12b}$$

(Hereafter we shall drop the boldface notation for m, J, and q which are $n \times n$ matrices.) Beals and Coifman prove that there exists a unique solution $m(x,\varsigma)$ bounded for all x, meromorphic in ς off the real ς-axis, with m normalized:

$$\lim_{\varsigma\to\infty} m(x,\varsigma) = I. \tag{12c}$$

For example, the integral equation governing $m_+(x,\varsigma)$, a function analytic for $\operatorname{Im}\varsigma > 0$ is given by:

$$m_+(x,\varsigma) = I + \int_{-\infty}^{x} dy\, e^{i(x-y)\varsigma\bar{J}}(\pi_0 + \pi_-)\,(qm_+)(y,\varsigma) + \int_{x}^{\infty} dy\, e^{i(x-y)\varsigma\bar{J}}\pi_+\,(qm_+)(y,\varsigma), \tag{12d}$$

where $\pi_\pm$ are strictly $+$: upper and $-$: lower triangular projection matrices, π_0 ia a pure diagonal projection matrix, and the *ad* action:

$$e^{i\alpha\bar{J}}q = e^{i\alpha J}qe^{-i\alpha J}.$$

For the inverse problem we need only call upon the well known formula:

$$m(x,\varsigma) = \frac{1}{2\pi i}\int\int_R dz' \wedge d\bar{z}' \frac{1}{z'-\varsigma}\frac{dm(x,z')}{d\bar{z}'} + \int_C dz' \frac{m(x,z')}{z'-\varsigma}, \tag{13}$$

where $z = z_R + iz_I$ and $dz \wedge d\bar{z} = -2i\, dz_R\, dz_I$. In order to implement (13) we take $R = R_\infty$ and $C = C_\infty$. Note that the normalization as $\varsigma \to \infty$ implies that the second term in (13) is I, and by direct calculation observe that

$$\frac{dm}{d\bar{z}} = (m_+ - m_-)\mu(z,\xi) + \sum_{j=1}^{N} \pi m_j \delta(z - z_j), \tag{12e}$$

where

$$\mu(z,\xi) := \begin{cases} 1, & z = \xi \text{ real}, \\ 0, & \text{otherwise}, \end{cases} \qquad m_j := m(x, z_j).$$

In (12*e*) we use

$$\begin{aligned} \frac{d}{d\bar{z}} &= \tfrac{1}{2}\left(\frac{\partial}{\partial z_R} + i\frac{\partial}{\partial z_I}\right), \\ \frac{d}{d\bar{z}}\left(\frac{1}{z - z_j}\right) &= \pi\delta(z - z_j). \end{aligned} \tag{12f}$$

Moreover we note that $(m_+ - m_-)(x,\xi)$ is also a solution of (12*b*) for $\varsigma = \xi$ real, hence we must have that

$$\begin{aligned} (m_+ - m_-)(x,\xi) &= m_-(x,\xi)\, W(x,\xi), \\ W(x,\xi) &= e^{ix\xi\bar{J}}V(\xi), \end{aligned} \tag{12g}$$

($V(\xi)$ is thhe scattering data) whereupon (13) yields

$$m(x,\varsigma) = I + \sum_{j=1}^{N} \frac{m_j(x)}{\varsigma - z_j} + \frac{1}{2\pi i}\int_{-\infty}^{\infty} d\xi\, \frac{m_-(x,\xi)\, W(x,\xi)}{\xi - \varsigma}. \tag{12h}$$

Note that (12*g*) is the RHBVP for the sectionally meromorphic functions m_+ and m_-, and (12*h*) is the corresponding Fredholm integral equation for the boundary values $m_\pm(x,\xi)$ obtained by taking the limit $\operatorname{Im}\varsigma \to \pm 0$.

An integral formula for the potential q is obtained from the usual asymptotic analysis ($\lim_{\varsigma\to\infty}$)

$$q = \left[J, \frac{1}{2\pi i}\int_{-\infty}^{\infty} d\xi\, m_-(x,\xi)\, W(x,\xi) - \sum_{j=1}^{N} m_j\right]. \tag{12i}$$

For nonlinear wave problems one only needs to supplement the above formulae with the proper time dependence of the scattering data. The requisite formula is given by

$$V(\xi, t) = e^{At}V(\xi, 0)e^{At}, \tag{12j}$$

where A is a diagonal matrix obtained from either the associated time equation (1*b*) or the nonlinear evolution equation itself.

The 2×2 scattering problem (*II*.2–*L*) is, of course, a special case of these more general results. The correspondence is given by

$$J = \begin{pmatrix} -1 & 0 \\ 0 & 1 \end{pmatrix}, \qquad q = \begin{pmatrix} 0 & q \\ r & 0 \end{pmatrix}, \qquad m_+ e^{iz\zeta J} = \begin{pmatrix} \phi_1 & \psi_1/a \\ \phi_2 & \psi_2/a \end{pmatrix}, \tag{12k}$$

where ϕ_1, ϕ_2, ψ_1, and ψ_2 are the components of ϕ and ψ respectively.

With regard to the scattering problem (*II*.5–*L*), this is a special case of the scalar problem discussed in detail by Beals [6]:

$$D^n_u = q_{n-2} D^{n-2} u + q_{n-3} D^{n-3} u + \cdots + q_1 Du + q_0 u + z^n u, \qquad D := \frac{1}{i}\frac{d}{dx}, \qquad q_j \to 0 \text{ sufficiently rapidly}\ . \tag{II.6–L}$$

By reducing (*II*.6–*L*) to a system, Beals shows that one can in fact construct suitable sectionally meromorphic functions and thereby obtain a RHBVP. The point $z = 0$ presents some technical difficulties which must be overcome in order to obtain the necessary analytic properties. Apart from this, a number of the results of the system [4] can be carried over. We will not discuss this analysis any further here however.

§6 Analytic problems associated with discrete scattering problems

Next we shall give the RHBVP's for the discrete scattering problems (*III*.1–*L*) and (*III*.2–*L*). For (*III*.1–*L*) define the scattering functions with the following asymptotic behaviors

$$n \to -\infty: \quad \phi_n(z) \sim z^{-n}, \qquad n \to \infty: \quad \begin{cases} \psi_n(z) \sim z^n, \\ \overline{\psi}_n(z) \sim z^{-n}, \end{cases} \tag{14a}$$

where we have taken $a_n \to 1/2$ and $b_n \to 0$ rapidly enough as $|n| \to \infty$, and $\lambda = \frac{1}{2}(z + z^{-1})$. Completeness of ψ_n and $\overline{\psi}_n$ implies:

$$\phi_n(z) = a(z)\overline{\psi}_n(z) + b(z)\psi_n(z), \qquad |z| = 1. \tag{14b}$$

By an analysis that essentially parallels that of the continuous problem the following functions are:

$$\begin{aligned} &\text{a.} \quad \text{Analytic for } |z| > 1: \quad M_n(z) := \phi_n(z)\, z^n, \qquad M_n(z) := \psi_n(z)\, z^{-n}, \qquad a(z) \\ &\text{b.} \quad \text{Analytic for } |z| < 1: \quad \overline{N}_n(z) := \overline{\psi}_n(z)\, z^n. \end{aligned} \tag{14c}$$

The symmetry condition is

$$\psi_n(z) = \overline{\psi}_n(z^{-1}) \qquad \text{or} \qquad N_n(z) = \overline{N}_n(z^{-1})\, z^{-2n}, \tag{14d}$$

whereupon the RHBVP associated with (*III*.1–*L*) is:

$$\frac{M_n(z)}{a(z)} = \overline{N}_n(z) + r(z) z^{-2n} \overline{N}_n(z^{-1}), \qquad |z| = 1, \tag{14e}$$

where $r(z) = b(z)/a(z)$, $\lim_{z\to\infty} M_n/a = 1$, and $\lim_{z\to\infty} \overline{N}_n = 1$. Nonlinear wave equations are solved by taking appropriate time dependencies for $r(z)$, *e.g.*, for the Toda lattice:

$$r(z,t) = r(z,0)e^{(z-1/z)t}, \tag{14f}$$

whereas other time dependencies give different nonlinear evolution equations. Similarly the zeros of $a(z)$, or the poles of the meromorphic function $M_n(z)/a(z)$, give rise to the soliton sector.

The analysis associated with (*III*.2–*L*) is similar. Taking $\lim_{|n|\to\infty} Q_n = 0$ and $\lim_{|n|\to\infty} R_n = 0$ rapidly enough, appropriate scattering functions are defined by the following boundary conditions:

$$\text{as } n \to -\infty: \quad \begin{cases} \phi_n \sim \begin{pmatrix} 1 \\ 0 \end{pmatrix} z^n, \\ \overline{\phi}_n \sim \begin{pmatrix} 0 \\ -1 \end{pmatrix} z^{-n}, \end{cases} \qquad \text{as } n \to \infty: \quad \begin{cases} \psi_n \sim \begin{pmatrix} 0 \\ 1 \end{pmatrix} z^{-n}, \\ \overline{\psi}_n \sim \begin{pmatrix} 1 \\ 0 \end{pmatrix} z^n. \end{cases} \tag{15a}$$

Completeness of these functions give:

$$\begin{aligned} \phi_n(z) &= a(z)\overline{\psi}_n(z) + b(z)\psi_n(z), \\ \overline{\phi}_n(z) &= -\overline{a}(z)\psi_n(z) + \overline{b}(z)\overline{\psi}_n(z). \end{aligned} \tag{15b}$$

Define: $M_n(z) := \phi_n(z)\,z^{-n}$, $\overline{M}_n(z) := \overline{\phi}_n(z)\,z^n$, $N_n(z) := \psi_n(z)\,z^n$, $\overline{N}_n(z) := \overline{\psi}_n(z)\,z^{-n}$. Analytic properties are given by:

$$\begin{aligned} &\text{Analytic for } |z| < 1: \quad && \overline{M}_n(z),\ \overline{N}_n(z),\ \overline{a}(z), \\ &\text{Analytic for } |z| > 1: \quad && M_n(z),\ N_n(z),\ a(z), \end{aligned} \tag{15c}$$

and the relevant RHBVP is:

$$\left.\begin{aligned} \frac{M_n(z)}{a(z)} &= \overline{N}_n(z) + r(z)z^{-2n}N_n(z), \\ \frac{\overline{M}_n(z)}{\overline{a}(z)} &= -N_n(z) + \overline{r}(z)z^{2n}\overline{N}_n(z), \end{aligned}\right. \qquad |z| = 1, \tag{15d}$$

where $r(z) = b(z)/a(z)$, $\overline{r}(z) = \overline{b}(z)/\overline{a}(z)$, and the eigenfunctions $\overline{M}_n$, $\overline{N}_n$ and $\overline{a}(z)$ all have constant asymptotic values (with respect to z) as $|z| \to \infty$, and similarly for M_n, N_n and $a(z)$ as $|z| \to 0$.

Nonlinear wave equations are obtained by taking suitable time dependencies for $r(z)$ and $\overline{r}(z)$:

$$\begin{aligned} &\text{For differential-difference equations:} \quad && r(z,t) = r(z,0)e^{i\omega_I(z)t}, \\ &\text{and for partial difference equations:} \quad && r(z,m) = r(z,0)(\omega_{II}(z))^m, \end{aligned} \tag{15e}$$

e.g., for (*III*.2): $\omega_I(z) = -(z^2 + 1/z^2 - 2)$, and for the Crank-Nicholson type partial difference scheme discussed below (*III*.2):

$$\omega_{II}(z) = \frac{1 - \frac{1}{2}i\sigma\,(z^2 + 1/z^2 - 2)}{1 + \frac{1}{2}i\sigma\,(z^2 + 1/z^2 - 2)}, \tag{15f}$$

where $\sigma = \Delta t/\Delta x^2$ (see also [1,2]).

§7 Analytic problem for ILW and B–O

Next we move on to summarize the situation for the ILW and B–O equations. The scattering parameter ς is parameterized by $\varsigma = \varsigma(k)$:

$$\varsigma = k + k\coth(2k\delta) - \frac{1}{2\delta}, \tag{16a}$$

and the relevant scattering functions have the following boundary conditions (hereafter only the functions W^+, analytic for $0 < \operatorname{Im} x < 2\delta$ are needed, and hence we shall drop the superscript $+$ on all the following):

$$x \to -\infty: \quad \begin{cases} M(x,\varsigma(k)) \sim 1, \\ \overline{M}(x,\varsigma(k)) \sim e^{2ikx}, \end{cases} \qquad x \to \infty: \quad \begin{cases} N(x,\varsigma(k)) \sim e^{2ikx}, \\ \overline{N}(x,\varsigma(k)) \sim 1. \end{cases} \tag{16b}$$

In [7], integral equations are developed for the functions in (16b) and it is shown that

i. the completeness relation is given by:

$$\frac{M(x,\varsigma)}{a(\varsigma)} = \overline{N}(x,\varsigma) + r(\varsigma)N(x,\varsigma), \qquad -\infty < \varsigma < \infty; \tag{16c}$$

ii. the analitic properties are:

$$\begin{aligned} &\text{analytic for } \operatorname{Im}\varsigma > 0: \quad M(x,\varsigma),\ a(\varsigma), \\ &\text{analytic for } \operatorname{Im}\varsigma < 0: \quad \overline{N}(x,\varsigma). \end{aligned}$$

iii. symmetry condition:

$$N(x,\varsigma(k)) = e^{2ik(x-i\delta)}\overline{N}(x,\varsigma(-k)). \tag{16d}$$

(These are RHBVP's with a *shift.*)

The ILW equation is solved by using the time dependence:

$$r(k,t) = r(k,0)\exp\left[-4ik(-k\coth(2k\delta) + \frac{1}{2\delta})t\right], \tag{16e}$$

and the potential $u(x) = u^+(x) + u^-(x)$, where $u^-(x) = (u^+(x))^*$ [$u(x)$ must be real], is obtained from

$$u^+(x) = \frac{1}{2\pi i}\int_{-1/2\delta}^{\infty} d\varsigma\, r(\varsigma)\, N(x,\varsigma) - i\sum_{j=1}^{N} c_j N_j, \tag{16f}$$

where $r(\varsigma) = 0$ for $\varsigma < -1/2\delta$, $N_j := N(x,\varsigma_j)$, and the norming constants $c_j = (\operatorname{Res} r(\varsigma))_{\varsigma=\varsigma_j}$ give rise to the solitons.

With regard to the B–O equation (*IV*.2) and its scattering problem (*IV*.2–*L*), we note that all of the results we will mention can either be obtained directly from [8] or via an intrincate limiting process from ILW [17] ($\lambda := 2k$). Nevertheless there is one rather substantial difference from the previous problems as will be seen below.

The relevant scattering functions (again only for the *plus* functions W^+) have the following asymptotic properties:

$$x \to -\infty: \quad \begin{cases} M(x,\lambda) \sim 1, \\ \overline{M}(x,\lambda) \sim e^{i\lambda x}, \end{cases} \qquad x \to \infty: \quad \begin{cases} N(x,\lambda) \sim e^{i\lambda x}, \\ \overline{N}(x,\lambda) \sim 1. \end{cases} \tag{17a}$$

The completeness relation is given by

$$M(x,\lambda) = \overline{N}(x,\lambda) + r(\lambda)\, N(x,\lambda), \qquad -\infty < \lambda < \infty, \tag{17b}$$

where $r(\lambda) = 0$ for $\lambda < 0$, and the eigenfunctions obey Fredholm integral equations. The solitons arise as poles of $M(x,\lambda)$ or $\overline{N}(x,\lambda)$ which are due to the existence of homogeneous solutions of these Fredholm equations. In particular M and $\overline{N}$ have the representations:

$$\begin{aligned} M(x,\lambda) &= 1 + \sum_{j=1}^{n} \frac{c_j \phi_j(x)}{\lambda - \lambda_j} + \eta_+(x,\lambda), \\ \overline{N}(x,\lambda) &= 1 + \sum_{j=1}^{n} \frac{\bar{c}_j \phi_j(x)}{\lambda - \lambda_j} + \bar{\eta}_-(x,\lambda), \end{aligned} \tag{17c}$$

where $\phi_j(x)$ obeys:

$$\begin{aligned} \phi_j(x) &= \int_{-\infty}^{\infty} dy\, G(x-y;\lambda_j)\, u(y)\, \phi_j(y), \\ G(x,\lambda_j) &= \frac{1}{2\pi} \int_0^{\infty} dp\, \frac{e^{ipx}}{p - \lambda_j}, \qquad \lambda_j < 0, \end{aligned} \tag{17d}$$

(*i.e.*, the homogeneous Fredholm equation referred to above), and η_+ and $\bar{\eta}_-$ are the boundary values of sectionally holomorphic functions with respect to the ray $0 < \lambda < \infty$. In order to convert (17*a*) into a RHBVP we must supplement it with a symmetry condition [8] which is given by:

$$\frac{\partial}{\partial \lambda}\left(N(x,\lambda) e^{-i\lambda x}\right) = F(\lambda) e^{-i\lambda x} \overline{N}(x,\lambda), \qquad \lambda > 0, \tag{17e}$$

$$x\phi_j(x) + \gamma_j \phi_j(x) = \left(\overline{N}(x,\lambda) - \frac{c_j \phi_j}{\lambda - \lambda_j}\right)_{\lambda=\lambda_j}, \qquad \lambda = \lambda_j < 0. \tag{17f}$$

The RHBVP comes from (17*b*) and (17*e*):

$$M(x,\lambda) = \overline{N}(x,\lambda) + \int_0^{\lambda} d\lambda'\, r(\lambda) F(\lambda') e^{ix(\lambda-\lambda')} \overline{N}(x,\lambda'). \tag{17g}$$

We see from (17*a*) and (17*g*) that:

i. The symmetry condition is *continuous* and not *discrete* as it was for the KdV and ILW scattering problems and,

ii. The RHBVP is *nonlocal* —a reflection of **i** above.

In fact this is the situation for many of the scattering problems associated with two spatial dimensional evolution equations.

We also remark that:

i. The scattering data is: $\{r(\lambda),\ F(\lambda),\ \gamma_j,\ \lambda_j\}$ (c_j is a fixed constant obtained by normalizing $\phi_j(x)$ appropriately).

ii. Equation (17g) can be converted into integral equations to determine $\overline{N}(x,\lambda)$ for $\lambda > 0$, and $\phi_j(x)$.

iii. To solve B–O, the time dependence of the scattering data is given by

$$\begin{aligned} r(\lambda,t) &= r(\lambda,0)e^{i\lambda^2 t}, \\ F(\lambda,t) &= F(\lambda,0)e^{-i\lambda^2 t}, \\ \gamma_j(t) &= 2\lambda_j t + i\gamma_{j,0}, \qquad \lambda_j = \text{ constant.} \end{aligned} \tag{17h}$$

iv. The potential can be obtained from: $u = u^+ + u^-$, where

$$u^+(x) = \frac{1}{2\pi i}\int_0^\infty d\lambda\, r(\lambda)N(x,\lambda) - \sum_{j=1}^{n} c_j\phi_j(x), \tag{17i}$$

and $u^-(x) = (u^+(x))^*$ assuming $u(x)$ is real.

v. The pure soliton solutions are obtained from

$$r(\lambda) = 0, \qquad \overline{N}(x,\lambda) = 1 + \sum_{l=1}^{n} \frac{c_l\phi_l}{\lambda - \lambda_l}$$

so that (17f) implies:

$$(x + 2\lambda_j t + i\gamma_{j,0})\phi_j(x) = 1 + \sum_{\substack{l=1 \\ l\neq j}}^{n} \frac{c_l\phi_l}{\lambda_j - \lambda_l}. \tag{17j}$$

Equation (17j) is a linear system, analogous to (7d) for KdV, only here the coefficients of the system are algebraic and not exponential. When $n = 1$:

$$\begin{aligned} \phi_1 &= \frac{1}{x + 2\lambda_1 t + i\gamma_{1,0}}, \\ u &= \frac{2\gamma_{1,0}}{(x - t/\gamma_{1,0})^2 + \gamma_{1,0}^2}. \end{aligned} \tag{17k}$$

vi. There exist very general direct linearizing transforms associated with the B–O equation which are analogous to those mentioned in connection with KdV, *e.g.*, (9a,b) (see also [18]). The point that we make here is that when a symmetry condition such as (17e) can not be isolated, nevertheless a linearization procedure still exists. In this case equation (17b) cannot be viewed as a RHBVP since it is *incomplete*, *i.e.*, for only two of the three functions do we have analytic or holomorphic properties. When this occurs we show[6] how the associated time operator (16) can be used to supplement the information at our disposal, *i.e.*, without analytic information on $N(x,\lambda)$ we go to the associated time equation ii above in order to develop a linearization.

[6]We have done this for B–O. For other equations the analysis is similar.

§8 Analytic problems for multidimensional systems

Next we comment upon the two-dimensional equations mentioned in (V). Since the contribution of Fokas and Ablowitz in these lectures [3] will go into more detail, we will only quote some of the results:

Associated with the KP equation we have $(V.1\text{–}L)$. Make the transformation

$$v(x,y;k) = m(x,y;k)\exp\left[i(kx + \frac{k^2}{\sigma}y)\right]. \tag{18a}$$

There are two cases:

a. **KPI** $\sigma = i$. In this case one may construct eigenfunctions which are sectionally meromorphic in the upper/lower half k-planes. The RHBVP is *nonlocal* as is the situation for B–O:

$$m_+(x,y;k) - m_-(x,y;k) = \int_{-\infty}^{\infty} dl\, m_-(x,y;l)\, f(k,l)\exp\left[i(l-k)x - i(l^2-k^2)y\right], \tag{18b}$$

where $f(k,l)$ is the scattering data which may be obtained explicitly in terms of an eigenfunction which depends on the two parameters k and l: $N(x,y;k,l)$. Schematically $M^{\pm}$ are obtained from integral equations of the form:

$$m_\pm(x,y;k) = 1 + \hat{G}_\pm(x,y;k)(um_\pm), \tag{18c}$$

and $N(x,y;k,l)$ from

$$N(x,y;k,l) = \exp\left[i(k-l)x - i(k^2-l^2)y\right] + \hat{G}_-(x,y;l)(uN). \tag{18d}$$

Underlying (18b) is the important continuous symmetry condition

$$\frac{\partial}{\partial l}\left(N(x,y;k,l)e^{i(lx-l^2y)}\right) = -\operatorname{sgn}(k-l)\, f(k,l)\, m_-(x,y;l)\exp\left[i(lx-l^2y)\right]. \tag{18e}$$

Remarks:

i. Poles of the functions $m_\pm(x,y;k)$ correspond to the lump type soliton solutions of KP, the spectral characterization of which depend on ***norming constants*** and suitable discrete eigenvalues.

ii. The time dependence of $f(k,l)$ is given by

$$f(k,l,t) = f(k,l,0)\exp\left[4i(l^3-k^3)t\right]. \tag{18f}$$

The norming constants evolve linearly in time and the discrete eigenvalues are fixed in time.

iii. There exist very general direct linearizing procedures for KP (see [9]).

b. **KPII** $\sigma = -1$. In this case the function $m(x,y;k)$ is bounded for all k and $\lim_{k\to\infty} m(x,y;k) = 1$, but is *nowhere analytic* in k. However the DBAR procedure discussed in connection with $(II.4\text{–}L)$[7] can be used. In particular in [10] we show that the structure of $\partial m/\partial\bar{k}$ is:

$$\frac{\partial m(x,y;k)}{\partial\bar{k}} = V(x,y;k_R,k_I)\, m(x,y;-\bar{k}), \tag{18g}$$

[7] See especially (13).

where $V(x,y;k_R,k_I)$ is scattering data, multiplied by a suitable exponential factor, which evolves simply in time. Equation (13) then yields an integral equation for the eigenfunctions:

$$m(x,y;k) = 1 + \frac{1}{2\pi i}\int\!\!\int_{R_\infty} dz \wedge d\bar{z}\,\frac{V(x,y;z_R,z_I)\,m(x,y;-\bar{z})}{z-k}, \tag{18h}$$

which now plays the role of the inverse problem. On the other hand (18g) is essentially a discrete symmetry condition for the function $\partial m(x,y;k)/\partial\bar{k}$.

It should be remarked that for sufficiently small initial data it can be proven that (18b) will always have a solution. In any event we do not expect homogeneous solutions in this situation.

Finally we briefly mention the analysis associated with the two-dimensional system (V.2–L). There are two important cases

i. J is a constant real diagonal matrix:

$$J = \mathrm{diag}(J_1, J_2, \ldots, J_n), \qquad J_1 > J_2 > \cdots > J_n.$$

ii. J is a constant purely imaginary diagonal matrix:

$$J = i\tilde{J} = i\,\mathrm{diag}(\tilde{J}_1, \tilde{J}_2, \ldots, \tilde{J}_n).$$

Make the transformation

$$v(x,y;k) = m(x,y;k)\exp\left[ik(Jx+y)\right], \tag{19a}$$

whereupon we find that m satisfies:

$$m_x = ik[J,m] + qm + Jm_y. \tag{19b}$$

For the case **i** there are functions which are analytic in the upper/lower half plane. Along the lines $\mathrm{Im}\,k = 0$ there is a RHBVP which is *nonlocal*. It has the form:

$$m_+(x,y;k) - m_-(x,y;k) = \int_{-\infty}^{\infty} dl\, m_-(x,y;l)\exp\left[il(Jx+y)\right] f(l,k)\exp\left[-ik(Jx+y)\right], \tag{19c}$$

where $f(l,k)$ is the scattering data, and it evolves in time when we relate (V.2–L) to a nonlinear evolution equation. For the case **ii**, the analogous function which is bounded in all space is *nowhere analytic*. In this situaton we use the same procedure as that of KPII, *i.e.*, the DBAR problem (13). It can be established that:[8]

$$\frac{\partial m(x,y;k)}{\partial\bar{k}} = \sum_{\substack{l=1\\ l\neq j}}^{n} T_{lj}\, m(x,y;k_R + i\frac{\tilde{J}_j}{\tilde{J}_l}k_I)\,\omega^{lj}, \tag{19d}$$

where $\omega^{lj} = \exp\left[i\Big(k_R(\tilde{J}_l-\tilde{J}_j)x + k_I(\tilde{J}_l-\tilde{J}_j)y/\tilde{J}_R\Big)\right]e^{lj}$, with $e^{lj}=0$ save for the lj^{th} entry in which case it is unity. The function T_{lj} plays the role of the scattering data, and evolves simply when we relate (V.2–L) to a nonlinear evolution equation. The inverse problem is complete when we use (13):

$$m(x,y;k) = I + \frac{1}{2\pi i}\int_{R_\infty}\!\!\int d\lambda\wedge d\bar{\lambda}\,\frac{1}{\lambda-k}\sum_{\substack{l=1\\ l\neq j}}^{n} T_{lj}\, m(x,y;\lambda_R + i\frac{\tilde{J}_j}{\tilde{J}_l}\lambda_I). \tag{19e}$$

This then completes the summary of our work on the direct and inverse scattering related to certain nonlinear evolution equations in one and two-spatial dimensions.

---

[8]For convenience we will assume no bound states.

This work was partially supported by the Air Force Office of Scientific Research under Grant Number 78-3674-D, the Office of Naval Research under Grant Number NOOO14-76-C-0867, and the National Science Foundation under Grant Number MCS-8202117.

References

[1] M. J. Ablowitz and H. Segur, *Solitons and the Inverse Scattering Transform.* SIAM studies in Applied Mathematics, 1981.

[2] T. R. Taha and M. J. Ablowitz: (i) On analytical and numerical aspects of certain nonlinear evolution equations. Part I: Analytical. I. N. S. #14 (preprint 1982); (ii) On analytical and numerical aspects of certain nonlinear evolution equations. Part II: numerical nonlinear Schrödinger equation. I. N. S. #15 (preprint 1982); (iii) On analytical and numerical aspects of certain nonlinear evolution equations. Part III: numerical Korteweg–de Vries equation. I. N. S. #16 (preprint 1982).

[3] A. S. Fokas and M. J. Ablowitz, Lectures on the inverse scattering transform for multidimensional $2+1$ problems. School of nonlinear phenomena, I. N. S. #28 (preprint, Dec 1982).

[4] R. Beals and R. Coifman: (i) Scattering, transformations spectrales, et équations d'évolution nonlineaire. Seminaire Goulaouic–Meyer–Schwartz, 1980-1981, exp. 22, École Polytechnique, Palaiseau; (ii) Scattering and inverse scattering for first order systems. (preprint); (iii) Scattering, transformations spectrales, et équations d'évolution Nonlineaire. Seminaire Goulaouic–Meyer–Schwartz, 1981-1982, exp. 21, École Polytechnique, Palaiseau.

[5] V. E. Zakharov and A. Shabat, Integration of thhe nonlinear equations of mathematical physics by the method of the inverse scattering problem. II. *Funct. Anal. Appl.* **13**, 166–174 (1979).

[6] R. Beals, The inverse problem for ordinary differential operators on the line. (Preprint).

[7] Y. Kodama, J. Satsuma, and M. J. Ablowitz, Nonlinear intermediate long wave equation: analysis and method of solution. *Phys. Rev. Lett.* **46**, 687–690 (1981); Y. Kodama, M. J. Ablowitz and J. Satsuma, Direct and inverse scattering problems of nonlinear intermediate long wave equations. *J. Math. Phys.* **23**, 564–576 (1982).

[8] A. S. Fokas and M. J. Ablowitz, The inverse scattering transform for the Benjamin–Ono equation —A pivot to multidimensional problems. *Stud. Appl. Math.* **68**, 1–10 (1983).

[9] A. S. Fokas and M. J. Ablowitz, On the Inverse Scattering and Direct Linearizing Transforms for the Kadomtsev–Petviashvili Equation. *Phys. Lett.* **94A**, 67–70 (1983).

[10] M. J. Ablowitz, D. Bar Yaacov, and A. S. Fokas, On the Inverse Scattering Transform for the Kadomtsev–Petviashvili Equation. I. N. S. #21, *Stud. Appl. Math.* (to be published).

[11] S. V. Manakov, The inverse scattering transform for the time-dependent Schrödinger equation and the Kadomtsev–Petviashvili equation. *Physica D* **3**, 420 (1981).

[12] (i)A. S. Fokas, On the inverse scattering of first order systems in the plane related to nonlinear multidimensional equations. I. N. S. #23 (preprint, Dec 1982); (ii) A. S. Fokas and M. J. Ablowitz, On the inverse scattering transform of multidimensional nonlinear equations related to first order systems in the plane. I. N. S. #24 (preprint, Jan 1983).

[13] D. J. Kaup, The inverse scattering solution for the full three-dimensional three-wave resonant interaction. *Physica D* **1**, 45–67 (1980).

[14] (i) N. I. Muskhelishvili, in*Singular Integral Equations.* J. Radok ed., Noordhoff, Groningen 1953; (ii) F. D. Gakhov, in *Boundary Value Problems*, I. N. Sneddon ed., Pergamon, New York, 1966, Vol. 85; (iii) N. P. Vekua, in *Systems of Singular Integral Equations*, J. H. Ferziger ed., Gordon and Breach, New York, 1967.

[15] A. S. Fokas and M. J. Ablowitz, Linearization of the Korteweg–de Vries and Painlevé II equations. *Phy. Rev. Lett.* **16**, 1096–1100 (1981).

[16] A. S. Fokas and M. J. Ablowitz, On a unified approach to transformations and elementary solutions of Pailevé equations. *J. Math. Phys.* **11**, 2033–2042 (1982).

[17] P. M. Santini, M. J. Ablowitz, and A. S. Fokas, On the limit from the intermediate long wave equation to the Benjamin–Ono equation. (Preprint, 1983).

[18] M. J. Ablowitz, A. S. Fokas, and R. L. Anderson, The direct linearizing transform and the Benjamin–Ono equation. *Phys. Lett.* **93A**, 375–378 (1983).

[19] A. S. Fokas and M. J. Ablowitz, On the initial value problem of the second Painlevé transcendent. I. N. S. #31 (preprint, 1983).

The Geometry of Complex Self-Dual Einstein Spaces

Charles P. Boyer

Instituto de Investigaciones
en Matemáticas Aplicadas y en Sistemas
Universidad Nacional Autónoma de México

Contents:

Introduction

In these lectures I would like to give a reasonably self-contained treatment of the differential geometry of complexified self-dual Einstein spaces, also known as ***H*-spaces** or ***Heavens***. I use the term ***self-dual*** not only because it has become more widely accepted, but also because, I believe, it is the more descriptive term. These lectures will focus on the interrelationship between two disparate approaches, namely, that of J. F. Plebański and collaborators [1–5], and that of R. Penrose and collaborators [6–9]. (See also [10].) A third approach due to E. T. Newman and collaborators will not be treated here, and the interested reader is referred to their comprehensive review article [11].

As the title indicates, our emphasis will be on the *geometry* of complexified self-dual Einstein spaces, and not on solution techniques —which is not to say that I believe that solution techniques are not important. On the contrary, one of the main goals of the geometric approach is to give the *"general solution"* of the self-dual Einstein equations in a form that makes further study of the properties of these spaces straightforward. For explicit solutions obtained by conventional techniques in partial differential equations, the reader is referred to [1–3, 5, 12, 13], while for those obtained by twistor methods the reader may consult [7, 14–17].

An important ingredient of these lectures —recently obtained in collaboration with J. F. Plebański [18]— is the use of an infinite hierarchy of conservation laws to deduce Penrose's curved twistor construction [6]. This result is obtained by setting up symplectic geometry on an infinite-dimensional manifold, and then realizing self-dual structures as certain maximal isotropic submanifolds. In this description, the twistor construction is equivalent to a certain twisted canonical transformation.

The methods used here offer several new insights:

1. Since we avoid the use of infinitesimal deformation theory, our methods may be more amenable to developing a global theory which would entail a study of the global behavior of the maximal isotropic submanifolds.

2. Group-theoretical methods may make it possible to generate new solutions from old ones, and

3. The appearance of an infinite number of conservation laws provides a closer connection with other nonlinear partial differential equations of mathematical physics and applied mathematics.

Indeed, we have heard from several Conference participants about similar techniques involving the soliton evolution equations from Mark Ablowitz and Athanasios Fokas, the stationary and symmetric Einstein equations from Frederick Ernst and Javier Chinea, and the self-dual Yang-Mills equations and two-dimensional chiral models from Ling-Lie Chau. A common picture seems to be emerging from these equations; namely, that an infinite hierarchy of conservation laws exists which can be used to convert the problem of solving the nonlinear partial differential equations into a Riemann-Hilbert-type problem, entailing patching together holomorphic data. In fact, the twistor construction is just this. Totally geodesic null two-surfaces play an important role in self-dual Einstein spaces. The problem of solving such differential equations becomes *encoded* in the holomorphic structure of the space of null two-surfaces. A similar result is true for the self-dual Yang-Mills equations [19–21].

Recently this procedure has been generalized to the full Yang-Mills [22–24] and supersymmetric Yang-Mills equations [22,25] by considering the space of null geodesics. An advantage of null geodesics over null geodesic surfaces is that the former exist in any space, so it appears that the general procedure works on any complexified Riemannian (local) manifold [7,26–28]. A disadvantage is that this procedure is an approximation in that it works on a finite jet, *i.e.*, up to a finite order in certain power-series expansion. Supersymmetric theories on supermanifolds appear to overcome this disadvantage [22,25]. Our approach using conservation laws may be relevant to such general theories.

It would be interesting to see how this procedure of encoding partial differential equations into the holomorphic structure of the space of null geodesics relates to solving the full Einstein equations. Perhaps a starting point would be the one-sided algebraically degenerate spaces of Plebański and Robinson [29].

Most of the material presented in these lectures was developed in collaboration with my friend Jerzy Plebański. I would like to take this opportunity to thank him. I would also like to thank Dan Finley for some stimulating discussions.

§1 A review of self-dual Einstein spaces

Let us recall the definition of complex self-dual Einstein spaces, or $\mathcal{H}$-spaces. Let M be a complex manifold with a *complex metric* g, *i.e.*, a nondegenerate holomorphic symmetric bilinear two-form of type (2,0) on the tangent space at each point of M. If M has complex dimension four, then the Levi-Civita connection on M with respect to g is torsion-free and has its values in the Lie algebra $so(4,\mathbb{C}) \simeq sl(2,\mathbb{C}) \oplus sl(2,\mathbb{C})$. Thus the corresponding curvature two-form R on M also has its values in $sl(2,\mathbb{C}) \oplus sl(2,\mathbb{C})$. Let R_+ and R_- denote the natural projections of R onto the first and second copies of $sl(2,\mathbb{C})$, respectively. Then, M is called an ***$\mathcal{H}$-space*** if either R_+ of R_- vanishes. The curvature form R is called ***self-dual*** (resp. ***anti-self-dual***) if R_- (resp. R_+) vanishes. Self-dual and anti-self-dual curvatures are equivalent under a change of orientation and we shall henceforth restrict our considerations to the case $R_- = 0$, in which case we refer to M as a self-dual manifold.

Let us look at this in more detail. A choice of local orthonormal frame sets up an isomorphism between the tangent space T_pM and a standard complex vector space W of complex dimension four. We shall use the techniques of spinor calculus [30,31]. Write $W = V_+ \otimes V_-$, where $V_\pm$ are vector spaces over $\mathbb{C}$ of complex dimension two, and endow $V_\pm$ with symplectic structures $\epsilon_\pm$. If we choose a basis e_A for V_+ and $e_{\dot A}$ for V_-, then $e_A \otimes e_{\dot A}$ is a basis for W, where $A, \dot A = 1,2$. Denote $\epsilon_{AB} = \epsilon_+(e_A, e_B)$ and $\epsilon_{\dot A \dot B} = \epsilon_-(e_{\dot A}, e_{\dot B})$ with the convention $\epsilon_{12} = -\epsilon_{21} \doteq \epsilon_{\dot 1 \dot 2} = -\epsilon_{\dot 2 \dot 1} = 1$. Elements of V_+ are called ***undotted spinors*** and those of V_-, ***dotted spinors***. We can use $\epsilon_\pm$ to identify $V_\pm$ with the dual $V_\pm^*$. In the index notation this is done by the standard raising and lowering procedure, *e.g.*, $\xi_A = \epsilon_{AB}\xi^B$ and we employ the summation convention over repeated indices.

We shall assume now that M has a spin structure and we denote by V_+M the bundle of self-dual spinors on M and by V_-M the bundle of anti-self-dual spinors. The reason for this terminology will become clear shortly. Then the holomorphic tangent bundle is the tensor product bundle $TM \simeq V_+M \otimes V_-M$. Let $\{\theta^a\}$ $a = 1,2,3,,4$, denote a local orthonormal coframe on M so that the metric is

$$g = \delta_{ab}\theta^a\theta^b,$$

where δ_{ab} is the Kronecker delta. We can write this in spinor language by defining the spinor coframe

$$\theta^{A\dot A} = \frac{1}{\sqrt{2}}\begin{pmatrix} \theta^3 + i\theta^4 & \theta^1 + i\theta^2 \\ -\theta^1 + i\theta^2 & \theta^3 - i\theta^4 \end{pmatrix},$$

so that the metric becomes

$$g = \epsilon_{AB}\epsilon_{\dot A \dot B}\theta^{A\dot A}\theta^{B\dot B} = 2\det\theta^{A\dot A}. \tag{1.1}$$

Notice also that $\delta = \epsilon_+ \otimes \epsilon_-$. This is the form of the metric which we shall use.

Let $\Lambda^2 M$ denote the bundle of two-forms on M. Since M has dimension four, the Hodge star operator $*$ maps two-forms into twwo-forms. So if we normalize so that $*^2 = 1$, we can consider the eigenspaces $\Lambda^2_{\pm}$ with eigenvalues ± 1 under $*$. Thus there is a natural splitting

$$\Lambda^2 M \simeq \Lambda^2_+ M \oplus \Lambda^2_- M. \tag{1.2}$$

Moreover,, since by using the metric, two-forms can be identified with antisymmetric endomorphisms of TM, this splitting corresponds to the Lie algebra isomorphism

$$\mathfrak{so}(4,\mathbb{C}) \simeq \mathfrak{sl}(2,\mathbb{C}) \oplus \mathfrak{sl}(2,\mathbb{C}).$$

Thus, the curvature two-form R is a section of $\Lambda^2_{\pm} M$ if and only if $R_{\mp} = 0$. It follows that an alternative definition of a self-dual manifold is that $*R = R$. Now, $\Lambda^2_+ M$ (resp. $\Lambda^2_- M$) is called the ***bundle of self-dual*** (resp. ***anti-self-dual***) ***two-forms***. Writing the tangent bundle as the tensor product of spin bundles then gives

$$\Lambda^2_{\pm} M \simeq S^2(V_{\pm} M), \tag{1.3}$$

where S^2 denotes the symmetric product.

The Cartan structure equations in spinor language become

$$\begin{aligned} d\theta^{A\dot A} + \Gamma^A{}_B \wedge \theta^{B\dot A} + \Gamma^{\dot A}{}_{\dot B} \wedge \theta^{A\dot B} &= 0, \\ d\Gamma^A{}_B + \Gamma^A{}_C \wedge \Gamma^C{}_B &= R^A{}_B, \\ d\Gamma^{\dot A}{}_{\dot B} + \Gamma^{\dot A}{}_{\dot C} \wedge \Gamma^{\dot C}{}_{\dot B} &= R^{\dot A}{}_{\dot B}. \end{aligned} \tag{1.4}$$

We identify the first copy of $\mathfrak{sl}(2,\mathbb{C})$ with S^2V_+ and the second copy with S^2V_-. The spinor coframe $\{\theta^{A\dot A}\}$ induces a local basis for $\Lambda^2 M$. In terms of the splitting (1.2) and isomorphism (1.3), we write

$$\theta^{A\dot A} \wedge \theta^{B\dot B} = \epsilon^{\dot A\dot B} S_{AB} + \epsilon^{AB} S^{\dot A\dot B},$$

so that

$$\begin{aligned} S^{AB} &= \tfrac{1}{2}\epsilon_{\dot A\dot B}\theta^{A\dot A} \wedge \theta^{B\dot B}, \\ S^{\dot A\dot B} &= \tfrac{1}{2}\epsilon_{AB}\theta^{A\dot A} \wedge \theta^{B\dot B}, \end{aligned} \tag{1.5}$$

determine a local basis for $\Lambda^2_{\pm} M$, respectively.

Decomposing the curvature, we have

$$\begin{aligned} R^A{}_B &= C^A{}_{BCD} S^{CD} + \tfrac{1}{6} R S^A{}_B + C^A{}_{B\dot C\dot D} S^{\dot C\dot D}, \\ R^{\dot A}{}_{\dot B} &= C^{\dot A}{}_{\dot B\dot C\dot D} S^{\dot C\dot D} + \tfrac{1}{6} R S^{\dot A}{}_{\dot B} + C^{\dot A}{}_{\dot B CD} S^{CD}, \end{aligned} \tag{1.6}$$

where C_{ABCD} (resp. $C_{\dot A\dot B\dot C\dot D}$) are the components of the self-dual (resp. anti-self-dual) Weyl tensor, $C_{\dot A\dot B CD} = C_{CD\dot A\dot B}$ are the components of the traceless Ricci tensor, and R is the scalar curvature. Equations (1.6) exhibit explicitly the decomposition

$$\begin{aligned} \Lambda^2 W \otimes S^2 V_{\pm} &\simeq S^2V_+ \otimes S^2V_{\pm} \oplus S^2V_- \otimes S^2V_{\pm} \\ &\simeq S^4V_{\pm} \oplus S^0V_{\pm} \oplus S^2V_{\mp} \otimes S^2V_{\pm}. \end{aligned}$$

Now suppose M is self-dual, then $R^{\dot{A}}{}_{\dot{B}} = 0$, and by a straightforward application of the Frobenius theorem to the last of Eqs. (1.4), there is a local spinor coframe $\theta^{A\dot{A}}$ such that the connection coefficients $\Gamma^{\dot{A}}{}_{\dot{B}}$ vanish. We also have from (1.6) that $C_{\dot{A}\dot{B}\dot{C}\dot{D}} = C_{\dot{A}\dot{B}CA} = R = 0$. Thus the Cartan equations reduce to

$$\begin{aligned} d\theta^{A\dot{A}} + \Gamma^{A}{}_{B} \wedge \theta^{B\dot{A}} &= 0, \\ d\Gamma^{A}{}_{B} + \Gamma^{A}{}_{C} \wedge \Gamma^{C}{}_{B} &= C^{A}{}_{BCD} S^{CD}. \end{aligned} \tag{1.7}$$

We see immediately that a self-dual manifold is Ricci-flat.

We shall assume that the group $SO(4,\mathbb{C}) \simeq SL(2,\mathbb{C}) \times SL(2,\mathbb{C})/\mathbb{Z}_2$ of the orthogonal frame bundle can be reduced to the subgroup $SL(2,\mathbb{C}) \times \{e\}/\mathbb{Z}_2$, where e is the identity in $SL(2,\mathbb{C})$. In this case we can deduce from an application of the Frobenius theorem to the first of Eqs. (1.7) that the holomorphic tangent bundle TM of M splits into the direct sum of two integrable subbundles E_1 and E_2. Let us assume for simplicity of discussion that the complementary foliations determined by E_1 and E_2 are regular, *i.e.*, there are complex manifolds M_2 and $\tilde{M}_2$ of complex dimension two and globally defined submersions

$$\begin{array}{ccc} & M & \\ {}^{p}\swarrow & & \searrow^{\tilde{p}} \\ M_2 & & \tilde{M}_2 . \end{array} \tag{1.8}$$

In this case E_1 and E_2 can be taken as $p^* T(M_2)$ and $\tilde{p}^* T(\tilde{M}_2)$, respectively.[1]

For any such local product structure (not necessarily self-dual) the leaves of each foliation are totally geodesic null surfaces with respect to the complex metric g on M. In terms of local complex coordinates $\{q^A\}$ and $\{\tilde{q}^A\}$ on M_2 and $\tilde{M}_2$, respectively, we can write (although by abuse of notation) the metric g on M locally as

$$g = 2\Phi_{AB}\, dq^A\, d\tilde{q}^B, \tag{1.9}$$

for some local holomorphic functions Φ_{AB} with $\Phi^{AB}\Phi_{AB} \neq 0$. In addition, if M is self-dual, then (1.7) can be used to deduce the existence of a local holomorphic function Ω such that

$$\Phi_{AB} = \Omega_{q^A \tilde{q}^B}, \tag{1.10}$$

where the subscripts q^A, $\tilde{q}^B$ denote partial differentiation. This equation can be used to define a nondegenerate antisymmetric two-form Ω_0 on M. In local coordinates

$$\Omega_0 = \tfrac{1}{2}\Omega_{q^A \tilde{q}^B}\, dq^A \wedge d\tilde{q}^B \tag{1.11}$$

is independent of the choice of coordinates compatible with the local product structure, and consequently defines a global nondegenerate two-form on M. Furthermore, one easily sees that Ω_0 is closed under exterior differentiation, so Ω_0 defines a complex symplectic structure on M. Now on a local product structure d splits as $d = \partial + \tilde{\partial}$; hence (1.11) can be written more succintly as

$$\Omega_0 = \tfrac{1}{2}\partial\tilde{\partial}\Omega. \tag{1.11$'$}$$

---

[1]It appears that this regularity assumption can be dropped by using symplectic vector bundles, and many of our results would still hold; however, the language would become more cumbersome.

The reader will notice the complete analogy with Kähler manifolds. In fact, the structure given by Eqs. (1.9)–(1.11) is just the complexification of a Kähler structure. Indeed, a complex manifold M with a complex metric g is called a ***complexified Kähler manifold*** (or **CK-*manifold***) if about every point there is a local function Ω and local coordinates $(q^A, \tilde{q}^A)$ such that (1.9) and (1.10) are satisfied. In particular, a self-dual manifold is a CK-manifold.

The remaining conditions for a manifold to be self-dual involve the bundle of anti-self-dual two-forms $\Lambda^2_- M$. From (1.3), $\Lambda^2_- M$ is a complex vector bundle of fiber dimension three, associated to the principal spin bundle with group $SL(2,\mathbb{C})$ (dotted indices). Since we have assumed that this group can be reduced to the identity group, it follows that the two-forms $S^{\dot{A}\dot{B}}$ of (1.5) form a global basis for $\Lambda^2_- M$. Now it can be shown [1] from (1.4) that necessary and sufficient conditions for M to be self-dual are that $S^{\dot{A}\dot{B}}$ be closed under exterior differentiation. Then with an appropriate choice of coframe we can arrange

$$S^{\dot{A}\dot{B}} = \begin{pmatrix} \omega & \Omega_0 \\ \Omega_0 & \tilde{\omega} \end{pmatrix},$$

where

$$\begin{aligned} \omega &:= \tfrac{1}{2} dq^A \wedge dq_A, \\ \tilde{\omega} &:= \tfrac{1}{2} d\tilde{q}^A \wedge d\tilde{q}_A, \end{aligned} \tag{1.13}$$

and after an appropriate normalization these three closed two-forms are seen to satisfy the quadratic relation

$$\omega \wedge \tilde{\omega} + 2\Omega_0^2 = 0. \tag{1.14}$$

This equation is the intrinsic form of Plebański's equation previously given in local coordinates [1,3,5] by

$$\tfrac{1}{2}\Omega_{q^A\tilde{q}^B}\Omega_{q_A\tilde{q}_B} = 1. \tag{1.14$'$}$$

Thus a self-dual structure on M can be described by the following data: Every point of M has a neighborhood U and a C^∞-function Ω on U such that (1.9), (1.10) and (1.14)$'$ are satisfied. We denote the set of all self-dual structures on M by $\mathcal{H}(M)$.

Notice that in order that (1.14) be intrinsic we must demand that the pseudogroup of coordinate transformations preserve the local product structure as well as the product $\omega \wedge \tilde{\omega}$. Thus the allowed coordinate transformations have the form

$$\begin{aligned} q'^A &= q'^A(q^B), \\ \tilde{q}'^A &= \tilde{q}'^A(\tilde{q}^B), \end{aligned}$$

with constant Jacobian determinants Δ and Δ^{-1}, respectively. Actually, in order to describe the infinite hierarchy intrinsically as we shall do, we shall further demand that $\Delta = 1$. Hence, ω and $\tilde{\omega}$ fix complex symplectic structures on M_2 and $\tilde{M}_2$, respectively.

Recall that the local product structure depended on a choice of dotted spinor frame. Any other such spinor frame differs from the original by a dotted $SL(2,\mathbb{C})$ gauge transformation. Since such a transformation must preserve the vanishing of the anti-self-dual components of the connection, it must be constant. Following Penrose [6], we introduce the constant spinor $\pi_{\dot{A}}$ and define

$$A = \pi_{\dot{A}}\pi_{\dot{B}} S^{\dot{A}\dot{B}}.$$

Then, (1.14) becomes

$$A^2 = 0 \qquad \text{for all spinors } \pi_{\dot A}. \tag{1.15}$$

Thus if we think of $\{\pi_{\dot A}\}$ as homogeneous coordinates on the complex projective plane $\mathbf{P}^1$, we have a projective plane's worth of closed two-forms satisfying (1.15); hence a $\mathbf{P}^1$'s worth of totally geodesic null two-surfaces passing through every point of M. Our local product structure picks out just two transversal null two-surfaces through each point; however, there is a pair of transversal totally geodesic null two-surfaces for every pair of distinct points of $\mathbf{P}^1$.

More generally, Penrose [6,7] showed that given any complex Riemannian manifold, necessary and sufficient conditions that through every point of M there pass a complex projective plane's worth of locally defined totally geodesic null two-surfaces, are that the anti-self-dual (or self-dual) components of $C_{\dot A\dot B\dot C\dot D}$ of the Weyl conformal tensor vanish. Following Atiyah, Hitchen, and Singer [20] (see also [15]), we formulate this somewhat differently. Let M be any complex Riemannian manifold and consider the anti-self-dual projective spinor bundle $PV\_M \longrightarrow M$, whose fibers are projective planes $\mathbf{P}^1$ with homogeneous coordinates $\pi_{\dot A}$ and whose total space has complex dimension five. Now let us define $\eta^A = \theta^{A\dot A}\pi_{\dot A}$, $\varsigma^A = \theta^{A\dot A}\kappa_{\dot A}$, with $\kappa^{\dot A}\pi_{\dot A} = 1$. In homogeneous coordinates the holomorphic differential of the second kind on $\mathbf{P}^1$ is $\pi^{\dot A} d\pi_{\dot A}$. The twisted one-form $\theta = \pi^{\dot A} d\pi_{\dot A} + \pi^{\dot A}\pi^{\dot B}\Gamma_{\dot A\dot B}$ is globally defined on PV_M. Under a change of frame by $(\ell, m) \in SL(2,\mathbf{C}) \times SL(2,\mathbf{C})$, θ is invariant and $\eta^A \mapsto \ell^A{}_B\eta^B$, so (η^A, θ) define a subbundle $E^* \subset T^* PV\_M$. The theorem says that E^* is integrable if and only if $C_{\dot A\dot B\dot C\dot D} = 0$. To prove this we compute the exterior derivative

$$d\eta^A = -\Gamma^A{}_B \wedge \eta^B - \theta \wedge \varsigma^A,$$

$$d\theta \equiv \pi_{\dot A}\pi_{\dot B}\pi_{\dot C}\pi_{\dot D} C_{\dot A\dot B\dot C\dot D}(\tfrac{1}{2}\varsigma^A \wedge \varsigma_A) \mod I,$$

where I denotes the ideal generated by $\{\eta^A, \theta\}$. Since this must hold for all $\pi^{\dot A}$ the result follows from Frobenius' theorem.

Thus by Frobenius' theorem there is a foliation of PV_M with two-dimensional leaves. Let us assume that this foliation is regular, that is, if $\mathcal{T}$ denotes the space of leaves, then $\mathcal{T}$ has a complex manifold structure such that the quotient projection is a holomorphic submersion[2]

$$\begin{array}{ccccc} & & PV\_M & & \\ & \overset{\alpha}{\swarrow} & & \overset{\beta}{\searrow} & \\ \mathcal{T} & & & & M. \end{array} \tag{1.16}$$

For $p \in M$, the fiber $\beta^{-1}(p)$ is a projective line $\mathbf{P}^1$ and $\alpha \circ \beta^{-1}(p)$ is thus an immersed holomorphic curve of $\mathcal{T}$. Hence, points of M correspond to holomorphic curves of $\mathcal{T}$. Conversely, if $p \in \mathcal{T}$, then $\alpha^{-1}(p)$ is a null two-surface in PV_M and this projects to a two-surface in M. Thus, points of $\mathcal{T}$ correspond to null two-surfaces of M. This is the *Penrose correspondence*.

A well known special case of (1.16) is when M is complexified, compactified Minkowski space which is the same as the complex Grassmannian of two planes in $\mathbf{C}^4$, $G(2,4)$. In this case PV_M is the complex flag manifold F_{12} consisting of the set of all nested subspaces $L_1 \subset L_2 \subset \mathbf{C}^4$, where L_j has complex dimension j. In this case, $\mathcal{T}$ is the complex projective plane $\mathbf{P}^3$ (*c.f.* Wells [31] for more details). Penrose's original point of view was to localize this and then use deformation theory [32].

[2] This assumption on the foliation places restrictions on the topology of M.

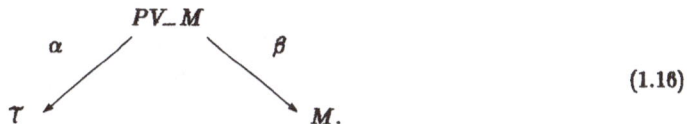

§2 A hierarchy of conservation laws

In [3], J. F. Plebański and I showed that there is a hierarchy of ***conservation laws*** associated with every self-dual manifold. This was done by rewriting the differential equation (1.14)′ as a closed differential ideal of exterior forms and then introducing new variables to prolong the ideal. In this way we obtain new closed Pfaffian one-forms whose existence are a consequence of the original ideal. Here we shall be somewhat more prosaic and briefly describe these conservation laws in terms of differential equations.

Notice that (1.14)′ can be rewritten as[3]

$$\partial_{\tilde{q}_B}(\Omega_{q_A}\Omega_{q^A\tilde{q}^B} - \tilde{q}_B) = 0, \tag{2.1a}$$

or

$$\partial_{q_B}(\Omega_{\tilde{q}_B}\Omega_{q^A\tilde{q}^B} - q_A) = 0. \tag{2.1b}$$

These equations have the form of a complexified conservation law. They are the local integrability conditions for the existence of potentials Σ and $\tilde{\Sigma}$ such that

$$\begin{aligned}\Omega_{q_A}\Omega_{q^A\tilde{q}^B} - \tilde{q}_B &= \Sigma_{\tilde{q}^B},\\ \Omega_{\tilde{q}_B}\Omega_{q^A\tilde{q}^B} - q_A &= \tilde{\Sigma}_{q^A}.\end{aligned} \tag{2.3}$$

Written in terms of an ideal of Pfaffian one-forms, these equations are equivalent to

$$\begin{aligned} d\Sigma - \tfrac{1}{2}p_A\,dp^A - s_A\,dq^A + \tfrac{1}{2}\tilde{q}_A\,d\tilde{q}^A &= 0,\\ d\Omega - p_A\,dq^A + \tilde{p}_A\,d\tilde{q}^A &= 0,\\ d\tilde{\Sigma} - \tfrac{1}{2}\tilde{p}_A\,d\tilde{p}^A - \tilde{s}_A\,d\tilde{q}^A + \tfrac{1}{2}q_A\,dq^A &= 0.\end{aligned} \tag{2.4}$$

Notice that we have introduced two new spinor variables s_A and $\tilde{s}_A$. Their meaning will become clear in a subsequent section. We also mention that we only really need the top two or bottom two equations to describe (1.14)′.

The next pair of conservation laws is not so transparent, *viz.*,

$$\begin{aligned}\partial_{\tilde{q}_A}\big(\tfrac{1}{2}\Omega_{q^B\tilde{q}^A}\Omega_{q_C q_B}\Omega_{q^C} + \Omega_{q^B\tilde{q}^A}\Sigma_{q_B}\big) &= 0,\\ \partial_{q_A}\big(\tfrac{1}{2}\Omega_{\tilde{q}^B q^A}\Omega_{\tilde{q}_C\tilde{q}_B}\Omega_{\tilde{q}^C} + \Omega_{\tilde{q}^B q^A}\tilde{\Sigma}_{\tilde{q}_B}\big) &= 0.\end{aligned} \tag{2.5}$$

These equations imply the local existence of a pair of new potentials Θ and $\tilde{\Theta}$, such that

$$\begin{aligned}\Omega_{q^B\tilde{q}^A}\big(\tfrac{1}{2}\Omega_{q^C}\Omega_{q_C q_B} + \Sigma_{q_B}\big) &= \Theta_{\tilde{q}^A},\\ \Omega_{\tilde{q}^B q^A}\big(\tfrac{1}{2}\Omega_{\tilde{q}^C}\Omega_{\tilde{q}_C\tilde{q}_B} + \tilde{\Sigma}_{q_B}\big) &= \tilde{\Theta}_{q^A}.\end{aligned} \tag{2.6}$$

If we write these equations as Pfaffians, we obtain

$$\begin{aligned} d\Theta - s_A\,dp^A - r_A\,dq^A &= 0,\\ d\tilde{\Theta} - \tilde{s}_A\,d\tilde{p}^A - \tilde{r}_A\,d\tilde{q}^A &= 0.\end{aligned} \tag{2.7}$$

---

[3]By a conservation law we mean any equation which can be written in the form $\partial_{q^A}Q^{AB_1\cdots B_k} = 0$.

Again we have introduced a pair of new spinor variables r_A and $\tilde{r}_A$. These equations suggest the use of (q^A, p^A) or $(\tilde{q}^A, \tilde{p}^A)$ as new independent variables. In fact, Plebański had already done so in [1], and showed that these new potentials must satisfy

$$\begin{aligned} &\tfrac{1}{2}\Theta_{p_A p_B}\Theta_{p^A p^B} + \Theta_{p_A q^A} = 0, \\ &\tfrac{1}{2}\tilde{\Theta}_{\tilde{p}_A \tilde{p}_B}\tilde{\Theta}_{\tilde{p}^A \tilde{p}^B} + \tilde{\Theta}_{\tilde{p}_A \tilde{q}^A} = 0. \end{aligned} \tag{2.8}$$

These equations have proved to be very efficient in obtaining explicit self-dual Einstein metrics [5].

Equations (2.8) can again be written as conservation laws:

$$\begin{aligned} &\partial_{p_A}\left(\tfrac{1}{2}\Theta_{p_B}\Theta_{p^A p^B} + \Theta_{q^A}\right) = 0, \\ &\partial_{\tilde{p}_A}\left(\tfrac{1}{2}\tilde{\Theta}_{\tilde{p}_B}\tilde{\Theta}_{\tilde{p}^A \tilde{p}^B} + \tilde{\Theta}_{\tilde{q}^A}\right) = 0, \end{aligned} \tag{2.9}$$

and thus imply the existence of new potentials Λ and $\tilde{\Lambda}$ satisfying

$$\begin{aligned} &\tfrac{1}{2}\Theta_{p_B}\Theta_{p^A p^B} + \Theta_{q^A} = \Lambda_{p^A}, \\ &\tfrac{1}{2}\tilde{\Theta}_{\tilde{p}_B}\tilde{\Theta}_{\tilde{p}^A \tilde{p}^B} + \tilde{\Theta}_{\tilde{q}^A} = \tilde{\Lambda}_{\tilde{p}^A}. \end{aligned} \tag{2.10}$$

We showed in [3] that this process continues by explicitly computing seven such pairs of conservation laws and their corresponding potentials. At that time we conjectured that this hierarchy was infinite. Recently [18], Plebański and I used the calculus of lifts to higher order tangent bundles to prove that there are indeed an infinite number of conservation laws, and then used these laws to derive Penrose's curved twistor construction. The remainder of these lectures will be devoted to a description of this process.

§3 Preliminaries on symplectic geometry

In this Section we shall collect together some well-known facts concerning symplectic geometry. Our basic references will be two articles of Weinstein [33,34] and the text of Guillemin and Sternberg [35].

As mentioned in the Introduction, we shall be interested in complex symplectic structures, so we shall work over the ground field $\mathsf{F} = \mathsf{R}$ or C. Let V be a vector space over F. A ***symplectic structure*** on V is a nondegenerate antisymmetric F-linear map $\omega : V \times V \longrightarrow \mathsf{F}$. We will also refer to ω as a ***symplectic two-form*** on V and to the pair (V, ω) as a ***symplectic vector space***. Now let M be a real C^∞-manifold (complex manifold) and ω a two-form on M. Then, the pair (M, ω) is a ***symplectic manifold*** if ω_p is a symplectic two-form on $T_p(M)$ for every $p \in M$, where $T_p(M)$ denotes the tangent space (holomorphic tangent space) at $p \in M$.

A subspace W of a symplectic vector space (V, ω) is called ***isotropic*** if $\omega(w_1, w_2) = 0$ for all $w_1, w_2 \in W$. It is called ***lagrangian*** if there is another subspace W' such that $V = W \oplus W'$. If V is finite-dimensional, then the lagrangian subspaces are just the maximal isotropic subspaces with $\dim_{\mathsf{F}} W = \frac{1}{2}\dim_{\mathsf{F}} V$. A submanifold N of a symplectic manifold (M, ω) is called ***isotropic*** (***lagrangian***) if the tangent space $T_p N$ is an isotropic (lagrangian) subspace of $T_p(M)$ for all $p \in N$.

Let $N \xrightarrow{\pi} M$ be a fibration. A submanifold $\iota : L \longrightarrow N$ is ***horizontal*** if $\pi \circ \iota$ is an embedding onto an open subset S of M [23]. If this is the case, then it can be shown that there exists a smooth section $\sigma : S \longrightarrow N$ such that $\sigma(S) = \iota(L)$. Notice that $\iota(L)$ intersects $\pi^{-1}(S)$ transversally. We shall be interested in horizontal isotropic and lagrangian submanifolds of certain fibrations. The previous result shows that such submanifolds can be given by sections of π over S.

The standard example of a symplectic manifold is the cotangent bundle T^*M of a smooth manifold M. It, however, has more structure, namely, a canonical one-form θ defined by $\sigma^*\theta = \sigma$ for any section σ of $\pi : T^*M \longrightarrow M$. The symplectic two-form ω on T^*M is given by $\omega = -d\theta$. Then, a horizontal submanifold of T^*M is lagrangian if and only if the corresponding section $\sigma : S \longrightarrow T^*M\,|_S$ is a closed one-form (*viz.*, $0 = \sigma^*\omega = -\sigma^* d\theta = -d\sigma^*\theta = -d\sigma$).

For a general symplectic manifold N, ω is not necessarily exact. However, there are a neighborhood U about each point and a one-form τ such that $\omega\,|_U = d\tau$. Now suppose $N \xrightarrow{\pi} M$ is a fibration and P is a horizontal isotropic submanifold of N. If σ is the corresponding section of N over the open subset S, then on $S \cap \pi(U)$, $\sigma^*\tau$ is a closed one-form. Thus, locally we can represent horizontal isotropic submanifolds by the graph of closed one-forms.

It is a well known result [33–35] that any symplectic manifold locally looks like a cotangent bundle (*c.f.* [34, Corollary 6.2]). It is thus expedient to introduce coordinates which reflect this local equivalence. Let (P, ω) be a symplectic manifold, M any manifold, and $U \subset P$ an open set such that $f : U \longrightarrow t^*M$ is an open embedding with $f^*d\theta = -\omega$, where θ is the canonical one-form on T^*M. Therefore, we can coordinatize U by pulling back the standard coordinates on T^*M. We refer to these coordinates on $U \subset P$ as ***cotangent coordinates***. As an example, consider the $\mathbb{C}$K-manifold M with complex symplectic form Ω_0 given by (1.11). We have the submersion $\rho : M \longrightarrow M_2$. Let q^A be coordinates on M_2 and (q^A, p_A) the standard coordinates on T^*M_2. Let $U \times \tilde{U} \subset M$ be a coordinate neighborhood in M with coordinates $(q^A, \tilde{q}^A)$ (abusing notation). We construct a diffeomorphism $f : U \times \tilde{U} \longrightarrow T^*U$ by sending $(q^A, \tilde{q}^A)$ into $\left(q^A, p_A = \frac{1}{2}\Omega_{q^A}(q, \tilde{q})\right)$ where Ω is the local function defined in (1.10). Then $f^*\Omega_0 = dq^A \wedge dp_A = -d\theta$.

To end this Section we want to mention that on any symplectic manifold (M, ω) there is a canonical isomorphism between TM and T^*M by sending the tangent vector v into the covector $\omega(v, \cdot)$. In this way we identify T^*M_2 and TM_2. Notice also that for a two-dimensional (complex) manifold, a (complex) symplectic structure on M_2 coincides with a (complex) volume structure on M_2.

§4 Higher-order tangent bundles

For any complex manifold M we can define the rth order holomorphic tangent bundle T^rM to be the set of r-jets of holomorphic curves from the origin in $\mathbb{C}$ to anywhere in M. Thus if $\psi : \mathbb{C} \longrightarrow M$ is a holomorphic curve and $\{q^A\}$ are local coordinates in M about $\psi(0)$, then writing

$$\begin{aligned} \psi^A(z) &= \sum_{k=0}^{r} p_k^A z^k + O^{r+1}(z), \\ p_k^A &= \frac{1}{k!}\frac{d^k\psi^A(z)}{dz^k}\bigg|_{z=0}, \end{aligned} \tag{4.1}$$

we determine standard coordinates $\{p_k^A\}_{k=0}^r$ in T^rM associated with the coordinates $q^A = p_0^A$ on M with the usual abuse of notation. Now, there are natural projection maps

$$\longrightarrow T^rM \xrightarrow{\pi^r_{r-1}} T^{r-1}M \longrightarrow \cdots \longrightarrow T^1M \longrightarrow T^0M, \tag{4.2}$$

where T^1M is identified with the ordinary tangent bundle TM, and T^0M with M itself. Hence, from the previous notation we have $p_1^A = p^A$. We define the infinite tangent bundle $T^\infty M$ as the projective limit of the sequence (4.2). Hence $T^\infty M$ is a closed subspace of the infinite product $\prod T^rM$ and a smooth infinite-dimensional manifold in the sense of Bernshtein and Rozenfel'd [36] (*i.e.*, locally it looks like $\mathsf{C}^\infty = \text{proj. lim.}\mathsf{C}^n$).

As an example consider $M = \mathsf{C}$, then $T^\infty\mathsf{C}$ can be identified with the ring of formal power series $\mathsf{C}[\![t]\!]$. The topology on $T^\infty\mathsf{C}$ is that induced from the infinite product. If we write a formal power series as

$$\sum_{k=0}^{\infty} a_k t^k, \qquad a_k \in \mathsf{C},$$

then neighborhoods U_j of zero take the form

$$U_j = \left\{ a_i \in \mathsf{C} : \begin{array}{c} |a_i| < \epsilon_i \quad \text{for } i = 0, 1, \ldots, j, \text{ and} \\ a_i \text{ arbitrary for } i = j+1, j+2, \ldots \end{array} \right\}.$$

In order to do calculus on T^rM we introduce the concept of lifts of functions, vector fields, forms, etc. from M to T^rM. The reader is referred to [37] for detailed proofs of statements made here. If $f : M \longrightarrow \mathsf{F}$ is a smooth function, then we define new functions $f^{(\lambda)} : T^rM \longrightarrow \mathsf{F}$, $\lambda = 0, 1, \ldots, r$ by

$$f^{(\lambda)}(j_r \circ \psi(0)) = \frac{1}{\lambda!} \frac{d^\lambda f \circ \psi}{dt^\lambda}\bigg|_{t=0}, \tag{4.3}$$

where $j_r(\psi)$ denotes the r-jet of ψ. We refer to $f^{(\lambda)}$ as the λth lift of f to T^rM. The sequence of numbers $\{f^{(0)}(j_r \circ \psi(0)), \ldots, f^{(r)}(j_r \circ \psi(0))\}$ describes the r-jet of f along ψ. Similarly, if X is a smooth vector field on M and f is a smooth function, then we define a vector field $X^{(\lambda)}$ on T^rM by

$$X^{(\lambda)} f^{(\mu)} = \begin{cases} (Xf)^{(\lambda+\mu-r)}, & \text{if } \lambda + \mu \geq r, \\ 0, & \text{if } \lambda + \mu < r. \end{cases} \tag{4.4}$$

It is easily checked that lifts are F-linear whether acting on functions or on vector fields. In addition, if ω is a smooth one-form on M, then we define a one-form $\omega^{(\lambda)}$ on T^rM by

$$\omega^{(\lambda)}(X^{(\mu)}) = \begin{cases} \omega(X)^{(\lambda+\mu-r)}, & \text{if } \lambda + \mu \geq r, \\ 0, & \text{if } \lambda + \mu < r. \end{cases} \tag{4.5}$$

We can extend this definition to the full exterior bundle ΛT^rM by the F-linearity and the formula

$$(\omega_1 \wedge \omega_2)^{(\lambda)} = \sum_{\mu=0}^{\lambda} \omega_1{}^{(\mu)} \wedge \omega_2{}^{(\lambda-\mu)}. \tag{4.6}$$

We refer to $X^{(\lambda)}$ and $\omega^{(\lambda)}$ as the λth lift of X and ω, respectively.

We mention some particularly useful properties of lifts. For the standard coordinates p_k^A on T^rM we have

$$(q^A)^{(\lambda)} = (p_0^A)^{(\lambda)} = p_\lambda^A, \qquad k = 0, 1, \ldots, r.$$

If we write a one-form ω in local coordinates with $q^A = p_0^A$,

$$\omega = \sum \omega_A \, dq^A,$$

then

$$\omega^{(\lambda)} = \sum_{\mu=0}^{\lambda} \omega_A{}^{(\mu)}\, dp^A_{\lambda-\mu}. \tag{4.7}$$

Lifts commute with exterior derivatives, in particular $(dq^A)^{(\lambda)} = dp^A_\lambda$. to make the correspondence with the notation of § 2, we notice that

$$p^A_0 = q^A,\ p^A_1 = p^A,\ p^A_2 = s^A,\ p^A_3 = r^A,$$

and similarly for $\tilde{p}^A_k$.

From the lifts one can reconstruct a particular representative of the equivalence class of r-jets. For example, if $f \in C^\infty M$ and $\psi : \mathsf{F} \longrightarrow M$ is a curve, then we can lift ψ to $T^r M$ and

$$\sum_{k=0}^{r} f^{(k)}(j_r \circ \psi(0))\, t^k$$

represents the r-jet $j_r f$ of f along ψ. In particular, we will be interested in the infinite jets of functions, vector fields, differential forms, and maps. They are defined to be the projective limits of the correponding finite jets. This gives a map from *objects* on M to *objects* on $T^\infty M$.

We shall make use of the following result whose proof is easy and thus omitted:

Proposition 1. *Let (M,ω) be a symplectic manifold, then $(T^r M,\omega^{(r)})$ is a symplectic manifold.*

§5 Self-dual Einstein spaces as isotropic and lagrangian submanifolds

In this Section we shall represent self-dual Einstein spaces as certain lagrangian and isotropic submanifolds of higher–order tangent bundles.

In what follows I assume that M is a regular local product manifold as in (1.8), and that M_2 and $\tilde{M}_2$ are symplectic manifolds with symplectic two-forms ω and $\tilde{\omega}$ respectively.[4] Then, M is a symplectic manifold with symplectic form $\rho^*\omega - \tilde{\rho}^*\tilde{\omega}$. The diagram (1.8) induces a splitting of the higher order tangent bundles, *viz.*,

$$T^r M = \rho^* T^r M_2 \oplus \tilde{\rho}^* T^r \tilde{M}_2. \tag{5.1}$$

By Proposition 1, $T^r M_2$ and $T^r \tilde{M}_2$ are symplectic manifolds with symplectic two-forms $\omega^{(r)}$ and $\tilde{\omega}^{(r)}$ respectively. Thus $T^r M$ is a symplectic manifold with symplectic form $\rho^*\omega^{(r)} - \tilde{\rho}^*\tilde{\omega}^{(r)}$.

For $r = 2$ we are interested in a slightly twisted situation. Consider the fibration $\rho^* T^2 M_2 \longrightarrow M$. The fiber dimension is four, so $\rho^* T^2 M_2$ is an eight-dimensional complex manifold. Furthermore, it is symplectic, with symplectic form $\rho^*\omega^{(2)} - \tilde{\rho}^*\tilde{\omega}^{(0)}$. From now on, for notational convenience, we will drop the pullback notation on forms and just write $\omega^{(2)}$ instead of $\rho^*\omega^{(2)}$. Similarly, $\omega^{(0)} - \tilde{\omega}^{(2)}$ is a symplectic form on $\tilde{\rho}^* T^2 \tilde{M}_2$. We also have the fibration $TM \longrightarrow M$ with symplectic form $\omega^{(1)} - \tilde{\omega}^{(1)}$. As mentioned previously, we identify $T^* M$ and TM. Under this identification, horizontal lagrangian submanifolds of TM represented by divergence-free vector fields can be also represented by closed one-forms.

Let us consider $T^2 M$. We have the following fibrations:

[4] Hereafter a self-dual structure will imply ω and $\tilde{\omega}$ fixed.

$$\begin{array}{ccccc} & & T^2M & & \\ & \overset{\tilde{\rho}^2}{\swarrow} & \downarrow \pi_1^2 & \overset{\rho^2}{\searrow} & \\ \tilde{\rho}^*\, T^2\tilde{M}_2 & & TM & & \rho^*\, T^2M_2. \end{array} \tag{5.2}$$

In all of the following considerations we will restrict ourselves to the right portion of this diagram, that is, to the fibration given by π_1^2 and ρ^2. It can be shown that all results could be equivalently formulated in terms of the fibrations π_1^2 and $\tilde{\rho}^2$.

Let N be a horizontal submanifold of $T^2M \longrightarrow M$. As discussed in § 3, any such submanifold can be equivalently given in terms of holomorphic sections $\sigma : S \longrightarrow T^2M$ over the open set $S \subset M$ determined by $\iota(N) = \sigma(S)$, where ι denotes the natural injection $\iota : N \longrightarrow T^2M$. Notice that N can be considered as a horizontal submanifold of both TM and $\rho^*\, T^2M_2$ by the composition maps $\pi_1^2 \circ \iota$ and $\rho^2 \circ \iota$, respectively. Furthermore, owing to the structure of T^2M, there are holomorphic sections σ_1 and σ_2 of TM and $\rho^*\, T^2M_2$, respectively, such that $\pi_1^2 \circ \iota(N) = \sigma_1(S)$ and $\rho^2 \circ \iota(N) = \sigma_2(S)$, or equivalently $\sigma_1 = \pi_1^2 \circ \sigma$ and $\sigma_2 = \rho^2 \circ \sigma$. Let $\mathcal{L}(S)$ denote the set of all holomorphic sections of T^2M such that $\pi_1^2 \circ \iota(N)$ and $\rho^2 \circ \iota(N)$ are *horizontal lagrangian submanifolds* of TM and $\rho^*\, T^2M_2$, respectively. Equivalently, $\mathcal{L}(S)$ is defined by the equations

$$\begin{aligned} \sigma_1^*(\omega^{(1)} - \tilde{\omega}^{(1)}) &= 0, \\ \sigma_2^*(\omega^{(2)} - \tilde{\omega}^{(0)}) &= 0. \end{aligned} \tag{5.3}$$

Let us see how an element $\sigma \in \mathcal{L}(S)$ determines a self-dual structure on S, that is, an element of $\mathcal{H}(S)$. By assumption M, and therefore S has a regular local product structure with (1.8) satisfied. Furthermore, (M_2, ω) and $(\tilde{M}_2, \tilde{\omega})$ are symplectic manifolds. Define a CK-structure on S by setting

$$\Omega_0 = \tfrac{1}{2}\sigma_1^*\, \omega^{(1)}. \tag{5.4}$$

This is clearly a closed two-form. Moreover, if θ_1 denotes the canonical one-fform on $\rho^*\, TM_3 \simeq \rho^*\, T^*\, M_2$, then $\sigma_1^*\, \omega^{(1)} = -\sigma_1^*\, d\theta_1 = -d\sigma_1$, and from our example at the end of § 3, it follows that Ω_0 locally has the form (1.11). It remains to show that Ω_0 thus defined has rank four. Let us assume this to be the case for the moment and inquire about the consequences of the second equations (5.3). It is clear that $\rho^*\, \omega - \tilde{\rho}^*\, \tilde{\omega}$ restricted to S defines a complex symplectic structure on S. We denote this two-form simply by $\omega - \tilde{\omega}$. We need to show that (1.14) is satisfied. Consider

$$0 = (\omega \wedge \omega)^{(2)} = 2\omega^{(2)} \wedge \omega^{(0)} + \omega^{(1)} \wedge \omega^{(1)}$$

as a two-form identity on T^2M. (We are abusing notation slightly by writing $\omega^{(1)}$ instead of $\pi_1^{2*}\, \omega^{(1)}$, but this should cause no confusion.) Now on T^2M the second of equations (5.3) implies

$$\sigma^*\, \omega^{(2)} = \tilde{\omega},$$

and from the above identity we have

$$\omega \wedge \tilde{\omega} = -\tfrac{1}{2}\sigma^*\, (\omega^{(1)} \wedge \omega^{(1)}) = -2\Omega_0^2,$$

verifying (1.14). It remains to show that if (5.3) holds, $\sigma^*\, \omega^{(1)}$ necessarily has rank four. If the rank of $\sigma^*\, \omega^{(1)}$ were less than four, then we would have $\omega^{(1)} \wedge \omega^{(1)} = 0$. But (5.3) would imply $\omega \wedge \tilde{\omega} = 0$, which contradicts the fact that M is symplectic.

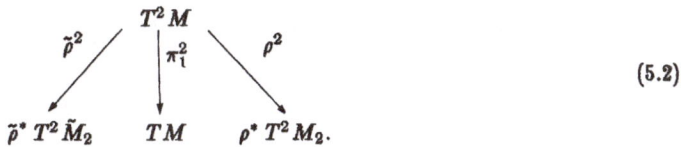

Conversely, suppose we are given a self-dual structure on S. Then S has a CK-structure with Ω_0 satisfying (1.14). Every one-form on S defines a horizontal submanifold of TM. Now suppose that the Dolbeault cohomology class $[\Omega_0] = 0$ in $H^2(S)$, then $\Omega_0 = -\frac{1}{2}d\sigma$ for some holomorphic one-form σ on S. Furthermore, this implies that if θ is the canonical one-form on $T^* M_2$ pulled back to $T^* M$, then $\sigma^* \theta = \theta$. But then $\Omega_0 = \frac{1}{2}\sigma^* \omega^{(1)}$, and the first of equations (5.3) is satisfied. Now extend σ to a section of $T^2 M$, then, from (1.14),

$$\sigma^* (\omega^{(2)} - \tilde{\omega}^{(0)}) \wedge \omega = 0.$$

Choose σ along the fibers of $T^2 M \longrightarrow TM$ such that the second of Eqs. (5.3) is satisfied. We have arrived at

Theorem 1. *Let $\iota : N \longrightarrow T^2 M$ be a horizontal submanifold with $\pi \circ \iota(N) = S$ and let $\sigma : S \longrightarrow T^2 M$ be its corresponding section. If $\pi_1^2 \circ \iota(N)$ and $\rho^2 \circ \iota(N)$ are lagrangian submanifolds of TM and $\rho^* T^2 M_2$, respectively, then S with the CK form defined by (5.4) is a self-dual manifold. Conversely, suppose Ω_0 defines a self-dual structure on S and that the cohomology class $[\Omega_0]$ vanishes, then there is a choice of σ along the fibers of $T^2 M \longrightarrow TM$ such that (5.3) are satisfied and Ω_0 is given by (5.4).*

Remark. This theorem can be globalized to the case $[\Omega_0] \neq 0$ by considering locally defined sections $\sigma_\alpha : U_\alpha \longrightarrow T^* M \simeq TM$ and then constructing global equivalence classes $[\sigma_\alpha]$, where $\sigma_\alpha \sim \sigma_\beta$ if on $U_\alpha \cap U_\beta$, $\sigma_\alpha = \sigma_\beta + dF_{\alpha\beta}$, where $\partial\tilde{\partial} F_{\alpha\beta} = 0$. Hereafter, we assume $[\Omega_0] = 0$.

We now consider the higher order structures. We shall be interested in the bundle

$$\rho^* T^r M_2 \longrightarrow M$$

for $r \geq 3$. By Proposition 1, $\omega^{(r)}$ defines a symplectic structure on $T^r M$, thus pulled back to $\rho^* T^r M_2$, $\omega^{(r)}$ is a presymplectic form (degenerate two-form). We call a holomorphic section $\sigma : S \longrightarrow T^r M |_S$ *$\mathcal{H}$-admissible* if the appropriate projections annihilate $\omega^{(1)} - \tilde{\omega}^{(1)}$ and $\omega^{(2)} - \tilde{\omega}^{(0)}$, *i.e.*, if σ generates a self-dual structure on S. We have

Theorem 2. *Let $\sigma : S \longrightarrow T^r S$ be an $\mathcal{H}$-admissible holomorphic section. Then, there is a choice of σ along the fibers of $T^r S \longrightarrow T^2 S$ such that $\sigma^* \omega^{(k)} = 0$ for all $k = 3, 4, \ldots, r$ and for all integers r.*

Proof. By induction on k, consider $k = 3$, and consider the kth lift of $\omega \wedge \omega$. By (4.6),

$$(\omega \wedge \omega)^{(3)} = 2[\omega^{(3)} \wedge \omega^{(0)} + \omega^{(2)} \wedge \omega^{(1)}].$$

We have, using Theorem 1,

$$\begin{aligned} \sigma^* (\omega^{(3)} \wedge \omega^{(0)}) &= -\sigma^* (\omega^{(2)} \wedge \omega^{(1)}) \\ &= \sigma^* \tilde{\omega}^{(0)} \wedge \sigma^* \tilde{\omega}^{(1)} \\ &= \sigma^* (\tilde{\omega}^{(0)} \wedge \tilde{\omega}^{(1)}) = 0. \end{aligned}$$

So we can choose σ along the fibers of $T^3 S \longrightarrow T^2 S$ such that $\sigma^* \omega^{(3)} = 0$. Now assume that we can choose a σ such that $\sigma^* \omega^{(j)} = 0$ for all $j = 3, 4, \ldots, k$. Then again by (4.6),

$$0 = (\omega \wedge \omega)^{(k+1)} = \sum_{j=0}^{k+1} \omega^{(j)} \wedge \omega^{(k+1-j)}.$$

Applying σ^* and using the induction hypothesis this implies $2\sigma^* (\omega^{(k+1)} \wedge \omega^{(0)}) = 0$ if $k > 3$. But if $k = 3$, then we have $2\sigma^* (\omega^{(4)} \wedge \omega^{(0)}) + \sigma^* (\omega^{(2)} \wedge \omega^{(2)}) = 0$. But again $\sigma^* (\omega^{(2)} \wedge \omega^{(2)}) = \tilde{\omega} \wedge \tilde{\omega} = 0$. Hence, by a choice of σ along the fibers of $T^{k+1} S \longrightarrow T^k S$, we have $\sigma^* \omega^{(k+1)} = 0$ for all $k \geq 3$. The same argument clearly works for $\tilde{\omega}^{(k)}$. *Q. E. D.*

We also have the following converse:

Proposition 2. *Let $r \geq 4$. If a section $\sigma : S \longrightarrow T^r S$ annihilates $\omega^{(3)}$ and $\omega^{(4)}$, then it is $\mathcal{H}$-admissible.*

Let us now define a holomorphic family of two-forms on $T^r M$ as follows:

$$\omega^r(t) = \sum_{k=0}^{r} \left[-\tilde{\omega}^{(k)}\, t^{-(k-1)} + \omega^{(k)}\, t^{(k-1)} \right], \tag{5.4}$$

for $t \in \mathbf{C} - \{0\}$.

Proposition 3. *For all $t \in \mathbf{C} - \{0\}$, and all $r = 0, 1, \ldots$ $\omega^r(t)$ defines a symplectic structure on $T^r M$.*

Proof. The tangent space to $T^r M$ at a point is spanned by $\{\partial_{p_k^A}, \partial_{\tilde{p}_k^A}\}$, $A = 1, 2$, $k = 0, 1, \ldots, r$. Since $\omega(t)$ is the sum of a two-form on $T^r M_2$ and a two-form on $T^r M_2$, and there is a symmetry $\tilde{\omega}(1/t) = \omega(t)$, it suffices to show that

$$\omega_2(t) = \sum_{k=0}^{r} \omega^{(k)}\, t^{(k-1)}$$

defines a symplectic structure on $T^r M_2$. We must show that $\omega_2(v, \partial_{p_k^A}) = 0$ for all $A = 1, 2$, $k = 0, 1, \ldots, r$ implies $v = 0$. Now $\omega_2(v, \partial_{p_k^A}) = 0$ implies the polynomial equations

$$\sum_{k=0}^{r} \omega^{(k)}(v, \partial_{p_j^A})\, t^k = 0.$$

Notice that for $j = r$ the equation is simply

$$\omega^{(r)}(v, \partial_{p_r^A}) = 0,$$

since $\omega^{(k)}(v, \partial_{p_j^A}) = 0$ for all $j > k$. This equation implies by (4.7) that v is tangent to fibers of the fibration $T^r M \longrightarrow M$. If we can show that v must be tangent to the fibers of the fibration $T^r M \longrightarrow T^s M$ for any $s = 0, 1, \ldots, r$, then, by taking $s = r$ we must have $v = 0$ and we are done. It thus suffices by induction to prove that if v is tangent to the fibers of $T^r M \longrightarrow T^s M$ it is necessarily tangent to the fibers of $T^r M \longrightarrow T^{s+1} M$. Consider

$$\sum_{k=0}^{r} \omega^{(k)}\, (v, \partial_{p_{r-s-1}^A})\, t^k = 0.$$

Now $\omega^{(k)}\, (v, \partial_{p_{r-s-1}^A})\, t^k = 0$ for all $k < r - s - 1$, and from (4.6) and (4.7), $\partial_{p_{r-s-1}^A}$ is paired in $\omega^{(k)}$ only to $dp_{A,k-r+s+1}$. But by the induction hypothesis, v is spanned by $\{\partial_{p_{s+1}^A}, \ldots, \partial_{p_r^A}\}$. Thus we must have

$$\omega^{(r)}\, (v, \partial_{p_{r-s-1}^A}) = 0;$$

that is, v is tangent to the fibers of $T^r M \longrightarrow T^{s+1} M$. *Q. E. D.*

We now have an easy corollary of Theorems 1 and 2:

Corollary. *Suppose $\sigma : S \longrightarrow T^r M$ is a holomorphic section and that there is a neighborhood of zero $I \subset \mathbf{C}$ such that $\sigma^* \omega^r(t) = 0$ for all $t \in I - \{0\}$ and any $r \geq 2$, then $\Omega_0 = \frac{1}{2}\sigma^* \omega^{(1)}$ defines a self-dual structure on S. Conversely, let Ω_0 define a self-dual structure on S, then there are a choice of holomorphic section σ and a neighborhood I of $0 \in \mathbf{C}$ such that $\sigma^* \omega^r(t) = 0$ for all $t \in I - \{0\}$ and all $r \geq 1$.*

In summary, we have constructed a holomorphic family of symplectic structures on each of the higher-order tangent bundles $T^r M$. Fix $t \in \mathbf{C} - \{0\}$ and let N be a horizontal isotropic submanifold of $T^r M$ with respect to $\omega^r(t)$. If we now allow t to vary, the horizontal submanifold may no longer be isotropic with repect to the new symplectic structure. The self-dual spaces are precisely those horizontal submanifolds which remain isotropic under such a deformation of the symplectic structure. We will also find it convenient to view the holomorphic family of two-forms $\omega^r(t)$ as a *generating two-form* for the hierarchy of two-forms $\omega^{(1)} - \tilde{\omega}^{(1)}$, $\omega^{(2)} - \tilde{\omega}^{(0)}$, $\omega^{(3)}$, $\omega^{(4)}$, ... plus their tilded counterparts.

§6 Symplectic geometry on the space of holomorphic curves

The space $T^\infty M$ can be though of as the space of parametrized formal curves on M in the sense of formal power series. The splitting (5.1) for each r induces aa splitting in the projective limit

$$T^\infty M = \rho^* T^\infty M_2 \oplus \tilde{\rho}^* T^\infty \tilde{M}_2. \tag{6.1}$$

Formal curves ψ on M_2 and $\tilde{\psi}$ on $\tilde{M}_2$ pull back to give a formal curve $(\psi, \tilde{\psi})$ on M.

Now on the projective limit $T^\infty M$ smooth or holomorphic p-forms are in a neighborhood $U \subset T^\infty M$ the pullbacks of p-forms defined on the finite jets $T^r M$ for some r. Thus, for example, a holomorphic or smooth two-form can never define a symplectic structure on $T^\infty M$. However, if (M, ω) is a symplectic manifold, Proposition 1 defines a symplectic structure on $T^r M$ for each r with two-form $\omega^{(r)}$. Thus, we can form the formal sum

$$\omega(t) = \sum_{p=0}^{\infty} \pi_k^* \omega^{(k)} t^k, \tag{6.2}$$

where $\pi_k : T^\infty M \longrightarrow T^k M$ is the natural projection onto the kth order tangent bundle. We can think of such a formal sum as a ***formal symplectic two-form*** on $T^\infty M$. This will provide $T^\infty M$ with a ***formal symplectic structure.***

Actually, we want a slightly more general situation. Let us consider antisymmetric bilinear maps from $T_p T^\infty M \times T_p T^\infty M$ into the ring of formal Laurent series $\mathbf{C}[[t, t^{-1}]]$. Alternatively, if $\Lambda^2 T^\infty M^*$ denotes the bundle of holomorphic two-forms on $T^\infty M$, then we consider the bundle $\Lambda^2 T^\infty M^* \otimes \mathbf{C}[[t, t^{-1}]]$. Sections of this bundle are referred to as ***formal two-forms*** on $T^\infty M$. We will say that a formal two-form ω is ***nondegenerate at*** $p \in T^\infty M$ if $\omega(u, v) = 0$ for all $v \in T_p T^\infty M$ implies $u = 0$. A formal two-form which is nondegenerate at every point $p \in T^\infty M$ is said to be ***nondegenerate.*** A pair $(T^\infty M, \omega)$, where ω is a holomorphic section of $\Lambda^2 T^\infty M^* \otimes \mathbf{C}[[t, t^{-1}]]$ is a closed nondegenerate formal two-form, is called a ***formal symplectic manifold.***

Let us now consider the formal sum

$$\omega(t) = \lim_{r \to \infty} \pi_r^* \omega^r(t) := \sum_{k=0}^{\infty} \left[\pi_k^* \omega^{(k)} t^{k-1} - \pi_k^* \tilde{\omega}^{(k)} t^{-k+1} \right]. \tag{6.3}$$

Proposition 4. *With ω defined by (6.2), $(T^\infty M, \omega)$ is a formal symplectic manifold.*

Proof. We only check the nondegeneracy. Notice that $\Lambda^2 T^\infty M^* \otimes \mathbf{C}[\![t, t^{-1}]\!]$ has a natural grading inherited from the natural grading of $\mathbf{C}[\![t, t^{-1}]\!]$. For $k \in \mathbf{Z}$, let ω_k denote the coefficient in $\Lambda^2 T^\infty M$ of the Laurent series of $\omega(t)$, then it follows from (6.3) that

$$\pi^*_{r+1}\, \omega^{r+1}(t) = \sum_{k=-r-1}^{r+1} \omega_k\, t^k. \tag{6.4}$$

Now, if $\omega(t)(u, v) = 0$ for all $v \in T_p T^\infty M$, then $\omega_k(u, v) = 0$ for all v and k, and (6.4) implies $\omega^r(t)(\pi_* u, \pi_* v) = 0$. Since π_{r*} is surjective and by Proposition 3 $\omega^r(t)$ is nondegenerate for all r, we have that $\pi_{r*} u = 0$ for each positive integer r. This implies that $u = 0$. *Q. E. D.*

We now have the following result as an immediate consequence of Theorems 1 and 2, and Proposition 4:

Theorem 3. *Consider the formal symplectic manifold $\big(T^\infty M, \omega(t)\big)$ with $\omega(t)$ given by (6.2). Let σ be a holomorphic section of $T^\infty M$ over the open submanifold $S \subset M$. Then, $\Omega_0 = \frac{1}{2}\sigma_1^*\, \omega^{(1)}$ defines a self-dual structure on S if and only if there is a choice of σ along the fibers of $T^\infty M \longrightarrow TM$ such that $\sigma^*\, \omega(t) = 0$.*

Let $\mathcal{L}(S)$ denote the subspace of formal holomorphic sections σ of $T^\infty M$ over S which satisfy $\sigma^*\, \omega(t) = 0$, *i.e.*, the space of horizontal maximal isotropic submanifolds[5] Usually such submanifolds are represented locally by local generating functions. This holds also in our infinite dimensional case if we allow the generating functions to have values in $\mathbf{C}[\![t, t^{-1}]\!]$. Before discussing local generating functions we briefly consider some cohomology on $T^\infty M$. Let $H^i(T^\infty M)$ denote the Dolbeault cohomology groups on $T^\infty M$ [36], and $H^i(M)$ these groups on M. From the fact that lifts are intrinsic one verifies

Proposition 5. *Suppose the Dolbeault cohomology classes $[\omega]$ and $[\tilde{\omega}]$ vanish in $H^2(M)$, then $[\omega(t)]$ vanishes in $H^2(T^\infty M)$.*

Hereafter we assume that $[\omega] = [\tilde{\omega}] = 0$. Thus there is a section $\tau(t)$ of $\Lambda^1(T^\infty M) \otimes \mathbf{C}[\![t, t^{-1}]\!]$ such that $\omega(t) = d\tau(t)$. Of course, $\tau(t)$ is not unique, for the substitution $\tau(t) \longrightarrow \tau(t) + d\lambda(t)$ for some $\lambda \in C^\infty(T^\infty M) \otimes \mathbf{C}[\![t, t^{-1}]\!]$ does not alter $\omega(t)$. A section $\sigma : S \longrightarrow T^\infty S$ is in $\mathcal{L}(S)$ if and only if the holomorphic section $\sigma^*\, \tau(t)$ of $\Lambda^1(S) \otimes \mathbf{C}[\![t, t^{-1}]\!]$ is closed. Assuming that S is such that $[\sigma^*\, \tau(t)] \in H^1(S)$ vanishes, we obtain a *generating function* $\Phi(t) \in C^\infty(S) \otimes \mathbf{C}[\![t, t^{-1}]\!]$ such that

$$\sigma^*\, \tau(t) = d\Phi(t). \tag{6.5}$$

With an appropiate choice of $\tau(t)$, the coefficients of t^0, t^{-1}, t^1, t^{-2}, t^2 in this equation are precisely equations (2.4) and (2.7).

Now consider the set LocDiff$T^\infty M$ of local diffeomorphisms of $T^\infty M$. This forms a pseudogroup on $T^\infty M$. We define the subpseudogroup ConSym$T^\infty M$ of ***conformal symplectic local diffeomorphisms*** to be the set of all $\phi \in$ LocDiff$T^\infty M$ which satisfy

$$\phi^*\, \omega(t) = \oint_C M(t, s)\, \omega(s)\, ds,$$

[5]For formal symplectic manifolds, maximal isotropic does not imply lagrangian. These maximal isotropic submanifolds have dimension four, yet lagrangian submanifolds are infinite-dimensional.

where C is any closed contour containing the origin and $M(t,s)$ is a formal Laurent series of holomorphic functions on $T^\infty M$ which satisfy the closure condition

$$\oint_C dM(t,s) \wedge \omega(s) = 0.$$

Here the integral is to be understood formally. It is easily seen that ConSym$T^\infty M\mid_S$ acts on $\mathcal{L}(S)$. It is conjectuired that ConSym$T^\infty S$ acts transitively there. If this were to be the case, then any two self-dual structures on S could be obtained one from the other by a member of ConSym$T^\infty M$. Work on this problem is currently in progress.

§7 The curved twistor construction

Our approach to the twistor construction is based on the classical symplectic fact that horizontal lagrangian submanifolds of a product of symplectic manifolds $M \times M$ can be identified locally with the graph of canonical transformations. Suppose (M,ω) is symplectic, then, so is (M,ω^-), where $\omega := -\omega$. Let $\pi_i : M \times M \longrightarrow M$ denote the ith factor, $i = 1,2$, then $(M \times M, \pi_1^*\omega + \pi_2^*\omega^-)$ is a symplectic manifold. Furthermore, a local diffeomorphism $\phi : M \longrightarrow M$ is a canonical transformation, *i.e.*, $\phi^*\omega = \omega$ if and only if

$$(\mathrm{gr}\phi)^*(\pi_1^*\omega + \pi_2^*\omega^-) = 0,$$

that is, the graph of ϕ is a horizontal lagrangian submanifold.

Now consider the fomal symplectic manifold $(T^\infty M, \omega(t))$. Suppose M_2 and $\tilde{M}_2$ are diffeomorphic and $M = M_2 \times \tilde{M}_2$ so that $T^\infty M \simeq T^\infty M_2 \times T^\infty M_2$. The formal symplectic two-form $\omega_2(t)$ on $T^\infty M_2$ can be viewed as a presymplectc form on $T^\infty M_2 \times \mathbf{C}^*$, where we write $\mathbf{C}^* = \mathbf{C} - \{0\}$. Thus we consider holomorphic maps $\hat{F} : T^\infty M_2 \times \mathbf{C}^* \longrightarrow T^\infty M_2 \times \mathbf{C}^*$ of the form $\hat{F} = (F, I)$, where $I : \mathbf{C}^* \longrightarrow \mathbf{C}^*$ is defined by $I(t) = t^{-1}$ and F satisfies

$$F^*\omega_2(t^{-1}) = t^{-2}\omega_2(t). \tag{7.1}$$

It follows from similar reasoning as above that locally gr$F$ can be identified with a local holomorphic section $\sigma : M \longrightarrow T^\infty M$ which annihilates $\omega(t)$.

In order to proceed further we will briefly recall [38] some basic geometric facts concerning line bundles over the complex projective plane $\mathbf{P}^1$. Since $\mathbf{P}^1$ is the space of complex lines through the origin in $\mathbf{C}^2$, there is a natural line bundle over $\mathbf{P}^1$ obtained by associating to each point $p \in \mathbf{P}^1$ the line it describes in $\mathbf{C}^2$. This is called the ***universal*** or ***tautological bundle*** on $\mathbf{P}^1$. We will denote it by $O(-1)$.[6] Coordinates (z_1, z_2) on $\mathbf{C}^2$ are called homogeneous coordinates on $\mathbf{P}^1$. They are to be considered as holomorphic sections of the dual bundle $O(1)$ to $O(-1)$. $O(1)$ is called the ***hyperplane bundle*** on $\mathbf{P}^1$. We form new line bundles $O(n)$ by taking tensor products, *i.e.*, $O(2) = O(1) \otimes O(1)$. If $O(0)$ denotes the trivial bundle on $\mathbf{P}^1 \times \mathbf{C}$, then the set of line bundles $\{O(n) : n \in \mathbf{Z}\}$ on $\mathbf{P}^1$ forms an abelian group (called the *Picard* group) under tensor product. Global holomorphic sections of $O(n)$ exist only if $n \geq 0$. For $n = 0$ they are the constant functions;[7] for $n > 0$ they are the homogeneous polynomials of degree n. We remark that if $M \xrightarrow{\nu} \mathbf{P}^1$ is a holomorphic submersion of a complex manifold M onto $\mathbf{P}^1$, then there are naturally defined bundles $\nu^* O(n)$ on M.

---

[6] More precisely, this notation is commonly used to denote the sheaf of germs of sections of the universal bundle.

[7] This is Liouville's theorem.

Let us now construct certain three-dimensional complex manifolds $\mathcal{T}$ as follows: Cover $\mathcal{T}$ with two coordinate neighborhoods $\mathcal{N}$ and $\tilde{\mathcal{N}}$ with holomorphic coordinates (t, z^A) and $(s, \tilde{z}^A)$, respectively. On $\mathcal{N} \cap \tilde{\mathcal{N}}$ define the transition functions

$$\begin{aligned} s &= t^{-1}, \\ \tilde{z}^A &= F^A(z^B, t), \end{aligned} \tag{7.2}$$

where F^A are holomorphic functions. We thus obtain a holomorphic foliation of $\mathcal{T}$ which we assume to be regular, *i.e.*, there is a global holomorphic submersion $\nu : \mathcal{T} \longrightarrow \mathsf{P}^1$. In order to make a connection with Eq. (7.1) we need to give $\mathcal{T}$ more structure. To do so consider the natural injection $\nu^* : T^*_{\nu(p)}\mathsf{P}^1 \longrightarrow T^*_p\mathcal{T}$ for $p \in \mathcal{T}$, and let Q^* denote the quotient bundle on $\mathcal{T}$. We require that on $\mathcal{T}$ there exist a certain global twisted closed two-form, more precisely, a global section of $\Lambda^2 Q^* \otimes \nu^* O(2)$ which is closed under exterior differentiation. This twisted two-form is denoted by μ. On $\mathcal{N}$ in local coordinates,

$$\mu = \tfrac{1}{2} dz^A \wedge dz_A \qquad \text{on } \mathcal{N}. \tag{7.3}$$

The global existence of μ then implies that the transition functions must satisfy

$$\tfrac{1}{2}\frac{\partial F^A}{\partial z^B}\frac{\partial F_A}{\partial z_B} = t^{-2}. \tag{7.4}$$

To make contact with Eq. (7.1), consider a local holomorphic section from $U_0 := \nu(\mathcal{N}) \subset \mathsf{C} \longrightarrow \mathcal{N}$ given by $t \mapsto \big(t, z^A = \psi^A(t)\big)$, and similarly a local holomorphic section $s \mapsto \big(s, \tilde{z}^A = \tilde{\psi}^A(s)\big)$ from $U_\infty := \nu(\tilde{\mathcal{N}})$ to $\tilde{\mathcal{N}}$. In $U_0 \cap U_\infty$ we have $\tilde{\psi}^A(t^{-1}) = F^A\big(\psi^B(t), t\big)$, where F^A satisfies (7.4). Thus F determines a global holomorphic section ψ of $\nu : \mathcal{T} \longrightarrow \mathsf{P}^1$. Denote by $\Gamma(\mathcal{T})$ the set of global holomorphic sections whose normal subbundle is isomorphic to $O(1) \oplus O(1)$, *i.e.*, compact holomorphic curves in $\mathcal{T}$. This determines an element of $\mathcal{L}(S)$ as follows: If we evaluate μ on $\psi \in \Gamma(\mathcal{T})$ we have on $\mathcal{N}$:

$$\mu \,|_{\psi(t)} = \tfrac{1}{2} d\psi^A(t) \wedge d\psi_A(t) = \omega_2(t).$$

Then (7.4) implies (7.1) and we take $\sigma = \mathrm{gr} F$.

Conversely, let $\sigma \in \mathcal{L}(S)$. Locally on S we can write $\sigma = \big(\psi^A(t), \tilde{\psi}^A(t)\big)$. Let us assume that ψ^A converges in some open disc U_0 containing 0 and that $\tilde{\psi}^A$ converges in an open disc U_∞ containing ∞, with $U_0 \cap U_\infty \neq \emptyset$. Then $t \mapsto \psi^A(t)$ and $s \mapsto \tilde{\psi}^A(s)$ define local holomorphic sections of $\mathcal{T}$ which by (7.1) patch together properly to define a global holomorphic section $\psi \in \Gamma(\mathcal{T})$. This is Penrose's curved twistor construction [6].[8] The nature (locally) of the solutions of the partial differential equation (1.14)′ is encoded in the global holomorphic structure of $\mathcal{T}$. Notice that the sections ψ are parametrized by the points of S. Furthermore, the points of $\mathcal{T}$ describe null two-sufaces of S, so $\mathcal{T}$ can be identified with $\mathcal{T}$ of the diagram (1.16). Thus there are $\mathcal{T}$'s worth (three-parameter family) of null two-surfaces in S obtained by fixing (t, q^A), say.

We have not addressed the problem of convergence of the local sections of $\psi^A(t)$ and $\tilde{\psi}^A(s)$. It should be possible to prove convergence if the metric is *"close enough"* to the flat metric. However, it is not clear whether in general the appropriate convergence can be obtained even if we localize in S. This brings us to a discrepancy between the twistor construction and the more conventional differential equations approach. By the twistor construction the *"general solution"* is given by two

[8] For this construction with cosmological constant, see [39].

holomorphic functions [F^A of (7.2)] of three complex variables with one differential constraint, given by (7.4); yet the *"general solution"* of Eq. (1.14)′ depends on two *arbitrary* holomorphic functions of three complex variables [3]. I am thus not convinced that the general self-dual metric on a sufficiently small neighborhood of $\mathbf{C}^4$ can be obtained by twistor methods.

I shall now give two simple examples of the twistor construction. For more sophisticated examples, see Ward's paper [14].

Example 1. Projective lines. We take the transition functions (7.2) satisfying (7.4) as $\tilde{z}^A = t^{-1} z^A$, so that the local sections satisfy

$$\tilde{\psi}^A(t^{-1}) = t^{-1}\psi^A(t).$$

It follows that the holomorphic curves are projective lines

$$\begin{aligned} \psi^A(t) &= q^A + \tilde{q}^A t \qquad \text{on } U_0 \\ \tilde{\psi}^A(t^{-1}) &= \tilde{q}^A + q^A t^{-1} \qquad \text{on } U_\infty \end{aligned}$$

Using some gauge freedom we obtain

$$\Omega = \epsilon_{AB} q^A \tilde{q}^B.$$

This represents flat space. In this case $\mathcal{T}$ is the total space of the holomorphic bundle $O(1) \oplus O(1)$ on $\mathbf{P}^1$. It can be shown that if F^A are any linear functions of z^B satisfying (7.4), then the metric is flat.

Example 2. Complex *pp*-waves. Consider the Jacobian

$$\frac{\partial F^A}{\partial z^B} = \begin{pmatrix} t^{-1} & 0 \\ F'(t^{-1}, z', t) & t^{-1} \end{pmatrix},$$

where F' is an arbitrary holomorphic function of its arguments. For local sections we have

$$\begin{aligned} \tilde{\psi}^1(t^{-1}) &= t^{-1}\psi^1(t), \\ \tilde{\psi}^2(t^{-1}) &= t^{-1}\psi^2(t) + F(t^{-1}\psi^1(t), t), \end{aligned}$$

where F is the integral of F' with respect to the first argument. The first equation again gives a projective line. Using this in the second equation and expanding F in a double power series we obtain from the coefficient of t^0 (after gauging),

$$\Omega = \epsilon_{AB} q^A \tilde{q}^B + G(q^1, \tilde{q}^1),$$

where G is an arbitrary holomorphic functon determined from F. This yields the metric for the complex *pp* waves [1,14,16]. In this case convergence of the holomorphic curves can be assured by appropriate choice of F.

In Ward's paper [14], he constructs three classes of self-dual metrics by twistor methods. The first is the complex *pp* waves of the previous example, the second is a class of metrics with one Killing vector field, given by solutions of the three-dimensional Laplace equation, which had been obtained previously by us [3], and independently by Hawking [41]. The third class, however, has not been obtained by more conventional techniques. The general procedure involved, given the holomorphic functions F^A on $U_0 \cap U_\infty$ satisfying (7.4), to split F^A into a sum of functions, one holomorphic about 0 and the other holomorphic about ∞. This splitting procedure can be explicitly implemented through the use of contour integration. However, in its proper algebraic setting this amounts to finding trivial cocycles [F^A] in certain sheaf cohomology groups. These cocycles can be used to construct solutions of the massless field equatons –*linear equations*. Thus one succeeds in *linearizing* the theory. It would be interesting to see what relationship, if any, this procedure has with other linearization techinques such as the inverse scattering transform [42,43].

We have recently found a group-theoretical formulation of the curved twistor construction which leads to a nonlinear superposition principle for the nonlinear graviton. See Ref. [18].

References

[1] J. F. Plebański, Some solutions of complex Einstein equations. *J. Math. Phys.* **16**, 2395–2402 (1975).

[2] J. D. Finley III and J. F. Plebański, Further heavenly metrics and their symmetries. *J. Math. Phys.* **17**, 585–596 (1976).

[3] C. P. Boyer and J. F. Plebański, Heavens and their integral manifolds. *J. Math. Phys.* **18**, 1022–1031 (1977).

[4] C. P. Boyer and J. F. Plebański, General relativity and *G*-structures. I. General theory and algebraically degenerate spaces. *Rep. Math. Phys.* **14**, 111–145 (1978).

[5] C. P. Boyer, J. D. Finley III, and J. F. Plebański, Complex general relativity, $\mathcal{H}$ and $\mathcal{HH}$ spaces —a survey of one approach. In *General Relativity and Gravitation*,Vol. 2, A. Held ed., Plenum Press, 1980, pp. 241–281.

[6] R. Penrose, Nonlinear gravitons and curved twistor theory. *GRG* **7**, 31–52 (1976).

[7] R. Penrose and R. S. Ward, Twistors for flat and curved space-time. In *General Relativity and Gravitation*,Vol. 2, A. Held ed., Plenum Press, 1980, pp. 283–328.

[8] R. O. Hansen, E. T. Newman, R. Penrose, and K. P. Tod, The metric and curvature properties of $\mathcal{H}$-space. Oxford University preprint.

[9] R. S. Ward, The self-dual Yang–Mills and Einstein equations. In *Complex Manifold Techniques in Theoretical Physics*, D. E. Lerner and P. D. Sommers eds., Pitman, 1979, pp. 12–34.

[10] E. J. Flaherty, *Hermitian and Kählerian Geometry in Relativity*. Springer Verlag, 1976.

[11] M. Ko, M. Ludvigsen, E. T. Newman, and K. P. Tod, The theory of $\mathcal{H}$-space. *Physics Reports* **71**, 51–139 (1981).

[12] J. D. Finley III and J. F. Plebański, The classification of all $\mathcal{H}$-spaces admitting a Killing vector. *J. Math. Phys.* **20**, 1938–1945 (1979).

[13] J. D. Finley III and J. F. Plebański, All algebraically degenerate $\mathcal{H}$-spaces via $\mathcal{HH}$-spaces. *J. Math. Phys.* **22**, 667–674 (1981).

[14] R. S. Ward, A class of self–dual solutions of Einstein's equations. *Proc. R. Soc. London* **A363**, 289–295 (1978).

[15] K. P. Tod and R. S. Ward, Self-dual metrics with self-dual Killing vectors. *Proc. R. Soc. London* **A368**, 411–427 (1979).

[16] W. D. Curtis, D. E. Lerner, and F. R. Miller, Complex *pp* waves and the nonlinear graviton construction. *J. Math. Phys.* **19**, 2024–2027 (1978).

[17] M. G. Eastwood, R. Penrose, and R. O. Wells, Cohomology and massless fields. *Commun. Math. Phys.* **78**, 305–351 (1981).

[18] C. P. Boyer and J. F. Plebański, An infinite hierarchy of conservation laws and nonlinear superposition principles for self-dual Einstein spaces. Preprint: Comunicaciones Técnicas IIMAS (1983).

[19] R. S. Ward, On self-dual gauge fields. *Phys. Lett.* **61A**, 81–82 (1977).

[20] M. F. Atiyah and R. S. Ward, Instantons and algebraic geometry, *Commun. Math. Phys.* **55**, 117-124 (1977).

[21] M. F. Atiyah, N. J. Hitchin, and I. M. Singer, Self-duality in four-dimensional Riemannian geometry. *Proc. R. Soc. London* **A362**, 425–461 (1978).

[22] E. Witten, An interpretation of classical Yang–Mills theory. *Phys. Lett.* **77B**, 394–398 (1978).

[23] J. Isenberg, P. B. Yasskin, and P. S. Green, Non-self-dual gauge fields. *Phys. Lett.* **78B**, 462–464 (1978).

[24] J. Isenberg and P. B. Yasskin, Twistor description of non-self-dual Yang–Mills fields. In *Complex Manifold Techniques in Theoretical Physics*, D. E. Lerner and P. D. Sommers eds., Pitman, 1979, pp. 180–206.

[25] Yu. I. Manin, Flag superpsaces and supersymmetric Yang–Mills equations. *preprint.*

[26] C. R. LeBrun, Spaces of complex geodesics and related structures. Oxford University thesis (1980).

[27] C. R. LeBrun, The first formal neighbourhood of ambitwistor space for curved space-time. Preprint IHES/M/81/54.

[28] Yu. I. Manin and I. B. Penkov, Null geodesics of complex Einstein spaces. *J. Funct. Anal. Appl.* **16**, 64–66 (1982).

[29] J. F. Plebański and I. Robinson, Left-degenerate vacuum metrics. *Phys. Rev. Lett.* **37**, 493–495 (1976).

[30] R. Penrose, Structure of space-time. In *Batelle Rencontres 1967*, C. M. de Witt and J. A. Wheeler, eds., Benjamin, 1968, pp. 121–235.

[31] R. O. Wells, Complex manifolds and mathematical physics. *Bull. Amer. Math. Soc. (new series)* **1**, 296–336 (1979).

[32] K. Kodaira, On stability of complex submanifolds of complex manifolds. *Amer. J. Math.* **85**, 79–94 (1963).

[33] A. Weinstein, Lagrangian submanifolds and hamiltonian systems. *Ann. Math.* **98**, 377–410 (1973).

[34] A. Weinstein, Symplectic manifolds and their lagrangian submanifolds. *Adv. Math.* **6**, 329–346 (1971).

[35] V. Guillemin and S. Sternberg, *Geometric Asymptotics.* Mathematical Surveys **14**, American Mathematical Society, 1977.

[36] I. N. Bernshtein and B. I. Rozenfel'd, Homogeneous spaces of infinite-dimensional Lie algebras and characteristic classes of foliations. *Russ. Math. Surv.* , 107–141 (19).

[37] K. Yano and S. Ishihara, *Tangent and Cotangent Bundles.* Marcel Dekker, 1973.

[38] R. O. Wells, *Differential Analysis on Complex Manifolds.* Springer Verlag, 1980.

[39] R. S. Ward, Self-dual space-times with cosmological constant. *Commun. Math. Phys.* **78**, 1–17 (1980).

[40] E. T. Newman, J. R. Porter, and K. P. Tod, Twistor surfaces and right-flat spaces. *GRG* **9**, 1129–1142 (1978).

[41] S. Hawking, Gravitational instantons. *Phys. Lett.* **60A**, 81–83 (1977).

[42] M. J. Ablowitz, D. J. Kaup, A. C. Newell, and H. Segur, The inverse scattering transform —Fourier analysis for nonlinear problems. *Studies in Applied Mathematics* **53**, 249–315 (1974).

[43] I. M. Krichever and S. P. Novikov, Holomorphic bundles over algebraic curves and nonlinear equations. *Russ. Math. Surv.* **35**, 53–79 (1980).

Integrable Dynamical Systems and Related Mathematical Results

Francesco Calogero

Dipartimento di Fisica
Università di Roma
Rome, Italy

Contents:

Introduction

The first part of these lecture notes presents an overview of results obtained over the last decade on finite-dimensional integrable dynamical systems. Since this topic has witnessed a research boom in this period, it is not possible to cover completely the field in the present framework, but the interested reader may perhaps pursue such a goal starting from the presentation given here and following up the references quoted. The main thread of this presentation is the investigation of solvable systems interpretable as (one-dimensional, classical) many-body problems with interparticle pair forces.

A spin-off of these studies has been a set of results on polynomials, differential operators, matrices, and singular integral equations which belong perhaps more to pure mathematics than to mathematical physics or applied mathematics. Central to these developments is the correspondence which exists between, on one side, a variable x and the corresponding differential operator d/dx, and on the other side, the two matrices $\mathbf{X}$ and $\mathbf{Z}$, of order n, defined in terms of n arbitrary (different) numbers x_j, by the formulae

$$\mathbf{X} = \mathrm{diag}(x_j), \quad X_{j,k} = \delta_{j,k} x_j, \tag{1}$$

$$Z_{j,k} = \begin{cases} \dfrac{1}{x_j - x_k}, & j \neq k, \\ \displaystyle\sum_{m=1}^{n}{}' \dfrac{1}{x_j - x_m}, & j = k. \end{cases} \tag{2}$$

These developments are outlined in the second part of these notes. However, since the first part has already overfilled the alloted space, the presentation of this second part will be extremely terse, being merely intended as a guide to the literature, which is indeed so recent that it does not justify yet an attempt for a comprehensive presentation.

In the following all equations are numbered progressively within each section; equation (1) of Section 2.1 is referred to as (1) within § 2.1, and as (2.1-1) elsewhere.

Chapter 1

Integrable Dynamical Systems

Introduction

Consider the hamiltonian system of $2n$ coupled (generally nonlinear) Ordinary Differential Equations (ODEs):

$$\dot{q}_j = \frac{\partial H(\mathbf{q},\mathbf{p})}{\partial p_j}, \qquad j = 1, 2, \ldots, n, \tag{1a}$$

$$\dot{p}_j = -\frac{\partial H(\mathbf{q},\mathbf{p})}{\partial q_j}, \qquad j = 1, 2, \ldots, n. \tag{1b}$$

Here the hamiltonian $H(\mathbf{q},\mathbf{p}) \equiv H(q_1, q_2, \ldots, q_n, p_1, p_2, \ldots, p_n)$ is a (given) function of the n lagrangian coordinates $q_j \equiv q_j(t)$ and of the n canonical momenta $p_j \equiv p_j(t)$, and of course the dot indicates differentiation with respect to the time t. A type of hamiltonian we shall be particularly interested in, especially in the following § 1.1, is that appropriate to describe the one-dimensional classical n-body problem of n point-like unit mass particles moving in the presence of an external potential $v(q)$ and interacting pairwise via the potential $V(q_j - q_k)$:

$$H(\mathbf{q},\mathbf{p}) = \tfrac{1}{2}\sum_{j=1}^{n} p_j^2 + \sum_{j=1}^{n} v(q_j) + \sum_{j>k=1}^{n} V(q_j - q_k). \tag{2}$$

The corresponding equations of motion read

$$\dot{q}_j = p_j, \qquad j = 1, 2, \ldots, n, \tag{3a}$$

$$\dot{p}_j = -v'(q_j) - {\sum_{k=1}^{n}}' V'(q_j - q_k), \qquad j = 1, 2, \ldots, n. \tag{3b}$$

In the last equation, the prime appended to a function denotes of course differentiation (with respect to the argument of the function), while the prime appended to the sum signifies omission of the term with $k = j$. This notation will be used throughout.

Insertion of (3a) in (3b) yields Newton's law of dynamics,

$$\ddot{q}_j = -v'(q_j) - \sum_{k=1}^{n}{}' V'(q_j - q_k), \qquad j = 1, 2, \ldots, n. \tag{4}$$

In writing (2), (3b), and (4) we have implicitly assumed that the two-body potential depends only on the distance between the particles, and not on their orientation, *i.e.*, the interparticle force is equal in strength and opposite in direction for the two particles, irrespective of whether a particle is to the right or to the left; this corresponds to the so-called *principle of action and reaction*, and implies of course

$$V(x) = V(-x), \qquad V'(x) = -V'(-x). \tag{5}$$

A quantity $I_m(\mathbf{q},\mathbf{p})$, having the property of remaining constant when the coordinates $q_j(t)$ and the momenta $p_j(t)$ evolve according to (3),

$$\dot{I}_m(\mathbf{q},\mathbf{p}) = 0, \tag{6}$$

is termed a ***constant of motion***. We have appended an index m to accommodate the possibility that there be several such quantities. See below.

It is easily seen, by direct computation using (3), that the time evolution of any function $F(\mathbf{q},\mathbf{p})$ of the coordinates $q_j(t)$ and the momenta $p_j(t)$ obeys the equation

$$\dot{F}(\mathbf{q},\mathbf{p}) = \{F(\mathbf{q},\mathbf{p}), H(\mathbf{q},\mathbf{p})\}, \tag{7}$$

where the *Poisson bracket* $\{A, B\}$ of two functions $A(\mathbf{q},\mathbf{p})$ and $B(\mathbf{q},\mathbf{p})$ is defined by the formula

$$\{A(\mathbf{q},\mathbf{p}), B(\mathbf{q},\mathbf{p})\} \equiv \sum_{j=1}^{n} \left(\frac{\partial A}{\partial q_j}\frac{\partial B}{\partial p_j} - \frac{\partial A}{\partial p_j}\frac{\partial B}{\partial q_j} \right). \tag{8}$$

Thus any constant of the motion is characterized by the property of having a vanishing Poisson bracket with the hamiltonian,

$$\{H(\mathbf{q},\mathbf{p}), I_m(\mathbf{q},\mathbf{p})\} = 0. \tag{9}$$

Since the Poisson bracket of any function with itself obviously vanishes [see (8)], the hamiltonian $H(\mathbf{q},\mathbf{p})$ is itself a constant of the motion,

$$\dot{H}(\mathbf{q},\mathbf{p}) = 0, \tag{10}$$

and this implies the *conservation of energy*. If in the n-body problem characterized by the hamiltonian (2) there is no external potential,

$$v(q) = 0, \tag{11}$$

it is immediately seen [summing (3b) over j and noting that the double sum in the right-hand side vanishes due to the antisymmetry of the summand, see (5)], that

$$I_1 \equiv \sum_{j=1}^{n} p_j(t) \tag{12}$$

provides a second constant of the motion: *conservation of momentum*. The index "1" may be associated with the property of I_1 to be of first order in the momenta p_j; accordingly, it is usual —see (2)— to identify the hamiltonian H with I_2:

$$I_2 \equiv H(\mathbf{q},\mathbf{p}). \tag{13}$$

The two constants I_1 and I_2 are obviously independent. If a hamiltonian system such as (1) or (3) admits n constants of the motion $I_m(\mathbf{q},\mathbf{p})$, $m=1,2,\ldots,n$, which are independent, namely none of them may be expressed as a function of the others, and are moreover *in involution*, namely the Poisson bracket between any pair of them vanishes:

$$\{I_j(\mathbf{q},\mathbf{p}), I_k(\mathbf{q},\mathbf{p})\} = 0, \qquad j,k=1,2,\ldots,n, \tag{14}$$

then the hamiltonian system is by definition ***integrable*** (or ***completely integrable***).

Clearly, for $n=1$, a hamiltonian system is always integrable, and for $n=2$ the system (3) is certainly integrable if (5) and (11) hold. But for $n \geq 3$ the integrable hamiltonian systems are exceptional rather than generic.

The motion of nonintegrable hamiltonian systems is generally extremely complicated (see, for instance [H1980]). The motion of integrable hamiltonian systems, on which we concentrate in the following, is much simpler; although a complete treatment of the relevant theory exceeds our present scope, let us outline some arguments in support of such a statement. Let us recall the essential elements of the theory of canonical transformations, which provide one route to the determination of the trajectories of a hamiltonian system. Consider the change of variables from the $2n$ hamiltonian coordinates q_j and p_j to $2n$ new coordinates $Q_j(\mathbf{q},\mathbf{p})$ and $P_j(\mathbf{q},\mathbf{p})$; assume this change of variables to be nonsingular (invertible, one-to-one), a necessary and sufficient condition for this being that the jacobian determinant of the new coordinates relative to the old ones (or viceversa) do not vanish:

$$\begin{vmatrix} \frac{\partial Q_1}{\partial q_1} \cdots \frac{\partial Q_1}{\partial q_n} & \frac{\partial Q_1}{\partial p_1} \cdots \frac{\partial Q_1}{\partial p_n} \\ \vdots \qquad \vdots & \vdots \qquad \vdots \\ \frac{\partial Q_n}{\partial q_1} \cdots \frac{\partial Q_n}{\partial q_n} & \frac{\partial Q_n}{\partial p_1} \cdots \frac{\partial Q_n}{\partial p_n} \\ \frac{\partial P_1}{\partial q_1} \cdots \frac{\partial P_1}{\partial q_n} & \frac{\partial P_1}{\partial p_1} \cdots \frac{\partial P_1}{\partial p_n} \\ \vdots \qquad \vdots & \vdots \qquad \vdots \\ \frac{\partial P_n}{\partial q_1} \cdots \frac{\partial P_n}{\partial q_n} & \frac{\partial P_n}{\partial p_1} \cdots \frac{\partial P_n}{\partial p_n} \end{vmatrix} \neq 0. \tag{15}$$

Assume moreover that the Poisson brackets of the new coordinates (relative to the old coordinates, or viceversa) have the properties

$$\{Q_j(\mathbf{q},\mathbf{p}), Q_k(\mathbf{q},\mathbf{p})\} = 0, \qquad j,k=1,2,\ldots,n, \tag{16a}$$
$$\{P_j(\mathbf{q},\mathbf{p}), P_k(\mathbf{q},\mathbf{p})\} = 0, \qquad j,k=1,2,\ldots,n, \tag{16b}$$
$$\{Q_j(\mathbf{q},\mathbf{p}), P_k(\mathbf{q},\mathbf{p})\} = \delta_{j,k}, \qquad j,k=1,2,\ldots,n. \tag{16c}$$

Then the change of variables is, by definition, a *canonical transformation.* Note incidentally that the identity transformation $Q_j = q_j$, $P_j = p_j$, $j=1,2,\ldots,n$, is a canonical transformation, since (8) indeed implies

$$\{q_j, q_k\} = 0 = \{p_j, p_k\}, \quad \{q_j, p_k\} = \delta_{j,k}, \qquad j,k=1,2,\ldots,n. \tag{17}$$

The main property of canonical transformations is to preserve the structure of Hamilton's equations, namely,

$$\dot{Q}_j = \frac{\partial K(\mathbf{Q},\mathbf{P})}{\partial P_j}, \qquad j = 1,2,\ldots,n, \tag{18a}$$

$$\dot{P}_j = -\frac{\partial K(\mathbf{Q},\mathbf{P})}{\partial Q_j}, \qquad j = 1,2,\ldots,n, \tag{18b}$$

where

$$K(\mathbf{Q},\mathbf{P}) = H(\mathbf{q},\mathbf{p}), \tag{19}$$

as can be easily verified using (16).

Consider now an integrable hamiltonian system, and introduce a canonical transformation from the variables $(\mathbf{q},\mathbf{p})$ to the variables $(\mathbf{Q},\mathbf{P})$, such that the new canonical momenta P_j coincide with the n constants of motion I_j:

$$P_j(\mathbf{q},\mathbf{p}) = I_j(\mathbf{q},\mathbf{p}), \qquad j = 1,2,\ldots,n. \tag{20}$$

[Note that the condition (14) is required to make this equation consistent with (16b).] Then (6) and (18b) imply that the new hamiltonian K must be independent of the new lagrangian coordinates Q_j,

$$K(\mathbf{Q},\mathbf{P}) = K(\mathbf{P}) = K(I_m), \tag{21}$$

and the equations of motion (18a) are immediately integrated

$$Q_j(t) = Q_j(0) + V_j t, \tag{22}$$

with

$$V_j = \frac{\partial K(I_1, I_2, \ldots, I_n)}{\partial I_j}, \qquad j = 1,2,\ldots,n. \tag{23}$$

Note that the time independence of the *velocities* V_j is implied by the time independence of the constants of motion I_j. Indeed, if the constants of motion I_j are chosen so that one of them coincides with the hamiltonian itself, say $I_2 = H = K$ [see (13)], then all the V_j's vanish except V_2:

$$V_j = 0, \qquad j = 0,1,3,4,\ldots,n. \tag{24}$$

The quantities $Q_1, Q_3, Q_4, \ldots, Q_n$ then provide $n-1$ additional constants of motion.

For an integrable system, thus, the problem of the motion can be solved by identifying n explicit constants of the motion $I_j(\mathbf{q},\mathbf{p}), j = 1,2,\ldots,n$, by choosing them as new canonical momenta —see (20)— and by identifying the corresponding lagrangian coordinates Q_j. (There exists in principle a standard procedure to perform this last step; this may, however, turn out to be impractical.) Once the quantities $Q_j(\mathbf{q},\mathbf{p})$ and $P_j(\mathbf{q},\mathbf{p})$ have been determined, the complete explicit solution of the problem is reduced to inverting these functions, namely, to express the quantitites q_j and p_j in terms of $\mathbf{Q}$ and $\mathbf{P}$:

$$q_j = q_j(\mathbf{Q},\mathbf{P}),\ p_j = p_j(\mathbf{Q},\mathbf{P}), \qquad j = 1,2,\ldots,n; \tag{25}$$

since the time evolution of the Q_j's and the P_j's is trivially simple, see (20) [with (6)] and (22) [possibly with (24)].

In the following sections, various techniques are discussed to identify, solve and/or investigate integrable dynamical systems. We concentrate on systems which can be identified as many-body problems, giving rise to equations of motion such as (4), or at least similar to it. Our treatment covers, in a more compact and coherent —if perhaps less complete— fashion, the material already reviewed in [C1978], [C1980], and [C1980a]; for a more complete treatment, see [OP1981].

§1.1 The Lax technique. The technique of Olshanetsky and Perelomov. Examples

Consider a system of coupled ODEs, such as Hamilton's equations (1.0–1) for some given hamiltonian $H(\mathbf{q},\mathbf{p})$; assume that two matrices of order n, $\mathbf{L}$ and $\mathbf{A}$, can be found [as functions of the hamiltonian coordinates, $\mathbf{L}=\mathbf{L}(\mathbf{q},\mathbf{p})$, $\mathbf{A}=\mathbf{A}(\mathbf{q},\mathbf{p})$], such that the equations of motion (1.0–1) can be cast in the matrix *Lax* (or *Heisenberg*) form

$$\dot{\mathbf{L}} = i[\mathbf{L},\mathbf{A}], \tag{1}$$

where the symbol $[\mathbf{A},\mathbf{B}]$ denotes the commutator,

$$[\mathbf{A},\mathbf{B}] \equiv \mathbf{AB}-\mathbf{BA}. \tag{2}$$

It can then be generally concluded that the system is integrable, since it is easy to prove that the n eigenvalues of $\mathbf{L}$ are time independent, providing therefore n constants of the motion. The proof goes as follows: let $\lambda^{(m)}$, $\mathbf{v}^{(m)}$, and $\mathbf{u}^{(m)}$ indicate the eigenvalues of $\mathbf{L}$ and the corresponding right and left eigenvectors:

$$\mathbf{L}\,\mathbf{v}^{(m)} = \lambda^{(m)}\mathbf{v}^{(m)}, \qquad m=1,2,\ldots,n, \tag{3}$$

$$\mathbf{u}^{(m)}\mathbf{L} \equiv \mathbf{L}^{\top}\mathbf{u}^{(m)} = \lambda^{(m)}\mathbf{u}^{(m)}, \qquad m=1,2,\ldots,n, \tag{4}$$

$$\delta_{\ell,m} = (\mathbf{u}^{(\ell)},\mathbf{v}^{(m)}) \equiv \sum_{j=1}^{n} u_j{}^{(\ell)} v_j{}^{(m)}. \tag{5}$$

Then of course

$$\lambda^{(m)} = (\mathbf{u}^{(m)},\mathbf{L}\mathbf{v}^{(m)}) = (\mathbf{u}^{(m)}\mathbf{L},\mathbf{v}^{(m)}), \qquad m=1,2,\ldots,n, \tag{6}$$

and therefore

$$\dot{\lambda}^{(m)} = (\mathbf{u}^{(m)},\dot{\mathbf{L}}\,\mathbf{v}^{(m)}) + (\dot{\mathbf{u}}^{(m)},\mathbf{L}\,\mathbf{v}^{(m)}) + (\mathbf{u}^{(m)}\mathbf{L},\dot{\mathbf{v}}^{(m)}) \tag{7a}$$

$$= i(\mathbf{u}^{(m)},[\mathbf{L},\mathbf{A}]\mathbf{v}^{(m)}) + \lambda^{(m)}[(\dot{\mathbf{u}}^{(m)},\mathbf{v}^{(m)}) + (\mathbf{u}^{(m)},\dot{\mathbf{v}}^{(m)})] \tag{7b}$$

$$= i[(\mathbf{u}^{(m)}\mathbf{L},\mathbf{A}\mathbf{v}^{(m)}) - (\mathbf{u}^{(m)},\mathbf{A}\mathbf{L}\,\mathbf{v}^{(m)})] + \lambda^{(m)}\frac{d(\mathbf{u}^{(m)},\mathbf{v}^{(m)})}{dt} \tag{7c}$$

$$= 0. \tag{7d}$$

To get (7b), equations (1), (3), and (4) have been used; (7c) coincides obviously with (7b), and (7d) follows from (7c) using (3), (4), and (5). Note that this proof requires $\mathbf{L}$ to be neither hermitian nor symmetric.

To make sure that the system under consideration is indeed completely integrable it must moreover be ascertained that the n eigenvalues $\lambda^{(m)}(\mathbf{q},\mathbf{p})$ are independent and that they are in involution. This has to be checked in each case.

It is often more convenient to focus attention, not on the n eigenvalues $\lambda^{(m)}$ of $\mathbf{L}$, but on the n symmetric invariants J_m of the matrix $\mathbf{L}$, defined by the formula

$$\det[\mathbf{L}+\lambda\mathbf{1}] = \lambda^n + \sum_{m=1}^{n} J_m\lambda^{n-m} = \prod_{m=1}^{n}[\lambda+\lambda^{(m)}], \tag{8}$$

or on the n traces of the powers of $\mathbf{L}$, defined by the formula

$$T_m = \frac{1}{m}\operatorname{tr}[\mathbf{L}^m] = \frac{1}{m}\sum_{j=1}^{n}[\lambda^{(j)}]^m, \qquad m=1,2,\ldots,n. \tag{9}$$

Indeed, these quantities are generally expressible more readily in terms of the matrix elements $L_{j,k}(\mathbf{q},\mathbf{p})$, and therefore also in terms of the hamiltonian coordinates q_j and p_j; the time independence of the eigenvalues guarantees that the quantities J_m and T_m —see (8) and (9)— are time independent as well.

The matrix equation (1) corresponds to n^2 scalar first order ODEs, while the equations of motion (1.0–1) number $2n$. Thus, for $n > 1$ it is in general not possible to cast the hamiltonian equations of motion (1.0–1) in the Lax form (1); indeed, the generic hamiltonian problem is not integrable for $n > 1$. We now show that there do however exist nontrivial cases when the hamiltonian equations of motion (1.0–1) can indeed be cast in the Lax form (1) and, moreover, the corresponding hamiltonian has the form (2), being therefore interpretable as a one-dimensional n-body problem.

A convenient *ansatz* for the matrices $\mathbf{L}$ and $\mathbf{A}$ reads ([M1975], [C1975])

$$L_{j,k} = \begin{cases} p_j & j = k, \\ \alpha(q_j - q_k) & j \neq k, \end{cases} \tag{10}$$

$$A_{j,k} = i \begin{cases} \sum_{\ell=1}^{n}{}' \beta(q_j - q_\ell), & j = k, \\ \gamma(q_j - q_k), & j \neq k, \end{cases} \tag{11}$$

where $\alpha(x)$, $\beta(x)$, and $\gamma(x)$ are functions to be determined. It is seen, after a little algebra, that (1) is equivalent to (1.0–3) with

$$v(x) = 0, \tag{12}$$

$$V(x) = \alpha(x)\,\alpha(-x) + \text{constant}, \tag{13}$$

provided the functions $\alpha(x)$, $\beta(x)$, and $\gamma(x)$ satisfy the equations

$$\gamma(x) = -\alpha'(x), \tag{14}$$

$$\beta(-x) = \beta(x), \tag{15}$$

$$\alpha'(x)\alpha(y) - \alpha(x)\alpha'(y) = \alpha(x+y)[\beta(x) - \beta(y)]. \tag{16}$$

Obviously, it is the functional equation (16) which constitutes the more important constraint. In fact, it can be shown ([C1976],[OP1981]) that the more general solution for the potential $V(x)$ consistent with (13), (15), and (16) is

$$V(x) = g^2 a^2 P(ax|\omega,\omega') + \text{constant}, \tag{17}$$

where g, a, ω, and ω' are constants and $P(z|\omega,\omega')$ is the Weierstrass function

$$P(z|\omega,\omega') \equiv z^{-2} + \sum_{\ell,m}{}' \left[\frac{1}{(z - 2\ell\omega - 2m\omega')^2} - \frac{1}{(2\ell\omega + 2m\omega')^2} \right]. \tag{18}$$

Here the sum extends over all (positive and negative) integers, excluding the single —and singular— term with $\ell = m = 0$. The prime appended to the sum is a reminder of this.

The function $\beta(x)$ is given, moreover —up to an additive constant which is clearly irrelevant— by

$$\beta(x) = V(x)/c \tag{19}$$

with the constant c defined by

$$c = \lim_{x \to 0} [x\alpha(x)]. \tag{20}$$

As for the function $\alpha(x)$, three different expressions [all producing the same potential (17) via (13)] read

$$\alpha_1(x) = iga' / \operatorname{sn}(a'x), \tag{21a}$$
$$\alpha_2(x) = iga' \operatorname{dn}(a'x) / \operatorname{sn}(a'x), \tag{21b}$$
$$\alpha_3(x) = iga' \operatorname{cn}(a'x) / \operatorname{sn}(a'x). \tag{21c}$$

Here sn, cn, and dn are Jacobian elliptic functions, and

$$a' = a\sqrt{e_1 - e_3}, \tag{22}$$

where we are using the notation of § 13.16 in [HTF1953]. Note that the potential (13) is automatically even —consistent with (1.0–5).

Let us recall that the n-body system characterized by (1.0–2) with (12)–(13), possesses the two constants of motion (1.0–12) and (1.0–13), namely, total momentum and energy. Indeed, it follows from (10) [see (1.0–2), (12), (13), (8), (9), (1.0–12), and (1.0–13)] that

$$J_1 = T_1 = I_1 = \sum_{j=1}^{n} p_j, \tag{23a}$$
$$J_2 = \tfrac{1}{2} I_1^2 - I_2 = \tfrac{1}{2} \Big(\sum_{j=1}^{n} p_j \Big)^2 - H, \tag{23b}$$
$$T_2 = I_2 = H. \tag{23c}$$

To obtain the last two formulae, we have set to zero the irrelevant constant in the right-hand side of (13).

In order that the potential $V(x)$ be real —see (17)— one of the two constants ω, ω' must be real and the other imaginary. They cannot both be real or both imaginary, or else the sum in the right-hand side of (18) diverges. In the following it is convenient to concentrate on the limiting cases when one or both of these constants diverge; then the potential $V(x)$ vanishes as $x \to \pm\infty$ and is nonsingular for real $x \neq 0$, thus allowing a more straightforward interpretation in terms of a "physical" n-body problem. In particular, for $\omega = i\omega' = \infty$,

$$V(x) = g^2 / x^2, \tag{24a}$$
$$\alpha(x) = ig/x, \tag{24b}$$

and for $\omega = \infty$, $\omega' = i\pi/2$,

$$V(x) = g^2 a^2 / \sinh^2(ax), \tag{25a}$$
$$\alpha(x) = iga / \sinh(ax). \tag{25b}$$

[The last formula corresponds to (21a) and (21b), while (21c) would give the equivalent —but less convenient— expression $\alpha(x) = iga \coth(ax)$; we have also set to zero the irrelevant constant in the right-hand side of (13), in order that $V(x)$ vanish asymptotically.]

Let us focus attention on the two-body potential (25), namely, on the n-body problem characterized by the hamiltonian

$$H = \frac{1}{2} \sum_{j=1}^{n} p_j^2 + g^2 a^2 \sum_{j>k=1}^{n} \frac{1}{\sinh^2[a(q_j - q_k)]}. \tag{26}$$

The typical "physical" phenomenon is then a scattering process corresponding to the asymptotic formulae

$$q_j(t) \rightarrow p_j(\pm\infty)t + x_j^{(\pm)} \quad \text{as} \quad t \rightarrow \pm\infty. \tag{27}$$

It is intrinsic to the definition of the scattering process, and required by consistency with (27) which describes the asymptotic free motion prevailing when all particles are widely separated, that the initial and final velocities satisfy the inequalities

$$p_j(-\infty) > p_{j+1}(-\infty), \qquad j = 1, 2, \ldots, n-1, \tag{28a}$$

$$p_j(+\infty) < p_{j+1}(+\infty), \qquad j = 1, 2, \ldots, n-1, \tag{28b}$$

provided the particles on the line are labeled so that

$$q_j(t) < q_{j+1}(t), \qquad j = 1, 2, \ldots, n-1. \tag{29}$$

Note that this ordering does not change throughout the motion since the particles are prevented from crossing each other by the singular and repulsive two-body interaction.

The solution of the scattering problem consists in predicting the final values $p_j(+\infty)$ and $x_j^{(+)}$ corresponding to given initial values $p_j(-\infty)$ and $x_j^{(-)}$, $j = 1, 2, \ldots, n$. The first half of this task turns out to be extremely simple. Indeed, let us recall that the eigenvalues of the Lax matrix

$$L_{j,k}(t) = \begin{cases} p_j(t), & j = k, \\ iga/\sinh\{a[q_j(t) - q_k(t)]\}, & j \neq k, \end{cases} \tag{30}$$

are time independent. But since the particles separate asymptotically, the off-diagonal part of $\mathbf{L}(t)$ vanishes as $t \rightarrow \pm\infty$:

$$L_{j,k}(\pm\infty) = \delta_{j,k} p_j(\pm\infty). \tag{31}$$

Thus the coincidence of the set of eigenvalues of $\mathbf{L}(-\infty)$ with the set of eigenvalues of $\mathbf{L}(+\infty)$ yields

$$\{p_j(+\infty),\ j = 1, 2, \ldots, n\} = \{p_j(-\infty),\ j = 1, 2, \ldots, n\}; \tag{32}$$

and this formula, together with (28), implies

$$p_j(+\infty) = p_{n-j+1}(-\infty), \qquad j = 1, 2, \ldots, n. \tag{33}$$

Thus the first particle gets finally the initial momentum of the last particle, the second gets the initial momentum of the next-to-last, and so on. (If n is odd, the central particle recovers at the end the same momentum it had initially.) Note that this result is independent of the value of the constants g and a characterizing the strength and range of the two-body potential, and it is also independent of the values of the n constants $x_j^{(-)}$ characterizing the incoming configuration, although of course the actual trajectories of the particles for finite time **do** depend on the values of these parameters![1]

[1]The result (33) was first proven in [C1975]. For $a = 0$ (so that $V(x) = g^2/x^2$), this property was first discovered in the quantal case, for $n = 3$ by C. Marchioro [M1970] and then for arbitrary n in [C1971]; see ahead. The quantal results implied of course the validity of the same outcome in the classical case, but a proof in the classical context was given, for arbitrary n, only in 1975 by J. Moser in the important paper [M1975], where the Lax technique, originally introduced in the context of nonlinear integrable partial differential equations in [L1968], was applied for the first time in the context of the classical n-body problem with pair interactions. It had been previously applied to the classical n-body Toda problem, which features only nearest-neighbour interactions, by H. Flaschka in [F1974] and [F1974a], and by S. V. Manakov in [Man1974]. For $n = 3$ (and always $a = 0$), (33) had been obtained in the classical context by C. Marchioro in 1971 (unpublished) and by D. C. Khandekar and S. V. Lawande [KL1972] via the explicit solution of the equations of motion; but this problem had been actually solved by C. Jacobi a century earlier in [J1866].

An exact formula yielding the values of the quantities $x_j^{(+)}$ can also be obtained, for instance, by carrying the argument given above one step further, *i.e.* keeping also the corrections to the leading terms as $t \to \pm\infty$. The result reads

$$x_j^{(+)} = x_{n-j+1}^{(-)} + \sum_{k=1}^{n}{}' \Delta[p_j(-\infty) - p_k(-\infty)], \qquad j = 1, 2, \ldots, n, \tag{34}$$

where Δ is the two-body shift, namely

$$\Delta[p_1(-\infty) - p_2(-\infty)] = x_1^{(+)} - x_2^{(-)} \quad \text{for} \quad n = 2. \tag{35}$$

This quantity is, of course, easily evaluated:

$$\Delta(p) = \operatorname{sign}(p) \frac{2}{a} \ln(1 + 4\frac{g^2 a^2}{p^2}). \tag{36}$$

Note that both (33) and (34) are precisely the results which would be obtained if the scattering process were a sequence of separate two-body collisions. Note moreover that the shift $\Delta(p)$ vanishes for $a = 0$; thus in the special case of the inverse-square two-body interaction $[V(x) = g^2/x^2$, corresponding to $a = 0$ —see (26)], (34) becomes closely analogous to (33),

$$x_j^{(+)} = x_{n-j+1}^{(-)}, \quad j = 1, 2, \ldots, n, \qquad [V(x) = g^2/x^2]. \tag{37}$$

By an amusing trick, it is possible to extend the n-body problem characterized by the hamiltonian (26) which describes n equal particles on the line interacting via the singular repulsive pair potential

$$V_e(x) = \frac{g^2 a^2}{\sinh^2(ax)} \tag{38a}$$

(where the subscript e stands for *equal*), to the problem of $n_1 + n_2 = n$ unit-mass particles on the line, with n_1 particles of one type and n_2 particles of another type, and the singular repulsive pair potential (38a) acting between equal particles, but the nonsingular attractive pair potential

$$V_d(x) = -\frac{g^2 a^2}{\cosh^2(ax)} \tag{38b}$$

acting between different particles. This is formally achieved by shifting the coordinates of the first n_1 particles (or the last n_2) by the imaginary amount $\frac{1}{2} i\pi/a$. Note that different particles can then cross each other, and they can form (stable) bound states, so that it is then possible to consider scattering processes with two-body bound states in the initial and final configuration, in addition to particles of either kind. There also exist m-body bound states with $m > 2$; for instance, the static symmetrical configuration of three particles, two of one kind sitting at a distance $d = (2a)^{-1} \operatorname{arccosh} 2$ to the right and left of one particle of the other kind[2]

It is actually possible to determine the trajectories $q_j(t)$ for all time in almost explicit form, by the technique due to Olshanetsky and Perelomov [OP1976], which is described here in the simpler case

[2] It can be shown that this configuration is not stable, tending to break up into single particles and two-body bound states. See [OR1978].

of the inverse-square potential (24a).[3] We thus focus on the hamiltonian

$$H = \tfrac{1}{2} \sum_{j=1}^{n} p_j^2 + g^2 \sum_{j>k=1}^{n} \frac{1}{(q_j - q_k)^2} \tag{39}$$

and the corresponding equations of motion

$$\ddot{q}_j = 2g^2 \sum_{k=1}^{n}{}' \frac{1}{(q_j - q_k)^3}, \qquad j = 1, 2, \ldots, n. \tag{40}$$

Let the matrices $\mathbf{Q}$ and $\mathbf{Y}$, of order n, be defined by the equations

$$\mathbf{Q}(t) = \mathrm{diag}[q_j(t)], \tag{41}$$

$$\mathbf{Y}(t) = \mathbf{U}(t)\,\mathbf{Q}(t)\,\mathbf{U}^{-1}(t), \tag{42}$$

where we reserve for a later time to make an appropriate choice for $\mathbf{U}(t)$. Note that this definition implies that the particle coordinates $q_j(t)$ are the eigenvalues of $\mathbf{Y}(t)$. We now t-differentiate $\mathbf{Y}$, and write

$$\dot{\mathbf{Y}}(t) = \mathbf{U}(t)\,\mathbf{L}(t)\,\mathbf{U}^{-1}(t), \tag{43}$$

having made the two positions

$$\mathbf{M}(t) = \frac{d\mathbf{U}^{-1}(t)}{dt}\mathbf{U}(t) = -\mathbf{U}^{-1}(t)\,\dot{\mathbf{U}}(t), \tag{44}$$

$$\mathbf{L}(t) = \dot{\mathbf{Q}}(t) + [\mathbf{Q}(t), \mathbf{M}(t)]. \tag{45}$$

A second differentiation yields

$$\ddot{\mathbf{Y}}(t) = \mathbf{U}(t)\{\dot{\mathbf{L}}(t) + [\mathbf{L}(t), \mathbf{M}(t)]\}\,\mathbf{U}^{-1}(t). \tag{46a}$$

For notational convenience we also rewrite (46a) in the form

$$\ddot{\mathbf{Y}}(t) = \mathbf{U}(t)\{\dot{\mathbf{L}}(t) - i[\mathbf{L}(t), \mathbf{A}(t)]\}\,\mathbf{U}^{-1}(t), \tag{46b}$$

having set

$$\mathbf{M}(t) = -i\mathbf{A}(t). \tag{47}$$

At this point a specific choice is made for the matrix $\mathbf{M}(t)$, or equivalently $\mathbf{A}(t)$:

$$A_{j,k}(t) = g \begin{cases} \displaystyle\sum_{\ell=1}^{n}{}' \frac{1}{(q_j - q_\ell)^2}, & j = k, \\[2ex] \displaystyle\frac{-1}{(q_j - q_k)^2}, & j \neq k. \end{cases} \tag{48}$$

[The attentive reader may notice that this definition of $A_{j,k}(t)$ coincides with (11) via (14), (19), (20), and (24a).] This of course implies, via (44), (47) and the initial condition[4]

$$\mathbf{U}(t=0) = \mathbf{U}^{-1}(t=0) = \mathbf{1} \tag{49}$$

[3]For the analogous treatment in the more general case (26), see [OP1981].

[4]This is hereafter assumed to hold for convenience —see ahead.

that the matrix **U** is determined, but there shall be no need to evaluate it. Note that (49) implies, via (42) and (43), the two convenient formulae

$$\mathbf{Y}(0) = \mathbf{Q}(0), \tag{50a}$$
$$\dot{\mathbf{Y}}(0) = \mathbf{L}(0). \tag{50b}$$

The choice (48) of **A** [and correspondingly for **M** —see (47)] yields, as can be readily verified, the following expression for **L** via (45):

$$L_{j,k}(t) = \begin{cases} \dot{q}_j(t), & j = k, \\ \frac{ig}{q_j(t) - q_k(t)}, & j \neq k. \end{cases} \tag{51}$$

[Again, the coincidence of this expression with (10) is easily checked, via (1.0–3a), (14), (19), (20), and (24a).] Moreover, from (48), (51), and (46b), one obtains the formula

$$\ddot{\mathbf{Y}}(t) = \mathbf{U}(t)\,\mathbf{D}(t)\,\mathbf{U}^{-1}(t), \tag{52}$$

where the diagonal matrix **D** is defined by

$$\mathbf{D}(t) = \mathrm{diag}\left(\ddot{q}_j(t) - 2g^2 \sum_{k=1}^{n}{}' \frac{1}{[q_j(t) - q_k(t)]^3} \right). \tag{53}$$

One may therefore conclude that, if the coordinates $q_j(t)$ evolve according to (40), then

$$\ddot{\mathbf{Y}}(t) = 0. \tag{54}$$

But this matrix differential equation is trivially integrated and, together with (50) and (51), it yields

$$Y_{j,k}(t) = \begin{cases} q_j(0) + \dot{q}_j(0)t, & j = k, \\ \frac{igt}{q_j(0) - q_k(0)} & j \neq k. \end{cases} \tag{55}$$

Let us emphasize that this is a completely explicit expression of the matrix $\mathbf{Y}(t)$ as a function of time and in terms of the initial positions $q_j(0)$ and velocities $\dot{q}_j(0)$ of the particles; while the trajectories $q_j(t)$ of the particles for all time are just the eigenvalues of **Y**,

$$\{q_j(t),\ j = 1, 2, \ldots, n\} = \text{eigenvalues of } \mathbf{Y}(t). \tag{56}$$

[See (41) and (42).] Thus, the two formulae (55) and (56) provide a fairly explicit solution of the initial value problem for the n-body system characterized by the hamiltonian (39).

It is easy to recover from these formulae the results (27) with (33) and (37). In fact, from (52) with (53), and (42) with (41), there immediately follows a more general result, namely, the assertion that the solutions $q_j(t)$ of the equation

$$\ddot{q}_j(t) = f(q_j) + 2g^2 \sum_{k=1}^{n}{}' \frac{1}{[q_j(t) - q_k(t)]^3}, \qquad j = 1, 2, \ldots, n, \tag{57}$$

coincide with the n eigenvalues of the matrix $\mathbf{Y}(t)$ solution of the second-order matrix differential equation

$$\ddot{\mathbf{Y}} = f(\mathbf{Y}), \tag{58}$$

supplemented by the initial conditions (50), with (51). [Note that while $\mathbf{Y}(0)$ is diagonal, $\dot{\mathbf{Y}}(0)$ is not.] Of course, the equations of motion (57) correspond to the hamiltonian (1.0–2) with (24a) and [see (1.0–3)]

$$f(x) = -v'(x). \tag{59}$$

The second-order matrix differential equation (58) can be explicitly solved for

$$f(\mathbf{Y}) = -\omega^2\mathbf{Y} + \phi(t)\mathbf{1}, \tag{60}$$

corresponding to the n-body hamiltonian

$$H = \frac{1}{2}\sum_{j=1}^{n}[p_j^2 + \omega^2 q_j^2 - 2\phi(t)q_j] + g^2 \sum_{j>k=1}^{n} \frac{1}{(q_j - q_k)^2}, \tag{61}$$

and to the equations of motion

$$\ddot{q}_j + \omega^2 q_j = \phi(t) + 2g^2 \sum_{k=1}^{n}{}' \frac{1}{(q_j - q_k)^3}, \quad j = 1, 2, \ldots, n. \tag{62}$$

Hence one immediately concludes that the trajectories $q_j(t)$ of this problem coincide with the eigenvalues of the matrix $\mathbf{Y}$ explicitly given as a function of time and of the initial positions $q_j(0)$ and velocities $\dot{q}_j(0)$ by the formula

$$Y_{j,k}(t) = \begin{cases} q_j(0)\cos(\omega t) + \dot{q}_j(0)\dfrac{\sin(\omega t)}{\omega} + \Phi(t), & j = k, \\ \dfrac{ig}{q_j(0) - q_k(0)}\,\dfrac{\sin(\omega t)}{\omega}, & j \neq k, \end{cases} \tag{63}$$

where

$$\Phi(t) \equiv \int_0^t dt'\,\phi(t)\,\frac{\sin[\omega(t - t')]}{\omega}. \tag{64}$$

Several remarks are now appropriate

Remark 1. The presence of the term proportional to $\phi(t)$ in (61) and (62) affects the trajectories $q_j(t)$ in a trivial way; indeed, if $q_j^{(0)}(t)$ indicate the solutions of the problem with $\phi(t) = 0$ and $q_j(t)$ the solutions with $\phi(t) \neq 0$ [and with the same initial conditions $q_j(0) = q_j^{(0)}(0)$, $\dot{q}_j(0) = \dot{q}_j^{(0)}(0)$], there holds the formula

$$q_j(t) = q_j^{(0)}(t) + \Phi(t), \qquad j = 1, 2, \ldots, n, \tag{65}$$

with $\Phi(t)$ defined by (64) above. This result is implied by the formula

$$\mathbf{Y}(t) = \mathbf{Y}^{(0)}(t) + \Phi(t)\mathbf{1} \tag{66}$$

—see (63). The latter is closely related to the following remark.

Remark 2. Let

$$q(t) = \frac{1}{n}\sum_{j=1}^{n} q_j(t), \tag{67}$$

to indicate the center-of-mass coordinate of the n-body system evolving according to the equations of motion (62). Then, clearly,

$$\ddot{q}(t) + \omega^2 q(t) = \phi(t), \tag{68}$$

implying

$$q(t) = q(0)\cos(\omega t) + \dot{q}(0)\frac{\sin(\omega t)}{\omega} + \Phi(t). \tag{69}$$

On the other hand there hold the identities

$$\sum_{j=1}^{n}(q_j - q)^2 = \frac{1}{n}\sum_{j>k=1}^{n}(q_j - q_k)^2 = \sum_{j=1}^{n} q_j^2 - nq^2. \tag{70}$$

One can therefore rewrite the hamiltonian (61) in the separated form

$$H = H_{CM} + \overline{H}, \tag{71}$$

$$H_{CM}(\mathbf{q},\mathbf{p}) = \tfrac{1}{2}n[p^2 + \omega^2 q^2 - 2q\phi(t)], \tag{72}$$

$$\overline{H}(\overline{\mathbf{q}},\overline{\mathbf{p}}) = \tfrac{1}{2}\sum_{j=1}^{n}\overline{p}_j^2 + \sum_{j>k=1}^{n}\left(\frac{\omega^2(\overline{q}_j - \overline{q}_k)^2}{2n} + \frac{g^2}{(\overline{q}_j - \overline{q}_k)^2}\right), \tag{73}$$

where

$$\overline{q}_j(t) = q_j(t) - q(t), \tag{74a}$$

$$\overline{p}_j(t) = \dot{\overline{q}}_j(t) = \dot{q}_j(t) - \dot{q}(t) = p_j(t) - p(t), \tag{74b}$$

$$p(t) = \dot{q}(t). \tag{74c}$$

The coordinates relative to the center-of-mass, $\overline{q}_j(t)$, satisfy then the equations of motion

$$\ddot{\overline{q}}_j = \sum_{k=1}^{n}\left(\frac{\omega^2(\overline{q}_j - \overline{q}_k)}{n} + \frac{2g^2}{(\overline{q}_j - \overline{q}_k)^3}\right) \tag{75}$$

and the constraint

$$\sum_{j=1}^{n}\overline{q}_j = 0, \tag{76}$$

which is clearly consistent with them. The conclusion of this remark is the essential equivalence of the hamiltonians (61) and (73), and of the equations of motion (62) and (75)–(76).

As the above remarks imply, the additional generality due to the presence of the ϕ term in (61) and (62) is trivial and therefore uninteresting. Hereafter we thus assume that ϕ [and therefore also Φ —see (64)] vanish:

$$\phi(t) = 0, \quad \Phi(t) = 0. \tag{77}$$

Remark 3. Clearly, the matrix $\mathbf{Y}$ defined by (63) with (77) is periodic, with period

$$T = 2\pi/\omega, \tag{78}$$

$$\mathbf{Y}(t+T) = \mathbf{Y}(t). \tag{79}$$

Therefore, all trajectories $q_j(t)$ solutions of (62) [with (77)] have the same property

$$q_j(t+T) = q_j(t), \qquad j = 1, 2, \ldots, n. \tag{80}$$

Note that this result holds independently of the initial conditions $q_j(0)$, $\dot{q}_j(0)$.[5]

Remark 4. Comparison of (63) [with (77)] and (55) suggests the following neat result [P1978]: the solutions $q_j(t)$ of the n-body problem characterized by the hamiltonian

$$H = \tfrac{1}{2}\sum_{j=1}^{n} p_j^2 + g^2 \sum_{j>k=1}^{n} \frac{1}{(q_j - q_k)^2} \tag{39}$$

and by the equations of motion

$$\ddot{q}_j = 2g^2 {\sum_{k=1}^{n}}' \frac{1}{(q_j - q_k)^3}, \qquad j = 1, 2, \ldots, n, \tag{40}$$

are in one-to-one correspondence with the solutions $r_j(t)$ of the n-body problem characterized by the hamiltonian

$$H = \tfrac{1}{2}\sum_{j=1}^{n} (p_j^2 + \omega^2 r_j^2) + g^2 \sum_{j>k=1}^{n} \frac{1}{(r_j - r_k)^2} \tag{81}$$

and by the equations of motion

$$\ddot{r}_j + \omega^2 r_j = 2g^2 {\sum_{k=1}^{n}}' \frac{1}{(r_j - r_k)^3}, \qquad j = 1, 2, \ldots, n, \tag{82}$$

via the simple relation

$$r_j(t) = \cos(\omega t)\, q_j\left(\frac{\tan(\omega t)}{\omega}\right), \qquad j = 1, 2, \ldots, n, \tag{83a}$$

$$q_j(t) = \sqrt{1 + (\omega t)^2}\, r_j\left(\frac{\arctan(\omega t)}{\omega}\right), \qquad j = 1, 2, \ldots, n. \tag{83b}$$

It is indeed trivial to verify this result directly from the equations of motion (40) and (82). One may thereby note that the relationships (83) hold, more generally, also for the solutions $q_j(t)$ and $r_j(t)$ of the (generally nonintegrable) problems with different coupling constants $g_{j,k}$ acting between different

[5] This property had been conjectured in 1971 on the basis of the solutions of the corresponding quantal problem in [C1971]. See ahead. It was first proven in the classical case in 1976, in [A1977] and [A1978]; an elementary proof is implied in the following Remark.

particle pairs:

$$H = \tfrac{1}{2}\sum_{j=1}^{n} p_j^2 + \sum_{j>k=1}^{n} \frac{g_{j,k}^2}{(q_j - q_k)^2}, \tag{84a}$$

$$\ddot{q}_j = 2\sum_{k=1}^{n}{}' \frac{g_{j,k}^2}{(q_j - q_k)^3}, \qquad j = 1, 2, \dots, n; \tag{84b}$$

$$H = \tfrac{1}{2}\sum_{j=1}^{n} (p_j^2 + \omega^2 r_j^2) + \sum_{j>k=1}^{n} \frac{g_{j,k}^2}{(r_j - r_k)^2}, \tag{85a}$$

$$\ddot{r}_j + \omega^2 r_j = 2\sum_{k=1}^{n}{}' \frac{g_{j,k}^2}{(r_j - r_k)^3}, \qquad j = 1, 2, \dots, n. \tag{85b}$$

Note that the relationships (83) map the *bounded* trajectories for the problem with $\omega \neq 0$ into the *unbounded* trajectories corresponding to the problem with $\omega = 0$. It appears moreover obvious from (83a) that the trajectories $r_j(t)$ are periodic with period $T = 2\pi/\omega$; but this conclusion holds true only if the functions $q_j(t)$ are algerbraic in their argument, as it is implied by the results given above [see, for instance (55) and (56)] for the case with equal particles, *i.e.*, $g_{j,k} = g$; while this is presumably not so in the (nonintegrable) case with different coupling constants $g_{j,k}$, in which case indeed the trajectories $r_j(t)$ are presumably not periodic.

Clearly, the problem (85) admits a unique equilibrium configuration

$$r_j(t) = \bar{r}_j, \quad \dot{r}_j(t) = 0, \quad j = 1, 2, \dots, n, \tag{86}$$

where the n (real) numbers $\bar{r}_j$ are the solutions of the system of n nonlinear algebraic equations

$$\omega^2 \bar{r}_j = 2\sum_{k=1}^{n}{}' \frac{g_{j,k}^2}{(\bar{r}_j - \bar{r}_k)^3}, \quad j = 1, 2, \dots, n. \tag{87}$$

Hence, via (83b), one infers that the n-body problem (84) admits the (*similarity*) solution

$$q_j(t) = \sqrt{1 + (\omega t)^2}\, \bar{r}_j, \qquad j = 1, 2, \dots, n. \tag{88}$$

In the special (integrable) case with equal particles ($g_{j,k} = g$), (87) reads

$$\omega^2 \bar{r}_j = 2g^2 \sum_{k=1}^{n}{}' \frac{1}{(\bar{r}_j - \bar{r}_k)^3}, \quad j = 1, 2, \dots, n, \tag{89}$$

and the quantities $\bar{r}_j$ are then conveniently rescaled setting

$$\bar{r}_j = \sqrt{g/\omega}\, \bar{x}_j, \qquad j = 1, 2, \dots, n, \tag{90}$$

so that the n numbers $\bar{x}_j$ are determined by the equations

$$\bar{x}_j = 2\sum_{k=1}^{n}{}' \frac{1}{(\bar{x}_j - \bar{x}_k)^3}, \quad j = 1, 2, \dots, n. \tag{91}$$

It so happens that these numbers are also solutions of the set of nonlinear equations

$$\bar{x}_j = \sum_{k=1}^{n}{}' \frac{1}{\bar{x}_j - \bar{x}_k}, \quad j = 1, 2, \dots, n, \tag{92}$$

and that they moreover coincide with the n zeros of the Hermite polynomial of order n :

$$H_n(\bar{x}_j) = 0, \qquad j = 1, 2, \dots, n. \tag{93}$$

These properties are proven and further discussed in § 1.3 and § 1.4.

The *quantal* n-body problem corresponding to the hamiltonians (39) and (81) have also been solved; we outline here the main results[6]

Let us first consider the $\omega = 0$ case (39). In units where $\hbar = m = 1$, the quantal hamiltonian reads

$$H = -\tfrac{1}{2}\sum_{j=1}^{n}\frac{\partial^2}{\partial x_j^2} + g^2\sum_{j>k=1}^{n}\frac{1}{(x_j - x_k)^2}, \tag{94}$$

and the corresponding stationary Schrödinger equation,

$$H\Psi(\mathbf{x}) = E\Psi(\mathbf{x}), \tag{95}$$

need be considered only in the sector of configuration space characterized by the inequalities

$$x_j < x_{j+1}, \qquad j = 1, 2, \ldots, n-1. \tag{96}$$

On the (finite) boundaries of this sector the wave function Ψ vanishes due to the repulsive and singular nature of the interaction:

$$\Psi(\mathbf{x}) = 0 \quad \text{if} \quad x_j = x_{j+1}, \qquad j = 1, 2, \ldots, n-1. \tag{97}$$

The scattering process in the sector (96) is associated with a wave function Ψ which behaves asymptotically as a superposition of an incoming and outgoing wave,

$$\Psi(\mathbf{x}) \approx \Psi_{\text{in}}(\mathbf{x}) + \Psi_{\text{out}}(\mathbf{x}), \tag{98}$$

with

$$\Psi_{\text{in}}(\mathbf{x}) = \exp\Big(i\sum_{j=1}^{n} p_j x_j\Big) \tag{99}$$

and

$$\Psi_{\text{out}}(\mathbf{x}) = \int dp'_1\cdots dp'_n \Big(\prod_{j=2}^{n}\theta(p'_j - p'_{j-1})\Big) S(\mathbf{p},\mathbf{p}')\exp\Big(i\sum_{j=1}^{n} p'_j x_j\Big). \tag{100}$$

The initial momenta p_j satisfy the inequalities

$$p_j > p_{j+1}, \qquad j = 1, 2, \ldots, n-1, \tag{101}$$

and the constraints

$$\tfrac{1}{2}\sum_{j=1}^{n} p_j^2 = E. \tag{102}$$

The fact that the outgoing momenta satisfy the conjugate inequalities

$$p'_j < p'_{j+1}, \qquad j = 1, 2, \ldots, n-1, \tag{103}$$

is explicitly enforced by the θ-functions in the integrand in the right-hand side of (100). The function S (S-*matrix*) has the general form

$$S(\mathbf{p},\mathbf{p}') = \delta\Big(\sum_{j=1}^{n} p_j^2 - \sum_{j=1}^{n} {p'_j}^2\Big)\,\delta\Big(\sum_{j=1}^{n} p_j - \sum_{j=1}^{n} p'_j\Big)\,\overline{S}(\mathbf{p},\mathbf{p}'), \tag{104}$$

[6] We refer the reader to [C1971], [CMR1975], [OP1977], and [OP1983] for more details.

reflecting energy and momentum conservation.

Thus far the treatment has been fairly general, describing the setup of a generic scattering process (with repulsive interparticle potentials, singular at zero separation). For the specific hamiltonian (94) and the corresponding Schrödinger equation (95) it is in fact possible to compute exactly the S-matrix, which turns out to have the extremely simple form

$$S(\mathbf{p},\mathbf{p}') = e^{-i\pi\lambda} \prod_{j=1}^{n} \delta(p'_j - p_{n-j+1}), \tag{105}$$

so that (100) becomes

$$\Psi_{\text{out}}(\mathbf{x}) = e^{-i\pi\lambda} \exp\left(i \sum_{j=1}^{n} p_{n-j+1} x_j \right). \tag{106}$$

The phase λ can also be explicitly computed:

$$\lambda = \tfrac{1}{2}(n-3) + \tfrac{1}{2}n(n-1)\alpha, \tag{107}$$

$$\alpha = \tfrac{1}{2}(1+\sqrt{1+4g^2}). \tag{108a}$$

The formula (105) corresponds of course to the property (33) for the corresponding classical problem. It can moreover be interpreted —say in the context of classical wave propagation in n dimensional space— by considering the n coordinates x_j as the components of a single point in n-dimensional space. Then, the potential

$$W(\mathbf{x}) = g^2 \sum_{j>k=1}^{n} \frac{1}{(x_j - x_k)^2}, \tag{109}$$

in the hamiltonian (94), appears as a (highly non-spherically-symmetric) function of the n-vector $\mathbf{x}$; the result we have just described is interpreted as the property of this potential to reflect an incoming plane wave with wave vector $\mathbf{p}$ [in the sector (96), and with the components of $\mathbf{p}$ satisfying the condition (101) which indeed qualifies $\mathbf{p}$ as the appropriate wave vector for an *incoming* wave in this sector] into an outgoing plane wave with wave vector $\mathbf{p}'$ [whose components are related to those of $\mathbf{p}$ by the rule $p'_j = p_{n-j+1}$, $j = 1, 2, \ldots, n$, implying $p'_j < p'_{j+1}$, so that $\mathbf{p}'$ qualifies as the wave vector of an outgoing plane wave in the sector (96)]. The remarkable feature of the potential (109) is of course to reflect an incoming plane wave into an outgoing plane wave —with no diffraction!

Let us turn next to the quantal problem with $\omega \neq 0$, whose hamiltonian reads

$$H = -\frac{\hbar^2}{2m} \sum_{j=1}^{n} \frac{\partial^2}{\partial x_j^2} + \tfrac{1}{2}\omega^2 \sum_{j=1}^{n} x_j^2 + g^2 \sum_{j>k=1}^{n} \frac{1}{(x_j - x_k)^2}. \tag{110}$$

Note that we have kept here the dimensional constants m and $\hbar$; we have moreover written this hamiltonian in non-translation-invariant version; the results given above [see in particular (61) and (71)–(73)] allow us to relate trivially the results given below to those for the translation-invariant version, where the external oscillator potential $\frac{1}{2}\omega^2 \sum_{j=1}^{n} x_j^2$ si replaced by the pair oscillator potential $\frac{1}{2}(\omega^2/n) \sum_{j>k=1}^{n} (x_j - x_k)^2$.

The Schrödinger equation reads now

$$H\Psi_\nu(\mathbf{x}) = E_\nu \Psi_\nu(\mathbf{x}); \tag{111}$$

again it need be considered only in the sector (96) of configuration space, in which boundaries again $\Psi_{\nu}(\mathbf{x})$ vanishes [see (97); actually now the eigenfunctions $\Psi_{\nu}(\mathbf{x})$ must vanish also at infinity, namely as $|\mathbf{x}|$ diverges, and this requirement causes the quantisation of the spectrum].

The n-vector ν has nonnegative integral components,

$$\nu = \{\nu_1, \nu_2, \ldots, \nu_n\}, \quad \nu_j = 0, 1, 2, \ldots; \tag{112}$$

to each value of ν there corresponds one eigenfunction [in the sector (96)], and the corresponding eigenvalue is given by the explicit formula

$$E_{\nu} = \frac{\hbar\omega}{\sqrt{m}}\left(\frac{3}{2} + \lambda + \sum_{j=1}^{n} j\nu_j\right), \tag{113}$$

with λ defined by (107), with (108a) being replaced now by

$$\alpha = \tfrac{1}{2}(1 + \sqrt{1 + 4mg^2\hbar^{-2}}). \tag{108b}$$

These results are independent of the statistics that the particles satisfy (Bose, Fermi, or Boltzmann); this is physically reasonable, since the particles, although all identical, are in fact distinguishable by their ordering on the line, which cannot change due to the singularity of the pair potential. Any assumption about their nature (bosons, fermions, or distinguishable) has indeed no effect on the energy spectrum (113), but merely determines the rule whereby the eigenfunction $\Psi_{\nu}(\mathbf{x})$ should be continued outside of the sector (96); in the boson (resp. fermion) case, by requiring that it be completely symmetric (resp. antisymmetric) in the particle coordinates x_j; in the distinguishable case, $n!$ linearly independent eigenfunctions can be envisaged, for instance, each of them nonvanishing only in one of the $n!$ sectors of configuration space which correspond to the $n!$ different orders of the particles on the line. That such choices are all consistent is implied by (97).

These considerations apply for $g^2 > 0$. As $g \to 0$ [implying $\alpha \to 1$, see (108)] the spectrum and eigenfunctions [in the sector (96)] of the $g^2 > 0$ problem go over into the spectrum and eigenfunctions, in the same sector, of the $g^2 = 0$ *pure oscillator* problem *with Fermi statistics*, because the boundary conditions (97) are preserved in the limit. The spectrum of the $g^2 = 0$ problem *with Bose statistics* is instead given by (113) and (107) with $\alpha = 0$. Note that these results imply that, up to a constant shift of all energy levels, the spectrum —including the multiplicities— of the pure oscillator Bose and Fermi problem, as well as that of the $g^2 > 0$ problem (with any statistics, up to a trivial $n!$ multiplicity factor in the distinguishable case) coincide.[7]

There is an amusing application of the results we have just described, which has yielded the value of a previously unknown multiple integral, namely:

$$\int_{-\infty}^{+\infty} dx_1 \cdots dx_n \exp\left(-\tfrac{1}{2}\omega^2 \sum_{j=1}^{n} x_j^2 - g^2 \sum_{j>k=1}^{n} \frac{1}{(x_j - x_j)^2}\right) = (2\pi)^{n/2}\omega^{-n} \exp[-\tfrac{1}{2}n(n-1)|g\omega|]. \tag{114}$$

Indeed, this multiple integral is proportional to the partition function Z_C for the classical hamiltonian (61) with (77) [or, equivalently, (81)]. On the other hand, the partition function Z_Q for the corresponding quantum hamiltonian (110) is easily computed from the spectrum (113). But in the limit $\hbar \to 0$, Z_Q must go over into Z_C, and this yields (114).[8]

---

[7] The spectrum E_{ν} is obviously equally spaced; this remark motivated the conjecture in [C1971] that the corresponding classical problem be completely periodic. This has, of course, turned out to be correct —as we saw above.

[8] This argument has been presented in mathematically rigorous fashion by G. Gallavotti and C. Marchioro in [GM1975].

Let us now return to the context of classical dynamics, to mention tersely some other developments, referring to the literature for more detailed treatments.[9] Several interesting results are obtained from the consideration of special configurations of the n-body systems considered above (possibly with $n = \infty$), configurations which have the property that they are preserved through time evolution. Let us briefly review various possibilities.

It is actually possible to derive the whole class of problems characterized by the hamiltonian (1.0–2) with (12) and (17) from the basic problem where (17) is replaced by (24a). For simplicity, let us confine our treatment to an outline of how the problem with the two-body potential (25a) can be generated by the problem with the two-body potential (24a). Consider indeed the latter problem with n particles having initially (real) coordinates $q_j(0)$ and velocities $\dot{q}_j(0)$, $j = 1, 2, \ldots, n$, to each of which is associated an infinite number of other particles having initially the (complex) coordinates $q_j(0) + i\pi k/a$, with $k = \pm 1, \pm 2, \pm 3$, and so on, all having velocities $\dot{q}_j(0)$. It is easily seen that such a configuration is preserved throughout time, so that it is sufficient to consider only the motion of the n particles with real coordinates $q_j(t)$, $j = 1, 2, \ldots, n$, each of which always carries along its infinite retinue of associates, which move along lines parallel to the real axis, remaining always exactly above or below $q_j(t)$ in the complex plane. On the other hand, the dynamics of $q_j(t)$ is not affected by the presence of its associates, whose forces neatly balance off in pairs. The force exerted on $q_j(t)$ by $q_k(t)$ and all its associates, meanwhile, is precisely the force that would be exerted by $q_k(t)$ alone if the interparticle potential were (25a) rather than (24a), since

$$\sum_{k=-\infty}^{+\infty} \frac{1}{(x + i\pi k/a)^2} = \frac{a^2}{\sinh^2(ax)}. \tag{115}$$

This formula applies to compare the potentials; the formula for the forces is obtained by differentiation.

Another amusing development emerges from the consideration of the many-body problem with particles of two types: the potential (38a) acting between equal particles and the potential (38b) acting between different particles.[10] Consider then the case of $2n$ particles, the first n of one type and the remaining n of the other. For obvious symmetry reasons, a configuration characterized initially by the condition

$$q_{n+j} = q_j, \quad \dot{q}_{n+j} = \dot{q}_j, \qquad j = 1, 2, \ldots, n, \tag{116}$$

is mantained throughout the motion, describing the evolution of n tightly bound two-body bound states. It is then sufficient to consider the evolution of the first n coordinates q_j; indeed, one might hope to generate in this manner a novel integrable n-body problem. But this novel problem is merely the original one (up to rescaling), since the interaction potential felt by the particle q_j due to the two (different) particles located at the position q_k reads, with $x = |q_j - q_k|$,

$$V_e(x) + V_d(x) = \left(\frac{2ag}{\sinh(2ax)}\right)^2 \tag{117}$$

—see (38).

Thus, rather than discovering a new integrable problem, one finds in this manner a new technique to deal with the old problem characterized by the interaction potential (25a). In particular, it

[9] Solvable n-body problems in two-dimensional space can be invented by complexifying the models considered above; see [C1976a] and [C1978]. For some integrable n-body problems analogous to those considered above but involving three-body forces, see [W1974], [CN1974], and [OP1981].

[10] Actually, the same approach could also be applied to the more general problem with four types of particles, which can be evinced from the many-body problem with the interparticle potential (17) by the same sort of trick which has yielded the problem with two types of particles and the pair interactions (38) out of the many-body model with the pair interaction (25a). For simplicity, however, we treat here only the simpler case with two types of particles with interactions (38).

is possible to introduce in this manner **L** and **A** Lax matrices of order $2n$ rather than n; and since the procedure can be repeated, also matrices of order $4n$, $8n$, and so on. Of course, whichever be the order of the Lax matrix **L**, only n independent constants of motion can be obtained out of its eigenvalues; the fact that only n eigenvalues of **L** are independent, even if **L** has order $2n$, $4n$, $8n$, etc., is related to the validity of formulae relating hyperbolic functions of argument $x/2$, $x/4$, $x/8$, etc., to hyperbolic functions of argument x. In the more general case associated with the Weierstrass potential (17), the relevant relations involve the Jacobi elliptic functions rather than hyperbolic ones.[11]

A third development connected with the consideration of special configurations is mainly of interest because it exhibits a remarkable relationship with the root systems associated with semisimple Lie algebras.[12] Consider any n-body system described by the hamiltonian (1.0–2), with (1.0–5), and moreover with

$$v(x) = v(-x). \tag{118}$$

Let $n = 2m + \nu$ and assume that the coordinates q_j and the momenta $p_j = \dot{q}_j$ satisfy the conditions

$$q_{m+j} = -q_j,\ p_{m+j} = -p_j,\ j = 1, 2, \ldots, m; \quad q_j = p_j = 0,\ j = 2m+1, 2m+2, \ldots, 2m+\nu. \tag{119}$$

Clearly, such a configuration is compatible with time evolution, namely if it exists at any one time (say, at $t = 0$), it holds for all time. One can then forget about the last $m + \nu$ particles; the last ν are in fact just fixed at the origin, and the positions of the other m are completely determined via (119) by the positions of the first m particles. Thus, the subcase of this n-body problem corresponding to the configuration (119) yields a novel *m-body problem*, characterized (up to an irrelevant constant which may be ignored even if it should be divergent) by the hamiltonian

$$H = \tfrac{1}{2}\sum_{j=1}^{m} p_j^2 + \sum_{j=1}^{m}[v(q_j) + V(2q_j) + \nu V(q_j)] + \sum_{j>k=1}^{m}[V(q_j - q_k) + V(q_j + q_k)]. \tag{120}$$

Unfortunately, the possibility to interpret the dynamical system characterized by this hamiltonian as an m-body problem is somewhat marred by the appearance of the sum (rather than the difference) of the particle coordinates as argument of the last term in the right-hand side. On the other hand it is generally unnecessary to require that ν be an integer.[13]

We end this section with some hitherto unpublished results, which are so easily obtained using the technique of Olshanetsky and Perelomov [see the treatment following (40)] to justify the compact presentation given below, without proofs.

Consider the n-body problem characterized by the equations of motion

$$\ddot{q}_j + \omega^2 q_j = \frac{\dot{g}}{g}\dot{q}_j + 2g^2 \sum_{k=1}^{n}{}' \frac{1}{(q_j - q_k)^3}, \qquad j = 1, 2, \ldots, n, \tag{121}$$

where we are now assuming that g, as well as ω, are (given) time-dependent quantities:

$$g = g(t), \qquad \omega = \omega(t). \tag{122}$$

---

[11] For a (marginally) more detailed presentation, the interested reader is referred to [C1976b] and [C1978].

[12] We refer to the review paper [OP1981] and to the literature quoted there for these mathematical developments, confining our treatment here to a mere outline of the basic idea.

[13] Let us again emphasize that the interest of this last development is the relation which can be established with properties of semisimple Lie algebras, as in [OP1981]. There are other group theoretical considerations which are also connected to some of the results described above; for these we refer to the original papers: [P1971], [G1975], [BR1977], [KKS1978], [B1980], and the review paper [OP1981].

[An additional term $\phi(t)$ could be added to the right-hand side of (121), but since such an addition affects only the motion of the center of mass of the system, we forsake such a trivial extension.]

The solutions $q_j(t)$ of these equations of motion, characterized by the initial data $q_j(0)$ and $\dot{q}_j(0)$, coincide with the eigenvalues of the matrix $\mathbf{Y}(t)$ defined by the differential equation

$$\ddot{\mathbf{Y}} - \frac{\dot{g}}{g}\dot{\mathbf{Y}} + \omega^2\mathbf{Y} = 0, \tag{123}$$

with the initial conditions (50) [with (41) and (51)]. This matrix $\mathbf{Y}(t)$ is therefore given by the formula

$$Y_{j,k}(t) = \begin{cases} q_j(0)y_0(t) + \dot{q}_j(0)y_1(t), & j = k, \\ \dfrac{ig(t)y_1(t)}{q_j(0) - q_k(0)}, & j \neq k, \end{cases} \tag{124}$$

where the two functions $y_s(t)$, $s = 0,1$ are the two solutions of the second-order scalar ordinary differential equation

$$\ddot{y}(t) - \frac{\dot{g}(t)}{g(t)}\dot{y}(t) + \omega^2(t)y(t) = 0, \tag{125}$$

characterized by the initial conditions

$$y_0(0) = 1, \qquad \dot{y}_0(0) = 0, \tag{126a}$$

$$y_1(0) = 0, \qquad \dot{y}_1(0) = 1. \tag{126b}$$

The differential equation (125) can be solved explicitly if

$$\omega^2(t) = \ddot{f}(t) - \dot{f}^2(t) - \frac{\dot{g}(t)}{g(t)}\dot{f}(t) + \omega_0^2 g^2(t)e^{4f(t)}, \tag{127}$$

where $f(t)$ is an arbitrary function and ω_0 is a constant (possibly vanishing). The general solution then reads

$$y(t) = \frac{a}{\omega_0}e^{-f(t)}\sin\left(\omega_0\int_{t_0}^{t} dt'\, g(t')\, e^{2f(t')}\right), \tag{128}$$

with a and t_0 arbitrary constants.

Let us display three especially simple cases. The first corresponds to

$$f(t) = 0; \tag{129a}$$

then

$$\omega(t) = \omega_0 g(t) \tag{129b}$$

and

$$y_0(t) = \cos\left(\omega_0\int_0^t dt'\, g(t')\right), \tag{129c}$$

$$y_1(t) = \frac{1}{\omega_0 g(0)}\sin\left(\omega_0\int_0^t dt'\, g(t')\right). \tag{120d}$$

In particular, if $\omega_0 = 0$ [implying $\omega(t) = 0$], $y_0(t) = 1$ and $y_1(t) = \int_0^t dt' \, g(t')/g(0)$; thus the eigenvalues of

$$Y_{j,k}(t) = \begin{cases} q_j(0) + \dfrac{\dot q_j(0)}{g(0)} \displaystyle\int_0^t dt' \, g(t'), & j = k, \\ i \dfrac{g(t)}{g(0)} \dfrac{1}{q_j(0) - q_k(0)} \displaystyle\int_0^t dt' \, g(t'), & j \neq k, \end{cases} \tag{130}$$

provide the solutions $q_j(t)$ of the equations of motion

$$\ddot q_j(t) = \frac{\dot g(t)}{g(t)} \dot q_j(t) + 2g^2(t) \sum_{k=1}^{n}{}' \frac{1}{[q_j(t) - q_k(t)]^3}, \qquad j = 1, 2, \ldots, n, \tag{131}$$

for any arbitrary choice of the function $g(t)$ [such that the right-hand side of (130) makes sense]. In fact, this result could have been obtained directly from (40), (55), and (56), by a redefinition of the time variable.

The second case corresponds to

$$f(t) = -\ln\left(\int_{t_1}^t dt' \, g(t')\right), \qquad e^{-f(t)} = \int_{t_1}^t dt' \, g(t'), \tag{132a}$$

where t_1 is an arbitrary constant. Then,

$$\omega(t) = \omega_0 g(t) \left(\int_{t_1}^t dt' \, g(t')\right)^{-2}. \tag{132b}$$

It is easily seen that, for $\omega_0 = 0$ [implying $\omega(t) = 0$], the results reported above are re-obtained.

The third case corresponds to

$$f(t) = -\tfrac{1}{2} \ln[g(t)], \qquad e^{-2f(t)} = g(t). \tag{133a}$$

Then,

$$\omega^2(t) = \omega_0^2 - \frac{1}{2} \frac{\ddot g(t)}{g(t)} + \frac{3}{4} \frac{\dot g^2(t)}{g^2(t)}. \tag{133b}$$

Note that in all these cases the choice of $g(t)$ remains free (except for the requirement that no singularity arises), while $\omega(t)$ is determined once $g(t)$ (and ω_0) have been assigned. If g is time independent, $g(t) = g$, the first and third cases reduce back to the standard case with constant ω, while the second case yields

$$\omega(t) = \frac{\omega_0}{g(t + \tau)^2}. \tag{134}$$

[We have set $t_1 = -\tau$, with the implication that τ be positive, so that $\omega(t)$ be nonsingular for $t \geq 0$.] The corresponding formulae for $y_0(t)$ and $y_1(t)$ read

$$y_0(t) = -\sqrt{1 + \frac{1}{a^2}} \left(1 + \frac{t}{\tau}\right) \sin\left(\frac{at}{t + \tau} - \arctan a\right), \tag{135a}$$

$$y_1(t) = -\frac{t + \tau}{g} \sin\left(\frac{at}{t + \tau}\right), \tag{135b}$$

$$a = \frac{\omega_0}{g\tau}. \tag{135c}$$

These formulae provide thus, via (124) and (56), the solution of the initial value problem for the n-body system characterized by the equations of motion

$$\ddot{q}_j + \frac{a^2\tau^2}{(t+\tau)^4} q_j = 2g^2 \sum_{k=1}^{n}{}' \frac{1}{(q_j - q_k)^3}, \qquad j = 1, 2, \ldots, n. \tag{136}$$

As a final example let us consider the case with time independent $\omega = \omega_0$ and with $g(t) = g_0 e^{-\alpha t}$, so that the equations of motion read

$$\ddot{q}_j + \omega_0^2 q_j = -\alpha \dot{q}_j + 2g_0 e^{-2\alpha t} \sum_{k=1}^{n}{}' \frac{1}{(q_j - q_k)^3}, \qquad j = 1, 2, \ldots, n, \tag{137}$$

with ω_0, α and g_0 (real) constants. The solutions $q_j(t)$ are then the eigenvalues of the matrix $\mathbf{Y}$, see (124), with [see (133); or solve directly (125)]

$$y_0(t) = \frac{\omega_0}{\tilde{\omega}} e^{-\alpha t/2} \sin\left(\tilde{\omega} t + \arctan \frac{2\tilde{\omega}}{\alpha}\right), \tag{138a}$$

$$y_1(t) = e^{-\alpha t/2} \frac{\sin(\tilde{\omega} t)}{\tilde{\omega}}, \tag{138b}$$

where

$$\tilde{\omega} = \sqrt{\omega_0^2 - \alpha^2/4}. \tag{138c}$$

These formulae apply even if $\tilde{\omega}$ is imaginary, or if it vanishes. Of course, if $\alpha > 0$, they imply that

$$q_j(t) \to 0, \quad j = 1, 2, \ldots, n, \quad \text{as } t \to +\infty; \tag{139}$$

a result which could have been immediately inferred from the equations of motion (137).

§1.2 Motion of the poles of special solutions of nonlinear partial differential equations. Examples

Let[14] us take as starting point for our discussion, the celebrated Korteweg–de Vries equation,

$$u_t + u_{xxx} - 6u_x u = 0, \qquad u \equiv u(x, t); \tag{1}$$

and let us investigate the possibility that this nonlinear evolution partial differential equation admits (possibly complex) solutions which are, for all time, rational in x. It is then easily seen that such solutions must have the form

$$u(x, t) = 2 \sum_{j=1}^{n} \frac{1}{[x - x_j(t)]^2}. \tag{2}$$

Moreover, the quantities $x_j(t)$ must satisfy the n constraints

$$\sum_{k=1}^{n}{}' \frac{1}{[x_j(t) - x_k(t)]^3} = 0, \qquad j = 1, 2, \ldots, n, \tag{3}$$

[14]The material in this and the following Sections is largely based on [C1978a]. See also [C1978].

and evolve in time according to the equation

$$\dot{x}_j(t) = -12 \sum_{k=1}^{n}{}' \frac{1}{[x_j(t) - x_k(t)]^2}, \qquad j = 1, 2, \ldots, n. \tag{4}$$

It can be shown (see Appendix 2.2.A) that the constraint (3) is compatible with the time evolution (4), namely that if (3) holds at any one time (say, at $t = 0$), and the quantities $x_j(t)$ evolve according to (4), then (3) holds for all time. Moreover, (3) and (4) imply (see the Appendix of § 1.2)

$$\ddot{x}_j(t) = 288 \sum_{k=1}^{n}{}' \frac{1}{[x_j(t) - x_k(t)]^5}, \qquad j = 1, 2, \ldots, n; \tag{5}$$

thus any solution of (3) and (4) provides a (special) solution of the n-body problem characterized by the hamiltonian

$$H(\mathbf{x}, \mathbf{p}) = \tfrac{1}{2} \sum_{j=1}^{n} p_j^2 + 72 \sum_{j>k=1}^{n} \frac{1}{(x_j - x_k)^4}. \tag{6}$$

The two-body potential which appears here,

$$V(x) = g^2/x^4, \tag{7}$$

is sometimes referred to as the (one-dimensional) *Maxwell* potential. Note that the value of the coupling constant g^2 can be modified by rescaling. There is no reason to expect the dynamical system (5) to be integrable for $n \geq 3$; indeed, no other solutions of this problem are known, besides the rather unphysical ones implied by the results we have just described (see also below).

The constraint (3) is trivially satisfied for $n = 1$, in which case (4) yields simply

$$\dot{x}_1(t) = 0, \qquad x_1(t) = x_1. \tag{8}$$

Indeed, $u(x,t) = 2/(x - x_1)^2$ is a (rather uninteresting) solution of (1).

Another case when (3) is obviously verified —for symmetry reasons— corresponds to the configuration (with $n = \infty$)

$$x_j(t) = \xi(t) + aj + b, \qquad j = 0, \pm1, \pm2, \pm3, \ldots, \tag{9a}$$

with a and b constants. It is convenient to set

$$a = i\pi/p, \qquad b = i\pi/2p; \tag{9b}$$

then (4) yields

$$\dot{\xi}(t) = \frac{24p^2}{\pi^2} \sum_{k=1}^{\infty} \frac{1}{k^2} = 4p^2, \tag{10}$$

and correspondingly (2) becomes

$$u(x,t) = \frac{-2p^2}{\cosh^2\{p[x - \xi(t)]\}}. \tag{11}$$

In this manner, thus, one has merely recovered the single-soliton solution of the KdV equation. Since we have allowed an infinite number of poles, this is of course not a *rational* solution of the KdV equation;

indeed, by taking an infinite square lattice of poles, one could have similarly obtained the solution of the KdV equation representing a ***cnoidal wave***, in which the hyperbolic function which appears in the single-soliton solution (11), is replaced by an appropriate Jacobi elliptic function.

More interesting, but less simple, are the cases with n finite and larger than unity. It is obvious that if the quantities x_j are all real, the constraint (3) cannot be satisfied [indeed, for the two values of j which correspond to the largest and the smallest x_j's, all the terms in the sum (3) have then the same sign]. It is less obvious, but nevertheless true (see [AMM1977]), that by allowing the x_j's to be complex, (3) can be satisfied if, and only if,

$$n = \tfrac{1}{2}m(m+1), \qquad m = 1, 2, 3, \ldots, \tag{12a}$$

namely for the values

$$n = 1,\ 3,\ 6,\ 10, \ldots. \tag{12b}$$

For instance, for $n = 3$, the set

$$x_j = y - ze^{2\pi ij/3}, \qquad j = 1,\ 2,\ 3, \tag{13a}$$

satisfies (3), as can be easily verified; and it is also easily seen that (4) implies the following time evolution for the quantities y and z:

$$\dot{y}(t) = 0, \tag{13b}$$

$$z(t) = \sqrt[3]{[z(0)]^3 + 12t}. \tag{13c}$$

A remarkable feature of the time evolution characterized by (3) and (4) is its relationship to the integrable n-body problem characterized by the hamiltonian (1.1–39).[15] These results are extremely interesting from a mathematical point of view, since they establish a connection between two integrable problems —the one-dimensional n-body system with inverse-square pair potentials and the KdV nonlinear PDE— each of which has played a seminal role in the developments which have recently occurred in the fields of finite-dimensional integrable dynamical systems (see § 1.1) and nonlinear PDEs integrable via the spectral transform technique. See, for instance, [CD1982].

The interest of these results as a source of solvable one-dimensional many-body problems is, on the other hand, marginal; not only because they generate problems already known to be solvable, but especially because they saddle these problems with unphysical constraints. For example, we have just seen that in the KdV case there are restrictions on the number n of *particles* and, moreover, the initial positions and velocities are constrained by (3) and (4), and these imply that these quantities cannot all be real.

There are some cases when these constraints are less stringent, but it appears ([C1978a]) that the only case when the motion of the poles of rational solutions of a nonlinear PDE can be identified with a *sort of* n-body problem without any constraint on the number of particles, nor on their initial positions and velocities, is associated with the (elementary) nonlinear PDE

$$\phi_t + \phi_x + \alpha\phi + \phi^2 = 0, \qquad \phi \equiv \phi(x, t). \tag{14}$$

[Hereafter we assume α to be a real constant, although the treatment would also apply if α were a t-dependent complex quantity.] We devote the rest of this Section to a succint analysis of this problem,

[15] For this and related developments, the interested reader is referred to the literature: [AMM1977], [CC1977], [AM1978], [C1978a],[K1978], [K1980], and [OP1981].

even though the same results can also be obtained more straightforwardly using techniques described in the following § 1.3.

The PDE (14) is trivially solved through noting that

$$\psi(x,t) = \frac{1}{\phi(x,t)} \tag{15}$$

satisfies a linear first-order PDE. The general solution reads

$$\phi(x,t) = \frac{\phi_0(x-t)}{e^{\alpha t} + \phi_0(x-t)[e^{\alpha t}-1]/\alpha}, \tag{16}$$

where of course

$$\phi_0(x) \equiv \phi(x,0). \tag{17}$$

It is plain, on the other hand, that any solution of (14) which is rational in x has the structure

$$\phi(x,t) = \sum_{j=1}^{n} \frac{r_j(t)}{x - x_j(t)}, \tag{18}$$

and that the residues $r_j(t)$ and poles $x_j(t)$ satisfy the evolution equations

$$\dot{x}_j = 1 - r_j, \qquad j = 1,2,\ldots,n, \tag{19a}$$

$$\dot{r}_j = -\alpha r_j - 2r_j \sum_{k=1}^{n}{}' \frac{r_k}{x_j - x_k}, \qquad j = 1,2,\ldots,n, \tag{19b}$$

implying

$$\ddot{x}_j = \alpha(1-\dot{x}_j) + 2(1-\dot{x}_j)\sum_{k=1}^{n}{}' \frac{1-\dot{x}_k}{x_j - x_k}, \qquad j = 1,2,\ldots,n. \tag{20}$$

Note that there are now no restrictions on the initial positions $x_j(0)$ or velocities $\dot{x}_j(0)$ of the n *particles* whose evolution is characterized by the *equations of motion* (20), nor is there any restriction on the number n, which can be any positive integer. But if the equations of motion (20) are interpreted in terms of Newton's law for the motion of n unit-mass particles on the line interacting via a pair force, then one sees that this force depends not only on the interparticle distance, but also on the velocities of the particles, *i.e.*, we have *velocity-dependent* forces.

The fact that the solutions $x_j(t)$ of (20) are the poles of (18) implies, via (16)–(18), that these quantities coincide with the n roots of the equation in x:

$$\sum_{j=1}^{n} \frac{1-\dot{x}_j(0)}{x - t - x_j(0)} = -\frac{\alpha}{1+e^{-\alpha t}}. \tag{21}$$

Thus, the solution of the initial value problem for the system of nonlinear coupled second-order ODEs (20) is reduced to the determination of the n roots of the single algebraic equation (21).

It is actually convenient to perform the change of variables

$$y_j(t) = x_j(t) - t, \qquad j = 1,2,\ldots,n, \tag{22}$$

and to interpret the quantities $y_j(t)$ —rather than the $x_j(t)$— as the coordinates of n particles on the line. The corresponding equations of motion read then

$$\ddot{y}_j = -\alpha \dot{y}_j + 2\dot{y}_j \sum_{k=1}^{n}{}' \frac{\dot{y}_k}{y_j - y_k}, \quad j = 1, 2, \ldots, n, \tag{23}$$

and the solutions $y_j(t)$ of these equations are the n roots of the algebraic equation in y,

$$\sum_{j=1}^{n} \frac{\dot{y}_j(0)}{y - y_j(0)} = \frac{\alpha}{1 - e^{-\alpha t}}. \tag{24}$$

The behaviour of the n roots of this equation as a function of time is conveniently analyzed by drawing a graph of the function of y represented by the left-hand side, and then noticing that the roots $y_j(t)$ are the values of y at which this graph crosses the horizontal line representing the right-hand side. Let us focus attention, for simplicity, on the case with all the velocities positive and with the particles labeled in increasing order from left to right:

$$\dot{y}_j > 0, \qquad j = 1, 2, \ldots, n, \tag{25a}$$
$$y_j < y_{j+1}, \qquad j = 1, 2, \ldots, n-1; \tag{25b}$$

note that these conditions are guaranteed to hold for all time if they hold at one time, say at $t = 0$ [this follows by inspection from (23)], and moreover, they are sufficient to exclude the occurrence of two-body collapse [since the singular two-body force is repulsive if (25a) holds].

It is now easy to analyse the motion. In particular, one finds the following asymptotic results: for $\alpha > 0$,

$$\lim_{t \to +\infty} y_j(t) = b_j(\alpha), \qquad j = 1, 2, \ldots, n, \tag{26a}$$
$$y_1(t) = -e^{-\alpha t} \frac{v}{\alpha} [1 + O(e^{\alpha t})] \quad \text{as} \quad t \to -\infty, \tag{26b}$$
$$\lim_{t \to -\infty} y_j(t) = a_{j-1}, \qquad j = 2, 3, \ldots, n. \tag{26c}$$

Here,

$$v \equiv \sum_{j=1}^{n} \dot{y}_j(0), \tag{27}$$

the n quantities $b_j(\alpha)$ are the n (real) roots, increasingly ordered, of the algebraic equation in b,

$$\sum_{j=1}^{n} \frac{\dot{y}_j(0)}{b - y_j(0)} = \alpha, \tag{28}$$

and the $n-1$ quantities a_j are the $n-1$ (real) roots, increasingly ordered, or the algebraic equation in a,

$$\sum_{j=1}^{n} \frac{\dot{y}_j(0)}{a - y_j(0)} = 0. \tag{29}$$

Note that the result (26a) is consistent with the presence of the *braking* effect represented, for $\alpha > 0$, by the first term in the right-hand side of (23).

For $\alpha = 0$ there is no braking, and the asymptotic results are accordingly modified as follows:

$$y_1(t) = vt + a_0 + O(|t|^{-1}) \qquad \text{as} \quad t \to -\infty, \tag{30a}$$
$$\lim_{t\to-\infty} y_j(t) = a_{j-1}, \qquad j = 2, 3, \ldots, n, \tag{30b}$$
$$\lim_{t\to+\infty} y_j(t) = a_j, \qquad j = 1, 2, \ldots, n-1, \tag{30c}$$
$$y_n(t) = vt + a_0 + O(t^{-1}) \qquad \text{as} \quad t \to +\infty, \tag{30d}$$

with the quantities a_j and v defined in (27) and (29), and

$$a_0 = \frac{1}{v}\sum_{j=1}^{n} \dot{y}_j(0)\, y_j(0). \tag{31}$$

These results (whose *physical* description is remarkably neat) follow of course from an analysis analogous to that described above, except for the fact that (24) reads now

$$\sum_{j=1}^{n} \frac{\dot{y}_j(0)}{y - y_j(0)} = \frac{1}{t}. \tag{32}$$

Consistently with the title of this Section, the *n-body problems* (20) [or, equivalently, (23)] have been identified with the motion of the poles of rational solutions of the explicitly solvable nonlinear PDE (14). But in fact this nonlinear PDE can be transformed into a linear PDE by the trivial change of dependent variable (15); hence the investigation of the motion of the n *poles* of solutions of (14) which are rational in x is tantamount to the study of the motion of the n *zeros* of solutions of a *linear* PDE, which are in fact polynomial in x. This suggests a more systematic investigation of the latter problem, which turns out in fact to be rather fruitful as a source of solvable dynamical systems resembling classical one-dimensional n-body problems. This approach is treated in the following § 1.3.

Appendix. Proof of (1.2–4) and (1.2–5)

Let

$$z_j(t) \equiv {\sum_{k=1}^{n}}' \frac{1}{[x_j(t) - x_k(t)]^3}, \qquad j = 1, 2, \ldots, n. \tag{1}$$

We now prove that (1.2–3), namely

$$\dot{x}_j(t) = c {\sum_{k=1}^{n}}' \frac{1}{[x_j(t) - x_k(t)]^2}, \qquad j = 1, 2, \ldots, n, \tag{2}$$

implies

$$\dot{z}_j(t) = 2c\left(z_j^2(t) + {\sum_{k=1}^{n}}' \frac{z_k(t)}{[x_j(t) - x_k(t)]^3}\right), \qquad j = 1, 2, \ldots, n, \tag{3}$$

and

$$\ddot{x}_j(t) = 2c^2 {\sum_{k=1}^{n}}' \left(\frac{1}{[x_j(t) - x_k(t)]^5} + \frac{z_k(t)}{[x_j(t) - x_k(t)]^2}\right), \qquad j = 1, 2, \ldots, n. \tag{4}$$

From these two formulae it is then immediate to infer that

$$z_j(0) = 0, \qquad j = 1, 2, \ldots, n \tag{5}$$

implies

$$z_j(t) = 0, \qquad j = 1, 2, \ldots, n \tag{6}$$

and

$$\ddot{x}_j(t) = 2c^2 \sum_{k=1}^{n}{}' \frac{1}{[x_j(t) - x_k(t)]^5}, \qquad j = 1, 2, \ldots, n; \tag{7}$$

these are the results to be proven in this Appendix.

The proofs of (3) and (4) are manipulative, and we indicate below the relevant steps. The symbol Σ'' indicates the double sum over k and l from 1 to n, with $k \neq l$, $k \neq j$, and $l \neq j$. We thus write:

$$\dot{z}_j = -3c \sum_{k=1}^{n}{}' (x_j - x_k)^{-4} \left(\sum_{l=1}^{n}{}' (x_j - x_l)^{-2} - \sum_{l=1}^{n}{}' (x_k - x_l)^{-2} \right) \tag{8a}$$

$$= -3c \sum{}'' (x_j - x_k)^{-4} [(x_j - x_l)^{-2} - (x_k - x_l)^{-2}] \tag{8b}$$

$$= -3c \sum{}'' (x_j - x_k)^{-4} [(x_j - x_l)^{-1} - (x_k - x_l)^{-1}][(x_j - x_l)^{-1} + (x_k - x_l)^{-1}] \tag{8c}$$

$$= 3c \sum{}'' (x_j - x_k)^{-3} [(x_j - x_l)^{-2}(x_k - x_l)^{-1} + (x_j - x_l)^{-1}(x_k - x_l)^{-2}] \tag{8d}$$

$$\begin{aligned} = 3c \sum{}'' (x_j - x_k)^{-3} [&(x_j - x_l)^{-2}(x_k - x_l)^{-1} - (x_j - x_l)^{-3} + (x_j - x_l)^{-3} \\ &+ (x_j - x_l)^{-1}(x_k - x_l)^{-2} - (x_k - x_l)^{-3} + (x_k - x_l)^{-3}] \end{aligned} \tag{8e}$$

$$= 3c \left(w_j + z_j^2 + \sum_{k=1}^{n}{}' z_k (x_j - x_k)^{-3} \right); \tag{8f}$$

$$\begin{aligned} w_j = \sum{}'' (x_j - x_k)^{-3} \big(&(x_j - x_l)^{-2} [(x_k - x_l)^{-1} - (x_j - x_l)^{-1}] \\ &+ (x_k - x_l)^{-2} [(x_j - x_l)^{-1} - (x_k - x_l)^{-1}] \big) \end{aligned} \tag{9a}$$

$$= \sum{}'' (x_j - x_k)^{-2} [(x_j - x_l)^{-3}(x_k - x_l)^{-1} - (x_k - x_l)^{-3}(x_j - x_l)^{-1}] \tag{9b}$$

$$= \sum{}'' (x_j - x_k)^{-2} (x_j - x_l)^{-2} (x_k - x_l)^{-2} [(x_k - x_l)(x_j - x_l)^{-1} - (x_j - x_l)(x_k - x_l)^{-1}] \tag{9c}$$

$$\begin{aligned} = \tfrac{1}{2} \sum{}'' &(x_j - x_k)^{-2} (x_j - x_l)^{-2} (x_k - x_l)^{-2} \\ &\times \big((x_k - x_l)[(x_j - x_l)^{-1} - (x_j - x_k)^{-1}] - (x_k - x_l)^{-1} [(x_j - x_l) - (x_j - x_k)] \big) \end{aligned} \tag{9d}$$

$$= -\tfrac{1}{2} \sum{}'' [(x_j - x_k)^{-3}(x_j - x_l)^{-3} + (x_j - x_k)^{-2}(x_j - x_l)^{-2}(x_k - x_l)^{-2}] \tag{9e}$$

$$= -\tfrac{1}{2} z_j^2 + \tfrac{1}{2} \sum_{k=1}^{n}{}' (x_j - x_k)^{-6} - \tfrac{1}{2} v_j; \tag{9f}$$

$$
\begin{aligned}
v_j &= \sum{}'' (x_j - x_k)^{-2}(x_j - x_l)^{-2}(x_k - x_l)^{-2} && (10a)\\
&= \sum{}'' [(x_j - x_k)^{-1} - (x_j - x_l)^{-1}]^2 (x_k - x_l)^{-4} && (10b)\\
&= 2\sum{}'' [(x_j - x_k)^{-2} - (x_j - x_k)^{-1}(x_j - x_l)^{-1}](x_k - x_l)^{-4} && (10c)\\
&= 2\sum{}'' (x_j - x_k)^{-2}(x_j - x_l)^{-1}(x_k - x_l)^{-3} && (10d)\\
&= 2\sum{}'' (x_j - x_k)^{-2}\left([(x_j - x_l)^{-1} - (x_j - x_k)^{-1}] + (x_j - x_k)^{-1}\right)(x_k - x_l)^{-3} && (10e)\\
&= 2\sum{}'' [-(x_j - x_k)^{-3}(x_j - x_l)^{-1}(x_k - x_l)^{-2} + (x_j - x_k)^{-3}(x_k - x_l)^{-3}] && (10f)\\
&= -2\sum{}'' (x_j - x_k)^{-2}(x_j - x_l)^{-2}(x_k - x_l)^{-1}[(x_j - x_k)^{-1} + (x_k - x_l)^{-1}]\\
&\quad + 2\sum_{k=1}^{n}{}' (x_j - x_k)^{-6} + 2\sum_{k=1}^{n}{}' z_k (x_j - x_k)^{-3} && (10g)\\
&= 2\sum_{k=1}^{n}{}' (x_j - x_k)^{-6} + 2\sum_{k=1}^{n}{}' z_k (x_j - x_k)^{-3} - 2v_j - u_j\,; && (10h)
\end{aligned}
$$

$$
\begin{aligned}
u_j &= 2\sum{}'' (x_j - x_k)^{-3}(x_j - x_l)^{-2}(x_k - x_l)^{-1} && (11a)\\
&= \sum{}'' (x_j - x_k)^{-2}(x_j - x_l)^{-2}(x_k - x_l)^{-1}[(x_j - x_k)^{-1} - (x_j - x_l)^{-1}] && (11b)\\
&= \sum{}'' (x_j - x_k)^{-3}(x_j - x_l)^{-3} && (11c)\\
&= z_j^2 - \sum_{k=1}^{n}{}' (x_j - x_k)^{-6}\,; && (11d)
\end{aligned}
$$

$$
\begin{aligned}
v_j &= \sum_{k=1}^{n}{}' (x_j - x_k)^{-6} - \frac{1}{3} z_j^2 + \frac{2}{3}\sum_{k=1}^{n}{}' z_k (x_j - x_k)^{-3}\,; && (12)\\
w_j &= -\frac{1}{3}\left(z_j^2 + \sum_{k=1}^{n}{}' z_k (x_j - x_k)^{-3}\right); && (13)\\
\dot{z}_j &= 2c\left(z_j^2 + \sum_{k=1}^{n}{}' z_k (x_j - x_k)^{-3}\right). && \textit{Q.E.D.}\quad (14)
\end{aligned}
$$

$$
\begin{aligned}
\ddot{x}_j &= -2c^2 \sum_{k=1}^{n}{}' (x_j - x_k)^{-3}\left(\sum_{l=1}^{n}{}' (x_j - x_l)^{-2} - \sum_{l=1}^{n}{}' (x_k - x_l)^{-2}\right) && (15a)\\
&= -2c^2 z_j \sum_{l=1}^{n}{}' (x_j - x_l)^{-2} + 2c^2 \sum_{k=1}^{n}{}' (x_j - x_k)^{-5} + 2c^2 a_j\,; && (15b)
\end{aligned}
$$

$$a_j = \sum{}'' (x_j - x_k)^{-3}(x_k - x_l)^{-2} \tag{16a}$$
$$= \sum{}'' (x_j - x_k)^{-1}[(x_j - x_k)^{-1} + (x_k - x_l)^{-1}]^2 (x_j - x_l)^{-2} \tag{16b}$$
$$= \sum{}'' [(x_j - x_k)^{-3}(x_j - x_l)^{-2} + (x_j - x_k)^{-1}(x_k - x_l)^{-2}(x_j - x_l)^{-2}] \tag{16c}$$
$$= \sum{}'' ((x_j - x_k)^{-3}(x_j - x_l)^{-2} + [(x_j - x_k)^{-1} - (x_j - x_l)^{-1}](x_k - x_l)^{-3}(x_j - x_l)^{-1}) \tag{16d}$$
$$= \sum{}'' [(x_j - x_k)^{-3} - (x_k - x_l)^{-3}](x_j - x_l)^{-2} \tag{16e}$$
$$= \sum_{l=1}^{n}{}' (z_j + z_l)(x_j - x_l)^{-2}\ ; \tag{16f}$$

$$\ddot{x}_j = 2c^2 \sum_{k=1}^{n}{}' [(x_j - x_k)^{-5} + z_k(x_j - x_k)^{-2}]\,. \qquad Q.E.D. \tag{17}$$

§1.3 Motion of the zeros of special solutions of linear partial differential equations. Examples

Let[16] $p_n(x,t)$ be a polynomial of degree n in x, whose n zeros $x_j(t)$ are, of course, functions of t :

$$p_n(x,t) = \prod_{j=1}^{n} [x - x_j(t)]. \tag{1}$$

Let then

$$p_{n-1}{}^{(j)}(x,t) = \frac{p_n(x,t)}{x - x_j(t)} = \prod_{\substack{k=1 \\ k \neq j}}^{n} [x - x_k(t)], \quad j = 1, 2, \ldots, n. \tag{2}$$

Thus $p_{n-1}{}^{(j)}(x,t)$ is a polynomial of degree $n-1$ in x. It can be easily shown that there exists a (non-unique) weight $w(x,t)$ which renders these polynomials orthogonal:

$$\int_a^b dx\, w(x,t) p_{n-1}{}^{(j)}(x,t)\, p_{n-1}{}^{(k)}(x,t) = 0 \quad \text{if} \quad j \neq k,\ j,k = 1, 2, \ldots, n. \tag{3}$$

(This result, however, will not be used in this Section.)

Through differentiating the logarithm of (1), and some algebra, it is easy to establish the following formula:

$$Dp_n(x,t) = \sum_{j=1}^{n} p_{n-1}{}^{(j)}(x,t) R_j, \tag{4}$$

[16]The material in this Section is largely based on the results of [C1978a], a subset of results being presented here. See also [C1978].

where D is the differential operator

$$
\begin{aligned}
D \equiv\ & [A_0 + A_1 x + A_2 x^2 + A_3 x^3] \frac{\partial^2}{\partial x^2} \\
& + [B_0 + B_1 x - 2(n-1) A_3 x^2] \frac{\partial}{\partial x} + C \frac{\partial^2}{\partial t^2} \\
& + [E - (n-1) D_2 x] \frac{\partial}{\partial t} + [D_0 + D_1 x + D_2 x^2] \frac{\partial^2}{\partial x \partial t} \\
& - [n(n-1)(A_2 - A_3 x) + n B_1],
\end{aligned}
\tag{5}
$$

and R_j is given, in terms of the zeros $x_j(t)$, by the formula

$$
\begin{aligned}
R_j \equiv\ & -C \ddot{x}_j - E \dot{x}_j + B_0 + B_1 x_j + \sum_{k=1}^{n}{}' \frac{2(A_0 + A_1 x_j + A_2 x_j^2 + A_3 x_j^2 x_k)}{x_j - x_k} \\
& + \sum_{k=1}^{n}{}' \frac{2C \dot{x}_j \dot{x}_k - (D_0 + D_1 x_j)(\dot{x}_j + \dot{x}_k) - D_2 x_j (\dot{x}_j x_k + \dot{x}_k x_j)}{x_j - x_k} .
\end{aligned}
\tag{6}
$$

Here the quantities $A_0, A_1, A_2, A_3, B_0, B_1, C, D_0, D_1, D_2$, and E are arbitrary, but of course independent of x and of the index j; they could depend arbitrarily on t, but for simplicity in the following we assume they are constant.

It is now plain that, for any given choice of all these constants, the set of n *nonlinear* ODEs

$$R_j = 0, \qquad j = 1, 2, \ldots, n, \tag{7}$$

implies that $p_n(x,t)$ satisfies the *linear* PDE

$$D p_n(x,t) = 0. \tag{8}$$

Thus the initial value problem for the (nonlinear) ***dynamical system*** (7) with (6) [*i.e.*, to determine $x(t)$ for given $x(0)$ and $\dot{x}(t)$], is reduced to solving the (linear!) PDE (8), with initial conditions

$$p_n(x,0) = \prod_{j=1}^{n} [x - x_j(0)], \tag{9a}$$

$$\frac{\partial p_n(x,0)}{\partial t} = - p_n(x,0) \sum_{j=1}^{n} \frac{\dot{x}_j(0)}{x - x_j(0)} ; \tag{9b}$$

and then to identifying the $x_j(t)$'s with the n zeros of the polynomial $p_n(x,t)$:

$$p_n(x_j(t), t) = 0, \qquad j = 1, 2, \ldots, n. \tag{10}$$

A related formula that follows directly from (1), reads

$$\dot{x}_j(t) = - \frac{\partial p_n(x,t)}{\partial t} \Big/ \frac{\partial p_n(x,t)}{\partial x} \Big|_{x = x_j(t)} , \qquad j = 1, 2, \ldots, n. \tag{11}$$

The dynamical system (7) with (6) is fairly general;[17] it can be interpreted as a classical one-dimensional n-body problem through identifying the zeros $x_j(t)$ with the coordinates of the particles, and

[17] Actually, even more general systems can be treated similarly: see [C1978a].

of course $\dot{x}_j(t)$ and $\ddot{x}_j(t)$ with the corresponding velocities and accelerations. Note that the technique of solution we have just outlined implies no restrictions on n nor on the initial values of the positions and velocities of the particles. On the other hand the term that reads as a pair force in the context of the n-body model interpretation, contains generally also "velocity-dependent" contributions, which cannot be eliminated through special choices of the various constants, at least as long as the equations of motion (7) with (6) are second-order in time, *i.e.*, as long as the constant C in (6) does not vanish.

In the following we discuss tersely only some examples of the many models one obtains from (7) and (6) through making specific choices for the values of the constants appearing in (6), and correspondingly in (5). Of course, interesting models appear only by setting to zero several of the constants which appear in (6) and (5); models with too many nonvanishing constants are clearly too general for a useful discussion. The diligent reader may profit from analyzing on his own, along the lines set by the examples discussed below, some model of his own choice. [In particular, it is easily seen that the equations of motion (1.2–20) correspond to (7) and (6) with $A_0 = C = 1$, $D_0 = 2$, $B_0 = E = \alpha$, and all the other constants set to zero, while (1.2–23) correspond to (7) and (6) with $C = 1$, $E = \alpha$, and all the others set to zero.]

Let us begin by discussing a simple model with first-order equations, namely

$$\dot{x}_j = \alpha x_j + \beta \sum_{k=1}^{n}{}' \frac{1}{x_j - x_k}, \qquad j = 1, 2, \ldots, n. \tag{12}$$

This corresponds to (7) and (6) with $E = 1$, $B_1 = \alpha$, $A_0 = \frac{1}{2}\beta$, and all the other constants set to zero. Before discussing the solution of the *dynamical system* described by these *equations of motion* via the technique introduced above, let us interject some remarks.

We note first of all that (12) imply the second-order equations of motion

$$\ddot{x}_j = \alpha^2 x_j - \beta^2 \sum_{k=1}^{n}{}' \frac{1}{(x_j - x_k)^3}, \qquad j = 1, 2, \ldots, n. \tag{13}$$

[For a proof, see the Appendix A to this Section.] Thus, for $\alpha = \pm i\omega$ and $\beta = \pm ig$, (13) reproduces exactly the equations of motion (1.1–82). But the solutions of (12) correspond of course only to a subset of the solutions of (1.1–82), namely those characterized by initial velocities given by (12) in terms of the initial positions. (Note, moreover, that to real positions there correspond imaginary velocities; thus generally these solutions are complex.)

Next, we note that if the $x_j(t)$'s satisfy (12), the quantities $y_j(t)$ defined by

$$y_j(t) = \sqrt{-2\alpha t}\, x_j(\tau), \qquad j = 1, 2, \ldots, n, \tag{14a}$$

$$\tau = \ln(-2\alpha t)/(-2\alpha), \tag{14b}$$

satisfy the equations of motion

$$\dot{y}_j = \beta \sum_{k=1}^{n}{}' \frac{1}{y_j - y_k}, \qquad j = 1, 2, \ldots, n. \tag{15}$$

On the other hand, it is obviously possible to change the constants α and β in (12) by a trivial rescaling of the dependent and independent variables. Thus we set hereafter, for simplicity —and motivated by (13)— $\alpha = -i$, $\beta = i$, so that (12) becomes

$$\dot{x}_j = i\left(-x_j + \sum_{k=1}^{n}{}' \frac{1}{x_j - x_k}\right), j = 1, 2, \ldots, n. \tag{16}$$

Let us now indicate how these equations of motion can be solved. The differential operator D corresponding to (16) —see above— reads now

$$D = \frac{\partial}{\partial t} + i\left(\frac{1}{2}\frac{\partial^2}{\partial x^2} - x\frac{\partial}{\partial x} + n\right). \tag{17}$$

It is therefore convenient to introduce, in addition to (1), a second representation of the polynomial $p_n(x, t)$, reading

$$p_n(x, t) = 2^{-n}H_n(x) + \sum_{m=1}^{n} a_m(t) H_{n-m}(x). \tag{18}$$

Here $H_m(x)$ indicates the Hermite polynomial of order m, *i.e.*, the polynomial of order m which satisfies the ODE

$$\left(\frac{1}{2}\frac{d^2}{dx^2} - x\frac{d}{dx} + m\right)H_m(x) = 0, \qquad m = 0, 1, 2, \ldots, \tag{19a}$$

(see, for instance, [HTF1953]) and is normalized by the condition

$$\lim_{x\to\infty} (2x)^{-m} H_m(x) = 1, \qquad m = 0, 1, 2, \ldots . \tag{19b}$$

Note the consistency of this last equation with (1) and (18).

The simultaneous validity of (1) and (18) induces a bijective mapping between the n *particle coordinates* $x_j(t)$ and the n *collective coordinates* $a_m(t)$. The usefulness of this mapping is due to the simplicity of the time evolution of the coordinates $a_m(t)$. Indeed, (8) and (17)–(19a) imply that the $a_m(t)$'s satisfy *uncoupled linear ODEs,* namely

$$\dot{a}_m(t) + ima_m(t) = 0, \qquad m = 1, 2, \ldots, n, \tag{20a}$$

and these equations are immediately solved:

$$a_m(t) = a_m(0)\exp(-imt), \qquad m = 1, 2, \ldots, n. \tag{20b}$$

It is clear from (16) that the system we are considering admits an equilibrium configuration

$$x_j(t) = \bar{x}_j, \quad \dot{x}_j(t) = 0, \qquad j = 1, 2, \ldots, n, \tag{21}$$

with the quantities $\bar{x}_j$ characterized by the system of n nonlinear algebraic equations

$$\bar{x}_j = \sum_{k=1}^{n}{}' \frac{1}{\bar{x}_j - \bar{x}_k}, \qquad j = 1, 2, \ldots, n. \tag{22}$$

On the other hand the equilibrium configuration corresponds to a time-independent polynomial p_n —see (1)— hence for this configuration,

$$a_m(t) = 0, \qquad m = 1, 2, \ldots, n. \tag{23}$$

—see (18) and (20b). Thus in this case the polynomial p_n is proportional to the Hermite polynomial of order n, and this identifies the n numbers $\bar{x}_j$ with the n zeros of the Hermite polynomial of order n,

$$H_n(\bar{x}_j) = 0, \qquad j = 1, 2, \ldots, n. \tag{24}$$

The fact that the n zeros of the Hermite polynomial of order n coincide with the n solutions of the system of nonlinear algebraic equations (22) is not new; it was discovered by Stieltjes almost a century ago —see, for instance [Sz1939]. The fact that these n zeros satisfy as well the equations

$$\bar{x}_j = 2 \sum_{k=1}^{n}{}' \frac{1}{(\bar{x}_j - \bar{x}_k)^3}, \qquad j = 1, 2, \ldots, n, \tag{25}$$

is implied by (13) with $\alpha = -i$, $\beta = i$. This result is more recent and appears in [C1977b] and [C1977c].[18]

Rather than discussing in more detail the solution of (16), or of other first-order systems of ODEs contained in (7) with (6) and $C = 0$, let us consider an analogous example with equations of motion of second order. Specifically, let us look at the *n-body problem* characterized by the equations of motion

$$\ddot{x}_j + x_j = \sum_{k=1}^{n}{}' \frac{1 + 2\dot{x}_j \dot{x}_k}{x_j - x_k}, \qquad j = 1, 2, \ldots, n. \tag{26}$$

This corresponds to (7) and (6) with $C = 1$, $B_1 = -1$, $A_0 = \frac{1}{2}$, and all the other constants set to zero. Thus in the corresponding linear PDE (8) the differential operator (5) reads

$$D = \frac{\partial^2}{\partial t^2} + \frac{1}{2}\frac{\partial^2}{\partial x^2} - x\frac{\partial}{\partial x} + n. \tag{27}$$

It is therefore convenient to introduce, in addition to (1), the representation (8) for the polynomial p_n which, to avoid any confusion, is now rewritten as follows:

$$p_n(x, t) = 2^{-n} H_n(x) + \sum_{m=1}^{n} b_m(t) H_{n-m}(x). \tag{28}$$

Again this induces a bijective mapping between the n particle coordinates $x_j(t)$ and the n collective coordinates $b_m(t)$. The term ***particle coordinates*** is now quite appropriate, since the equations of motion (26) look now indeed as Newton's equations for n unit-mass particles moving on a straight line, being attracted to the origin by an elastic linear force, and interacting pairwise with a velocity-dependent force inversely proportional to the interparticle distance. Note that the pair force is singular; for small velocites it is repulsive, and therefore prevents the particles from crossing each other. But if the velocities of two particles are opposite and sufficiently large, the force becomes attractive and it may therefore cause the collision (*collapse*) of two contiguous particles, at which point the equations (26) become singular. The natural continuation of the solutions beyond this singularity is complex —see below.

As for the term ***collective coordinates***, it is also now well justified. Indeed, it is easily seen that the first of these coordinates, $b_1(t)$, is proportional to the center of mass,

$$b_1(t) = -2^{n-1} \sum_{j=1}^{n} x_j(t). \tag{29}$$

---

[18] Additional results on the n zeros of the Hermite polynomial of order n stem from investigating the behaviour of the system (16) in the neighborhood of its equilibrium configuration; these results were also obtained, essentially in this manner, fairly recently: in [C1977] and [C1978a]. They can be also proven in other ways, and are discussed in the following § 1.4.

The following results —see (30)— might suggest terming the b_m's ***normal mode coordinates***, were it not for the fact that the normal modes refer generally to the eigenmodes of the *linearized* equations describing the *small* oscillations of a system in the neighborhood of its equilibrium configuration, whereas here, via the nonlinear transformation between the particle coordinates x_j and the collective coordinates b_m induced by (1) and (28), one is solving *exactly* the nonlinear equations of motion (26). See below.

As in the preceding case, the merit of the collective coordinates $b_m(t)$ rests on the simplicity of their time evolution. Indeed, (8), (27), (28), and (19a) imply

$$\ddot{b}_m(t) + m b_m(t) = 0, \qquad m = 1, 2, \ldots, n, \tag{30a}$$

and these equations are immediately solved by the explicit formula

$$b_m(t) = b_m(0)\cos(\sqrt{m}\,t) + \dot{b}_m(0)\,\frac{\sin(\sqrt{m}\,t)}{\sqrt{m}}, \qquad m = 1, 2, \ldots, n. \tag{30b}$$

Note that these results imply that the generic solution of the equations of motion (26) is not periodic, but rather multiply-periodic, generally with incommensurate periods. Also note that the system characterized by the equations of motion (26) is clearly hamiltonian —although it is not easy to write it in hamiltonian form— since the transformation from the n coordinates $x_j(t)$ to the n coordinates $b_m(t)$ via (1) and (28) is obviously canonical —as any transformation to new coordinates which depends only on the old coordinates and not on the old momenta.

It is obvious that, as in the case discussed above, the system (28) admits the equilibrium configuration (21) corresponding to

$$b_m(t) = 0, \qquad m = 1, 2, \ldots, n, \tag{31}$$

with (22) and (24). This will be further discussed in the following § 1.4.

In some cases the solutions of (26) can be displayed in rather explicit form. Take for instance the case characterized by initial conditions $x_j(0)$ such that

$$H_n(x_j(0)) = \alpha, \qquad j = 1, 2, \ldots, n, \tag{32a}$$

and by the initial velocities

$$\dot{x}_j(0) = \frac{\beta}{H_n'(x_j(0))}, \qquad j = 1, 2, \ldots, n, \tag{32b}$$

with α and β two arbitrary real constants (not too large in modulus; see below). Then,

$$b_m(0) = -2^{-n}\alpha\delta_{n,m}, \qquad m = 1, 2, \ldots, n, \tag{33a}$$

$$\dot{b}_m(0) = -2^{-n}\beta\delta_{n,m}, \qquad m = 1, 2, \ldots, n. \tag{33b}$$

Thus in this case,

$$b_m(t) = -2^{-n}B\sin(\sqrt{n}\,t + \theta)\,\delta_{m,n}, \qquad m = 1, 2, \ldots, n, \tag{34}$$

with

$$B = \sqrt{\alpha^2 + \beta^2/n}, \tag{35a}$$

$$\theta = \arctan(\beta/\sqrt{n}\,\alpha), \tag{35b}$$

and the coordinates $x_j(t)$ for all time are the n roots of the simple equation in x,

$$H_n(x) = B\sin(\sqrt{n}\,t + \theta). \tag{36}$$

This result follows of course from the fact that (1) and (28) in this case yield

$$p_n(x) = \prod_{j=1}^{n}[x - x_j(t)] = 2^{-n}H_n(x) + b_n(t), \tag{37}$$

which incidentally provides the more convenient starting point to derive (32a) and (32b).[19] The analysis of the time evolution of the n roots of (36) lends itself to a transparent analysis, by drawing the graph of the Hermite polynomial $H_n(x)$ and considering its intersection with a horizontal straight line which oscillates periodically, as indicated by the right-hand side of (36). Note that a necessary and sufficient condition to guarantee that the n roots of (36) be real for all time is that the modulus of B not exceed the smallest relative maximum of the modulus of $H_n(x)$ for real x.

Another interesting initial condition is

$$x_j(0) = \bar{x}_j, \qquad j = 1, 2, \ldots, n, \tag{38a}$$

$$\dot{x}_j(0) = \delta_{j,k}\,\dot{x}_k(0), \qquad j = 1, 2, \ldots, n; \tag{38b}$$

it describes the system initially in its equilibrium configuration (thus the $\bar{x}_j$'s are the n zeros of the Hermite polynomial of order n), with only the kth particle moving. The corresponding *collective coordinates* $b_m(t)$ are then given by (30b) with

$$b_m(0) = 0, \qquad m = 1, 2, \ldots, n, \tag{39a}$$

$$\dot{b}_m(0) = -\dot{x}_k(0)\,\frac{2^m(n-1)!}{(n-m)!}\,\frac{H_{n-m}(\bar{x}_k)}{H_{n-1}(\bar{x}_k)}, \qquad m = 1, 2, \ldots, n. \tag{39b}$$

For a proof, see Appendix B of this Section.

Exercise 1. Solve the dynamical system characterized by the equations of motion

$$\ddot{x}_j + x_j = -2\alpha\dot{x}_j + \sum_{k=1}^{n}{}' \frac{1 + 2\dot{x}_j\dot{x}_k}{x_j - x_k}, \qquad \alpha > 0, \quad j = 1, 2, \ldots, n. \tag{40}$$

In particular, prove that

$$\lim_{t\to+\infty} x_j(t) = \bar{x}_j, \qquad j = 1, 2, \ldots, n, \tag{41}$$

where the $\bar{x}_j$'s are the n zeros of the Hermite polynomial of order n (for any initial data which do not lead to a collision; and indeed, even if there are collisions, for any reasonable continuation of the trajectories beyond the singularity).

Exercise 2. Discuss the two-dimensional real system characterized by the equations of motion

$$\ddot{\vec{r}}_j + \vec{r}_j = -2\alpha\dot{\vec{r}}_j + \sum_{k=1}^{n}{}' \frac{\vec{\varphi}^{\,j,k}}{|\vec{r}_j - \vec{r}_k|^2}, \qquad j = 1, 2, \ldots, n, \tag{42}$$

where $\vec{r}_j \equiv (x_j, y_j)$ are n two-dimensional (time-dependent) vectors and the x and y components of the vectors $\vec{\varphi}^{\,j,k}$ are defined as follows:

$$\varphi_x^{j,k} = (x_j - x_k)[1 + 2(\dot{x}_j\dot{x}_k - \dot{y}_j\dot{y}_k)] + 2(y_j - y_k)(\dot{x}_j\dot{y}_k + \dot{x}_k\dot{y}_j), \tag{43a}$$

$$\varphi_y^{j,k} = -(y_j - y_k)[1 + 2(\dot{x}_j\dot{x}_k - \dot{y}_j\dot{y}_k)] + 2(x_j - x_k)(\dot{x}_j\dot{y}_k + \dot{x}_k\dot{y}_j). \tag{43b}$$

Hint: $z_j = x_j + iy_j$.

[19] Note, incidentally, that Eq. (3.12b) of [C1980a] is misprinted.

Appendix A. Proof of (1.3–13)

Differentiation of (1.3–12) with respect to time, yields

$$\ddot{x}_j = \alpha \dot{x}_j - \beta \sum_{k=1}^{n}{}' \frac{\dot{x}_j - \dot{x}_k}{(x_j - x_k)^2}, \qquad j = 1, 2, \ldots, n. \tag{1}$$

Using (1.3–12) again in the right-hand side yields

$$\ddot{x}_j = \alpha^2 x_j - \beta^2 \sum_{k=1}^{n}{}' \frac{1}{(x_j - x_k)^2} \left(\sum_{l=1}^{n}{}' \frac{1}{x_j - x_l} - \sum_{l=1}^{n}{}' \frac{1}{x_k - x_l} \right). \tag{2}$$

By separating out, in the last two sums, the terms with $l = k$ and with $l = j$, it is immediate to see that (2) yields (1.3–13), with the additional term in the right-hand side,

$$Z_j \equiv -\beta^2 \sum_{\substack{k,l=1 \\ k \neq l \neq j \neq k}}^{n} \frac{1}{(x_j - x_k)^2} \left(\frac{1}{x_j - x_l} - \frac{1}{x_k - x_l} \right). \tag{3}$$

Thus the thesis is proven if Z_j is shown to vanish identically. Indeed, by combining the last two terms, one obtains

$$Z_j = \beta^2 \sum_{\substack{k,l=1 \\ k \neq l \neq j \neq k}}^{n} \frac{1}{(x_j - x_k)(x_j - x_l)(x_k - x_l)}, \tag{4}$$

and this clearly vanishes since the summand is antisymmetrical in the two dummy indices l and k.

Q.E.D.

This result was first obtained in [CC1977].

Appendix B. Proof of (1.3–39a) and (1.3–39b).

The basic formula is of course the relation

$$\prod_{j=1}^{n} [x - x_j(t)] = 2^{-n} H_n(x) + \sum_{m=1}^{n} b_m(t) H_{n-m}(x), \tag{1a}$$

implying

$$\sum_{j=1}^{n} \dot{x}_j(t) \prod_{\substack{k=1 \\ k \neq j}}^{n} [x - x_k(t)] = -\sum_{m=1}^{n} \dot{b}_m(t) H_{n-m}(x). \tag{1b}$$

Thus, for $t = 0$ and $x_j(0) = \bar{x}_j$ with

$$H_n(\bar{x}_j) = 0, \tag{2}$$

(1a) becomes

$$2^{-n} H_n(x) = 2^{-n} H_n(x) + \sum_{m=1}^{n} b_m(0) H_{n-m}(x), \tag{3a}$$

while if in addition $\dot{x}_j(0) = \dot{x}_k(0)\delta_{j,k}$, (1b) yields

$$\dot{x}_k(0) \prod_{\substack{j=1 \\ j \neq k}}^{n} (x - \bar{x}_j) = -\sum_{m=1}^{n} \dot{b}_m(0) H_{n-m}(x). \tag{3b}$$

Thus (1.3–39a) is an immediate consequence of (3a), while (3b) yields, for $x = \bar{x}_j$,

$$\sum_{m=1}^{n} \dot{b}_m(0) H_{n-m}(\bar{x}_j) = -2^{-n} \delta_{j,k} \dot{x}_k H'_n(\bar{x}_k), \qquad j = 1, 2, \ldots, n. \tag{4}$$

Thus the validity of (39b) corresponds to the formula

$$\sum_{m=1}^{n} \frac{2^m (n-1)!}{(n-m)!} H_{n-m}(\bar{x}_k) H_{n-m}(\bar{x}_j) = \delta_{j,k} H_{n-1}(\bar{x}_k) H'_n(\bar{x}_k), \quad j, k = 1, 2, \ldots, n. \tag{5}$$

But this is a consequence of the standard formula

$$\sum_{m=1}^{n} \frac{2^m (n-1)!}{(n-m)!} H_{n-m}(x) H_{n-m}(y) = \frac{H_n(x) H_{n-1}(y) - H_{n-1}(x) H_n(y)}{x - y}. \tag{6}$$

This is obtained, for instance, from Eq. (10.13–11) of [HTF1953] by replacing n by $n-1$ and the dummy index m by $n-m$. Indeed, for $x = \bar{x}_j$ and $y = \bar{x}_k$ with $j \neq k$, the right-hand side of this formula vanishes —see (2)— while for $x = y$, (6) becomes

$$\sum_{m=1}^{n} \frac{2^m (n-1)!}{(n-m)!} [H_{n-m}(x)]^2 = H'_n(x) H_{n-1}(x) - H'_{n-1}(x) H_n(x), \tag{7}$$

and this, using again (2), reproduces (5) for $j = k$ when $x = \bar{x}_j$.

§1.4 Investigation of certain integrable dynamical systems near equilibrium. Properties of the zeros of the classical polynomials. Remarkable properties of certain matrices

In the preceding Section we have seen that the dynamical system characterized by the equations of motion [see (1.3–16)]

$$\dot{x}_j = i\left(-x_j + \sum_{k=1}^{n}{}' \frac{1}{x_j - x_k}\right), \qquad j = 1, 2, \ldots, n, \tag{1}$$

can be solved by relating the n coordinates x_j to the n coordinates a_m via the nonlinear mapping induced by the polynomial identity [see (1.3–1) and (1.3–18)]

$$\prod_{j=1}^{n} [x - x_j(t)] = 2^{-n} H_n(x) + \sum_{m=1}^{n} a_m(t) H_{n-m}(x); \tag{2}$$

for indeed, if the n coordinates $x_j(t)$ evolve according to (1), the n coordinates a_m evolve according to the decoupled linear equations [see (1.3–20a)]

$$\dot{a}_m + i m a_m = 0, \qquad m = 1, 2, \ldots, n, \tag{3}$$

and these equations are easily solved [see (1.3–20b)]. We also noted that this implies that the n numbers $\bar{x}_j$ corresponding to the equilibrium configuration of the system (1),

$$x_j(t) = \bar{x}_j, \quad \dot{x}_j(t) = 0, \qquad j = 1, 2, \ldots, n, \tag{4}$$

besides being characterized by the n algebraic relations

$$\bar{x}_j = \sum_{k=1}^{n}{}' \frac{1}{\bar{x}_j - \bar{x}_k}, \qquad j = 1, 2, \ldots, n, \tag{5}$$

correspond to the n zeros of the Hermite polynomial of order n:

$$H_n(\bar{x}_j) = 0, \qquad j = 1, 2, \ldots, n. \tag{6}$$

We now look at the system (1) not only at equilibrium, but near equilibrium, and in this manner we derive some additional properties of these numbers, *i.e.*, some novel properties of the zeros of Hermite polynomials.

Let us investigate the small oscillations of the system (1) in the neighborhood of its equilibrium configuration (4). To this end we set

$$x_j(t) = \bar{x}_j + \epsilon \xi_j(t), \qquad j = 1, 2, \ldots, n, \tag{7}$$

treat ϵ as a small parameter, and thereby linearize the equations of motion (1). One thus obtains, in vector notation, the linear equation

$$\dot{\boldsymbol{\xi}}(t) + i(\mathbf{1} + \mathbf{A})\boldsymbol{\xi}(t) = \mathbf{0}, \tag{8}$$

where of course the vector $\boldsymbol{\xi}(t)$ has n components $\xi_j(t)$, $\mathbf{1}$ is the unit matrix of order n, and the symmetric matrix $\mathbf{A}$ is defined as follows:

$$A_{j,k} = \begin{cases} \displaystyle\sum_{l=1}^{n}{}' \frac{1}{(\bar{x}_j - \bar{x}_l)^2}, & j = k, \\ \displaystyle\frac{-1}{(\bar{x}_j - \bar{x}_k)^2}, & j \neq k. \end{cases} \tag{9}$$

Thus the n circular frequencies of the small oscillations of the system (1) in the neighborhood of its equilibrium configuration coincide with the n eigenvalues of the matrix $\mathbf{1} + \mathbf{A}$. But it follows from (3) and the relationship between the $x_j(t)$'s and the $a_m(t)$'s that these frequencies are just the first n positive integers, from 1 to n. One arrives thus at the following conclusion, which we now state as a

Theorem.[20] *The matrix* $\mathbf{A}$*, defined in terms of the* n *zeros of the Hermite polynomial of order* n *by (9), has the first* n *nonnegative integers, from* 0 *to* $n-1$*, as eigenvalues:*

$$\mathbf{A}\,\mathbf{v}^{(H)(m)} = (m-1)\mathbf{v}^{(H)(m)}, \qquad m = 1, 2, \ldots, n. \tag{10}$$

The result we have just reported suggested (see [C1977] and [C1978a]) the following

[20] This result was originally proven, essentially in this manner, in [C1978a] (see also [C1977]); it can also be proven by other techniques, see [ABCOP1979], [C1981], and ahead. The proof given here can be made more precise by inserting (7) in (2) and treating ϵ as a small parameter; indeed, in this manner it is possible to find the explicit form of the eigenvectors $\mathbf{v}^{(H)(m)}$. See (10). The diligent reader will try to do this; the result may be verified looking in [ABCOP1979] or [C1981].

Conjecture. *Let the matrix* $\mathbf{A}$*, of order* n*, be defined, in terms of* n *a priori arbitrary numbers* x_j*, by the formula*

$$A_{j,k} = \begin{cases} \sum_{l=1}^{n}{}' \dfrac{1}{(x_j - x_l)^2}, & j = k, \\ \dfrac{-1}{(x_j - x_k)^2}, & j \neq k, \end{cases} \tag{11}$$

and require this matrix to have the first n *nonnegative integers, from* 0 *to* $n-1$*, as eigenvalues. Then the* n *numbers* x_j *coincide, up to a common translation, with the* n *zeros of the Hermite polynomial of order* n*,*

$$H_n(x_j + x_0) = 0, \qquad j = 1, 2, \ldots, n. \tag{12}$$

But this conjecture is valid only for $n = 2$ and $n = 3$; it is false for $n = 4$ (see [C1982a]) and presumably, *a fortiori*, for $n > 4$.

It may be of interest in this connection to mention that the set of n algebraic relations (5) do characterize the n zeros of the Hermite polynomial of order n, namely, the n numbers x_j satisfy the n algebraic relations

$$x_j = \sum_{k=1}^{n}{}' \frac{1}{x_j - x_k}, \qquad j = 1, 2, \ldots, n, \tag{5}$$

if and only if they coincide with the n zeros of the Hermite polynomial of order n,

$$H_n(x_j) = 0, \qquad j = 1, 2, \ldots, n. \tag{14}$$

For proofs, see [Sz1939], [S1978], and [Mu1978].

There exist analogous results for all the classical polynomials. Let us merely report here the relevant formulae from [ABCOP1979]:[21]

Laguerre polynomials:

$$L_n^{\alpha}(y_j) = 0, \qquad j = 1, 2, \ldots, n, \tag{15a}$$

$$\sum_{k=1}^{n}{}' \frac{1}{y_j - y_k} = \tfrac{1}{2}\left(1 - \frac{1+\alpha}{y_j}\right), \qquad j = 1, 2, \ldots, n, \tag{15b}$$

$$B_{j,k} = \begin{cases} \sum_{l=1}^{n}{}' \frac{y_l}{(y_j - y_l)^2}, & j = k, \\ -\frac{y_k}{(y_j - y_k)^2}, & j \neq k, \end{cases} \tag{15c}$$

$$\mathbf{B}\,\mathbf{v}^{(L)(m)} = \tfrac{1}{2}(m-1)\mathbf{v}^{(L)(m)}, \qquad m = 1, 2, \ldots, n. \tag{15d}$$

Jacobi polynomials:

$$P_n^{(\alpha,\beta)}(z_j) = 0, \qquad j = 1, 2, \ldots, n, \tag{16a}$$

$$\sum_{k=1}^{n}{}' \frac{1}{z_j - z_k} = \tfrac{1}{2}\,\frac{\alpha - \beta + (\alpha + \beta + 2)/z_j}{1 - z_j^2}, \qquad j = 1, 2, \ldots, n, \tag{16b}$$

$$C_{j,k} = \begin{cases} \sum_{l=1}^{n}{}' \frac{1 - z_l^2}{(z_j - z_l)^2}, & j = k, \\ -\frac{1 - z_k^2}{(z_j - z_k)^2}, & j \neq k, \end{cases} \tag{16c}$$

$$\mathbf{C}\,\mathbf{v}^{(J)(m)} = \tfrac{1}{2}(m-1)(2n - m + \alpha + \beta)\mathbf{v}^{(J)(m)}, \qquad m = 1, 2, \ldots, n. \tag{16d}$$

---

[21] For the definition of the classical polynomials see [HTF1953]

The result mentioned above for the zeros of Hermite polynomials could as well have been derived from an analysis of the dynamical system characterized by the equations of motion (1.3–26), namely

$$\ddot{x}_j + x_j = \sum_{k=1}^{n}{}' \frac{1 + 2\dot{x}_j \dot{x}_k}{x_j - x_k}, \qquad j = 1, 2, \ldots, n, \tag{17}$$

rather than (1); this is left as an exercise for the diligent reader (see [C1978a]). Note that the equilibrium configuration, as well as the small oscillations around equilibrium for the integrable system (17), coincide with those of the dynamical system defined by the hamiltonian

$$H(\mathbf{q}, \mathbf{p}) = \tfrac{1}{2} \sum_{j=1}^{n} (p_j^2 + q_j^2) - \sum_{j>k=1}^{n} \ln(q_j - q_k), \tag{18}$$

whose equations of motion read

$$\ddot{q}_j + q_j = \sum_{k=1}^{n}{}' \frac{1}{q_j - q_k}, \quad j = 1, 2, \ldots, n. \tag{19}$$

Indeed, in both cases the equilibrium configuration corresponds to the zeros of the Hermite polynomial of order n, and the circular frequencies of the small oscillations around equilibrium are the square roots of the first n positive integers, from 1 to n. [See (1.3–30b), or note that the square of these frequencies are the eigenvalues of $\mathbf{1} + \mathbf{A}$; see (9) and (10).]

It is also of interest to note that the system characterized by the hamiltonian

$$H(\mathbf{q}, \mathbf{p}) = \tfrac{1}{2} \sum_{j=1}^{n} (p_j^2 + q_j^2) + \sum_{j>k=1}^{n} \frac{1}{(q_j - q_k)^2} \tag{20}$$

has again the equilibrium configuration

$$q_j(t) = \bar{x}_j, \quad \dot{q}_j(t) = 0, \qquad j = 1, 2, \ldots, n, \tag{21}$$

while the circular frequencies of the small oscillations for this system are the first n positive integers, so that the small oscillations are always completely periodic (as indeed they must be, since any motion of the system characterized by the hamiltonian (20) is periodic; see § 2.1). Indeed, the relationship between the system characterized by the hamiltonians (18) and (20) is an instance of application of the following result [P1978a]:

Theorem. (Perelomov) *Let two hamiltonian systems be characterized by the two hamiltonians*

$$H^{(s)}(\mathbf{q}, \mathbf{p}) = \tfrac{1}{2} \sum_{j=1}^{n} p_j^2 + V^{(s)}(\mathbf{q}), \qquad s = 1, 2, \tag{22}$$

with

$$V^{(2)}(\mathbf{q}) = \tfrac{1}{2} \sum_{j=1}^{n} \left(\frac{\partial V^{(1)}(\mathbf{q})}{\partial q_j} \right)^2 + \text{constant}. \tag{23}$$

Then, if $\bar{\mathbf{q}}$ *is an equilibrium configuration for system* **1**, *it is an equilibrium configuration for system* **2** *as well; moreover, if* ${\omega_m}^{(s)}$, $m = 1, 2, \ldots, n$, $s = 1, 2$, *indicate the circular frequencies of the small oscillations around this equilibrium configuration for the two systems, there holds the relationship*

$${\omega_m}^{(2)} = \left({\omega_m}^{(1)}\right)^2, \qquad m = 1, 2, \ldots, n. \tag{24}$$

The proof of this theorem is straightforward. The equations of motion corresponding to (22) read

$$\ddot{q}_j = -\frac{\partial V^{(s)}(\mathbf{q})}{\partial q_j}, \qquad j=1,2,\ldots,n, \quad s=1,2. \tag{25}$$

Thus the equilibrium configurations $\overline{\mathbf{q}}^{(s)}$ are characterized by the equations

$$\left.\frac{\partial V^{(s)}(\mathbf{q})}{\partial q_j}\right|_{\mathbf{q}=\overline{\mathbf{q}}^{(s)}} = 0, \qquad j=1,2,\ldots,n, \quad s=1,2. \tag{26}$$

Moreover, the small oscillations around the equilibrium configurations are characterized by the equations

$$\mathbf{q}^{(s)}(t) = \overline{\mathbf{q}}^{(s)}(t) + \epsilon\boldsymbol{\eta}^{(s)}(t), \qquad s=1,2, \tag{27}$$

$$\ddot{\boldsymbol{\eta}}^{(s)}(t) + \mathbf{M}^{(s)}\boldsymbol{\eta}^{(s)}(t) = 0, \qquad s=1,2, \tag{28}$$

$$M_{j,k}{}^{(s)} = \left.\frac{\partial^2 V^{(s)}(\mathbf{q})}{\partial q_j \partial q_k}\right|_{\mathbf{q}=\overline{\mathbf{q}}^{(s)}}, \qquad s=1,2, \tag{29}$$

the square of the circular eigenfrequencies $\omega_m{}^{(s)}$ being the n eigenvalues of the symmetric matrix $\mathbf{M}$. But (23) implies

$$\frac{\partial V^{(2)}(\mathbf{q})}{\partial q_j} = \sum_{l=1}^{n} \frac{\partial^2 V^{(1)}(\mathbf{q})}{\partial q_j \partial q_l}\frac{\partial V^{(1)}(\mathbf{q})}{\partial q_l} \tag{30}$$

and

$$\frac{\partial^2 V^{(2)}(\mathbf{q})}{\partial q_j \partial q_k} = \sum_{l=1}^{n} \frac{\partial^2 V^{(1)}(\mathbf{q})}{\partial q_j \partial q_l}\frac{\partial^2 V^{(1)}(\mathbf{q})}{\partial q_l \partial q_k} + \sum_{l=1}^{n} \frac{\partial^3 V^{(1)}(\mathbf{q})}{\partial q_j \partial q_k \partial q_l}\frac{\partial V^{(1)}(\mathbf{q})}{\partial q_l}. \tag{31}$$

Thus if $\overline{\mathbf{q}}^{(1)} = \overline{\mathbf{q}}$ is an equilibrium configuration for system **1**, namely if (26) holds for $s=1$ and $\overline{\mathbf{q}}^{(1)} = \overline{\mathbf{q}}$, then (30) implies that $\overline{\mathbf{q}}$ is also an equilibrium configuration for system **2**, namely (26) also holds for $s=2$ with $\overline{\mathbf{q}}^{(2)} = \overline{\mathbf{q}}$; moreover, (31) implies, via (26) and (29), the matrix formula

$$\mathbf{M}^{(2)} = \left(\mathbf{M}^{(1)}\right)^2. \tag{32}$$

This last formula implies of course (24), thereby completing the proof of the theorem; note that it also implies that the eigenmodes of the small oscillations are the same for the two systems.

The applicability of Perelomov's theorem to the case mentioned above obtains because the two definitions [see (18), (20), and (22)]

$$V^{(1)}(\mathbf{q}) = \tfrac{1}{2}\sum_{j=1}^{n} q_j^2 - \sum_{j>k=1}^{n} \ln(q_j - q_k) \tag{33a}$$

and

$$V^{(2)}(\mathbf{q}) = \tfrac{1}{2}\sum_{j=1}^{n} q_j^2 + \sum_{j>k=1}^{n} \frac{1}{(q_j - q_k)^2}, \tag{33b}$$

imply

$$V^{(2)}(\mathbf{q}) = \tfrac{1}{2}\sum_{j=1}^{n}\left[\frac{\partial V^{(1)}(\mathbf{q})}{\partial q_j}\right]^2 + \tfrac{1}{2}n(n-1). \tag{33c}$$

This is proven in the Appendix to this Section.

Let us note, in connection with Perelomov's theorem, that while (23) implies that any equilibrium configuration for system **1** is also an equilibrium configuration for system **2**, the converse need not be true. For instance, in the example discussed above, the equilibrium configuration for the system characterized by the hamiltonian (22) with (33a) ($s = 1$) is uniquely given, up to permutations, by the formula $\bar{q}_j^{(1)} = \bar{x}_j$, where the quantities $\bar{x}_j$ are characterized by the n algebraic equations (5) or, equivalently, by the requirement to be the zeros of the Hermite polynomial of order n —see (6). Of course this configuration, $\bar{q}_j^{(2)} = \bar{x}_j$, is at equilibrium also for the system characterized by the hamiltonian (22) with (33b) ($s = 2$); indeed, it is easy to verify that (5), *i.e.*

$$\bar{x}_j = \sum_{k=1}^{n}{}' \frac{1}{\bar{x}_j - \bar{x}_k}, \qquad j = 1, 2, \ldots, n, \tag{34a}$$

implies

$$\bar{x}_j = 2 \sum_{k=1}^{n}{}' \frac{1}{(\bar{x}_j - \bar{x}_k)^3}, \qquad j = 1, 2, \ldots, n. \tag{34b}$$

(This is of course implied by the results of the Appendix to this Section; for a direct proof, see [C1977c].) But the converse is obviously not true; for instance $\bar{x}'_j = i\bar{x}_j$ clearly satisfies (34b) but not (34a). Thus $\bar{q}_j^{(2)} = \bar{x}'_j = i\bar{x}_j$ provides an equilibrium configuration for system **2** but not for system **1**. It is, however, quite obvious on physical grounds that, *in the real domain*, the equilibrium configuration for system **2** [*i.e.*, for (22) with (34b); namely, for the system discussed in § 1.1] is unique, and it therefore coincides with that for system **1** (namely, with the n zeros of the Hermite polynomial of order n).

Let us end this Section by investigating, in a manner analogous to that employed above, the behaviour around equilibrium of another dynamical system. The main purpose of this exercise is the derivation of some remarkable results, consisting again in the identification of the eigenvalues of certain matrices which are now constructed with arbitrary numbers, rather than with the zeros of the classical polynomials. Consider the system characterized by the equations of motion

$$\ddot{x}_j = \sum_{k=1}^{n}{}' \frac{2\dot{x}_j \dot{x}_k - x_j(\dot{x}_j + \dot{x}_k)}{x_j - x_k}, \qquad j = 1, 2, \ldots, n; \tag{35}$$

this can be solved by the method described in § 1.3. The relevant formulae read

$$p_n(x,t) = \prod_{j=1}^{n} [x - x_j(t)] = x^n + \sum_{m=1}^{n} a_m(t) x^{m-1}, \tag{36}$$

$$\left(\frac{\partial^2}{\partial t^2} + x \frac{\partial^2}{\partial x \partial t}\right) p_n(x,t) = 0, \tag{37}$$

$$\ddot{a}_m + (m-1)\dot{a}_m = 0, \qquad m = 1, 2, \ldots, n. \tag{38}$$

Let us now consider an *equilibrium configuration* for this system; any set of n numbers x_j will do —see (35). Then we investigate the small oscillations of the system around this equilibrium configuration, setting

$$x_j(t) = x_j + \epsilon \xi_j(t), \qquad j = 1, 2, \ldots, n \tag{39}$$

in (35), treating ϵ as a small parameter and thereby obtaining the linearized equation

$$\ddot{\boldsymbol{\xi}}(t) + \mathbf{N}\, \dot{\boldsymbol{\xi}}(t) = 0, \tag{40}$$

which we have written in vector form. The matrix **N**, or order n, is defined in terms of the n arbitrary numbers x_j through the formula

$$N_{j,k} = \begin{cases} x_j \sum_{l=1}^{n}{}' \dfrac{1}{x_j - x_l}, & j = k, \\ \dfrac{x_j}{x_j - x_k}, & j \neq k. \end{cases} \tag{41}$$

A comparison of (40) with (38) implies that the eigenvalues of the matrix **N** must coincide with the numbers $m-1, m = 1, 2, \ldots, n$. One has thus obtained the following:

Theorem[22] *The matrix* **N**, *of order n, defined in terms of the n arbitrary numbers x_j by (41), has the first n nonnegative integers, from 0 to $n-1$, as eigenvalues:*

$$\mathbf{N}\,\mathbf{v}^{(m)} = (m-1)\mathbf{v}^{(m)}, \qquad m = 1, 2, \ldots, n. \tag{42}$$

Let us end indicating how the theorem given above for the zeros of Hermite polynomials [see (10)] can be derived directly from the results mentioned in the footnote. Analogous arguments apply to the results for the other classical polynomials [see (15) and (16)]. the starting point is the observation that these results imply[see § 2.1] that another matrix of order n, constructed in terms of the $n+1$ *arbitrary* numbers $x_p, p = 0, 1, \ldots, n$, and having the first n nonnegative integers, from 0 to $n-1$, as its eigenvalues, reads

$$\mathcal{A} = (n-1)\mathbf{1} + \tfrac{1}{2}\mathbf{Z}^2 + x_0\mathbf{Z} - \mathbf{N}, \tag{43}$$

with **N** defined by (41) and **Z** defined by

$$Z_{j,k} = \begin{cases} \sum_{l=1}^{n}{}' \dfrac{1}{x_j - x_l}, & j = k, \\ \dfrac{1}{x_j - x_k}, & j \neq k. \end{cases} \tag{44}$$

It can be moreover easily shown, by explicit computation that

$$\mathcal{A} = \mathbf{A} - \mathbf{B}, \tag{45}$$

where

$$A_{j,k} = \begin{cases} \sum_{l=1}^{n}{}' \dfrac{1}{(x_j - x_l)^2}, & j = k, \\ \dfrac{-1}{(x_j - x_k)^2}, & j \neq k, \end{cases} \tag{46}$$

$$B_{j,k} = \begin{cases} \sum_{l=1}^{n}{}' \dfrac{y_l}{x_j - x_l}, & j = k, \\ \dfrac{y_j}{x_j - x_k}, & j \neq k, \end{cases} \tag{47}$$

$$y_j = x_j - x_0 - \sum_{k=1}^{n}{}' \frac{1}{x_j - x_k}, \qquad j = 1, 2, \ldots, n. \tag{48}$$

---

[22] This result was first proven essentially in this manner in [ABCOP1979]; it can also be proven by other techniques, see [ABCOP1979] and [C1981]. The diligent reader may try to refine the proof given here by inserting (39) in (36), treating ϵ as a small parameter, and in this manner he may also obtain the explicit form of the eigenvectors $\mathbf{v}^{(m)}$ —see (42). The derivation of this result, as outlined here, makes it appear as rather surprising. In fact, it has been shown in [C1980c], [C1980d], [C1981], [C1981a], [BC1981], [C1982], and [C1982b], that this result can be set into an overall framework, where it appears as a trivial instance, rather than as an exceptional case; these developments are outlined in the following Chapter 2.

These results hold for any arbitrary choice of the numbers $x_p, p = 0, 1, \ldots, n$ —with the obvious restriction $x_j \neq x_k$, $j, k = 1, 2, \ldots, n$, $j \neq k$. For the special choice

$$x_j = x_0 + \overline{x}_j, \tag{49}$$

where the $\overline{x}_j$'s are the n zeros of the Hermite polynomial of order n [see (6)], the matrix **B** vanishes identically [see (5), (48), and (47)]. Then **A** coincides with $\mathcal{A}$ [see (45)], and it is thereby once more proven that the matrix (9) has the first n nonnegative integers, from 0 to $n-1$, as its eigenvalues.

Appendix. Proof of (1.4–33c)

$$\frac{1}{2}\sum_{j=1}^{n}\left(\frac{\partial V^{(1)}(\mathbf{q})}{\partial q_j}\right)^2 = \frac{1}{2}\sum_{j=1}^{n}\left(q_j + {\sum_{k=1}^{n}}' \frac{1}{q_j - q_k}\right)^2 \tag{1a}$$

$$= \frac{1}{2}\sum_{j=1}^{n} q_j^2 - \alpha + \beta, \tag{1b}$$

where

$$\alpha = \sum_{j=1}^{n} q_j {\sum_{k=1}^{n}}' \frac{1}{q_j - q_k}, \tag{2}$$

$$\beta = \frac{1}{2}\sum_{j=1}^{n}{\sum_{k=1}^{n}}' \frac{1}{q_j - q_k}{\sum_{l=1}^{n}}' \frac{1}{q_j - q_l}. \tag{3}$$

To prove (1.4–33c), we now show that

$$\alpha = \frac{1}{2}n(n-1), \tag{4}$$

$$\beta = {\sum_{j>k=1}^{n}}' \frac{1}{(q_j - q_k)^2}. \tag{5}$$

The proof of (4) is easy:

$$\alpha = \sum_{j=1}^{n}{\sum_{k=1}^{n}}' \frac{q_j}{q_j - q_k} \tag{6a}$$

$$= n(n-1) + \sum_{j=1}^{n}{\sum_{k=1}^{n}}' \frac{q_k}{q_j - q_k} \tag{6b}$$

$$= n(n-1) - \alpha. \qquad \textit{Q.E.D.} \tag{6c}$$

[Eq. (6b) is obtained replacing the q_j in the numerator by $q_j - q_k + q_k$; (6c) is obtained exchanging the dummy indices j and k in the sum, and this of course implies (4).]

The proof of (5) is also fairly easy:

$$\beta = \tfrac{1}{2} \sum_{j=1}^{n} \sum_{\substack{k=1 \\ k \neq j}}^{n} \sum_{\substack{l=1 \\ l \neq j}}^{n} \frac{1}{(q_j - q_k)(q_j - q_l)} \tag{7a}$$

$$= \tfrac{1}{2} \sum_{j=1}^{n} \sum_{k=1}^{n}{}' \frac{1}{(q_j - q_k)^2} + \tfrac{1}{2} \sum_{\substack{j,k,l=1 \\ j \neq k \neq l \neq j}}^{n} \frac{1}{(q_j - q_k)(q_j - q_l)} \tag{7b}$$

$$= \sum_{j>k=1}^{n} \frac{1}{(q_j - q_k)^2} + \tfrac{1}{2}\gamma, \tag{7c}$$

where

$$\gamma = \sum_{\substack{j,k,l=1 \\ j \neq k \neq l \neq j}}^{n} \frac{1}{(q_j - q_k)(q_j - q_l)} . \tag{8}$$

Thus (5) holds if γ vanishes. But clearly one can rewrite (8) in the form

$$\gamma = \sum_{\substack{j,k,l=1 \\ j \neq k \neq l \neq j}}^{n} \left(\frac{1}{q_j - q_k} - \frac{1}{q_j - q_l} \right) \frac{1}{q_k - q_l} , \tag{9}$$

and this, together with (8), and the fact that the dummy indices can be exchanged, implies

$$\gamma = -2\gamma, \qquad \text{namely,} \qquad \gamma = 0. \tag{10}$$

Chapter 2

Matrices, Differential Operators, Polynomials, Singular Integral Equations

Introduction

In § 1.4 we have obtained some mathematical results from the analysis of the behaviour of certain solvable dynamical systems in the neighborhood of their equilibrium configurations. For instance, in this manner it was found that the matrix $\mathbf{A}$, of order n,

$$A_{j,k} = \begin{cases} \sum_{m=1}^{n}{}' \frac{1}{(\bar{x}_j - \bar{x}_m)^2}, & j = k, \\ \frac{-1}{(\bar{x}_j - \bar{x}_k)^2}, & j \neq k, \end{cases} \tag{1}$$

has the first n nonnegative integers as its eigenvalues if the n numbers $\bar{x}_j$ are the n zeros of the Hermite polynomial of order n,

$$H_n(\bar{x}_j) = 0, \quad j = 1, 2, \ldots, n; \tag{2}$$

and the matrix $\mathbf{N}$, of order n,

$$N_{j,k} = \begin{cases} x_j \sum_{m=1}^{n}{}' \frac{1}{x_j - x_m}, & j = k, \\ \frac{x_j}{x_j - x_k}, & j \neq k, \end{cases} \tag{3}$$

has the first n nonnegative integers as its eigenvalues, for any arbitrary choice of the n numbers x_j (all different). These results, although arrived at in the manner indicated in § 1.4, can be better understood by placing them in a general framework, that has nothing to do with integrable dynamical systems, but pertains rather to the theory of linear differential operators, polynomials and polynomial interpolation, and finite-dimensional matrices. This approach, and the corresponding results, are outlined in the following Sections of this Chapter, whose presentation is however, for the reasons indicated in the Introduction, extremely terse, being essentially limited to the display of the main formulae, with little explanation and no proofs. The reader who wishes to pursue the matter in more detail may do so through the references given in each Section.

§2.1 Matrix representations of x and d/dx

Let x_j be n arbitrary numbers (all different), and define in terms of them the two matrices $\mathbf{X}$ and $\mathbf{Z}$, of order n, as follows:

$$\mathbf{X} = \mathrm{diag}(x_j), \quad X_{j,k} = \delta_{j,k} x_j, \tag{1}$$

$$Z_{j,k} = \begin{cases} \sum_{m=1}^{n}{}' \dfrac{1}{x_j - x_m}, & j = k, \\ \dfrac{1}{x_j - x_k}, & j \neq k. \end{cases} \tag{2}$$

It is then possible to define a (non unique) weight $w(x)$ and a set of n polynomials (of degree $n-1$) $p_{n-1}^{(j)}(x)$, such that [C1981]

$$\int dx\, w(x) \tilde{p}_{n-1}^{(j)}(x) p_{n-1}^{(k)} = \delta_{j,k}, \tag{3}$$

$$\int dx\, w(x) \tilde{p}_{n-1}^{(j)}(x)\, x\, p_{n-1}^{(k)} = X_{j,k}, \tag{4}$$

$$\int dx\, w(x) \tilde{p}_{n-1}^{(j)}(x) \frac{d}{dx} p_{n-1}^{(k)} = Z_{j,k}, \tag{5}$$

with

$$p_{n-1}^{(j)}(x) = k_n \prod_{\substack{k=1 \\ k\neq j}}^{n} (x - x_k) = \frac{p_n(x)}{x - x_j}, \tag{6}$$

$$\tilde{p}_{n-1}^{(j)}(x) = C_j p_{n-1}^{(j)}(x). \tag{7}$$

A (non-unique) realization of the weight $w(x)$ is that corresponding to the set of $n+1$ orthogonal polynomials $p_m(x)$, of degree $m = 0, 1, 2, \ldots, n$ [C1981]:

$$\int dx\, w(x) p_l(x) p_m(x) = \delta_{l,m}; \quad l, m = 0, 1, 2, \ldots, n. \tag{8}$$

Note that the last polynomial of this set has, consistently with (6), the n numbers x_j as its zeros:

$$p_n(x) = k_n \prod_{j=1}^{n} (x - x_j). \tag{9}$$

The constants C_j are defined, in terms of this polynomial set, by the formulæ

$$C_j = k_{n-1} p_{n-1}(x_j) / p_n'(x_j), \tag{10a}$$

$$= [k_{n-1} p_{n-1}(x_j)]^2 \lambda_j, \tag{10b}$$

$$= [p_n'(x_j)]^{-2} \lambda_j^{-1}, \tag{10c}$$

where

$$k_m = \lim_{x\to\infty} \frac{p_m(x)}{x^m}, \quad m = n, n-1, \ldots, \tag{11}$$

and the λ_j's are the Christoffel numbers associated with the set of orthogonal polynomials $p_m(x)$ (see Section 10.4 of [HTF1953]).

These results [C1981] may also be obtained in the framework of the Lagrangian polynomial interpolation of a function and its derivatives [C1982b]. This yields moreover an alternative definition of the weight $w(x)$, namely [C1982b]

$$w(x) = \sum_{j=1}^{n} \left[p_{n-1}^{(j)}(x_j)\right]^{-2} \delta(x - x_j) \tag{12a}$$

$$= \sum_{j=1}^{n} \left[p'_{n-1}(x_j)\right]^{-2} \delta(x - x_j), \tag{12b}$$

which is consistent with (3)–(7) with

$$C_j = 1, \quad j = 1, 2, \ldots, n. \tag{12c}$$

As a consequence of these results, every formula for differential operators which holds in the functional space spanned by polynomials of degree less than n has, in the finite-dimensional vector space of dimension n, a direct counterpart, which may be obtained via the rule $x \mapsto \mathbf{X}$, $d/dx \mapsto \mathbf{Z}$. More precisely:

Lemma *If there holds, for the differential operator*

$$\mathcal{F} = \sum_m f_m(x) \frac{d^m}{dx^m}, \tag{13}$$

where $f_m(x)$ is a polynomial in x of degree m (or less), the equation

$$\mathcal{F} p_q(x) = 0, \quad q < n, \tag{14}$$

where $p_q(x)$ is a polynomial of degree q (less than n),

$$p_q(x) = \sum_{r=0}^{q} a_r x^r, \tag{15}$$

then there holds also the vector formula

$$\mathbf{F} \boldsymbol{\pi} = \mathbf{0}, \tag{16}$$

with the matrix $\mathbf{F}$, of order n, defined by

$$\mathbf{F} = \sum_m f_m(\mathbf{X}) \mathbf{Z}^m, \tag{17}$$

and the n-vector $\boldsymbol{\pi}$ defined by

$$\boldsymbol{\pi} = p_q(\mathbf{X}) \mathbf{v}, \tag{18}$$

$$v_j = \frac{k_n}{p_{n-1}^{(j)}(x_j)} = \frac{k_n}{p'_n(x_j)} = \prod_{k=1}^{n}{}' \frac{1}{x_j - x_k}. \tag{19}$$

Thus, for instance, the fact (see § 1.4 and the Introduction to this Chapter) that the matrix

$$\mathbf{N} = \mathbf{X}\,\mathbf{Z} \tag{20}$$

has eigenvalues $0, 1, 2, \ldots, n-1$,

$$\mathbf{N}\mathbf{v}^{(m)} = (m-1)\mathbf{v}^{(m)}, \qquad m = 1, 2, \ldots, n, \tag{21}$$
$$\mathbf{v}^{(m)} = \mathbf{X}^{m-1}\mathbf{v}, \qquad m = 1, 2, \ldots, n, \tag{22}$$

is an immediate consequence of the trivial differential formula

$$x\frac{d}{dx}x^{m-1} = (m-1)x^{m-1}. \tag{23}$$

Similarly, the fact that the matrix $\mathbf{A}$ (see (1.3–43)) has eigenvalues $0, 1, 2, \ldots, n-1$,

$$\mathbf{A}\mathbf{v}^{(H)(m)} = (n-m)\mathbf{v}^{(H)(m)}, \qquad m = 1, 2, \ldots, n, \tag{24}$$
$$\mathbf{v}^{(H)(m)} = H_{m-1}(\mathbf{X} - x_0\mathbf{1})\mathbf{v}, \qquad m = 1, 2, \ldots, n, \tag{25}$$

is an immediate consequence of the formula[1]

$$\left[n - 1 + \frac{1}{2}\frac{d^2}{dx^2} + (x_0 - x)\frac{d}{dx}\right]H_{m-1}(x - x_0) = (n-m)H_{m-1}(x - x_0). \tag{26}$$

Note that the explicit form of the eigenvectors is now also exhibited [see (22), (25), (18), and (19)].

It is clear that this technique provides an ample possibility to construct explicit matrices, defined in terms of n, or more, arbitrary parameters (or, as subcases, of n or more numbers defined by some special property), whose spectrum and eigenvectors are completely, or partly, known [C1981]. One of the possible applications of this technique is to test computer programs to diagonalize large matrices. An even more powerful implementation of this technique is implied by the results of the following § 2.3.

§2.2 Algebra of the matrices X and Z

In this Section we report a collection of algebraic results for the matrices **X**, **Z**, and **N**, defined by (2.1–1), (2.1–2), and (2.1–20):[2]

$$[\mathbf{Z}, \mathbf{X}] = \mathbf{1} - \mathbf{J}, \tag{1}$$
$$(\mathbf{1})_{j,k} = \delta_{j,k}, \tag{2}$$
$$J_{j,k} = 1, \quad \mathbf{J}^2 = n\mathbf{J}, \quad \mathbf{J}\,\mathbf{Z} = \mathbf{0}; \tag{3}$$

$$p_n(x) = k_n \prod_{j=1}^{n}(x - x_j) = \sum_{m=0}^{n} a_m x^m, \quad a_n = k_n; \tag{4}$$

[1]See (10.13.-12) of [HTF1953].

[2]For proofs and additional results, see [C1980b], [C1980c], [C1981], and [C1981a].

$$v_j = \frac{k_n}{p'_{n-1}(x_k)} = \frac{k_n}{p_{n-1}{}^{(j)}(x_k)} = \prod_{k=1}^{n}{}' \frac{1}{x_j - x_k}, \tag{5}$$

$$\mathbf{v}^{(m)} = \mathbf{X}^{m-1}\mathbf{v}, \quad v_j{}^{(m)} = x_j^{m-1} \prod_{k=1}^{n}{}' \frac{1}{x_j - x_k}, \quad \mathbf{v}^{(1)} = \mathbf{v}; \tag{6}$$

$$\begin{aligned}
\mathbf{J}\,\mathbf{v}^{(m)} &= \mathbf{0}, & & m = 1,2,\ldots,n-1, & &(7a)\\
\mathbf{J}\,\mathbf{v}^{(n)} &= \mathbf{u}, \quad u_j = 1, & & & &(7b)\\
\mathbf{X}\,\mathbf{v}^{(m)} &= \mathbf{v}^{(m+1)}, & & m = 1,2,\ldots,n-1, & &(8)\\
\mathbf{Z}\,\mathbf{v}^{(m)} &= (m-1)\mathbf{v}^{(m-1)}, & & m = 1,2,\ldots,n, & &(9)\\
\mathbf{N}\,\mathbf{v}^{(m)} &= (m-1)\mathbf{v}^{(m)}, & & m = 1,2,\ldots,n; & &(10)
\end{aligned}$$

$$u_j = 1, \tag{11}$$

$$\mathbf{u}^{(m)} = \frac{1}{k_n} \sum_{l=m}^{n} a_l \mathbf{X}^{l-m}\mathbf{u}, \qquad m = 1,2,\ldots,n, \tag{12a}$$

$$\mathbf{u}^{(n)} = \mathbf{u}, \tag{12b}$$

$$\begin{aligned}
\mathbf{u}^{(m)}\mathbf{Z} \equiv \mathbf{Z}^{\top}\mathbf{u}^{(m)} &= m\mathbf{u}^{(m+1)}, & & m = 1,2,\ldots,n-1, & &(13a)\\
\mathbf{u}^{(n)}\mathbf{Z} = \mathbf{u}\,\mathbf{Z} \equiv \mathbf{Z}^{\top}\mathbf{u} &= \mathbf{0}, & & & &(13b)\\
\mathbf{u}^{(m)}\mathbf{N} \equiv \mathbf{N}^{\top}\mathbf{u}^{(m)} &= (m-1)\mathbf{u}^{(m)}, & & m = 1,2,\ldots,n; & &(14)
\end{aligned}$$

$$V_{j,k} = v_j{}^{(k)} = x_j^{k-1} \prod_{m=1}^{n}{}' \frac{1}{x_j - x_m}, \tag{15}$$

$$U_{j,k} = u_k{}^{(j)} = \frac{1}{k_n} \sum_{m=j}^{n} a_m x_k^{m-1}, \tag{16}$$

$$\mathbf{U}\mathbf{V} = \mathbf{V}\mathbf{U} = \mathbf{1}, \tag{17}$$

$$\det \mathbf{V} = \prod_{j>k=1}^{n} \frac{1}{x_j - x_k}, \tag{18a}$$

$$\det \mathbf{U} = \prod_{j>k=1}^{n} (x_j - x_k), \tag{18b}$$

$$\begin{aligned}
\mathbf{U}\,\mathbf{N}\,\mathbf{V} &= \mathrm{diag}(j-1), & & & &(19)\\
(\mathbf{U}\,\mathbf{Z}\,\mathbf{V})_{j,k} &= j\delta_{j,k-1}, & & & &(20)\\
(\mathbf{U}\mathbf{X}\mathbf{V})_{j,k} &= \delta_{j,k+1}, & & j = 1,2,\ldots,n, \quad k = 1,2,\ldots,n-1, & &(21a)\\
(\mathbf{U}\mathbf{X}\mathbf{V})_{j,k} &= -a_{j-1}/k_n, & & j = 1,2,\ldots,n; & &(21b)
\end{aligned}$$

$$\frac{\partial \mathbf{N}}{\partial x_j} = [\mathbf{M}^{(j)}, \mathbf{N}], \tag{22}$$

$$M_{j,k}{}^{(m)} = Z_{m,k}(\delta_{j,m} - \delta_{j,k}), \quad j,k,m = 1,2,\ldots,n, \tag{23}$$

$$\dot{\mathbf{N}} = [\mathbf{M}, \mathbf{N}], \tag{24}$$

$$\mathbf{M} = \dot{\mathbf{V}}\,\mathbf{U} = -\mathbf{V}\,\dot{\mathbf{U}} = \dot{\mathbf{X}}\,\mathbf{Z} - \mathrm{diag}(\dot{\mathbf{x}}\,\mathbf{Z}); \tag{25}$$

$$\mathbf{N}(\mathbf{y}) = \mathbf{W}(\mathbf{y},\mathbf{x})\mathbf{N}(\mathbf{x})[\mathbf{W}(\mathbf{y},\mathbf{x})]^{-1}, \tag{26}$$

$$\mathbf{W}(\mathbf{y},\mathbf{x}) = \mathbf{V}(\mathbf{y})\mathbf{U}(\mathbf{x}), \tag{27a}$$

$$W_{j,k}(\mathbf{y},\mathbf{x}) = \frac{y_j - x_j}{y_j - x_k} \prod_{\substack{m=1 \\ m\neq j}}^{n} \frac{y_j - x_m}{y_j - y_m}, \tag{27b}$$

$$[\mathbf{W}(\mathbf{y},\mathbf{x})]^{-1} = \mathbf{W}(\mathbf{x},\mathbf{y}). \tag{28}$$

In (24) and (25) one assumes that the n numbers x_j depend arbitrarily on a parameter t (*time*), and indicates by a superimposed dot the corresponding derivative. Note that (24) has the form of a Lax equation (see § 1.1); but it has, of course, no dynamical content (*i.e.* it is an identity).

In the last four equations, two sets of arbitrary numbers enter, which are indicated by the two n-vectors $\mathbf{x}$ and $\mathbf{y}$; the corresponding notation should be self-explanatory. Note that the explicit representation (27b) (from [C1980c]) is nontrivial, as indeed the fact that the matrix $\mathbf{W}(\mathbf{y},\mathbf{x})$ defined by this explicit formula in terms of the two arbitrary vectors $\mathbf{x}$ and $\mathbf{y}$ satisfies (28). Indeed, this last formula could be used to test computer programs for inverting large matrices.

§2.3 Explicit realizations of the algebra of raising and lowering operators. Construction of matrices with known eigenvalues and eigenvectors. Representations of the classical polynomials

Consider a functional space spanned by the denumerable set of linearly independent functions $\phi_m(x)$, $m = 1,2,3,\ldots$, and assume that there exist two (*lowering* and *raising*) operators, $\mathcal{L}$ and $\mathcal{R}$, acting in this space as follows:

$$\mathcal{L}\phi_m(x) = (m-1)\phi_{m-1}(x), \qquad m = 1,2,\ldots, \tag{1a}$$

$$\mathcal{R}\phi_m(x) = \phi_{m+1}(x), \qquad m = 1,2,\ldots. \tag{1b}$$

Note that (1a) implies that $\mathcal{L}$ annihilates the first of the basis functions, $\mathcal{L}\phi_1(x) = 0$.

These formulae imply that the *number* operator $\mathcal{N}$,

$$\mathcal{N} = \mathcal{R}\mathcal{L}, \tag{2a}$$

has the $\phi_m(x)$'s as eigenfunctions, with integral eigenvalues,

$$\mathcal{N}\phi_m(x) = (m-1)\phi_m(x), \qquad m = 1,2,\ldots, \tag{2b}$$

and that there hold the commutation rules

$$[\mathcal{L},\mathcal{N}]=\mathcal{L},\qquad [\mathcal{R},\mathcal{N}]=-\mathcal{R},\qquad [\mathcal{L},\mathcal{R}]=1. \tag{3}$$

Explicit realizations of this algebra are, for instance

$$\mathcal{R}=\frac{d}{dx},\qquad \mathcal{L}=x,\qquad \phi_m(x)=x^{m-1}, \tag{4a}$$
$$\mathcal{R}=\frac{1}{2}\frac{d}{dx},\qquad \mathcal{L}=2x-\frac{d}{dx},\qquad \phi_m(x)=H_{m-1}(x). \tag{4b}$$

Note moreover that (4b) implies that, if $\mathcal{L}$ and $\mathcal{R}$ are realizations of this algebra, then

$$\mathcal{L}^{(H)}=\tfrac{1}{2}\mathcal{L},\qquad \mathcal{R}^{(H)}=2\mathcal{R}-\mathcal{L} \tag{5}$$

are also realizations, and so on. Similarly, if the differential operator

$$\mathcal{F}=\sum_m f_m\frac{d^m}{dx^m} \tag{6a}$$

has a given spectrum, then the operator

$$\tilde{\mathcal{F}}=\sum_m f_m(\mathcal{R})\mathcal{L}^m \tag{6b}$$

has the same spectrum, for any realization of the operators $\mathcal{R}$ and $\mathcal{L}$; and so on.

Consider instead an n-dimensional vector space spanned by the n linearly independent vectors $\mathbf{v}^{(m)}$, $m=1,2,\dots,n$; and assume that there exist two matrices of order n, $\mathbf{L}$ and $\tilde{\mathbf{R}}$, which act on these vectors as follows:

$$\mathbf{L}\,\mathbf{v}^{(m)}=(m-1)\mathbf{v}^{(m-1)},\qquad m=1,2,\dots,n, \tag{7a}$$
$$\tilde{\mathbf{R}}\,\mathbf{v}^{(m)}=\mathbf{v}^{(m+1)},\qquad m=1,2,\dots,n-1. \tag{7b}$$

Note that (7a) implies that $\mathbf{L}$ annihilates the first of the basis vectors, $\mathbf{L}\,\mathbf{v}^{(1)}=\mathbf{0}$, while, writing (7b), we have omited to indicate the action of $\tilde{\mathbf{R}}$ on the last basis vector: $\tilde{\mathbf{R}}\,\mathbf{v}^{(n)}=\,?$ Indeed, the superimposed tilde on $\mathbf{R}$ has been introduced as a reminder of this fact; while we indicate by $\mathbf{R}$ a matrix which, in addition to having the property (7b), annihilates the highest vector $\mathbf{v}^{(n)}$:

$$\mathbf{R}\,\mathbf{v}^{(m)}=\begin{cases}\mathbf{v}^{(m+1)}, & m=1,2,\dots,n-1,\\ 0, & m=n.\end{cases}$$

Given a matrix $\tilde{\mathbf{R}}$, it is easy to construct the corresponding matrix $\mathbf{R}$, using the matrix $\mathbf{P}$ which projects on the highest vector,

$$\mathbf{P}\,\mathbf{v}^{(m)}=\delta_{m,n}\mathbf{v}^{(m)},\qquad m=1,2,\dots,n; \tag{8}$$

since clearly (7b) and (8) imply (7c) with

$$\mathbf{R}=\tilde{\mathbf{R}}(\mathbf{1}-\mathbf{P}). \tag{9}$$

These matrices, together with the matrix

$$\mathbf{N}=\mathbf{R}\mathbf{L}=\tilde{\mathbf{R}}\mathbf{L}, \tag{10}$$

satisfy the following additional relations:

$$\mathbf{N}\,\mathbf{v}^{(m)} = (m-1)\mathbf{v}^{(m)}, \qquad m = 1, 2, \ldots, n, \tag{11}$$

$$\mathbf{P}^2 = \mathbf{P}, \qquad \mathbf{P\,L} = \mathbf{RP} = \mathbf{0}, \qquad \mathbf{P\,N} = \mathbf{NP} = (n-1)\mathbf{P}, \tag{12}$$

$$[\mathbf{L}, \mathbf{N}] = \mathbf{L}, \qquad [\mathbf{R}, \mathbf{N}] = -\mathbf{R}, \tag{13}$$

$$\mathbf{L\,R} = \mathbf{N} + \mathbf{1} - n\mathbf{P}, \qquad [\mathbf{L}, \mathbf{R}] = \mathbf{1} - n\mathbf{P}. \tag{14}$$

There hold, moreover, the following formulae involving the set of n vectors $\mathbf{u}^{(m)}$, orthonormal to $\mathbf{v}^{(m)}$:

$$(\mathbf{u}^{(l)}, \mathbf{v}^{(m)}) = \delta_{l,m} = \sum_{j=1}^{n} u_j{}^{(l)} v_j{}^{(m)}, \tag{15a}$$

$$U_{j,k} = u_k{}^{(j)}, \qquad V_{j,k} = v_j{}^{(k)}, \qquad \mathbf{UV} = \mathbf{VU} = \mathbf{1}; \tag{15b}$$

$$\mathbf{P} = \mathbf{v}^{(n)} \otimes \mathbf{u}^{(n)}, \qquad P_{j,k} = v_j{}^{(n)} u_k{}^{(n)}, \tag{16}$$

$$\mathbf{N} = \sum_{m=1}^{n} (m-1)\mathbf{v}^{(m)} \otimes \mathbf{u}^{(m)}, \qquad N_{j,k} = \sum_{m=1}^{n} (m-1) v_j{}^{(m)} u_k{}^{(m)}, \tag{17a}$$

$$\mathbf{L} = \sum_{m=2}^{n} (m-1)\mathbf{v}^{(m-1)} \otimes \mathbf{u}^{(m)}, \qquad L_{j,k} = \sum_{m=2}^{n} (m-1) v_j{}^{(m-1)} u_k{}^{(m)}, \tag{17b}$$

$$\mathbf{R} = \sum_{m=1}^{n-1} \mathbf{v}^{(m+1)} \otimes \mathbf{u}^{(m)}, \qquad R_{j,k} = \sum_{m=1}^{n-1} v_j{}^{(m+1)} u_k{}^{(m)}, \tag{17c}$$

$$\mathbf{U\,N\,V} = \mathrm{diag}(j-1), \qquad (\mathbf{U\,N\,V})_{j,k} = \delta_{j,k}(j-1), \tag{18a}$$

$$(\mathbf{U\,L\,V})_{j,k} = j\,\delta_{j,k-1}, \qquad (\mathbf{U\,R\,V})_{j,k} = \delta_{j,k+1}. \tag{18b}$$

The main result[3] which we now note is that an *explicit realization* of the equations from (7a) to (18b) is provided by the formulae of the preceding § 2.2 [including the definitions of the basis vectors $\mathbf{v}^{(m)}$ and $\mathbf{u}^{(m)}$], with the matrices $\mathbf{X}$ and $\mathbf{Z}$ defined by (2.1–1) and (2.1–2):

$$\mathbf{L} = \mathbf{Z}, \qquad \tilde{\mathbf{R}} = \mathbf{X}, \qquad \mathbf{R} = \mathbf{X}(\mathbf{1} - \mathbf{P}), \tag{19}$$

$$P_{j,k} = x_j^{n-1} \prod_{m=1}^{n}{}' \frac{1}{x_j - x_m}, \tag{20}$$

$$R_{j,k} = \delta_{j,k} x_j - x_j^n \prod_{m=1}^{n}{}' \frac{1}{x_j - x_m}. \tag{21}$$

It is moreover plain how other explicit finite-dimensional realizations of this algebra can be obtained. For instance, one such realization is provided —see (5)— by 5607

$$\mathbf{L}^{(H)} = \tfrac{1}{2}\,\mathbf{Z}, \quad \tilde{\mathbf{R}}^{(H)} = 2\mathbf{X} - \mathbf{Z}, \quad \mathbf{N}^{(H)} = \tilde{\mathbf{R}}^{(H)}\mathbf{L}^{(H)} = \mathbf{XZ} - \tfrac{1}{2}\,\mathbf{Z}^2, \tag{22}$$

[3]See [C1982] and [BC1981].

always with $\mathbf{X}$ and $\mathbf{Z}$ defined by (2.1–1) and (2.1–2); and the corresponding (right) eigenvectors read

$$\mathbf{v}^{(H)(m)} = H_{m-1}(\mathbf{X})\mathbf{v}, \qquad m = 1, 2, \ldots, n, \tag{23}$$

with $\mathbf{v}$ defined by (2.2–5):

$$\mathbf{N}^{(H)}\mathbf{v}^{(H)(m)} = (m-1)\mathbf{v}^{(H)(m)}, \qquad m = 1, 2, \ldots, n, \tag{24}$$
$$\mathbf{L}^{(H)}\mathbf{v}^{(H)(m)} = (m-1)\mathbf{v}^{(H)(m-1)}, \qquad m = 1, 2, \ldots, n, \tag{25a}$$
$$\tilde{\mathbf{R}}^{(H)}\mathbf{v}^{(H)(m)} = \mathbf{v}^{(H)(m+1)}, \qquad m = 1, 2, \ldots, n-1, \tag{25b}$$
$$\tilde{\mathbf{R}}^{(H)}\mathbf{v}^{(H)(n)} = H_n(\mathbf{X})\mathbf{v}. \tag{25c}$$

These formulae apply for an arbitrary choice of the n numbers x_j. If, on the other hand, the special choice $x_j = \bar{x}_j$ is made, with $\bar{x}_j$ the zeros of the Hermite polynomial of order n, then the right-hand side of the last equation above, (25c), vanishes, so that in this case the tilde over $\mathbf{R}$ may be dropped [Compare (25c) with (7c)].

These results yield clearly a more flexible version of the Lemma of § 2.1, since they imply the possibility of a correspondence between the operators d/dx and x in a functional space spanned by polynomials, not only with the finite-dimensional matrices $\mathbf{Z}$ and $\mathbf{X}$, but more generally with any realization of the matrices $\mathbf{L}$ and $\tilde{\mathbf{R}}$ satisfying the algebra described above. This provides a powerful tool to construct explicit matrices, defined in terms of n or more arbitrary numbers, or of the zeros of special functions, whose eigenvectors can be given in explicit form together with their eigenvalues; the latter being expressed simply in terms of integers or in terms of the zeros of special functions. The interested reader will find a more explicit analysis and several examples in [C1981], [BC1981], and [C1982]. Here we merely mention one spinoff of these results, namely the following representation of Hermite polynomials (from [BC1981]):

$$H_n(x) = 2^n \det[x\mathbf{1} - \mathbf{M}^{(H)}(\lambda)], \tag{26}$$

where

$$\mathbf{M}^{(H)}(\lambda) = \frac{1}{\sqrt{2}}\left(\lambda\mathbf{L} + \lambda^{-1}\mathbf{R}\right). \tag{27}$$

Here λ is arbitrary, and the two matrices $\mathbf{R}$ and $\mathbf{L}$, of order n, are defined according to the above treatment: for instance, in terms of n arbitrary numbers x_j, by (21), (19) and (2.1–2); or, in terms of the n zeros $\bar{x}_j$ of the Hermite polynomial $H_n(x)$, by (22) with the tilde on $\mathbf{R}$ omitted [see the remarks after (25c); note that in this case the validity of (26) becomes trivial if, in (27), $\lambda = \sqrt{2}$]. Analogous representations exist for all the classical polynomials.

§2.4 On some singular integral equations

In the previous Sections, techniques have been outlined to construct explicit vector equations. The idea that we would like to convey here is the possibility (only partially explored until now) to obtain results for singular integral equations by letting n, the order of these vectors and matrices, diverge to infinity.

Let us outline briefly one example. We have seen in § 1.4 that the matrix $\mathbf{A}$, defined in terms of the n zeros $\bar{x}_j$ of the Hermite polynomial of order n,

$$H_n(\bar{x}_j) = 0, \tag{1}$$

by the formula

$$A_{j,k}=\begin{cases}\displaystyle\sum_{m=1}^{n}{}'\frac{1}{(\bar{x}_j-\bar{x}_m)^2}, & j=k,\\ \displaystyle\frac{-1}{(\bar{x}_j-\bar{x}_k)^2}, & j\neq k,\end{cases} \tag{2}$$

has the first n nonnegative integers as eigenvalues:

$$\mathbf{A}\,\mathbf{v}^{(m)}=(m-1)\mathbf{v}^{(m)},\qquad m=1,2,\ldots,n, \tag{3a}$$

$$\sum_{k=1}^{n}\frac{v_j{}^{(m)}-v_k{}^{(m)}}{(\bar{x}_j-\bar{x}_k)^2}=(m-1)v_j{}^{(m)},\qquad j,m=1,2,\ldots,n. \tag{3b}$$

Taking the limit of (3b) as $n\to\infty$, it was found in [CP1978] that the singular integral operator $\mathcal{A}$, defined by

$$\mathcal{A}f(x)\equiv\frac{P}{\pi}\int_{-1}^{1}dy\,a(y)\,\frac{f(x)-f(y)}{(x-y)^2}, \tag{4a}$$

$$a(y)=\sqrt{1-y^2}, \tag{4b}$$

has the integral eigenvalues

$$\mathcal{A}U_m(x)=mU_m(x),\qquad m=0,1,2,\ldots. \tag{5}$$

It was moreover noted in [C1979] that the eigenfunctions $U_m(x)$ are the Chebyshev polynomials of the second kind (see [HTF1953]):

$$U_m(\cos\theta)=\frac{\sin[(m+1)\theta]}{\sin\theta}. \tag{6}$$

It was also shown in [C1979] that the singular integral operator (4a) has integer eigenvalues and polynomial eigenfunctions if and only if the weight $a(y)$ has the (more general) form

$$a(y)=\sqrt{1-y^2}\left(1+\frac{\alpha}{1-y}+\frac{\beta}{1+y}\right). \tag{4c}$$

The eigenfunctions and spectrum in this more general case were investigated in considerable detail in [A1979] and [C1979a].

References

[A1977] M. Adler, Some finite-dimensional integrable systems. *Commun. Math. Phys.* **55**, 195–230 (1977).

[A1978] M. Adler, Some finite-dimensional integrable systems. In [FG1978], pp. 237–244.

[A1979] D. Atkinson, Inversion of an equation of Calogero. In [S1979], pp. 267–270.

[ABCOP1979] S. Ahmed, M. Bruschi, F. Calogero, M. A. Olshanetsky, and A. M. Perelomov, Properties of the zeros of the classical polynomials and of the Bessel functions. *Nuovo Cimento* **49B**, 173–199 (1979).

[AM1978] M. Adler and J. Moser, On a class of polynomials connected with the KdV equation. *Commun. Math. Phys.* **61**, 1–30 (1978).

[AMM1977] H. Airault, H. P. McKean, and J. Moser, Rational and elliptic solutions of the Korteweg-de Vries equation and a related many-body problem. *Comm. Pure Appl. Math.* **30**, 95–148 (1977).

[B1980] G. Barucchi, Graphs and an exactly solvable N-body problem in one dimension.

[Ba1978] A. O. Barut (ed.), *Nonlinear Equations in Physics and Mathematics* . Reidel, 1978.

[BB1980] C. Bardos and D. Bessis (eds.), *Bifurcation Phenomena in Mathematical Physics and Related Topics.* Reidel, 1980.

[BC1981] M. Bruschi and F. Calogero, Finite dimensional matrix representations of the operator of differentiation through the algebra of raising and lowering operators: general properties and explicit examples. *Nuovo Cimento* **62B**, 337–351 (1981).

[BR1977] G. Barucchi and T. Regge, Conformal properties of a class of exactly solvable N-body problems in space dimension one. *J. Math. Phys.* **18**, 1149–1153 (1977).

[C1971] F. Calogero, Solution of the one-dimensional N-body problem with quadratic and/or inversely quadratic pair potentials. *J. Math. Phys.* **12**, 419–436 (1971).

[C1975] F. Calogero, Exactly solvable one-dimensional many-body problems. *Lett. Nuovo Cimento* **13**, 411–416 (1975).

[C1976] F. Calogero, On a functional equation connected with integrable many-body problems. *Lett. Nuovo Cimento* **16**, 77-80 (1976).

[C1976a] F. Calogero, Exactly solvable two-dimensional many-body problems. *Lett. Nuovo Cimento* **16**, 35–38 (1976).

[C1976b] F. Calogero, A sequence of Lax matrices for certain integrable hamiltonian systems. *Lett. Nuovo Cimento* **16**, 22-24 (1976).

[C1977] F. Calogero, On the zeros of the classical polynomials. *Lett. Nuovo Cimento* **19**, 505–508 (1977).

[C1977b] F. Calogero, Equilibrium configuration of the one-dimensional n-body problem with quadratic and inversely quadratic pair potentials. *Lett. Nuovo Cimento* **20**, 251–253 (1977).

[C1977c] F. Calogero, On the zeros of Hermite polynomials. *Lett. Nuovo Cimento* **20**, 489–490 (1977).

[C1978] F. Calogero, Integrable many-body problems. In [Ba1978], pp. 3–53.

[C1978a] F. Calogero, Motions of poles and zeros of special solutions of nonlinear and linear partial differential equations and related "solvable" many-body problems. *Nuovo Cimento* **43B**, 177–241 (1978).

[C1979] F. Calogero, Singular integral operators with integral eigenvalues and polynomial eigenfunctions. *Nuovo Cimento* **51B**, 1–14 (1979); **53B**, 463 (1979).

[C1979a] F. Calogero, Integral representations and generating function for the polynomials $U_n^{(\alpha,\beta)}(x)$. *Lett. Nuovo Cimento* **24**, 595–600 (1979).

[C1980] F. Calogero, Solvable many-body problems and related mathematical findings (and conjectures). In [BB1980], pp. 371–384.

[C1980a] F. Calogero, Integrable many-body problems and related mathematical results. In [Co1981], pp. 151–164.

[C1980b] F. Calogero, Isospectral matrices and polynomials. *Nuovo Cimento* **58B**, 169–180 (1980).

[C1980c] F. Calogero, Finite transformations of certain isospectral matrices. *Lett. Nuovo Cimento* **28**, 502–504 (1980).

[C1981] F. Calogero, Matrices, differential operators and polynomials. *J. Math. Phys.* **22**, 919–932 (1981).

[C1981a] F. Calogero, Additional identities for certain isospectral matrices. *Lett. Nuovo Cimento* **30**, 342–344 (1981).

[C1982] F. Calogero, Isospectral matrices and classical polynomials. *Linear Alg. Appl.* **44**, 55–60 (1982).

[C1982a] F. Calogero, Disproof of a conjecture. *Lett. Nuovo Cimento* **35**, 181–185 (1982).

[C1982b] F. Calogero, Lagrangian interpolation and differentiation. *Lett. Nuovo Cimento* **35**, 273–278 (1982).

[CC1977] D. V. Chudnovky and G. V. Chudnovsky, Pole expansions of nonlinear partial differential equations. *Nuovo Cimento* **40B**, 339–353 (1977).

[CD1982] F. Calogero and A. Degasperis, *Spectral Transform and Solitons: Tools to Solve and Investigate Nonlinear Evolution Equations.* Volume 1. North Holland, 1982.

[CM1974] F. Calogero and C. Marchioro, Exact solution of a one-dimensional three-body scattering problem with two-body and/or three-body inverse-square potentials. *J. Math. Phys.* **15**, 1425–1430 (1974).

[CMR1975] F. Calogero, C. Marchioro, and O. Ragnisco, Exact solution of the classical and quantal one-dimensional many-body problems with the two-body potential $g^2a^2/\sinh^2(ax)$. *Lett. Nuovo Cimento* **13**, 383–387 (1975).

[Co1981] E. G. D. Cohen (ed.), *Fundamental Problems in Statistical Mechanics, V*, Proceedings of the 1980 Enschede Summer School, North Holland, 1981.

[CP1978] F. Calogero and A. M. Perelomov, Asymptotic density of the zeros of Hermite polynomials of diverging order, and related properties of certain singular integral operators. *Lett. Nuovo Cimento* **23**, 650–652 (1978).

[F1974] H. Flaschka, The Toda lattice. I. *Phys. Rev.* **B9**, 1924–1925 (1974).

[F1974a] H. Flaschka, On the Toda lattice. II. *Prog. Theor. Phys.* **51**, 703–716 (1974).

[FG1978] H. Flaschka and D. W. McLaughlin (eds.), *Proceedings of the conference on the Theory and Application of Solitons* (Tucson, Jan. 1976). *Rocky Mountain J. Math.* **8**, No. 1 and 2, 1978.

[G1975] P. J. L. Gambardella, Exact results in quantum many-body systems of interacting particles in many dimensions with $SU(1,1)$ as the dynamical group. *J. Math. Phys.* **16**, 1172–1187 (1975).

[GM1975] G. Gallavotti and C. Marchioro, On the calculation of an integral. *J. Math. Anal. Appl.* **44**, 661–675 (1975).

[H1980] R. Helleman, Self-generated chaotic behaviour in nonlinear mechanics. In [Co1981], pp. 165–233.

[HTF1953] *Higher Transcendental Functions*, Vol. 2, (compiled by the staff of the Bateman manuscript project, A. Erdèlyi, director), Mc Graw-Hill, 1953.

[J1866] C. Jacobi, Problema trium corporum mutis attractionibus cubus distantiarum inverse proportionalibus recta linea se moventium. In *Gesammelte Werke*, Vol. 4, Berlin, 1866.

[K1978] I. M. Krichever, Rational solutions of the Kadomtsev-Petviashvili equation and many-body problems. *Func. Anal. Appl.* **12**, 59–61 (1978). [Original Russian reference: *Funk. Anal ego Pril.* **12**, 76–78 (1978).]

[K1980] I. M. Krichever, Elliptic solutions of the Kadomtsev-Petviashvili equation and integrable systems of particles. *Func. Anal. Appl.* **14**, 282–290 (1980). [Original Russian reference: *Funk. Anal ego Pril.* **14**, 45–54 (1980).]

[KKS1978] D. Každan, B. Kostant, and S. Sternberg, Hamiltonian group actions and dynamical systems of the Calogero type. *Comm. Pure Appl. Math.* **31**, 481–507 (1978).

[KL1972] D. C. Khandekar and S. V. Lawande, Solution of a one-dimensional three-body problem in classical mechanics. *Amer J. Phys.* **40**, 458–462 (1972).

[L1968] P. D. Lax, Integrals of nonlinear equations of evolution and solitary waves. *Commun. Pure Appl. Math.* **21**, 467–490 (1968).

[M1970] C. Marchioro, Solution of a three-body scattering problem in one dimension. *J. Math. Phys.* **11**, 2193–2196 (1970).

[M1975] J. Moser, Three integrable hamiltonian systems connected with isospectral deformations. *Adv. in Math.* **16**, 197–220 (1975).

[Man1974] S. V. Manakov, Complete integrability and stochastization of discrete dynamical systems. *Sov. Phys. JETP* **40**, 269–274 (1974).

[Mu1978] M. E. Muldoon, An infinite system of equations characterizing the zeros of Bessel functions. *Lett. Nuovo Cimento* **23**, 447–448 (1978).

[OP1976] M. A. Olshanetsky and A. M. Perelomov, Explicit solution of the Calogero model in the classical case and geodesic flows of zero curvature. *Lett. Nuovo Cimento* **16**, 333–339 (1976).

[OP1977] M. A. Olshanetsky and A. M. Perelomov, Quantum completely integrable systems connected with semi-simple Lie algebras. *Lett. Math. Phys.* **2**, 7–13 (1977).

[OP1981] M. A. Olshanetsky and A. M. Perelomov, Classical integrable finite-dimensional systems related to Lie algebras. *Physics Reports* **71**, 313–400 (1981).

[OP1983] M. A. Olshanetsky and A. M. Perelomov, Quantal integrable systems related to Lie algebras. *Physics Reports* (to be published).

[OP1978] M. A. Olshanetsky and V.-B. K. Rogov, Bound states in completely integrable systems with two types of particles. *Ann. Inst. H. Poincaré* **29**, 169–177 (1978).

[P1971] A. M. Perelomov, Algebraic approach to the solution of a one-dimensional model of N interacting particles. *Teor. Mat. Fiz.* **6**, 364 (1971).

[P1978] A. M. Perelomov, The simple relation between certain dynamical systems. *Commun. Math. Phys.* **63**, 9–11 (1978).

[P1978a] A. M. Perelomov, Equilibrium configuration and small oscillations of some dynamical systems. *Ann. Inst. H. Poincaré* **A28**, 407–415 (1978).

[S1978] P. C. Sabatier, On the solutions of an infinite system of nonlinear equations. *Lett. Nuovo Cimento* **21**, 41–44 (1978).

[S1979] P. C. Sabatier (ed.), *Problèmes inverses, évolution non linéaire.* Comptes rendus de la rencontre R.C. P. 264 (Montpellier, Oct. 1979), Editions du CNRS, Paris, 1980.

[Sz1939] G. Szegő, *Orthogonal Polynomials.* AMS Coll. Publ. **23**, Providence, R. I., 1939.

[W1974] J. Wolfes, On the three-body linear problem with three-body interaction. *J. Math. Phys.* **15**, 1420–1424 (1974).

Chiral Fields, Self-Dual Yang–Mills Fields as Integrable Systems, and the Role of the Kac–Moody Algebra*

Ling–Lie Chau

Physics Departament
Brookhaven National Laboratory
Upton, New York, USA

Contents:

* Authored under Contract No. DE–AC02–76CH00016 with the U. S. Department of Energy.

Introduction

It has become increasingly clear that, besides its mathematical beauty, the Yang–Mills theory [1–4] may provide the key to our understanding of strong interactions. With the recent experimental observations of gluon jets [5], the ideas of non-abelian gauge theory for strong interactions is brought one step further to reality. Despite many interesting theoretical and phenomenological observations, like confinement [6], asymptotic freedom [7], and Quantum Chromodynamics (QCD) pertubative studies [8], the non-abelian gauge theory is far from fully solved.

In the past ten years or so, powerful mathematical tools have been developed in completely solving many two-dimensional nonlinear differential equations [9]. The characteristics of these so-called integrable systems are the existence of the corresponding linear systems, Bianchi–Bäcklund transformations (BT), obtaining infinite number of conservation laws, soliton solutions, and even the construction of an S–matrix [10]. It is our intention to see whether these powerful techniques can be applied to solve the Yang–Mills theory. From our work of the past three years or so, we have found that, amazingly, the Self-Dual Yang–Mills (SDYM) fields in the J-formulation possess many characteristics of those integrable systems [11–23]. It thus provides a beautiful mathematical system in four dimensions. Due to the extreme similarity in appearance between the SDYM equation and the two-dimensional chiral fields [24–36], many fruitful results have come about by investigating both systems hand in hand.

In §1 we discuss the chiral fields of their integrable characteristics: Parametric BT [24], local conservation laws [24], Riccati equations and the linear systems they lead to [25–27], nonlocal conservation laws [28,29] and their relations to the linear systems [30,15,16], and the recent development of Kac–Moody algebra [31–35].

In §2 we discuss the SDYM fields and the properties of a parametric BT [13], nonlocal conservation laws [14] and their relation to the linear systems [12,15–19], and the derivation of Kac–Moody algebra [20–23], especially with the explicit use of the derivation of the Riemann–Hilbert (RH) method.

I have given quite a few talks on the subject [15,16,18–23]. In order to give an overall view, previously elaborated subjects will be only briefly mentioned. What have been emphasized are: Riccati equations for the chiral fields and their relation to linear systems; using the RH method to derive the transformation that had led to the derivation of the Kac–Moody algebra in the SDYM equations.

§1 Chiral fields

Chiral theories are theories with geometric constraints. They have rich a geometric structure and are relevant to certain physical systems. In particle physics, these models are studied because of their similarity (asymptotic freedom, instantons, etc.) with four-dimensional Yang–Mills fields. Further, it has recently been shown that the classical Yang–Mills theory can be formulated as chiral fields in the loop space. Later in §2 we shall demonstrate that, when properly formulated in what we call the J-formulation, the SDYM equations in four dimensions have strikingly similar appearance as the chiral equation in two dimensions.

The principal chiral fields $g(\varsigma, \eta)$ of group $SU(N)$ are $n \times n$ matrix fields, which have the following Lagrangian density and constraints:

$$L = \mathrm{Tr}(\partial_\varsigma g)(\partial_\eta g^+), \qquad \text{with } g^+ g = g g^+ = I. \tag{1.1}$$

Defining

$$A_\varsigma \equiv g^+ \partial_\varsigma g, \qquad A_\eta \equiv g^+ \partial_\eta g, \tag{1.2}$$

the equation of motion obtained from (1.1) is:

$$\partial_\varsigma A_\eta + \partial_\eta A_\varsigma = 0. \tag{1.3}$$

Here we use the light-cone variables $\varsigma \equiv x_0 + x_1$ and $\eta \equiv x_0 - x_1$. Notice here that A_ς and A_η form the algebra of *SU(N)*, and that they can be considered as pure gauge potentials due to $\partial_\eta A_\varsigma - \partial_\varsigma A_\eta - [A_\varsigma, A_\eta] = 0$. Equation (1.3) has the appearance of a continuity equation. Equation (1.2) characterizes the most important properties of the system, *i.e.*, curvatureless gauge potential. These properties are shared by many nonlinear differential equations including the properly formulated SDYM equations, as we shall discuss later in §2.

1.1 Parametric Bianchi–Bäcklund transformations and local conservation laws

The BT we have constructed [24] for the principal chiral fields were:

$$g'^{+}\partial_\varsigma g' - g^{+}\partial_\varsigma g = \partial_\varsigma(g^{+}g'), \tag{1.1.1}$$
$$g'^{+}\partial_\eta g' - g^{+}\partial_\eta g = -\partial_\eta(g^{+}g'), \tag{1.1.2}$$

with the constraint

$$g^{+}g' + g'^{+}g = 2\beta I, \qquad \text{where } \beta \leq 1, \tag{1.1.3}$$

and

$$g'^{+}g' = g^{+}g = 1. \tag{1.1.4}$$

Here, β is a constant parameter. It is easy to show that (1.1.1) and (1.1.2) are the BT: $(1.1.1)_{,\eta} + (1.1.2)_{,\varsigma} \Rightarrow \partial_\eta A'_\varsigma + \partial_\varsigma A'_\eta - \partial_\eta A_\varsigma - \partial_\varsigma A_\eta = 0$. So A'_ς and A'_η satisfy the equation of motion (1.2) if A_ς and A_η do. The BT can be rewritten as:

$$2(1-\beta)\partial_\varsigma(g'+g) = (g'-g)\left[(\partial_\varsigma g^{+})g' + g'^{+}(\partial_\varsigma g)\right], \tag{1.1.5}$$

$$2(1+\beta)\partial_\eta(g'-g) = -(g'+g)\left[(\partial_\eta g^{+})g' + g'^{+}(\partial_\eta g)\right]. \tag{1.1.6}$$

Then, incorporating the equation of motion (1.3), a continuity-like equation could be obtained, namely

$$(1-\beta)\partial_\varsigma\left\{\mathrm{Tr}\left[(\partial_\eta g^{+})g' + g'^{+}(\partial_\eta g)\right]\right\} + (1+\beta)\partial_\eta\left\{\mathrm{Tr}\left[(\partial_\varsigma g^{+})g' + g'^{+}(\partial_\varsigma g)\right]\right\} = 0. \tag{1.1.7}$$

Using the procedure of expanding around $\beta = 1$, local conservation laws can be derived. The reader is referred to Refs. [24] and [15–16] for details.

1.2 Riccati equation ⇔ Linear systems

We can rewrite the BT equations (1.1.1) and (1.1.2) in the form of Riccati equations:

$$g' \times (1.1.1) \Rightarrow \partial_\varsigma g' - g'g^{+}\partial_\varsigma g = \lambda g'(\partial_\varsigma g^{+})g' + \lambda g'g^{+}\partial_\varsigma g'. \tag{1.2.1}$$

For the last term

$$\begin{aligned}\lambda g' g^+ \partial_\varsigma g' &= 2\beta \partial_\varsigma g' - g g'^+ \partial_\varsigma g', && \text{using (1.1.3)}\\ &= 2\beta \partial_\varsigma g' - g[g^+ \partial_\varsigma g + \partial_\varsigma (g^+ g')], && \text{using (1.1.1)}\\ &= 2\beta \partial_\varsigma g' - \partial_\varsigma g - g(\partial_\varsigma g^+) g' + \partial_\varsigma g'.\end{aligned} \tag{1.2.2}$$

Substituting (1.2.2) into (1.2.3) we obtain the Riccati equation from the ς-BT equation:

$$2(1-\beta)\partial_\varsigma g' - g'(\partial_\varsigma g^+) g' - g'(g^+ \partial_\varsigma g) + (g \partial_\varsigma g^+) g' + \partial_\varsigma g = 0. \tag{1.2.3}$$

Defining

$$\Gamma \equiv g^+ g', \tag{1.2.4}$$

equation (1.2.3) can be written in the following way:

$$2(1-\beta)\partial_\varsigma \Gamma + \Gamma A_\varsigma \Gamma - \Gamma A_\varsigma + (1-2\beta) A_\varsigma \Gamma - A_\varsigma = 0. \tag{1.2.5}$$

Similarly, for the η-BT equation (1.1.2),

$$2(1+\beta)\partial_\eta \Gamma - \Gamma A_\eta \Gamma - \Gamma A_\eta + (1+2\beta) A_\eta \Gamma - A_\eta = 0. \tag{1.2.6}$$

Here some general comments are in order on the relation between the Riccati equation and the corresponding linear system. Using the notation of Levin [37], consider two Riccati equations,

$$\partial_\varsigma \Gamma + \Gamma G_3^\varsigma \Gamma + \Gamma G_4^\varsigma - G_1^\varsigma \Gamma - G_2^\varsigma = 0, \tag{1.2.7}$$

$$\partial_\eta \Gamma + \Gamma G_3^\eta \Gamma + \Gamma G_4^\eta - G_1^\eta \Gamma - G_2^\eta = 0. \tag{1.2.8}$$

One can show that the sufficient conditions for the integrability of these two Riccati equations,

$$\partial_\eta \partial_\varsigma \Gamma - \partial_\varsigma \partial_\eta \Gamma = 0, \tag{1.2.9}$$

are the same as the integrability conditions for the following linear system:

$$\frac{dM}{d\varsigma} = G^\varsigma M, \tag{1.2.10}$$

$$\frac{dM}{d\eta} = G^\eta M, \tag{1.2.11}$$

i.e.,

$$\partial_\eta G^\varsigma - \partial_\varsigma G^\eta + [G^\varsigma, G^\eta] = 0, \tag{1.2.12}$$

where

$$G^\varsigma = \begin{pmatrix} G_1^\varsigma & G_2^\varsigma \\ G_3^\varsigma & G_4^\varsigma \end{pmatrix}, \qquad G^\eta = \begin{pmatrix} G_1^\eta & G_2^\eta \\ G_3^\eta & G_4^\eta \end{pmatrix}, \qquad M = \begin{pmatrix} M_1 & M_2 \\ M_3 & M_4 \end{pmatrix}, \tag{1.2.13}$$

with G_i^ς, G_i^η, and M_i also matrices. Conversely, as discussed in the paper of Levin [37], given the linear system (1.2.10–11) one can show that a Γ defined as

$$\Gamma \equiv (M_1 \Gamma_0 + M_2)(M_3 \Gamma_0 + M_4)^{-1}, \tag{1.2.14}$$

Γ_0 being a constant matrix, satisfies the Riccati equations (1.2.7). Following this discussion, we can construct the corresponding linear system for the Riccati equations (1.2.7–8) which we obtained from the BT (1.1.1–2):

$$\frac{dM}{d\varsigma} = \frac{1}{2(1-\beta)} A_\varsigma \otimes \begin{pmatrix} -(1-2\beta) & -1 \\ +1 & -1 \end{pmatrix} M \equiv \frac{1}{2(1-\beta)} A_\varsigma \otimes C^\varsigma M, \tag{1.2.15}$$

$$\frac{dM}{d\eta} = \frac{1}{2(1+\beta)} A_\eta \otimes \begin{pmatrix} -(1+2\beta) & +1 \\ -1 & -1 \end{pmatrix} M \equiv \frac{1}{2(1+\beta)} A_\eta \otimes C^\eta M, \tag{1.2.16}$$

i.e.,

$$G^\varsigma = \frac{1}{2(1-\beta)} A_\varsigma \otimes \begin{pmatrix} -(1-2\beta) & -1 \\ +1 & -1 \end{pmatrix} \equiv \frac{1}{2(1-\beta)} A_\varsigma \otimes C^\varsigma M, \tag{1.2.17}$$

$$G^\eta = \frac{1}{2(1+\beta)} A_\eta \otimes \begin{pmatrix} -(1+2\beta) & +1 \\ -1 & -1 \end{pmatrix} \equiv \frac{1}{2(1+\beta)} A_\eta \otimes C^\eta M. \tag{1.2.18}$$

Here we ask whether we can simultaneously diagonalize the constant matrices C^ς and C^η, so that the linear system reduces to that of Zakharov and Mikhaĭlov [30].

First we diagonalize C^ς, *i.e.*,

$$UC^\varsigma U^{-1} = \frac{1}{2(1-\beta)} \begin{pmatrix} \lambda_+ & 0 \\ 0 & \lambda_- \end{pmatrix},$$

where $\lambda_\pm$ are the eigenvalues of the matrix C^ς,

$$\lambda_\pm = -(1-\beta) \pm i\sqrt{1-\beta^2}, \tag{1.2.19}$$

and

$$U = -\frac{1}{2i\sqrt{1-\beta^2}} \begin{pmatrix} -1 & \beta + i\sqrt{1-\beta^2} \\ 1 & -\beta + i\sqrt{1-\beta^2} \end{pmatrix}, \qquad U^{-1} = \begin{pmatrix} \beta - i\sqrt{1-\beta^2} & \beta + i\sqrt{1-\beta^2} \\ 1 & 1 \end{pmatrix}. \tag{1.2.20}$$

The miraculous thing is that the same similarity transformation also diagonalizes C^η, *i.e.*,

$$UC^\eta U^{-1} = \frac{1}{2i\sqrt{1-\beta^2}} \begin{pmatrix} \lambda_+ & 0 \\ 0 & -\lambda_- \end{pmatrix}. \tag{1.2.21}$$

Therefore, the original $2n \times 2n$ matrix linear equations (1.2.15–16) can be written as two sets of $n \times n$ matrix linear equations,

$$\frac{dM_1}{d\varsigma} = \frac{\lambda_+}{2(1-\beta)} A_\varsigma M_1, \tag{1.2.22}$$

$$\frac{dM_1}{d\eta} = \frac{i\lambda_+}{2\sqrt{1-\beta^2}} A_\eta M_1; \tag{1.2.23}$$

and

$$\frac{dM_3}{d\varsigma} = \frac{\lambda_-}{2(1-\beta)} A_\varsigma M_3, \tag{1.2.24}$$

$$\frac{dM_3}{d\eta} = -\frac{\lambda_-}{2\sqrt{1-\beta^2}} A_\eta M_3. \tag{1.2.25}$$

The reason and implication for this possibility of simultaneous diagonalization is not clear. I did this part of study in 1979 [25]. Recently the Montréal school [26] has explored in the direction of geometric meaning of the superposition principle for nonlinear systems.

1.3 Nonlocal conservation laws ⇔ Linear system

Besides local conservation laws, the chiral fields also have nonlocal conservation laws. The existence of such nonlocal currents for the σ-model was first obtained by Lüscher and Pohlmeyer [28]. Here I shall demonstrate it using the method of Brezin *et al.* [29]. As I mentioned before, the equation of motion (1.3) is like a continuity equation. So let us consider A_ς and A_η to be the first currents, *i.e.*,

$$V_\varsigma^{(1)} \equiv A_\varsigma = \partial_\varsigma \chi^{(1)}, \qquad V_\eta^{(1)} \equiv A_\eta = -\partial_\eta \chi^{(1)}. \tag{1.3.1}$$

Such $\chi^{(1)}$ exists and can be obtained from the A's by integration because of the equation of motion (1.3). The higher currents are then obtained by an iteration procedure. Suppose the nth currents $V_\varsigma^{(n)}$ and $V_\eta^{(n)}$ exist, *i.e.*,

$$\partial_\eta V_\varsigma^{(n)} + \partial_\varsigma V_\eta^{(n)} = 0; \qquad V_\varsigma^{(n)} = \partial_\varsigma \chi^{(n)}, \quad V_\eta^{(n)} = -\partial_\eta \chi^{(n)}. \tag{1.3.2}$$

Then, the $(n+1)^{\text{th}}$ currents can be constructed from $\chi^{(n)}$ by

$$V_\varsigma^{(n+1)} = D_\varsigma \chi^{(n)}, \qquad V_\eta^{(n+1)} = D_\eta \chi^{(n)}, \tag{1.3.3}$$

where

$$D_\varsigma \equiv \partial_\varsigma + A_\varsigma, \qquad D_\eta \equiv \partial_\eta + A_\eta. \tag{1.3.4}$$

Using the curvatureless equation (1.2) and the equation of motion (1.3), it is easy to show that

$$\partial_\eta V_\varsigma^{(n+1)} + \partial_\varsigma V_\eta^{(n+1)} = 0. \tag{1.3.5}$$

Therefore the $(n+1)^{\text{th}}$ currents constructed from (1.3) are conserved. To obtain the $(n+1)^{\text{th}}$ charge, we need integration and differentiation of lower charges at different points in ς and η:

$$\chi^{(n+1)} = \int_\varsigma d\varsigma'\, D_{\varsigma'} \chi^{(n)} = \int_\varsigma d\varsigma'\, D_{\varsigma'} \Big[\int_{\varsigma'} d\varsigma''\, D_{\varsigma''} \chi^{(n-1)} \Big] = \cdots \tag{1.3.6}$$

Thus the term ***nonlocal*** is used for these conservation laws.

In the quantum mechanical version, Lüscher [28] showed that in the σ-model these nonconservation laws also imply no particle production, which is the basis for constructing the S-matrix [10]. Thus far, however, the physical origin and meaning of these nonlocal currents for the chiral field have not yet been exploited in general. This is one of the leading challenges in the field.

Now we want to show how to obtain the ***linearized*** equations, sometimes called the inverse scattering equations, for the chiral equations. These equations were known [30]. Here we would like to demonstrate the method which was introduced in [15], and which will be used for the SDYM fields in §2.

From (1.3.2) and (1.3.3), we obtain

$$\partial_\varsigma \chi^{(n)} = D_\varsigma \chi^{(n-1)}, \tag{1.3.7}$$
$$\partial_\eta \chi^{(n)} = D_\eta \chi^{(n-1)}. \tag{1.3.8}$$

Multiplying (1.3.7) by L^n, where L is an arbitrary constant, and summing over n, we obtain:

$$\sum_{n=1}^{\infty} L^n \partial_\varsigma \chi^{(n)} = \sum_{n=1}^{\infty} L^n D_\varsigma \chi^{(n-1)}. \tag{1.3.9}$$

Equation (1.3.9) can be rewritten as

$$\partial_\varsigma \sum_{n=0}^{\infty} L^n \chi^{(n)} = L D_\varsigma \sum_{n=0}^{\infty} L^n \chi^{(n)}, \tag{1.3.10}$$

where the sum on the left hand side of (1.3.9) can be extended to $n = 0$ because $\chi^{(0)} = 1$. Now define

$$\psi \equiv \sum_{n=0}^{\infty} L^n \chi^{(n)}, \tag{1.3.11}$$

which is a function of ς, η, and L. Equation (1.3.10) then becomes

$$\partial_\varsigma \psi = L D_\varsigma \psi. \tag{1.3.12}$$

By a similar procedure we obtain

$$\partial_\eta \psi = -L D_\eta \psi. \tag{1.3.13}$$

To claim that (1.3.12) and (1.3.13) are the linearized equations for the chiral fields, we need to show that the integrability condition on ψ from (1.3.12) and (1.3.13) implies the chiral field. Equations (1.3.12) and (1.3.13) can be rewritten as

$$\partial_\varsigma \psi = \frac{L}{(1-L)} A_\varsigma \psi, \tag{1.3.14}$$

$$-\partial_\eta \psi = \frac{L}{(1+L)} A_\eta \psi. \tag{1.3.15}$$

Now $\partial_\eta(1.3.14) + \partial_\varsigma(1.3.15) \Rightarrow$

$$\frac{L}{(1-L)} [(\partial_\eta A_\varsigma)\psi + A_\varsigma \partial_\eta \psi] + \frac{L}{(1+L)} [(\partial_\varsigma A_\eta)\psi + A_\eta \partial_\varsigma \psi] = 0. \tag{1.3.16}$$

Using (1.3.14) and (1.3.15) in (1.3.16), and after simple manipulations, one obtains

$$\{\partial_\eta A_\varsigma + \partial_\varsigma A_\eta + L(\partial_\eta A_\varsigma - \partial_\varsigma A_\eta - [A_\varsigma, A_\eta])\}\psi = 0. \tag{1.3.17}$$

We see that the integrability of ψ for arbitrary L implies

$$\partial_\eta A_\varsigma - \partial_\varsigma A_\eta [A_\varsigma, A_\eta] = 0, \qquad \text{and } \partial_\eta A_\varsigma + \partial_\varsigma A_\eta = 0.$$

These are just the conditions of curvatureless of the gauge potential A, and the continuity-like equation of (1.2) and (1.3). Notice that if we define $\lambda = L^{-1}$, (1.3.14) and (1.3.15) are just the inverse scattering equations of Zakharov and Mikhaĭlov [30] for the chiral fields. Thus we see that the existence of conserved nonlocal currents is closely related to the fact that there is an arbitrary parameter in the linearized equations. This aspect should be further analyzed in order to shed light on the meaning of those nonlocal currents.

Lastly we note that, in the x_1–x_2 coordinate, (1.3.14) and (1.3.15) become

$$[\partial_\mu - A_\mu - L^{-1} \epsilon_{\mu\nu} \partial_\nu]\, \psi(x, L) = 0, \qquad \mu = 1, 2. \tag{1.3.18}$$

Parallel developments can be carried out for the super-chiral field equations [31].

§2 Self-dual Yang–Mills fields

2.1 The J-formulation of the self-dual Yang–Mills field in complexified E^4 space

In the complexified E^4 space,

$$\sqrt{2}\,y = x_1 + ix_2, \quad \sqrt{2}\,\bar{y} = x_1 - ix_2, \quad \sqrt{2}\,z = x_3 - ix_4, \quad \sqrt{2}\,\bar{z} = x_3 + ix_4,$$

the SDYM equations $F_{\mu\nu} = \frac{1}{2}\epsilon_{\mu\nu\rho\sigma}F_{\rho\sigma}$ are [11]:

$$F_{yz} = 0 = F_{\bar{y}\bar{z}}, \qquad F_{y\bar{y}} + F_{z\bar{z}} = 0. \tag{2.1.1}$$

The first two equations imply that the gauge potential A_μ can always be written in the following form:

$$A_y = D^{-1}\partial_y D, \quad A_z = D^{-1}\partial_z D, \quad A_{\bar{y}} = \overline{D}^{-1}\partial_{\bar{y}}\overline{D}, \quad A_{\bar{z}} = \overline{D}^{-1}\partial_{\bar{z}}\overline{D}. \tag{2.1.2}$$

For the gauge group *SL(N,C)*, $\det D = \det \overline{D} = 1$. For real *SU(N)*-potentials A_μ, one can show that D and $\overline{D}$ are related, in real coordinate space, as $D^\dagger \doteq \overline{D}^{-1}$. We define a matrix J by

$$J \equiv D\overline{D}^{-1}, \tag{2.1.3}$$

which can be shown to be gauge invariant, and $\det J = 1$ for the gauge group *SL(N,C)*; J can be made hermitian $J^\dagger \doteq J$ in the real coordinate space for real *SU(N)* gauge fields. Now the SDYM equations can be written, using

$$B_y \equiv J^{-1}\partial_y J, \qquad B_z \equiv J^{-1}\partial_z J, \tag{2.1.4}$$

as

$$\partial_{\bar{y}}B_y + \partial_{\bar{z}}B_z = 0. \tag{1.2.5}$$

This we call the *left-SDYM-J equation*. Equivalently, the SDYM equation can be written, using

$$\hat{B}_{\bar{y}} \equiv J\partial_{\bar{y}}J^{-1}, \qquad \hat{B}_{\bar{z}} \equiv J\partial_{\bar{z}}J^{-1}, \tag{2.1.6}$$

as

$$\partial_y\hat{B}_{\bar{y}} + \partial_z\hat{B}_{\bar{z}} = 0, \tag{2.1.7}$$

which we call the *right-SDYM-J equation*.

2.2 Two-parameter Bianchi–Bäcklund transformations

One can easily show that the following transformation is a Bianchi–Bäcklund transformation [13] for both sets of equations (2.1.4–5) and (2.1.6–7):

$$J^{-1}\partial_y J - J'^{-1}\partial_y J' = \lambda'\partial_{\bar{z}}(J^{-1}J'), \tag{2.2.1}$$
$$J^{-1}\partial_z J - J'^{-1}\partial_z J' = -\lambda'\partial_{\bar{y}}(J^{-1}J'), \tag{2.2.2}$$

i.e., if J satisfies (2.1.4) and (2.1.5) so does J'. For hermitian J and J', they can be shown to satisfy the algebraic constraint $J'J^{-1} - J'^{-1}J = BT$, and $\lambda' = e^{i\alpha}$, where α and β are real. The reader is referred to [13] for a detailed discussion.

2.3 Infinite nonlocal conservation laws

Consider B_y and B_z of (2.1.4) and (2.1.5) being the first conserved currents,

$$V_y^{(1)} \equiv B_y = \partial_{\bar{z}}\chi^{(1)}, \qquad V_z^{(1)} \equiv B_z = -\partial_{\bar{y}}\chi^{(1)}, \tag{2.3.1}$$

where $\chi^{(1)}$ exists because of (2.1.4–7). From these first currents we can generate an infinite number of them by the following iterative procedure:

$$V_y^{(n+1)} = D_y\chi^{(n)} = \partial_{\bar{z}}\chi^{(n+1)}, \qquad V_z^{(n+1)} = D_z\chi^{(n)} = -\partial_{\bar{y}}\chi^{(n+1)}. \tag{2.3.2}$$

Such generated currents $V_y^{(n+1)}$ and $V_z^{(n+1)}$ can be shown [14] to be conserved and to satisfy (2.2.2). The $\chi^{(n+1)}$ can be regarded as conserved charges. Similar discussions can be made for the right-SDYM-J equations.

2.4 The linear system for the self-dual Yang–Mills fields in the J-formulation

From these infinite nonlocal conservation laws, using a method given in [5], we can obtain the following linear differential equations

$$\partial_{\bar{z}}\chi = \lambda D_y\chi \equiv \lambda(\partial_y + J^{-1}\partial_y J)\chi, \tag{1.4.1}$$

$$-\partial_{\bar{y}}\chi = \lambda D_z\chi \equiv \lambda(\partial_z + J^{-1}\partial_z J)\chi, \tag{1.4.2}$$

as already demonstrated for the chiral fields in §**1.3**. The integrability of this equations gives the left-SDYM equations of motion (2.1.4) and (2.1.5). Similarly, for the right-SDYM equations (2.1.6) and (2.1.7) we have

$$\partial_z\hat{\chi} = -\frac{1}{\lambda}\hat{D}_{\bar{y}}\hat{\chi} = -\frac{1}{\lambda}(\partial_{\bar{y}} + J\partial_{\bar{y}}J^{-1})\hat{\chi}, \tag{2.4.3}$$

$$\partial_y\hat{\chi} = \frac{1}{\lambda}\hat{D}_{\bar{z}}\hat{\chi} = \frac{1}{\lambda}(\partial_{\bar{z}} + J\partial_{\bar{z}}J^{-1})\hat{\chi}. \tag{2.4.4}$$

Now we shall discuss some properties of the solutions of (2.4.1) and (2.4.2). If χ_1 and χ_2 are both solutions, we find, through the following manipulation,

$$\begin{aligned}(\partial_{\bar{z}} - \lambda\partial_y)(\chi_1^{-1}\chi_2) &= (\chi_1^{-1}\chi_1)\left[(\partial_{\bar{z}} - \lambda\partial_y)\chi_1^{-1}\right]\chi_2 + \chi_1^{-1}(\partial_{\bar{z}} - \lambda\partial_y)\chi_2 \\ &= -\chi_1^{-1}[(\partial_{\bar{z}} - \lambda\partial_y)\chi_1]\chi_1^{-1}\chi_2 + \chi_1^{-1}\lambda B_y\chi_2 \\ &= -\chi_1^{-1}\lambda B_y\chi_1\chi_1^{-1}\chi_2 + \chi_1^{-1}\lambda B_y\chi_2 = 0.\end{aligned} \tag{2.4.5}$$

Similarly, we have

$$(\partial_{\bar{y}} + \lambda\partial_z)(\chi_1^{-1}\chi_2) = 0,$$

and therefore

$$\chi_1^{-1}\chi_2 = A(\lambda\bar{z} + y, \lambda\bar{y} - z, \lambda), \tag{2.4.6}$$

which is an arbitrary matrix function of the variables $\lambda\bar{z} + y$, $\lambda\bar{y} - z$, and λ.

Another property is that $\chi^{-1}(\lambda)$ and $\chi^{+}(-1/\overline{\lambda})J$ satisfy the same set of equations. We can see this through the following arrangements. For χ satisfying (2.4.1) and (2.4.2), χ^{-1} satisfies the following corresponding equations:

$$\chi^{-1}(2.4.1)\chi \Rightarrow \begin{cases} \chi^{-1}[(\partial_{\overline{z}}-\lambda\partial_y)\chi]\chi^{-1} = \lambda\chi^{-1}J^{-1}J_y, \text{ or} \\ \qquad -(\partial_{\overline{z}}-\lambda\partial_y)\chi^{-1} = \lambda\chi^{-1}J^{-1}J_y. \end{cases} \tag{2.4.7}$$

Similarly,

$$\chi^{-1}(2.4.2)\chi \Rightarrow \begin{cases} \chi^{-1}[(\partial_{\overline{y}}+\lambda\partial_z)\chi]\chi^{-1} = \lambda\chi^{-1}J^{-1}J_z, \text{ or} \\ \qquad (\partial_{\overline{y}}-\lambda\partial_z)\chi^{-1} = \lambda\chi^{-1}J^{-1}J_z. \end{cases} \tag{2.4.8}$$

Now we take the hermitian conjugate of (2.4.1) and (2.4.2):

$$(2.4.1)^{+} \Rightarrow (\partial_z-\overline{\lambda}\partial_{\overline{y}})\chi^{+} = \overline{\lambda}\chi^{+}J_{,\overline{y}}J^{-1},$$

where $J^{+}=J$ is used, or

$$\begin{aligned} [(\partial_z-\lambda\partial_{\overline{y}})\chi^{+}]J-\overline{\lambda}\chi^{+}J_{,z} &= 0, \text{ or} \\ (\partial_z-\overline{\lambda}\partial_{\overline{y}})(\chi^{+}J)-(\chi^{+}J)J^{-1}J_{,z} &= 0. \end{aligned} \tag{2.4.9}$$

Similarly,

$$(2.4.2)^{+} \Rightarrow -(-\overline{\lambda}\partial_{\overline{z}}-\partial_y)(\chi^{+}J)-(\chi^{-1}J)J^{-1}J_{,y} = 0. \tag{2.4.10}$$

Comparing (2.4.9) with (2.4.8), and (2.4.10) with (2.4.7), we see that $\chi^{-1}(\lambda)$ and $\chi^{+}(-1/\overline{\lambda})J$ satisfy the same equations. From (2.4.6), we have

$$\chi^{+}(-1/\overline{\lambda})J = \chi^{-1}(\lambda)A(\lambda\overline{z}+y,\lambda\overline{y}-z,\lambda). \tag{2.4.11}$$

Picking $A=1$ is a very special choice.

It is an interesting fact that the χ's of these linear equations can be solved for a given J from those $(5n-4)$-parameter instanton solutions, see Refs. [18,19].

2.4.1 Connection with the Belavian–Zakharov linear systems

We want to emphasize that the linear systems (2.4.1) and (2.4.2) derived here, although related, are quite different from those obtained by Belavian and Zakharov [12]. Ours are here closely related to nonlocal conservation laws, and are useful for deriving the Kac–Moody algebra —as we shall see in §**2.5**. The ones by Belavian and Zakharov are closely related to the CP^3 geometry, as we shall discuss later in this section.

Now we want to express (2.4.1) and (2.4.2) in terms of the potentials A_u, with $u=y,\overline{y},z,\overline{z}$. The A_u's are expressed in terms of D and $\overline{D}$ matrices in (2.1.2). From (2.1.3), $J\equiv D\overline{D}^{-1}$,

$$\begin{aligned} J^{-1}J_{,y} &= \overline{D}(D^{-1}D_y-\overline{D}^{-1}\overline{D}_{,y})\overline{D}^{-1} \\ &= \overline{D}(A_y-\overline{D}^{-1}D_{,y})\overline{D}^{-1}, \end{aligned} \tag{2.4.12}$$

and

$$\begin{aligned} \partial_y\chi &= \partial_y(\overline{D}\overline{D}^{-1}\chi) \\ &= \overline{D}\overline{D}^{-1}(\partial_y\overline{D})\overline{D}^{-1}\chi+\overline{D}\partial_y(\overline{D}^{-1}\chi), \end{aligned} \tag{2.4.13}$$

$$\begin{aligned}\partial_{\bar{z}}\chi &= \partial_{\bar{z}}(\overline{D}\,\overline{D}^{-1}\chi)\\ &= \overline{D}\,\overline{D}^{-1}(\partial_{\bar{z}}\overline{D})\overline{D}^{-1}\chi + \overline{D}\partial_{\bar{z}}(\overline{D}^{-1}\chi),\\ &= \overline{D}A_{\bar{z}}(\overline{D}^{-1}\chi) + \overline{D}\partial_{\bar{z}}(\overline{D}^{-1}\chi).\end{aligned} \tag{2.4.14}$$

Substituting (2.4.13) and (2.4.14) into one of the linear equations (2.4.1), $(\partial_{\bar{z}} - \lambda\partial_y)\chi = \lambda J^{-1}J_{,y}\chi$, we obtain

$$A_{\bar{z}}(\overline{D}^{-1}\chi) + \partial_{\bar{z}}(\overline{D}^{-1}\chi) - \lambda\partial_y(\overline{D}^{-1}\chi) = \lambda A_y(\overline{D}^{-1}\chi). \tag{2.4.15}$$

Now define

$$\psi = D^{-1}\chi, \tag{2.4.16}$$

then (2.4.15) becomes

$$(\lambda A_y - A_{\bar{z}})\psi = (-\lambda\partial_y + \partial_{\bar{z}})\psi. \tag{2.4.17}$$

For the other linear equation, (2.4.2), we obtain

$$(\lambda A_z + A_{\bar{y}})\psi = -(\lambda\partial_z + \partial_{\bar{y}})\psi. \tag{2.4.18}$$

As for χ, (2.4.6) and (2.4.11), we can derive the corresponding relations for ψ:

$$\psi_1^{-1}\psi_2 = B(\lambda\bar{z} + y, \lambda\bar{y} - z, \lambda), \tag{2.4.19}$$

where ψ_1 and ψ_2 are both solutions of (2.4.17) and (2.4.18), and B is an arbitrary matrix function of the indicated variables.. After operations similar to those in (2.4.7) and (2.4.10), we find that $\psi^{-1}(\lambda)$ and $\psi^{+}(-1/\bar{\lambda})$ satisfy the same set of equations. Therefore, from (2.4.19) we obtain

$$\psi^{+}(-1/\bar{\lambda})\,\psi(\lambda) = B(\lambda\bar{z} + y, \lambda\bar{y} - z, \lambda). \tag{2.4.20}$$

2.4.2 Geometric construction of the linear system

In §**2**, the SDYM equations were viewed as two curvatureless conditions $f_{yz} = 0$ and $f_{\bar{y}\bar{z}} = 0$ on the yz and the $\bar{y}\bar{z}$-planes, plus a third constraining equation $f_{y\bar{y}} + f_{z\bar{z}} = 0$; see (2.1.1). Actually, there is an infinite number of such planes. All those planes passing through a given point can be characterized by a free complex parameter λ such that the three equations are encompassed in only one [18]:

$$F_{(y-\lambda\bar{z})\,(z+\lambda\bar{y})} = 0. \tag{2.4.21}$$

Since (2.4.21) holds for all values of λ, the λ^0 term gives $f_{yz} = 0$; the λ^2 term gives $f_{\bar{z}\bar{y}} = 0$; and the λ^1 term gives $f_{y\bar{y}} + f_{z\bar{z}} = 0$. From (2.4.21), which can be interpreted as curvatureless in the complex variables $y - \lambda\bar{z}$ and $z + \lambda\bar{y}$, the potentials must be the following form:

$$\begin{cases} A_{(y-\lambda\bar{z})} = \psi\partial_{(y-\lambda\bar{z})}\psi^{-1},\\ A_{(z+\lambda\bar{y})} = \psi\partial_{(z+\lambda\bar{y})}\psi^{-1},\end{cases} \tag{2.4.22}$$

or equivalently

$$\begin{cases}\lambda A_y - A_{\bar{z}} = [-(\lambda\partial_y - \partial_{\bar{z}})\psi]\psi^{-1},\\ \lambda A_z + A_{\bar{y}} = [-(\lambda\partial_z + \partial_{\bar{y}})\psi]\psi^{-1},\end{cases} \tag{2.4.23}$$

and finally

$$\begin{cases}(\lambda A_y - A_{\bar{z}})\psi = -(\lambda\partial_y - \partial_{\bar{z}})\psi,\\(\lambda A_z + A_{\bar{y}})\psi = -(\lambda\partial_z - \partial_{\bar{y}})\psi.\end{cases}$$

Since the gauge potentials are traceless, $\det\psi = 1$. These are precisely the linearized equations of Belavian and Zakharov [12].

With this elementary introduction we can now appreciate the way Atiyah and Ward [3] formulated the problem using more sophisticated language. Any point x in four-complex-dimensional euclidean space can be expressed as a two-by-two complex matrix

$$x = x_4 + i\vec{x}\cdot\vec{\sigma} = \begin{pmatrix} x_4 + ix_3 & x_2 + ix_1 \\ -x_2 + ix_1 & x_4 - ix_3 \end{pmatrix} = i\begin{pmatrix} z & \bar{y} \\ y & -\bar{z} \end{pmatrix}. \tag{2.4.24}$$

Note that

$$\det x = \sum_{i=1}^{4} x_i^2. \tag{2.4.25}$$

For two given complex spinors

$$\pi \equiv \begin{pmatrix}\pi_1\\ \pi_2\end{pmatrix}, \qquad \omega \equiv \begin{pmatrix}\omega_1\\ \omega_2\end{pmatrix}, \tag{2.4.26}$$

where π_1, π_2, ω_1, and ω_2 are complex numbers,

$$x\pi = \omega \tag{2.4.27}$$

defines a null plane. It is very easy to show that if $x_1, x_2 \in x$, then $(x_1 - x_2)\pi = 0$, and thus $\det(x_1 - x_2) = 0$, or $(x_1 - x_2)$ has null lenght. By straightforward calculation one can show that any area on such a null plane is either self-dual or anti-self-dual:

$$d\sigma_{\mu\nu} = \pm\epsilon_{\mu\nu\alpha\beta}d\sigma_{\alpha\beta}. \tag{2.4.28}$$

On such planes, anti-self-dual or self-dual fields, respectively, are curvatureless,

$$\begin{aligned} f_{\mu\nu}d\sigma_{\mu\nu} &= f_{\mu\nu}(\pm\epsilon_{\mu\nu\alpha\beta}d\sigma_{\alpha\beta}) \\ &= -f_{\alpha\beta}d\sigma_{\alpha\beta}, \qquad \text{if } f_{\mu\nu}\epsilon_{\mu\nu\alpha\beta} = \mp f_{\alpha\beta}; \end{aligned} \tag{2.4.29}$$

therefore, $f_{\mu\nu} = 0$ on the plane.

Since π and ω, or $c\pi$ and $c\omega$, specify the same set of x, the null plane x is specified by three complex numbers, thus CP^3. Now, let us choose

$$\pi = \begin{pmatrix}1\\ \pi_2/\pi_1\end{pmatrix} = \begin{pmatrix}1\\ -\lambda\end{pmatrix}, \qquad \text{for } \pi_1 \neq 0, \text{ or } \lambda \neq \infty. \tag{2.4.30}$$

Note that we need another description for $\pi_2 \neq 0$,

$$\pi = \begin{pmatrix}\pi_2/\pi_1\\ 1\end{pmatrix} = \begin{pmatrix}-1/\lambda\\ 1\end{pmatrix}. \tag{2.4.31}$$

On the null plane $(dx)\pi = 0$, *i.e.*,

$$\begin{pmatrix} dz & d\bar{y} \\ dy & -d\bar{z} \end{pmatrix}\begin{pmatrix}1\\ -\lambda\end{pmatrix} = 0, \quad \text{or} \quad \begin{pmatrix} dz - \lambda d\bar{y} \\ dy + \lambda d\bar{z} \end{pmatrix} = 0. \tag{2.4.32}$$

Since the potential is curvatureless on such a plane, the phase factor ψ can be integrated out:

$$A_\alpha dx_\alpha = \psi_1(\lambda, x)\partial_\alpha \psi_1^{-1}(\alpha, x)dx_\alpha, \qquad \text{for } \pi_1 \neq 0 \text{ or good at } \lambda = 0, \tag{2.4.33}$$
$$= \psi_2(\lambda, x)\partial_\alpha \psi_2^{-1}(\alpha, x)dx_\alpha, \qquad \text{for } \pi_2 \neq 0 \text{ or good at } \lambda = \infty. \tag{2.4.34}$$

Using (2.4.32) and expanding out (2.4.33) and (2.4.34), we obtain

$$\lambda A_z + A_{\bar{y}} = \psi_1(\lambda\partial_z + \partial_{\bar{y}})\psi_1^{-1}, \tag{2.4.35}$$
$$-\lambda A_y + A_{\bar{z}} = \psi_1(-\lambda\partial_y + \partial_{\bar{z}})\psi_1^{-1}, \tag{2.4.36}$$

and $\psi_2(\lambda, x)$ satisfies the same equations. These are just the linear equations (2.4.17) and (2.4.18). The two functions (also called *sections*), $\psi_1(\lambda, x)$ and $\psi_2(\lambda, x)$, in the overlap region of the two patches in CP^3, are related by a transition function $B(z-\lambda\bar{y}, y+\lambda\bar{z}, \lambda)$. Now $\psi_1(\lambda, x) = \psi_2(\lambda, x)B(z-\lambda\bar{y}, y+\lambda\bar{z}, \lambda)$, or $B(z-\lambda\bar{y}, y+\lambda\bar{z}, \lambda) = \psi_2^{-1}(\lambda, x)\psi_1(\lambda, x)$, which is just (2.4.19); $\psi_1(\lambda, x)$ has good properties at $\lambda = 0$, and $\psi_2(\lambda, x)$ at $\lambda = \infty$. This formulation of Atiyah and Ward to solve the SDYM equation was to find an appropriate B and split it into ψ_1 and ψ_2 with appropriate analyticity properties in λ. The observation is extremely beautiful. As we know, however, the instantons were not actually found in this way.

2.5 The Riemann–Hilbert transform and the Kac–Moody algebra for the self-dual Yang–Mills fields

2.5.1 The case of the *SL(N,C)* SDYM fields

We introduce the following two infinitesimal parametric transformations for the J-field [20]:

$$\delta_\alpha(\lambda)J = \alpha_a\delta_a(\lambda)J = -J\chi(\lambda)T_\alpha\chi(\lambda)^{-1} = \sum_{m=0}^{\infty}\lambda^m\alpha_a\delta_a^{(m)}J, \tag{2.5.1}$$

$$\hat{\delta}_\alpha(-\frac{1}{\lambda})J = \alpha_a\hat{\delta}_a(-\frac{1}{\lambda})J = \chi(-\frac{1}{\lambda})T_\alpha\chi(-\frac{1}{\lambda})^{-1}J = \sum_{m=0}^{\infty}\lambda^m\alpha_a\hat{\delta}_a^{(m)}J, \tag{2.5.2}$$

where $T_\alpha \equiv \alpha^a T_a$, the α^a's are infinitesimal parameters, and the T_a's are traceless anti-hermitian matrices satisfying $[T_a, Tb] = C_{ab}^c T_c$, with C_{ab}^c the structure constants of *su(N)*. For complex α, the T_α's span the Lie algebra *sl(N,C)*, while for real α the T_α's span the Lie algebra *su(N)*. Using (2.4.1–4), it is easy to show that $J+\delta_\alpha J$ and $J+\hat{\delta}_\alpha J$ satisfy the self duality equations (2.1.4), (2.1.5), and (2.1.6), (2.1.7), respectively. Moreover, it follows from $\mathrm{Tr}(T_a) = 0$ that $\det(J+\delta_\alpha J) = 1 = \det(J+\hat{\delta}_\alpha J)$. These transformations are therefore infinitesimal Bäcklund transformations. We can actually show that they satisfy the same Bäcklund transformations constructed previously in [13], and given in §**2.2**.

Historically, the transformations (2.5.1) and (2.5.2) were first found by guesswork [20]. We can now show that they can be derived from the RH transform. I shall derive the infinitesimal transformation (2.5.1) and (2.5.2) from this transform.

For the linear system (2.4.1) and (2.4.2), we pick the boundary condition

$$\chi(\lambda = 0, y, \bar{y}, z, \bar{z}) = I; \tag{1.5.3}$$

differentiating the linear system with respect to λ, we obtain

$$\partial_{\bar{z}}\dot{\chi} = D_y\chi + \lambda D_y\dot{\chi}, \tag{2.5.4}$$

for $\lambda = 0$,

$$\partial_{\bar{z}}\dot{\chi}(\lambda = 0) = \partial_y \chi(\lambda = 0) + B_y \chi(\lambda = 0). \tag{2.5.5}$$

Using the boundary condition (2.5.3) we obtain

$$\partial_{\bar{z}}\dot{\chi}(\lambda = 0) = B_y. \tag{2.5.6}$$

Therefore, the ***potential*** B_y can be calculated if we now χ. Similarly,

$$-\partial_{\bar{y}}\dot{\chi}(\lambda = 0) = B_z. \tag{2.5.7}$$

The RH transform provides a method for generating a new χ from a given one: first one selects a closed contour C in the λ-plane, with C_+ being the region inside the contour and C_- the region outside, such that $\chi(\lambda)$ is analytic on $C_+ \supset C_-$. We pick a group element $U(\lambda)$, $U(\lambda) \supset C$, and define

$$H(\lambda) \equiv \chi(\lambda)\, U(\lambda)\, \chi(\lambda)^{-1}. \tag{2.5.8}$$

The main task now is the RH problem of constructing two functions $Y_\pm(\lambda)$ for a given $H(\lambda)$, which are analytic in $C_\pm$ and continuous on C, such that

$$H(\lambda) = Y_- Y_+^{-1} \qquad \text{with } Y_+(\lambda = 0) = I. \tag{2.5.9}$$

Then one can show that a new $\chi'(\lambda)$ can be constructed as follows:

$$\chi'(\lambda) = \begin{cases} Y_+(\lambda) \quad U(\lambda) \\ Y_-(\lambda) \quad \chi(\lambda) \quad U(\lambda)^{-1} \end{cases} \qquad \text{in } C_+, \tag{2.5.10}$$

and the new potential

$$B_{y'} = \partial_{\bar{z}}\dot{\chi}'(\lambda = 0), \qquad i.e., \tag{2.5.11}$$

$$\delta B_y \equiv B_{y'} - B_y = \partial_{\bar{z}}\left[\dot{\chi}'(\lambda = 0) - \dot{\chi}(\lambda = 0)\right] \equiv \partial_{\bar{z}}\delta\dot{\chi}(\lambda = 0), \tag{2.5.12}$$

and

$$B_{z'} = -\partial_{\bar{y}}\dot{\chi}'(\lambda = 0), \qquad i.e., \tag{2.5.13}$$

$$\delta B_z \equiv B_{z'} - B_z = -\partial_{\bar{y}}\left[\dot{\chi}'(\lambda = 0) - \dot{\chi}(\lambda = 0)\right] \equiv -\partial_{\bar{y}}\delta\dot{\chi}(\lambda = 0). \tag{2.5.14}$$

An integral transformation relating χ' to χ can be derived:

$$\chi'(\lambda) + \frac{\lambda}{2\pi i}\int \frac{d\lambda'}{\lambda'(\lambda' - \lambda)}\chi'(\lambda')[U(\lambda') - 1]\chi'(\lambda')^{-1}\chi(\lambda) = \chi(\lambda), \tag{2.5.15}$$

which can be shown to be an identity by using the following equations:

$$\int_C d\lambda' \frac{1}{\lambda'(\lambda' - \lambda)} = 0, \qquad \text{for } \lambda \subset C_+, \tag{2.5.16}$$

and

$$\int_C d\lambda' \frac{Y_-(\lambda')}{\lambda'(\lambda' - \lambda)} = 0, \qquad \text{for } \lambda \subset C_-, \tag{2.5.17}$$

since $Y_-(\lambda)$ is analytic in C_-. From (2.5.15), we obtain

$$\delta\chi(\lambda) \equiv \chi'(\lambda) - \chi(\lambda) \equiv d(\lambda)\chi(\lambda), \tag{2.5.18}$$

and

$$d(\lambda) = -\frac{\lambda}{2\pi i}\int_C \frac{d\lambda'}{\lambda'(\lambda'-\lambda)}\chi'(\lambda')[U(\lambda')-1]\chi'(\lambda')^{-1}. \tag{2.5.19}$$

Let us now consider an infinitesimal RH transformation

$$v_\alpha(\lambda) \equiv U_\alpha(\lambda) - 1 = \alpha^a T_a \lambda^{-k}, \tag{2.5.20}$$

where the α^a are infinitesimal parameters, k is an arbitrary given integer, and the T_a's are the generators of the Lie algebra,

$$[T_a, T_b] = C^c_{ab} T_c. \tag{2.5.21}$$

Substituting (2.5.20) into (2.5.19), considering

$$\chi(\lambda, x) T_a \chi^{-1}(\lambda, x) = \sum_{m=0}^{\infty} \lambda^m T_a^{(m)}(x), \tag{2.5.22}$$

and using the following identities

$$\frac{\lambda}{\lambda'(\lambda'-1)} = \sum_{n=0}^{\infty} \frac{\lambda^{n+1}}{\lambda'^{n+2}}, \tag{2.5.23}$$

$$\frac{1}{2\pi i}\int_C d\lambda' \lambda'^{-k} = \delta_{k1}, \tag{2.5.24}$$

we find

$$\begin{aligned} d_\alpha^{(k)}(\lambda) &\equiv -\frac{\lambda}{2\pi i}\int_C \frac{d\lambda'}{\lambda'(\lambda'-1)}\chi'(\lambda')[U(\lambda')-1]\chi'(\lambda')^{-1} \\ &= -\sum_{n=0}^{\infty} \alpha^a T_a^{(k+n+1)}(x)\,\lambda^{n+1}. \end{aligned} \tag{2.5.25}$$

We thus obtain

$$\delta_\alpha^{(k)}\chi(\lambda) = d^k(\lambda)\chi(\lambda) = -\alpha^a \sum_{n=0}^{\infty} \lambda^{n+1} T_a^{(k+n+1)}(x)\chi(\lambda). \tag{2.5.26}$$

The corresponding changes in the potentials are

$$\delta_\alpha^{(k)} B_y(x) = \partial_{\bar{z}} \delta_a^{(k)} \dot{\chi}(\lambda = 0, x) = -\alpha^a \partial_{\bar{z}} T_a^{(k+1)}(x), \tag{2.5.27}$$

$$\delta_\alpha^{(k)} B_z(x) = -\partial_{\bar{y}} \delta_a^{(k)} \dot{\chi}(\lambda = 0, x) = \alpha^a \partial_{\bar{y}} T_a^{(k+1)}(x), \tag{2.5.28}$$

and

$$\delta_\alpha B_y(x) \equiv \sum_{k=0}^{\infty} \lambda^k \delta_\alpha^k B_y = -\frac{1}{\lambda}\partial_{\bar{z}}[\chi(\lambda) T_\alpha \chi^{-1}(\lambda)] = -D_y[\chi(\lambda) T_\alpha \chi^{-1}(\lambda)], \tag{2.5.29}$$

using the linear equation (2.4.1), and

$$\delta_\alpha B_z(x) \equiv \sum_{k=0}^{\infty} \lambda^k \delta_\alpha^k B_z = \frac{1}{\lambda}\partial_{\bar y}[\chi(\lambda)T_\alpha\chi^{-1}(\lambda)] = -D_z[\chi(\lambda)T_\alpha\chi^{-1}(\lambda)], \tag{2.5.30}$$

where $D_u \equiv \partial_u + [B_u, \cdot]$, for $u = y, z$. To relate this change in the potential to that of J, we observe that

$$\delta B_y = \delta(J^{-1}\partial_y J) = D_y(J^{-1}\delta J), \tag{2.5.31}$$

after simple manipulations, and

$$\delta B_z = \delta(J^{-1}\partial_z J) = D_z(J^{-1}\delta J). \tag{2.5.32}$$

Comparing with those obtained from the RH method, (2.5.29) and (2.5.30), we find the transformation

$$J^{-1}\delta_\alpha J = -\chi T_\alpha \chi^{-1}, \tag{2.5.33}$$

i.e., the same as the one we first guessed in our paper [20].

After lengthy calculations, from (2.5.1) and (2.5.2) we can derive the following infinite algebraic relations [20,22]

$$[\Delta_a^{(m)}, \Delta_b^{(n)}]J = -C_{ab}^c \Delta_c^{(m+n)} J, \qquad -\infty \leq m, n \leq \infty, \tag{2.5.34}$$

where $\Delta_a^{(m)} = \delta_a^{(m)}$ for $m > 0$, $\Delta_a^{(m)} = \delta_a^{(0)} + \hat{\delta}_a^{(0)}$ for $m = 0$, and $\Delta_a^{(m)} = (-)^{(m)}\hat{\delta}_a^{(-m)}$ for $m < 0$. This is the now-well known Kac–Moody algebra $sl(N, C) \times C(\lambda, \lambda^{-1})$. The important point is that it lacks the center of the algebra, which is of the form $c\delta_{ab}\delta_{k,-\lambda}$, where c is a constant.

Since the indices m and n in (2.5.34) cover all integers, we can condense this equation into a single commutator in the complementary variables θ and θ'. Multiplying both sides of (2.5.34) by $e^{im\theta}e^{in\theta'}$, and summing with the definition

$$Q(\theta) \equiv \sum_{m=-\infty}^{\infty} e^{im\theta} Q_a^{(m)},$$

we obtain

$$[Q_a(\theta), Q_b(\theta')] = C_{ab}^c Q_c(\theta)\,\delta(\theta - \theta'), \tag{2.5.35}$$

where the variable θ can be identified (for unimodular λ) as $e^{i\theta} = \lambda$, the CP^3 parameter.

2.5.2 The case of real *SU(N)* self-dual Yang–Mills fields

For J hermitian and α_a real, we see that δJ and $\hat{\delta} J$ give new $J' \equiv J + \delta J$ and $\hat{J}' \equiv J = \hat{\delta} J$, respectively, with $\det J' = 1 = \det \hat{J}'$; but J' and $\hat{J}'$ are not hermitian. From the condition $\hat{\chi}^{-1}(\lambda) = \chi^\dagger(-1/\lambda)$ we can easily show that $[\delta(-1/\lambda)J]^\dagger = \delta(\overline{\lambda})J$. Therefore, we can form two hermitian transformations

$$\overset{(+)}{\delta}{}_a(\lambda)J \equiv \delta_a(\lambda)J + \delta_a(-1/\lambda)J \quad = \sum_{k=0}^{\infty} \lambda^k \overset{(+)}{\delta}{}_a^{(k)} J, \tag{2.5.36}$$

$$\overset{(-)}{\delta}{}_a(\lambda)J \equiv i[\delta_a(\lambda)J - \delta_a(-1/\lambda)J] \quad = \sum_{k=0}^{\infty} \lambda^k \overset{(-)}{\delta}{}_a^{(k)} J, \tag{2.5.37}$$

where J is restricted to be hermitian. After lengthy derivation we find the algebra

$$\begin{aligned}
&[\overset{(+)}{d}{}_a^{(m)}, \overset{(+)}{d}{}_b^{(n)}] = C_{ab}^c \overset{(+)}{d}{}_c^{(m+n)}, \\
&[\overset{(-)}{d}{}_a^{(m)}, \overset{(-)}{d}{}_b^{(n)}] = C_{ab}^c \overset{(-)}{d}{}_c^{(m+n)}, \qquad 0 < m, n < \infty, \\
&[\overset{(-)}{d}{}_a^{(m)}, \overset{(-)}{d}{}_b^{(n)}] = C_{ab}^c \sum_{\ell=0}^{m+n} a_\ell \overset{(+)}{d}{}_c^{\ell},
\end{aligned} \tag{2.5.38}$$

where

$$\overset{(+)}{d}{}_a^{(0)} \equiv \overset{(+)}{\delta}{}_a^{(0)}, \quad \overset{(\pm)}{d}{}_a^{(1)} \equiv \overset{(\pm)}{\delta}{}_a^{(1)}, \qquad \text{and} \quad [\overset{(\pm)}{d}{}_a^{(m)}, \overset{(\pm)}{d}{}_b^{(1)}] \equiv C_{ab}^c \overset{(\pm)}{d}{}_c^{(m+1)},$$

and the coefficients a_λ are completely determined in the calculation. So we see that $\overset{(+)}{d}{}_a^{(m)}$ and $\overset{(-)}{d}{}_b^{(n)}$ form a symmetric space-like algebra over $su(N) \times R(\lambda)$ [22].

The existence of such an infinite-dimensional algebra in the four-dimensional SDYM theory certainly is a very nice addition to the list of total-integrability characteristics of the theory. However, the full implications of the algebra are still to be discovered. Recently, the infinite-dimensional algebra of the Toda lattice has been used to construct an S-matrix for the theory [38]. The hope is that a similar development can be achieved for the four-dimensional Yang–Mills field.

References

[1] C. N. Yang and R. L. Mills, *Phys. Rev.* **96**, 191 (1954); C. N. Yang, *Phys. Rev. Lett.* **33**, 445 (1970).

[2] T. T. Wu and C. N. Yang, *Phys. Rev.* **D12**, 3845 (1975); C. H. Gu and C. N. Yang, *Scientia Sinica* **18**, 483 (1975); C. N. Yang, *Proceedings of Sixth Hawaiian Topical Conference in Particle Physics*, University of Hawaii Press, 1975; T. T. Wu and C. N. Yang, *Nucl. Phys.* **B107**, 1 (1976); C. N. Yang, *Monopoles and Fiber Bundles. In understanding of the fundamental constituents of matter*, A. Zichichi ed., 1976.

[3] A. Belavian, A. Polyakov, A. Schwartz, and Y. Tyupkin, *Phys. Lett.* **59B**, 85 (1975); M. F. Atiyah, V. G. Drinfield, N. J. Hitchin, and Yu. I. Manin, *Phys. Lett.* **65A**, 185 (1978); M. F. Atiyah and R. S. Ward, *Comm. Math. Phys.* **55**, 117 (1977).

[4] A. M. Polyakov, *JETP Lett.* **20**, 194 (1974); G. 't Hooft, *Nucl. Phys.* **B79**, 276 (1974); M. K. Prasad and C. M. Sommerfield, *Phys. Rev. Lett.* **35**, 760 (1975); E. B. Bogomolny, *Sov. J. Nucl. Phys.* **24**, 861 (1976).

[5] *In e^+e^- reactions see*: MARK J Collaboration, D. P. Barber *et al.*, *Phys. Rev. Lett.* **43**, 830 (1979); TASSO Collaboration, R. Brandelik *et al.*, *Phys. Lett.* **86B**, 243 (1979); PLUTO Collaboration, C. Berger *et al.*, *Phys. Lett.* **86B**, 418 (1979); JADE Collaboration, W. Bartel *et al.*, *DESY* (1979), preprint 79/80. *In hadronic reactions see*: Spectrometer Collaboration at ISR, CERN, T. Akesson *et al.*, *Phys. Lett.* **118B**, 185 (1982); Spectrometer Collaboration at ISR, CERN, T. Akesson *et al.*, *Phys. Lett.* **118B**, 193 (1982); UA2 Collaboration at CERN $\bar{p}p$ Collider, M. Banner *et al.*, *Phys. Lett.* **119B**, 203 (1982); UA1 Collaboration at CERN $\bar{p}p$ Collider, G. Arnison *et al.*, *Phys. Lett.* **123B**, 115 (1983).

[6] K. Wilson, *Phys. Rev.* **D10**, 2445 (1974); A. M. Polyakov, *Phys. Lett.* **59B**, 82 (1975); G. 't Hooft, *Phys. Rev.* **D14**, 3432 (1976). *For recent work, see*: M. Creutz, *Phys. Rev. Lett.* **43**, XXXX (1979); M. Creutz, L. Jacobs, and C. Rabbi, *Phys. Rev.* **D20**, 1915 (1979); *ibid.*, *Phys. Rep.* (to be published).

[7] G. 't Hooft, *Nucl. Phys.* **B33**, 173 (1971), and private communication; H. D. Politzer, *Phys. Rev. Lett.* **30**, 1346 (1973); D. J. Gross and F. A. Wilczek, *Phys. Rev. Lett.* **30**, 1343 (1973).

[8] H. D. Politzer, *Nucl. Phys.* **B129**, 301 (1977).

[9] See *Bäcklund transformations, the inverse scattering method, solitons and their applications*, Vanderbilt University, 1974. R. M. Miura, ed., Lectures notes in mathematics, Springer–Verlag, and references therein.

[10] A. B. Zamolodchikov, *Comm. Math. Phys.* **55**, 183 (1977); A. B. B. Zamolodchikov and A. B. Zamolodchikov, *Nucl. Phys.* **B133**, 525 (1978).

[11] C. N. Yang, *Phys. Rev. Lett.* **38**, 1377 (1977); S. Ward, *Phys. Lett.* **61A**, 81 (1977); Y. Brihaye, D. B. Fairlie, J. Nuyts, and R. F. Yates, *J. Math. Phys.* **19** 2528 (1978).

[12] A. A. Belavian and V. E. Zakharov, *Phys. Lett.* **73B**, 53 (1978).

[13] M. K. Prasad, A. Sinha, and L.–L. Chau Wang, *Phys. Rev. Lett.* **43**, 750 (1979).

[14] M. K. Prasad, A. Sinha, and L.–L. Chau Wang, *Phys. Lett.* **87B**, 237 (1979).

[15] L.–L. Chau Wang, *Proceedings of the Guanzhou (Canton) Conference on Theoretical Particle Physics*, 1980, p. 1082.

[16] L.–L. Chau Wang, *Proceedings of the International School of Subnuclear Physics. The high energy limit*, Z. Zichichi ed., 1983, p. 249; *ibid.*, *Proceedings of the International Workshop on High Energy Physics*, Protvino, Serpukhov, USSR 1980, p. 402.

[17] K. Pohlmeyer, *Comm. Math. Phys.* **72**, 37 (1980).

[18] L.–L. Chau Wang, *Lectures at the 18th Winter School of Theoretical Physics*, Karpacz, Poland, 1981.

[19] L.–L. Chau Wang, M. K. Prasad, and A. Sinha, *Phys. Rev.* **D23**, 2321 (1981) *ibid.*, *Phys. Rev.* **D24**, 1574 (1981).

[20] L.–L. Chau, M. L. Ge, and Y. S. Wu, *Phys. Rev.* **D25**, 1086 (1982).

[21] K. Ueno and N. Nakamura, *Phys. Lett.* **117**, 208 (1982).

[22] L.–L. Chau and Y. S. Wu, *Phys. Rev.* **D26**, 3581 (1983); L. L. Chau, M. L. Ge, A. Sinha, and Y. S. Wu, *Phys. Lett.* **121B**, 391 (1983).

[23] L.–L. Chau, *Invited Talk at the 11th International Colloquium on Group Theoretical Methods in Physics*, Boğaciçi University, Istambul, Turkey, 1982.

[24] A. T. Ogielskĭ, M. K. Prasad, A. Sinha, and L.–L. Chau Wang, *Phys. Lett.* **91B**, 387 (1980). For $\beta = 0$, there is the Bäcklund transformations in: K. Pohlmeyer, *Comm. Math. Phys.* **46**, 207 (1976); I. V. Cheredrik, *Teor. Mat. Fiz.* **38**, 120 (1979).

[25] L.–L. Chau, (1979) (unpublished).

[26] P. Winternitz, *Lie Groups and Solutions of Nonlinear Differential Equations* in these Proceedings.

[27] J. Harnad, Y. Saint-Aubin, and S. Shnider, Université de Monréal, preprint CRMA-1074, 1982.

[28] M. Lüscher and K. Pohlmeyer, *Nucl. Phys.* **B137**, 44 (1978); M. Lüscher, *Nucl. Phys.* **B135**, 1 (1978).

[29] E. Brezin, C. Itzykson, J. Zinn-Justin, and J. B. Zuber, *Phys. Lett.* **82B**, 442 (1979).

[30] V. E. Zakharov and A. V. Mikhaĭlov, *Sov. Phys. JETP* **47**, 1017 (1978).

[31] Z. Popowicz and L.–L. Chau Wang, *Phys. Lett.* **98B**, 253 (1981).

[32] L. Dolan and A. Roos, *Phys. Rev.* **D22**, 20 (1980).

[33] B. Y. Hou, Yale preprint YTP80-29, 1980 (unpublished); B. Y. Hou, M. L. Ge, and Y. S. Wu, *Phys. Rev.* **D24**, 2238 (1981); M. L. Ge and Y. S. Wu, *Phys. Lett.* **108B**, 411 (1982).

[34] L. Dolan, *Phys. Rev. Lett.* **47**, 1371 (1981).

[35] C. Devchand and D. B. Fairlie, *Nucl. Phys.* **B194**, 232 (1982).

[36] Y. S. Wu, *IAS* (preprint 1982).

[37] J. J. Levin, *Bull. Amer. Math. Soc.* **10**, 519 (1959).

[38] B. Kostant, *Adv. Math.* **34**, 195 (1979); *ibid.*, *Inv. Math.* **48**, 101 (1978); M. Adler, *Inv. Math.* **50**, 219 (1979); A. G. Reyman and M. A. Semenov-Tian-Shansky, *Inv. Math.* **54**, 81 (1979).

The Homogeneous Hilbert Problem, the Geroch Conjecture, and a New Nine-Parameter Solution of the Einstein–Maxwell Equations*

Frederick J. Ernst

Department of Physics
Illinois Institute of Technology, USA

Contents:

* Research supported in part by National Science Foundation grants PHY–79–08627 and PHY–82–05608

§1 Introduction

If, in the twenty-first century, one is able to look upon nonlinear partial differential equations with as much confidence as one looks upon linear partial differential equations, it is likely that the source of that confidence will be our own perseverance in attempting to cope with such nonlinear systems as occur in the general theory of relativity, the theory of self-dual Yang–Mills fields, and in hydrodynamics.

When in 1968 I proposed [1] that, of all the unsolved problems of general relativity, one might do well to concentrate one's effort upon the problem of finding all stationary axially symmetric solutions of the Einstein vacuum and electrovac[1] field equations, my motivations were more mathematical than physical. Indeed, even though the mathematical problem is now essentially solved, the problem of adjoining to such *exterior* fields corresponding *interior* fields satisfying the Einstein equations, with the appropriate stress-energy tensor to obtain global solutions, remains quite intractable.

One striking concurrence of facts gave me hope that the mathematical problem presented by the exterior field might be solvable; namely:

i. I found that the Einstein field equation could be replaced by a single complex potential equation[2]

$$(\mathrm{Re}\,E) * d * dE = *(dE * dE), \tag{1}$$

in the vacuum case, and by the pair of equations

$$\begin{aligned}(\mathrm{Re}\,E + M^*M) * d * dE &= *[(dE + 2M^*dM) * dE]\,,\\ (\mathrm{Re}\,E + M^*M) * d * dM &= *[(dE + 2M^*dM) * dM]\,,\end{aligned} \tag{2}$$

in the electrovac case, and

ii. furthermore, that the Kerr and Kerr–Newman blackhole metrics satisfied these equations in a **trivial** manner, while these metrics were by no means obvious solutions of the original Einstein and Einstein–Maxwell equations, respectively.

In recent years, equations very similar to Eq. (1) have been encountered in such diverse areas of mathematical physics as the Heisenberg model of a ferromagnet, the chiral model, and the theory of self-dual Yang–Mills fields. The present School and Workshop is one manifestation of the growing awareness that these diverse fields are united by common mathematical structures.

In the gravitation case there are really two problems, the mathematical description of which are almost identical; one is the stationary axially symmetric problem of describing the spacetime structure outside a body which is rotating uniformly about a symmetry axis. This problem is characterized by two commuting Killing vector fields, one of which is spacelike and the other timelike, while the second

[1] The term 'electrovac' refers to solutions of the coupled Einstein–Maxwell field equations.

[2] Here I have employed the coordinate-independent language of differential forms. In particular, the symbol (*) designates the *duality operator* in Euclidean 3-space.

problem is that of cylindrical waves, where both Killing vector fields are spacelike. There is, of course, an important difference in these two problems, for the first involves an elliptic system of differential equations while the latter involves a hyperbolic system. Thus, whereas one may show that the E- and M-potentials in the former problem are holomorphic in the complex extensions of the non-ignorable coordinates, such a conclusion is impossible in the case where one deals with a hyperbolic system.

One case in point is a conjecture made by R. Geroch [2] in 1972, and proved by I. Hauser [3] and myself in 1980. The proof of the conjecture depended very much upon the elliptic character of the equations with which we were dealing, a property without which we could not have inferred the holomorphy upon which our proof depended so much. Indeed, it might well be argued that in the case of the cylindrical wave problem, it is precisely the non-holomorphic E- and M-potentials which are most interesting to consider.

The equations we are discussing have been found to possess an extremely rich structure, admitting both Bäcklund transformations[3] and Kinnersley–Chitre (K–C) transformations [4]. The latter constitute an infinite parameter group. This group was only vaguely defined until we completed our proof of the Geroch conjecture, *viz.*, that essentially all stationary axially symmetric vacuum solutions may be generated through the action of a group of transformations upon any given solution such as Minkowski space. In fact, it is extremely simple to generalize our proof of the Geroch conjecture to show that essentially all stationary axially symmetric electrovac fields may be obtained from Minkowski space through the action of the K–C group. Here, the qualification *essentially* refers to the fact that our proof depends upon the existence of the complex potentials E and M in some open neighborhood on the symmetry axis. We do not consider solutions which are everywhere singular on the axis.

I should like to mention that our way of understanding the group was obtained only after mastering the infinite-hierarchy approach of Kinnersley and Chitre. Later we learned how to obtain our results more directly and elegantly without the intervention of the infinite hierarchy of potentials, but this in no way diminishes the historical importance of the studies of Kinnersley and Chitre.

To explain our proof of the Geroch conjecture or its electrovac generalization, I must identify the group itself and the potential which characterizes any given spacetime, and upon which the group acts to produce one solution from another. For simplicity of exposition I shall restrict attention to the vacuum case initially. However, the application which I shall describe later involves electrovac fields, so you will in the end see that the vacuum formalism generalizes in a very simple way.

§2 The Geroch group

After Hauser and I published our proof of the Geroch conjecture, the way in which we conceived of the Geroch group evolved. The objective of the present lectures, however, is primarily to facilitate your reading of the published papers. Therefore, I shall identify the group in the same way as we did in those papers, namely, in terms of representations which involve the choice of a complex plane contour we call L. I shall not talk about the contourless description of the group which becomes possible when one uses the Hauser–Ernst gauge consistently. I think that this will be described in a forthcoming paper by I. Hauser. We employed a family of *representations* K_L of the Geroch group, one for each smooth contour L surrounding the origin in the complex plane C, and symmetric with respect to the real axis. K_L is the set of all holomorphic 2×2 complex matrix functions u such that the domain of u is a subregion of C, such that for all t in this region

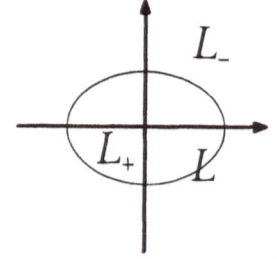

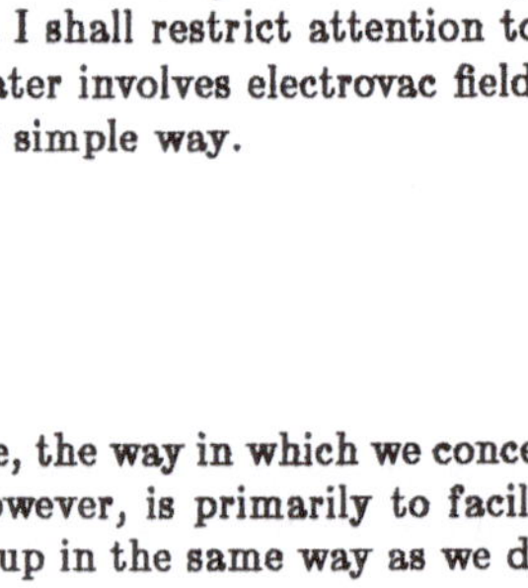

[3]Bäcklund transformations for these equations were studied by B. K. Harrison, G. Neugebauer, and C. Cosgrove.

$$u(t)^\dagger \epsilon u(t) = \epsilon, \qquad \det u(t) = 1, \qquad \epsilon = \begin{pmatrix} 0 & 1 \\ -1 & 0 \end{pmatrix},$$

that L is contained in this domain of u, and that

$$\begin{pmatrix} t^{-1} & 0 \\ 0 & 0 \end{pmatrix} u(t) \begin{pmatrix} t & 0 \\ 0 & 1 \end{pmatrix}$$

is holomorphic in $L + L_-$ (including at infinity). Other representations are possible, but this is the one actually used in the published papers.

§3 The H – and F –potentials

Starting with a commuting pair of Killing vector fields $\mathbf{X}_a$ and their respective coforms X_a, we introduce the self-dual two-forms $W_b = -2P\, dX_b$, where P is a projection operator and the coefficient -2 was inserted for historical reasons. Then it is easy to show that the contraction $\mathbf{X}_a W_b =: dH_{ab}$ is exact differential in the vacuum case. Given any stationary axially symmetric vacuum spacetime, it is a straightforward procedure to evaluate the 2×2 H –potential. (The time-time component is just the E –potential.) For a given H –potential, the linear equation

$$dF(t) = \gamma(t)\Omega\, F(t), \qquad \Omega = i\epsilon,$$

where $F(0) = \Omega$ and $\dot{F}(0) = H$, can always be solved. Here,

$$\gamma(t) := t[(1 - 2tz) - 2t\rho *]^{-1} dH,$$

where ρ and z are Weyl canonical coordinates, and t is an arbitrary complex parameter (in our paper we spell out all domains very carefully). It is always possible to choose the gauge so that $F(t)$ is holomorphic everywhere except for branch points, at $t = 1/[2(z \pm i\rho)]$, connected by a suitable cut. (In our paper we also discuss certain algebraic constraints which may be imposed upon $F(t)$, but here I am merely trying to sketch the ideas and not overwhelm you with details.)

§4 The homogeneous Hilbert problem approach to Kinnersley–Chitre transformations

Suppose $F_0(t)$ is known for some seed metric, *e.g.*, Minkowski space, and $u(t)$ has been selected from K_L. Then

$$u(t)\, F_0(t)^{-1}$$

has singularities of both $u(t)$ and of $F_0(t)$. The homogeneous Hilbert problem (HHP) consists of finding 2×2 matrix fields $X_-(t)$ and $F(t)$ such that

$$X_-(t) = F(t)\, u(t)\, F_0(t)^{-1}, \qquad F(0) = \Omega,$$

where $F(t)$ is holomorphic in $L + L_+$ and $X_-(t)$ is holomorphic in $L + L_-$ (including $t = \infty$). (Such Hilbert problems occur in other areas of physics, *e.g.*, partial wave analysis, where the HHP is the basis of the N over D methods.)

We have established that this procedure gives rise to a new $F(t)$ potential from which a new metric may be constructed in a straightforward manner. One defines the new H–potential by

$$H := \dot{F}(0),$$

and then

$$h_{ab} = \mathbf{X}_a \cdot \mathbf{X}_b = -\operatorname{Re} H_{ab}$$

gives (most of) the metric tensor. The only nontrivial step is solving the HHP, and this is a *linear* problem which is solvable in principle.

§5 Solving the homogeneous Hilbert problem

A number of methods have already been explored for solving the HHP:

i. A Cauchy type integral equation,

$$\int_L ds \, \frac{F(s)\,u(s)\,F_0(t)^{-1}}{s(s-t)} = 0, \qquad t \in L_+, \quad z \in \Delta_L \text{ (defined in the paper).}$$

ii. A matrix nonsigular Fredholm integral equation;

$$F(t) - \tfrac{1}{2\pi i} \int_L ds\, F(s)\, K(s,t) = F_0(t), \qquad t \in L, \quad z \in \Delta_L,$$

where

$$K(s,t) := \frac{t}{s(s-t)} \Big[M(s,t) - u(s)\, M(s,t)\, u(t)^{-1} \Big], \qquad M(s,t) := F_0(s)^{-1} F_0(t).$$

(This approach was used in the proof of Geroch conjecture.)

iii. Single-variable nonsingular Fredholm equation (unpublished work).

iv. Direct inspection combined with shrewd guesswork.

§6 The axis relation

A powerful tool for determining which K–C transformation to use in order to obtain one spacetime from another is provided by the axis relation

$$-iE = \frac{-iE_0 u^3{}_3(1/2z) + 2zu^3{}_4(1/2z)}{-iE_0(1/2z)u^4{}_3(1/2z) + u^4{}_4(1/2z)},$$

where $E = E(z,0)$, and $E_0 = E_0(z,0)$.

§7 The Geroch conjecture

To prove the Geroch conjecture, we took $E_0 = 1$ and any other E_g, and used the axis relation to identify a $u(t)$. For example, if on the axis

$$E_g(z,0) = e^{-2b(1/2z)} + ia(1/2z),$$

then we may select

$$u(t) = \begin{pmatrix} e^{-b(t)} & ta(t)\, e^{b(t)} \\ 0 & e^{b(t)} \end{pmatrix}.$$

It is then only necessary to prove that the E –potential obtained by solving the HHP for this $u(t)$ and the Minkowski space $F_0(t)$ is in fact E_g. This we did in our paper.

§8 Nine-parameter electrovac Tomimatsu–Sato generalization

As an example of the HHP approach to generating new solutions from old ones, I have with two students, Y. Chen and D. S. Guo, worked out [5] a natural electrovac generalization of the simplest Tomimatsu–Sato (T–S) vacuum metric. The calculations, which involved the evaluation of a fair number of determinants, were performed entirely upon a 64k 4MHz home computer with a Z80 CPU using a newly developed symbolic evaluation program [6] written in the MuLISP[4] language.

The specific K–C transformation we employed was that introduced by Cosgrove [7], who showed that the (beyond extreme) charged Kerr–NUT solution results from applying his transformation once to Minkowski space. Using the Hauser–Ernst τ-plane formulation[5] of the HHP, which we now much prefer to the original t-plane formulation, we may identify the Cosgrove transformation by

$$v(\tau) = \exp\,[J\eta(\tau)],$$

where

$$J = \tfrac{i}{3}I + \tfrac{2}{E}hh^{\dagger}Q,$$

$$Q = \begin{pmatrix} 0 & 1 & 0 \\ -1 & 0 & 0 \\ 0 & 0 & -\frac{i}{2} \end{pmatrix}, \qquad E = 2ih^{\dagger}Qh,$$

$$h = \begin{pmatrix} h_1 \\ h_2 \\ h_3 \end{pmatrix} = \text{complex constant matrix},$$

and

$$\eta(\tau) = -i\ln\left(\frac{\tau - K}{\tau - K^*}\right).$$

Note the appearance of 3×3 matrices in the electrovac problem where we had 2×2 matrices in the vacuum problem.

We employed the Cosgrove transformation twice in succession, upon Minkowski space, as the seed metric. The resulting complex potentials,

$$E = E_{\text{seed}} - \frac{2iN}{D} \qquad \text{and} \qquad M = M_{\text{seed}} - \frac{2N'}{D},$$

[4] MuLISP is available for approximately $200 from the Soft Warehouse, P.O. Box 11174, Honolulu, Hawai 96828-0174.

[5] $\tau = 1/2t$. I believe I. Hauser will soon publish a paper in which the merits of the τ-plane formulation of the vacuum problem are described. In the electrovac case, the advantages are even more impressive.

where

$$\begin{aligned} N =&\Delta_{456}\Delta_{32}R_1(K_1-K_1^*)(K_1-K_2^*)(K_1^*-K_2)\\ &+\Delta_{423}\Delta_{65}R_1(K_1-K_1^*)(K_2-K_2^*)(K_2-K_1^*)\\ &+\Delta_{523}\Delta_{61}R_4(K_1-K_1^*)(K_2^*-K_1)(K_2-K_2^*)\\ &+\Delta_{623}\Delta_{51}R_4(K_1-K_1^*)(K_2-K_2^*)(K_1-K_2^*)\\ &+\Delta_{123}\Delta_{65}R_4(K_1-K_2)(K_1-K_2^*)(K_2-K_2^*) \end{aligned}$$

and

$$\begin{aligned} D =&\Delta_{423}\Delta_{156}(K_2-K_1)(K_2^*-K_1^*)\\ &+\Delta_{523}\Delta_{164}(K_2^*-K_1)(K_2-K_1^*)\\ &+\Delta_{623}\Delta_{145}(K_2^*-K_1)(K_2-K_1^*) \end{aligned}$$

were published by Guo and Ernst [8].

The expression for N' was like the expression for N except that primes appeared on the two-index deltas. The three-index deltas were determinants

$$\Delta_{ijk} = \det(V_i\, V_j\, V_k)$$

of certain eigenvectors

$$V_i = \begin{pmatrix} Q_i \\ R_i \\ S_i \end{pmatrix},$$

while the two-index deltas were subdeterminants

$$\Delta_{ij} = \begin{vmatrix} R_i & R_j \\ S_i & S_j \end{vmatrix}, \qquad \text{and} \qquad \Delta'_{ij} = \begin{vmatrix} Q_i & Q_j \\ R_i & R_j \end{vmatrix}.$$

For any given choice of seed metric, the eigenvectors V_1 through V_6 can be determined in terms of the locations K_1 and K_2 of the τ-plane singularities and the complex constants h_1, h_2, h_3 and h_4, h_5, h_6 of the respective Cosgrove transformations by a straightforward procedure. If P_0 is the P-potential of the seed metric and

$$\gamma_1(\tau) = P_0(\tau)\, J_1\, P_0(\tau)^{-1},$$

then V_1 is the unique eigenvector of $\gamma_1(K_1^*)$ corresponding to the nondegenerate eigenvalue $-\frac{2}{3}i$, while V_2 and V_3 are linearly independent eigenvectors of $\gamma_1(K_1)$ corresponding to the doubly-degenerate eigenvalue $+\frac{1}{3}i$.

For Minkowski space one finds that the eigenvectors V_1 through V_3 may be selected as follows:

$$Q_1 = -i\Big[R_2^* r(K^*) - R_1(K^* - z)\Big], \qquad R_1 = h_1 + ih_2, \qquad S_1 = 2ir(K^*)h_3,$$

$$Q_2 = -i\Big[R_1^* r(K) - R_2(K - z)\Big], \qquad R_2 = h_1^* + ih_2^*, \qquad S_2 = 0,$$

$$Q_3 = h_3\Big[R_2^* r(K) - R_1(K - z)\Big], \qquad R_3 = ih_3^* R_1, \qquad S_3 = 2r(K)\Big[R_1^* R_1 - R_2^* R_2\Big],$$

where

$$r(\tau) = \sqrt{(z-\tau)^2 + \rho^2}.$$

The eigenvectors V_4 through V_6 may be selected to have the same form as V_1 through V_3, except that K_1 is replaced by K_2 and h_1 through h_3 are replaced, respectively, by h'_4 through h'_6, which components are given as

$$h' = \begin{pmatrix} h'_4 \\ h'_5 \\ h'_6 \end{pmatrix} = \exp[-J_1\eta(K_2^*)]\begin{pmatrix} h_4 \\ h_5 \\ h_6 \end{pmatrix}.$$

To obtain the T–S generalizations we took a limit of the solution described above, letting K_2 and K_1 tend to a common limit K and (h'_4, h'_5, h'_6) tend toward (h_1, h_2, h_3). Following this we were able to introduce oblate spheroidal coordinates and a parametrization which is analogous to that of the original T–S solution.

To specify our nine-parameter electrovac generalization of the simplest T–S solution we shall write E and M in the form

$$E = \frac{U-W}{U+W}, \qquad M = \frac{V}{U+W},$$

where

$$U = e^{i\gamma}\Big(U_1(x^4-1) + U_2(y^4-1) + U_3xy(x^2+y^2) + U_4(x^2+y^2)^2$$
$$+ U_5xy(-x^2+y^2-2) + U_6(-x^2+y^2+2x^2y^2)\Big),$$

$$V = V_1x(x^2+1) + V_2y(1-y^2) + V_3x(x^2+y^2) + V_4y(x^2+y^2),$$

and

$$W = W_1x(x^2+1) + W_2y(1-y^2) + W_3x(x^2+y^2) + W_4y(x^2+y^2),$$

where the constant coefficients are given by

$$U_1 = p^2 + \tfrac{1}{2}h^*h - \tfrac{1}{2}h^*h(q-p)^2\Big[1 - \tfrac{1}{2}h^*h + 2(\sigma^2+\tau^2)\Big],$$
$$U_2 = -q^2 + \tfrac{1}{2}h^*h + \tfrac{1}{2}h^*h(q-p)^2\Big[1 - \tfrac{1}{2}h^*h + 2(\sigma^2+\tau^2)\Big],$$
$$U_3 = 2ipq + ih^*h(q-p)^2\Big[1 - \tfrac{1}{2}h^*h + 2(\sigma^2+\tau^2)\Big],$$
$$U_4 = -(\alpha^2+\beta^2) + h^*h\Big[(\alpha+\sigma)^2 + (\beta+\tau)^2\Big],$$
$$U_5 = 2i\beta(1-h^*h) - 2ih^*h\tau,$$
$$U_6 = 2i\alpha(1-h^*h) - 2ih^*h\sigma,$$

$$
\begin{aligned}
V_1 &= h^*\Big[2p + (q-p)(h^*h + 2i\sigma + 2\tau)\Big],\\
V_2 &= ih^*\Big[-2q + (q-p)(h^*h + 2i\sigma + 2\tau)\Big],\\
V_3 &= -2ih^*[\alpha p - i(\beta + i\sigma + \tau)q]\\
&\qquad + h^*(q-p)\Big[2(\alpha + i\beta)(\sigma - i\tau) + h^*h(-i\alpha + \beta - i\sigma + \tau)\Big],\\
V_4 &= -2ih^*[i\alpha q + (\beta + i\sigma + \tau)p]\\
&\qquad - ih^*(q-p)\Big[2(\alpha + i\beta)(\sigma - i\tau) + h^*h(-i\alpha + \beta - i\sigma + \tau)\Big],
\end{aligned}
$$

and

$$
\begin{aligned}
W_1 &= 2p + h^*h(q-p)(1 + 2i\sigma + 2\sigma),\\
W_2 &= -2iq + ih^*h(q-p)(1 + 2i\sigma + 2\sigma),\\
W_3 &= -2i(\alpha p - i\beta q) + h^*h(q-p)[-i\alpha + \beta - 2i\sigma + 2(\sigma^2 + \tau^2) + 2(\alpha + i\beta)(\sigma - i\tau)],\\
W_4 &= -2i(i\alpha q + \beta p) - ih^*h(q-p)[-i\alpha + \beta - 2i\sigma + 2(\sigma^2 + \tau^2) + 2(\alpha + i\beta)(\sigma - i\tau)].
\end{aligned}
$$

It will be noticed that when $h = 0$ this reduces to the five-parameter vacuum solution of Kinnersley and Chitre ($q^2 - p^2 = 1$ case). Furthermore, when $\alpha = \beta = \sigma = \tau = 0$, our solution reduces to the charged T–S solution of Ernst. A similar procedure will, when applied to the N-fold Cosgrove solution worked out by Wang, Guo, and Wu [9], result in rational funtion electrovac generalizations of the higher Kinnersley–Chitre–Tomimatsu–Sato solutions.

References

[1] F. J. Ernst, New formulation of the axially symmetric gravitational field problem. *Phys. Rev.* **167**, 1175–1178; . *ibid.* New formulation of the axially symmetric gravitational field problem. II. *Phys. Rev.* **168**, 1415–1417 (1968).

[2] R. Geroch, A method for generating new solutions of Einstein's equations, II. *J. Math. Phys.* **13**, 394–404 (1972).

[3] I. Hauser and F. J. Ernst, Proof of a Geroch conjecture. J. Math. Phys. **22**, 1051–1063 (1981).

[4] W. Kinnersley and D. M. Chitre, Symmetries of the stationary Einstein–Maxwell field equations. *J. Math. Phys.* **18**, 1538–1542 (1977); *ibid.* Symmetries of the stationary Einstein–Maxwell field equations. II. *J. Math. Phys.* **19**, 1926–1931; *ibid.* Symmetries of the stationary Einstein–Maxwell equations. IV. transformations which preserve asymptotic flatness. *J. Math. Phys.* **19**, 2037–2042 (1978).

[5] Y. Chen, D. S. Guo, and F. J. Ernst, Charged spinning mass field involving rational functions. *J. Math. Phys.* (June 1983.)

[6] F. J. Ernst, *Symbolic Microcomputer Calculations in General Relativity and Differential Geometry.* Includes user's manual and source listings of all programs (five dollars). Also available on IBM-3740 formatted SSSD 8 inch disk (fifteen dollars).

[7] C. Cosgrove, Bäcklund tansformations in the Hauser–Ernst formalism for stationary axisymmetric spacetimes. *J. Math. Phys.* **22**, 2624–2639 (1981).

[8] D. S. Guo and F. J. Ernst, Electrovac generalization of Neugebauer's $N = 2$ solution of the Einstein vacuum field equations. *J. Math. Phys.* **23**, 1359–1363 (1982).

[9] S. K. Wang, H. Y. Guo, and K. Wu, The N-fold Kerr family and charged Kerr family solutions. *Preprint* AS-ITP-82-025, Academia Sinica Institute of Theoretical Physics, August 1982.

The Inverse Scattering Transform for Multidimensional (2+1) Problems*

Athanassios S. Fokas
and
Mark J. Ablowitz

Department of Mathematics and Computer Science
Clarkson College of Technology
Potsdam, New York, USA

Contents:

* This article consists of expanded material of five lectures presented by one of us (A. S. Fokas) at the School and Workshop on Nonlinear Phenomena.

§1 Introduction

This article is mainly concerned with the question of whether it is possible to extend the IST formalism to:

i. solve inverse problems in the plane, *i.e.*, given appropriate scattering data reconstruct the potential $u(x, y)$;

ii. solve the initial value problem associated with certain evolution equations in two spatial and one temporal dimension, *i.e.*, given $u(x, y; 0)$, and assuming it is sufficiently decaying at infinity, find $u(x, y; t)$.

It is shown here that the answer to the above question is affirmative. Furthermore, a unified method is presented for solving the above problems. The exposition of this method is the main goal of the present article. However, in order to put this work into perspective, various related aspects in $1 + 1$ (*i.e.*, one spatial and one temporal) dimensional problems are discussed.

Due to the work of several Soviet scientists[1] it became clear that the IST in $1 + 1$ is intimately related to de so-called Riemann–Hilbert boundary value problem (RHBVP). Recently Beals and Coifman [2], in their elegant and significant treatment of the IST of first order systems related to $1+1$ problems, have indicated that the RH problem should be viewed as a special case of a so-called $\overline{\partial}$ (DBAR) problem. More recently, the work of the authors and Bar Yaacov on the IST for $2 + 1$ (*i.e.*, two spatial and one temporal) problems seems to indicate that the inverse problems in multidimensions cannot, in general, be solved in terms of RHBVP's. The role of the RH problem is now necessarily played by a DBAR problem (we note that in certain cases in $2+1$, *e.g.*, KPI, the RH problem formulation is still adequate).

At this point we would like to note that RH problems have also been useful in connection with two other recent significant discoveries:

i. The integration of the Ernst eqaution (the static axisymmetric reduction of Einstein's equations in a vacuum). In particular the RH problem seems indispensible with regards to the proof of the Geroch conjecture [3] (see Frederick Ernst's contribution in *these Proceedings*).

ii. The integration of the self-dual Yang–Mills (SDYM) equations in four-dimensional Euclidean space, in particular with respect to the Atiyah–Ward construction [4].

The SDYM equations as a result of their special structure (although defining a four-dimensional model) have many properties similar to those of two-dimensional problems (see L. L. Chau's contribution in *this proceedings*). Motivated from the above discussion we expect that the DBAR problem will also be useful for the exact integration of various models in both the fields of relativity as well as that of particle physics. Furthermore, we also expect these ideas to be useful for multidimensional differential-difference and purely difference equations.

Associated with a given *exactly solvable* equation there exists at least two interrelated aspects of fundamental interest:

1. The development of a method of solution.

---

[1]For example Zakharov, Shabat, Manakov, Mikhailov, and Belavin [1].

2. The investigation of various *algebraic* properties of the equation, which includes:

i. finding the hierarchy of all equations which have similar properties to the given equation, *e.g.*, they are solved by the same eigenvalue problem and/or have the same conserved quantities;

ii. finding a set of infinitely many conserved quantities of the equation and examining their Poisson bracket (commutator) relations;

iii. establishing the Hamiltonian (or bi-Hamiltonian) and action-angle formulations of the equations as well as that of every member of its hierarchy;

iv. finding Bäcklund transformations and their superposition laws (Bianchi identities);

v. investigating the geometric as well as the group-theoretic origin of the algebraic properties;

vi. classifying the equation into a more general hierarchy using for example some Kac–Moody algebra classification. For discussion of some of the above aspects see [5–15] for $1+1$ equations and [16–19] for $2+1$ equations.

In this article we are concerned with the first aspect, *i.e.*, *the development of a method of solution.* In connection with this we note two rather different problems.

1.1 Find a method for generating particular solutions of the given equation

There exist, to our knowledge, the following methods:

i. Appropriate use of Bäcklund transformations [20,21].

ii. The bilinear approach of Hirota [20,22].

iii. The more general τ-function approach of Sato–Miwa–Jimbo–Date–Kashiwara [19] related to Kac–Moody algebras.

iv. The so-called *dressing method* of Zakharov–Shabat [23], *i.e.*, exploit the fact that there exist linear integral equations of the Gelfand–Levitan–Marchenko (GLM) type, which have the property that their solutions also solve certain nonlinear equations in $1+1$ and $2+1$.

v. The *RH direct method* of Zakharov–Shabat–Mikhailov [24], *i.e.*, exploit the fact that there exist local matrix RH problems which can be directly used for generating solutions of certain nonlinear equations in $1+1$. Recently Manakov [25] has extended this method to equations in $2+1$ by using nonlocal RH problems.

vi. The direct linearizing method proposed by the authors [26] and extended recently by Santini and the authors [27]. This method is closely related to the perturbation approach of Rosales [28] and consists essentially of summing up his series about an arbitrary state. It exploits the fact that there exist rather general linear integral equations, involving an arbitrary measure-contour, which can be used to linearize certain nonlinear equations in $1+1$ and $2+1$ (for details see §2). The interrelations of the direct methods are discussed in [27].

1.2 Find a method for solving general initial value problems

Such a method, in one form or another, is a generalization of the IST. There exist two types of initial value problems which are of fundamental interest:

i. The initial value problem for potentials decaying at infinity (for reference in $1+1$ see §3).

ii. The initial value problem for periodic potentials [1,20].

The above discussion of methods of solution is summarized in the following

(1) Method of Solutions

1.1. Generating Particular Solutions

i. Bäcklund transforms

ii. Hirota's bilinear approach

iii. τ-function approach of Sato *et al.*

iv. *Dressing method*

v. *RH direct method*

vi. Direct linearization

1.2. Solving General Initial Value Problems

i. IST for potentials decaying at infinity

ii. IST for periodic potential.

We would like to know that the authors and R. L. Anderson, in their first effort to solve the initial value problem associated with the Benjamin–Ono (B–O) equation, introduced another method, which they called the direct linearizing transform [29]. This method goes a step further than **1.1.vi** and involves solving an additional linear integral equation for obtaining the necessary scattering data. However, the authors latter discovered [30] that this additional complication may be avoided provided one uncovers certain additional analytic structure of the underlying Jost eigenfunctions (see also the contributions of M. J. Ablowitz and A. S. Fokas in *this proceedings*).

This paper is primarily concerned with **1.2.i** above for problems in $2+1$ [we note that **1.2.ii** for problems in $2+1$ is still, to our knowledge, open]. However, for completeness we also:

i. elaborate on **1.1.vi**;

ii. review the main eigenvalue problems related to IST in $1+1$;

iii. remark on several RH problems;

iv. illustrate the use of RH problems with regards to IST in $1+1$, by using the Korteweg–de Vries (KdV) as an example.

§2 Direct linearization in $1+1$ and $2+1$

In this section we establish the existence of rather general linear integral equations, involving an arbitrary measure-contour, which linearize certain equations.

Proposition 2.1 (Fokas and Ablowitz [26], Rosales [28]).

Let $\phi(k;x,t)$ be a solution of the following linear integral equation in k,

$$\phi(k;x,t)+ie^{i(kx+k^3t)}\int_L d\varsigma(l)\,\frac{\phi(l;x,t)}{l+k}=e^{i(kx+k^3t)}, \tag{2.1}$$

where the contour L and measure $d\varsigma(l)$ are essentially arbitrary. Assume that the homogeneous integral equation corresponding to (2.1) has only the zero solution. Then

$$u=-\frac{\partial}{\partial x}\int_L d\varsigma(k)\,\phi(k;x,t) \tag{2.2}$$

solves the KdV equation

$$u_t+6uu_x+u_{xxx}=0. \tag{2.3}$$

Remarks

i. The above result can be motivated in two different ways. The first one due to Rosales consists of looking for a perturbation solution of (2.3) around $u = 0$. The second is related to the RH treatment of the IST of the KdV (see §5).

ii. The solution $\phi(k; x, t)$ of (2.1) also solves the Schrödinger eigenvalue problem

$$\phi_{xx} + u\phi - ik\phi_x = 0. \tag{2.4}$$

Let $\phi = \psi e^{i(kx+k^3t)/2}$ and (2.4) reduces to the usual Schrödinger equation

$$\psi_{xx} + ((\tfrac{1}{2}k)^2 + u)\psi = 0. \tag{2.5}$$

iii. If the nonlinearity is absent the equation (2.1) yields $\overline{\phi} = e^{i(kx+k^3t)}$. Hence equation (2.2) implies

$$\overline{u} = -\frac{\partial}{\partial x}\int_L d\varsigma(k)\, e^{i(kx+k^3t)}.$$

In fact this is the most *general* solution (Ehrenpreis principle) of the linear equation:

$$\overline{u}_t + \overline{u}_{xxx} = 0.$$

iv. Equation (2.1) with a *specific* choice of the measure-contour reduces to the so-called k-space equation recently introduced by R. G. Newton [31], and is shown to be equivalent to the GLM equation associated with KdV. As it was pointed out above the linear limit of (2.1) yields the *general solution* of the underlying linear equation $u_t + u_{xxx} = 0$. This should be contrasted with the linear limit of the GLM equation which yields only those decaying solutions of $u_t + u_{xxx} = 0$ which are obtained through the Fourier transform. In this sense equation (2.1) can be thought of as the analogue of a generalized transform for solving a nonlinear partial differential equation (PDE) in the same way that the GLM equation corresponds to the Fourier transform.

v. There exist, similar to (2.1), linear integral equations associated with various other nonlinear equations, see [32].

vi. In [26] it was explicitly demonstrated that (2.1) characterizes solutions which are not contained in the GLM equation. This was achieved by considering the self-similar reduction of the KdV

$$\frac{d^3u}{dx^3} + 6u\frac{du}{dx} - \left(2u + x\frac{du}{dx}\right) = 0. \tag{2.6}$$

Using (2.1), a three-parameter family of solutions of (2.6) was found. However, the question of whether or not this is the general solution of (2.6) was let open. Equation (2.6) is related [33,34] to Painlevé II equation

$$\frac{d^2v}{dx^2} - xv - 2v^3 = \alpha. \tag{2.7}$$

We have recently shown that the solution of (2.6) obtained in [26] corresponds only to a specific range of α $(0 \leq \alpha < 1)$. In connection with (2.6) we are motivated to look for a more general linear integral equation which also linearizes the KdV.

Proposition 2.2 (Santini, Ablowitz, and Fokas [27]).

Let $\phi(k;x,t)$ be a solution of

$$\phi(k;x,t) + ie^{i(kx+k^3t)} \int_L d\varsigma(l) \frac{g(k,l,x,t)}{k+l} \phi(l;x,t) = h(k,x,t)e^{i(kx+k^3t)}, \tag{2.8}$$

where $h(k,x,t)$ is a solution of the coupled system

$$\begin{cases} h_{xx} + ikh_x + u_0 h = 0, & (2.9a) \\ h_t + (2u_0 - k^2)h_x + (iku_0 - u_{0_x})h = 0, & (2.9b) \end{cases}$$

$u_0(x,t)$ is any solution of the KdV equation (2.3), and $g(k,l,x,t)$ is defined in terms of $h(k,x,t)$ by

$$g(k,l,x,t) := h(k,x,t)\, h(l,x,t) + \frac{2i}{k-l}\left[h(k,x,t)\, h_x(l,x,t) - h(l,x,t)\, h_x(k,x,t)\right]. \tag{2.10}$$

Assuming that the homogeneous integral equation corresponding to (2.8) has only the zero solution, then

$$u(x,t) = u_0(x,t) - 2\frac{\partial}{\partial x}\int_L d\varsigma(k)\, \phi(k,x,t)\, h(k,x,t) \tag{2.11}$$

solves the KdV equation (2.3).

Remarks

i. In the special case of $u_0 = 0$, equation (2.8) reduces to (2.1). Similarly if $u_0 = -2/x^2$ the above linearization reduces to the one given in [35].

ii. There exist linear integral equations similar to (2.8) associated with various other nonlinear equations [27].

iii. The linearization expressed by (2.8) is closely related to the *RH direct* method. This is discussed in [27].

We now present the analogue of (2.1) associated with KPI.

Proposition 2.3 (Fokas and Ablowitz [36]).

Let $\mu(x,y,t,k)$ be a solution of the linear integral equation

$$\mu(x,y,t,k) + i\iint_L d\varsigma(l,\nu)\, w(x,y,t,k,l,\nu)\, \mu(x,y,t,l) = v(x,y,t,k), \tag{2.12}$$

where the contour L and measure $d\varsigma(l,\nu)$ are essentially arbitrary, v is any solution of

$$\begin{cases} iv_y + v_{xx} + 2ikv_x = 0, & (2.13a) \\ v_t + 4v_{xxx} + 12ikv_{xx} - 12k^2 v_x = 0, & (2.13b) \end{cases}$$

and w is given by

$$w(x,y,t,k,l,\nu) = \exp\left[\hat{\theta}(x,y,t,l,k,\nu)\right]\left[\int_\alpha^x d\xi\, v(\xi,y,t,k)e^{i(k-\nu)\xi} + B(y,t,k,l,\nu)e^{i(k-\nu)\alpha}\right], \tag{2.14}$$

where $\hat{\theta}(x,y,t,l,k,\nu) := i(l-k)x - i(l^2-\nu^2)y + 4i(l^3-\nu^3)t$ and $B = B(y,t,k,l,\nu)$ is defined in terms of v through

$$B_y + i(\nu^2 - k^2)B = iv_x(\alpha) - (k+\nu)v(\alpha), \qquad v(\alpha) := v(\alpha, y, t, k), \tag{2.15a}$$
$$B_t + 4i(k^3 - \nu^3)B = -4\left[v_{xx}(\alpha) + i(\nu + 2k)v_x(\alpha) - (k^2 + \nu^2 + \nu k)v(\alpha)\right]. \tag{2.15b}$$

Assume that the homogeneous solution corresponding to (2.12) has only the zero solution. Then

$$u(x,y,t) = 2i\frac{\partial}{\partial x}\iint_L d\varsigma(l,\nu)\,\mu(x,y,t,l)\exp[\hat{\theta}(x,y,t,l,\nu,\nu)], \tag{2.16}$$

solves the KPI equation

$$(u_t + 6uu_x + u_{xxx})_x = 3u_{yy}. \tag{2.17}$$

Remarks

i. If μ solves (2.12) and u is defined in terms of μ via (2.16), then μ also solves the Lax pair associated with KPI:

$$\begin{cases} i\mu_y + \mu_{xx} + 2ik\mu_x = -u\mu, & (2.18a)\\ \mu_t + 4\mu_{xxx} + 12ik\mu_{xx} - 12k^2\mu_x + 6u\mu_x & \\ \qquad + 3iku\mu + 3iu\mu_x + 3u_x\mu - 3i\left(\int_{-\infty}^{x} dx'\, u_y\right)\mu = 0. & (2.18b)\end{cases}$$

ii. If $\partial_t = 0$ and $l = \nu$, the linearization expressed by (2.12) reduces to the linearization of the B–O equation

$$u_y + 2uu_x + \frac{1}{\pi}\frac{\partial^2}{\partial x^2}⨍_{-\infty}^{\infty} d\xi\,\frac{u(\xi,y)}{\xi - x} = 0$$

presented in [37]. This is consistent with the fact that equation (2.18a) is related to the y-part of the Lax pair of the B–O equation.

iii. If $\partial_y = 0$ then (2.12) reduces to the linearization of the KdV equation given by Proposition 2.1.

§3 Review of the main eigenvalue problems related to the inverse scattering transform in 1+1

A necessary condition for a given nonlinear equation to be solvable via an IST formalism is the existence of a nontrivial pair of equations, the so-called Lax pair [38], such that the original equation can be viewed as the solvability condition of this Lax pair of equations. Provided that this is the case, then most of the analysis of the IST is carried out on the time-independent part of the Lax pair. In this sense, the investigation of the IST for a given equation is essentially tantamount to the investigation of the inverse problem of the underlying time-independent part of its Lax pair.

The IST was discovered by Gardner, Green, Kruskal, and Miura [39], who were able to relate the celebrated KdV equation (2.3) to the classical time-independent Schrödinger equation[2]

$$\psi_{xx} + \left(u(x,t) + k^2\right)\psi = 0. \tag{3.1}$$

[2]The time-dependent part of the Lax pair of the KdV is given by $\psi_t = u_x\psi + (4k^2 + 2u)\psi_x$.

The inverse problem associated with (3.1) for potentials decaying as $|x| \to \infty$ has been investigated in [40].

The next eigenvalue problem to receive much attention was the so-called Ablowitz–Kaup–Newell–Segur (AKNS) scattering problem [41]:[3]

$$\psi_x = \lambda \begin{pmatrix} -i & 0 \\ 0 & i \end{pmatrix} \psi + \begin{pmatrix} 0 & q_1(x,t) \\ q_2(x,t) & 0 \end{pmatrix} \psi, \tag{3.2}$$

where $\psi(x,t,\lambda)$ is a two-vector. Equation (3.2) is the time-independent part of the Lax pair of several physically significant equations. In particular (3.2) can be used for the exact integration of the following equations [43]

$$\begin{aligned} q_1 = \mp\bar{q}_2 :\quad & i(q_2)_t = (q_2)_{xx} \pm 2q_2^2\bar{q}_2, && \text{(nonlinear Schrödinger)} \\ q_1 = \pm q_2 :\quad & (q_2)_t + (q_2)_{xxx} \pm 6q_2^2(q_2)_x = 0, && \text{(modified KdV)} \\ q_1 = -q_2 = -\tfrac{1}{2}u_x :\quad & u_{xt} = \sin u. && \text{(sine-Gordon)} \end{aligned}$$

A third order generalization of (3.1) has been considered by Kaup [44] and Deift, Trubowitz, and Tomei [45]. This eigenvalue problem is related to the Boussinesq equation. Also a 3×3 generalization of (3.2) has been considered by Kaup [46] and Zakharov, and Manakov [47] in connection with the so-called three-wave resonant interaction.

Gel'fand and Dikii [7i] proposed an nth order generalization of (3.1),

$$D^n\psi + q_{n-2}D^{n-2}\psi + q_{n-3}D^{n-3}\psi + \cdots + q_0\psi = \lambda^n\psi, \qquad D := i\frac{\partial}{\partial x}, \tag{3.3}$$

and investigated many of the *algebraic* properties of the equations solvable by (3.3). Symes [48] obtained the recursion operator generating all equations solvable by the Gel'fand–Dikii operator. Similarly Ablowitz and Haberman [49] proposed an $n \times n$ generalization of (3.2),

$$\psi_x = \lambda J\psi + Q(x,t)\psi, \tag{3.4}$$

where J is a constant diagonal matrix and Q is an off-diagonal matrix, and established that (3.4) can be used to solve several physically important equations. Many investigators (*e.g.*, [50,8ii]) obtained the recursion operator generating all equations solvable by (3.4).

However, the question of solving the inverse problems associated with (3.3) and (3.4) remained open for a rather long time. This, in our opinion, was mainly due to the fact that the solution of the above problem required a deeper understanding of the analytic properties of the underlying Jost eigenfunctions. The inverse problems associated with both (3.3) and (3.4) have been recently solved by Beals [51] and Beals and Coifman [2] respectively (see also Caudrey [52]).

The eigenvalue problems (3.1) and (3.2), and their generalizations (3.3) and (3.4) are, in our opinion, the main differential problems which have been used in connection with the IST in $1+1$. There exist several variants of the above problems, which however should be solvable by some simple variation of the procedure used for solving (3.1–4). Such variants include systems of Schrödinger eigenvalue problems [53,54], eigenvalue problems with more complicated dependence on the eigenvalues [55–57], etc.

So far we have limited our exposition to differential equations. However, it has recently be discovered that there exist certain classes of singular integro-differential equations which are also solvable by the

[3] This problem was first introduced as a natural generalization of the eigenvalue problem used by Zakharov and Shabat [42] in connection with the so-called nonlinear Schrödinger equation.

IST. Kodama, Ablowitz, and Satsuma [58] have implemented the IST needed for the exact integration of the so-called intermediate long wave equation (ILWE) [58]. Here we only comment on a certain singular limit of the ILWE: the so-called B–O equation

$$u_t + 2uu_x + Hu_{xx} = 0, \qquad (Hv)(x) := \frac{1}{\pi} \int_{-\infty}^{\infty} d\xi \, \frac{v(\xi)}{\xi - x}.$$

The associated linear eigenvalue problem, obtained by Bock and Kruskal [59] and Nakamura [60], turns out to be a differential RH problem in the extended x-complex plane,

$$i\psi_x^+ + \lambda(\psi^+ - \psi^-) = -u\psi^+, \tag{3.5}$$

where ψ^+ (ψ^-) denotes a function analytic in the upper (lower) half x-complex plane. The inverse problem associated with (3.5) has been considered by the authors in [30]. This eigenvalue problem, in addition of being interesting in its own right, also provides a pivot to a multidimensional IST. This follows from the fact that many of the novel aspects of (3.5) are also present in the IST for KPI (for further details see the contribution of M. J. Ablowitz and A. S. Fokas in *this proceedings*). Proper generalization of (3.5) [corresponding to (3.2–4)] are still open (although some progress has been made in [6]).

§4 Remarks on Riemann–Hilbert boundary value problems

RH problems have played (explicitly or implicitly) an important role in soliton theory. In this section we review some basic elements of RH theory, as well as the most important RH problems encountered in soliton theory.

4.1 Classical Riemann-Hilbert problems for Hölder functions

We first introduce some definitions and results needed in the sequel.

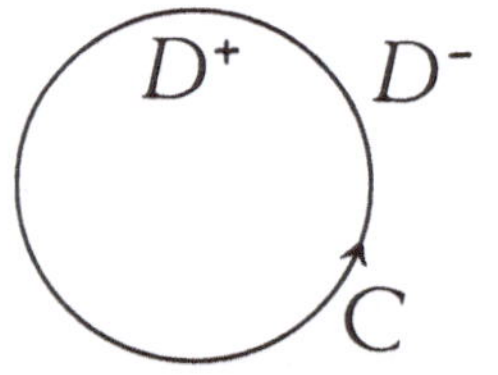

1. Let C be a simple smooth, closed contour dividing the complex z-plane into two regions D^+ and D^- (the positive direction of C will be taken as that for which D^+ is on the left; more than one contour could be considered but for simplicity here we consider only one). A scalar function $\phi(z)$ defined in the entire plane, except for points on C, will be called ***sectionally holomorphic*** if:

i. the function $\phi(z)$ is holomorphic in each of the regions D^+ and D^- except, perhaps, at $z = \infty$;

ii. the function $\phi(z)$ is sectionally continuous with respect to C, *i.e.*, as z approaches any point t on C along any path which lies wholly in either D^+ or D^-, the function $\phi(z)$ approaches a definite limiting value $\phi^+(t)$ or $\phi^-(t)$ respectively.

It then follows according to a theorem of Painlevé [62] that $\phi(z)$ is continuous in the closed region $D^+ + C$ if it is assigned the value $\phi^+(t)$ on C. Similarly for the region $D^- + C$.

2. A function $\phi(t)$ defined on C will be called ***Hölder*** if for any two arbitrary points on C

$$|\phi(t_2) - \phi(t_1)| < A|t_2 - t_1|^\lambda, \qquad A > 0, \quad 0 < \lambda \leq 1. \tag{4.1}$$

(If $\lambda = 1$ then the Hölder condition becomes the Lipschitz condition.)

3. The *index of a function $\phi(t)$ with respect to C* is the increment of its argument, in transversing the curve in the positive direction, divided by 2π, *i.e.*,

$$\operatorname{ind}\phi(t) := \frac{1}{2\pi}\,[\arg\phi(t)]_C = \frac{1}{2\pi i}\,[\ln\phi(t)]_C = \frac{1}{2\pi i}\int_C d(\ln\phi(t)). \tag{4.2}$$

4. The sectionally holomorphic function $\phi(z)$ has a *finite degree at infinity* if for some finite integer m $\lim_{z\to\infty}\phi(z)/|z|^m = 0$. Then

$$\phi(z) \sim c_k z^k + O(z^{k-1}) \qquad \text{as} \qquad z\to\infty. \tag{4.3}$$

5. The following result, due to Plemelj, is of fundamental importance in the theory of RH problems:

   Assume that $f(t)$ is Hölder on C. Then the function

$$\phi(z) := \frac{1}{2\pi i}\int_C d\tau\,\frac{f(\tau)}{\tau - z} \tag{4.4a}$$

   is a sectionally holomorphic function vanishing at infinity. Furthermore the limits $\phi^+(t)$ and $\phi^-(t)$ as t approaches C are given by

$$\phi^{\pm}(t) = \pm\tfrac{1}{2}f(t) + \frac{1}{2\pi i}\int_C d\tau\,\frac{f(\tau)}{\tau - t}, \tag{4.4b$^\pm$}$$

   where principal value integrals are assumed if needed.

   An intermediate consequence of the *Plemelj formulae* (4.4b)$^\pm$ is that any function $f(t)$, Hölder on C, can be written as

$$\phi^+(t) - \phi^-(t) = f(t), \tag{4.5}$$

   where $\phi^\pm(t)$ are the boundary values of the sectionally holomorphic function $\phi(z)$, uniquely defined in terms of $f(t)$ [assuming $\phi(z) \sim O(1/z)$ as $z\to\infty$].

6. All above notions are easily extended to vector functions $\boldsymbol{\phi}(z)$. So for example the degree of $\boldsymbol{\phi}(z)$ at infinity is the highest degree of any of its components. The Plemelj's formulae are also valid for vector functions.

Having introduced the above notions, we can now formulate the simplest RH problem.

4.1.1 Scalar homogeneous Riemann–Hilbert problem

The scalar homogeneous RH problem is defined as follows:

Given a contour C, and a function $G(t)$ which is Hölder on C with $G(t) \neq 0$ on C, find a sectionally holomorphic function $\phi(z)$, with finite degree at infinity, such that

$$\phi^+(t) = G(t)\,\phi^-(t) \qquad \text{on } C, \tag{4.6}$$

where $\phi^\pm(t)$ are the boundary values of $\phi(z)$ on C.

The solution of (4.6) can be given in closed form. This solution is essentially characterized by the index of $G(t)$: Let

$$k := \operatorname{ind} G(t). \tag{4.7}$$

Then the sectionally holomorphic function $X(z)$ defined by

$$X(z) := \begin{cases} e^{\Gamma(z)}, & z \text{ in } D^+, \\ z^{-k} e^{\Gamma(z)}, & z \text{ in } D^-, \end{cases} \qquad \Gamma(z) := \frac{1}{2\pi i} \int_C d\tau \frac{\ln\left[\tau^{-k} G(\tau)\right]}{\tau - z}, \tag{4.8}$$

is the *fundamental solution* of (4.6). Any solution $\phi(z)$ of (4.6) (with finite degree at infinity) is given by

$$\phi(z) = X(z)\, P(z), \tag{4.9}$$

where $P(z)$ is an arbitrary polynomial of z.

Remarks

i. The above results can be easily obtained by taking the ln of (4.6) and then using Plemelj's formulae. The introduction of the term τ^{-k} in (4.8) makes the function $\ln\left[\tau^{-k} G(\tau)\right]$ Hölder on C.

ii. RH problems are intimately related to singular integral equations. For this and others reasons one is interested in *finding all solutions* of (4.6) *which vanish at infinity*. Equations (4.8) and (4.9) imply that:

   a. if $k > 0$ then there exist k linearly independent solutions of (4.6) vanishing at infinity;

   b. if $k \leq 0$ then there exists no solution of (4.6) vanishing at infinity.

iii. *Inhomogeneous RH problems*

$$\phi^+(t) = g(t)\, \phi^-(t) + F(t), \tag{4.10}$$

where $F(t)$ is Hölder on C, are simply solved in terms of homogeneous RH problems. In particular replacing $G(t)$ by $X^+(t)/X^-(t)$ in (4.10) and using Plemelj's formula it follows that

$$\phi(z) = X(z)\left[\psi(z) + P(z)\right], \qquad \psi(z) := \frac{1}{2\pi i} \int_C d\tau \frac{F(\tau)}{X^+(\tau)\,(\tau - z)}. \tag{4.11}$$

Hence, if one is looking for solutions vanishing at infinity it follows that:

   a. if $k > 0$ then there exist k linearly independent solutions given by (4.11) with $P(z) = P_{k-1}(z)$;

   b. if $k = 0$ then there exists a unique solution $X(z)\psi(z)$;

   c. if $k < 0$ then there exists a unique solution $X(z)\psi(z)$ if and only if the following orthogonality conditions are valid:

$$\int_C d\tau \frac{F(\tau)\, \tau^{n-1}}{X^+(\tau)}, \qquad n = 1, 2, \ldots, -k.$$

4.1.2 Vector homogeneous Riemann–Hilbert problems

The vector homogeneous RH problem is defined as follows:

Given a contour C, and a $n \times n$ matrix $G(t)$ which is Hölder and nonsingular on C (i.e., $\{G\}_{ij}$ are Hölder and $\det G(t) \neq 0$ on C), find a sectionally holomorphic vector function $\boldsymbol{\phi}(z)$, with finite degree at infinity, such that

$$\boldsymbol{\phi}^+(t) = G(t)\boldsymbol{\phi}^-(t). \tag{4.12}$$

The solution of (4.12) cannot, in general, be found in closed form. It is characterized through a system of linear Fredholm integral equations. However, in contrast to the scalar case, where the existence or nonexistence of solutions is a priori determined in terms of the index, the existence and uniqueness of solutions of (4.12) must be investigated in terms of certain Fredholm equations. This difficulty arises from the fact that the solutions of (4.12) depend on the individual indices $k_1, \ldots, k_n$ which *cannot be a priori calculated.* Only their sum,

$$k = k_1 + \cdots + k_n = \operatorname{ind} \det G(t), \tag{4.13}$$

is a priori known.

Solving (4.12) means finding a *fundamental solution matrix* $X(z)$ composed of solution vectors $\mathbf{X}_1(z), \ldots, \mathbf{X}_n(z)$; the individual index k_l is related to the behavior of $\mathbf{X}_l(z)$ as $z \to \infty$. Suppose we are looking for solutions of (4.12) whose degree at infinity does not exceed r. Among these solutions there exist some with the lowest possible degree $-k_1$. Let $\mathbf{X}_1$ denote a solution with degree $-k_1$. From the remaining solutions consider all those which cannot be obtained from $\mathbf{X}_1$ by $\boldsymbol{\phi}(z) = P(z)\mathbf{X}_1(z)$, where $P(z)$ is any polynomial. Among these solutions pick one with the lowest possible degree. Call $\mathbf{X}_2$ and $-k_2$ the solution and degree respectively, etc. It can be shown that the solution matrix $X(z)$ constructed this way has the following two properties:

i. $\det X(z) \neq 0$ for all finite z, and

ii. $\det(z^{k_1}\mathbf{X}^1, \ldots, z^{k_n}\mathbf{X}^n) \neq 0$ at infinity.

Furthermore, any solution of (4.12) is given by

$$\boldsymbol{\phi}(z) = X(z)\mathbf{P}(z), \tag{4.14}$$

where $\mathbf{P}(z)$ is a polynomial vector.

The above nonconstructive approach for determining $X(z)$ can be turned into a constructive one provided that k_l, $l = 1, \ldots, n$ are known. For example, suppose that $k_1 = k_2 = \cdots = k_n = 0$ (a necessary but not sufficient condition, for this is $\operatorname{ind}\det G(t) = 0$). Then using (4.12) and

$$\tfrac{1}{2}\mathbf{X}_l^-(t) + \frac{1}{2\pi i}\int_C d\tau\, \frac{\mathbf{X}_l^-(\tau)}{\tau - t} - \mathbf{I}_l = 0 \tag{4.15}$$

(which is a necessary and sufficient condition for $\mathbf{X}_l^-(z)$ to be holomorphic in D^- and tend to the unit l^{th} vector $\mathbf{I}_l$ at infinity), it follows that $\mathbf{X}_l^-$ is found from the vector Fredholm integral equation

$$\mathbf{X}_l^-(t) - \frac{1}{2\pi i}\int_C d\tau\, \frac{G^{-1}(t)\,G(\tau) - I}{\tau - t}\mathbf{X}^-(\tau) = \mathbf{I}_l, \qquad l = 1, \ldots, n. \tag{4.16}$$

Conversely if the linear Fredholm integral equation (4.16) have *unique* solutions $\mathbf{X}_l^-$, $l = 1, \ldots, n$, then all the individual indices k_l, $l = 1, \ldots, n$ are zero. To find k_l, in general one has to investigate integral equations of the type (4.16) (with some arbitrary forcing instead of $\mathbf{I}_l$).

Remarks

i. To establish the above results one introduces several RH problems related to (4.12) —so-called adjoint, accompanying or associate— and then appropriately uses these problems and the known results of Fredholm theory (in particular one makes extensive use of Fredholm's alternative theorem).

ii. The results expressed by remarks **ii** and **iii** of the scalar case are easily generalized to the vector case [for example $G(t) = X^+(t)/X^-(t)$].

4.1.3 A note on the history of Riemann–Hilbert problems

The RHBVP[4] was first introduced by Riemann [63] in connection with the so-called ***Riemann monodromy*** (RM) problem. This problem is defined roughly as follows:

*Given branch points $\{a_1, \ldots, a_n\}$in the complex plane and $m \times m$ matrices $M_1, \ldots, M_n$ characterizing the **monodromy group** of $a_1, \ldots, a_n$ find all linear ordinary differential equations (ODE's)*

$$\frac{dy}{dx} = \sum_{i=1}^{n} \frac{A_i}{x - a_i} y, \qquad A_i : \quad m \times m \text{ matrices} \tag{4.17}$$

which possess the monodromy group generated by M_i, $i = 1, \ldots, n$.

The connection between the RH and RM problems as well as their implications to soliton theory is truely fascinating [64][5]. The RM problem was reduced to a RH problem by Hilbert [67], Plemelj [68], and Birkoff [69]. In this way one obtains a vector RH problem with *discontinuous coefficients* (see below). Actually Plemelj has extensively used the results on RH problems to study the RM problem. Parenthetically we mention that Lappo Danilevskij studied the RM problem in terms of hyperlogarithm functions [70]. More recently Sato *et al.* [66], using the equivalence between the RM problem and the so-called Schlesinger equations, studied the RM problem using classical operators of field theory.

Riemann made no attempt to solve the problem formulated by himself. The first solution of the homogeneous scalar RH problem (4.6) was given by Hilbert in 1905 in terms of a Fredholm integral equation (Picard in 1927 proceeded also along similar lines). Plemelj in 1907 gave the first closed form solution of (4.6) in the case that $k = 0$. Carleman in 1932 solved a particular singular integral equation (as it was pointed out earlier, singular integral equations are related to RH problems). Gakhov in 1938 gave the full solution of (4.6). The vector RH problem (4.12) was considered by Plemelj, Gakhov, Muskhelishvili, and exaustively by Vekua. One should also mention that the work of Wiener and Hopf is also related to RH problems.

Discontinuous RH problems, *i.e.*, problems where $G(t)$ is Hölder on L except on a finite number of points where it is discontinuous, have also been extensively considered in the literature. A complete solution of the scalar discontinuous RH problem was given in 1941 independently by Gakhov and Muskhelishvili. These authors used two rather different methods; Gakhov's method was generalized by Vekua to the vector discontinuous RH problem[6]

4.2 Some results of Krein and Gohberg

After the classical work on RH problems for Hölder functions reviewed in §4.1, many investigators have considered RH problems for functions in various other spaces. An up to date review of these results can be found in [74]. Here we only mention some results taken from the papers of Krein [75] and Gohberg [76]. These results:

i. serve to illustrate further the notion of index;

ii. contain the only, to our knowledge, result providing an *a priori* determination of all individual indices[7]

---

[4]Sometimes referred to in the literature as the ***Hilbert*** or ***Riemann*** problem.

[5]For the role of the RM problem in soliton theory see [78,65,66].

[6]A complete bibliography on the above work is contained in the books of Gakhov [71], Muskhelishvili [72], and Vekua [73].

[7]We thank P. Deift for bringing this result to our attention as well as for pointing out to us its relevance to the solution of the inverse problem of the Schrödinger equation [77].

We first introduce some notation. Let L_1 denote the normed linear ring of all measurable absolutely integrable functions with norm $\|f\|_{L_1} := \int_{-\infty}^{\infty} dt\,|f(t)|$. Let $F(\lambda)$ be the Fourier transform of $f(t)$, where $f(t) \in L_1$ and R^0 the ring of $F(\lambda)$. Let R denote the ring of $F(\lambda)$, where $F(\lambda)$ denotes the Fourier transform of $a\delta(t) + f(t)$ and a any complex number, *i.e.*, R is the set of all functions of the form

$$a + \int_{-\infty}^{\infty} dt\, f(t) e^{i\lambda t}, \qquad f(t) \in L_1. \tag{4.18}$$

Here R_+ is the subset of R consisting of all functions of the form $a + \int_0^{\infty} dt\, f(t)e^{i\lambda t}$. It is clear that functions in R_+ are analytically continuable in the upper half λ-plane π_+. Similar considerations apply to R_- and π_-.

Gohberg and Krein [76] were interested in the solution of

$$\chi(t) - \int_0^{\infty} ds\, k(t-s)\chi(s) = f(t), \tag{4.19}$$

where $\chi(t)$ and $f(t)$ are vectors, and $k(t)$ is an $n \times n$ matrix with elements in L_1. Assume for concreteness that $f(t) \in L_{1(n\times 1)}(0,\infty)$ and seek solutions $\chi(t) \in L_{1(n\times 1)}(0,\infty)$. The solution of (4.19) is intimately related to the vector RH problem

$$\phi^+(\lambda) = G(\lambda)\phi^-(\lambda), \qquad G(\lambda) \in R_{(n\times n)}, \quad \det G(\lambda) \neq 0, \quad \lambda \in L:\ \{-\infty,\infty\}. \tag{4.20}$$

We note that (4.20) is not a special case of (4.12) since $G(\lambda)$ is continuous but, in general, not Hölder. Gohberg and Krein have developed a beautiful theory for analyzing (4.19) and (4.20). Here we only touch upon this theory; in particular we concentrate only on some of the results concerning (4.20). For completeness we first review the scalar case —corresponding to (4.19) being a scalar equation.

4.2.1 Scalar factorization problem

Assume that the scalar function $G(\lambda) \in R$ is defined on the closed line $L:\ \{-\infty,\infty\}$ and $G \neq 0 \in L$. The basic factorization problem associated with G, which as we shall see is related to a RH problem, can be formulated as follows: Find $G_\pm \in R_\pm$ such that

$$G = G_+G_-, \qquad G_\pm(\infty) = 1. \tag{4.21}$$

We call a factorization ***proper*** if either $G_+ \neq 0 \in \pi_+$ or $G_- \neq 0 \in \pi_-$. We call a factorization ***canonical*** if both $G_\pm \neq 0 \in \pi_\pm$.

Krein [75], using fundamental theorems of Wiener and Levy, has proven that:

1. The function $G(\lambda)$ (specified above) admits a canonical factorization if and only if

$$G(\lambda) \neq 0 \qquad \text{and} \qquad \operatorname{ind} G = 0.$$

If it admits a canonical factorization then this is its only proper one.

2. Suppose that

$$G \neq 0 \qquad \text{and} \qquad \operatorname{ind} G = k \neq 0.$$

i. If $k > 0$ then there exists a choice of points $\alpha_1, \ldots, \alpha_m \in \pi_+$ and integers $p_1, \ldots, p_m$ satisfying $p_1 + \cdots + p_m = k$, such that there exists a function $G_+ \in R_+$ with zeros at $\alpha_1, \ldots, \alpha_m$ and no other zeros in π_+. Furthermore, there exists a function $G_- \in R$ such that $G = G_+G_-$ and G_- has no zeros in π_-.

ii. If $k < 0$ analogous statements are valid (of course now $p_1 + \cdots + p_m = -k$).

Remarks

i. Equation (4.21) can be written as

$$G_+ = G(G_-)^{-1}, \qquad G_\pm(\infty) = 1. \tag{4.22}$$

Then assuming, for the sake of comparison, that G is also Hölder, equation (4.22) becomes a RH problem with the *jump* given along the real axis and the boundary condition of boundness at infinity. Suppose that $k > 0$, then G_- has no zeros in π_- and hence $(G_-)^{-1} \in R_-$. Thus $\phi^+(\lambda) = G_+(\lambda)$, $\phi^-(\lambda) = (G_-(\lambda))^{-1}$. However, if $k < 0$ then $(G_-)^{-1}$ is not holomorphic in π_- and the RH (4.22) has no solution bounded at infinity. This is consistent with the scalar RH theory.

ii. The functions G_+ and G_- clearly admit the following representations

$$G_+(\lambda) = 1 + \int_0^\infty dt\, e^{i\lambda t}\gamma_+(t), \qquad \lambda \in \pi_+,$$
$$G_-(\lambda) = 1 + \int_0^\infty dt\, e^{-i\lambda t}\gamma_-(t), \qquad \lambda \in \pi_-,$$

where $\gamma_\pm \in L_1$. The above formulae are useful in establishing various properties of $G_\pm$, however they have the disadvantage that they are not constructive. Krein [75] also gives constructive formulae for $G_\pm$ which coincide with the ones obtained via RH theory.

4.2.2 A vector Riemann–Hilbert problem with all its indices positive

A fundamental result of [76] is a factorization theorem for matrices $G(\lambda) \in R_{(n\times n)}$. In preparation for this result we first consider a vector RH problem with all its indices positive.

Consider the adjoint of the homogeneous equation corresponding to (4.19)

$$\boldsymbol{\psi}(t) - \int_0^\infty ds\, k'(t-s)\,\boldsymbol{\psi}(s) = 0. \tag{4.23}$$

Equations (4.19) and (4.23) are intimately related to the RH problem

$$\mathbf{F}_-(\lambda) = \big(I - K(\lambda)\big)\mathbf{F}_+(\lambda), \qquad \mathbf{F}_\pm \in R^\pm_{(n\times 1)}, \tag{4.24}$$

where $K(\lambda)$ is the Fourier transform of $k(t)$ [recall that $k(t) \in L_{1(n\times n)}$]. If (4.23) has only the trivial solution then all the individual indices of the RH problem (4.24) are positive. Actually, one may prove the following:

Assume that (4.23) has only the trivial solution. Then

1. The RH problem (4.24) has n solutions in $R^+_{(n\times 1)}$ whose values at infinity are linearly independent.

2. The multiplicity of any zero λ_0, $\operatorname{Im}\lambda_0 \geq 0$, of $\mathbf{F}_+(\lambda) \in R^+_{(n\times 1)} \leq \alpha$, where α is the number of linearly independent solutions of the homogeneous version of (4.19).

   In the above case one may also introduce the notion of a *standard solution matrix* which is analogous to the fundamental solution matrix introduced in §4.1. This matrix F is defined as follows:

   i. F_+ solves (4.24);

   ii. $\det F_+(\infty) \neq 0$;

iii. Assume that the jth column vector of F_+ has a zero at $\lambda = i$ of multiplicity k_j. Choose the columns of F_+ in such an order that $k_1 \geq k_2 \geq \cdots \geq k_n$. Define $k = (k_1, k_2, \ldots, k_n)$ to be the index of (4.24). Because of the result (2) above, k is bounded and hence from all solution matrices in $R^+_{(n\times n)}$ there must be one with the greatest index. This is the standard solution matrix.

The following results are proven in [76] about the above standard solution matrix.

3. Let $F_+(\lambda)$ be the standard solution matrix of (4.24). Then $\det F_+(\lambda) \neq 0$ for every $\lambda \in \pi_+$, with the exception of $\lambda = i$ where the multiplicity of zeros of $\det F_+(\lambda)$ is equal to $\sum_{j=1}^n k_j$. Furthermore,

$$\sum_{j=1}^{n} k_j = -\operatorname{ind}\det(I - K(\lambda)) \tag{4.25}$$

and hence $\det F_-(\lambda) \neq 0$ for every $\lambda \in \pi_-$.

4. Let $F_+(\lambda) = (\phi_1, \ldots, \phi_j, \ldots, \phi_n)$ be the standard solution matrix of (4.24). Then every solution of (4.24) has the form

$$\mathbf{F}(\lambda) = \sum_{j=1}^{n} \left(a_{j0} + a_{j1}(\lambda - i)^{-1} + \cdots + a_{jk_j}(\lambda - i)^{-k_j}\right)\phi_j(\lambda).$$

4.2.3 A matrix factorization theorem

1. *Every nonsingular matrix $G(\lambda) \in R_{(n\times n)}$ possesses the following factorization:*

$$G(\lambda) = G_+(\lambda)\,\Delta(\lambda)\,G_-(\lambda), \tag{4.26}$$

where $G_\pm \in R^\pm_{(n\times n)}$, $\det G_\pm(\lambda) \neq 0$ for $\lambda \in \pi_\pm$, and $\Delta(\lambda) = \operatorname{diag}\left(\left(\frac{\lambda - i}{\lambda + i}\right)^{k_1}, \ldots, \left(\frac{\lambda - i}{\lambda + i}\right)^{k_n}\right)$ for some integers $k_1 \geq k_2 \geq \cdots \geq k_n$.

2. *If a matrix possesses another factorization $G = \tilde{G}_+(\lambda)\,\tilde{\Delta}(\lambda)\,\tilde{G}_-(\lambda)$ then $\tilde{\Delta}(\lambda) = \Delta(\lambda)$, $\tilde{G}_+(\lambda) = G_+A(\lambda)$, and $\tilde{G}_-(\lambda) = A^{-1}G_-(\lambda)$ for some nonsingular matrix A.*

3. $$\sum_{j=1}^{n} k_j = \operatorname{ind}\det G(\lambda). \tag{4.27}$$

4.2.4 A theorem about indices

Let m^* denote the Hermitian conjugate of m, *i.e.*, $m^* = \overline{m'}$, where the prime denotes the transpose and the overline the complex conjugate. Define the real and imaginary parts of a matrix through

$$m_R := \tfrac{1}{2}(m + m^*), \qquad m_I := \tfrac{1}{2}(m - m^*). \tag{4.28}$$

A matrix m is definite if $\xi^* m \xi$ is real and has only one sign for all $\xi \neq 0$. The following result is valid [76]:

If the real or imaginary part of $m(\lambda) \in R_{(n\times n)}$ is definite and $\det m(\lambda) \neq 0$ then all its individual indices are zero.

The above result is a simple corollary of the matrix factorization theorem presented in §4.2.3.

4.3 Riemann–Hilbert problems appearing in inverse scattering transforms

The following types of RH problems have been used in connection with the exact integration of nonlinear equations.

1. **Scalar RH problems with a "shift"**, *e.g.*, the RH problem associated with the KdV is of the form

$$\psi^+(k) = \psi^+(-k) + C(k)\psi^+(k), \qquad k \text{ real}. \tag{4.29}$$

It is clear that $\psi^+(-k)$ is a *minus* function, hence by considering (4.29) together with the equation obtained from (4.29) by letting $k \mapsto -k$, we obtain a standard vector RH problem. The RH problem associated with the so-called ILWE is also of a *shift* type.

2. **Vector RH problems**, *e.g.*, the RH problem associated with AKNS or with the $n \times n$ of the AKNS (3.4) where all enties of J are either real or imaginary.

3. **Vector RH problems on complicated contours**, *e.g.*, the general case of the eigenvalue problem (3.4). In this case the RH problem is defined along rays which are determined by J.

4. **Discontinuous vector RH problems on complicated contours**, *e.g.*, the RH problem associated with the initial value problem of Painlevé II equation [78]. In this case the matrix $G(k)$ is discontinuous at the origin and the jump conditions are defined on a set of rays.

5. **Scalar differential RH problems**, *e.g.*, the RH problem associated with the eigenvalue problem (time-independent part of the Lax pair) or B–O [see (3.5)].

6. **Nonlocal scalar and vector RH problems**. The inverse problem of B–O and of KPI give rise to scalar nonlocal RH problemss. Similarly the inverse problem associated with hyperbolic systems in the plane gives rise to nonlocal vector RH problems of the form

$$\mu^+(k) = \mu^-(k) + \int_{-\infty}^{\infty} dl\, \mu^-(l) F(l,k).$$

A class of generalized operator RH problems is considered in [79].

§5 Inverse scattering transform for the Korteweg–de Vries equation formulated as a Riemann–Hilbert problem

Recall that the Lax pair of the KdV is given by

$$\begin{cases} v_{xx} + \left(u(x,t) + k^2\right)v = 0, & (5.1) \\ v_t = (u_x + \nu)v + (4k^2 - 2u)v_x, & (5.2) \end{cases}$$

where ν is an arbitrary constant parameter. As it was pointed out earlier most of the analysis of the IST is carried out on (5.1). Hence, for the time being we suppress the time dependence.

5.1 Bounded eigenfunctions and their relationship

Let ϕ and $\overline{\phi}$ denote *left* eigenfunctions, while ψ and $\overline{\psi}$ denote *right* ones. They are specified by the following asymptotic behavior,

$$\begin{aligned} &\phi \to e^{-ikx}, \quad \overline{\phi} \to e^{ikx} \qquad \text{as } x \to -\infty, \\ &\psi \to e^{ikx}, \quad \overline{\psi} \to e^{-ikx} \qquad \text{as } x \to \infty, \end{aligned} \tag{5.3}$$

i.e., ϕ, for example, is that solution of (5.1) which also tends to e^{-ikx} as $x \to -\infty$ (it turns out that $\overline{\phi}$ will not be needed in the sequel).

Using the linear independence property of the solutions of the second order ODE (5.1), we obtain the scattering equation

$$\phi(k,x) = a(k)\,\overline{\psi}(k,x) + b(k)\,\psi(k,x). \tag{5.4}$$

It turns out that it is more convenient to work with eigenfunctions normalized to unity at infinity, hence let

$$M := \phi e^{ikx}, \qquad N := \psi e^{-ikx}, \qquad \overline{N} := \overline{\psi} e^{ikx}. \tag{5.5}$$

Equations (5.4) and (5.5) imply

$$\frac{M(k,x)}{a(k)} = \overline{N}(k,x) + \rho(k)e^{2ikx}N(k,x), \qquad \rho(k) := \frac{b(k)}{a(k)}. \tag{5.6}$$

Equation (5.6) is the central equation associated with the inverse problem of (5.1). Our goal is to view (5.6) as a RHBVP in the k-complex plane, where:

i. x is a parameter;

ii. a and ρ are given;

iii. M, $\overline{N}$, and N are to be found.

In order to achieve our goal we need:

i. to establish a symmetry relationship between N and $\overline{N}$;

ii. to find the analytic properties of M, N, a, and ρ with respect to k.

5.1.1 A "symmetry" relationship

Under the discrete transformation $k \mapsto -k$, equation (5.1) remains invariant and the boundary condition associated with ψ goes to that of $\overline{\psi}$. Hence $\psi(k,x) = \overline{\psi}(-k,x)$ which implies

$$\overline{N}(k,x) = N(-k,x). \tag{5.7}$$

5.1.2 The analytic properties of M, N, and a

The analytic properties of M and N are most easily found by characterizing M and N in terms of linear integral equations. Let us work, for example, with M. Since ϕ solves (5.1), it follows that M solves

$$M_{xx} - 2ikM_x = -uM. \tag{5.8}$$

We seek a bounded solution of (5.8) in the form

$$M(k,x) = 1 + \int_{-\infty}^{\infty} d\xi\, G(x,\xi,k)\,u(\xi)\,M(k,\xi), \tag{5.9}$$

where $G(x, \xi, k)$ is the Green's function solving $G_{xx} - 2ikG_x = -\delta(x - \xi)$, i.e.,

$$G(x, \xi, k) = \frac{1}{2\pi} \int_C dP \, \frac{e^{iP(x-\xi)}}{P(P - 2k)}, \tag{5.10}$$

where C is an appropriate contour. This contour is chosen from the requirement that $G = 0$ for $\xi > x$, since $\lim_{x \to -\infty} M = 1$ [therefore equation (5.9) should only involve an integral from $-\infty$ to x]. Thus C must be taken as

$$C: \quad -\infty \underset{\underset{P=0}{\smile}}{\rule{5cm}{0.4pt}} \underset{\underset{P=2k}{\smile}}{\rule{5cm}{0.4pt}} \infty, \tag{5.11}$$

and

$$G(x, \xi, k) = \begin{cases} 0, & \xi > x, \\ \dfrac{i}{2k}(-1 + e^{2ik(x-\xi)}), & \xi < x. \end{cases} \tag{5.12}$$

Hence

$$M(k, x) = 1 + \frac{i}{2k} \int_{-\infty}^{x} d\xi \, (-1 + e^{2ik(x-\xi)}) u(\xi) M(k, \xi). \tag{5.13}$$

Equation (5.13), for an appropriate small norm of u, is a Volterra integral equation. Furthermore its kernel is a $(+)$ function with respect to k, i.e., its kernel is holomorphic in the upper half of the k-complex plane. *Hence its solution $M(k, x)$ is also a $(+)$ function* with respect to k. Similarly, one may establish that N satisfies

$$N(k, x) = 1 + \frac{i}{2k} \int_{x}^{\infty} d\xi \, (-1 + e^{-2ik(x-\xi)}) u(\xi) N(k, \xi), \tag{5.14}$$

which implies that $N(k, x)$ is also a $(+)$ function with respect to k.

To establish the analytic properties of $a(k)$ we use the following integral representations of $a(k)$ and $b(k)$:

$$\begin{aligned} a(k) &= 1 - \frac{i}{2k} \int_{-\infty}^{\infty} d\xi \, u(\xi) M(k, \xi) \\ b(k) &= \frac{i}{2k} \int_{-\infty}^{\infty} d\xi \, u(\xi) M(k, \xi) e^{-2ik\xi}. \end{aligned} \tag{5.15}$$

From the above representation it follows that $a(k)$ *is a $(+)$ function* [while $b(k)$ cannot, in general, be analytically continued off real k].

To derive the above results let $\Delta(k, x) := M(k, x) - a(k) N(-k, x)$. Then using (5.13) and (5.14) it follows that Δ satisfies

$$\begin{aligned} \Delta(k, x) = 1 - a(k) &+ \frac{i}{2k} \int_{-\infty}^{\infty} d\xi \, (-1 + e^{2ik(x-\xi)}) u(\xi) M(k, \xi) \\ &- \frac{i}{2k} \int_{x}^{\infty} d\xi \, (-1 + e^{2ik(x-\xi)}) u(\xi) \Delta(k, \xi). \end{aligned} \tag{5.16}$$

On the other hand, using (5.6), $\Delta = b(k) e^{2ikx} N(k, x)$ and therefore it satisfies [using (5.14)]

$$\Delta(k, x) = b(k) e^{2ikx} - \frac{i}{2k} \int_{x}^{\infty} d\xi \, (-1 + e^{2ik(x-\xi)}) u(\xi) \Delta(k, \xi). \tag{5.17}$$

Comparing (5.16) and (5.17) one finds (5.15).

Let us now return to equation (5.6) with $\overline{N}(k,x) = N(-k,x)$;

$$\frac{M(k,x)}{a(k)} = N(-k,x) + \rho(k)e^{2ikx}N(k,x). \tag{5.18}$$

It can be shown that $a(k)$ may have simple zeros $k_1, \ldots, k_n$ in the upper half k-complex plane. Hence in general M/a will be meromorphic in the upper half k-plane. So let

$$\frac{M(k,x)}{a(k)} = \mathcal{M}(k,x) + \sum_{j=1}^{n} \frac{A_j(x)}{k-k_j}, \tag{5.19}$$

where $\mathcal{M}(k,x)$ is a (+) function in k. Integrating Eq. (5.19) around k_j it follows that $A_j(x) = C_j\, e^{2ik_jx} \times N(k_j,x)$. Hence equation (5.18) yields

$$\mathcal{M}(k,x) = N(-k,x) + \sum_{j=1}^{n} \frac{C_j\, e^{2ik_jx} N(k_j,x)}{k-k_j} + \rho(k)e^{2ikx}N(k,x). \tag{5.20}$$

5.2 The inverse problem

Equation (5.20) defines a RH problem in terms of the scattering data $\{k_j, C_j\}_{j=1}^{n}$ and $\rho(k)$. Actually, by letting $k \mapsto -k$ in (5.20) one obtains a system of RH problems for the (+) and (−) vectors

$$\begin{pmatrix} \mathcal{M}(k,x) \\ N(k,x) \end{pmatrix}, \qquad \begin{pmatrix} \mathcal{M}(-k,x) \\ N(-k,x) \end{pmatrix}.$$

It is interesting that one can very easily establish the uniqueness and existence of the solution to this system of RH problems. This follows by calling upon the theorem of Gohberg and Krein [76] (see §4.2) about definite matrices, to prove that the relevant individual indices are zero.

Equation (5.20) yields the following integral equation for $N(k,x)$:

$$N(k,x) - \frac{1}{2\pi i}\int_{-\infty}^{\infty} dl\, \frac{\rho(l)e^{2ilx}N(l,x)}{l+k+i0} = 1 - \sum_{j=1}^{n} \frac{C_j\, e^{2ik_jx}}{k+k_j} N(k_j,x). \tag{5.21}$$

The above equation is obtained by taking the (−) projection of (5.20) and then letting $k \mapsto -k$.

Finally the potential $u(x)$ can be reconstructed from the following expression:

$$u = -2\frac{\partial}{\partial x}\left[\frac{1}{2\pi}\int_{-\infty}^{\infty} dl\, \rho(l)e^{2ilx}N(l,x) - i\sum_{j=1}^{n} C_j\, e^{2ik_jx}N(k_j,x)\right]. \tag{5.22}$$

Equation (5.22) follows by considering the large k asymptotics of (5.21) and then using [see (5.14)]

$$\lim_{k\to\infty} N(k,x) = 1 - \frac{i}{2k}\int_x^{\infty} d\xi\, u(\xi).$$

Let us summarize the solution of the inverse problem: Given the scattering data $\{k_j, C_j\}_{j=1}^{n}$ and $\rho(k)$, equation (5.21) yields $N(k,x)$. Then equation (5.22) implies $u(x)$.

5.3 The time dependence

Assume now that u depends on t. Then the scattering data will also depend on t. This dependence will be in general complicated. However, if u evolves in t according to the KdV, then this dependence is very simple:

$$\left\{\begin{array}{l}\dfrac{\partial k_j}{\partial t}=0\\ C_j(t)=C_j(0)e^{8ik_j^3t}\end{array}\right\}_{j=1}^{n}, \qquad \rho(k,t)=\rho(k,0)e^{8ik^3t}. \tag{5.23}$$

Using equations (5.23) and the solution of the inverse problem associated with (5.1), one may easily solve the initial value problem of the KdV: Given $u(x,0)$ compute the initial scattering data k_j, $C_j(0)$, and $\rho(k,0)$. Then use (5.23) to compute k_j, $C_j(t)$, and $\rho(k,t)$. Then use (5.21) to obtain $N(k,x,t)$ and finally use (5.22) to obtain $u(x,t)$.

Equations (5.23) can be derived as follows: By considering (5.2) with $v=\phi$ as $x\to-\infty$ it follows that $\nu=4ik^3$. Also (5.4), as $x\to\infty$, implies

$$\phi\sim ae^{-ikx}+be^{ikx}. \tag{5.24}$$

Finally using (5.24) in (5.2) [with $\nu=4ik^3$] as $x\to\infty$, one obtains

$$a_t=0, \qquad b_t=8ik^3b.$$

Hence $\rho(k,t)=\rho(k,0)e^{8ik^3t}$. The evolution of $C_j(t)$ is obtained in a similar way.

Remarks

i. Pure soliton solutions correspond to $\rho=0$. Hence they are characterized through the following linear system of algebraic equations [see (5.21)]:

$$N(k_j,x)=1-\sum_{l=1}^{n}\frac{C_l\,e^{2ik_lx}N(k_l,x)}{k_j+k_l}, \tag{5.25}$$

$$u=-2i\frac{\partial}{\partial x}\sum_{j=1}^{n}C_j\,e^{2ik_jx}N(k_j,x). \tag{5.26}$$

ii. The connection with the direct linearization for the KdV given in §2 [see (2.1) and (2.2)] is established by considering (5.21) and (5.22): Equation (5.21) can be written as

$$N(k,x)-\frac{1}{2\pi i}\int_{L_1}d\varsigma(l)\,\frac{e^{2ilx}N(l,x)}{l+k+i\epsilon}=1, \tag{5.27}$$

where

$$d\varsigma(l)=\left\{\begin{array}{ll}\rho(l)dl, & \text{for } l \text{ real}\\ -2\pi iC_j\delta(l-k_j)dl, & \text{for } l \text{ imaginary.}\end{array}\right. \tag{5.28}$$

Equation (2.1) follows by considering (5.27) with an *arbitrary* measure-contour —and making a simple exponential transformation. Similarly, equation (2.2) follows from (5.22).

iii. The GLM equation can also be derived from (5.27). Assuming

$$N(k,x)=1+\int_x^{\infty}ds\,K(x,s)e^{ik(s-x)}, \tag{5.29}$$

substituting (5.29) in (5.27) and operating with $\frac{1}{2\pi}\int_{-\infty}^{\infty} dk\, e^{ik(x-y)}$ one obtains

$$K(x,y)+F(x+y)+\int_x^{\infty} ds\, K(x,s)\, F(s+y)=0, \qquad y>x, \tag{5.30}$$

where

$$F(x):=\frac{1}{2\pi}\int_{L_1} d\varsigma(k)\, e^{ikx}. \tag{5.31}$$

Similarly equation (5.22) yields

$$u(x)=2\frac{\partial}{\partial x}K(x,x). \tag{5.32}$$

§6 The inverse scattering transform for $2+1$ dimensional problems

It is well known [20] that, just as in one spatial dimension, many physically important $2+1$ nonlinear equations are related to Lax pairs, *i.e.*, to pairs of linear systems the solvability condition of which implies the given equation.

In particular certain two-dimensional analogues of the KdV, the so-called KPI and KPII equations [80],

$$(u_t+6uu_x+u_{xxx})_x=-3\sigma^2 u_{yy}, \qquad \sigma=i \text{ or } -1,$$

have been related to [81],

$$\sigma\psi_y+\psi_{xx}+(u+\lambda)\psi=0, \qquad \sigma=i \text{ or } -1 \tag{6.1}$$

(KPI corresponds to $\sigma=i$ and KPII to $\sigma=-1$). Equation (6.1) can be thought of as a two-dimensional generalization of (3.1) (note that in the case of $\sigma=i$, equation (6.1) becomes the *time*-dependent Schrödinger equation).

Similarly Ablowitz and Haberman [49] proposed a two-dimensional generalization of (3.4) in the form

$$\psi_x=\lambda B\psi+Q\psi+J\psi_y, \tag{6.2}$$

where B is a constant diagonal matrix, J is a constant diagonal matrix either purely real (hyperbolic case) or purely imaginary (elliptic case) and $Q=Q(x,y)$ is an off-diagonal matrix containing the potentials. Furthermore, several investigators have established that many physically interesting $2+1$ evolution equations are related to (6.2); among them the Davey–Stuartson (DS) equation [82] with or without surface tension (the DS equation is the long wave limit of the Benney–Roskes equation [83]), the N-wave interaction in $2+1$, and the modified KPI and modified KPII.

However, the question of finding a suitable scheme for solving the initial value problems of such equations remained largely open for a rather long time. In this regard we mention that some progress had been made in connection with KPI [84,85] and with the three-wave interaction [86]. However, it was not clear from this work that a viable unified scheme to handle these and other equations could be obtained. In particular in Manakov's treatment of KPI:

i. the usual IST had to be supplemented with solving an additional pair of GLM-type equations in scattering space;

ii. the lump solutions (algebraically decaying solutions) were excluded.

We point out that the manifestation of lumps is one of the novel aspects of IST in 2+1.[8] Similarly Kaup's treatment of the three-wave interaction exploits crucially the existence of characteristic coordinates. Furthermore, what is perhaps more important, the IST has been considered, so far, within the framework of RH problems —local in 1+1, nonlocal in KPI. However, this framework seems inadequate for handling other types of two-dimensional problems, *e.g.*, KPII.

We have recently developed a method for

i. solving certain inverse problems in the plane;

ii. solving initial value problems (for initial data decaying at infinity) of certain nonlinear equations in 2 + 1. This method has emerged from our treatment of several concrete two-spatial dimensional problems. Our program of study began with the B–O equation [30] which, although is a one-dimensional equation, has all the essential features of a two-dimensional problem (this is a consequence of its nonlocal character). Equation (6.1) with $\sigma = i$ and the related KPI were considered in [36,87]. Equation (6.1) with $\sigma = -1$ and KPII were considered by Bar Yaacov and the authors in [88]. We stress that the treatment of KPII was of crucial importance in the development of our method, since it was the first case to discover the inadequacy of the RH problem formulation of the IST. The hyperbolic as well as the elliptic versions of (6.2) have been considered in [89–91].

We note that in all of the above problems we make some assumptions about the compactness of certain operators, the existence or nonexistence of eigenvalues, etc. Some of these assumptions can be directly justified by assuming appropriate smothness of $u(x,y)$ and that a certain norm of $u(x,y)$ is sufficiently small. Definitely, a rigorous investigation of these questions is needed. We have recently, together with Bar Yaacov, undertaken such an investigation.

Generally speaking our method involves the following steps:

i. Define an eigenfunction $\mu(x,y,k)$ which is bounded for all complex values of the spectral parameter k, and which is appropriately normalized. This eigenfunction is usually defined in terms of a Fredholm linear integral equation of the second type and it may have different representations in different sections of the complex k-plane. The above Fredholm integral equation may have homogeneous solutions. These homogeneous modes, corresponding to discrete eigenfunctions, give rise to lumps, *i.e.*, algebraically decaying solitons.

ii. Compute $\partial\mu/\partial\bar{k}$. This is in general expressed in terms of some other bounded eigenfunction, which we call $N(x,y,l,k)$, and appropriate scattering data.

iii. Employ a suitable ***symmetry*** relationship between N and μ to express $\partial\mu/\partial\bar{k}$ in terms of μ and appropriate scattering data. This symmetry relation may be ***discrete***, *e.g.*, KPII, or differential, *e.g.*, KPI.

iv. Use the following extension of Cauchy's formula [92],

$$\mu(x,y,k) = \frac{1}{2\pi i}\iint_R dz \wedge d\bar{z}\,\frac{1}{z-k}\,\frac{\partial\mu(x,y,z)}{\partial\bar{z}} + \frac{1}{2\pi i}\int_C dz\,\frac{\mu(x,y,z)}{z-k}, \tag{6.3}$$

to find a linear integral equation for $\mu(x,y,k)$ in k. This equation is uniquely defined in terms of the above mentioned scattering data. Typically the second integral is the identity as μ is normalized to unity for large z.

v. Calculate the potential $u(x,y)$ directly from the solution of the inverse problem —typically given by integrals over the $\mu(x,y,k)$ and the scattering data.

---

[8] For a further discussion of Manakov's results on KPI see [87].

vi. In order to solve the initial value problem of some related nonlinear evolution equation, one needs only to find the evolution of the scattering data. This is straightforward, and in all concrete cases the scattering data evolves simply and the initial scattering data is expressed in terms of $u(x,y,0)$. Hence equation (6.3) is uniquely defined in terms of the initial data.

We note that in some problems, *e.g.*, B–O and KPI, $\mu(k)$ is sectionally meromorphic, *i.e.*, it is holomorphic, modulo poles, in regions of the complex plane separated by certain contours and it has a jump across these contours. In these cases $\partial\mu/\partial\bar{k}$ is zero everywhere except on the pole locations and on the above contours, and equation (6.3) reduces to a RH problem.

In what follows we shall illustrate the above method by recalling the main steps involved with the inverse problems of equations (6.1) and (6.2). Proofs, in general, will be omitted since they can be found in cited papers.

§7 On the inverse scattering transform of the "time"-dependent Schrödinger equation and Kadomtsev–Petviashvili I equation

In this section we present the results of Fokas and Ablowitz contained in references [36] and [87]. We consider the inverse problem associated with

$$i\mu_y + \mu_{xx} + 2ik\mu_x = -u\mu. \tag{7.1}$$

We assume that $u(x,y) \to 0$ rapidly enough for large x and y. Equation (7.1) is obtained from the well known *time*-dependent Schrödinger equation

$$i\psi_y + \psi_{xx} + (u+\lambda)\psi = 0, \tag{7.2}$$

by letting $\lambda = 0$ (this is without loss of generality since λ can be scaled out), and then defining $\psi = \mu e^{i(kx-k^2y)}$.

7.1 Bounded eigenfunctions and their relationship

We first consider step **i** of the method of §6, *i.e.*, we introduce an eigenfunction $\mu(x,y,k)$ which solves (7.1), is bounded for all complex values of $k = k_R + ik_I$, and tends to unity as $k \to \infty$. Such an eigenfunction is given by

$$\mu(x,y,k) = \begin{cases} \mu^+(x,y,k), & k_I \geq 0, \\ \mu^-(x,y,k), & k_I \leq 0, \end{cases} \tag{7.3}$$

where μ^+ and μ^- satisfy the following linear Fredholm integral equations

$$\mu^\pm(x,y,k) = 1 + [g^\pm_{k,u}\,\mu^\pm(\cdot,\cdot,k)](x,y), \tag{7.4$^\pm$}$$

where

$$\begin{aligned}[g^+_{k,u}f](x,y) := \frac{i}{2\pi}\left(-\int_y^\infty d\eta \int_0^\infty dm \int_{-\infty}^\infty d\xi + \int_{-\infty}^y d\eta \int_{-\infty}^0 dm \int_{-\infty}^\infty d\xi\right) \\ \times \exp\left[im(x-\xi) - im(m+2k)(y-\eta)\right] u(\xi,\eta)\, f(\xi,\eta),\end{aligned} \tag{7.5a}$$

and

$$[g_{k,u}^{-}f](x,y) := \frac{i}{2\pi}\left(-\int_{y}^{\infty} d\eta \int_{-\infty}^{0} dm \int_{-\infty}^{\infty} d\xi + \int_{-\infty}^{y} d\eta \int_{0}^{\infty} dm \int_{-\infty}^{\infty} d\xi\right) \times \exp\left[im(x-\xi) - im(m+2k)(y-\eta)\right] u(\xi,\eta)\, f(\xi,\eta). \tag{7.5b}$$

It is clear that the kernel of equation $(7.4)^{+}$ is a $(+)$ function with respect to k, *i.e.*, it may be analytically continued in the upper half k-plane. Similarly the kernel of equation $(7.4)^{-}$ is a $(-)$ function.

There exist several ways for deriving equations $(7.4)^{\pm}$. For example, by formally taking the x-Fourier transform of (7.1) it follows that

$$\begin{aligned}\mu(x,y,k) &= \frac{1}{2\pi}\int_{-\infty}^{\infty} dm\, A(k,m) \exp\left[i(m-k)x - i(m^2-k^2)y\right] \\ &+ \frac{i}{2\pi}\int_{-\infty}^{y} d\eta \int_{-\infty}^{\infty} d\xi \int_{-\infty}^{\infty} dm\, u(\xi,\eta)\,\mu(\xi,\eta) \exp\left[im(x-\xi) - im(m+2k)(y-\eta)\right],\end{aligned} \tag{7.6}$$

where $A(k,m)$ is an arbitrary function of k and m, and the first integral in (7.6) represents a *homogeneous* solution of (7.1) —corresponding to $u = 0$. Equations $(7.4)^{\pm}$ easily follow from (7.6) by splitting the second integral and properly choosing $A(k,m)$, so that boundedness for all complex values of k is achieved. Alternatively, equations $(7.4)^{\pm}$ can also be obtained by using a Green's function approach: A representation of a solution of (7.1) is given by

$$\begin{aligned}\mu(x,y,k) &= \int_{-\infty}^{\infty} dm\, A(k,m) \exp\left[i(m-k)x - i(m^2-k^2)y\right] \\ &+ \int_{-\infty}^{\infty} d\xi \int_{-\infty}^{\infty} d\eta\, G(x,y,\xi,\eta,k)\, u(\xi,\eta)\,\mu(\xi,\eta,k),\end{aligned} \tag{7.7}$$

where the Green's function G satisfies

$$iG_y + G_{xx} + 2ikG_x = -\delta(x-\xi)\,\delta(y-\eta). \tag{7.8}$$

We seek bounded solutions of (7.8). A solution of (7.8) is

$$\begin{aligned}G(x-\xi, y-\eta, k) &= \frac{i}{2\pi}\int_{-\infty}^{\infty} dm\, e^{im(x-\xi)} g(y-\eta,k,m) \\ g(y-\eta,k,m) &= \frac{1}{2\pi i}\int_{-\infty}^{\infty} dq \frac{e^{iq(y-\eta)}}{q+(m^2+2km)}\end{aligned} \tag{7.9}$$

It is clear that if $k = k_R + ik_I$, g will have a jump across $k_I = 0$. Let g^{+} and g^{-} denote g for $k_I = 0^{\pm}$ respectively. These functions can be analytically continued into all $k_I > 0$ and $k_I < 0$ respectively. Then

$$\begin{aligned}g^{+} &= e^{-im(m+2k)(y-\eta)}\begin{cases}-H(\eta-y), & m>0,\\ H(y-\eta), & m<0,\end{cases} \\ g^{-} &= e^{-im(m+2k)(y-\eta)}\begin{cases}H(y-\eta), & m>0,\\ -H(\eta-y), & m<0,\end{cases}\end{aligned} \tag{7.10}$$

where $H(x)$ denotes the usual Heaviside function, *i.e.*, $H(x) = 0$ for $x < 0$ and $H(x) = 1$ for $x > 0$. Equations $(7.4)^{\pm}$ are obtained from (7.7) by taking 1 as the homogeneous solution and $G^{\pm}$ as the Green's function.

We have elaborated on the derivation of $(7.4)^{\pm}$ because all bounded eigenfunctions associated with (6.1) and (6.2) can be obtained in a similar way.

Fredholm's theory implies that equations $(7.4)^{\pm}$ have solutions meromorphic in k. Hence $\mu(x,y,k)$ *is a sectionally meromorphic function of k with some jump across $k_I = 0$, i.e.,* $\partial\mu/\partial\bar{k}$ is zero everywhere except at the pole location and along $k_I = 0$. To compute this jump one must evaluate $\mu^+(x,y,k) - \mu^-(x,y,k)$ for $k = k_R$. Using $(7.4)^{\pm}$ it follows that

$$\begin{aligned} \mu^+(x,y,k) - \mu^-(x,y,k) &= \int_{-\infty}^{\infty} dl\, T(k,l)\, N(x,y,l,k), \\ T(k,l) &:= \frac{i}{2\pi}\operatorname{sgn}(k-l)\int_{-\infty}^{\infty} d\xi \int_{-\infty}^{\infty} d\eta\, u(\xi,\eta)\,\mu^+(\xi,\eta,k)e^{\theta(\xi,\eta,k,l)}, \qquad k,l \text{ real}, \end{aligned} \tag{7.11}$$

where $\theta(x,y,l,k) := i(l-k)x - i(l^2-k^2)y$. In (7.11) the eigenfunction $N(x,y,l,k)$ solves (7.1) in k and also satisfies the integral equation

$$N(x,y,l,k) = e^{\theta(x,y,l,k)} + [g^-_{\bar{k},u} N(\cdot,\cdot,l,k)](x,y), \qquad k,l \text{ real}. \tag{7.12}$$

[Equation (7.12) is obtained from (7.7) by taking $e^{\theta(x,y,l,k)}$ as the homogeneous solution and G^- as the Green's function.]

Step **iii** of the method of §6 involves finding a *symmetry* relationship between $N(x,y,l,k)$ and $\mu(x,y,k)$. This crucial relationship turns out to be

$$\frac{\partial}{\partial k}\left(N(x,y,l,k)e^{i(kx-k^2y)}\right) = -F(l,k)\,\mu^-(x,y,k)e^{i(kx-k^2y)}, \qquad l,k \text{ real} \tag{7.13}$$

or, in integrated form

$$N(x,y,l,k) = \mu^-(x,y,l)e^{\theta(x,y,l,k)} - \int_l^k dp\, F(l,p)\,\mu^-(x,y,p)e^{\theta(x,y,p,k)}, \tag{7.14a}$$

where

$$F(l,p) = \frac{i}{2\pi}\int_{-\infty}^{\infty} d\xi \int_{-\infty}^{\infty} d\eta\, u(\xi,\eta)\,N(\xi,\eta,l,p), \qquad k,l,p \text{ real}. \tag{7.14b}$$

To derive (7.13) use $(7.4)^-$ and (7.12). Then (7.14a) follows by using $N(x,y,k,k) = \mu^-(x,y,k)$.

If equations $(7.4)^{\pm}$ have homogeneous solutions then one needs to establish a relationship between bounded and discrete (homogeneous) eigenfunctions. Equations $(7.4)^{\pm}$ are Fredholm equations of the second type which we assume to be regular (actually by defining $M^{\pm} := \mu^{\pm}\sqrt{|u|}$ it follows that $M^{\pm}$ satisfy integral equations with weakly singular kernels and hence Fredholm theory applies). Let $\phi_j^+(x,y)$ and $\phi_j^-(x,y)$ denote their homogeneous solutions corresponding to eigenvalues k_j^+ and k_j^- respectively, where $\operatorname{Im} k_j^+ > 0$ and $\operatorname{Im} k_j^- < 0$. We assume that there exists a finite number of such eigenvalues and that they are all simple. (Actually if u is real, one can show that $k_j^- = \overline{(k_j^+)}$, where the overline denotes complex conjugate.) Then Fredholm theory implies that μ^+ and μ^- admit the following representations,

$$\mu^{\pm}(x,y,k) = 1 + \sum_{j=1}^{n} \frac{C_j^{\pm}\phi_j^{\pm}(x,y)}{k - k_j^{\pm}} + \tilde{\mu}^{\pm}(x,y,k), \tag{7.15$^{\pm}$}$$

where $\tilde{\mu}^+$ and $\tilde{\mu}^-$ are (+) and (−) functions, respectively, with respect to k ($C_j^{\pm}$ are introduced only for the purpose of normalization).

The following important relations are valid,

$$\lim_{k\to k_j^\mp}\left(\mu^\pm - \frac{C_j^\mp \phi_j^\mp}{k-k_j^\mp}\right) = (x-2k_j^\mp y+\gamma_j^\mp)\phi_j^\mp, \tag{7.16$^\pm$}$$

where $\gamma_j^\mp$ are constants and $C_j^\mp = i$ if $\phi_j^\mp$ are normalized by $(x-2k_j^\mp y)\phi_j^\mp \to 1$ as $\sqrt{x^2+y^2}\to\infty$. The constants $\gamma_j^\mp$ are fixed by the asymptotics of equations (7.15)$^\pm$ in the neighborhood of $k=k_j^\mp$.

To derive equations (7.16)$^\pm$, formulate integral equations for $\tilde{\mu}^\pm(x,y,k)$ and then take their limit as $k\to k_j^\mp$. In this way (7.16)$^\pm$ are obtained if and only if

$$1 \mp \frac{1}{2\pi}\int_{-\infty}^{\infty} d\xi \int_{-\infty}^{\infty} d\eta\, u(\xi,\eta)\,\phi_j^\mp(\xi,\eta) = 0. \tag{7.17$^\pm$}$$

Equations (7.17)$^\pm$ should also follow from the homogeneous versions of (7.4)$^\pm$

7.2 Scattering equation and the inverse problem

Using (7.14) in (7.11) one obtains the scattering equation, *i.e.*, a relationship connecting the jump of $\mu(x,y,k)$ across k_I with $\mu(x,y,k)$, and the scattering data:

$$\mu^+(x,y,k) = \mu^-(x,y,k) + \int_{-\infty}^{\infty} dl\, f(k,l) e^{\theta(x,y,l,k)} \mu^-(x,y,l), \qquad k,l \text{ real}, \tag{7.18}$$

where

$$f(k,l) = \tfrac{i}{2\pi}\operatorname{sgn}(k-l)\int_{-\infty}^{\infty} d\xi \int_{-\infty}^{\infty} d\eta\, u(\xi,\eta)\, N(\xi,\eta,k,l), \qquad k,l \text{ real}. \tag{7.19}$$

Equation (7.18) is the central equation associated with the solution of the inverse problem of (7.1). Equation (7.18), together with (7.15)$^\pm$ and (7.16)$^\pm$, defines a RHBVP for the sectionally meromorphic function $\mu(x,y,k)$. Its solution is expressed through the following linear integral equations

$$\mu^-(k) - i\sum_{l=1}^{n}\left(\frac{\phi_l^+}{k-k_l^+}+\frac{\phi_l^-}{k-k_l^-}\right) - \frac{1}{2\pi i}\int_{-\infty}^{\infty}\int_{-\infty}^{\infty} dl\, d\nu\, \frac{f(\nu,l)e^{\theta(x,y,l,\nu)}\mu^-(l)}{\nu-k+i0} = 1, \qquad k \text{ real} \tag{7.20}$$

$$(x-2k_j^\mp y+\gamma_j^\mp)\phi_j^\mp - i\sum_{l=1}^{n}\left(\frac{\phi_l^+}{k_j^\mp-k_l^+}+\frac{\phi_l^-}{k_j^\mp-k_l^-}\right) - \frac{1}{2\pi i}\int_{-\infty}^{\infty}\int_{-\infty}^{\infty} dl\, d\nu\, \frac{f(\nu,l)e^{\theta(x,y,l,\nu)}\mu^-(l)}{\nu-k_j^\mp} = 1, \tag{7.21$^\pm$}$$

where $\mu^-(k) := \mu^-(x,y,k)$ and $\sum$ means summation from $l=1$ to n unless any of the denominators vanishes, *i.e.*, it omits those terms.

Equations (7.20) and (7.21)$^\pm$ were derived in [36,87] by splitting equation (7.18) into its (+) and (−) parts. Here we use an alternative approach in order to illustrate (6.3). Recall that

$$\mu(k) = \begin{cases} 1+\tilde{\mu}^+(k) + i\displaystyle\sum_{j=1}^{n}\frac{\phi_j^+}{k-k_j^+}, & k_I \geq 0, \\ 1+\tilde{\mu}^-(k) + i\displaystyle\sum_{j=1}^{n}\frac{\phi_j^-}{k-k_j^-}, & k_I \leq 0. \end{cases} \tag{7.22}$$

Hence, using

$$\frac{\partial}{\partial \bar{k}} \tilde{\mu}^{\pm}(k) = 0, \qquad \frac{\partial}{\partial \bar{k}}\left(\frac{1}{k-k_0}\right) = \pi\,\delta(k-k_0),$$

equation (7.22) yields

$$\frac{\partial \mu}{\partial \bar{k}} = \begin{cases} \pi i \sum_{j=1}^{n} \phi_j^+ \delta(k-k_j^+), & k_I > 0, \\ \mu^+(k) - \mu^-(k), & k_I = 0, \\ \pi i \sum_{j=1}^{n} \phi_j^- \delta(k-k_j^-), & k_I < 0. \end{cases} \tag{7.23}$$

Thus, using (7.18), (7.23), and $dz \wedge d\bar{z} = -2i\, dz_R dz_I$, equation (6.3) yields

$$\begin{aligned} \mu(k) - \frac{1}{2\pi i}\int_{-\infty}^{\infty} \frac{dz_R}{z_R - k}\int_{-\infty}^{\infty} dl\, f(z_R, l)\, \mu^-(l) \exp\left[i(l-z_R)x - i(l^2 - z_R^2)y\right] \\ -i\sum_{l=1}^{n}\left(\frac{\phi_l^+}{k-k_l^+} + \frac{\phi_l^-}{k-k_l^-}\right) = 1. \end{aligned} \tag{7.24}$$

Taking the limits $k \to k_R - i0$ and $k \to k_j^{\pm}$ in equation (7.24), equations (7.20) and (7.21)$^{\pm}$ follow.

Equations (7.20) and (7.21)$^{\pm}$ define $\mu^-(k)$ and $\{\phi_j^+, \phi_j^-\}_{j=1}^{n}$ in terms of the *scattering data*

$$\{k_j^{\pm}, \gamma_j^{\pm}\}_{j=1}^{n}, \qquad f(k,l). \tag{7.25}$$

Step **v** of the method involves calculating the potential $u(x,y)$ in terms of μ^- and $\{\phi_j^+, \phi_j^-\}_{j=1}^{n}$. By taking the large k asymptotics of (7.20) and (7.1) it follows that

$$u(x,y) = \frac{\partial}{\partial x}\left\{2\sum_{j=1}^{n}(\phi_j^+(x,y) + \phi_j^-(x,y)) + \frac{1}{\pi}\int_{-\infty}^{\infty} dl \int_{-\infty}^{\infty} dk\, f(k,l) e^{\theta(x,y,l,k)} \mu^-(x,y,l)\right\} \tag{7.26}$$

7.3 The initial value problem for Kadomstev–Petviashvili I

Assume now that u also depends on t, *i.e.*, $u = u(x,y,t)$. Then the scattering data (7.25) will depend, in general, on t in a complicated way. However, if u evolves according to certain evolution equations, *e.g.*, KPI or any member of its hierarchy, then this dependence is simple. In particular if u satisfies KPI (2.17) then

$$\frac{\partial k_j^{\pm}}{\partial t} = 0, \qquad \gamma_j^{\pm}(t) = 12(k_j^{\pm})^2 t + \gamma_j^{\pm}(0), \qquad f(k,l,t) = f(k,l,0)e^{4i(l^3-k^3)t}. \tag{7.27}$$

At $t = 0$, $u(x,y,0)$ is given and hence the scattering data (7.25) at $t = 0$ can be calculated. For example

$$f(k,l,0) = \frac{i}{2\pi}\,\mathrm{sgn}(k-l)\int_{-\infty}^{\infty} d\xi \int_{-\infty}^{\infty} d\eta\, u(\xi,\eta,0)\, N(\xi,\eta,0,k,l),$$

where $N(\xi,\eta,0,k,l)$ can be obtained from (7.12). Using (7.27), (7.21)$^{\pm}$ yield $\mu^{-}(x,y,t,k)$ and $\phi_j^{\mp}(x,y,t)$. Finally, equation (7.26) determines $u(x,y,t)$.

From the above discussion it follows that the solution of the inverse problem of (7.1) immediately yields the solution of the initial value problem of KPI, provided one establishes (7.27). This can be done as follows: [for convenience we work with $\psi(x,y,k)$ instead of $\mu(x,y,k)$] KPI can be written as the compatibility condition of (7.2) and

$$L_k\psi := \left[\partial_t + 4\partial_x^3 + 6u\partial_x + 3\left(u_x - i\int_{-\infty}^{x} dx'\, u_y\right) + \alpha(k)\right]\psi = 0, \tag{7.28}$$

where $\alpha(k)$ is an arbitrary function of k. Since $\mu \sim 1$ for large k, then $\psi \sim e^{i(kx-k^2y)}$ and hence $\alpha(k) = 4ik^3$. Equation (7.18) implies that

$$\psi^{+}(k) = \psi^{-}(k) + \int_{-\infty}^{\infty} dl\, \psi^{-}(l)\, f(k,l), \tag{7.29}$$

where $\psi^{\pm}(k) := \psi^{\pm}(x,y,t,k)$. Applying the linear operator L_k to (7.29) it follows that $f_t = 4i(l^3 - k^3)f$. The evolution of $\gamma_j^{\mp}$ is found in a similar way.

Remarks

i. The pure lump solutions correspond to $f(k,l,0) = 0$ and are then characterized by

$$u(x,y,t) = 2\frac{\partial}{\partial x}\sum_{j=1}^{n}\left(\phi_j^{+}(x,y,t) + \phi_j^{-}(x,y,t)\right), \tag{7.30}$$

where $\phi_j^{\mp}(x,y,t)$ satisfy the following system of *linear algebraic* equations

$$\left[x - 2k_j^{\mp}y + 12(k_j^{\mp})^2t + \gamma_j^{\mp}(0)\right]\phi_j^{\mp} - i\sum_{l=1}^{n}\left(\frac{\phi_l^{+}}{k_j^{\mp} - k_l^{+}} + \frac{\phi_l^{-}}{k_j^{\mp} - k_l^{-}}\right) = 1. \tag{7.31$^{\pm}$}$$

In particular the 1-lump is given by

$$u = 2\partial_x^2 \ln\left[(x' + p_1y')^2 + p_2^2y'^2 + \frac{1}{p_2^2}\right], \tag{7.32}$$

where

$$x' = x - 3(p_1^2 + p_2^2)t - x_0, \qquad y' = y + 6p_1t - y_0.$$

ii. In the case that $\phi_j^{\mp} = 0$, using

$$i\int_{-\infty}^{x} d\xi\, e^{i\xi(k-\nu-i\epsilon)} = \frac{e^{ix(k-\nu)}}{k-\nu-i\epsilon},$$

multiplying equation (7.20) by $f(k,k')e^{\theta(x,y,k,k')}$, and integrating over $dkdk'$ one directly obtains a GLM-type equation.

§8 On the inverse scattering transform of a certain diffusion equation and Kadomtsev–Petviashvili II equation

This section gives an account of the results of Ablowitz, Bar Yaacov and Fokas in [88]. Consider

$$-\mu_y + \mu_{xx} + 2ik\mu_x = -u\mu, \tag{8.1}$$

which is related to

$$-\psi_y + \psi_{xx} + (u+\lambda)\psi = 0, \tag{8.2}$$

in the same way that (7.1) is related to (7.2).

8.1 Bounded eigenfunctions and their relationship

As before we introduce an eigenfunction $\mu(x,y,k)$ which solves (8.1), is bounded for all complex values of $k = k_R + ik_I$, and tends to unity as $k \to \infty$. This eigenfunction has two different representations according to whether $k_R > 0$ or $k_R < 0$:

$$\mu(x,y,k) = 1 + [g_{k_R,k_I,u}\,\mu(\cdot,\cdot,k)](x,y), \tag{8.3}$$

where

$$[g_{k_R,k_I,u}\,f](x,y) := \frac{i}{2\pi}\left(-\int_y^{\infty} d\eta \int_{-\infty}^{\infty} d\xi \int_{-2k_R}^{0} dm + \int_{-\infty}^{y} d\eta \int_{-\infty}^{\infty} d\xi \left\{\int_0^{\infty} dm + \int_{-\infty}^{-2k_R} dm\right\}\right)$$
$$\times \exp\left[im(x-\xi) - im(m+2k)(y-\eta)\right] u(\xi,\eta)\, f(\xi,\eta), \quad k_R > 0, \tag{8.4}$$

and for $k_R < 0$ the integrals with respect to m are replaced by $\int_0^{-2k_R} dm$ and $\{\int_{-\infty}^{0} dm + \int_{-2k_R}^{-\infty} dm\}$ respectively.

We note that:

i. $\mu(x,y,k)$ does not have a jump across $k_R = 0$ since $g_{0+,k_I,u}\,f = g_{0-,k_I,u}\,f$;

ii. equation (8.3), and hence μ, *depends explicitly on* k_R. Thus the eigenfunction μ defined by (8.3), although bounded for all complex values of k, is analytic nowhere with respect to k, and $\partial\mu/\partial\bar{k} \neq 0$. We emphasize the explicit dependence on k_R by writing $q_{k_R,k_I,u}$ instead of $g_{k,u}$. However, for convenience of notation we still write $\mu(x,y,k)$ instead of $\mu(x,y,k_R,k_I)$.

The second step of the method of §6 is to evaluate the ***departure from holomorphicity*** of μ, *i.e.*, $\partial\mu/\partial\bar{k}$. Equation (8.3) implies

$$\frac{\partial\mu(x,y,k)}{\partial\bar{k}} = F(k_R,k_I)\,N(x,y,k_R,k_I), \tag{8.5}$$

where

$$F(k_R,k_I) := -\frac{1}{4\pi}\,\mathrm{sgn}(k_R)\int_{-\infty}^{\infty} d\xi \int_{-\infty}^{\infty} d\eta\, e^{i(2k_R\xi - 4k_Ik_R\eta)} u(\xi\eta)\,\mu(\xi,\eta,k). \tag{8.6}$$

In equation (8.6) N is a solution of (8.1) satisfying also

$$N(x,y,k_R,k_I) = e^{-i(2k_Rx - 4k_Ik_Ry)} + [g_{k_R,k_I,u}\,N(\cdot,\cdot,k_R,k_I)](x,y). \tag{8.7}$$

[Clearly (8.5) and (8.7) are the analogues of (7.11) and (7.12).]

In order to view (8.5) as a DBAR problem one needs to relate N and μ. In this case one finds the following *discrete* symmetry relationship:

$$N(x,y,k_R,k_I) = \mu(x,y,-\bar{k})e^{-i(2k_Rx-4k_Ik_Ry)}. \tag{8.8}$$

We assume that equation (8.3) has no homogeneous solutions. Hence equation (8.5) and (8.8) are sufficient for obtaining the solution of the inverse problem.

8.2 Scattering equation and the inverse problem

Using (8.8) and (8.5) it follows that

$$\frac{\partial \mu(x,y,k)}{\partial \bar{k}} = F(k_R,k_I)\,\mu(x,y,-\bar{k})e^{-i(2k_Rx-4k_Ik_Ry)}. \tag{8.9}$$

Equation (8.9) defines a DBAR problem in k, for the bounded function μ, in terms of the scattering data $F(k_R,k_I)$ defined by (8.6). Its solution is given by [see (6.3)]

$$\mu(x,y,k) = 1 + \frac{1}{2\pi i}\iint_{R_\infty} \frac{dz \wedge d\bar{z}}{z-k} F(z_R,z_I)e^{-i(2z_Rx-4z_Iz_Ry)}\mu(x,y,-\bar{z}), \tag{8.10}$$

where R_∞ is the entire z-complex plane and $dz \wedge d\bar{z} = -2i\, dz_R dz_I$.

Once $\mu(x,y,k)$ is found, the potential $u(x,y)$ is easily calculated from

$$u(x,y) = \frac{1}{\pi}\frac{\partial}{\partial x}\iint_{R_\infty} dz \wedge d\bar{z}\, F(z_R,z_I)e^{-i(2z_Rx-4z_Rz_Iy)}\mu(x,y,-\bar{z}). \tag{8.11}$$

Equation (8.11) again follows from the large k asymptotics of (8.10) and (8.1).

8.3 The initial value problem of Kadomtsev–Petviashvili II

To solve the initial value problem of KPII one has only to supplement equations (8.10) and (8.11) with the evolution of the scattering data $F(k_R,k_I)$ (see the discussion in §7.3). This evolution is obtained as follows: KPII is the compatibility condition of (8.2) and $\hat{L}_k\psi = 0$, where $\hat{L}_k$ is defined from (7.28) by changing $-i\int_{-\infty}^{x} dx'\, u_y$ to $\int_{-\infty}^{x} dx'\, u_y$. Equation (8.9) implies

$$\frac{d\psi(k)}{d\bar{k}} = \psi(-\bar{k})\,F(k_R,k_I). \tag{8.12}$$

Applying the linear operator $\hat{L}_k$ to (8.12) it follows that $F_t = -4i(k^3+\bar{k}^3)F$. Hence

$$F(k_R,k_I,t) = F(k_R,k_I,0)e^{-4i(k^3+\bar{k}^3)t}. \tag{8.13}$$

§9 On the inverse scattering transform of hyperbolic systems in the plane, n-wave, and the DSI

This section gives an account of the results of Fokas and Ablowitz in [89] and [91]. Consider the inverse problem of the hyperbolic system

$$\mu_x = i\kappa \hat{J}\mu + q\mu + J\mu_y, \qquad \hat{J}f := Jf - fJ, \tag{9.1}$$

where $\mu(x, y, \kappa)$ is an nth order matrix, J is a constant real diagonal matrix with elements $J_1 > J_2 > \cdots > J_n$, and $q = q(x, y)$ is an nth order off-diagonal matrix containing the potentials $q_{ij}(x, y)$. We assume that $q_{ij}(x, y) \to 0$ rapidly enough for large x and y. Equation (9.1) is obtained from the well known [49] equation

$$\psi_x = i\lambda B\psi + q\psi + J\psi_y, \tag{9.2}$$

by taking $B = 0$ (this is without loss of generality since all the equations solvable by (9.2) are independent of B [49,18]), and using $\psi = \mu e^{i\kappa(Jx+Iy)}$ where I denotes the unit diagonal matrix.

9.1 Bounded eigenfunctions

Let $\pi_0\mu$, $\pi_+\mu$, and $\pi_-\mu$ denote the diagonal, strictly upper diagonal, and strictly lower diagonal parts of the matrix μ. A solution of (9.1) bounded for all complex values of $\kappa = \kappa_R + i\kappa_I$ and tending to unity as $\kappa \to \infty$ is given by

$$\mu(x, y, \kappa) = \begin{cases} \mu^+(x, y, \kappa), & \kappa_I \geq 0, \\ \mu^-(x, y, \kappa), & \kappa_I \leq 0, \end{cases} \tag{9.3}$$

where $\mu^{\pm}(x, y, \kappa)$ satisfy the following linear integral equations

$$\begin{aligned} \mu^{\pm}(x, y, \kappa) = 1 &+ \frac{1}{2\pi}\int_{-\infty}^{x} d\xi\, E_{x-\xi,J}\, e^{i\kappa(x-\xi)\hat{J}}(\pi_0 + \pi_{\pm})q(\xi, \eta)\,\mu^{\pm}(\xi, \eta, \kappa) \\ &- \frac{1}{2\pi}\int_{x}^{\infty} d\xi\, E_{x-\xi,J}\, e^{i\kappa(x-\xi)\hat{J}}\pi_{\mp}\, q(\xi, \eta)\,\mu^{\pm}(\xi, \eta, \kappa). \end{aligned} \tag{9.4$^\pm$}$$

In equations $(9.4)^{\pm}$ the linear operator $E_{x-\xi,J}$ is defined by

$$(E_{x-\xi,J}f(\cdot))(x-\xi, y) := \int_{-\infty}^{\infty} d\eta \int_{-\infty}^{\infty} dm\, e^{im(x-\xi)J + im(y-\eta)} f(\eta) = f(y + (x-\xi)J), \tag{9.5}$$

where $f(y + (x-\xi)J)$ denotes the matrix obtained from $f(x)$ by evaluating the lth row of the matrix $f(x)$ at $y + (x-\xi)J_l$. Also from the definition of J it follows that

$$e^{\hat{J}}f = e^J f e^{-J}. \tag{9.6}$$

Equations $(9.4)^{\pm}$ can be derived by using a Green's function approach similar to the one used in [87,88]. Alternatively, one may use a Fourier transform in the y direction.

Assuming that the linear integral equations $(9.4)^{\pm}$ have no homogeneous solutions it follows that μ^+ and μ^- are holomorphic functions of κ, for $\kappa_I > 0$ and $\kappa_I < 0$ respectively. Hence the function $\mu(x, y, \kappa)$ defined by (9.3) is a *sectionally holomorphic function of* κ *having a jump across* $\kappa_I = 0$. Thus $\partial\mu/\partial\bar{\kappa} = 0$ for all κ, with $\kappa_I \neq 0$, and $\partial\mu/\partial\bar{\kappa} = \mu^+(x, y, \kappa) - \mu^-(x, y, \kappa)$ for $\kappa = \kappa_R$. Rather than following in detail the method of §6, *i.e.*, computing $\mu^+ - \mu^-$ in terms of some other bounded eigenfunction N and then establishing a *symmetry* condition relating μ and N, we find it more convenient to obtain directly a scattering equation.

9.2 Scattering equation and the solution of the inverse problem

Using $(9.4)^{\pm}$ it can be shown that

$$\mu^{-}(x,y,\kappa) - \mu^{-}(x,y,\kappa) = \int_{-\infty}^{\infty} dl\, \mu^{-}(x,y,l)\, e^{ilJx+ily} f(l,\kappa)\, e^{-i\kappa Jx - i\kappa y}, \qquad \kappa = \kappa_R. \tag{9.7}$$

In equation (9.7) the scattering data $f(l,\kappa)$ satisfies

$$f(l,k) - \int_{-\infty}^{\infty} dm\, T_{+}(l,m) f(m,\kappa) = T_{+}(l,\kappa) - T_{-}(l,\kappa), \qquad l,k \text{ real}, \tag{9.8}$$

where

$$T_{\pm}(l,\kappa) := \frac{1}{2\pi}\int_{-\infty}^{\infty} d\xi \int_{-\infty}^{\infty} d\eta\, e^{-ilJ\xi - il\eta}\, \pi_{\pm} q(\xi,\eta)\, \mu^{\mp}(\xi,\eta,\kappa) e^{i\kappa\xi J + i\kappa\eta}, \qquad l,k \text{ real}. \qquad (9.9)^{\pm}$$

Before indicating how the above equations can be derived we note the remarkable fact that equation (9.8) can be solved in *closed form*. This is because its kernel is strictly upper triangular.

Example

Suppose that $n = 2$, *i.e.*,

$$q(x,y) = \begin{pmatrix} 0 & q_1(x,y) \\ q_2(x,y) & 0 \end{pmatrix}, \qquad T_{+}(l,\kappa) = \begin{pmatrix} 0 & T_{+12}(l,\kappa) \\ 0 & 0 \end{pmatrix}, \qquad T_{-}(l,\kappa) = \begin{pmatrix} 0 & 0 \\ T_{-21}(l,\kappa) & 0 \end{pmatrix},$$

where

$$T_{+12}(l,\kappa) := \frac{1}{2\pi}\int_{-\infty}^{\infty} d\xi \int_{-\infty}^{\infty} d\eta\, \{q(\xi,\eta)\, \mu^{-}(\xi,\eta,\kappa)\}_{12}\, e^{i\kappa\xi J_2 + i\kappa\eta - il\xi J_1 - il\eta},$$

$$T_{-21}(l,\kappa) := \frac{1}{2\pi}\int_{-\infty}^{\infty} d\xi \int_{-\infty}^{\infty} d\eta\, \{q(\xi,\eta)\, \mu^{+}(\xi,\eta,\kappa)\}_{21}\, e^{i\kappa\xi J_1 + i\kappa\eta - il\xi J_2 - il\eta}.$$

Then equation (9.8) implies

$$f_{22}(l,\kappa) = 0, \qquad f_{21}(l,\kappa) = -T_{+12}(l,\kappa),$$

$$f_{12}(l,\kappa) = T_{-12}(l,\kappa), \qquad f_{11}(l,\kappa) = -\int_{-\infty}^{\infty} dm\, T_{+12}(l,m)\, T_{-21}(m,\kappa).$$

Equation (9.8) can be solved in a similar manner for any n.

Equation (9.7) is the central equation associated with the inverse problem of (9.1). It defines a nonlocal RH problem in the complex κ-plane for the sectionally holomorphic matrix function $\mu(x,y,\kappa)$. By taking the *minus* projection of (9.7) it follows that $\mu^{-}(x,y,\kappa)$ solves the following linear integral equation

$$\mu^{-}(x,y,\kappa) + \frac{1}{2\pi i}\int_{-\infty}^{\infty} dl \int_{-\infty}^{\infty} d\nu\, \frac{\mu^{-}(x,y,l) e^{ilJx} f(l,\nu) e^{-i\nu Jx + i(l-\nu)y}}{\nu - \kappa + i0} = I. \tag{9.10}$$

Alternatively, equation (9.10) can be derived by using (6.3).

Equation (9.10) uniquely defines $\mu^{-}(x,y,\kappa)$ in terms of the scattering data $f(l,\nu)$. Once $\mu^{-}(x,y,\kappa)$ is found, the potential $q(x,y)$ is easily obtained:

$$q(x,y) = -\frac{1}{2\pi}\hat{J}\int_{-\infty}^{\infty} dl \int_{-\infty}^{\infty} d\nu\, \mu^{-}(x,y,l)\, e^{ilJx} f(l,\nu)\, e^{-i\nu Jx + i(l-\nu)y}. \tag{9.11}$$

9.3 On the initial value problem of n-wave interaction and DSI

A specific evolution equation for $q(x, y, t)$ implies a specific Lax pair, which in turn implies a unique evolution of the associated scattering data. Let us illustrate how this evolution can be determined by considering two different forms for the time-dependent part of the Lax pair whose time-independent part is given by (9.2) —with $B = 0$.

9.3.1 Lax pairs containing the n-wave interaction

Consider the Lax pair

$$\begin{cases} \psi_x = J\psi_y + q\psi, \\ \psi_t = A_1\psi + A_2\psi_y, \end{cases} \tag{9.12}$$

and assume that

$$A_1 \to 0, \quad A_2 \to A_{20} \qquad \text{for large } x, y, \tag{9.13}$$

where A_{20} is a constant real diagonal matrix. Hence for large x and y the eigenfunction $\psi \to \exp[i\kappa(Jx + y + A_{20}t)]$. However, the IST was based on an eigenfunction normalized to I. Thus, let $\psi = \mu \exp[i\kappa(Jx + y \ +A_{20}t)]$, (9.12) yield (9.1), and

$$\mu_t = i\kappa(A_2\mu - \mu A_{20}) + A_1\mu + A_2\mu_y. \tag{9.14}$$

Applying the operator $L_\kappa := \partial_t - i\kappa A_2 - A_1 - A_2\partial_y$ to the scattering equation (9.7) and using (9.14), it follows that

$$f(l, \kappa, t) = \exp[iltA_{20}]\, f(l, \kappa, 0) \exp[-i\kappa t A_{20}]. \tag{9.15}$$

The n-wave interaction equations are [49]

$$(q_{ij})_t = \alpha_{ij}(q_{ij})_x + \beta_{ij}(q_{ij})_y + \sum_{\substack{\kappa=1\\ \kappa\neq j}}^{n} (\alpha_{i\kappa} - \alpha_{\kappa j})\, q_{i\kappa}\, q_{\kappa j}, \qquad i, j, \kappa = 1, \ldots, n, \tag{9.16}$$

where α_{ij} and β_{ij} are real constants related to the x and y components of the underlying group velocities. Equations (9.16) are the compatibility equations for the Lax pair (9.12) with A_1 and A_2 defined by

$$\{A_1\}_{ij} = \alpha_{ij}q_{ij}, \qquad \{A_1\}_{ii} = 0, \qquad A_2 = \mathrm{diag}(C_1, \ldots, C_n), \tag{9.17}$$

where J_l and C_l are given in terms of α_{ij} and β_{ij} via

$$\alpha_{ij} = \frac{C_i - C_j}{J_i - J_j}, \qquad \beta_{ij} = C_i - J_i\alpha_{ij}. \tag{9.18}$$

Hence the initial value problem of (9.16) can be solved via (9.10) and (9.11); in these equations the initial scattering data is found from (9.8) and $(9.9)^{\pm}$, and $f(l, \kappa, t)$ is found from (9.15) with $A_{20} = A_2$.

9.3.2 Lax pairs containing DSI

Consider the Lax pair

$$\begin{cases} \psi_x = q\psi + J\psi_y, \\ \psi_t = A_1\psi + A_2\psi_y + A_3\psi_{yy}, \end{cases} \tag{9.19}$$

and assume that

$$A_1 \to 0, \quad A_2 \to 0, \quad A_3 \to A_{30} \qquad \text{for large } x, y, \tag{9.20}$$

where A_{30} is a purely imaginary diagonal matrix. Letting $\psi = \mu \exp[i\kappa(Jx + \psi + i\kappa A_{30}t)]$, equations (9.19) yield (9.1) and

$$\mu_t = A_3\mu_{yy} + A_2\mu_y + A_1\mu + (i\kappa)^2(A_3\mu - \mu A_{30}) + 2i\kappa A_3\mu_y + i\kappa A_2\mu. \tag{9.21}$$

The evolution of the scattering data associated with (9.19) is given by

$$f(l,\kappa,t) = \exp\left[(il)^2 tA_{30}\right] f(l,\kappa,0) \exp\left[-(i\kappa)^2 tA_{30}\right]. \tag{9.22}$$

We call DSI the following set of equations:

$$\begin{aligned} iQ_t + \tfrac{1}{2}(Q_{xx} + Q_{yy}) &= -\sigma|Q|^2 Q + \phi Q, \\ \phi_{xx} - \phi_{yy} &= 2\sigma\left(|Q|^2\right)_{xx}, \qquad \sigma = \pm 1. \end{aligned} \tag{9.23}$$

Equations (9.23) are the compatibility conditions of (9.19) with

$$J_1 = 1, \quad J_2 = -1, \qquad A_2 = iq, \quad A_3 = \operatorname{diag}(i, -i), \tag{9.24}$$

$$A_1 = \begin{pmatrix} A_{11} & A_{12} \\ A_{21} & A_{22} \end{pmatrix}, \qquad q = \begin{pmatrix} 0 & Q \\ \sigma\overline{Q} & 0 \end{pmatrix}, \tag{9.25}$$

where the entries of A_1 are defined by

$$\begin{aligned} A_{12} &= \tfrac{1}{2}i(Q_x + Q_y), \\ A_{21} &= \tfrac{1}{2}i\sigma(-\overline{Q}_x + \overline{Q}_y), \\ A_{11_x} - A_{11_y} &= -\tfrac{1}{2}i\sigma\left[(Q\overline{Q})_x + (Q\overline{Q})_y\right], \\ A_{22_x} + A_{22_y} &= \tfrac{1}{2}i\sigma\left[(Q\overline{Q})_x - (Q\overline{Q})_y\right]. \end{aligned} \tag{9.26}$$

Hence (assuming that A_{11} and A_{22} tend to zero for large x and y) the initial value problem of (9.23) can be solved via (9.10) and (9.11). In these equations $J = \operatorname{diag}(1, -1)$; furthermore the initial scattering data $f(l, \kappa, 0)$ is determined from (9.8) and (9.9)±, and $f(l, \kappa, t)$ is found from (9.22) where $A_{30} = A_3 = \operatorname{diag}(i, -i)$.

§10 Elliptic systems in the plane, DSII

In this final section we give an account of the results of Fokas and Ablowitz in [89] and [91]. Consider the inverse problem associated with the elliptic system

$$\mu_x = i\kappa\hat{J}\mu + q\mu - iJ\mu_y, \tag{10.1}$$

where $\mu(x, y, \kappa)$ is an nth order matrix, J is a constant real diagonal matrix with all its entries different from each other, and $q(x, y)$ is an nth order off-diagonal matrix containing the potentials $q_{ij}(x, y)$. We assume that $q_{ij}(x, y) \to 0$ rapidly enough for large x and y. Equation (10.1) is obtained from the well known equation $\psi_x = q\psi - iJ\psi_y$ [49], through the transformation $\psi = \mu e^{i\kappa Jx - \kappa y}$.

10.1 Bounded eigenfunctions and their relations

For the solution of the inverse problem associated with (10.1) we follow very closely the steps outlined in §6. We first consider step **i**, *i.e.*, we introduce an eigenfunction $\mu(x,y,\kappa)$ which solves (10.1), is bounded for all complex values of κ, and tends to unity as $\kappa \to \infty$; such an eigenfunction satisfies the following linear integral equation:

$$\mu(x,y,\kappa) = I + \left(G_{\kappa_R,\kappa_I,q}\,\mu(\cdot,\cdot,\kappa)\right)(x,y). \tag{10.2}$$

The operator $G_{\kappa_R,\kappa_I,q}$ is a linear integral matrix operator defined by the following: Let $f(x,y)$ be some $n\times n$ matrix, then the ij^{th} entry of the operator $G_{\kappa_R,\kappa_I,q}$ applying on $f(x,y)$ is

$$\begin{aligned}\{(G_{\kappa_R,\kappa_I,q}f)(x,y)\}_{ij} &:= \frac{1}{2\pi}\left(\int_{-\infty}^{x} d\xi \int_{-\infty}^{C_{ij}\kappa_I} dm \int_{-\infty}^{\infty} d\eta - \int_{x}^{\infty} d\xi \int_{C_{ij}\kappa_I}^{\infty} dm \int_{-\infty}^{\infty} d\eta\right)\\ &\times\left\{\exp\left[(mJ+i\kappa\hat{J})(x-\xi)+im(y-\eta)\right]q(\xi,\eta)\,f(\xi,\eta)\right\}_{ij}, \qquad J_i>0,\end{aligned} \tag{10.3}$$

where $C_{ij} := (J_i - J_j)/J_i$ and, for $J_i < 0$, the integrals with respect to dm are replaced by $\int_{C_{ij}\kappa_I}^{\infty} dm$ and $\int_{-\infty}^{C_{ij}\kappa_I} dm$ respectively; f is a matrix and $\{f\}_{ij}$ denotes its ij^{th} entry. Hence

$$\begin{aligned}&\left\{\exp[(mJ+i\kappa\hat{J})(x-\xi)+im(y-\eta)]\,q(\xi,\eta)\,f(\xi,\eta)\right\}_{ij}\\ &\quad= \exp\{[mJ_i+i\kappa(J_i-J_j)](x-\xi)+im(y-\eta)\}\{q(\xi,\eta)\,f(\xi,\eta)\}_{ij}.\end{aligned} \tag{10.4}$$

Sometimes it will be convenient to work with the column vectors of the matrix μ. Letting $\mu = (\boldsymbol{\mu}^1,\ldots,\boldsymbol{\mu}^j,\ldots,\boldsymbol{\mu}^n)$ it follows from (10.2) that $\boldsymbol{\mu}^j$ satisfies

$$\boldsymbol{\mu}^j(x,y) = \mathbf{I}^j + \left(\mathbf{g}^j_{\kappa_R,\kappa_I,q}\,\boldsymbol{\mu}^j(\cdot,\cdot,\kappa)\right)(x,y), \tag{10.5^j}$$

where $\mathbf{I}^j$ denotes the jth unit vector, and

$$\begin{aligned}\{(\mathbf{g}^j_{\kappa_R,\kappa_I,q}\boldsymbol{f})(x,y)\}_l &:= \frac{1}{2\pi}\left(\int_{-\infty}^{x} d\xi \int_{-\infty}^{C_{lj}\kappa_I} dm \int_{-\infty}^{\infty} d\eta - \int_{x}^{\infty} d\xi \int_{C_{lj}\kappa_I}^{\infty} dm \int_{-\infty}^{\infty} d\eta\right)\\ &\times \exp\{[mJ_l+i\kappa(J_l-J_i)](x-\xi)+im(y-\eta)\}\{q(\xi,\eta)\,\boldsymbol{f}(\xi,\eta)\}_l, \qquad J_l>0,\end{aligned} \tag{10.6}$$

[for $J_l < 0$ the integrals with respect to dm must be altered just as in (10.3)]. If $\boldsymbol{f}$ is a vector, $\{\boldsymbol{f}\}_l$ denotes its lth entry.

Equation (10.2) can be derived in a similar way as equation (9.4). Comparing (10.2) with (9.4) it follows that:

i. Equation (10.2), in contrast with (9.4), has no jump across $\kappa_I = 0$.

ii. Equation (10.2) *depends explicitly* on κ_I.

We emphazise this dependence by writing $G_{\kappa_R,\kappa_I,q}$ instead of $G_{\kappa,q}$. However, for simplicity of notation, we still write $\mu(x,y,\kappa)$ instead of $\mu(x,y,\kappa_R,\kappa_I)$. From the above it follows that the solution $\mu(x,y,\kappa)$ although bounded everywhere, *i.e.*, for all complex values of κ, is analytic nowhere with respect to κ, and $\partial\mu/\partial\bar{k} \neq 0$.

The *departure from holomorphicity* of $\mu(x,y,\kappa)$ is measurable by $\partial\mu/\partial\bar{k}$. Hence we are lead to step **ii** of §6, namely compute $\partial\mu/\partial\bar{k}$. Differentiating (10.2) with respect to $\bar{\kappa}$ it follows that for all κ which are not eigenvalues of (10.2)

$$\frac{\partial \mu(x,y,\kappa)}{\partial \bar{k}} = \Omega(x,y,\kappa_R,\kappa_I) + \left(G_{\kappa_R,\kappa_I,q}\frac{\partial \mu(\cdot,\cdot,\kappa)}{\partial \bar{k}}\right)(x,y), \qquad \kappa \notin R^*, \tag{10.7}$$

where the matrix Ω is defined by

$$\begin{aligned} &\{\Omega\}_{ii} = 0, \quad \{\Omega\}_{ij} = T_{ij}(\kappa_R,\kappa_I)\exp\left[\theta_{ij}(x,y,\kappa_R,\kappa_I)\right], \qquad i \neq j; \\ &\theta_{ij}(x,y,\kappa_R,\kappa_I) = iC_{ij}\left(J_i\kappa_R x + \kappa_I y\right), \end{aligned} \tag{10.8}$$

and T_{ij} is given by

$$T_{ij}(\kappa_R,\kappa_I) := \frac{i}{4\pi}\operatorname{sgn}(J_i)C_{ij}\int_{-\infty}^{\infty} d\xi \int_{-\infty}^{\infty} d\eta \, \{q(\xi,\eta)\,\mu(\xi,\eta,\kappa)\}_{ij} \exp\left[-\theta_{ij}(\xi,\eta,\kappa_R,\kappa_I)\right]. \tag{10.9}$$

[R^* denotes the set of points in the complex κ-plane for which (10.2) has a homogeneous solution.]

Equation (10.7) motivates the introduction of another bounded eigenfunction, which we call:

$$N(x,y,\kappa_R,\kappa_I) = w(x,y,\kappa_R,\kappa_I) + \left(G_{\kappa_R,\kappa_I,q}\, N(\cdot,\cdot,\kappa_R,\kappa_I)\right)(x,y), \tag{10.10}$$

where w is the matrix Ω with $T_{ij} = 1$, *i.e.*,

$$\{w\}_{ii} = 0, \quad w_{ij} = \exp\left[\theta_{ij}(x,y,\kappa_R,\kappa_I)\right], \qquad i \neq j. \tag{10.11}$$

To show that $N(x,y,\kappa_R,\kappa_I)$ also solves (10.1) one needs only to show that the w above satisfies the *homogeneous* version of (10.1), *i.e.*,

$$w_x = (i\kappa_R - \kappa_I)\hat{J}w - iJw_y;$$

this is straightforward.

Equations (10.8) and (10.11) imply

$$\Omega = \sum_{\substack{i,j=1 \\ i\neq j}}^{n} T_{ij}w^{ij}, \tag{10.12}$$

where w^{ij} is a matrix with zeros everywhere, except at its ij^{th} entry,

$$\{w^{ij}\}_{l\nu} = 0, \quad i \neq l \text{ and/or } j \neq \nu; \qquad \{w^{ij}\}_{ij} = e^{\theta_{ij}}. \tag{10.13}$$

Hence, equation (10.7) implies

$$\frac{\partial \mu}{\partial \bar{\kappa}} = \sum_{\substack{i,j=1 \\ i\neq j}}^{n} T_{ij}\, N^{ij}, \tag{10.14}$$

where the matrix N^{ij} is defined by

$$N^{ij}(x,y,\kappa_R,\kappa_I) = w^{ij}(x,y) + \left(G_{\kappa_R,\kappa_I,q}N^{ij}(\cdot,\cdot,\kappa_R,\kappa_I)\right)(x,y). \tag{10.15}$$

Step iii of the method of §6 consists of finding a relationship between $\mu(x,y,\kappa)$ and $N^{ij}(x,y,\kappa_R,\kappa_I)$. This crucial relationship is as follows:

$$N^{ij}(x,y,\kappa_R,\kappa_I) = \mu(x,y,\kappa_R + i\frac{J_j}{J_i}\kappa_I)\, w^{ij}(x,y,\kappa_R,\kappa_I). \tag{10.16}$$

To derive (10.16) one uses the fact that the operator $\mathbf{g}^j_{\kappa_R,\kappa_I,q}$ possesses the following property:

$$\left(\mathbf{g}^j_{\kappa_R,\kappa_I,q} \exp\left[\theta_{ij}(\cdot,\cdot,\kappa_R,\kappa_I) - \theta_{ij}(x,y,\kappa_R,\kappa_I)\right] f(\cdot,\cdot)\right)(x,y) = \left(\mathbf{g}^i_{\kappa_R,\kappa_I J_j/J_i,q} f(\cdot,\cdot)\right)(x,y). \tag{10.17}$$

This equation can be established by a simple change of variables.

Equations (10.14) and (10.15) imply

$$\frac{\partial\mu(x,y,\kappa)}{\partial\bar{\kappa}} = \sum_{\substack{i,j=1\\ i\neq j}}^{n} \mu(x,y,\kappa_R + i\frac{J_j}{J_i}\kappa_I)\,\Omega^{ij}(x,y,\kappa_R,\kappa_I), \qquad \kappa\notin R^*, \tag{10.18}$$

where the matrix Ω^{ij} has a nonzero element only at its ij^{th} entry,

$$\{\Omega^{ij}(x,y,\kappa_R,\kappa_I)\}_{\nu l} = \begin{cases} 0, & \nu\neq i \text{ and/or } l\neq j,\\ T_{ij}(\kappa_R,\kappa_I)\exp\left[\theta_{ij}(x,y,\kappa_R,\kappa_I)\right], & \nu = i,\ l = j. \end{cases} \tag{10.19}$$

Equation (10.18) is the basic equation needed for the solution of the inverse problem of (10.1). However, this equation is complete only if equation (10.2) has no homogeneous solutions. In the sequel we are going to obtain the additional relationship needed in the case that homogeneous modes exist. The following proposition is valid:

Proposition 4.1

Suppose that equation $(10.5)^i$ has a homogeneous solution $\boldsymbol{\phi}^i(x,y)$ at κ^i_R,κ^i_I. Then equation $(10.5)^j$, with $j = 1,\ldots,n$ and $i\neq j$, also has a homogeneous solution $\boldsymbol{\phi}^i(x,y)\exp[-\hat{\theta}_{ji}(x,y,\kappa^i_R,\kappa^i_I)]$ at the position $\kappa^i_R,\kappa^i_I J_i/J_j$, where $\hat{\theta}_{ji}(x,y,\kappa^i_R,\kappa^i_I) = \theta_{ji} + i\sigma_{ji}$, and σ_{ji} is a constant depending on κ^i.

The above result follows directly from equation (10.17) and the definition of $\boldsymbol{\phi}^i(x,y)$:

$$\boldsymbol{\phi}^i(x,y) = \left(\mathbf{g}^i_{\kappa^i_R,\kappa^i_I,q}\boldsymbol{\phi}^i(\cdot,\cdot)\right)(x,y). \tag{10.20}$$

The homogeneous solutions $\boldsymbol{\phi}^i$ give rise to lumps for the corresponding nonlinear equations in a way similar to that found in B–O and KPI. Suppose that $(10.5)^i$ has homogeneous solutions $\boldsymbol{\phi}^i_{l_i}$ at $\kappa^i_{l_i}$, where $l_i = 1,2,\ldots,\lambda_i$. Then all $\boldsymbol{\mu}^j$, with $j = 1,2,\ldots,n$ and $j\neq i$ will have singularities at $\kappa^i_{R l_i},\kappa^i_{I l_i} J_i/J_j$. Assuming these singularities are simple poles we obtain the following representation for the vector $\boldsymbol{\mu}^i$:

$$\begin{aligned}\boldsymbol{\mu}^i(x,y,\kappa) = \tilde{\boldsymbol{\mu}}^i(x,y,\kappa) &+ \sum_{l_1=1}^{\lambda_1} \frac{\boldsymbol{\phi}^1_{l_1}\exp[-\hat{\theta}_{i1}(x,y,\kappa^1_{R l_1},\kappa^1_{I l_1})]}{\kappa - \left(\kappa^1_{R l_1} + i\kappa^1_{I l_1}\frac{J_1}{J_i}\right)} + \cdots\\ &+ \sum_{l_i=1}^{\lambda_i}\frac{\boldsymbol{\phi}^i_{l_i}}{\kappa-\kappa^i_{l_i}} + \cdots + \sum_{l_n=1}^{\lambda_n}\frac{\boldsymbol{\phi}^n_{l_n}\exp[-\hat{\theta}_{in}(x,y,\kappa^n_{R l_n},\kappa^n_{I l_n})]}{\kappa - \left(\kappa^n_{R l_n} + i\kappa^n_{I l_n}\frac{J_n}{J_i}\right)},\end{aligned} \tag{10.21a}$$

where

$$\hat{\theta}_{ij}(x,y,\kappa^i_{R_{l_i}},\kappa^i_{I_{l_i}}) = \theta_{ij}(x,y,\kappa^i_{R_{l_i}},\kappa^i_{I_{l_i}}) + (\sigma_{ij})^i_{l_i}. \tag{10.21b}$$

The above representation takes the concise form

$$\boldsymbol{\mu}^i(x,y,\kappa) = \tilde{\boldsymbol{\mu}}^i(x,y,\kappa) + \mathbf{s}^i(x,y,\kappa), \tag{10.22}$$

where $\tilde{\boldsymbol{\mu}}^i$ is the nonsingular part of $\boldsymbol{\mu}^i$, and $\mathbf{s}^i$ represents the singular portion of $\boldsymbol{\mu}^i$:

$$\mathbf{s}^i(x,y,\kappa) = \sum_{j=1}^{n}\sum_{l_j=1}^{\lambda_j} \frac{\boldsymbol{\phi}^j_{l_j}\exp[-\hat{\theta}_{ij}(x,y,\kappa^j_{R_{l_j}},\kappa^j_{I_{l_j}})]}{\kappa - \left(\kappa^j_{R_{l_j}} + i\kappa^j_{I_{l_j}}\frac{J_j}{J_i}\right)} \tag{10.23}$$

Equations (10.18), (10.22), and (10.23) imply

$$\frac{\partial\mu(x,y,\kappa)}{\partial\bar{k}} = \begin{cases} \displaystyle\sum_{\substack{i,j=1\\ i\neq j}}^{n} \mu(x,y,\kappa_R + i\frac{J_j}{J_i}\kappa_I)\,\Omega^{ij}(x,y,\kappa_R,\kappa_I), & \kappa\notin R^*, \\ \Delta(x,y,\kappa), & \kappa\in R^*, \end{cases} \tag{10.24}$$

where R^* is the set of eigenvalues of (10.2) and Δ is the matrix $\Delta(x,y,\kappa) = \left(\boldsymbol{\Delta}^1(x,y,\kappa),\ldots,\boldsymbol{\Delta}^n(x,y,\kappa)\right)$ with

$$\boldsymbol{\Delta}^i(x,y,\kappa) = \pi\sum_{j=1}^{n}\sum_{l_j=1}^{\lambda_j}\boldsymbol{\phi}^j_{l_j}\exp[-\hat{\theta}_{ij}(x,y,\kappa^j_{R_{l_j}},\kappa^j_{I_{l_j}})]\,\delta\left[\kappa - \left(\kappa^j_{R_{l_j}} + i\kappa^j_{I_{l_j}}\frac{J_j}{J_i}\right)\right]. \tag{25}$$

To solve the DBAR problem associated with (10.24) one has further to establish a relationship between $\boldsymbol{\mu}^i$ and $\boldsymbol{\phi}^i$. This relationship is as follows:

$$\lim_{\kappa\to\kappa^i_{l_i}}\left[\boldsymbol{\mu}^i(x,y,\kappa) - \frac{\boldsymbol{\phi}^i_{l_i}(x,y)}{\kappa-\kappa^i_{l_i}}\right] = \boldsymbol{\phi}^i_{l_i}(-iJ_ix + y + \gamma^i_{l_i}), \qquad l_i = 1,2,\ldots,\lambda_i, \tag{10.26}$$

where $\boldsymbol{\phi}^i_{l_i}$ is normalized by $\lim[\boldsymbol{\phi}^i_{l_i}(J_ix+iy)] \to i$ for large x and y, and the constant $\gamma^i_{l_i}$ can be fixed from asymptotics. Equation (10.26) is valid provided that

$$\begin{aligned} &1 - \frac{\operatorname{sgn}(J_i)}{2\pi i}\int_{-\infty}^{\infty}d\xi\int_{-\infty}^{\infty}d\eta\,\{q(\xi,\eta)\,\boldsymbol{\phi}^i_{l_i}(\xi,\eta)\}_i = 0, \\ &\int_{-\infty}^{\infty}d\xi\int_{-\infty}^{\infty}d\eta\,\exp[-\theta_{li}(\xi,\eta,\kappa^i_{R_{l_i}},\kappa^i_{I_{l_i}})]\{q(\xi,\eta)\,\boldsymbol{\phi}^i_{l_i}(\xi,\eta)\}_l = 0, \end{aligned} \qquad l = 1,2,\ldots,n, \quad l\neq i. \tag{10.27}$$

Equations (10.26) and (10.27) can be derived following similar steps to those used in B–O and KPI.

10.2 The Inverse Problem

The basic equations formally expressing the solution of the inverse problem of (10.1) is the matrix equation

$$\mu(x,y,\kappa) - \left(T_{x,y,\Omega}\,\mu(x,y,\cdot)\right)(\kappa) - s(x,y,\kappa) = 1 \tag{10.28}$$

and the following set of vector equations

$$(-ixJ_1 + y + \gamma^1_{l_1})\boldsymbol{\phi}^1_{l_1} - \hat{\mathbf{T}}^1_{l_1}(x,y) - \hat{\mathbf{s}}^1_{l_1}(x,y) = \mathbf{I}^1, \qquad l_1 = 1,2,\ldots,\lambda_1, \tag{10.29$^1_{l_1}$}$$

$$\ldots$$

$$(-ixJ_n + y + \gamma^n_{l_n})\boldsymbol{\phi}^n_{l_n} - \hat{\mathbf{T}}^n_{l_n}(x,y) - \hat{\mathbf{s}}^n_{l_n}(x,y) = \mathbf{I}^n, \qquad l_n = 1,2,\ldots,\lambda_n. \tag{10.29$^n_{l_n}$}$$

In (10.28) the linear operator $T_{x,y,\Omega}$ is defined by

$$\big(T_{x,y,\Omega}f(\cdot)\big)(\kappa) := \frac{1}{2\pi i}\iint_{R_\infty} dz \wedge d\bar{z}\,\frac{1}{z-\kappa}\sum_{\substack{i,j=1\\ i\neq j}}^{n} f(z_R + i\frac{J_j}{J_i}z_I)\,\Omega^{ij}(x,y,z_R,z_I), \tag{10.30}$$

[R_∞ is the entire z-complex plane, $dz \wedge d\bar{z} = -2i\,dz_R\,dz_I$] and the matrix $S(x,y,\kappa)$ has columns $\mathbf{s}^i$, $i = 1,2,\ldots,n$ which are defined by (10.23). In (10.29) the vectors $\hat{\mathbf{T}}^i_{l_i}$ and $\hat{\mathbf{s}}^i_{l_i}$ are defined by

$$\begin{aligned}\hat{\mathbf{T}}^i_{l_i}(x,y) &:= \{(T_{x,y,\Omega}\,\mu(x,y,\cdot))(\kappa^i_{l_i})\}_i,\\ \hat{\mathbf{s}}^i_{l_i} &:= \sum_{j=1}^{n}\sum_{l_j=1}^{\lambda_j}\frac{\boldsymbol{\phi}^j_{l_j}\exp[-\hat{\theta}_{ij}(x,y,\kappa^j_{R_{l_j}},\kappa^j_{I_{l_j}})]}{\kappa^i_{l_i} - \left(\kappa^j_{l_j} + i\kappa^j_{l_j}\frac{J_j}{J_i}\right)},\end{aligned} \tag{10.31}$$

where $\{f\}_i$ denotes the ith column of the matrix f, and $\sum_{l_i=1}^{\lambda_i}$ denotes the sum from $l_i = 1$ to $l_i = \lambda_i$ unless any of the denominators vanishes. Equations (10.29) can be written in the concise form

$$\phi(x,y)(-ixJ + y\mathbf{I} + \Gamma) - \hat{T}(x,y) - \hat{s}(x,y) = I, \tag{10.32}$$

where the columns of ϕ, $\hat{T}$, and $\hat{s}$ are $\boldsymbol{\phi}^i_{l_i}$, $\hat{\mathbf{T}}^i_{l_i}$, and $\hat{\mathbf{s}}^i_{l_i}$ respectively, and Γ is a diagonal matrix with elements $\gamma^i_{l_i}$

The linear integral equations (10.28) and $(10.29)^i_{l_i}$ define the functions

$$\mu(x,y,\kappa), \quad \{\phi^i_{l_i}\}^{\lambda_i}_{l_i=1}, \qquad i = 1,2,\ldots,n$$

in terms of the *scattering data*

$$\{\kappa^i_{l_i}, \gamma^i_{l_i}, (\sigma_{ij})^i_{l_i}\}^{\lambda_i}_{l_i=1}, \quad i = 1,2,\ldots,n; \qquad T_{ij}(\kappa_R,\kappa_I), \quad \begin{matrix} i,j = 1,2,\ldots,n\\ i\neq j\end{matrix} \tag{10.33}$$

Given the scattering data (10.33), equations (10.28) and $(10,29)^i_{l_i}$ yield the functions $\mu(x,y,\kappa)$ and $\boldsymbol{\phi}^i_{l_i}$. Then the potential $q(x,y)$ can be reconstructed from

$$\begin{aligned} q(x,y) = \hat{J}\Bigg\{&\frac{1}{2\pi}\iint_{R_\infty} dz \wedge d\bar{z}\sum_{\substack{i,j=1\\ i\neq j}}^{n}\mu(x,y,z_R + i\frac{J_j}{J_i}z_I)\,\Omega^{ij}(x,y,z_R,z_I)\\ &- i\sum_{j=1}^{n}\Bigg\{\sum_{l_j=1}^{\lambda_j}\boldsymbol{\phi}^j_{l_j}\exp[-\hat{\theta}_{1j}(x,y,\kappa^j_{R_{l_j}},\kappa^j_{I_{l_j}})] + \cdots\\ &+ \sum_{l_j=1}^{\lambda_j}\boldsymbol{\phi}^j_{l_j}\exp[-\theta_{nj}(x,y,\kappa^j_{R_{l_j}},\kappa^j_{I_{l_j}})]\Bigg\}\Bigg\}. \end{aligned} \tag{10.34}$$

The above equations can be obtained by using (6.3), (10.24), and (10,26).

We note that pure lump solutions correspond to $\Omega^{ij} = 0$ and hence are obtained via the *linear system of algebraic equations* $(10.29)^i_{l_i}$, with $\Omega^{ij} = 0$.

10.3 On the initial value problem of DSII

Consider the Lax pair

$$\begin{cases} \psi_x = q\psi - iJ\psi_y, \\ \psi_t = A_1\psi + A_2\psi_y + a_3\psi_{yy}, \end{cases} \tag{10.35}$$

and assume (9.20) (J and A_{30} are real and purely imaginary diagonal matrices respectively). Letting $\psi = \mu \exp[\kappa(iJx - y + \kappa A_{30}t)]$, equations (10.35) yield (10.1) and

$$\mu_t = A_3\mu_{yy} + A_2\mu_y + A_1\mu + \kappa^2(A_3\mu - \mu A_{30}) - 2\kappa A_3\mu_y - \kappa A_2\mu.$$

The evolution of the scattering data Ω^{ij} —of (10.19)— associated with (10.35) is given by

$$\Omega^{ij}(x, y, \kappa_R, \kappa_I, t) = \exp(\hat{\kappa}^2 A_{30}t)\,\Omega^{ij}(x, y, \kappa_R, \kappa_I, 0)\exp(-\kappa^2 A_{30}t), \qquad \hat{\kappa} := \kappa_R + i\frac{J_j}{J_i}\kappa_I. \tag{10.36}$$

To derive equation (10.36) apply the linear operator L_κ,

$$L_\kappa := \partial_t - A_3\partial_y^2 - A_2\partial_y - A_1 - \kappa^2 A_3 + 2\kappa A_3\partial_y + \kappa A_2$$

on equation (10.18). Similarly one can establish that

$$\begin{aligned} \frac{\partial}{\partial t}\gamma^i_{l_i} &= -2\kappa^i_{l_i}A_{30_i}, \\ \frac{\partial}{\partial t}(\sigma_{ji})^i_{l_i} &= A_{30_j}\left[2i\kappa^i_{R_{l_i}}\kappa^i_{I_{l_i}}\left(1 - \frac{J_i}{J_j}\right) - \left(1 - \frac{J_i}{J_j}\right)^2 (\kappa^i_{I_{l_i}})^2\right]. \end{aligned} \tag{10.37}$$

We call DSII the following set of equations:

$$\begin{aligned} iQ_t + \tfrac{1}{2}(Q_{yy} - Q_{xx}) &= \sigma|Q|^2Q + \phi Q, \\ \phi_{xx} + \phi_{yy} &= -2\sigma(|Q|^2)_{xx}. \end{aligned} \tag{10.38}$$

Equations (10.38) are the compatibility conditions of (10.35) with

$$J_1 = 1, \quad J_2 = -1, \qquad A_2 = -q, \quad A_3 = \mathrm{diag}(i, -i) \tag{10.39}$$

and A_1 and q are defined by (9.25) with

$$\begin{aligned} A_{12} &= -\tfrac{1}{2}(iQ_x + Q_y), \\ A_{21} &= -\tfrac{1}{2}\sigma(-i\overline{Q}_x + \overline{Q}_y), \\ A_{11_x} + iA_{11_y} &= \tfrac{1}{2}\sigma\left[i(Q\overline{Q})_x + (Q\overline{Q})_y\right], \\ A_{22_x} - iA_{22_y} &= \tfrac{1}{2}\sigma\left[-i(Q\overline{Q})_x + (Q\overline{Q})_y\right]. \end{aligned} \tag{10.40}$$

Hence (assuming that A_{11} and A_{22} tend to zero for large x and y) the initial value problem of (10.38) can be solved via (10.28), (10.29), and (10.34). In these equations $J = \mathrm{diag}(1, -1)$; furthermore, the evolution of the scattering data (10.33) is determined from equations (10.36) and (10.37), where $A_{30} = \mathrm{diag}(i, -i)$.

This work was partially supported by the Office of Naval Research under Grant Number NOOO14-76-C-0867, the National Science Foundation under Grant Number MCS-8202117, and the Air Force Office of Scientific Research under Grant Number 78-3674-D.

References

[1] V. E. Zakharov, S. V. Manakov, S. P. Novikov, and L. P. Pitayevskiĭ, *Theory of Solitons. The Method of the Inverse Scattering Problem*, Nauka, Moscow (in Russian), 1980.

[2] R. Beals and R. R. Coifman: (i) Scattering, transformations spectrales, et équations d'évolution non lineaires, Seminaire Goulaoric–Meyer–Schwartz, 1980–1981, exp. 22, École Polytechnique, Palaiseau; (ii) Scattering and inverse scattering for first order systems (preprint); (iii) Scattering, transformations spectrales, et équations d'évolution non lineaires, Seminaire Goulaoric–Meyer–Schwartz, 1981–1982, exp. 21, École Polytechnique, Palaiseau.

[3] (i) I. Hauser and F. J. Ernst, A homogeneous Hilbert problem for the Kinnersley–Chitre transformations. *J. Math. Phys.* **21**, 1126–1140 (1980); (ii) A homogeneous Hilbert problem for the Kinnersley–Chitre transformations of electrovac space-times. *ibid.* **21**, 1418–1422 (1980); (iii) Proof of a Geroch conjecture. *ibid.* **22**, 1051 (1981); (iv) V. A. Belinskiĭ and V. E. Zakharov, Integration of the Einstein equations by the inverse scattering method and calculation of the exact soliton solution. *Sov. Phys. JETP* **48**, 985–993 (1978); (v) C. Cosgrove, Bäcklund transformations in the Hauser–Ernst formalism for stationary axisymmetric space-times. *J. Math. Phys.* **22**, 2624–2639 (1981).

[4] (i) M. F. Atiyah and R. S. Ward, instantons and algebraic geometry. *Commun. Math. Phys.* **55**, 117–124 (1977); (ii) M. F. Atiyah, N. J. Hitchin, V. G. Drinfield, and Manin I. Yu., Construction of instantons. *Phys. Lett.* **65A**, 185–187 (1978).

[5] P. J. Olver, Evolution equations possessing infinetely many symmetries *J. Math. Phys.* **18**, 1212–1215 (1977); *Math. Proc. Camb. Phil. Soc.* **88**, 71 (1980).

[6] F. Magri, A simple model of the integrable hamiltonian equation. *J. Math. Phys.* **19**, 1156–1162 (1978); *ibid.* Lecture Notes in Physics #120, p. 233, M. Boiti, F. Pipinelli, and G. Soliani, eds. Springer Verlag (1978).

[7] (i) I. M. Gel'fand and L. A. Dikii, Resolvents and hamiltonian systems. *Funct. Anal. Appl.* **11**, 93–104 (1977); (ii) I. M. Gel'fand and I. Ya. Dorfman, *Funct. Anal. Appl.* **13**, 248 (1979); *ibid.* **14** (1980).

[8] (i) A. S. Fokas, A symmetry approach to exactly solvable evolution equations. *J. Math. Phys.* **21**, 1318 (1980);

(ii) A. S. Fokas and R. L. Anderson, On the use of isospectral eigenvalue problems for obtaining hereditary symmetries for hamiltonian systems. *J. Math. Phys.* **23**, 1066 (1982);

(iii) A. S. Fokas and B. Fuchssteiner, Bäcklund transformations for hereditary symmetries. *Nonlinear Anal. TMA* **5**, 423 (1981);

(iv) *ibid.*, On the structure of symplectic operators and hereditary symmetries. *Lett. Nuovo Cimento* **28**, 299 (1980);

(v) *ibid.*, The hierarchy of the Benjamin–Ono equation. *Phys. Lett.* **86A**, 341 (1981);

(vi) B. Fuchssteiner and A. S. Fokas, Symplectic structures, their Bäcklund transformations and hereditary symmetries. *Physica* **4D**, 47 (1981);

(vii) B. Fuchssteiner, *Nonlinear Anal. TMA* **3**, 849 (1979); *ibid. Prog. Theor. Phys.* **65**, 861 (1981); *ibid.* **68** (1982); *ibid. Lett. Math. Phys.* **4**, 1977 (1980).

[9] A. P. Fordy and J. Gibbons, Factorization of operators. II. *J. Math. Phys.* **22**, 1170 (1981); *ibid.*, Integrable nonlinear Klein–Gordon equations and Toda lattices. *Commun. Math. Phys.* **77**, 21 (1980).

[10] (i) B. A. Kupershmidt and G. Wilson, *Invent. Math.* **62**, 403 (1981); *ibid. Commun. Math. Phys.* **81**, 189 (1981); (ii) Y. Kosmann–Schwartzbach, Hamiltonian systems of fibered manifolds. *Lett. Math. Phys.* **5**, 229–237 (1981).

[11] B. G. Konopelchen'ko, Transformation properties of the integrable evolution equations. *Phys. Lett.* **75A**, 447 (1980); *ibid.* **79A**, 39 (1980); *ibid.* **95B**, 83 (1980); *ibid.* **100B**, 254–260 (1981).

[12] G. Z. Tu, Commutativity theorem of partial differential equations. *Commun. Math. Phys.* **77**, 289–297 (1980).

[13] (i) C. S. Gardner, The Korteweg–de Vries equation and generalizations. IV. The Korteweg–de Vries equation as a hamiltonian system. *J. Math. Phys.* **12**, 1548–1551 (1971); (ii) V. E. Zakharov and L. D. Fadeev, Korteweg–de Vries equation, a completely integrable hamiltonian system. *Funct. Anal. Appl.* **5**, 280–287 (1971).

[14] H. D. Wahlquist and F. B. Estabrook, Prolongation structures and nonlinear evolution equations *J. Math. Phys.* **16**, 1–7 (1975) *ibid.* **17**, 1293–1297 (1976); see also the contribution of these authors in *Nonlinear Equations Solvable Via the IST*, F. Calogero ed. (1977).

[15] L. Abellanas and A. J. Galindo, *J. Math. Phys.* **20**, 1239 (1979); *ibid. J. Math. Phys.***24**, 504 (1983); *ibid. Lett. Math. Phys.* **6**, 391 (1982).

[16] W. Oevel and B. Fuchssteiner, Explicit formulas for symmetries and conservation laws of the Kadomtsev–Petviashvili equation. *Phys. Lett.* **88A**, 323–327 (1982).

[17] H. H. Chen, J. C. Lee, and J. E. Lin, to appear in *Nonlinear Waves* (preprint 1982).

[18] B. G. Konopelchen'ko, The two-dimensional matrix spectral problem: General structure of the integrable equations and their Bäcklund transformations. *Phys. Lett.* **86A**, 346–350 (1981); *ibid. Commun. Math. Phys.* (1982); *ibid. J. Phys. A.* **15** (1982).

[19] M. Kashiwara and T. Miwa, *Proc. Japan Acad.* **57A**, 342 (1981); E. Date, M. Kashiwara, and T. Miwa, *Proc. Japan Acad.* **57A**, 387 (1981); M. Jimbo, E. Date, M. Kashiwara, and T. Miwa, *J. Phys. Soc. of Japan* **50**, 3806 (1981); and RIMS preprints 359 and 360.

[20] M. J. Ablowitz and H. Segur, *Solitons and the Inverse Scattering Transform*, SIAM Studies in Applied Mathematics #4, 1981.

[21] H. H. Chen, A Bäcklund transformation in two dimensions. *J. Math. Phys.* **16**, 2382–2384 (1975); D. V. Chudnovski, *J. Math. Phys.* **20**, 2416 (1979).

[22] J. Satsuma, *N*-soliton solution of the two-dimensional Korteweg–de Vries equation. *J. Phys. Soc. of Japan* **40**, 286–290 (1976); J. Satsuma and M. J. Ablowitz, Two-dimensional lumps in nonlinear dispersive systems. *J. Math. Phys.* **20**, 1496 (1979); A. Nakamura, *Phys. Rev. Lett.* **46**, 751 (1981); *ibid. Phys. Lett.* **88A**, 55 (1982); *ibid. J. Math. Phys.* **22**, 2456 (1981); *ibid. J. Phys. Soc. of Japan* **50**, 2469 (1981) *ibid.* **51**, 19 (1981).

[23] V. E. Zakharov and P. B. Shabat, A scheme for integrating the nonlinear equations cf mathematical physics by the method of the inverse scattering problem. I. *Funct. Anal. Appl.* **8**, 226–235 (1974).

[24] (i) V. E. Zakharov and P. B. Shabat, Integration of the nonlinear equations of mathematical physics by the method of the inverse scattering problem. II. *Funct. Anal. Appl.* **13**, 166-174 (1979); (ii) V. E. Zakharov and A. V. Mikhaĭlov, Relativistically invariant two-dimensional models of field theory which are integrable by means of the inverse scattering problem method. *Sov. Phys. JETP* **47**, 1071–1027 (1978).

[25] S. V. Manakov (private communication).

[26] A. S. Fokas and M. J. Ablowitz, On a linearization of the Korteweg–de Vries and Painlevé II equations. *Phys. Rev. Lett.* **47**, 1096 (1981).

[27] P. Santini, M. J. Ablowitz, and A. S. Fokas, The Direct Linearization and the RH Direct Method. (Preprint).

[28] R. Rosales, Exact solutions of some nonlinear evolution equations. *Stud. Appl. Math.* **59**, 117–151 (1978).

[29] M. J. Ablowitz, A. S. Fokas, and R. L. Anderson, The direct linearizing transform and the Benjamin–Ono equation. *Phys. Lett. A.* **93**, 375 (1983).

[30] A. S. Fokas and M. J. Ablowitz, The inverse scattering transform for the Benjamin–Ono —a pivot to multidimensional problems. *Stud. Appl. Math.* **68**, 1–10 (1983).

[31] R. G. Newton, *J. Math. Phys.* **21**, 493 (1980); *ibid. Geophys. J. R. Astr. Soc.* **65**, 191 (1981).

[32] (i) G. R. W. Quispel and H. W. Capel, *Phys. Lett.* **88A**, 371 (1981); *ibid* **85A**, 248 (1981); *ibid Physica* **110A**, 41 (1981); (ii) F. W. Nijhoff, J. Van der Linden, G. R. W. Quispel, H. W. Capel, and J. Velthuizen, *Physica A*, (1982).

[33] G. B. Whitham (private communication).

[34] A. S. Fokas and M. J. Ablowitz, On a unified approach to transformations and elementary solutions of Painlevé equations. *J. Math. Phys.* **23**, 2033–2042 (1982).

[35] A. S. Fokas and M. J. *Ablowitz, Direct Linearization of the Korteweg-de Vries Equation.* AIP Conference Proceedings #88, 237–241 (1982). M. Tabor and Y. M. Treve eds.

[36] A. S. Fokas and M. J. Ablowitz, On the inverse scattering and direct linearizing transforms for the Kadomtsev–Petviashvili equation. *Phys. Lett. A.* **94A**, 67–70 (1983).

[37] M. J. Ablowitz and A. S. Fokas, *A Direct Linearization Associated with the Benjamin–Ono equation.* Conference Proceedings #88, 229–236 (1982). M. Tabor and Y. M. Treve eds.

[38] P. D. Lax, Integrals of nonlinear equations of evolution and solitary waves. *Comm. Pure Appl. Math.* **21**, 467–490 (1968).

[39] C. S. Gardner, J. M. Greene, M. D. Kruskal, and R. M. Miura, The Korteweg–de Vries equation and generalizations. VI. Methods for exact solution. *Comm. Pure Appl. Math.* **27**, 97–133 (1974).

[40] P. Delft and E. Trubowitz, Inverse scattering on the line. *Comm. Pure Appl. Math.* **32**, 121–125 (1979).

[41] (i) M. J. Ablowitz, D. J. Kaup, A. C. Newel, and H. Segur, Method for solving the sine-Gordon equation. *Phys. Rev. Lett.* **30**, 1262–1264 (1973); (ii) *ibid.*, Nonlinear evolution equations of physical significance. *Phys. Rev. Lett.* **31**, 125–127 (1973); (iii) *ibid.*, The inverse scattering transform —Fourier analysis for nonlinear problems. *Stud. Appl. Math.* **53**, 249–315 (1974).

[42] V. E. Zakharov and P. B. Shabat, Exact theory of two-dimensional self-focusing and one-dimensional self-modulation of waves in nonlinear media. *Sov. Phys. JETP* **34**, 62-69 (1972).

[43] G. B. Whitham, *Linear and Nonlinear Waves.* Wiley-Interscience, 1974.

[44] D. J. Kaup, On the inverse scattering problem for cubic eigenvalue problems of the class $\psi_{xxx} + 6Q\psi_x + 6R\psi = \lambda\psi$. *Stud. Appl. Math.* **62**, 189 (1980).

[45] P. Delft, C. Tomei, and E. Trubowitz, *Commun. Pure Appl. Math.* **35**, 567 (1982).

[46] D. J. Kaup, The three-wave interaction —a nondispersive phenomenon. *Stud. Appl. Math.* **55**, 9–44 (1976).

[47] V. E. Zakharov and S. V. Manakov, The theory of resonance interaction of wave packets in nonlinear media. *Sov. Phys. JETP* **42**, 842–850 (1976).

[48] W. Symes, Relations among generalized Korteweg–de Vries systems. *J. Math. Phys.* **20**, 721–725 (1979).

[49] M. J. Ablowitz and R. Haberman, Nonlinear evolution equations —two and three dimensions. *Phys. Rev. Lett.* **35**, 1185–1188 (1975).

[50] A. C. Newell, The general structure of integrable evolution equations. *Proc. R. Soc. London* **A365**, 283–311 (1979).

[51] R. Beals, *The Inverse Problem for Ordinary Differential Operators on the Line*, Yale, 1982 (preprint).

[52] P. J. Caudrey, The inverse problem for a general $N \times N$ spectral equation. *Physica* **6D**, 51–66 (1982).

[53] (i) F. Calogero and A. Degasperis, Nonlinear evolution equations solvable by the inverse spectral transform. I and II. *Nuovo Cimento* **32B**, 201–242 (1976); (ii) *ibid.*, Nonlinear evolution equations solvable by the inverse spectral transform. *Nuovo Cimento* **39B**, 1–54 (1977).

[54] (i) M. Wadati, The modified Korteweg–de Vries equation. *J. Phys. Soc. of Japan* **32**, 1681 (1977); (ii) M. Wadati, H. Sanki, and K. Konno, Relationships among inverse methods, Bäcklund transformation and an infinite number of conservation laws. *Prog. Theor. Phys.* **53**, 419–436 (1975).

[55] M. Jaulent, Inverse scattering problems in absorbing media. *J. Math. Phys.* **17**, 1351–1360 (1976).

[56] D. J. Kaup and A. C. Newell, An exact solution for a derivative nonlinear Schrödinger equation. *J. Math. Phys.* **19**, 798–801 (1978); A. V. Mikhaĭlov, *Sov. Phys. JETP* **23**, 356 (1976).

[57] L. M. Alonso, Schrödinger spectral problems with energy-dependent potentials as sources of nonlinear Hamiltonian evolution equations. *J. Math. Phys.* **21**, 2342–2349 (1980).

[58] (i) Y. Kodama, J. Satsuma, and M. J. Ablowitz, Nonlinear intermediate long wave equation: analysis and method of solution. *Phys. Rev. Lett.* **46**, 687–690 (1981); (ii) Y. Kodama, M. J. Ablowitz, and J. Satsuma, Direct and inverse scattering problem of nonlinear intermediate long wave equations. *J. Math. Phys.* **23**, 564–576 (1982).

[59] T. L. Bock and M. D. Kruskal, A two-parameter Miura transformation of the Benjamin–Ono equation *Phys. Lett.* **74A**, 173–176 (1979).

[60] A. Nakamura, A direct method calculating periodic wave solutions to nonlinear evolution equations. I. Exact two-periodic wave solution. *J. Phys. Soc. of Japan* **47**, 1701–1705 (1979).

[61] J. Satsuma, T. Taha, and M. J. Ablowitz, On a Bäcklund Transformation and Scattering Problem for the Modified Intermediate Long Wave Equation. Preprint INS #12, May 1982.

[62] E. Hille, *ODE's in the Complex Domain.* Wiley-Interscience, 1976.

[63] B. Riemann, *Gesammelte Mathematische Werke.* Leipzig, 1982.

[64] D. V. Chudnovski, In *Bifurcation Phenomena in Mathematical Physics and Related Topics*, p. 385. C. Bardos and D. Bessis eds., D. Reidel, 1980.

[65] H. Flaschka and A. C. Newell, Monodromy and spectrum preserving deformations. *Comm. Math. Phys.* **76**, 65–116 (1980).

[66] (i) M. Sato, T. Miwa, and M. Jimbo, *RIMS, Kyoto Univ.* **14**, 223 (1977); *ibid.* **15**, 201 (1979); *ibid.* **15**, 871 (1979); (ii) M. Jimbo, T. Miwa, and K. Ueno, *Physica* **2D**, 306 (1981); M. Jimbo and T. Miwa, *Physica* **2D**, 407 (1981); *ibid. Physica* **4D**, 26 (1982).

[67] D. Hilbert, *Grundzüge der Integralgleichungen.* Leipzig, 1924.

[68] J. Plemelj, *Problems in the Sense of Riemann and Klein.* Wiley, 1964.

[69] G. D. Birkhoff, *Transactions of the AMS*, 436 (1909); *ibid.* 199 (1910).

[70] N. P. Erugin, *Linear Systems of Ordinary Differential Equations.* Academic Press, 1966.

[71] F. D. Gakhov, *Boundary Value Problems.* Pergamon, 1966.

[72] N. I. Muskhelishvili, *Singular Integral Equations.* Noordhoff, Groningen, 1953.

[73] N. P. Vekua, *Systems of Singular Integral Equations.* Gordon and Breach, 1967.

[74] (i) P. P. Zabreyko, A. I. Koshelev, M. A. Krasnoselskiĭ, S. G. Mikhlin, L. S. Rakovshchik, and V. Ya, *Integral Equations —a Reference Text.* Noordhoff Int., Leyden, 1975; (ii) S. Prossdorf, *Some Classes of Singular Equations.* North-Holland, 1978.

[75] M. G. Krein, *Uspekhi Mat. Nauk* **13**, 3 (1958).

[76] I. Gohberg and M. G. Krein, *Uspekhi Mat. Nauk* **13**, 2 (1958).

[77] P. Delft (private communication).

[78] A. S. Fokas and M. J. Ablowitz, On the initial value problem of Painlevé II. (Preprint).

[79] R. G. Newton, On a generalized Hilbert problem. *J. Math. Phys.* **23**, 2257–2265 (1982).

[80] B. B. Kadomtsev and V. I. Petviashvili, On the stability of solitary waves in weakly dispersive media. *Sov. Phys. Doklady* **15**, 539–541 (1970).

[81] V. Dryuma, *Sov. Phys. JETP Lett.* **19**, 387 (1974).

[82] A. Davey and K. Stewartson, On three-dimensional packets of surface waves. *Proc. Roy. Soc. London* **A338**, 101–110 (1974).

[83] D. J. Benney and G. J. Roskes, Wave instabilities. *Stud. Appl. Math.* **48**, 377–385 (1969).

[84] (i) V. E. Zakharov and S. V. Manakov, Soliton theory. *Sov. Sci. Phys. Rev.* **1**, 133–190 (1979); (ii) S. V. Manakov, The inverse scattering transform for the time-dependent Schrödinger equation and Kadomtsev–Petviashvili equation. *Physica* **3D**, 420–427 (1981).

[85] H. Segur, *Comments on IS for the Kadomtsev–Petviashvili Equation*, AIP Conference Proceedings #88, 211–228 (1982). La Jolla, 1981. M. Tabor and Y. M. Treve eds.

[86] (i) D. J. Kaup, The inverse scattering solution for the full three-dimensional three-wave resonant interaction. *Physica* **1D**, 45–67 (1980); (ii) H. Cornille, Solutions for the nonlinear three-wave equations in three-spatial dimensions. *J. Math. Phys.* **20**, 1653–1666 (1979); (iii) L. P. Niznik, *Ukranian Math. J.* **24**, 110 (1972).

[87] A. S. Fokas and M. J. Ablowitz, On the inverse scattering of the time dependent Schrödinger equation and the associated KPI equtaion, INS #22, Nov. 1982 (to appear in *Stud. Appl. Math.*).

[88] M. J. Ablowitz, D. Bar Yaacov, and A. S. Fokas, On the IST for KPII, INS #21, Nov. 1982 (to appear in *Stud. Appl. Math.*).

[89] A. S. Fokas, On the inverse scattering of first order systems in the plane related to nonlinear multidimensional equations, INS #23, Dec. 1982 (preprint).

[90] A. S. Fokas and M. J. Ablowitz, On a method of solution for a class of multidimensional problems. (Preprint).

[91] A. S. Fokas and M. J. Ablowitz, On the inverse scattering transform of multidimensional nonlinear equations related to first order systems in the plane. Preprint INS #24, Jan. 1983.

[92] L. Hörmander, *Complex Analysis in Several Variables.* Van Nostrand, 1966.

The Technique of Variable Separation for Partial Differential Equations

Willard Miller, Jr.

School of Mathematics
University of Minnesota, USA

Contents:

Introduction

In these lectures I will present the basic concepts in the theory of separation of variables for partial differential equations and explore some of the deep relations between variable separation and the generalized Lie symmetries of these equations. Historically, the theory of variable separation has been developed most intensively and proved most useful for two classes of partial differential equations: first order differential equations and linear equations of all orders. However, the concepts are clearly applicable to general nonlinear equations.

The primary use of variable separation is for the computation of explicit solutions of partial differential equations. The solutions can be calculated by solving ordinary differential equations (the separation equations). Many of the solutions obtained by this method prove so important that these functions are studied and tabulated in their own right: the special functions of mathematical physics. For Hamilton–Jacobi equations variable separation is used to obtain complete integrals, which in turn lead to explicit solutions of the associated Hamiltonian system.

Basically, a partial differential equation is (additively) separable in the independent variables $x_1, \ldots, x_n$ if the equation admits a nontrivial solution of the form $u = \sum_{i=1}^{n} S^{(i)}(x_i)$. One can also talk about product separation $v = \prod_{i=1}^{n} T^{(i)}(x_i)$ or more complicated types of separation such as $w = \tan\left[\sum_{i=1}^{n} S^{(i)}(x_i)\right]$. However, a change of dependent variable reduces these other types to additive separation, *e.g.*, $u = \ln v$ or $u = \arctan w$.

We shall see that there are many varieties of additive separation and that they can be classified. This should reduce some of the confusion concerning basic definitions that abounds in the theory of variable separation.

In §2 we shall review briefly the theory of generalized Lie symmetries of partial differential equations and show that the standard procedure for computing ***symmetry adapted*** solutions of these equations from a knowledge of the Lie symmetries is an example of variable separation of a particularly simple type.

In §3 we shall apply the theory of variable separation to two particularly simple and physically important problems: orthogonal variable separation for Hamilton–Jacobi and Helmholtz equations. This will lead us to the mathematics of Stäckel form. Then in §4 we will provide an intrinsic characterization of variable separation for these equations in terms of Lie symmetries.

We will conclude with an application of the ideas introduced earlier; an analysis of ***related evolution equations***.

These lectures are concerned with the method of variable separation itself; space does not permit a study of the properties of the separable solutions and the relationship between these properties and Lie symmetries. This relationship is explored in the author's monograph [1].

§1 The general concept of variable separation

We begin with the definition of additive separability for a partial differential equation

$$H(x_i, u, u_i, u_{ij}, u_{ijk}, \ldots) = E \tag{1.1}$$

in the coordinates $x_1, \ldots, x_n$. Here, u is the dependent variable, $u_i = \partial_{x_i} u$, $u_{ij} = \partial_{x_i}\partial_{x_j} u$, etc., where $1 \leq i, j, k, \ldots \leq n$, and E is a parameter. We assume (for convenience) that H is a polynomial in the variables u_i, u_{ij}, ... with coefficients which are real analytic functions of the variables x_i, u, all defined in a common domain $D \times J$, $D \subseteq R^n$ with $(0, \ldots, 0) \in D$, and J an open interval on the real line.

A solution of (1.1) is a function $u = S(x, E)$ defined and analytic for x in a nonzero domain $D' \subseteq D$ and E in an open interval $I \subseteq R$, such that substitution of this function into (1.1) renders (1.1) an identity for all $(x, E) \in D' \times I$. A *separable solution* is a solution of the form $u = \sum_{j=1}^{n} S^{(j)}(x_j, E)$.

Since for a separable solution $u_{ij} = 0$ for $i \neq j$, without loss of generality we can set all mixed partial derivatives identically equal to zero in (1.1) and obtain the simpler equation

$$H(x_i, u, u_i, u_{ii}, \ldots) = E. \tag{1.2}$$

For convenience we set $u_{i,1} \equiv u_i$, $u_{i,j+1} = \partial_{x_i} u_{i,j}$, $j = 1, 2, \ldots$ and define m_i to be the largest number ℓ such that $\partial_{u_{i,\ell}} H = H_{u_{i,\ell}} \not\equiv 0$. To avoid discussion of degenerate cases we require $m_i > 0$ for $i = 1, 2, \ldots, n$. Let D_i denote the total differentiation operators

$$D_i = \partial_{x_i} + u_{i,1}\partial_u + u_{i,2}\partial_{u_{i,1}} + \cdots + u_{i,m_i+1}\partial_{u_{i,m_i}} + \cdots. \tag{1.3}$$

If u is a separable solution of (1.2) such that $H_{u_{j,m_j}} \neq 0$ in some domain for all j, then $D_i H(x, u) = 0$ or

$$u_{i,m_i+1} = -\frac{\tilde{D}_i H}{H_{u_{i,m_i}}}, \qquad i = 1, 2, \ldots, n, \tag{1.4}$$

where

$$\tilde{D}_i = \partial_{x_i} + u_{i,1}\partial_u + \cdots + u_{i,m_i}\partial_{u_{i,m_i-1}}. \tag{1.5}$$

Clearly, u satisfies the integrability conditions $D_j u_{i,m_i+1} = 0$, $j \neq i$ or

$$\begin{aligned} H_{u_{i,m_i}} H_{u_{j,m_j}} (\tilde{D}_i \tilde{D}_j H) + H_{u_{i,m_i} u_{j,m_j}} (\tilde{D}_i H)(\tilde{D}_j H) \\ = H_{u_{j,m_j}} (\tilde{D}_i H)(\tilde{D}_j H_{u_{i,m_i}}) + H_{u_{i,m_i}} (\tilde{D}_j H)(\tilde{D}_i H_{u_{j,m_j}}). \end{aligned} \tag{1.6}$$

Note that this expression is a polynomial in the variables $u_{k,\ell}$. In general, (1.6) is a restriction both on the coefficients of H and the form of the particular separable solution u. However, there is an important special case where (1.6) is an identity in the dependent variables u, $u_{k,\ell}$. [Indeed, this case will occur if (1.2) admits so many separable solutions that for each $x^0 \in D$ and each set of real constants u^0, u_i^0, $u_{ii}^0, \ldots$, $i = 1, 2, \ldots, n$ satisfying $H(x^0, u^0, u_i^0, u_{ii}^0, \ldots) = E$, there is a separable solution $u(x)$ such that $u(x^0) = u^0$, $u_i(x^0) = u_i^0, \ldots$.] Then conditions (1.6) reduce to restrictions on the coefficients of H which are independent of the choice of separable solution. If (1.6) is an identity we say that $\{x_i\}$ is a regular *separable coordinate system* (for the equation $H = E$).

Suppose $\{x_i\}$ is a regular separable coordinate system and consider the equations

$$\begin{aligned} D_i v &= v_{i,1}, \\ D_i v_{j,1} &= \delta_{i,j} v_{j,2}, \\ D_i v_{j,m_j-1} &= \delta_{i,j} v_{j,m_j}, \\ D_i v_{j,m_j} &= -\delta_{i,j} \frac{\tilde{D}_j H(x, v)}{H_{u_{j,m_j}}(x, v)}, \qquad 1 \leq i,\ j \leq n. \end{aligned} \tag{1.7}$$

The integrability condition for this system of equations is $D_kD_iv_{j,m_j} = D_iD_kv_{j,m_j}$, equivalent to (1.6). Since (1.6) is satisfied identically, for $x^0 \in D$ and each set of constants $v^0_{i,j}$, $1 \leq i \leq n$, $0 \leq j \leq m_i$, such that $H_{u_{j,m_j}}(x^0, v^0) \neq 0$, there is a unique solution v of the system (1.7) such that $v(x^0) = v^0$, $D_iv(x^0) = v^0_{i,1}$, $D_iv_{i,j}(x^0) = v^0_{i,j+1}$ (see [2, Chapter 1]). Choose E such that $E = H(x^0, v^0)$. Then $u = v(x)$ is a separable solution of (1.2) such that $u(x^0) = v^0$ and $u_{i,j}(x^0) = v^0_{i,j}$, $1 \leq i \leq n$, $1 \leq j \leq m_i$. Indeed $D_{x_i}H(x, v) = 0$, so $H(x, v) = E$.

Theorem 1. *If $\{x_i\}$ is a regular separable system for the equation $H = E$, i.e., if equations (1.6) are satisfied identically, then for every set of $m_1 + m_2 + \cdots + m_n + 1$ constants $\{v^0, v^0_{i,j}\}$ with $H(x^0, v^0) = E$ and $H_{u_{j,m_j}}(x^0, v^0) \neq 0$, there is a unique separable solution u of $H(x, u) = E$ such that $u(x^0) = v^0$, $u_{i,j}(x^0) = v^0_{i,j}$, $1 \leq i \leq n$, $1 \leq j \leq m_i$.*

If equations (1.6) are not satisfied identically, separable solutions still may exist, but they will depend on fewer than $\sum_{i=0}^{n} m_i + 1$ parameters. This type of separation is ***nonregular***.

Example 1. $H = (x_1 + x_2)(u_{11} + u_{22}) - 2(u_1 + u_2)$. Equations (1.6) are satisfied identically so $\{x_1, x_2\}$ is a regular separable system. The general separable solution depends on five parameters and is given by

$$u = (\alpha x_1^3 + \beta x_1^2 + \gamma x_1 - \tfrac{1}{2}Ex_1) + (-\alpha x_2^3 + \beta x_2^2 - \gamma x_2 + \delta). \tag{1.8}$$

Example 2. $H = u_{11}^2 + u_1 + u_{22}$. Here we have $u_{111} = -\frac{1}{2}$ (provided $u_{11} \neq 0$) and $u_{222} = 0$ so equations (1.6) are satisfied identically and $\{x_1, x_2\}$, is a regular separable system. The general separable solution depends on five parameters:

$$u = \left(-\tfrac{1}{12}x_1^3 + \alpha x_1^2 + \beta x_1\right) + \left(\tfrac{1}{2}(E - 4\alpha^2 - \beta)x_2^2 + \gamma x_2 + \delta\right). \tag{1.9}$$

Example 3. $H = x_2u_{11} + x_1u_{22} + u_1 + u_2$. Equations (1.6) reduce to the requirement $u_{11} + u_{22} = 0$. The general separable solution depends on four parameters:

$$u = \left(\alpha x_1^2 + \beta x_1\right) + \left(-\alpha x_2^2 + (E - \beta)x_2 + \gamma\right). \tag{1.10}$$

This is a nonregular separable system.

Example 4. $H = (u_{11} + u_{22})/u$. Equations (1.6) are satisfied identically for $u \neq 0$. The general separable solution depends on five parameters:

$$u = \alpha \exp(x_1\sqrt{E}) + \beta \exp(-x_1\sqrt{E}) + \gamma \exp(x_2\sqrt{E}) + \delta \exp(-x_2\sqrt{E}) \tag{1.11}$$

for $E > 0$, with obvious modifications for $E \leq 0$.

Associated with any separable coordinate system, regular or not, there is a system of separation equations. Let $u = v(x) = \sum_{j=0}^{n} v^j(x_j)$ be a particular separable solution of the equation $H(x, u) = E$, uniquely determined by its initial conditions $u = v^0$, $u_{j,k} = v^0_{j,k}$ at the point $x = x^0$. Now fix i and set $x = (y_1, \ldots, y_{i-1}, x_i, y_{i+1}, \ldots, y_n)$ where we consider x_i as a variable and y_j as a parameter, $j \neq i$. Substituting $u = v$ into $H(x, u) = E$ we obtain an ordinary differential equation

$$H^{(i)}(x_i, u^i, v^j(y_j), y_j) = E, \qquad i = 1, 2, \ldots, n, \quad j \neq i, \tag{1.12}$$

for the function $u^i = v^i(x_i)$, an equation that depends on the parameters y_j, $v^j(y_j)$. Each such expression (1.12) is a ***separation equation*** for u^i. It is important to observe that if $u^i = \hat{v}^i(x_i)$ is any solution of the

i^{th} separation equation (1.12), valid for all values of the y_j, then

$$\hat{u}(x) = \sum_{j \neq i} v^j(x_j) + \hat{v}^i(x_i) \tag{1.13}$$

is a separable solution of (1.1).

To write the separation equations in a normal form, we solve for the highest derivative term u_{i,m_i} in (1.12) and obtain

$$u_{i,m_i} = F^{(i)}(x_i, u_{i,\ell}, v^j(y_j), y_j, E), \qquad 0 \le \ell < m_i, \quad j \neq i. \tag{1.14}$$

[Since H is a polynomial in the derivatives of u, there may be several distinct solutions (1.14). We choose that solution which corresponds to v_{i,m_i}.]

We say that the separation equations corresponding to the separable coordinates $\{x_i\}$ are *normal* if each of the functions $F^{(i)}$, $i = 1, 2, \ldots, n$ in (1.14) is independent of y_j, $j \neq i$, for each fixed choice of v^j. If the separation equations are normal there is a single equation for each unknown function $u^{(i)}$. The equations then take the form

$$u_{i,m_i} = \mathcal{F}^{(i)}(x_i, u_{i,\ell}, \lambda_1, \ldots, \lambda_q), \qquad i = 1, 2, \ldots, n, \tag{1.15}$$

where $1 \le q \le \sum_{i=0}^{n} m_i + 1$, $\lambda_k = \lambda_k(u_{j,s})$, $1 \le k \le q$, and the parameters λ_k are functionally independent as functions of the $\sum_{i=0}^{n} m_i + 1$ parameters $u^0_{i,\ell}$. We choose q to be minimal: The q n-vectors $\left(\frac{\partial \mathcal{F}^{(i)}}{\partial \lambda_k}(x_i, \lambda_j)\right)$, $k = 1, \ldots, q$ are linearly independent over the field of functions $f(\lambda_1, \ldots, \lambda_q)$.

Theorem 2. *If the separation equations are normal, then $q = n$ and for each set of solutions $u^{(i)}(x_i)$, $i = 1, 2, \ldots, n$, of the ordinary differential equations (1.15) the function $u = \sum_{i=0}^{n} u^{(i)}(x_i)$ is a separable solution of (1.2). All separable solutions of (1.2) arise in this manner.* See reference [3] for the proof.

Corollary 1. *If $\{x_i\}$ is a separable system with normal separation equations then it is a regular separable system.*

Note that Examples 2 and 4 have normal separation equations, while Examples 1 and 3 are nonnormal.

There is a similar theory of additive separation for a partial differential equation of the form (1.1) with $E = 0$, *i.e.*, an equation not depending on a parameter. We make the same assumptions on H as before and take the equation in the form (1.2) with $E = 0$:

$$H(x_i, u, u_i, u_{ii}) = 0. \tag{1.16}$$

Then a separable solution u of (1.16) must satisfy the integrability conditions (1.6). In case the integrability conditions are identities in the sense that there exist functions $P_{i,j}(x_k, u, u_{k,\ell})$, polynomials in $u_{k,\ell}$, such that

$$\begin{aligned}\mathcal{F}_{ij} \equiv\; & H_{u_{i,m_i}} H_{u_{j,m_j}} (\tilde{D}_i \tilde{D}_j H) + H_{u_{i,m_i}, u_{j,m_j}} (\tilde{D}_i H)(\tilde{D}_j H) \\ & - H_{u_{j,m_j}} (\tilde{D}_i H)(\tilde{D}_j H_{u_{i,m_i}}) - H_{u_{i,m_i}} (\tilde{D}_j H)(\tilde{D}_i H_{u_{j,m_j}}) \\ =\; & P_{i,j} H, \qquad i \neq j,\end{aligned} \tag{1.17}$$

we say that $\{x_k\}$ is a *regular separable coordinate* system for the equation $H = 0$.

Theorem 3. *If $\{x_k\}$ is a regular separable system for $H = 0$ then for every set of $m_1+m_2+\cdots+m_n+1$ constants $\{v^0, v^0_{i,j}\}$ with $H(x^0, v^0) = 0$ and $H_{u_{j,m_j}}(x^0, v^0) \not\equiv 0$, there is a unique separable solution u of $H(x, u) = 0$ such that $u(x^0) = v^0$, $u_{i,j}(x^0) = v^0_{i,j}$, $1 \le i \le n$, $1 \le j \le m_i$.*

Again we observe that if equations (1.17) are not satisfied identically, separable solutions still may exist but will depend on fewer that $\sum_{i=0}^n m_i$ independent parameters. This is ***nonregular*** separation. Examples 1–4 above for $E = 0$ are instances of regular and nonregular separation. Less trivial is

Example 5. $H = (x_2 - x_3)u_{11} + (x_3 - x_1)u_{22} + (x_1 - x_2)u_{33}$. Equations (1.17) are satisfied with $P_{i,j} \not\equiv 0$, so $\{x_k\}$ is a regular separable system for $H = 0$, through not for $H = E$. The general separable solution depends on six parameters and is given by

$$u = \tfrac{1}{6}\alpha(x_1^3 + x_2^3 + x_3^3) + \tfrac{1}{2}\beta(x_1^2 + x_2^2 + x_3^2) + \gamma_1 x_1 + \gamma_2 x_2 + \gamma_3 x_3 + \delta. \tag{1.18}$$

We can define the concepts of ***separation equation*** and ***normal*** separation for $H = 0$ in exact analogy with the definitions for $H = E$. The separation equations depend on q independent parameters and we have:

Theorem 4. *If the separation equations for $H = 0$ are normal in the coordinates $\{x_k\}$, then $q = n - 1$ and for each set of solutions $u^{(i)}(x_i)$, $i = 1, 2, \ldots, n$, of the separation equations, the function $u = \sum_{i=0}^n u^{(i)}(x_i)$ is a separable solution of $H = 0$.*

The proof is virtually the same as that of Theorem 2. The only difference is that, since $E = 0$, there is one less parameter in the separation equations. It is easy to check that the separation equations for Example 5 are normal:

$$u_{ii} + \alpha x_i + \beta = 0, \qquad i = 1, 2, 3.$$

§2 Generalized Lie symmetries

In this section we list the fundamental concepts and formulas from the Lie symmetry theory of partial differential equations that are essential for understanding the significance of variable separation. We limit ourselves here to an enumeration of basic results. For the detailed proofs and background material see the standard references [4,7].

In the following all functions are assumed to be locally real analytic. We consider local independent coordinates $x_1, \ldots, x_n$ and a dependent coordinate u. Given an assignment $u = f(\mathbf{x})$ we introduce the notation

$$u^J = \partial^J f(x), \qquad \partial^J = \frac{\partial^{|J|}}{\partial^{j_1}_{x_1}\partial^{j_2}_{x_2}\cdots\partial^{j_n}_{x_n}}, \tag{2.1}$$

to represent the kth order derivatives of f. Here $J = (j_1, \ldots, j_n)$ with each j_i a nonnegative integer and $|J| = j_1 + \ldots + j_n = k$. A ***generalized vector field*** on the $(n+1)$-dimensional space of independent and dependent variables is an expression of the form

$$\hat{Z} = \sum_{i=1}^n \xi_i(\mathbf{x}, u^J)\frac{\partial}{\partial x_i} + \varphi(\mathbf{x}, u^J)\frac{\partial}{\partial u}, \tag{2.2}$$

where ξ_i and φ depend on $\mathbf{x}$, u and finitely many derivates of u, ($|J| \le \ell < \infty$). Locally this field generates new coordinates $\mathbf{x}^*(\alpha)$, $u^*(\alpha)$ obtained by solving

$$\frac{\partial x_i^*(\alpha)}{\partial} = \xi_i(\mathbf{x}^*, u^{*J}), \qquad x_i^*(0) = x_i,$$
$$\frac{\partial u^*(\alpha)}{\partial \alpha} = \varphi(\mathbf{x}^*, u^{*J}), \qquad u^*(0) = u = f(\mathbf{x}).$$

If ξ_i and φ depend only on $\mathbf{x}$ and u, then $\hat{Z}$ is the generator of a Lie point transformation. In an obvious way $\hat{Z}$ can be prolonged to a vector field on the jet space with coordinates $(\mathbf{x}, u, u^J)$:

$$Z = \hat{Z} + \sum_{|K|>0} \varphi^K(x, u^J)\frac{\partial}{\partial u^K}, \tag{2.3}$$

where

$$\begin{aligned} \varphi^K &= D^K(\varphi - \sum_{i=1}^{n} u^i \xi_i) + \sum_{i=1}^{n} u^{J,i}\xi_i, \\ K &= (k_1, \ldots, k_n), \qquad J = (j_1, \ldots, j_n), \\ J, i &= (j_1, \ldots, j_{i-1}, j_i + 1, j_{i+1}, \ldots, j_n), \\ D^K &= D_1^{k_1} D_2^{k_2} \cdots D_n^{k_n}, \end{aligned} \tag{2.4}$$

and D_i is the ***total derivative***

$$D_i = \frac{\partial}{\partial x_i} + \sum_{|K|\geq 0} u^{K,i}\frac{\partial}{\partial u^K}. \tag{2.5}$$

[Although the sums in (2.3) and (2.5) are formally infinite, in practice we will apply these operators only to functions that depend on finitely many derivatives u^K.]

Let $H(\mathbf{x}, u^K) = 0$ be a partial differential equation for u. We say that the generalized vector field $\hat{Z}$ is a ***generalized symmetry*** operator for this equation provided

$$ZH(\mathbf{x}, u^K) = 0, \tag{2.6}$$

whenever $H(\mathbf{x}, u^K) = 0$, *i.e.*, ZH vanishes for all solutions u of $H = 0$. In the following we will make the technical hypothesis, almost always satisfied by partial differential equations of physical interest, that for each generalized symmetry operator $\hat{Z}$ there are a finite number of functions $\chi_J(\mathbf{x}, u^K)$ such that

$$ZH = \sum_J \chi_J D^J H. \tag{2.7}$$

See [5, page 135], for a discussion of this specialization of (2.6).

The ***characteristic*** ψ of a generalized vector field $\hat{Z}$ is defined by

$$\psi = \varphi - \sum_{i=1}^{n} u^i \xi_i. \tag{2.8}$$

Note that the generalized vector field $\hat{X}(\psi) = \psi\partial_u$ has the prolongation

$$X(\psi) = \psi\partial_u + \sum_{|K|>0} D^K\psi\partial_u K. \tag{2.9}$$

We call $\hat{X}(\psi)$ the ***standard representation*** of $\hat{Z}$. It follows from (2.3) and (2.4) that

$$Z = X(\psi) + \sum_{i=1}^{n} \xi_i D_i, \tag{2.10}$$

and from (2.7) that $\hat{Z}$ is a generalized symmetry for the equation $H = 0$ if and only if its standard representation $X(\psi) = \psi\partial_u$ is a generalized symmetry. Thus, from the viewpoint of Lie symmetries there is no loss of generality in restricting to operators of the form $\psi\partial_u$.

The ***commutator*** of two prolongations of standard representation operators is an operator of the same form:

$$[X(\psi_1), X(\psi_2)] \equiv X(\psi_1)X(\psi_2) - X(\psi_2)X(\psi_1) = X(\{\psi_1, \psi_2\}), \tag{2.11}$$

where

$$\{\psi_1, \psi_2\} = X(\psi_1)\psi_2 - X(\psi_2)\psi_1. \tag{2.12}$$

Note that the functions $\psi(\mathbf{x}, u^K)$ form an (infinite-dimensional) Lie algebra under the bracket $\{\cdot, \cdot\}$. Indeed,

$$\begin{gathered}\{\psi_1, \psi_2\} = -\{\psi_2, \psi_1\}\\ \{a_1\psi_1 + a_2\psi_2, \psi_3\} = a_1\{\psi_1, \psi_3\} + a_2\{\psi_2, \psi_3\}, \qquad a_i \in R,\\ \{\{\psi_1, \psi_2\}, \psi_3\} + \{\{\psi_2, \psi_3\}, \psi_1\} + \{\{\psi_3, \psi_1\}, \psi_2\} = 0.\end{gathered} \tag{2.13}$$

In terms of this bracket the symmetry condition (2.6)–(2.7) becomes

$$\{\psi, H\} = 0 \quad \text{whenever} \quad H = 0. \tag{2.14}$$

If ψ_1 and ψ_2 depend only on the variables x_i and u^i, then (2.12) reduces to the Poisson bracket:

$$\{\psi_1, \psi_2\} = \sum_{i=1}^{n} (\partial_{x_i}\psi_1\partial_{u^i}\psi_2 - \partial_{x_i}\psi_2\partial_{u^i}\psi_1). \tag{2.15}$$

If ψ_1 and ψ_2 are linear in the u^K,

$$\psi_i = \sum_K a_K{}^{(i)}(\mathbf{x})\, u^K, \qquad i = 1, 2,$$

then the bracket agrees with the usual operator commutator bracket:

$$\begin{aligned}\{\psi_1, \psi_2\} &= -[L_1, L_2]u = -(L_1L_2 - L_2L_1)u,\\ L_i &= \sum_K a_K{}^{(i)}(x)D^K, \qquad i = 1, 2.\end{aligned} \tag{2.16}$$

Let $\mathbf{y}$ and v be a new set of independent and dependent coordinates, related to $\mathbf{x}$ and u by a transformation of the type

$$\mathbf{y} = \mathbf{f}(\mathbf{x}), \qquad v = g(\mathbf{x}, u).$$

Then the operator $X(\psi)$, (2.9), in standard form with respect to the $\mathbf{x}$, u coordinates, transforms to the operator $X'(\psi')$ in standard form with respect to the $\mathbf{y}$,v coordinates:

$$X(\psi) = X'(\psi'), \qquad \psi' = \psi\, \partial_u g.$$

We are now ready to highlight the simplest point of contact between separation of variables methods and the symmetries of partial differential equation

$$H(\mathbf{x}, u^K) = 0. \tag{2.17}$$

Let $\hat{Z} \neq 0$ be an infinitesimal point symmetry for (2.17) which is ***projectable***, *i.e.*, of the form

$$\hat{Z} = \sum_{i=1}^{n} \xi_i(x)\frac{\partial}{\partial x_i} + \varphi(x,u)\frac{\partial}{\partial u}, \tag{2.18}$$

and, in addition, suppose not all ξ_i are zero. Then by Lie's theorem (see [2, pages 34, 49, and 50]) we can find new coordinates $s, y_2, \ldots, y_n, v$ such that

$$\mathbf{x} = A(x, \mathbf{y}), \qquad u = B(s, \mathbf{y}, v), \tag{2.19}$$

and, in the new coordinates

$$\hat{Z} = \frac{\partial}{\partial s}. \tag{2.20}$$

Since $\hat{Z}$ is a point symmetry, it follows from (2.7) (and very mild technical assumptions on H) that H must be of the form

$$H = H_1(s, \mathbf{y})H_2(\mathbf{y}, v^L), \tag{2.21}$$

where H_2 has nontrivial dependence on the v^L. Thus, each solution $v(x, \mathbf{y})$ of

$$H_2(y, v^L) = 0 \tag{2.22}$$

determines a solution u of $H(\mathbf{x}, u^K) = 0$. The equations $H = 0$ and $H_2 = 0$ are essentially equivalent. However, the second equation depends on s only through derivatives terms in v. Again, under mild technical assumptions on H, we can find solutions $v(\mathbf{y})$ of (2.22), *i.e.*, solutions such that $\partial_s v \equiv 0$. [If $\partial_v H_2 \equiv 0$ then we can find solutions $v(s, \mathbf{y}) = \lambda s + v'(\mathbf{y})$ where λ is a constant.] For $n = 2$, Eq.(2.22) reduces to an ordinary differential equation and the solutions of this equation yield ***symmetry adapted*** solutions of (2.17). Furthermore (s, y_2) is a separable coordinate system for (2.17), though not necessarily a regular separable system. In general we have reduced a partial differential equation in n variables to one in $n-1$ variables, and this provides an instance of partial separation.

As a simple example, consider the symmetry

$$\hat{Z} = t\partial_x + c\partial_t - \partial_u, \qquad c \neq 0 \tag{2.23}$$

for the Korteweg–de Vries equation

$$\partial_t u - \partial_{xxx} u - u\,\partial_x u = 0. \tag{2.24}$$

(The symmetry algebra for this equation can be found in many references, *e.g.*, [5].) The requirement $\hat{Z} = \partial_s$ leads to new coordinates (s, y, v) such that

$$t = cs, \qquad y = \tfrac{1}{2}cs^2 + x, \qquad v = -s + u.$$

In terms of the new coordinates, (2.24) becomes

$$\partial_s v - c\partial_{yyy} v - cv\partial_y v - 1 = 0. \tag{2.25}$$

The ordinary differential equation obtained by setting $\partial_s v \equiv 0$ is related to the first Painlevé transcendent, [5, page 107]. In §5 we shall give some additional examples of this approach to variable separation.

§3 Separability for Hamilton–Jacobi, Helmholtz, and Laplace equations

We now apply the results of §1 to determine the possible regular separable coordinate systems for the Hamilton–Jacobi equation on a pseudo-Riemannian manifold V^n

$$H(x^i, u_i) \equiv \sum_{\substack{i=1\\j=1}}^{n} g^{ij} u_i u_j = E, \qquad g^{ij} = g^{ji}, \tag{3.1}$$

where $u_i = \partial_i u$. In the local coordinate $\{x^i\}$ the metric on V^n is $ds^2 = \sum_{i,j=1}^n g_{ij}\, dx^i dx^j$ and $\sum_{j=1}^n g_{ij} g^{jk} = \delta_i^k$, $g = \det(g_{ij}) \neq 0$. (Here we change notation again to that of Eisenhart's book [8]. Recall that a solution $u(\mathbf{x}, \lambda_1, \ldots, \lambda_n)$, $\lambda_1 = E$ is a *complete integral* of (3.1) provided $\det(\partial_{x^i \lambda_j} u) \neq 0$, and that the knowledge of a complete integral enables us to solve the associated Hamiltonian system $\dot{p}_i = -\partial_{x^i} H(\mathbf{x}, \mathbf{p})$, $\dot{x}^i = \partial_{p_i} H(\mathbf{x}, \mathbf{p})$. Separation of variables is a powerful technique for obtaining explicit complete integrals for many Hamilton–Jacobi equations.) Initially we limit ourselves to ***orthogonal*** coordinates $\{x^i\}$, *i.e.*, coordinates for which $ds^2 = \sum_{i=1}^n H_i^2 (dx^i)^2$, so that $g_{ij} = 0$ if $i \neq j$. Thus (3.1) becomes

$$\sum_{i=1}^{n} H_i^{-2} u_i^2 = E, \tag{3.2}$$

and from the integrability conditions (1.6) we see that $\{x^i\}$ is a regular separable system if and only if

$$\partial_{jk} H_i^{-2} = \partial_j H_i^{-2} \partial_k \ln H_j^{-2} + \partial_k H_i^{-2} \partial_j \ln H_k^{-2}, \qquad j \neq k. \tag{3.3}$$

These are the standard Levi-Civita separability conditions and are well known to be equivalent to the requirement that the metric coefficients be in Stäckel form with respect to the coordinates $\{x^i\}$ (see [9,10]). That is, there exists an $n \times n$ matrix $s_{ji}(x^j)$, whose j^{th} row depends only on x^j, such that $S = \det(s_{ji}) \neq 0$, (a ***Stäckel matrix***), and

$$H_j^{-2} = \frac{s^{j1}}{S}, \tag{3.4}$$

where s^{j1} is the $(j,1)$ minor of (s_{ij}).

We will work out the proof of this fact and study the theory of Stäckel matrices. Given a Stäckel matrix, consider the set of ordinary differential equations (the separation equation)

$$u_i^2 + \sum_{j=1}^{n} \lambda_j s_{ij}(x^i) = 0, \qquad i = 1, \ldots, n, \tag{3.5}$$

where $\lambda_1 = -E, \lambda_2, \ldots, \lambda_n$ are parameters and $u_i = u_i(x^i)$. Multiplying the ith separation equation by $T^{1j} = s^{j1}/S$, where $T^{ji} = (S^{-1})^{ji}$, and summing on i, we obtain (3.2), where $H_j^{-2} = T^{1j}$. Assuming $T^{1j} \neq 0$ for all j, we see that $\{x^i\}$ is a regular separable orthogonal coordinate system for the Hamilton–Jacobi equation (3.1). Thus, the Stäckel form coefficients H_j^{-2} must satisfy the integrability conditions (3.3).

To prove the converse, suppose $s_{ji}(x^j)$ is a Stäckel matrix, consider the equation

$$\sum_{i=1}^{n} s_{ji}(x^j) T^{ik} = \delta_j^k \tag{3.6}$$

for the inverse of s_{ji}, and assume $T^{1j} \neq 0$ for any j. Clearly

$$\sum_{i=1}^{n} s_{ji}(x^j)\partial_q T^{ik} = 0, \qquad j \neq q,$$

where $\partial_q = \partial_{x^q}$. Thus there exists a function $f_q^{(k)}(\mathbf{x})$ such that

$$\partial_q T^{ik} = f_q^{(k)} T^{iq}, \qquad i = 1, \ldots, n. \tag{3.7}$$

Now introduce the nonvanishing functions $H_j^{-2} = T^{1j}$ and define the roots $\rho_k^{(i)}(\mathbf{x})$ by

$$T^{ik} = \rho_k^{(i)} H_k^{-2}. \tag{3.8}$$

Clearly $\rho_k^{(1)} = 1$ for $1 \leq k \leq n$. Substituting (3.8) into (3.7) we find

$$\partial_q \rho_k^{(i)} H_k^{-2} + \rho_k^{(i)} \partial_q H_k^{-2} = f_q^{(k)} \rho_q^{(i)} H_q^{-2}.$$

Setting $i = 1$ in this expression we find $\partial_q H_k^{-2} = f_q^{(k)} H_q^{-2}$. Thus

$$\partial_q \rho_k^{(i)} = (\rho_q^{(i)} - \rho_k^{(i)})\partial_q \ln H_k^{-2},$$

and we see that the system of equations

$$\partial_q \rho_k = (\rho_q - \rho_k)\partial_q \ln H_k^{-2}, \qquad q, k = 1, \ldots, n, \tag{3.9}$$

admits a set of n linearly independent vector solutions $(\rho_1^{(i)}, \ldots, \rho_n^{(i)})$, $i = 1, \ldots, n$. It follows that the integrability conditions $\partial_q(\partial_l \rho_k) = \partial_l(\partial_q \rho_k)$ for this system must be identically satisfied. As is easily verified, these conditions are precisely the separability conditions (3.3).

Conversely, suppose we are given n nonvanishing functions H_j^{-2} satisfying the separability conditions (3.3). Then the system of equations (3.9) has an n-dimensional solution space. Let $(\rho_1^{(i)}, \ldots, \rho_n^{(i)})$, $i = 1, \ldots, n$, be a basis for this space. Clearly $\rho_k = 1$, $1 \leq k \leq n$ is a solution, so we can choose this basis such that $\rho_k^{(1)} = 1$. Now we define the nonsingular $n \times n$ matrix (T^{ik}) by $T^{ik} = \rho_k^{(i)} H_k^{-2}$ and let (s_{jl}) be the inverse of this matrix:

$$\sum_{i=1}^{n} s_{ji} T^{ik} = \delta_j^k. \tag{3.10}$$

It follows from (3.9) that equation (3.7) holds where $f_q^{(k)} = H_q^2 \partial_q H_k^{-2}$. Differentiating both sides of (3.10) with respect to x^q, where $q \neq j$, and using (3.7), we find

$$\sum_{i=1}^{n} \partial_q s_{ji} T^{ik} = 0, \qquad q \neq j,$$

for all k. Thus $\partial_q s_{ji} = 0$ for $q \neq j$. It follows that $s_{ji} = s_{ji}(x^j)$ and (s_{jl}) is a Stäckel matrix. Thus the functions H_j^{-2} are in Stäckel form. This proves the equivalence of (3.3) and (3.4).

For the Hamilton–Jacobi equation with potential

$$\sum_{i=1}^{n} H_i^{-2} u_i^2 + V(\mathbf{x}) = E, \tag{3.11}$$

the results are similar. The integrability conditions reduce to (3.3) and

$$\partial_{ik}V - \partial_k \ln H_j^{-2}\partial_j V - \partial_j \ln H_k^{-2}\partial_k V = 0, \qquad j \neq k. \tag{3.12}$$

As shown in reference [11], this last condition means precisely that the potential function can be expressed in the form

$$V = \sum_{i=1}^{n} f^{(i)}(x^i) H_i^{-2}. \tag{3.13}$$

Again the separation equations are normal.

A regular orthogonal separable system for the Hamilton–Jacobi equation (3.2) with $E = 0$ is characterized by the integrability conditions

$$\partial_{jk}H_i^{-2} - \partial_j H_i^{-2}\partial_k \ln H_j^{-2} - \partial_k H_i^{-2}\partial_j \ln H_k^{-2} = \rho_{jk}(x) H_i^{-2} H_j^2 H_k^2, \qquad j \neq k, \tag{3.14}$$

for some functions ρ_{jk}. These equations are equivalent to

$$\partial_{jk} \ln K_i^{-2} + \partial_j \ln K_i^{-2}\partial_k \ln K_i^{-2} - \partial_j \ln K_i^{-2}\partial_k \ln K_j^{-2} - \partial_j \ln K_k^{-2}\partial_k \ln K_i^{-2} = 0, \tag{3.15}$$

for $K_i^{-2} = H_i^{-2}/H_1^{-2}$. Furthermore, as shown in reference [12], the equations are equivalent to the requirement that $H_i^{-2} = Q(\mathbf{x})\mathcal{H}_i^{-2}$, where the metric $ds^2 = \sum_{j=1}^{n} \mathcal{H}_j^2 (dx^j)^2$ is in Stäckel form. The separation equations have the appearance (3.5) with $E = 0$.

Next we study the problem of (multiplicative) separation of variables for the Helmholtz (or Schrödinger) equation

$$(\Delta + V(\mathbf{x}))\,\Psi(\mathbf{x}) = E\,\Psi(\mathbf{x}), \tag{3.16}$$

on the pseudo-Riemannian manifold V^n. Here

$$\Delta = \sum_{i,j=1}^{n} \frac{1}{\sqrt{g}}\,\partial_i(\sqrt{g}\, g^{ij}\partial_j) \tag{3.17}$$

is the Laplace–Beltrami operator on V^n, defined independent of local coordinates. To convert this product separation problem, $\Psi = \prod_{i=1}^{n}\Psi^{(i)}(x^i)$, to the standard additive separation form, we introduce the new dependent variable $u = \ln \Psi$. Further, we restrict ourselves to orthogonal separable systems

$$ds^2 = \sum_{i=1}^{n} H_i^2 (dx^i)^2.$$

Then (3.16) becomes

$$H \equiv \sum_{i=1}^{n} \left[H_i^{-2}(u_{ii} + u_i^2) + s_i u_i\right] + V = E, \tag{3.18}$$

where

$$s_i = \frac{1}{\sqrt{g}}\,\partial_i(\sqrt{g}\,H_i^{-2}), \qquad \sqrt{g} = H_1 H_2 \cdots H_r. \tag{3.19}$$

The integrability conditions (1.6) for regular separation lead to (3.3), (Stäckel form), upon comparison of the coefficients of u_i^2. Comparison of the coefficients of u_{ii} in (1.6) yields the ***Robertson condition***

$$\partial_{ij} \ln(\sqrt{g}\,H_i^{-2}) = 0, \qquad i \neq j. \tag{3.20}$$

Comparison of the constant terms in (1.6) yields the conditions (3.12) on the potential $V(\mathbf{x})$, *i.e.*, the potential must be expressable in the form (3.13) to permit separation. There are no additional consequences of the integrability conditions. Again the separation equations are normal and take the form

$$u_{ii} + u_i^2 + g_i(x^i)u_i + f_i(x^i) + \sum_{j=1}^{n} \lambda_j s_{ij}(x^i) = 0, \tag{3.21}$$

where $\lambda_1 = -E$.

It follows that every orthogonal coordinate system permitting product separation of the Helmholtz equation (3.16) corresponds to a Stäckel form; hence it permits additive separation of the Hamilton–Jacobi equation (3.1). Eisenhart has shown, [10], that the additional Robertson condition for product separation is equivalent to the requirement $R_{ij} = 0$ for $i \neq j$, where R_{ij} is the Ricci tensor of V^n expressed in the Stäckel coordinates $\{x^i\}$. It follows that the Robertson condition is automatically satisfied in Euclidean space, a space of constant curvature or any Einstein space.

More generally we can introduce the notion of R-separation for the Helmholtz equation (3.18) in orthogonal coordinates $\{x^i\}$. Here, R-separable solutions take the form $\Psi = \exp\big(R(\mathbf{x})\big)\prod_{i=1}^{n}\Psi^{(i)}(x^i) = e^R\Theta$, where $R(\mathbf{x})$ is a fixed function, independent of parameters. If $R \equiv 0$ we have ***separation***, and if $R(x) = \sum_{i=1}^{n} R^{(i)}(x^i)$ we have ***trivial*** R-separation. Otherwise the R-separation is ***nontrivial***. Writing $u = \ln\Theta = R - \ln\Psi$, we have the following generalization of (3.18):

$$H \equiv \sum_{i=1}^{n}\Big[H_i^{-2}(u_{ii} + u_i^2) + (2H_i^{-2}\partial_i R + s_i)u_i + H_i^{-2}(\partial_{ii}R + (\partial_i R)^2) + s_i\partial_i R\Big] + V = E. \tag{3.22}$$

Comparing the coefficients of u_i^2 in the integrability conditions (1.6) we again find that the metric $ds^2 = \sum_{i=1}^{n} H_i^2(dx^i)^2$ must be in Stäckel form. Comparison of the coefficients of u_{ii} yields

$$\partial_{ij}\big[2R + \ln(\sqrt{g}\,H_i^{-2})\big] = 0, \qquad i \neq j. \tag{3.23}$$

Finally, comparison of the constant terms in (1.6) and use of (3.23) leads to the requirement (3.12) for the ***modified potential***,

$$\tilde{V} = V - \tfrac{1}{2}\sum_{i=1}^{n} H_i^{-2}(\partial_i \ell_i + \tfrac{1}{2}\ell_i^2), \tag{3.24}$$

where

$$\ell_i = \partial_i \ln(\sqrt{g}\,H_i^{-2}) = \partial_i \ln\frac{\sqrt{g}}{S}. \tag{3.25}$$

We see that whenever $\tilde{V}$ satisfies (3.12), hence (3.13), equation (3.16) permits orthogonal R-separation with

$$R = -\tfrac{1}{2}\ln\frac{h}{S} + \sum_{i=1}^{n} L^{(i)}(x^i), \tag{3.26}$$

where the functions $L^{(i)}$ are arbitrary. Thus through appropiate choice of V, every additively separable coordinate system $\{x^i\}$ for the zero-potential Hamilton–Jacobi equation can occur as a multiplicatively separable system for the Helmholtz equation. In all these cases the separation equations are normal. Details are given in reference [13].

The question arises whether nontrivial R-separation occurs for $V = 0$. From (3.20), (3.23), and Eisenhart's formulation of Robertson's condition as $R_{ij} = 0$, $i \neq j$, we see that only trivial orthogonal

R-separation can occur in an Einstein space. However, as shown in references [13,14], nontrivial R-separation can occur for $V = 0$, even in conformally flat spaces. An example is

$$ds^2 = (x+y+z)[(x-y)(x-z)dx^2 + (y-z)(y-x)dy^2 + (z-x)(z-y)dz^2],$$
$$e^R = (x+y+z)^{-1/4}. \tag{3.27}$$

Finally, we take up orthogonal R-separation for the Laplace equation on V^n :

$$\Delta\Psi(\mathbf{x}) = 0. \tag{3.28}$$

Here the Laplace–Beltrami operator is given by (3.17). We are interested in solutions of the form $\Psi(\mathbf{x}) = \exp\big(R(\mathbf{x})\big)\,\Theta(\mathbf{x})$, where $\Theta(\mathbf{x}) = \prod_{i=1}^n \Psi^{(i)}(x^i)$ and the metric becomes $ds^2 = \sum_{i=1}^n H_i^{-2}(dx^i)^2$ in the coordinates $\{x^i\}$. Writing $u = \ln\Theta$ we can write (3.28) in the standard form

$$H \equiv \sum_{i=1}^n \Big[H_i^{-2}(u_{ii} + u_i^2) + (2H_i^{-2}\partial_i R + s_i)u_i + H_i^{-2}(\partial_{ii}R + (\partial_i R)^2) + s_i\partial_i R\Big] = 0, \tag{3.29}$$

where

$$s_i = \frac{1}{\sqrt{g}}\,\partial_i(\sqrt{g}\,H_i^{-2}), \qquad \sqrt{g} = H_1\cdots H_n. \tag{3.30}$$

We now substitute these expressions into the integrability conditions (1.17) to find the requirements that $\{x^j\}$ be a regular separable coordinate system. Equating the coefficients of u_i^2 we obtain the conditions (3.14), hence (3.15), on the metric components H_i^{-2}. Thus there exists a function $Q(\mathbf{x})$ such that $H_i^{-2} = Q\mathcal{H}_i^{-2}$, $i = 1, \ldots, n$, where the metric $\hat{ds}^2 = \sum_{i=1}^n \mathcal{H}_i^{-2}(dx^i)^2$ is in Stäckel form. Let $\big(s_{ij}(x^i)\big)$ be a Stäckel matrix associated with this form. Comparison of the coefficients of u_{ii} yields

$$\partial_{ij}\Big[2R + \ln\big(\frac{Q^{1-n/2}h}{S}\big)\Big] = 0, \qquad i \neq j, \tag{3.31}$$

where $h = \mathcal{H}_1\cdots\mathcal{H}_n$. Comparison of the constant terms in (1.17) and use of (3.31) leads to the remaining requirement that the ***potential***

$$\tilde{V} = \sum_{i=1}^n \mathcal{H}_i^{-2}(\partial_i \ell_i + \tfrac{1}{2}\ell_i^2), \qquad \ell_i = \partial_i \ln\big(\frac{Q^{1-n/2}h}{S}\big), \tag{3.32}$$

satisfies

$$\partial_{jk}\tilde{V} - \partial_k \ln \mathcal{H}_j^{-2}\partial_j\tilde{V} - \partial_j \ln\mathcal{H}_k^{-2}\partial_k\tilde{V} = 0, \qquad j \neq k; \tag{3.33}$$

hence $\tilde{V}$ is a ***Stäckel multiplier***:

$$\tilde{V} = \sum_{i=1}^n f^{(i)}(x^i)\mathcal{H}_i^{-2}. \tag{3.34}$$

If conditions (3.15) and (3.33) are satisfied then $\{x^j\}$ is R-separable with

$$R(\mathbf{x}) = -\tfrac{1}{2}\ln\big(\frac{Q^{1-n/2}h}{S}\big) + \sum_{i=1}^n L^{(i)}(x^i), \tag{3.35}$$

where the $L^{(i)}$ are arbitrary.

Because of space and time limitations we shall not treat here the general case of (possibly) nonorthogonal separation for the Hamilton–Jacobi, Helmholtz, and Laplace equations. The complete details of the general case can be found in [3], and references contained therein.

§4 Intrinsic characterization of variable separation

For the Hamilton–Jacobi, Helmholtz, and Laplace equations on V^n, introduced in the previous section, (R-) separable coordinate systems can always be characterized intrinsically, *i.e.*, in a coordinate-free manner. Consider first the problem of additive orthogonal separation for the Hamilton–Jacobi equation (3.1). Let H be the quadratic form

$$H = \sum_{\substack{i=1\\j=1}}^{n} g^{ij} u_i u_j, \qquad u_i = \partial_{y_i} u, \tag{4.1}$$

where $(g^{ij}(y))$ is the metric expressed in terms of the general coordinates $\{y^k\}$ and u is the dependent variable. Expression (4.1) is invariant under a change of independent coordinates. Now assume that (3.1) is separable in the orthogonal coordinates $\{x^i\}$. From the separation equations (3.5), (3.8), and (3.9) we are led to the quadratic forms

$$\mathcal{A}^\ell = \sum_{j=1}^{n} T^{\ell j} u_\ell^2 = \sum_{j=1}^{n} \rho_j^{(\ell)} H_j^{-2} u_j^2, \qquad \ell = 1, 2, \ldots, n, \tag{4.2}$$

which can easily be shown to have the following properties:

1. $\mathcal{A}^1 = H$.
2. The n element set $(\mathcal{A}^\ell)$ is linearly independent (as a set of quadratic forms).
3. $(\mathcal{A}^\ell)$ is in involution, *i.e.*, $\{\mathcal{A}^i, \mathcal{A}^j\} = 0$, and each $\mathcal{A}^i$ is a quadratic symmetry of equation (3.1).
4. The differentials of the separable coordinates, $\omega^j = dx^j$, constitute a simultaneous eigenbasis for the $(\mathcal{A}^\ell)$. [Here, ρ is a *root* of a quadratic form $\mathcal{A} = (a^{ij})$ with respect to the metric g^{ij} if $\det(a^{ij} - \rho g^{ij}) = 0$, and $\omega = \sum_\ell \lambda_\ell dx^\ell$ is an eigenform corresponding to ρ if $\omega \neq 0$ and $\sum_{j=1}^{n} (a^{ij} - \rho g^{ij})\lambda_j = 0$.]
5. $\mathcal{A}^\ell(\mathbf{x}, u_x) = -\lambda_\ell$ for each additively separable solution $u = \sum_{i=1}^{n} u^{(i)}(x^i, \boldsymbol{\lambda})$ of (3.1).

In [15] the authors proved the following converse of these statements:

Theorem 5. *Let $(\mathcal{A}^\ell)$, $\mathcal{A}^1 = H$, be a linearly independent set of n second order symmetric quadratic forms such that*

1. $\{\mathcal{A}^\ell, \mathcal{A}^m\} = 0, \quad 1 \le \ell, m \le n,$
2. *The $(\mathcal{A}^\ell)$ have a common eigenbasis $\{\omega^{(j)}\}$.*

Then there is a separable coordinate system $\{x^j\}$ for the Hamilton–Jacobi equation $H(\mathbf{y}, v_{\mathbf{y}}) = E$ on V^n such that $\omega^{(j)} = f^{(j)}(x)\, dx^j$ for some functions $f^{(j)}$. The separable solutions u are determined by $\mathcal{A}^\ell(\mathbf{x}, u_{\mathbf{x}}) = -\lambda_\ell$.

The main point of this theorem is that, under the required hypotheses the eigenforms $\omega^{(\ell)}$ of the quadratic forms a^{ij} are normalizable, *i.e.*, that up to multiplication by a nonzero function, $\omega^{(\ell)}$ is the differential of a coordinate. This fact, which is proved through use of the commutation relations $\{\mathcal{A}^\ell, \mathcal{A}^m\} = 0$ to verify appropriate integrability conditions, permits us to compute the coordinates directly from a knowledge of the symmetry operators.

As an example of the use of Theorem 5 we consider the Hamilton–Jacobi equation for two dimensional Minkowski space. In cartesian coordinates this equation is

$$H \equiv u_x^2 - u_t^2 = E. \tag{4.2}$$

The vector space of all symmetries of the form $a(x,t)u_x + b(x,t)u_y$ (Killing vectors) is easily shown to be closed under the bracket $\{\cdot,\cdot\}$; hence the symmetries form a Lie algebra. Furthermore we have

$\{\mathcal{H}, \mathcal{L}\} \equiv 0$ for each linear symmetry. The Lie algebra is three dimensional over the field of real scalars, with basis

$$\begin{array}{lll} \mathcal{L}_1 = u_x, & \mathcal{L}_2 = u_t, & \mathcal{L}_3 = tu_x + xu_t, \\ \{\mathcal{L}_1, \mathcal{L}_2\} = 0, & \{\mathcal{L}_3, \mathcal{L}_1\} = \mathcal{L}_2, & \{\mathcal{L}_3, \mathcal{L}_2\} = \mathcal{L}_1. \end{array} \tag{4.3}$$

It can be shown that every symmetry which is quadratic in the first derivatives of u (second order Killing tensor) is a polynomial in the linear symmetries $\mathcal{L}_i$. (This is true for all spaces of constant curvature, *e.g.*, [15].) Thus all candidates for variable separation can be built from the basis symmetries (4.3).

Consider, for example, the quadratic symmetry $\mathcal{A}^2 = 2\mathcal{L}_3\mathcal{L}_1$. With respect to cartesian coordinates, the corresponding symmetric quadratic forms are

$$\mathcal{A}^2 \sim \begin{pmatrix} 2t & x \\ x & 0 \end{pmatrix}, \qquad \mathcal{A}^1 = \mathcal{H} \sim \begin{pmatrix} 1 & 0 \\ 0 & -1 \end{pmatrix}.$$

Clearly, $\mathcal{A}^2$ has roots $\rho = t \pm \sqrt{t^2 - x^2}$ (assuming $t > |x|$) with a basis of eigenforms $\omega_1 = (t + \sqrt{t^2 - x^2})\, dx - x dt$, $\omega_2 = (t - \sqrt{t^2 - x^2})\, dx - x dt$. By Theorem 5, $\mathcal{A}^2$ does define a regular separable coordinate system $\{\xi^1, \xi^2\}$ for (4.2) and there exist functions f_i such that $d\xi^i = f_i\omega_i$, $i = 1, 2$. We find $f_1 = \left[\xi^2((\xi^2)^2 - (\xi^1)^2)\right]^{-1}$, $f_2 = -\left[\xi^1((\xi^2)^2 - (\xi^1)^2)\right]^{-1}$,

$$t = \tfrac{1}{2}((\xi^1)^2 + (\xi^2)^2), \qquad x = \xi^1\xi^2. \tag{4.4}$$

On the other hand the symmetry

$$\mathcal{A} = 2\mathcal{L}_3(\mathcal{L}_1 - \mathcal{L}_2) \tag{4.5}$$

has two equal roots and only one eigenform. Thus $\mathcal{A}$ cannot determine a separable coordinate system.

For manifolds of dimension $n \geq 3$ there is a second way that a system of $n - 1$ commuting symmetries may fail to determine separable coordinates: although each quadratic symmetry determines a basis of eigenforms, there is no basis of eigenforms for all symmetries simultaneously. See [15] for a simple example.

These results extend to R-separation of the Helmholtz equation on V^n,

$$\Delta\, \Psi(\mathbf{x}) = E\, \Psi(\mathbf{x}), \tag{4.6}$$

where in local coordinates

$$\Delta = \frac{1}{\sqrt{g}} \sum_{\substack{i=1\\j=1}}^{n} \partial_i(\sqrt{g}\, g^{ij}\partial_j). \tag{4.7}$$

Here a linear differential operator A on V^n is a ***symmetry operator*** for Δ if

$$[\Delta, A] \equiv \Delta A - A\Delta = 0. \tag{4.8}$$

[This agrees with the bracket (2.16).] Note that uniquely associated with every second order symmetry operator

$$A = \sum_{\substack{i=1\\j=1}}^{n} a^{ij}\partial_{ij} + \sum_{i=1}^{n} b^i\partial_i + c$$

in local coordinates $\{y^\ell\}$, is the second order Killing tensor

$$\mathcal{A} = \sum_{\substack{i=1\\j=1}}^{n} a^{ij}u_iu_j.$$

Indeed, $[\Delta, A] = 0$ implies $\{\mathcal{H}, \mathcal{A}\} = 0$, though the converse is false.

Theorem 6. [13,16] *Necessary and sufficient conditions for the existence of an orthogonal R-separable coordinate system $\{x^j\}$ for the Helmholtz equation $\Delta\Psi = E\Psi$ are that there exist n second order differential operators $A^1 = \Delta, A^2, \ldots, A^n$ such that*

1. $[A^i, A^j] = 0, \quad 1 \le i, j \le n.$
2. *The associated set of Killing tensors $(\mathcal{A}^i)$ is linearly independent.*
3. *There is a basis $\{\omega^{(j)} : 1 \le j \le n\}$ of simultaneous eigenforms for the $\mathcal{A}^i$.*

If these conditions are satisfied then there exist functions $f^j(\mathbf{x})$ such that $\omega^{(j)} = f^j\,dx^j$, $1 \le j \le n$. Theorem 6 follows from Theorem 5 through exploitation of the commutation relations **1**. Indeed, these relations can be used to show that the separation conditions for $\tilde{V}$ are valid.

Corollary 2. [13] *Suppose the second order differential operators $A^1 = \Delta, A^2, \ldots, A^n$ satisfy conditions* **1–3** *of Theorem 6, and in addition that they are in self-adjoint form with respect to the measure $dV = \sqrt{g}\,dy^1 \ldots dy^n$:*

$$A^\ell = \frac{1}{\sqrt{g}} \sum_{\substack{i=1\\j=1}}^{n} \partial_{y^i}(\sqrt{g}\,a^{ij}_{(\ell)}\partial_{y^j}) + c_\ell(\mathbf{y}), \qquad \ell = 1, 2, \ldots, n, \tag{4.9}$$

(a form which is independent of the choice of local coordinates $\{y^j\}$). Then the R-separable solutions $\Psi = e^R \prod_{i=1}^n \Psi^{(i)}(x^i)$ of $\Delta\Psi = E\Psi$ are characterized as the eigenfunctions of the A^ℓ:

$$A^\ell\Psi = -\lambda_\ell\Psi, \qquad \ell = 1, 2, \ldots, n, \tag{4.10}$$

where $\lambda_1 = -E$ and $\lambda_2, \ldots, \lambda_n$ are separation constants.

For the Hamilton-Jacobi equation

$$\mathcal{H}(x, \partial_i u) = 0, \tag{4.11}$$

i.e., $E = 0$, there is an analogous characterization of additive separation by conformal symmetries. A function $\mathcal{L}(x, \partial_i u)$ is said to be a ***conformal symmetry*** provided there is a function $q(x, \partial_i u)$ such that

$$\{\mathcal{L}, \mathcal{H}\} = q\mathcal{H}. \tag{4.12}$$

Theorem 7. [12] *Necessary and sufficient conditions for the existence of an orthogonal separable coordinate system $\{x^j\}$ for the Hamilton–Jacobi equation (4.11) are that there exist $n-1$ symmetric quadratic functions $\mathcal{B}^k = \sum_{i=1, j=1}^{n} b^{ij}_{(k)}(\mathbf{x})\,u_i u_j$, $2 \le k \le n$, such that*

1. *Each $\mathcal{B}^k$ is a conformal symmetry.*
2. $\{\mathcal{B}^i, \mathcal{B}^j\} = 0, \quad 2 \le i, j \le n.$
3. *The set $(\mathcal{H}, \mathcal{B}^k)$ is linearly independent (as n quadratic forms).*
4. *There is a basis $\{\omega^{(j)} : 1 \le j \le n\}$ of simultaneous eigenforms for the $\{\mathcal{B}^k\}$.*

If conditions **1–4** *are satisfied then there exist functions $f^j(\mathbf{x})$ such that $\omega^{(j)} = f^j dx^j$, $1 \le j \le n$.*

Again, these results extend to R-separation of the Laplace equation on V^n:

$$\Delta\Psi(\mathbf{x}) = 0. \tag{4.13}$$

A linear differential operator B on V^n is a ***conformal symmetry*** operator for Δ if there exists a linear differential operator C such that

$$[\Delta, B] \equiv \Delta B - B\Delta = C\Delta. \tag{4.14}$$

Uniquely associated with every second order conformal symmetry operator

$$B = \sum_{\substack{i=1\\j=1}}^{n} b^{ij}\partial_{ij} + \sum_{i=1}^{n} b^{i}\partial_i + c$$

(in local coordinates $\{y^\ell\}$) is the second order conformal symmetry

$$\mathcal{B} = \sum_{\substack{i=1\\j=1}}^{n} b^{ij} u_i u_j.$$

Theorem 8. [18] *Necessary and sufficient conditions for the existence of an orthogonal R-separable coordinate system $\{x^j\}$ for the Laplace equation (4.13) are that there exist $n-1$ second order differential operators $B^2, \ldots, B^n$ on V^n such that*

1. *Each B^k is a conformal symmetry operator.*
2. $[B^i, B^j] = 0, \quad 2 \le i,j \le n.$
3. *The set $(\mathcal{H}, \mathcal{B}^2, \ldots, \mathcal{B}^n)$ is linearly independent.*
4. *There is a basis $\{\omega^{(j)} : 1 \le j \le n\}$ of simultaneous eigenforms for the $\mathcal{B}^k$.*

If conditions **1–4** *are satisfied then there exist functions $f^j(\mathbf{x})$ such that $\omega^{(j)} = f^j(\mathbf{x})\,dx^j$, $j = 1, 2, \ldots, n$.*

The preceding theorems characterize orthogonal separation and R-separation for Hamilton–Jacobi, Helmholtz, and Laplace equations in terms of symmetries. There are similar results for nonorthogonal separation of these equations, though the results are more complicated to state and prove: see [11,17].

We can reach several important conclusions concerning variable separation and R-separation for Hamilton–Jacobi, Helmholtz, and Laplace (or wave) equations. First, one must recognise the intrinsic geometric nature of R-separation. The apparently technical conditions for R-separation are equivalent to the existence of an n-dimensional family of commuting symmetry operators which can be simultaneously diagonalized. In spaces, such as those of constant curvature, for which all symmetry operators can be constructed from the Lie symmetry algebra, all R-separation questions become problems in algebra [1,15].

Second, comparing Theorems 5 and 6, it is obvious that R-separation, not ordinary separation, for the Helmholtz equation is the natural analogy of additive separation for the Hamilton–Jacobi equation. Finally, we note the close relationship between variable separation and quantization theory. Corresponding to a separable system $\{x^j\}$ for the Hamilton–Jacobi equation we have an involutive family $\{\mathcal{A}^\ell\}$ of quadratic constants of the motion. The Helmholtz equation R-separates in these same coordinates if and only if second order operators $\{A^\ell\}$ can be found (with the pure second order terms in A^ℓ agreeing with those of $\mathcal{A}^\ell$) such that the A^ℓ pairwise commute.

§5 Related evolution equations

By an *evolution equation* we mean a partial differential equation of the form

$$(*) \qquad \Omega \equiv v_t - K(y^1, \ldots, y^n, v, v_{i_1 \ldots i_n}) = 0, \qquad (5.1)$$

where

$$v_t = \partial_t v, \qquad v_{i_1 \ldots i_n} = \partial_{y^1}^{i_1} \ldots \partial_{y^n}^{i_n} v,$$

and K depends on only a finite number $m > 0$ of the derivatives $v_{i_1 \ldots i_n}$. We assume that K is a real local analytic function of its $m+n+1$ variables, and for technical reasons, that it is a polynomial in the derivatives $v_{i_1 \ldots i_n}$. A ***solution*** of (5.1) is a function $v = v(t, y^1, \ldots, y^n)$, locally analytic in the variables $(t, \mathbf{y})$, such that (5.1) is well defined and identically satisfied for all $(t, \mathbf{y}) \in S$ where S is a nonempty open set in C^{n+1}. A second evolution equation

$$(+) \qquad \Phi \equiv u_s - J(x^1, \ldots, x^n, u, u_{i_1 \ldots i_n}) = 0, \tag{5.2}$$

is said to be a ***related*** to $(*)$ if there is a coordinate transformation

$$t = t(s, \mathbf{x}), \quad y^j = y^j(s, \mathbf{x}), \quad v = v(s, \mathbf{x}, u), \tag{5.3}$$

$j = 1, \ldots, n$, which maps $(*)$ to $(+)$. Here we assume that the Jacobian $\det(\frac{\partial(t, \mathbf{y})}{\partial(s, \mathbf{x})})$ is locally nonzero and $\frac{\partial v}{\partial u} \neq 0$. It is clear that an arbitrary transformation of the form

$$t = s, \qquad y^j = y^j(\mathbf{x}), \qquad v = v(\mathbf{x}, u), \tag{5.4}$$

will map $(*)$ to a related evolution equation, so we consider transformations of the form (5.4) to be trivial. Our interest is in determining all equivalence classes of evolution equations related to a given equation, where two evolution equations are equivalent if they are related by a trivial coordinate transformation.

Our basic observation is

Lemma 1. *Let*

$$\Omega \equiv v_t - K(\mathbf{y}, v, v_{i_1 \ldots i_n}) = 0 \tag{5.5}$$

be an evolution equation and

$$\Phi \equiv u_s - J(\mathbf{x}, u, u_{j_1 \ldots j_n}) = 0 \tag{5.6}$$

an evolution equation related to Ω *by means of the coordinate transformation*

$$t = T(s, \mathbf{x}), \quad y^j = Y^j(s, \mathbf{x}), \quad v = V(s, \mathbf{x}, u). \tag{5.7}$$

Then $X(f)$ *is a standard form point symmetry of* $\Omega = 0$*, where*

$$f = \partial_s V - \sum_{j=1}^{n} \partial_s Y^j \cdot v_{y^j} - \partial_s T \cdot K. \tag{5.8}$$

Proof. It is obvious from (5.6) that $\hat{Y} = \partial_s$ is a point symmetry operator for $\Phi = 0$, hence for $\Omega = 0$. From (2.6), (2.9) and (5.7) we see that Y corresponds to the standard form symmetry $X(f)$. *Q.E.D.*

The converse of Lemma 1 is false; there may exist point symmetries of $\Omega = 0$ that do not correspond to a related evolution equation. For example,

$$\Omega \equiv v_t - v_{y^1 y^1} - v_{y^2 y^2} = 0, \qquad \hat{Y} = \partial_{y^1}. \tag{5.9}$$

The following result isolates a special class of point symmetries that do correspond to related evolution equations.

Lemma 2. *Let*

$$Y = \tau(t)\partial_t + \sum_{j=1}^{n} \xi^j(t, \mathbf{y})\partial_{y^j} + \eta(t, \mathbf{y}, v)\partial_v \tag{5.10}$$

be a point symmetry operator for $\Omega = 0$, *where* $\tau \not\equiv 0$. *Then there exists a transformation to new coordinates* $(s, \mathbf{x}, u)$,

$$t = T(s), \qquad y^j = Y^j(s, \mathbf{x}), \qquad v = V(s, \mathbf{x}, u), \tag{5.11}$$

such that (in the new coordinates) $\hat{Y} = \partial_s$ *and the transformed equation can be expressed as an evolution equation*

$$\Phi \equiv u_s - J(\mathbf{x}, u, u_{j_1 \ldots j_n}) = 0.$$

The coordinate transformation is not unique because of the possibility of trivial transformations (5.4). We are identifying an equivalence class of related equations.

Proof. It follows directly from (5.10) and Lie's theorem [2, pp. 34, 49, and 50], that there exists a new coordinate system $(s, \mathbf{x}, u)$ such that the coordinate transformation takes the form (5.11) and $\hat{Y} = \partial_s$. Introducing the new coordinates into the equation $\Omega = 0$ we see that only the term v_t contributes a derivative of u with respect to s. Thus this equation can be rewritten as

$$\Phi \equiv u_s - J(s, \mathbf{x}, u, u_{j_1 \ldots j_n}) = 0.$$

However, since $\hat{Y} = \partial_s$ is a point symmetry operator for $\Phi = 0$, we must have $\partial_s J = 0$.

The form of the point symmetry operator (5.10) appears somewhat special. However, for a large class of evolution equations it is perfectly general. Given an evolution equation $\Omega = 0$, (5.1), we can express K as a polynomial in the derivatives $v_{i_1 \ldots i_n}$. We say that a given monomial $v_{i_1 \ldots i_n} v_{j_1 \ldots j_n} v_{\ell_1 \ldots \ell_n}$ has ***order*** $i_1 + i_2 + i_n + j_1 + \ldots + \ell_n$. In reference [18] we introduce the technical definition of ***nondegeneracy*** for K which means, roughly speaking, that for the monomials of highest order $m > 0$ it is not possible to choose new special coordinates $\mathbf{x}(\mathbf{y})$ such that the transformed monomials contain no derivatives with respect to one of the coordinates x^1. For example, with $n = m = 2$ the forms $v_{y^1 y^1} + v_{y^2 y^2}$, $v_{y^1}^2 + v_{y^2 y^2}$, and $y^1 v_{y^1 y^2}$, $(y^1 \neq 0)$ are nondegenerate while $v_{y^2}^2$ is degenerate. In reference [18] we prove:

Theorem 9. *Let*

$$\Omega \equiv v_t - K(\mathbf{y}, v, v_{i_1 \ldots i_n}) = 0 \tag{5.12}$$

be a nondegenerate evolution equation. There is a one-to-one correspondence between (equivalence classes of) nondegenerate evolution equations related to $\Omega = 0$ *and point symmetry operators for* $\Omega = 0$ *of the form*

$$\hat{Y} = \tau(t, \mathbf{y})\partial_t + \sum_{j=1}^{n} \xi^j(t, \mathbf{y})\partial_{y^j} + \eta(t, \mathbf{y}, v)\partial_v \tag{5.13}$$

with $\tau \neq 0$. *In fact, all such point symmetries have the property that* $\partial_{y^i}\tau = 0$, $i = 1, \ldots, n$.

An example of the application of this theorem is provided by the KdV equation (2.24) and the related equation (2.25).

We now modify these ideas to apply to the Hamilton–Jacobi equation

$$2\lambda u_t - g^{ij}(\mathbf{y})u_{y^i} u_{y^j} - 2\lambda \xi^i(\mathbf{y})u_{y^i} - \lambda^2 V(\mathbf{y}) = 0. \tag{5.14}$$

(For convenience, we adopt the Einstein summation convention for indices) Here ξ^i and V are given functions, λ is a parameter and (g^{ij}) is an $n \times n$ nonsingular matrix defining a metric on a pseudo-Riemannian space V^n. We can interpret (5.14) as the (time-dependent) Hamilton–Jacobi equation for a one-particle hamiltonian system on V^n with (velocity-dependent) potential.

Our interest is in transformations of (5.14) which map this equation into another equation of the same type:

$$2\lambda\tilde{u}_s - \tilde{g}^{ij}(\mathbf{x})u_{x^i}u_{x^j} - 2\lambda\tilde{\xi}^i(\mathbf{x})\tilde{u}_{x^i} - \lambda^2\tilde{V}(\mathbf{x}) = 0. \tag{5.15}$$

Here, λ is unchanged and $\tilde{g}^{ij}$ defines a metric on a pseudo-Riemannian space $\tilde{V}^n$. In terms of the variables determining (5.14), the allowable transformations are

$$t = T(s,\mathbf{x}), \qquad y^j = Y^j(s,\mathbf{x}), \qquad u = \tilde{u} + \lambda h(s,\mathbf{x}). \tag{5.16}$$

We must determine which of these transformations will map (5.14) into an equation of the form (5.15).

We claim that the allowable point symmetries for (5.14) should be those of the type

$$Y = \tau(t,\mathbf{y})\partial_t + Y^j(t,\mathbf{y})\partial_{y^j} + \lambda k(s,\mathbf{y})\partial_u \tag{5.17}$$

(It is straightforward to show that the space of all symmetries of this type forms a Lie algebra under the usual operator conmutator bracket).

Theorem 10. *There is a one-to-one correspondence between (equivalence classes of) Hamilton–Jacobi equations related to (5.14) and point symmetry operators $\hat{Y}$ for (5.14) of the form (5.17) with $\tau \neq 0$. All symmetries of the form (5.17) satisfy $\partial_{y^i}\tau = 0$, $i = 1, \ldots, n$.*

Proof. Suppose the Hamilton–Jacobi equation (5.15) is related to (5.14) via a transformation of the form (5.16). This means that we can obtain (5.15) by substituting the transformation (5.16) into (5.14) and solving for $\lambda\tilde{u}_s$ in the resulting expression. Since $(\tilde{g}^{ij})$ is nonsingular, the $n \times n$ matrix $(\frac{\partial x^i}{\partial y^j})$ must also be nonsingular. Furthermore, the coefficient of $u_s u_\ell$ in the resulting expression is $2g^{ij}\frac{\partial s}{\partial y^i}\frac{\partial x^\ell}{\partial y^i}$. Since this must vanish for each ℓ, we have $\frac{\partial s}{\partial y^i} = 0$, so $s = S(t)$ or $t = T(s)$ with $\partial_s T \neq 0$. Clearly, $\hat{Y} = \partial_s$ is a point symmetry of (5.15) which implies that $\hat{Y} = T'(s)\partial_t + \frac{\partial y^i}{\partial s}\partial_{y^i} + \lambda\frac{\partial h}{\partial s}\partial_u$ is a point symmetry of (5.14).

Conversely, suppose $\hat{Y}$, (5.17) is a point symmetry operator for (5.14) with $\tau \neq 0$. Then $X(\lambda k - \xi^j u_{y^j} - \tau u_t)$ is a standard form symmetry operator for this equation. This is possible only if $\partial_{y^i}\tau = 0$, $i = 1, \ldots, n$, [since $(g^{i\ell})$ is nonsingular]. By Lie's theorem we can introduce new coordinates $s, \mathbf{x}$ and a new dependent variable $\tilde{u}$ such that

$$\begin{gathered} \partial_s = \tau\partial_t + f^j\partial_{y^j}, \\ t = T(s), \qquad y^j = Y^j(s,\mathbf{x}), \qquad u = \tilde{u} + \lambda h(s,\mathbf{x}), \end{gathered} \tag{5.18}$$

where $\partial_s h = k$. Clearly, $\frac{\partial s}{\partial y^i} = 0$ and $\frac{\partial x^j}{\partial y^i}$ are nonsingular. Thus, substituting the new coordinates s, x, $\tilde{u}'$ into (5.14), we see that the coefficients of $\tilde{u}_s^2$ and $\tilde{u}_s\tilde{u}_i$ vanish in the resulting expression while the quadratic form $g^{ij}\frac{\partial x^\ell}{\partial y^i}\frac{\partial x^m}{\partial y^j}\tilde{u}_{x^\ell}\tilde{u}_{x^m}$ is nonsingular. The coefficient of $\tilde{u}_s$ is $\lambda\frac{\partial s}{\partial t} = \lambda f(s) \neq 0$ and, since $\hat{Y} = \partial_s + \lambda k\partial_u$ is a point symmetry, if we multiply all terms in our equation by $f^{-1}(s)$ we obtain an expression of the form (5.15), where the coefficient of each term is independent of s and $(\tilde{g}^{ij})$ is nonsingular.

Corollary 3. *If the Hamilton–Jacobi equations (5.14) and (5.15) are related by a transformation (5.16) then the tensors (g^{ij}) and $\pm(\tilde{g}^{ij})$ define metrics on the same pseudo-Riemannian manifold (There is a possible sign ambiguity).*

There is a similar theory of transformations that map a Schrödinger equation

$$2\lambda\partial_t\Psi - \frac{1}{\sqrt{g}}\partial_{y^i}(g^{ij}\sqrt{g}\,\partial_{y^j})\Psi - (2\lambda\xi^i + \rho^i)\partial_{y^i}\Psi - (\lambda^2V + \lambda U + W)\Psi = 0 \tag{5.19}$$

to another Schrödinger equation

$$2\lambda\partial_s\Theta - \frac{1}{\sqrt{\tilde{g}}}\partial_{x^i}(\tilde{g}^{ij}\sqrt{\tilde{g}}\,\partial_{x^j})\Theta - (2\lambda\tilde{\xi}^i + \tilde{\rho}^i)\partial_{x^i}\Theta - (\lambda^2\tilde{V} + \lambda\tilde{U} + \tilde{W})\Theta = 0. \tag{5.20}$$

Here, λ is a parameter which we can roughly identify with $1/i\hbar$ where $\hbar$ is Planck's constant [1], $\big(g^{ij}(\mathbf{y})\big)$ determines a metric on a pseudo-Riemannian n-dimensional manifold, and $g^{-1} = \det(g^{ij})$. The allowable coordinate transformations are of the form

$$\begin{gathered} t = T(s,\mathbf{x}), \qquad y^j = Y^j(s,\mathbf{x}), \\ \Psi(t,\mathbf{y}) = \exp\big(\lambda R^{(1)}(s,\mathbf{x}) + R^{(2)}(s,\mathbf{x})\big)\,\Theta(s,\mathbf{x}), \end{gathered} \tag{5.21}$$

where $\frac{\partial(t,\mathbf{y})}{\partial(s,\mathbf{x})}$ is nonsingular. The allowable point symmetries of (5.19) take the form

$$\hat{Y} = \tau(t,\mathbf{y})\partial_t + \gamma^j(t,\mathbf{y})\partial_{y^j} + \big(\lambda k(t,\mathbf{y}) + \ell(t,\mathbf{y})\big)\Psi\partial_\Psi. \tag{5.22}$$

The vector space of symmetries of this type is a Lie algebra under the usual operator commutator bracket; we call this the ***Lie symmetry algebra*** of (5.19).

Theorem 11. *There is a one-to-one correspondence between (equivalence classes of) Schrödinger equations related to (5.19) and allowable point symmetry operators $\hat{Y}$ for (5.19) of the form (5.22) with $\tau \neq 0$. All symmetries of the form (5.22) satisfy $\partial_{y^i}\tau = 0$, $i = 1, \ldots, n$.*

Corollary 4. *If the Schrödinger equations (5.19) and (5.20) are related by a transformation (5.21) then the tensors (g^{ij}) and $\pm(g^{ij})$ define metrics on the same pseudo-Riemannian manifold.*

The proof of these statements and the connection between the symmetry operator and the coordinate transformation is very similar to that of Theorem 10. Again the time coordinates s, t of related Schrödinger equations satisfy $t = T(s)$ where $T' = \tau$.

To provide further examples of related evolution equations we will consider one case in detail: the free particle Hamilton–Jacobi equation

$$2\lambda u_t = u_{y^1}^2 + u_{y^2}^2. \tag{5.23}$$

The symmetry algebra $\mathcal{G}$ of this equation (the Schrödinger algebra) is nine-dimensional, with basis

$$\begin{aligned} P_i &= \partial_{y^i}, \\ B_i &= -t\partial_{y^i} + \lambda y^i\partial_u, \qquad i = 1,2, \\ M &= y^1\partial_{y^2} - y^2\partial_{y^1}, \\ E &= 2\lambda\partial_u, \\ K_2 &= -t^2\partial_t - t(y^1\partial_{y^1} + y^2\partial_{y^2}) + \tfrac{1}{2}\lambda(y^1y^1 + y^2y^2)\partial_u, \\ K_{-2} &= \partial_t, \\ D &= y^1\partial_{y^1} + y^2\partial_{y^2} + 2t\partial_t. \end{aligned} \tag{5.24}$$

We will identify two related evolution equations if one can be transformed into the other through an action of the Schrödinger group [1, Chapter 2]. Thus to classify the possible related evolution equations we first determine a complete set of orbit representatives for one-dimensional subalgebras of $\mathcal{G}$ (under the adjoint action of the Schrödinger group). We then compute the related evolution equation, if any, associated with the orbit representative. A list of orbit representatives is given in [1, p. 124]. Our final results are presented in Table 1.

We express any related equation in terms of Cartesian coordinates:

$$2\lambda \tilde{u}_s = \tilde{u}_{x^1}^2 + \tilde{u}_{x^2}^2 + 2\lambda(\tilde{\xi}^1 \tilde{u}_{x^1} + \tilde{\xi}^2 \tilde{u}_{x^2}) + \lambda^2 \tilde{V}. \tag{5.25}$$

Thus the related equation can be determined by merely listing its associated potential.

Note that the operators corresponding to orbits 8–10 are not associated with an evolution equation since they contain no term in ∂_t. The analogous results for the free-particle Schrödinger equation are virtually identical with those presented here for the Hamilton–Jacobi equation.

The above results are closely related to the separation of variables problem for Hamilton–Jacobi and Schrödinger equations. We can see this by considering a single example: orbit 1 on Table 1. We look for solutions of the associated equation (5.25) such that $\tilde{u}_s = E$, a constant. The resulting equation,

$$\tilde{u}_{x^1}^2 + \tilde{u}_{x^2}^2 + \lambda^2(x^1x^1 + x^2x^2) = E, \tag{5.26}$$

is the Hamilton–Jacobi equation for the harmonic oscillator. It is well known, [1], that (5.26) separates in precisely three coordinate systems: cartesian coordinates $\{x^1, x^2\}$, polar coordinates $\{r, \theta\}$, and elliptic coordinates $\{\alpha, \beta\}$, where

$$x^1 = r\cos\theta, \qquad x^2 = r\sin\theta,$$

and

$$x^1 = d\cosh\alpha\,\cos\beta, \qquad x^2 = d\sinh\alpha\,\sin\beta,$$

where d is a positive constant. Transforming back to the free particle equation (5.23) via the transformation listed on Table 1, we see that each of these separable systems leads to an R-separable coordinate system for (5.23). Similarly, orbits 4 ($\beta = 0$), 6 and 7 yield R-separable coordinates for (5.23) and, indeed, all "interesting" R-separable coordinates for this equation arise from separable coordinates for the time-independent free particle, linear potential, harmonic oscillator, and repulsive oscillator cases. For the remaining cases the time independent equation does not separate in any coordinate system.

We note here a recent paper by Kumei and Bluman that uses generalized Lie symmetries to characterize those nonlinear differential equations that can be transformed to linear equations via 1 : 1 contact transformations, [20].

§6 Concluding remarks

We have demonstrated a close connection between symmetry and variable separation for linear equations and first order partial differential equations, as well as in the computation of symmetry-adapted solutions. However, the intrinsic characterization and connection with symmetry of separable coordinates for general nonlinear PDE's has yet to be established. This would be an interesting topic for future research.

In these notes we have not attacked the computationally difficult problem of computing for a fixed PDE, all the coordinate systems in which the equation admits variable separation. A brief review of the history of this problem and recent progress can be found in [3]. Of particular interest may be reference [21] where a simple graphical procedure is given for the construction of all separable coordinate systems for the Hamilton–Jacobi and Helmholtz equations on the n-sphere and in Euclidean n-space, for all n.

Operator	Coordinates	Potential
1. $K_{-2} - K_2$	$t = \tan s$ $y^i = x^i \sec s$ $u = \tilde{u} - \frac{1}{2}\lambda(x^1x^1 + x^2x^2)\tan s$	$\lambda(x^1x^1 + x^2x^2)$
2. $K_{-2} - K_2 + \beta M$	$t = \tan s$ $y^1 = (x^1 \sin\beta s + x^2 \cos\beta s)\sec s$ $y^2 = (-x^1 \cos\beta s + x^2 \sin\beta s)\sec s$ $u = \tilde{u} - \frac{1}{2}\lambda(x^1x^1 + x^2x^2)\tan s$	$2\lambda\beta(x^1\tilde{u}_{x^2} - x^2\tilde{u}_{x^1})$ $+ \lambda^2(x^1x^1 + x^2x^2)$
3. $K_{-2} - K_2 + M + \gamma B_1$	$t = \tan s$ $y^1 = x^1 \tan s + x^2 - \frac{1}{2}\gamma s \tan s + \frac{1}{2}\gamma$ $u = \tilde{u} - \frac{1}{2}\lambda[(x^1x^1 + x^2x^2)\tan s$ $-\gamma x^1 s \tan s - \gamma x^2 s$ $+\frac{1}{4}\gamma^2 s^2 \tan s - \frac{3}{4}\gamma^2 s]$	$2\lambda(x^1\tilde{u}_{x^2} - x^2\tilde{u}_{x^1} - \frac{1}{2}\gamma\tilde{u}_{x^1})$ $+ \lambda^2(x^1x^1 + x^2x^2 - \gamma x^2 - \frac{3}{4}\gamma^2)$
4. $D + \beta M$	$t = e^{2s}$ $y^1 = \sqrt{2}e^s(x^1 \cos\beta s - x^2 \sin\beta s)$ $y^2 = \sqrt{2}e^s(x^1 \sin\beta s - x^2 \cos\beta s)$ $u = \tilde{u}$	$2\lambda(x^1\tilde{u}_{x^1} - x^2\tilde{u}_{x^2})$ $+ 2\lambda\beta(x^1\tilde{u}_{x^2} - x^2\tilde{u}_{x^1})$
5. $-K_2 - M$	$t = -s^{-1}$ $y^1 = \frac{1}{2}x^1 \sin s + x^2 s^{-1} \cos s$ $y^2 = x^1 s^{-1} \cos s + x^2 s^{-1} \sin s$ $u = \tilde{u} + \frac{1}{2}\lambda s^{-1}(x^1x^1 + x^2x^2)$	$2\lambda(x^2\tilde{u}_{x^1} - x^1\tilde{u}_{x^2})$
6. $-K_2 - P_1$	$t = -s^{-1}$ $y^1 = -\frac{1}{2}s + x^1 s^{-1}$ $y^2 = x^2 s^{-1}$ $u = \tilde{u} + \frac{1}{2}\lambda[(x^1x^1 + x^2x^2)s^{-1}$ $-\frac{1}{12}s^3 + x^1 s]$	$-2\lambda^2 x^1$
7. K_{-2}	$t = s$ $y^j = x^j$ $u = \tilde{u}$	0
8. M	—	
9. $P_1 + B_2$	—	
10. P_1	—	

Evolution equations related to $2\lambda u_t = u_{y^1}^2 + u_{y^2}^2$.

References

[1] W. Miller, Jr., *Symmetry and Separation of Variables.* Addison-Wesley, 1977.

[2] L. P. Eisenhart, *Continuous Groups of Transformations.* Dover reprint, 1961.

[3] E. G. Kalnins and W. Miller, Jr., Intrinsic characterization of variable separation for the partial differential equations of mechanics. Symposium on Modern Developments in Analytical Mechanics (Torino, Italy, June 1982). (To appear.)

[4] P. J. Olver, Symmetry groups and group invariant solutions of partial differential equations. *J. Diff. Geom.* **14**, 497–542 (1979).

[5] P. J. Olver, *Applications of Lie Groups to Differential Equations.* Mathematical Institute Lecture Notes, Oxford, 1980.

[6] G. W. Bluman and J. D. Cole, *Similarity Methods for Differential Equations.* Applied Mathematical Series # 13, Springer-Verlag, 1974.

[7] L. V. Ovsjannikov, *Group Properties of Differential Equations.* Academic Press, 1982.

[8] L. P. Eisenhart, *Riemannian Geometry.* Princeton University Press, 2nd printing, 1949.

[9] T. W. Koornwinder, A precise definition of separation of variables. Proceedings of the Scheveningen Conference of Differential Equations. Lecture Notes in Mathematics, Springer-Verlag, 1980.

[10] L. P. Eisenhart, Separable systems of Stäckel. *Ann. Math.***35**, 284–305 (1934).

[11] E. G. Kalnins and W. Miller, Jr., Killing tensors and nonorthogonal variable separation for the Hamilton–Jacobi equation. *SIAM J. Math. Anal.* **12**, 617–638 (1981).

[12] E. G. Kalnins and W. Miller, Jr., Conformal Killing tensors and variable separation for the Hamilton–Jacobi equation. *SIAM J. Math. Anal.* (To appear.)

[13] E. G. Kalnins and W. Miller, Jr., The theory of orthogonal R-separation for Helmholtz equations. *Advances in Mathematics* (To appear.)

[14] E. G. Kalnins and W. Miller, Jr., Some Remarkable R-separable coordinate systems for the Helmholtz equation. *Lett. Math. Phys.* **4**, 469–474 (1980).

[15] E. G. Kalnins and W. Miller, Jr., Killing tensors and variable separation for Hamilton–Jacobi and Helmholtz equations. *SIAM J. Math. Anal.* **11**, 1011–1026 (1980).

[16] E. G. Kalnins and W. Miller, Jr., Intrinsic characterization of orthogonal R-separation for Laplace equations. (To appear.)

[17] E. G. Kalnins and W. Miller, Jr., The general theory of R-separation for Helmholtz equations. *J. Math. Phys.* (To appear.)

[18] E. G. Kalnins and W. Miller, Jr., Related evolution equations and Lie symmetries. (Submitted.)

[19] L. Landau and E. Lifshitz, *Quantum Mechanics, Non-Relativistic Theory.* (Translated from Russian.) Addison-Wesley, 1958.

[20] S. Kumei and G. W. Bluman, When nonlinear differential equations are equivalent to linear differential equations. *SIAM J. Appl. Math.* **42**, 1157–1173 (1982).

[21] E. G. Kalnins and W. Miller, Jr., Separation of variables on n-dimensional Riemannian manifolds. I. The n-sphere and Euclidean n-space. (To appear.)

Collectivity and Geometry

Marcos Moshinsky*

Instituto de Física
Universidad Nacional Autónoma de México

Abstract

The papers on which my talk at this School was based, were submitted for publication in the Journal of Mathematical Physics. Only a brief abstract will be published here.

We have examined the way in which collective behavior has been discussed in nuclei, and show that it can be related to the symplectic geometry of the many-body nucleon system when we introduce in it appropriate constraints. Specifically, the n-body system in d-dimensional space is associated with the symplectic group in $2dn$ dimensions, *i.e.*, $Sp(2dn)$. This group admits a subgroup $Sp(2d) \times O(n)$ and a reasonable definition of collectivity is obtained when the states are constrained to a definite irreducible representation of $O(n)$ —and thus also of $Sp(2d)$— and the collective Hamiltonian is restricted to the enveloping algebra of $Sp(2d)$.

The group $Sp(2d)$ admits the subgroup $Sp(2) \times O(d)$ and the mathematical problem is to determine the states characterized by definite irreducible representations of this chain of groups, as well as the matrix elements of the generators of $Sp(2d)$ with respect to these states. The problem has been solved for $d = 2$ in the second paper of the series and will be discussed for $d = 3$ in the third paper which is in preparation. For physical space, *i.e.*, $d = 3$, the Casimir operator of $Sp(2)$ goes into the Bohr–Mottelson collective Hamiltonian when we are constrained to a scalar representation of $O(n)$ and $n \gg 1$. The Casimir operator of the $O(2)$ subgroup of $Sp(2)$ goes into the Interacting Boson Model Hamiltonian. Other collective models are possible when different types of constraints are chosen.

In future papers of this series, the techniques developed in the third paper will be applied to the systematic discussion of collective effects in nuclear models.

---

* Member of the Instituto Nacional de Investigaciones Nucleares and El Colegio Nacional.

Some Physical Applications of Solitons

Harvey Segur

Aeronautical Research Associates of Princeton, Inc.
Princeton, New Jersey, USA

Contents:

---

* By Harvey Segur, Allan Finkel (Institute for Advanced Study, Princeton), and Hilda Philander (A.R.A.P., Princeton).

** H. Segur and J. L. Hammack, *J. Fluid Mech.* **118**, 285–304 (1982)

*** H. Segur, *Journal of Mathematical Physics* **24**, 1439–1443 (1983)

Chapter 1

Integrable Models of Shallow Water Waves*

Abstract

The Korteweg-de Vries and the Kadomtsev-Petviashvili (KP) equations both model the evolution of relatively long water of moderate amplitude as they propagate in shallow water. Both equations are completely integrable. In this paper we review the derivation of each equation as an approximate model of shallow water waves, and compare their solutions with some of the experimental observations of waves in shallow water. We also describe in detail the family of doubly periodic KP solutions. These are the natural two-dimensional generalizations of cnoidal waves in one-dimension; one way think of them as describing ***typical*** patterns of nonlinear, two-dimensional waves in shallow water.

§1.1 Introduction

Two equations that have been studied intensively in recent years are the Korteweg-de Vries (KdV; 1895) equation,

$$u_{\tau} + 6uu_{\chi} + u_{\chi\chi\chi} = 0 \,, \tag{1}$$

and a generalization of it due to Kadomtsev and Petviashvili (KP; 1970),

$$(u_{\tau} + 6uu_{\chi} + u_{\chi\chi\chi})_{\chi} + 3u_{\eta\eta} = 0 \,. \tag{2}$$

Most of this interest is due to the remarkable fact that each equation can be solved exactly as an initial-value problem by a method now known as the Inverse Scattering Transform. This method was first discovered for the KdV equation on $-\infty < \chi < \infty$, in the famous papers of Gardner, Greene, Kruskal and Miura (1967, 1974). The corresponding work for (1) on a periodic interval was published by several people during 1974–1976 [Novikov (1974), Dubrovin and Novikov (1974), Dubrovin (1975), Lax (1975), Its and Matveev (1975), McKean and van Moerbeke (1975), McKean and Trubowitz (1976), Dubrovin, Matveev and Novikov (1976)]. For the KP equation on $-\infty < \chi, \eta < \infty$, a method of solution was given very recently by Ablowitz, Bar Yaacov, and Fokas (1982); *cf.* Ablowitz and Fokas, *in these Proceedings*).

As we shall see below, it happens that both (1) and (2) also model the evolution of water waves of moderate amplitude as they propagate in one direction in relatively shallow water. In physical

* By Harvey Segur, Allan Finkel (Institute for Advanced Study, Princeton), and Hilda Philander (A.R.A.P., Princeton).

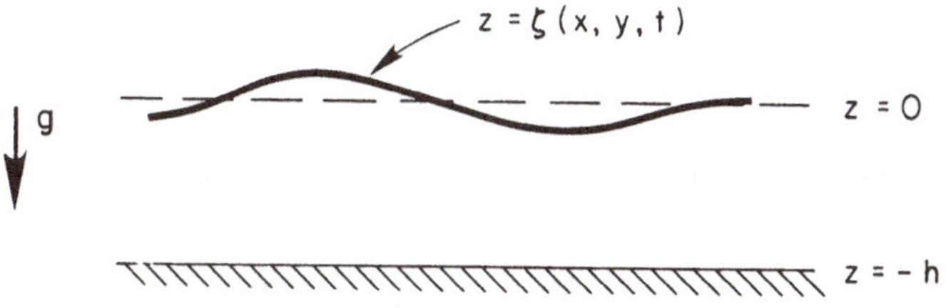

Figure 1.1. Physical configuration showing notation for Equations (3)–(6).

terms, the KdV equation arises if the waves are strictly one-dimensional (*i.e.*, one spatial dimension plus time), while the KP equation arises if they are only nearly one-dimensional. Because both equations are completely integrable, they provide very precise predictions about the evolution of water waves under appropiate conditions.

This paper has two objetives. The first is to examine the validity of (1) and (2) as models of water waves, by comparing their solutions with some of the available experimental data. The comparison given here will be brief, but much more detailed verifications of (1) have been made elsewhere (*e.g.*, Hammack and Segur, 1974, 1978). The second objetive is to describe in detail a special family of solutions of (2), which are periodic in each of two independent variables. It appears that these doubly-periodic solutions may have great practical importance as models of long water waves. Among other things, they describe:

i. the nonlinear interaction of two trains of finite amplitude waves in shallow water;

ii. the reflection of a train of finite-amplitude shallow-water waves from a vertical wall (by replacing an appropriate line of symmetry with the wall);

iii. the reflection of such a wave train by a change in bottom topography; or

iv. *typical* finite-amplitude, short-crested waves in shallow water. In this last sense, they are the natural generalizations to two dimensions of cnoidal waves in one dimension.

§1.2 Derivation of the equations

The classical problem of water waves is to find the irrotational motion of an inviscid, incompressible, homogeneous fluid, subject to a constant gravitational force (g). The fluid rests on a horizontal and impermeable bed of infinite extent at $z = -h$ and has a free surface at $z = \varsigma(x, y, t)$; see Figure 1.1. In this derivation we neglect the effects of surface tension at the free surface, although it can be included without difficulty (*e.g.*, see Ch. 4 of Ablowitz and Segur, 1981).

The fluid has a velocity potencial, ϕ, which satisfies

$$\nabla^2\phi = 0\,, \qquad -h < z < \varsigma(x, y, t)\,; \tag{3}$$

(irrotational motion of an incompressible fluid). It is subject to boundary conditions on the bottom, $z = -h$:

$$\phi_z = 0, \tag{4}$$

(impermeable bed); and along the free surface, $z = \varsigma$:

$$\frac{D\varsigma}{Dt} \equiv \varsigma_t + \phi_x\varsigma_x + \phi_y\varsigma_y = \phi_z \tag{5}$$

(kinematic condition);

$$\phi_t + g\varsigma + \tfrac{1}{2}|\nabla\phi|^2 = 0 \tag{6}$$

(dynamic condition).

Boundary conditions in (x, y) and initial conditions also are required. If the waves in question are isolated, then $\nabla\phi$ and ς should vanish as $(x^2 + y^2) \to \infty$. In other problems, periodic boundary conditions in x and in y may be relevant.

This problem, first posed by Stokes (1874), remains unsolved. To make further progress, we impose additional assumpions on the solutions of (3–6). The first such assumption is that the wave amplitudes should be small. If we interpret *small* to mean infinitesimal, then we may linearize (3–6) about $\nabla\phi = 0$, $\varsigma = 0$, and seek solutions of the linearized equations proportional to $\exp\{i(kx+my-wt)\}$ (*e.g.*, see Lamb, 1932, §§ 228, 266, 267). The result is the linearized dispersion relation,

$$\omega^2 = g\kappa \tanh \kappa h\,, \tag{7}$$

where $\kappa^2 = k^2 + m^2$. From this one computes the group velocity and shows that the linearized problem is dispersive at most wave numbers, but not as $\kappa \to 0$, (*i.e.*, long waves, or shallow water waves), where it is only weakly dispersive. Both (1) and (2) arise as models of the water wave problem in this weakly dispersive limit, $\kappa h \ll 1$.

To derive (1) or (2), we assume that:

A. wave amplitude are small,

$$\epsilon \equiv \varsigma_{\max}/h \ll 1\,;$$

B. the relevant length scale in the x-direction is much longer than the fluid depth (*i.e.*, shallow water waves),

$$(kh)^2 \ll 1\,;$$

C. either the motion is strictly one-dimensional (for KdV),

$$m = 0\,,$$

C′ or it is nearly one-dimensional (for KP)

$$\left(\frac{m}{k}\right)^2 \ll 1\,;$$

D. all of these effects balance,

$$\mathrm{KdV}: (kh)^2 = O(\epsilon), \qquad \mathrm{KP}: (kp)^2 = O(\epsilon) = O(mh)\,.$$

These assumptions imply a certain scaling of the original equations, [see (9) below], which may then be solved as a perturbation series, term-by-term in ϵ.[1]

At leading order, the equations are linear (from **A**), nondispersive (from **B**), and one-dimensional (from **C**). The solution is simply the solution of the linear, one-dimensional wave equation:

$$\varsigma(x, y, t) = \epsilon h[f(x - \sqrt{gh}\,t; y) + F(x + \sqrt{gh}\,t; y)] + O(\epsilon^2)\,. \tag{8}$$

[1]We omit the details here, but the entire derivation may be found in Chapter 4 of Ablowitz and Segur (1981), or elsewhere.

At this order, every wave has permanent form, not because it is a soliton but because we are solving the one-dimensional, linear wave equation.

When this perturbation expansion is carried to second order, the one-dimensional, linear wave equation has homogeneous (forcing) terms representing weak nonlinearity, weak dispersion and weak two-dimensionality. Each of these effects contributes to a secular term at second order. We may eliminate them by introducing a second slow time-scale (T):

$$r = \sqrt{\epsilon}\,(x - \sqrt{gh}\,t)/h\,, \qquad l = \sqrt{\epsilon}\,(x + \sqrt{gh}\,t)/h\,,$$

$$\eta = \epsilon y/h\,, \qquad T = (\epsilon)^{3/2}\sqrt{g/h}\,t\,, \tag{9}$$

and requiring that the right-running waves satisfy

$$(2f_T + 3ff_r + \tfrac{1}{3}f_{rrr})_r + f_{\eta\eta} = 0\,, \tag{10}$$

which may be rescaled to (2). In the one-dimensional case, $\partial_\eta \equiv 0$, and (10) leads to (1). The left-running waves must satisfy a similar equation.

At this order, no secular terms arise from interactions between left- and right-running waves provided that f and F are smooth, and that $\int f dr$, $\int F dl$ remain bounded. For periodic initial data, the latter requirement implies that

$$\oint f dr = 0\,, \qquad \oint F dl = 0\,, \tag{11}$$

where the intergrals are taken over one period. Because the original equations are galilean-invariant, (11) amounts only to a normalization.

Some observations about the physical meaning of (1) and (2) may be made at this point.

A. The time in (1) or (2) has the physical meaning of a slow time-scale. Concepts like solitons that are implied by these equations have physical meaning only on this long time-scale.

B. Our original equations, (3–6), are themselves only approximately correct, because they neglect real effects like viscosity. Thus, there is a short (linear, non-dispersive, one-dimensional), time-scale, on which all waves have permanent form, from (8). This is followed by a longer (KdV or KP) time-scale on which soliton interactions are relevant. This is then followed by an even longer time-scale on which other effects (like viscosity) become important.

C. The KdV and KP equations are fully nonlinear, but they model water waves only when the water waves are weakly nonlinear. Fully nonlinear water waves are known to break, as solutions of (1) or (2) do not. Similarly, (2) models water waves that are only weakly two-dimensional, although this restriction is not evident in (2) by itself.

D. Because (1) and (2) are first-order in time, neither descibes the interactions of left-with right-running waves. However, this is not because such interactions were not admitted, but because they are not important on the time-scale on which these equations apply. Their interaction is given by (8), to leading order.

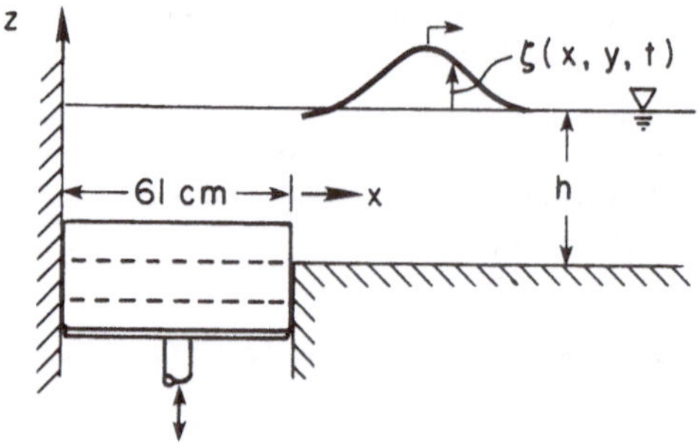

Figure 1.2. Schematic drawing of the wave generator used by Hammack and Segur (1974).

§1.3 Experimental evidence of solitons

The consequences of the theory of Inverse Scattering Transforms for the KdV equation are as follows. (Details may be found in Ablowitz and Segur, 1981, among other places.) Let $u_0(\chi)$ be any any smooth function on $(-\infty, \infty)$ that vanishes rapidly along with its derivatives as $\chi \to \pm\infty$. Then (1) has a unique solution that concides with $u_0(\chi)$ at $\tau = 0$. As $\tau \to \infty$, this solution evolves into N isolated solitons, ordered by amplitude, followed by an oscillatory wave-train that disperses in time (*radiation*). When the solitons have separated, each is given (locally) by

$$u(\chi, \tau) = 2k^2 \operatorname{sech}^2\{k(\chi - 4k^2\tau + \chi_0)\} . \tag{12}$$

The number of solitons (N), their amplitudes (k_j, $j = 1, \ldots, N$), and all the other details of the long-time solution can be found directly from $u_0(\chi)$. There is no need to advance (1) numerically in time.

Corresponding to (12) is a water wave, whose surface elevation is given by

$$\zeta(x,t) = \tfrac{4}{3}\epsilon k^2 h \operatorname{sech}^2\left\{\frac{\sqrt{\epsilon k}}{h}[x \pm \sqrt{gh}(1 + \tfrac{2}{3}\epsilon k^2)t + x_0]\right\} + O(\epsilon^2) . \tag{13}$$

Similarly, all the other consequences of (1) imply predictions about real water waves. Hammack and Segur (1974, 1978) tested several aspects of this theory by comparing it with laboratory experiments. Next we briefly reiterate some of their results, to give the reader an idea of the validity of the KdV equation as a model of long water waves of moderate amplitude. [Experimental comparisons of (1) with water waves also have been made by Zabusky and Galvin (1971), and by Weidman and Maxworthy (1978).]

The experiments of Hammack and Segur were conducted in a wave tank 31.6 m long, 61 cm deep and 39.4 cm wide. As shown schematically in Figure 1.2, the wave generator consisted of a rectangular piston located in the tank bed adjacent to the upstream end wall of the tank. The piston spanned the tank width, and was 61 cm long for the experiments we will discuss. The time-history of its vertical displacement was prescribed for each experiment.

Wave measurements were made during each experiment at several positions down the tank using parallel-wire resistance gauges. In the experiments described here the fluid depth (h) was 5 cm, and the waves were measured at $x/h = 0, 20, 180$ and 400, where $x = 0$ at the downstream edge of the piston.

Figure 1.3 shows a wave generated simply by raising the piston. The piston motion was fast enough that the shape of the wave at $x/h = 0$ is effectively the shape of the piston; (because of the

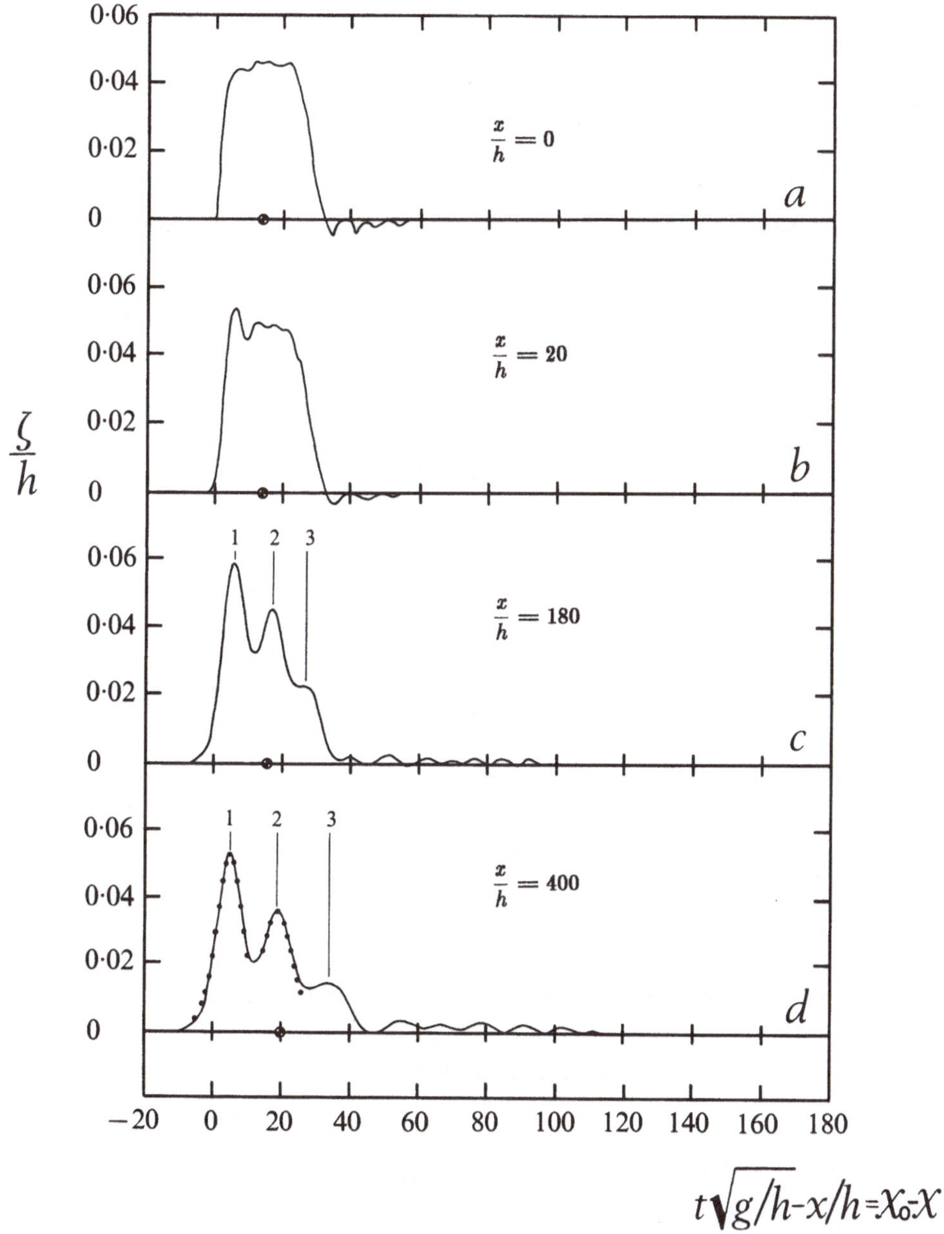

Figure 1.3. Evolution of surface waves, measured at four locations downstream of the wave maker. The front of each wave is to the left in the coordinate system used, which translates with speed $\sqrt{gh}$. ——, measured wave profiles; ·····, soliton profiles computed using (13) and the measured peak amplitude of each wave.

reflecting wall at the upstream end of the piston, the wave at $x = 0$ was actually twice as long as the piston and half as high as its displacement).

On a short time-scale, according to (8), this wave should simply translate with speed $\sqrt{gh}$. The wave measured at $x/h = 20$ (Figure 1.3b) fits this description approximately; its shape is basically that of the wave at $x = 0$. (The front of the wave is to the left in these figures, and a wave which translates with speed $\sqrt{gh}$ shows no horizontal displacement in succeeding frames).

That solitons emerge on a long time scale may be seen in Figures 1.3c and 1.3d. Solving the appropriate scattering problem with the wave measured at $x = 0$ as the potential yields three discrete eigenvalues, representing three solitons. These correspond to the three positive, more-or-less permanent waves seen at $x/h = 180$ and $x/h = 400$. According to (13), these waves all should move to the left in these figures, since their speeds all exceed $\sqrt{gh}$. That they do not is a measure of the effect of viscosity in these experiments.

Even so, we assert that these waves are solitons on the basis of their shapes. The entire profile of a single soliton is determined from (13) once its amplitude is known. The peak amplitudes of the first two waves in Figure 1.3d were measured and the dots in that figure represent evaluations of (13) based on those amplitudes. The agreement with the measured wave shapes is striking.

The results shown in Figure 1.3 suggest the following picture of long water waves moderate amplitude.

a. There is a short (linear) time-scale, during which the left- and right-running waves separate from each other.

b. There is a long (KdV) time-scale, during which the right- (or left-) running waves evolve in N solitons plus radiation.

c. There is an even longer viscous time-scale, during which the energy in these solitons is gradually dissipated. Because the KdV time-scale is shorter, however, the solitons continually readjust their shapes and speeds as they lose energy so that *locally*, as in Figure 1.3d, they look and act like solitons.

Several other experiments were performed in this series, to test other aspects of the KdV theory. We refer the reader to the original papers or to Chapter 4 of Ablowitz and Segur (1981) for more details.

Next we consider the KP equation, (2), as a model of nearly one-dimensional water waves of moderate amplitude, propagating in shallow water. Here the theory is still incomplete, and we are aware of no systematic experimental study. Consequently we are forced to discuss special solutions of (2), and fortuitous experimental observations.

Obviously, every KdV solution also solves KP. More generally, Satsuma (1976) showed that (2) admits an N-soliton solution, with the N solitons traveling in N different directions. The two-soliton formula is

$$u(\chi, \eta, \tau) = 2\partial_\chi^2 \ln f \,, \tag{14}$$

where

$$f = 1 + \exp\phi_1 + \exp\phi_2 + \exp(\phi_1 + \phi_2 + A) \,,$$

$$\phi_j = k_j(\chi + p_j\eta - c_j\tau) \,, \qquad c_j = k_j^2 + 3p_j^2 \,,$$

$$\exp A = \frac{(k_1 - k_2)^2 - (p_1 - p_2)^2}{(k_1 + k_2)^2 - (p_1 - p_2)^2} \,.$$

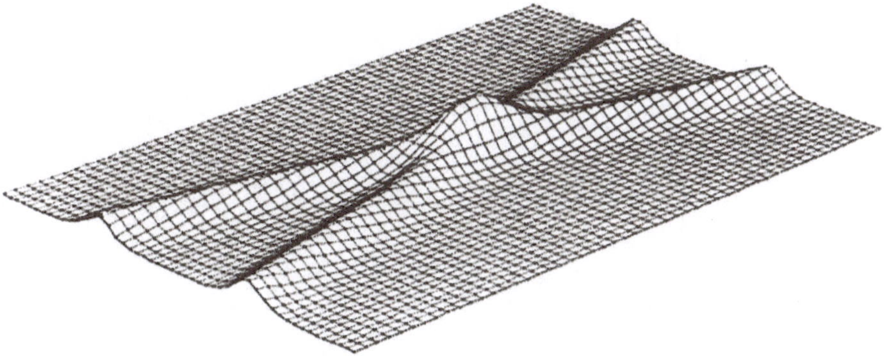

Figure 1.4. Two-soliton solution of the KP equation, with $k_1 = k_2 = 1$, $p_1 = -p_2 = 4$ in (14).

A typical solution is shown in Figure 1.4. Far from the interaction region, each wave is essentially a KdV soliton, but traveling at an angle to the χ-axis. The interaction of two waves is necessarily nonlinear, and a phase shift of each wave as a result of the interaction is evident. There are two phases in this solution and two spatial coordinates, so unless $p_1 = p_2$ in (14), there is a uniformly translating coordinate system in which this solution is stationary.

How well does this two-soliton solution of (2) predict water wave interactions? No quantitative experimental data is available but the photograph in Figure 1.5, of two long-crested waves interacting in shallow water off a beach in Oregon, certainly is suggestive. Each of two waves apparently is part of a train of periodic waves coming in from deep water, but their wavelengths seem to be long enough that each acts like a solitary wave in a shallow water. The comparison of Figures 1.4 and 1.5 is only quantitative, of course, but certainly it suggests that the KP equation might be as useful for weakly two-dimensional water waves as the KdV equation is for one-dimensional waves.

Because (2) only models weakly two-dimensional wave interactions, some oblique interactions are not predicted by it. Miles (1977a,b) examined oblique interactions of two solitary water waves of moderate amplitude, using a method based directly on (3–6) and which includes (14), the two-soltion solution of (2), as a special case. He found two possible types of interactions, depending on whether or not two-dimensionality dominated nonlinearity. If the angle between the two waves were large enough, each segment of one wave is affected by the other only for a short time, and the interaction is *weak*. These interactions are qualitatively like linear wave interactions, as shown in Figure 1.6a. They correspond to the limit $(p_1 - p_2)^2 \gg (k_1 \pm k_2)^2$ in (14). Alternatively, when the angle between the waves in small, then each segment of one wave feels the other wave for a long time, and the interactions is *strong*. Strong interactions are shown in Figures 1.4, 1.5, and 1.6b.

In the water wave problem, these strong interactions occurs only if the angle between the two waves does not exceed a certain critical value. Maxworthy (1980) found experimental evidence that such a critical angle exists, and that the interaction of two solitary waves of given amplitude changes from *weak* to *strong* as one moves through this critical angle. Johnson (1982) has examined this critical angle in more detail.

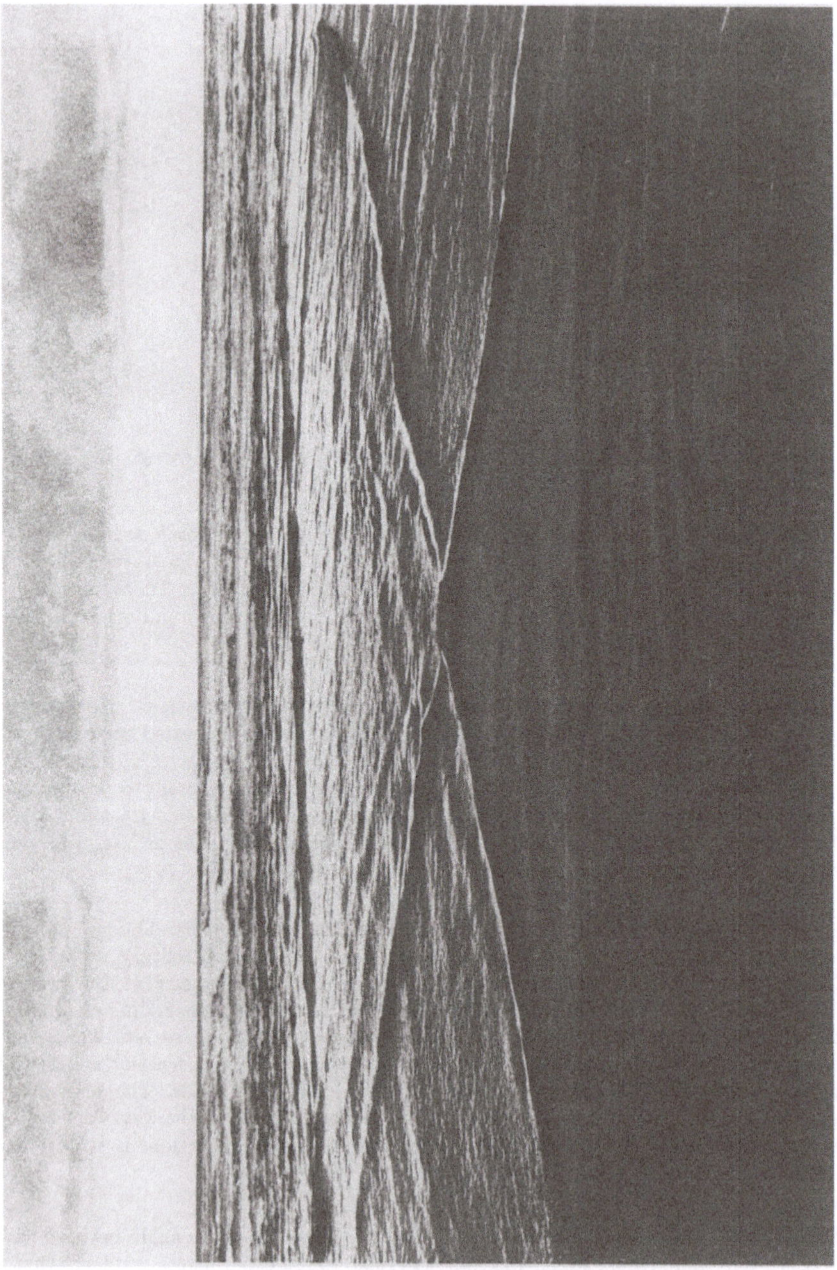

Figure 1.5. Oblique interaction of two waves in shallow water, observed off a beach in Oregon. (Photograph courtesy of T. Toedtemeier.)

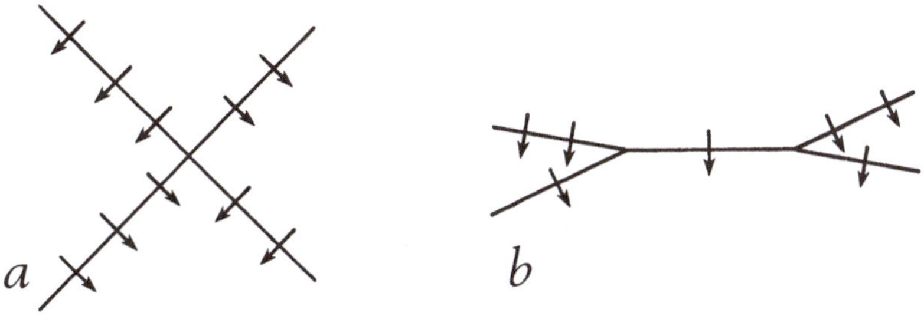

Figure 1.6. Two types of oblique interactions of KP solitons, showing the patterns of the wave crests. Arrows indicate direction of propagation.
(a) Weak interation corresponding to a large angle between the waves, or to small amplitude waves; *(b)* strong interaction, for a small angle between the waves or for larger aplitude waves.

§1.4 Periodic waves in shallow water

We turn now to the question of periodic waves of moderate amplitude in shallow water, and to periodic solutions of (1) and (2). The derivation of (1) or (2) given in §1.2 remains valid for periodic initial data, provided that the waves still satisfy assumptions (A–D) and the normalization condition, (11). In the case of one-dimensional periodic waves, $f(r)$ and $F(l)$ in (8) are periodic functions, and (1) becomes the appropiate evolution equation on $0 \le \chi \le L$, with periodic boundary conditions,

$$u(\chi + L, \tau) = u(\chi, \tau)\,, \tag{15a}$$

$$\int_0^L d\chi\, u = 0\,. \tag{15b}$$

The first non-trivial solution of the periodic KdV problem was given by Korteweg and de Vries (1895):

$$u(\chi, \tau) = 2k^2\nu^2 \mathrm{cn}^2[k(\chi - c\tau) + \chi_0;\, \nu] + u_0\,. \tag{16}$$

Here $\mathrm{cn}\,[\phi; \nu]$ is a Jacobian elliptic function with modulus ν $(0 \le \nu \le 1)$, whence the name ***cnoidal wave***; also,

$$c = 6u_0 - 4k^2\,(1 - 2\nu^2)\,,$$

$$kL = 2K(\nu)\,,$$

$$u_0 = -2k^2\left[\frac{E(\nu)}{K(\nu)} - 1 + \nu^2\right],$$

where $K(\nu)$, $E(\nu)$ are the complete elliptic integrals of the first and second kinds, respectively (*cf.* Byrd and Friedman, 1971). A typical cnoidal wave is shown in Figure 1.7. These waves reduce to infinitesimal, sinusoidal waves if $\nu \to 0$, and to solitons, (12), if $\nu \to 1$.

From one perspective, we may view cnoidal waves as the simplest solutions of the KdV problem with periodic boundary conditions. Viewed from a more physical perspective, (16) represents a periodic, shallow-water wave according to

$$\varsigma(x, t) = \tfrac{2}{3}\epsilon h\, u(\chi, \tau) + O(\epsilon^2)\,.$$

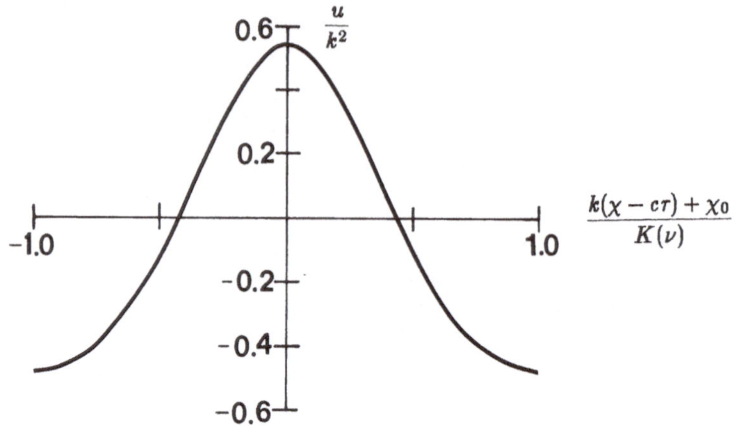

Figure 1.7. One period of cnoidal wave solution of the KdV equation with $\nu^2 = \frac{1}{2}$ in (16).

Thus, (16) defines a family of *typical* nonlinear waves in shallow water, from which one may extract *typical* values of waves speeds, forces, etc., for engineering purposes. This second viewpoint is common among ocean engineers, naval architects and others responsible for the design of large structures in relatively shallow water [2]

The practical value of this approach is evident: it provides realistic estimates of forces from waves of finite amplitude. Its major limitation also is evident: it is a one-dimensional theory. Water waves which impinge on a vertical wall obliquely, (*i.e.*, waves coming from almost any direction) are excluded from consideration. In particular, if a wave impinges on a wall of some structure obliqely, the point of interaction moves along the wall at a finite speed, and could conceivably excite a resonant frequency of the structure. Such wave-structure interaction cannot be predicted using only cnoidal waves, because the one-dimensional nature of cnoidal waves excludes this possibility.

Thus there is a practical need for a two-dimensional generalization of cnoidal waves. We discuss here the simplest meaninful generalization: doubly periodic solutions of the KP equation. As we will see, these solutions are periodic in each of two real, generally non-orthogonal directions, and except for degenerate cases, they are stationary in a uniformly translating cordinate system.

Without any calculations, one can almost guess how doubly periodic KP solutions might look. Because the cnoidal wave is a periodic generalization of one soliton, one should seek a doubly periodic generalization of a two-soliton solution. Thus, Figures 1.8 show a sketch of hypothetical *weak* and *strong* interactions of periodic waves, obtained simply by extending periodically the interactions of solitons shown in Figure 1.6. Note that with strong interactions, all of the waves are short-crested. As with Figure 1.6, one expects each of these conjectured interactions figures to be stationary in an appropriately translating coordinate system.

Preliminary experiments by Hammack (1980, unpublished) seem to confirm that some obliquely interacting, shallow-water waves form patterns like those shown in Figure 1.8b. The experiments were performed in a simple ripple tank (6.1 m long × 1.13 m wide) with a periodic wave maker across one

[2]*e.g.*, see the book on the engineering design of such structures by Sarpkaya and Isaacson, 1981.

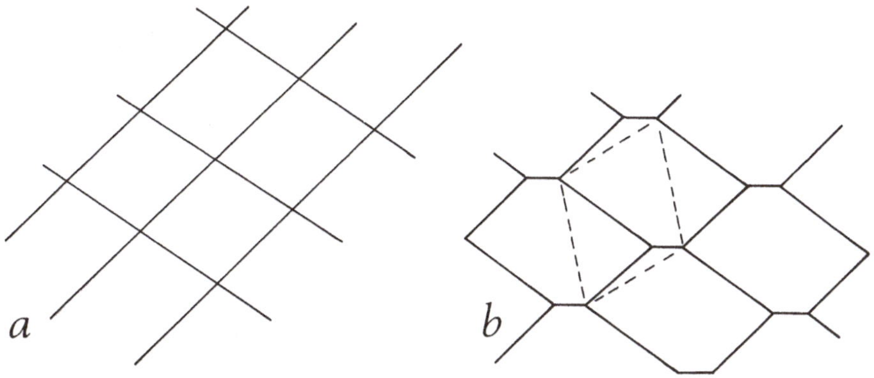

Figure 1.8. Hypothetical pattern of wave crests for periodic KP solutions. *(a)* Weak interaction, as in linear theory; *(b)* strong interaction. The dashed lines indicate a period parallelogram.

end of the tank.[3] In the experiment in question, a uniform train of one-dimensional, shallow-water was reflected obliquely by a uniform step, placed at 30^0 to the incident waves, as shown in Figure 1.9a. The water depth changed discontinuously across the step from 1.9 cm to 6.2 cm. The incident and reflected waves constituted two trains of obliquely interacting, finite-amplitude, shallow-water waves. Figure 1.9b shows the observed interaction pattern, in qualitative agreement with that in Figure 1.8b. Obviously, such a comparison of a hypothetical solution of (2) with a crude qualitative observation of water waves confirms nothing, but it suggests that doubly periodic KP solutions might have practical importance.

Now let us return from this flight of fancy to the question of doubly periodic solutions of (2). Using methods of algebraic geometry, Krichever (1976) first observed that (2) admits quasi-periodic solutions[4] of the form

$$u(\chi,\eta,\tau) = 2\partial_\chi^2 \ln \Theta(Z_1,\ldots,Z_N)\,, \tag{17a}$$

where $\Theta(Z_1,\ldots,Z_N)$ denotes Riemann's theta function and

$$Z_j = U_j\chi + V_j\eta + W_j\tau + Z_{j_0}\,, \qquad j = 1,\ldots,N\,. \tag{17b}$$

This idea has been pursued in a series of papers by Krichever and Novikov (see Krichever and Novikov, 1980, for a review). However, Riemann theta functions of several arguments are mathematically well-defined, but poorly understood, and is has been difficult to extract concrete information from this work to date.

Fortunately, recent work by Dubrovin (1981) on compact Riemann surfaces of low genus (1,2 and 3) has provided the information necessary to change (17) from an abstract symbol into a computationally effective tool for these low genera. We note that for theta functions, the genus of the Riemann surface equals the number of arguments in the theta function. Thus, genus 1 gives back the (single-argument) cnoidal waves discussed above; genus 2, corresponding to a theta function of two arguments, gives the intrinsically two-dimensional, doubly periodic generalizations of cnoidal waves that are of interest here. We will not consider solution of genus 3 in this paper.

---

[3]In a ripple tank, light shining through shallow water from below is focussed by wave crests and defocussed by troughs. The result is an instantaneous picture of the pattern of water waves in which the wave crests appear as bright lines.

[4]Technical definition: a quasi-periodic function with N arguments is periodic in each of its arguments separately.

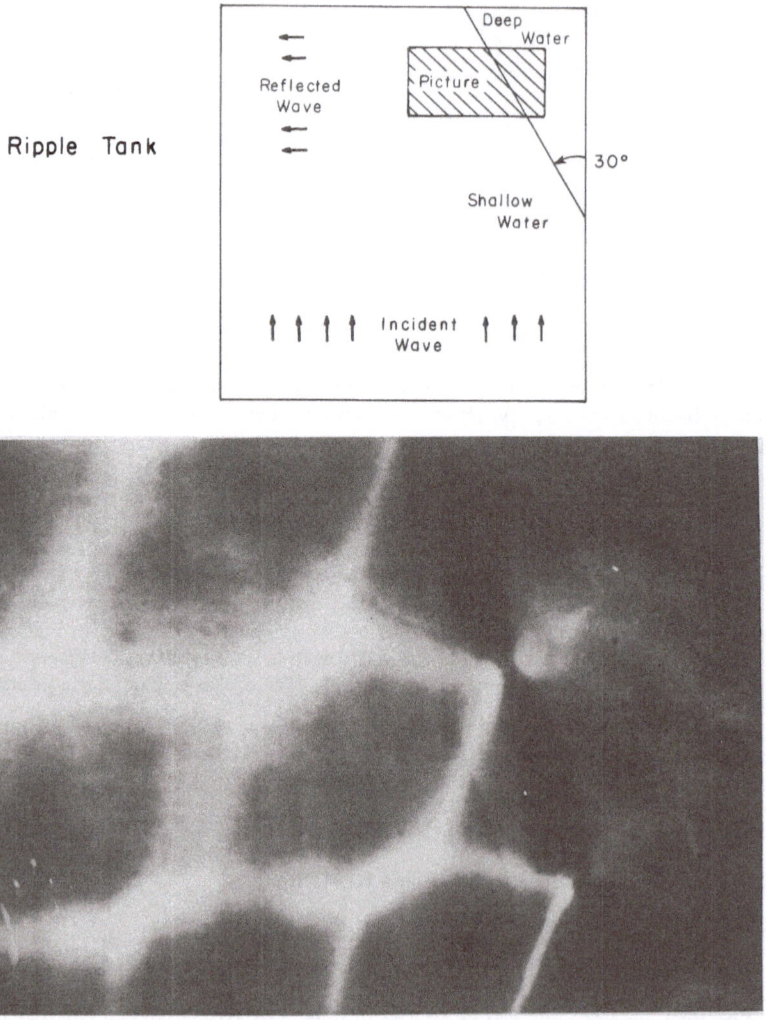

Figure 1.9. *(a)* **Experimental apparatus to create oblique interaction of trains of finite amplitude waves in shallow water.** *(b)* **Observed configuration of wave crests. (Photograph courtesy of J. L. Hammack.)**

Dubrovin's work enables one to solve the following problems:

i. Parameterize all possible real-value KP solutions that can be expressed in terms of a theta function of two arguments, as in (17).

ii. Give an effective algorithm to compute these solutions.

iii. Given the instantaneous values (*e.g.*, at $\tau = 0$) of a KP solution in the form of (17) with $N = 2$, describe this solution for all time.

Because these problems can be solved, one can solve (2) as an initial value problem, for the very restricted family of solutions related to theta functions of genus 2.

Before describing the solution of these three problems, we mention earlier work by Nakumura (1979) and Bryant (1982) on finding doubly periodic solutions of KP. Nakamura (1979) also used (17), but without the results of Dubrovin (1981) he was unable to give a complete solution of the problem. Bryant (1982) gave an algorithm to compute approximately specific doubly periodic KP solutions directly in terms of Fourier series. No claim of completeness was made in his work.

i. Parameterize the KP solutions of genus 2.

Note that (2) has several obvious symmetries: χ-, η- and τ-translations, two coordinate reflections ($\eta \to -\eta$; $\chi \to -\chi$, $\tau \to -\tau$), *coordinate rotation* ($\partial_x \to \partial_x, \partial_\eta \to \partial_\eta + \alpha\partial_\chi, \partial_\tau \to \partial_\tau - 6\alpha\partial_\eta - 3\alpha^2\partial_\chi$), galilean invariance ($u \to u + \beta, \chi \to \chi - 6\beta\tau, \eta \to \eta, \tau \to \tau$) and scaling ($u \to \lambda^{-2}u, \chi \to \lambda\chi, \eta \to \lambda^2\eta, \tau \to \lambda^3\tau$). All except scaling follow from corresponding symmetries in the original problem, (3–6); scaling follows from the arbitrary definition of ϵ in (9). Each of these symmetries introduces a free parameter into any KP solution. Our objective here is to identify the parameters that characterize the KP solutions of genus 2, beyond these automatic parameters. [According to this way of counting, a single KdV soliton (12) has no free parameters, a cnoidal wave (16) has one free parameter, while the two-soliton solutions of KP in (14) have two free parameters.]

Using methods of algebraic geometry, Dubrovin shows that each KP solution of genus 2 corresponds to a point on the surface of a compact Riemann surface of genus 2. (Topologically, this is a sphere with two handles.) This surface is identified uniquely by a 2×2 Riemann matrix (a symmetric matrix with negative definite real part); *i.e.*, by three complex parameters. A fourth complex parameter identifies the point on the Riemann surface. Thus each KP solution of genus 2 is identified by four parameters. We will see below that if these four parameters all are real and satisfy certain inequalities, then the resulting KP solution is real.

ii. Calculate KP solutions of genus 2.

Let

$$\mathbf{B} = \begin{pmatrix} b_{11} & b_{12} \\ b_{12} & b_{22} \end{pmatrix}, \quad \mathbf{m} = \begin{pmatrix} m_1 \\ m_2 \end{pmatrix}, \quad \mathbf{p} = \begin{pmatrix} p_1 \\ p_2 \end{pmatrix}, \quad \mathbf{Z} = \begin{pmatrix} Z_1 \\ Z_2 \end{pmatrix}, \tag{18}$$

where b_{ij} and Z_j are arbitrary complex numbers, $p_j = 0$ or $\frac{1}{2}$, and m_j are arbitrary real integers. A Riemann theta function is defined by a Fourier series of the form

$$\Theta(\mathbf{Z}) = \sum_{m_1=-\infty}^{\infty} \sum_{m_2=-\infty}^{\infty} \exp\{\tfrac{1}{2}\mathbf{m}\cdot\mathbf{B}\cdot\mathbf{m} + \mathbf{m}\cdot\mathbf{Z}\}, \tag{19a}$$

where

$$\mathbf{m}\cdot\mathbf{B}\cdot\mathbf{m} = \sum_{i,j=1}^{2} b_{ij} m_i m_j, \qquad \mathbf{m}\cdot\mathbf{Z} = \sum_{j=1}^{2} m_j Z_j. \tag{19b}$$

In the present application Z_j is given by (17b), and is pure imaginary. Then $\Theta(\mathbf{Z})$ is a real if $\mathbf{B}$ is a real matrix, which we now assume. $\mathbf{B}$ is negative definite if

$$b_{11} + b_{22} < 0, \tag{20a}$$

$$b_{11}b_{22} - b_{12}^2 > 0. \tag{20b}$$

Without loss of generality we may also assume

$$b_{22} \le b_{11} < 0. \tag{20c}$$

We want (17) to solve KP, with $N = 2$ and $\Theta(\mathbf{Z})$ defined by (19). It becomes necessary to define certain *theta-constants*, as follows.

$$\hat{\Theta}[\mathbf{p}] = \sum_{m_1=-\infty}^{\infty} \sum_{m_2=-\infty}^{\infty} \exp\{(\mathbf{m}+\mathbf{p}) \cdot \mathbf{B} \cdot (\mathbf{m}+\mathbf{p})\}\,, \tag{21a}$$

$$\hat{\Theta}_{ij}[\mathbf{p}] = \sum_{m_1=-\infty}^{\infty} \sum_{m_2=-\infty}^{\infty} (m_i+p_i)(m_j+m_j) \exp\{(\mathbf{m}+\mathbf{p}) \cdot \mathbf{B} \cdot (\mathbf{m}+\mathbf{p})\}\,, \tag{21b}$$

$$\hat{\Theta}_{ijkl}[\mathbf{p}] = \sum_{m_1} \sum_{m_2} (m_i+p_i)(m_j+p_j)(m_k+p_k)(m_l+p_l) \exp\{(\mathbf{m}+\mathbf{p}) \cdot \mathbf{B} \cdot (\mathbf{m}+\mathbf{p})\}\,, \tag{21c}$$

For fixed (i,j,k,l), each of these is a four-component vector, indexed by the four choices of $\mathbf{p}$.

Define the 4×4 real matrix $\mathbf{D}$ by

$$\mathbf{D} = (\hat{\Theta}_{11}[\mathbf{p}],\quad \hat{\Theta}_{12}[\mathbf{p}],\quad \hat{\Theta}_{22}[\mathbf{p}],\quad \hat{\Theta}[\mathbf{p}])\,. \tag{22}$$

The Riemann matrix $\mathbf{B}$ is said to be *indecomposable* if

$$\det \mathbf{D} \neq 0\,. \tag{23}$$

This condition assures that the theta function associated with $\mathbf{B}$ through (19) is honestly genus 2, and has not degenerated into genus 1. Beyond (20) and (23), there are no further restrictions on the real matrix $\mathbf{B}$, which completely characterizes the underlying Riemann surface.

Each KP solution of genus 2 corresponds to a point on the Riemann surface. We now select the point, by selecting (U_1, U_2) in (17b). Because of (23), $\mathbf{D}^{-1}$ exists. Denote the rows of this matrix by

$$\mathbf{D}^{-1} = \begin{pmatrix} d^{11}[\mathbf{p}] \\ d^{12}[\mathbf{p}] \\ d^{22}[\mathbf{p}] \\ d[\mathbf{p}] \end{pmatrix}. \tag{24}$$

For $\mathbf{U} = (U_1, U_2)$, define

$$\partial_U^4 \hat{\Theta}[\mathbf{p}] = \sum_{i,j,k,l=1}^{2} U_i U_j U_k U_l \hat{\Theta}_{ijkl}[\mathbf{p}]\,. \tag{25}$$

Also define, for $j \geq i$,

$$Q_{ij}(\mathbf{U}) = \sum_{\mathbf{p}} d^{ij}[\mathbf{p}]\, \partial_U^4 \hat{\Theta}[\mathbf{p}]\,. \tag{26}$$

Dubrovin[5] shows that with $N = 2$, (17) defines a KP solution if and only if $(\mathbf{U}, \mathbf{V}, \mathbf{W})$ are related by

$$\begin{aligned} U_1 W_1 + 4Q_{11}(\mathbf{U}) + 3V_1^2 &= 0\,, \\ U_2 W_2 + 4Q_{22}(\mathbf{U}) + 3V_2^2 &= 0\,, \\ U_1 W_2 + U_2 W_1 + 4Q_{12} + 6V_1 V_2 &= 0\,. \end{aligned} \tag{27}$$

[5]Dubrovin's $\mathbf{W}$ differs from ours by a factor of (-4), because of different scalings of (2).

Self-consistency of (27) implies

$$(U_1V_2 - U_2V_1)^2 + \tfrac{4}{3}P(U_1, U_2) = 0\,, \tag{28a}$$

where

$$P(U_1, U_2) = U_1^2 Q_{22}(\mathbf{U}) + U_2^2 Q_{11}(\mathbf{U}) - U_1 U_2 Q_{12}(\mathbf{U})\,. \tag{28b}$$

If $U_2 \neq 0$, the scaling invariance and coordinate rotation of (2) permit the normalization,

$$\mathbf{U} = (s, 1)\,, \qquad \mathbf{V} = (V_1, 0)\,. \tag{29}$$

Given (29), it is evident from (28a) that the six roots of

$$P(s, 1) = 0$$

give six solutions with $\mathbf{V} = 0$, *i.e.*, genus 2 solutions of the KdV equation. These solutions correspond to the six Weierstrass points of the Riemann surface. To obtain real-valued, two-dimensional solutions of KP, we need

$$P(s, 1) > 0 \quad \text{for real } s\,. \tag{30}$$

Then it follows from (27) that (17) defines a real-valued KP solution of genus 2 if (Z_1, Z_2) are defined using

$$\begin{aligned} \mathbf{U} &= i\,(s, 1)\,, \\ \mathbf{V} &= i\,\big(2\sqrt{\tfrac{1}{3}P(s,1)}, 0\big)\,, \\ \mathbf{W} &= 4i\,\big(-sQ_{22}(s,1) + Q_{12}(s,1), Q_{22}(s,1)\big)\,. \end{aligned} \tag{31}$$

There is one additional solution, corresponding to $U_2 = 0$. If

$$Q_{22}(1, 0) > 0\,, \tag{32}$$

then (17) defines a real-valued KP solution of genus 2 if (Z_1, Z_2) are defined using

$$\begin{aligned} \mathbf{U} &= i\,(1, 0)\,, \\ \mathbf{V} &= i\,\big(0, 2\sqrt{\tfrac{1}{3}Q_{22}(1,0)}\big)\,, \\ \mathbf{W} &= 4i\,\big(Q_{11}(1,0), Q_{12}(1,0)\big)\,. \end{aligned} \tag{33}$$

This KP solution is also a KdV solution if $Q_{22}(1, 0)$ vanishes.

The parameter s, which selects the point on the Riemann surface, has a simple geometric interpretation. The function $u(\chi, \eta, \tau)$ is defined on a fundamental period parallelogram, then repeated periodically in two directions (see Figure 1.8). It is easy to show that the area, A, of this parallelogram satisfies

$$(U_1V_2 - U_2V_1)^2 = \left(\frac{4\pi^2}{A}\right)^2\,. \tag{34}$$

According to (28a), for fixed scaling of (2), choosing a solution of (30) amounts to choosing the area of the period parallelogram.

To summarize, **B** contains three real parameters, which must satisfy (20) and (23). Then s is a real parameter that satisfies (30), or (32) in the special case. Every legitimate choice of these four real parameters produces a KP solution of genus 2, using (17) with either (31) or (33). Naturally, this four-parameter family of solutions may be generalized by using the symmetries of the KP equation discussed in problem **i.** All of the solutions with $U_1V_2 \neq U_2V_1$ that are produced in this way are: *(i)* real-valued; *(ii)* quasi-periodic functions of two variables (Z_1, Z_2); and *(iii)* stationary in time in some uniformly translating coordinate system. In this sense, they are the natural generalizations of cnoidal waves to two dimensions. We conjeture that no other KP solutions have these three properties, but we have not proven it.

Figures 1.10 show some KP solutions of genus 2, obtained by implementing this algorithm numerically. All of the solutions shown there correspond to the same Riemann matrix: $b_{11} = -1.72$, $b_{12} = 1.18$, $b_{22} = -3.55$. A phase shift of these short-crested waves, due to their nonlinear interactions, is evident in the figures.

iii. Given the instantaneous values of a KP solution of genus 2, describe this solution for all time.

By hypothesis, $u(\chi, \eta, \tau = 0)$ has the form (17) with $N = 2$. Because $u(\chi, \eta, 0)$ is given pointwise, we may measure **U** and **V** directly. If $U_1V_2 = U_2V_1$, then the solution in question is actually a KdV solution of genus 2, and it may be described using the well-established theory for the periodic KdV equation (*e.g.*, Dubrovin & Novikov, 1974, or Ch. 2.3 of Ablowitz & Segur, 1981). Thus we need to consider only the case in which

$$U_1V_2 \neq U_2V_1 \, , \tag{35}$$

so that the solution is non-trivially periodic in two spatial directions.

Every KP solution of genus 2 that satisfies (35) is stationary in a uniformly translating coordinate system, whose velocity is given by **W** in (17*b*). Once **W** is know, then the initial data in a period parallelogram plus **W** determine the solution for all time. It is not necessary to reconstruct **B**, the Riemann matrix, for KP solutions of genus 2 that satisfy (35).

Algebraic equations for **W** may be obtained in a variety of ways. The method presented here is valid either if $\mathbf{U}_2 = 0$ or if (U_1/U_2) is rational. This restriction is always satisfied in applications, where (U_1/U_2) is measured only to a finite accuracy. In problem **ii.**, it amounts to requiring that the parameter s be rational.

If (U_1/U_2) is rational (or if $U_2 = 0$), then $u(\chi, \eta, 0)$ is a strictly periodic function of χ, holding (η, τ) fixed. Denote this χ-period by L. Because $u(\chi, \eta, 0)$ has the form (17), it follows that for all (χ_0, η),

$$\int_0^L d\chi \, u(\chi + \chi_0, \eta, 0) = 0 \, . \tag{36}$$

Define

$$\phi(\chi, \eta; \chi_0) = \frac{1}{L} \int_0^L d\xi \cdot \xi u(\chi + \chi_0 + \xi, \eta) \, . \tag{37}$$

ϕ is the unique anti-derivative of u with the same periodicity as u, and with zero mean in χ; *i.e.*, ϕ also satisfies (36), and $\partial_\chi \phi = u$. The corresponding anti-derivative of ϕ may be defined in a similar way.

One equation for **W** may be obtained by multiplying (2) by u, and integrating over a period parallelogram (with area A). The result is

$$\int\!\!\int_A d\chi \, d\eta \, [u_\chi u_\tau + 6uu_\chi^2 - u_{\chi\chi}^2 + 3u_\eta^2] = 0 \, . \tag{38}$$

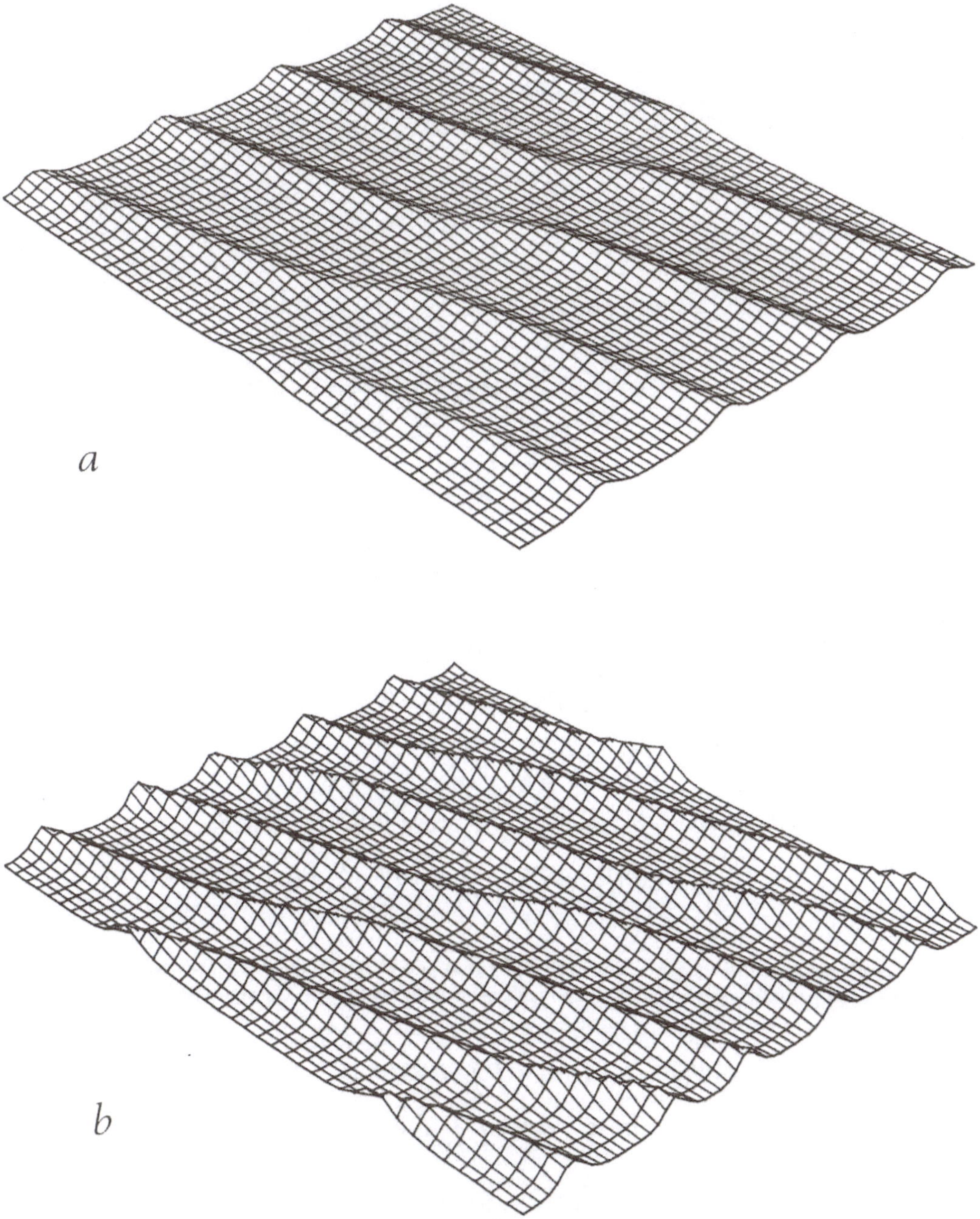

Figure 1.10. Sample of KP solutions of genus 2, corresponding to different points on the same Riemman surface. *(a)* $s = U_1/U_2 = 0.0$ in (29); *(b)* $s = 0.35$ (continues).

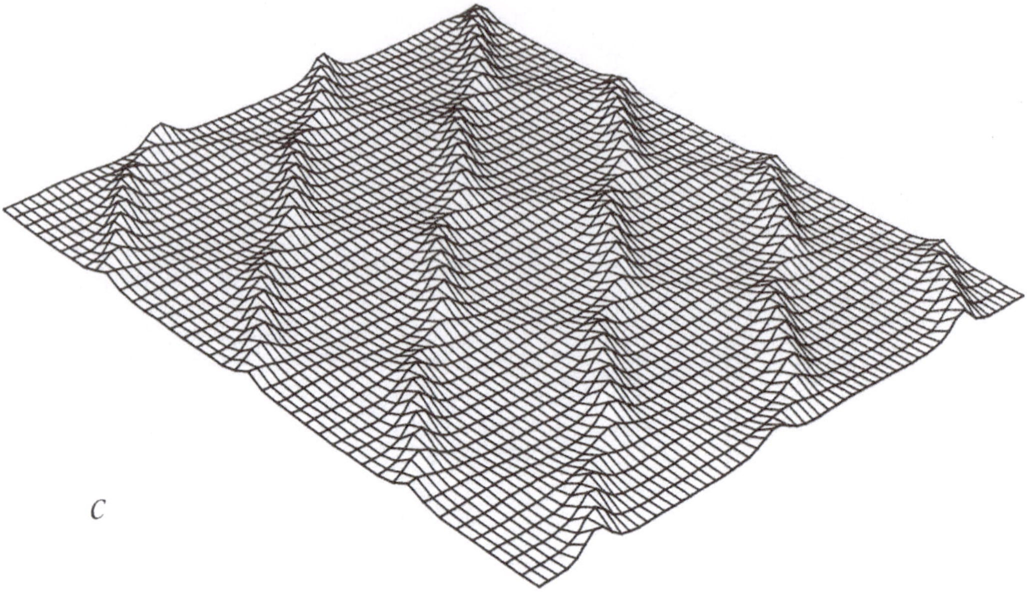

Figure 1.1 (Cont.) *(c)* $s = -0.70$.

By hypothesis, $u(\chi, \eta, \tau) = f(Z_1, Z_2)$, where Z_j is given by (17b). It follows that

$$(U_1 V_2 - U_2 V_1) u_\tau = (W_1 V_2 - W_2 V_1) u_\chi + (U_1 W_2 - U_2 W_1) u_\eta \,, \tag{39}$$

so (38) becomes

$$\begin{aligned} \int\!\!\int_A d\chi\, d\eta\, [V_2 u_\chi^2 - U_2 u_\chi u_\eta] W_1 + \int\!\!\int_A d\chi\, d\eta\, [U_1 u_\chi u_\eta - V_1 u_\chi^2] W_2 \\ + (U_1 V_2 - U_2 V_1) \int\!\!\int_A d\chi\, d\eta\, [6 u u_\chi^2 - u_{\chi\chi}^2 + 3 u_\eta^2] = 0 \,. \end{aligned} \tag{40}$$

To obtain a second equation, multiply (2) by that anti-derivative of ϕ (*i.e.*, the second integral of u) with the same periodicity as u and ϕ. The result after integrating over the same period parallelogram is:

$$\begin{aligned} \int\!\!\int_A d\chi\, d\eta\, [V_2 u^2 - U_2 u \phi_\eta] W_1 + \int\!\!\int_A d\chi\, d\eta\, [U_1 u \phi_\eta - V_1 u^2] W_2 \\ + (U_1 V_2 - U_2 V_1) \int\!\!\int_A d\chi\, d\eta\, [3u^3 - u_\chi^2 + 3\phi_\eta^2] = 0 \,. \end{aligned} \tag{41}$$

Equations (40) and (41) are two linear algebraic equations for (W_1, W_2). Whenever they are linearly independent, their common solution defines (W_1, W_2), which completes the mathematical specification of the KP solutions of genus 2. It is easy to show that (40) and (41) are linearly independent for small enough wave amplitudes, but we have not yet established this property in general. In the cases we have tested numerically, (40) and (41) are independent, and their common solution agrees with that in (31).

More generally, given any initial data, $f(\chi, \eta)$, such that

$$\lim_{L\to\infty} \frac{1}{L} \int_0^L d\chi\, f(\chi, \eta) = 0 , \tag{42}$$

one may seek the KP solution of genus 2 that best ***approximates*** $f(\chi, \eta)$ at $\tau = 0$, along with an appropriate measure of goodness of fit. It is evident that such approximation procedures will be necessary if these KP solutions of genus 2 are to become a practical tool in physical problems. However, we leave this and other questions of physical implications of these solutions for a future paper.

We are grateful to T. Toedtemeier and J. L. Hammack for their permission to use Figures 1.5 and 1.9, respectively. We acknowledge with pleasure several helpful conversations with R. I. Sykes concerning the algorithms in §1.4. The work in this Chapter was partially supported by the Office of Naval Research, and by NSF Grant #MCS-8108814 (A01).

References

M. J. Ablowitz, D. Bar Yaacov and A. S. Fokas, On the inverse scattering transform for the Kadomtsev–Petviashvili equation. Preprint, 1982.

M. J. Ablowitz and A. S. Fokas, in *these Procedings.*

M. J. Ablowitz and H. Segur, *Solitons and the Inverse Scattering Transform.* SIAM, Philadelphia, 1981.

P. J. Bryant, *J. Fluid Mech.* **115**, 525–532 (1982).

P. F. Byrd and M. D. Friedman, *Handbrook of Elliptic Integrals for Engineers and Scientists.* Spinger-Verlag, New York, 1971.

B. A. Dubrovin, *Funct. Anal. Appl.* **9**, 215–223 (1975).

B. A. Dubrovin, *Russian Math. Surveys* **36**, 11–92 (1981).

B. A. Dubrovin, V. B. Matveev, and S. P. Novikov, *Russian Math. Surveys* **31**, 59–146 (1976).

B. A. Dubrovin and S. P. Novikov, *Sov. Phys. JETP* **40**, 1058–1063 (1974).

C. S. Gardner, J. M. Greene, M. D. Kruskal, and R. M. Miura, *Phys. Rev. Lett.* **19**, 1095–1097 (1967).

C. S. Gardner, J. M. Greene, M. D. Kruskal, and R. M. Miura, *Commun. Pure Appl. Math.* **27**, 97–133 (1974).

J. L. Hammack and H. Segur, *J. Fluid Mech.* **65**, 289–314 (1974).

J. L. Hammack and H. Segur, *J. Fluid Mech.* **84**, 337–358 (1978).

A. R. Its and V. B. Matveev, *Funct. Anal. Appl.* **9**, 67ff (1975).

R. S. Johnson, *J. Fluid Mech.* **120**, 49–70 (1982).

B. B. Kadomtsev and V. I. Petviashvili, *Sov. Phys. Dokl.* **15**, 539–541 (1970).

D. J. Korteweg and G. de Vries, *Philos. Mag. Ser. 5*, **39**, 422–443 (1895).

I. M. Krichever, *Sov. Math. Dokl.* **17**, 394–397 (1976).

I. M. Krichever and S. P. Novikov, in *Sov. Scient. Rev.* §*C*, *Math. Phys. Rev.* **1**, ed. by S. P. Novikov, (1980).

H. Lamb, *Hydrodynamics.* Dover, New York, (1932).

P. D. Lax, *Comm. Pure. Appl. Math.* **28**, 141–188 (1975).

T. Maxworthy, *J. Fluid Mech.* **96**, 47–64 (1980).

H. P. McKean and E. Trubowitz, *Comm. Pure Appl. Math.* **29**, 143–226 (1976).

H. P. McKean and P. van Moerbeke, *Invent. Math.* **30**, 217ff (1975).

J. W. Miles, *J. Fluid Mech.* **79**, 157–169 (1977*a*).

J. W. Miles, *J. Fluid Mech.* **79**, 171–179 (1977*b*).

A. Nakumura, *J. Phys. Soc. Japan* **47**, 1701–1705 (1979).

S. P. Novikov, *Funct. Anal. Appl.* **8**, 236–246 (1974).

T. Sarpkaya and M. Isaacson, *Mechanics of Waves Forces on Offshore Structures.* Van Nostrand Reinhold Co., New York, 1981.

J. Satsuma, *J. Phys. Soc. Japan* **40**, 286–290 (1976).

G. G. Stokes, *Trans. Cambridge Phil. Soc.* **8**, 441–455 (1847).

P. D. Weidman and T. Maxworthy, *J. Fluid Mech.* **85**, 417–431 (1978).

N. J. Zabusky and C. J. Galvin, *J. Fluid Mech.* **47**, 811–824 (1971).

Chapter 2

Solition Models of Long Internal Waves**

Abstract

The Korteweg–de Vries (KdV) equation and the finite-depth equation of Joseph (1977) and Kubota, Ko & Dobbs (1978) both describe the evolution of long internal waves of small but finite amplitude, propagating in one direction. In this paper, theories are tested experimentally by comparing measured and theoretical soliton shapes. The KdV equation predicts the shapes of our measured solitons with remarkable accuracy, much better than does the finite-depth equation. When carried to second-order, the finite-depth theory becomes about as accurate as (first-order) KdV theory for our experiments. However, second-order corrections to the finite-depth theory also identify a bound on the range of validity of that entire expansion. This range turns out to be rather small; it includes only about half of the experiments reported by Koop & Butler (1981).

§2.1 Introduction

The evolution of long internal waves with small amplitudes in a stably stratified fluid is governed approximately by a linear wave equation with small but cumulative corrections due to weak nonlinearity, dispersion and dissipation, and possibly to a slowly varying background. Several theoretical models exist which include various combinations of these cumulative effects. The purpose of this paper is to test two of these theoretical models experimentally in order to obtain some notion of their accuracy and range of validity.

The two theoretical models that we consider are weakly nonlinear and weakly dispersive: the Korteweg–de Vries (KdV) (1895) equation.

$$f_\tau + 6ff_\chi + f_{\chi\chi\chi} = 0 \tag{1}$$

and an equation due to Joseph (1977) and to Kubota, Ko & Dobbs (1978),

$$f_\tau + ff_\chi + T[f_{\chi\chi}] = 0 \tag{2a}$$

where

$$T[f] = -\tfrac{1}{2} \int_{-\infty}^{\infty} f(y) \coth \tfrac{1}{2}\pi(\chi - y)\, dy \tag{2b}$$

and the integral is evaluted in the principal-value sense. The latter equation, which we call the ***finite-depth equation***, may be written in a variety of equivalent ways, of which (2) is perhaps the simplest. This list of

** H. Segur and J. L. Hammack, *J. Fluid Mech.* **118**, 285–304 (1982)

equation could logically include an equation proposed by Benjamin (1967) and later derived by Ono (1975),

$$f_\tau + f f_X - H[f_{XX}] = 0. \tag{3}$$

where $H[\,]$ is the Hilbert transform. However, we will not consider (3) because we have no experimental data in the range of parameters where (3) is valid.

It is well-known that the KdV equation describes the slow evolution of internal waves of fairly small amplitude that are long in comparison with the total fluid depth (see *e.g.* Benney 1966). However, this meaning of *long* is overly restrictive because it excludes internal waves whose wavelengths may be comparable or even less than the total fluid depth, but which are much longer that the thickness of an appropriate thin layer defined by the background density distribution. For example, Osborne & Burch (1980) have observed internal waves in the Andaman Sea that seem to behave like KdV solitons, even though their observed wavelengths are only comparable to the total water depth.

As we will see in §2, the derivation of (2) permits wavelengths comparable to the total fluid depth, provided only that they are much longer than the thickness of an appropriate thin layer. Thus, there is a sense in which (2) generalizes (1), and one might expect (2) to be at least as accurate as (1) in predicting experimental data.

We find the opposite to be true: for our data set, the predictions of (2) are always less accurate than those of (1). The finite-depth equations predicts the data accurately only in so far as it agrees with the KdV equation. This conclusion is based on our limited set of data, but Koop & Butler (1981) have reached subtantially the same conclusion on the basis of independent experiments.

As we will show, the resolution of this paradox is as follows. Both (1) and (2) are derived from Euler's equations of motion by comparable asymptotic expansions. In each case, the solution of the equation, (1) or (2), provides the dominant term in the asymptotic expansion. However, in terms of allowable wave amplitudes, the range of validity of (2) is rather small; in fact, it is much smaller than that of (1). This limited range is found by carrying the expansion that leads to (2) to the next order, and comparing the second-order theory with experimental data.

Thus, under somewhat different conditions, both (1) and (2) predict the slow evolution of long internal waves of small amplitude as they travel in one direction. However, waves that are long enough to satisfy the requirements for (2) are not necessarily long enough for (1). On the other hand, waves that have amplitudes small enough to satisfy the requirements for (1) may lie outside the range of validity of (2).

§2.2 Derivation of the equations

We consider a two-fluid configuration, in which a layer of lighter fluid overlies a layer of heavier fluid, resting on a horizontal impermeable bed in a constant gravitational field (see Figure 2.1). This is the simplest configuration that supports internal gravity waves, and it is adequate to model the waves observed in our experiments, in those Koop & Butler (1981), and in those of Obsborne & Burch (1980). It excludes higher vertical modes, including those studied by Kubota *et al.* (1978). Therefore, our results cannot be compared directly with theirs, although the derivations themselves may be compared.

The problem of finding the two-dimensional, infinitesimal, irrotational disturbances admitted by two stably stratified layers of incompressible fluid in a constant gravitational field was discussed by Lamb (1932, §231). The velocity potentials in the upper and lower fluids may be written as

$$\phi_1 \sim (A \sinh kz + B \cosh kz) \exp ik[x - c(k)t],$$
$$\phi_2 \sim D \cosh k(z + h_2) \exp ik[x - c(k)t].$$

Here $c(k)$ must satisfy the linear dispersion relation

$$\left(\frac{kc^2}{g}\right)^2 [1 + (1 - \Delta)T_1 T_2] - \frac{kc^2}{g}[T_1 + T_2] + \Delta T_1 T_2 = 0, \tag{4a}$$

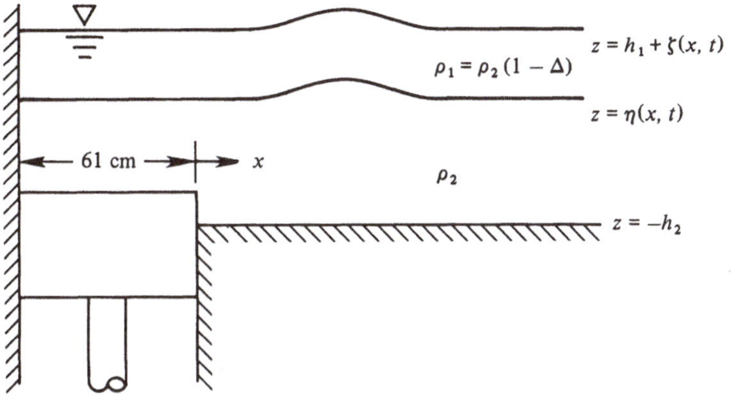

Figure 2.1. Piston moves up into a two-layer system, generating waves both at the free surface (ζ) and at the interface (η).

where

$$\Delta = \frac{\rho_2 - \rho_1}{\rho_2}, \tag{4b}$$

$$T_i = \tanh kh_i \qquad (i = 1, 2). \tag{4c}$$

In our experiments $\Delta \simeq 0.05$; therefore, we will use the Boussinesq approximation ($\Delta \to 0$, but $g\Delta$ finite) to simplify results. Some generalizations to arbitrary Δ ($0 \leq \Delta \leq 1$) are mentioned below. For small Δ, the roots of (4a) are

$$c_s^2(k) = \frac{g}{k} \tanh k(h_1 + h_2) + O(\Delta) \qquad \text{(surface waves)}, \tag{5a}$$

$$c_i^2(k) = \frac{g\Delta T_1 T_2}{k(T_1 + T_2)}[1 + O(\Delta)] \qquad \text{(internal waves)}. \tag{5b}$$

In the Boussinesq limit, these speeds always are distinct for a given k. Henceforth, we will let $\Delta \to 0$ (and $g \to \infty$), and retain only dominant terms.

A major difference between (1) and (2) is due to different aproximations of (5b). The KdV limit is obtained by letting $k(h_1 + h_2) \to 0$, so that

$$c_s^2(0) = g(h_1 + h_2), \tag{6a}$$

$$c_i^2(0) = \frac{g\Delta h_1 h_2}{h_1 + h_2} \equiv c_0^2. \tag{6b}$$

The dispersive term in (1) corresponds to the first correction to (6b) for small $k(h_1 + h_2)$. On the other hand, the finite-depth limit amounts to $h_2/h_1 \to 0$, $\quad kh_1$ finite, so that

$$c_i^2(0) = g\,\Delta h_2 \equiv \bar{c}^2. \tag{7}$$

The dispersive term in (2) corresponds to the first correction to (7) for small kh_2. Note that the speeds of the long infinitesimal waves in (6b) and (7) differ, unless we also require $h_2/h_1 \to 0$ in (6b).

2.2.1 The KdV equation

Aspects of KdV theory for long internal waves have been discussed by Keulegan (1953), Long (1956), Peters & Stoker (1960), Benney (1966) and others [*cf.* Miles (1979), and references cited therein]. We now outline briefly the derivation of (1) for the two-layer configuration shown in Figure 2.1. The basic assumptions are as follows.

i. The waves are long in comparison with total fluid depth:

$$k^2(h_1+h_2)^2 \ll 1,$$

where k^{-1} represents a characteristic horizontal wavelength.

ii. The waves are small, so that if $\bar{\eta}$ denotes a characteristic wave amplitude then

$$\bar{\eta}/(h_1+h_2) \ll 1.$$

iii. The two effects are in approximate balance:

$$\epsilon = \bar{\eta}/(h_1+h_2) = O[k^2(h_1+h_2)^2] \ll 1.$$

iv. Viscous effects are weaker than either of these. In addition, we assume throughout that the motion is two-dimensional,and that the fluid is incompresible.

It is consistent with these assumptions to define dimensionless (*) variables as follows

$$z^* = \frac{z}{h_1+h_2}, \quad x^* = \frac{\sqrt{\epsilon}x}{h_1+h_2}, \quad t^* = \sqrt{\frac{\epsilon g}{h_1+h_2}}\, t. \tag{8}$$

We also introduce a slow time variable,

$$\tau^* = \epsilon t^*. \tag{9}$$

The interface is defined by

$$\eta = \epsilon(h_1+h_2)\,\eta^*(x^*,t^*,\tau^*;\epsilon); \tag{10a}$$

if the upper surface is free, it is defined by

$$\varsigma = h_1 + \epsilon(h_1+h_2)\,\varsigma^*(x^*,t^*,\tau^*;\epsilon). \tag{10b}$$

There is a velocity potential in each layer. In the lower fluid, because of Laplace's equation and the boundary condition at $z=-h_2$, the potential has the formal expansion

$$\phi_2 = \phi_0(x^*,t^*,\tau^*;\epsilon) - \tfrac{1}{2}\epsilon^2\left[z^* + \frac{h_2}{h_1+h_2}\right]^2 \frac{\partial^2\phi_0}{\partial(x^*)^2} + O(\epsilon^4).$$

There is a corresponding expansion in the upper fluid. At the interface, the normal velocity and the pressure must be continuous, while the pressure mut vanish at the free surface. In addition, both the free surface and the interface satisfy kinematic conditions ($D\varsigma/Dt = D\eta/Dt = 0$). Finally, all motion should vanish as $|x| \to \infty$. In order to satisfy these conditions order-by-order in ϵ, it necessary to expand η^*, ς^*, and the velocity potentials. Thus, for example

$$\eta^* = \eta_1 + \epsilon\eta_2 + \cdots.$$

At the leading order (infinitely long waves of infinitesimal amplitude), the equations are hyperbolic and linear. An initial disturbance is decomposed into four wave modes:

$$\eta_1 = \sum_{j=1}^{4} f_j\left(\frac{x-c_j t}{\sqrt{h_1 h_2}}\right), \tag{11}$$

where $[c_j]$ are the four roots in (6). Because these speeds are so different, localized surface and internal waves quickly separate in space.

Weak nonlinear interactions and weak dispersive effects appear on the next scale $[\tau^* = O(1)]$, when the expansion is carried to the next order and secular terms are eliminated. There are no nonlinear interactions between modes, because they interact with each order for too short a time. However, each mode interacts with itself for a long time, and each four waves in (11) evolves according to its own KdV equation. Omitting details of the analysis, the dimensional equation for the internal wave mode is

$$\frac{1}{c_0}\frac{\partial \eta}{\partial t} + \frac{\partial \eta}{\partial x} + \frac{3}{2}\left(\frac{1}{h_2} - \frac{1}{h_1}\right)\eta\frac{\partial \eta}{\partial x} + \frac{1}{6}h_1 h_2 \frac{\partial^3 \eta}{\partial x^3} = 0, \tag{12}$$

where c_0 is given by (6b). [Recall that (12) is valid for $\Delta \ll 1$. The generalization to arbitrary Δ is given by Leone, Segur & Hammack (1982) if the upper surface is free, and by Djordjevic & Redekopp (1978) if the upper surface is rigid.] To reduce (12) to (1) let

$$\chi = \frac{x - c_0 t}{\sqrt{h_1 h_2}}, \quad \tau = \frac{1}{6}\sqrt{\frac{g\Delta}{h_1 + h_2}}\quad t, \quad f = \frac{3}{2}\left(\frac{1}{h_2} - \frac{1}{h_1}\right)\eta. \tag{13}$$

Then $f(\chi, t)$ satisfies

$$f_\tau + 6ff_\chi + f_{\chi\chi\chi} = 0.$$

We now state some of the consequences of KdV theory that may be tested experimentally. More details may be found in Segur (1973), Hammack & Segur (1974, 1978), and elsewhere.

i. The solution for KdV is

$$f(\chi, \tau) = 2\kappa^2 \operatorname{sech}^2\{\kappa(\chi - 4\kappa^2\tau\chi_0)\}, \tag{14a}$$

where κ, χ_0 are arbitrary constants. In dimensional terms,

$$\eta = \overline{\eta}\operatorname{sech}^2\{p(x - pvt - x_0)\}, \tag{14b}$$

where

$$p^2 = \frac{3(h_1 - h_2)\overline{\eta}}{4h_1^2 h_2^2}, \quad v = \sqrt{\frac{g\Delta h_1 h_2}{h_1 + h_2}}\left[1 + \left(\frac{1}{2}\frac{h_1 - h_2}{h_1 h_2}\right)\overline{\eta}\right]. \tag{14c}$$

These results were given first by Keulegan (1953), except for a misprint. Because $f \geq 0$, it follows that a internal soliton always thickens the thin layer. Thus, the soliton raises the interface if $h_1 > h_2$, and lowers it if $h_1 < h_2$.

ii. In our experiments, $\eta \geq 0$ initially. It follows that the internal wave evolves into solitons when $h_1 > h_2$, but not when $h_1 < h_2$.

iii. The nonlinear term in (12) vanishes if $h_1 = h_2$. We will examine this special case elsewhere (Segur & Hammack, Chapter 1 of *these Proceedings*). Here we assume $h_1 \neq h_2$.

iv. Arbitrary initial data that are smooth and localized will evolve into N solitons, ordered by amplitude, followed by a dispersive oscillatory tail. The number of solitons that emerge from $f(\chi, 0)$ is the number of zeros of

$$\frac{d^2\psi}{d\chi^2} + f(\chi, 0)\psi = 0, \quad \psi \to 1 \quad \text{as} \quad \chi \to -\infty. \tag{15}$$

2.2.2 The finite-depth equation

Kubota *et al.* (1978) derived (2) for cases in which the fluid is confined between two rigid walls and the background density distribution is continuous. The derivation we present here differs slightly from theirs because our background density distribution is discontinuous. A more important difference, however, is that our choice of small parameters ($\epsilon \ll 1$) differs from theirs. Both derivations lead to (2) but the scaling of the physical variables is somewhat different. For simplicity, we will use the Boussinesq approximation, and also replace the free surface with a rigid lid.

The assumptions underlying (2) are that:

i. there is a thin (lower) layer, $\epsilon = h_2/h_1 \ll 1$;

ii. the characteristic horizontal wavelenght is comparable to the depth of the thick layer, $kh_1 = O(1)$;

iii. wave amplitudes are small, $\overline{\eta}/h_2 \ll 1$;

iv. these two effects balance, $\overline{\eta}/h_2 = O(\epsilon)$;

v. viscous effects may be neglected.

Note that the assumption of long waves ($kh_2 \ll 1$) is implied by i and ii. Note further that assumption iv is not consistent with corresponding assumption for KdV. Consequently, we may expected the KdV limit of (2) to be somewhat singular.

The following scaling is consistent with tha assumptions listed above. In the lower layer, dimensionless (*) variables are

$$x^* = x/h_1, \quad z^* = z/h_2 = z/\epsilon h_1, \quad t^* = \overline{c}t/h_1, \quad \tau = \epsilon t^*. \tag{16a}$$

The wave speed $\overline{c}$ is to be determined. The interface is defined by

$$\eta = \epsilon h_2[\eta_1 + \epsilon\eta_2 + \epsilon^2\eta_3] + O(\epsilon^4). \tag{16b}$$

After satisfying Laplace's equation in the lower fluuid and the boundary condition at $z = -h_2$, one finds that the velocity in the lower fluid at the interface may be represented by

$$\begin{aligned} u &= \epsilon\overline{c}[u_1(x^*, t^*; \tau, \epsilon) + \epsilon u_2] + O(\epsilon^3), \\ w &= -\epsilon^2\overline{c}\left[\frac{\partial u_1}{\partial x^*} + \epsilon\frac{\partial u_2}{\partial x^*} + \epsilon\eta_1\frac{\partial u_1}{\partial x^*}\right] + O(\epsilon^4). \end{aligned} \tag{17}$$

Because no further terms will be needed in this expasion in order to derive (2), it will follow that the entire dispersive effect in (2) is due to the upper fluid. It will also turn out that the entire nonlinear effect is due to the lower fluid.

The velocity potential in the upper layer may be written as

$$\phi_1 = \frac{1}{2\pi}\int_{-\infty}^{\infty}\frac{\cosh k(h_1 - z)}{\sinh kh_1}\Psi(k, t^*, \tau)\exp ikx\, dk.$$

It is consistent with assumption ii to define

$$m = kh_1, \qquad x^* = x/h_1, \tag{18}$$

and to rewrite ϕ_1 in terms of these variables. Then if we expand Ψ as

$$\Psi = h_2^2\overline{c}(A_1 + \epsilon A_2 + \cdots), \tag{19}$$

the velocities in the upper fluid at the interface take the form

$$\begin{aligned} U &\sim \frac{i\epsilon^2\overline{c}}{2\pi}\int_{-\infty}^{\infty}(m\coth m)(A_1 + \epsilon A_2 + \cdots)\exp imx^*\, dm, \\ W &\sim -\frac{\epsilon^2\overline{c}}{2\pi}\int_{-\infty}^{\infty} m(A_1 + \epsilon A_2 + \cdots)\exp imx^*\, dm, \end{aligned} \tag{20}$$

There are two kinematic conditions at the interface:

$$\frac{\partial \eta}{\partial t} + u\frac{\partial \eta}{\partial x} = w, \tag{21a}$$

$$(U - u)\frac{\partial \eta}{\partial x} = W - w. \tag{21b}$$

Continuity of pressure at the interface may be written in terms of Bernoulli's law,

$$\frac{\partial \phi_1}{\partial t} + \tfrac{1}{2}(U^2 + W^2) - \frac{\partial \phi_2}{\partial t} - \tfrac{1}{2}(u^2 + w^2) - g\Delta\eta = \text{const.}, \tag{21c}$$

once we have made the Boussinesq approximation. Because (21c) is valid along the interface, its tangential derivate vanishes there. Subtituing (17) and (20) into these conditons, one finds that non-trivial solutions exist at leading order if

$$\bar{c}^2 = g\Delta h_2 \tag{22}$$

as anticipated by (7). [A major difference between this derivation and one following Kubota *et al.* (1978) is that their leading-order wave speed satisfies

$$\bar{c}^2 = g\Delta h_2(1 - \epsilon).$$

Thus, they retain a term at leading order that we regard as a higher-order effect. This difference in ordering persists at higher orders in the expansion.] In the present derivation, the solution of the leading-order equation is

$$\begin{aligned} \eta_1(x^*, t^*, \tau) &= f(r, \tau) + g(l, \tau), \\ u_1(x^*, t^*, \tau) &= f(r, \tau) - g(l, \tau), \\ A_1(x^*, t^*, \tau) &= i\hat{f}(m, \tau)\exp(-imt^*) - i\hat{g}(m, \tau)\exp imt^*, \end{aligned} \tag{23}$$

where

$$r = x^* - t^*, \quad l = x^* + t^*,$$

$(\hat{\cdot})$ denotes the Fourier transform (assumed to exist), and we have used the boundary condition that all motion ceases as $|x| \to \infty$.

Secular terms arise at the next order unless the right-going waves satisfy

$$2\frac{\partial f}{\partial \tau} + 3f\frac{\partial f}{\partial r} - \frac{\partial}{\partial r}\frac{1}{2\pi}\int (m \coth m)\hat{f}(m, \tau)\exp imr\, dm = 0, \tag{24a}$$

and the left-going waves satisfy a similar equation. This nonlinear evolution equation has several representations. Another is

$$2\frac{\partial f}{\partial \tau} + 3f\frac{\partial f}{\partial r} - \tfrac{1}{2}\frac{\partial^2}{\partial r^2} \fint f(y, \tau)\coth \tfrac{1}{2}\pi(r - y)\, dy = 0. \tag{24b}$$

which may be scaled to (2). In dimensional variables

$$(\bar{c})^{-1}\frac{\partial \eta}{\partial t} + \frac{\partial \eta}{\partial x} + \frac{3}{2}\frac{\eta}{h_2}\frac{\partial \eta}{\partial x} - \frac{h_2}{4h_1}\frac{\partial^2}{\partial x^2}\fint \coth\left(\tfrac{1}{2}\pi\frac{x - z}{h_1}\right)\eta\left(\frac{z}{h_1}\right) dz = 0, \tag{25}$$

where $\bar{c}$ is defined by (22).

In this derivation we have assumed that all functions vanish as $|x| \to \infty$. Periodic boundary conditions are more natural in many problems such as when the waves are generated by the periodic motion of the tides. In this case it is neccessary only to replace the Fourier integrals with Fourier sums. The analogue of (24) on $(-\pi, \pi)$

with periodic boundary conditions is

$$2\frac{\partial f}{\partial \tau} + 3f\frac{\partial f}{\partial r} - \frac{\partial}{\partial r}\frac{1}{2\pi}\sum_{-\infty}^{\infty} n \coth n \hat{f}_n(\tau) \exp inr = 0, \tag{26a}$$

where $[\hat{f}_n]$ are the coefficients of the Fourier-series representation of f. The sum in (26a) can be written as a convolution integral, with a kernel involving elliptic functions. The derivation of (26a) also requieres

$$\hat{f}_0(\tau) = \int_{-\pi}^{\pi} f(r,\tau)dr = 0. \tag{26b}$$

which amounts to normalization. The analysis of (26) is quite similar to that of (24), as discussed by Ablowitz *et al*, (1982).

Based on the work of Joseph & Egri (1978), Chen & Lee (1979), Satsuma, Ablowitz & Kodama (1979), and Kodama, Satsuma & Ablowitz (1981), we may assert that (24) is completely integrable, and that its solutions are qualitatively similar to those of the KdV equation. In particular, a soliton solution of (24) is

$$\tfrac{3}{2}f = \frac{\lambda \sin\lambda}{\cos\lambda + \cosh\lambda[r + \frac{1}{2}\tau\lambda\cot\lambda + r_0]}, \tag{27a}$$

where (λ, r_0) are arbitrary parameters except that $0 \leq \lambda \leq \pi$. In dimensional terms

$$\frac{\eta}{h_2} = \frac{\frac{3}{2}(h_2/h_1)\lambda\sin\lambda}{\cos\lambda + \cosh\Theta} \tag{27b}$$

$$\Theta = \lambda\frac{(x+x_0)}{h_1} - \frac{\lambda\bar{c}t}{h_1}\left[1 - \frac{1}{2}\frac{h_1}{h_2}\lambda\cot\lambda\right]. \tag{27c}$$

This reduces to a KdV soliton in the limit $\lambda \to 0$, h_2/h_1 fixed.

In order to carry this expansion to higher order, it is convenient to consider only right-going waves, so that $g(l,\tau) \equiv 0$ in (23). Then at second-order

$$\begin{aligned} \eta_2(r,\tau) &= f_2 \\ u_2 &= f_2 - \tfrac{1}{4}f^2 + \tfrac{1}{2}\frac{\partial}{\partial r}T[f], \\ \frac{i}{2\pi}\int A_2 \exp imx^* dm &= -f_2 - \tfrac{3}{4}f^2 - \tfrac{1}{2}\frac{\partial}{\partial r}T[f]. \end{aligned} \tag{28}$$

At second order $f_2(r,\tau)$ is free, but secular terms arise at third order unless

$$L[f_2] = \tfrac{1}{2}f\frac{\partial f}{\partial \tau} + \tfrac{3}{2}\frac{\partial}{\partial r\partial \tau}T[f] + \tfrac{1}{2}\frac{\partial}{\partial r}(f^3) - \frac{\partial}{\partial r}\left(f\frac{\partial}{\partial r}T[f]\right) - \tfrac{1}{3}\frac{\partial^3 f}{\partial r^3}, \tag{29}$$

where f is a solution of (24), and

$$L[v] = 2\frac{\partial v}{\partial \tau} + 3\frac{\partial}{\partial r}(fv) + \frac{\partial^2}{\partial r^2}T[v].$$

Because $L[v]$ is the linearization of (24), one finds easily that

$$L\left[\frac{\partial f}{\partial r}\right] = L\left[\frac{\partial f}{\partial \tau}\right] = 0, \tag{30a}$$

$$L\left[\tau\frac{\partial f}{\partial r}\right] = 2\frac{\partial f}{\partial r}, \tag{30b}$$

$$L[f] = \tfrac{3}{2}\frac{\partial}{\partial r}(f^2). \tag{30c}$$

We now seek specifically the second-order correction to a soliton (27) so that

$$\frac{\partial f}{\partial \tau} \sim -(c_1 + \epsilon c_2)\frac{\partial f_2}{\partial r}, \qquad \frac{\partial f_2}{\partial \tau} \sim -c_1\frac{\partial f_2}{\partial r},$$

and c_2 is to be determined. Moreover,

$$\frac{\partial}{\partial r}T[f] + \tfrac{3}{2}f^2 - 2c_1 f \sim 0,$$

$$\frac{\partial^2 f}{\partial r^2} = \lambda^2 f + 9c_1 f^2 - \tfrac{9}{2}f^3.$$

In this case, (29) reduces to

$$L[f_2] = \frac{\partial}{\partial r}\{(2c_2 - 3c_1^2 - \tfrac{1}{3}\lambda^2)f - 3c_1 f^2 + \tfrac{7}{2}f^3\}. \tag{31}$$

From (30a); this equation has no unique solution. From (30b), we must choose

$$2c_2 = 3c_1^2 + \tfrac{1}{3}\lambda^2, \tag{32}$$

in order to avoid unbounded growth in f_2. Thus the dimensional speed of a soliton to this order is

$$C \sim \bar{c}\left[1 - \tfrac{1}{2}\epsilon\lambda\cot\lambda + \tfrac{1}{2}\epsilon^2\lambda^2(\tfrac{3}{4}\cot^2\lambda + \tfrac{1}{3})\right], \tag{33}$$

where $\epsilon = h_2/h_1$, and λ is the parameter for the soliton.

In the small-amplitude limit, $\lambda \to 0$ and (33) reduces to the expansion of (6b) for small ϵ. At the other extreme, if $\lambda \to \pi$ then to this order of approximation,

$$C \sim \bar{c}\Big/\sqrt{1 - \frac{\epsilon\pi}{\pi - \lambda}},$$

Clearly

$$\lambda < \pi(1 - \epsilon) \tag{34}$$

is necessary for the validity of (33) as an asymptotic expansion. It is probably not sufficient. Joseph & Adams (1981) also carried this expansion to second order in a related problem. They suggest that their expansion is valid only if $C < 1.4c_*$, where c_* corresponds to C_0 defined by (6b). In this problem, their cutoff corresponds to about

$$\lambda < \pi(1 - 2\epsilon).$$

However, they give no clear criterion for this choice.

We now return to (31). Let $v(r)$ denote a particular solution of

$$L[v] = \frac{\partial}{\partial r}\{f^3\}. \tag{35}$$

We have found no analytic expression for $v(r)$, but a numerical procedure to solve (35) approximately is outlined in the appendix. A solution of (31), corresponding to $f[r - (c_1 + \epsilon c_2)\tau; \lambda]$ is

$$f_2 = -2c_1 f + \tfrac{7}{2}v, \tag{36}$$

or

$$\eta \sim \epsilon h_2[(1 - 2\epsilon c_1)f + \tfrac{7}{2}\epsilon v].$$

This is the expansion that the expression that we will compare with our data in §3.

In the experiments of Koop & Butler (1981), the dimesnsionless density difference $\Delta = 0.367$, and the Boussinesq approximation is valid. However, the analysis for arbitrary Δ follows nearly identical lines, and may simply state the main results.

With the thin layer on the bottom and a free surface on top, the leading-order wave speed is still given by (22). The evolution equation becomes

$$2\frac{\partial \tilde{f}}{\partial r} + 3\tilde{f}\frac{\partial \tilde{f}}{\partial r} + (1-\Delta)\frac{\partial^2}{\partial r^2}T[\tilde{f}] = 0, \tag{37}$$

instead of (24). For fixed λ, the relation between a soliton with $\Delta \neq 0, (\tilde{f}, \bar{c}_1)$ and the earlier results (f, c_1) is

$$\bar{c}_1(\lambda;\Delta) = (1-\Delta)c_1(\lambda), \quad \tilde{f} = (1-\Delta)f(r - \bar{c}_1\tau;\lambda). \tag{38}$$

The evolution equation for $\tilde{f}_2$ is somewhat complicated, but if $\tilde{f}$ is a soliton and $\tilde{f}_2$ is a permanent wave travelling with the same speed, then

$$\begin{aligned} \tilde{L}[\tilde{f}_2] &\equiv \frac{\partial}{\partial r}\{2\bar{c}_1\tilde{f}_2 + 3\tilde{f}\tilde{f}_2 + (1-\Delta)\frac{\partial}{\partial r}T[\tilde{f}_2]\} \\ &= \frac{\partial}{\partial r}\Bigg\{\Bigg[2\bar{c}_2 - 3\bar{c}_1^2 - \tfrac{1}{3}\lambda^2 + \Delta(1-\Delta)\Big(\frac{\lambda}{\sin\lambda}\Big)^2\Bigg]\tilde{f} \\ &\qquad - \frac{3\bar{c}}{(1-\Delta)^2}\tilde{f}^2 + \frac{7 - 17\Delta + 13\Delta^2}{2(1-\Delta)^2}\tilde{f}^3 - \frac{3\Delta}{2}\frac{\partial}{\partial r}T[\tilde{f}]\Bigg\}. \end{aligned} \tag{39}$$

The result is

$$\begin{aligned} 2\bar{c}_2 &= 3\bar{c}_1^2 + \tfrac{1}{3}\lambda^2 - \Delta(1-\Delta)\Big(\frac{\lambda}{\sin\lambda}\Big)^2, \\ \tilde{f}_2 &= -2\bar{c}_1\frac{(1+\Delta-\Delta^2)}{(1-\Delta)^2}\tilde{f} - \frac{3\Delta}{2(1-\Delta)}\tilde{f}^2 + [2(1-\Delta)^2 + \tfrac{3}{2}]\,v, \end{aligned} \tag{40}$$

where $(\tilde{f}, \bar{c}_1)$ satisfy (38), and v satisfies (35). Let $\bar{\eta}$ denote the maximum displacement of the interface. Then it follows from (40) that

$$\begin{aligned} \frac{h_1\bar{\eta}}{(1-\Delta)h_2^2} &= \frac{\frac{2}{3}\lambda\sin\lambda}{1+\cos\lambda}\Bigg[1 + \epsilon\lambda\cot\lambda\frac{1+\Delta-\Delta^2}{(1-\Delta)^2} - \frac{\epsilon\Delta\lambda\sin\lambda}{1+\cos\lambda}\Bigg] \\ &\quad + \epsilon\Bigg[2(1-\Delta) + \frac{3}{2(1-\Delta)}\Bigg]V_{\max} + O(\epsilon^2). \end{aligned} \tag{41}$$

Koop & Butler (1981) define an integral length scale by

$$\bar{\eta}\Lambda \equiv \int_0^\infty \eta\, dx. \tag{42}$$

It follows from the results above that

$$\frac{\bar{\eta}\Lambda}{h_2^2} = \frac{2\lambda}{3}(1-\Delta) + \tfrac{2}{3}\epsilon\lambda^2\cot\lambda + \epsilon\big(2(1-\Delta)^2 + \tfrac{3}{2}\big)I + O(\epsilon^2), \tag{43}$$

where

$$I = \int_0^\infty v(r)\, dr,$$

and must be found numerically.

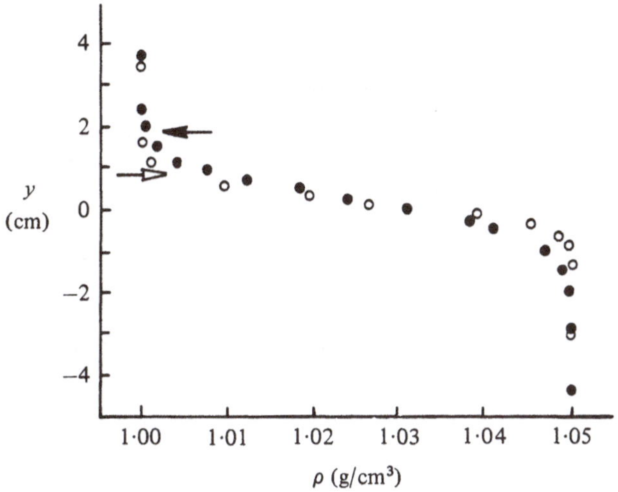

Figure 2.2. Typical density stratifications before (∘) and after (•) a set of experiments. Arrows indicate the position of the dye interface.

§2.3 Comparison with experiments

A series of experiments was performed in which long internal waves were generated by the vertical uplift of a rectangular piston at one end of a wave tank, as shown schematically in Figure 2.1. The tank was 30 m long, 60 cm deep, and 39.4 cm wide. The piston was 61 cm long and spanned the tank width. The time-displacement history of the piston was controlled by electrohydraulic-servo system, so that repeatable motions were easily obtained. Both the tank and wavemaker have been described in detail by Hammack (1972).

The stratified fluid used in these experiments consisted of a layer of fresh water overlying a layer of brine with $h_1 = 45$ cm, $h_2 = 5$ cm and $\Delta = 0.048$. Actual density stratifications were measured using a conductivity probe before, during, and after a set of experiments. In all cases the initial thickness region (pycnocline) was about 1 cm; experiments were terminated when the pycnocline thickness reached 2 cm. Typical beginning and ending stratifications near the pycnocline are shown in Figure 2.2. Because the density varied continuously with depth in the experiments, linear theory would predict an infinite set of internal wave modes. However, we are primarily interested in the lowest internal wave mode, for which the two-layer model discussed in §2 is applicable.

Detailed discussions of the tank-filling procedures and stratification properties are given by Hammack (1980), where it is demonstrated that the growth of the interfacial thickness was dominated by molecular diffusion. Hence, the shear layers produced by these long waves at the pycnocline were laminar and did not cause appreciable mixing.

Rapid uplift of the rectangular piston generated both surface and internal waves. Both initial waves were rectangular in shape, with a length of 122 cm (twice the piston length since the tank end wall adjacent to the piston acts as a plane of symmetry in linear, inviscid theory). The maximum amplitude of the surface wave was one-half the piston uplift, while the internal-wave amplitude was futher attentuated by the ratio $h_1/(h_1+h_2)$. (A detailed analysis and discussion of the generation is given by Hammack 1980). The faster surface wave separated quickly from the internal wave so that there was effectively no interaction. In order to prevent the surface wave from returning to the internal wave region, a vertical plate, located 18.8 m from the piston, was lowered carefully into the water after the passage of the surface wave. In addition to trapping the surface wave, the plate effectively lengthened the test section for the internal wave, since this wave eventually reflected off the plate and propagated back through the test section.

Internal wave measurements were obtained using a laser-optics-detector system described by Hammack (1980). Briefly, the light beam of a laser was converted to a uniform-width light sheet that was directed across the glass-walled tank. With the quiescent interface intercepting the light sheet and the brine dyed blue, subsequent displacements of the interface varied the amount of light traversing the tank cross section. (The dye inteface remained sharp and distinct throughout a set of experiments. It was always located near the top of the salinity interface, as shown in Figure 2.2.) The transmitted light was focused onto a photodetector, whose output voltage was recorded. The system was calibrated before each experiment to the determine the (nonlinear) correlation between output voltage of the photodetector and the vertical displacement of the interface. Only one gauge was avaible; hence the same experiment was repeated in order to measure time histories of the interfacial displacement at vaious positions along the tank. Both the repeatability of the piston motion and the quiescent conditons prior to an experiment were carefully monitored and assured for all reported data. The small changes in the background stratification between experiments led to slight changes in the phase speed of the internal waves.

Figure 2.3 shows the evolution of internal wave, as measured at seven downstream locations in the tank. The co-ordinate system in Figure 2.3 moves with the linearized wave speed c_0, and reverses the sense of direction of the wave. Thus the front of these waves is to the left in each record. Moreover, we have omitted te weak train of oscillatory waves that follows the dominant waves that are shown. The first five measurements were recorded ahead of the plate at $x/\sqrt{h_1 h_1} = 125$.

For this experiment the piston uplift was 2 cm., so we may take $k^{-1} = 122$ cm., and $\overline{\eta} = 1$, corresponding to the dominant lenght scale and maximum amplitude of the initial wave. Hence, $k^2(h_1 + h_2)^2 \sim 0.16$, $\overline{\eta}/(h_1 + h_2) \sim 0.02$, and we may apply KdV theory at least provisionally. Using either the first or second wave record as the potential in (15), one finds that according to KdV theory, this wave should evolve into two solitons, followed by a weak oscillatory wave train. Certainly this prediction is in qualatitive agreement with the wave records shown in Figure 2.3.

A more rigorous test of KdV theory is shown in Figure 2.4, where have plotted the lead wave from each of the last five records on a single graph. According to (14), the data from all these wave records should fall on a single curve once the peak amplitude at each station is known. The agreement in Figure 2.4 between the predicted wave shape and all these data is so good that we conclude that these are (locally) KdV solitons.

The major discrepacies between the predicted and observed wave shapes occur: (i) at the rear (*ie.* right) of each wave, where the influence of the trailing soliton becomes important; and (ii) in the data at $x/\sqrt{h_1 h_2} = 151$, just downstream of the reflecting plate, where the incident and reflected waves from this plate apparently are still superposed.

The amplitude of the soliton is slowly decaying as it propagates down the tank, presumably due to viscosity, but this decay is sufficiently slow that the wave continually readjusts its shape as it decays so that it is locally a KdV soliton. On the basis of a much larger set of experiments, Hammack & Segur (1974) found that this same description (locally KdV solitons, slowly attenuated by viscosity) also applies to the corresponding long surface waves. Leone, Segur & Hammack (1982) discuss the viscous decay of these waves.

Based on the remarkably good agreement in Figure 2.4 and the fact that initially $k^2(h_1 + h_2)^2$ exceeds $\overline{\eta}/(h_1 + h_2)$ by a factor of about 10, it is tempting to conclude that the KdV equation has a fairly large range of validity. This conclusion may well be correct, but it is not necessarily implied by Figure 2.4. While $k^2(h_1 + h_2)^2 \sim 0.16$ may describe then initial data, the evolving solitons have characteristics wavelenghts that are much longer, so that the required balance is achieved for the solitons. The stronger conclusion applies only if the equation predicts correctly from the initial data the wavelengths of the solitons that the emerge.

KdV theory predicts that all solitons should move somewhat faster than c_0, the linear long-wave speed, and therefore should move to the left in Figure 2.3. The larger, faster soliton does move somewhat to the left until the last station, but the smaller soliton clearly is moving to the right. Thus the observed

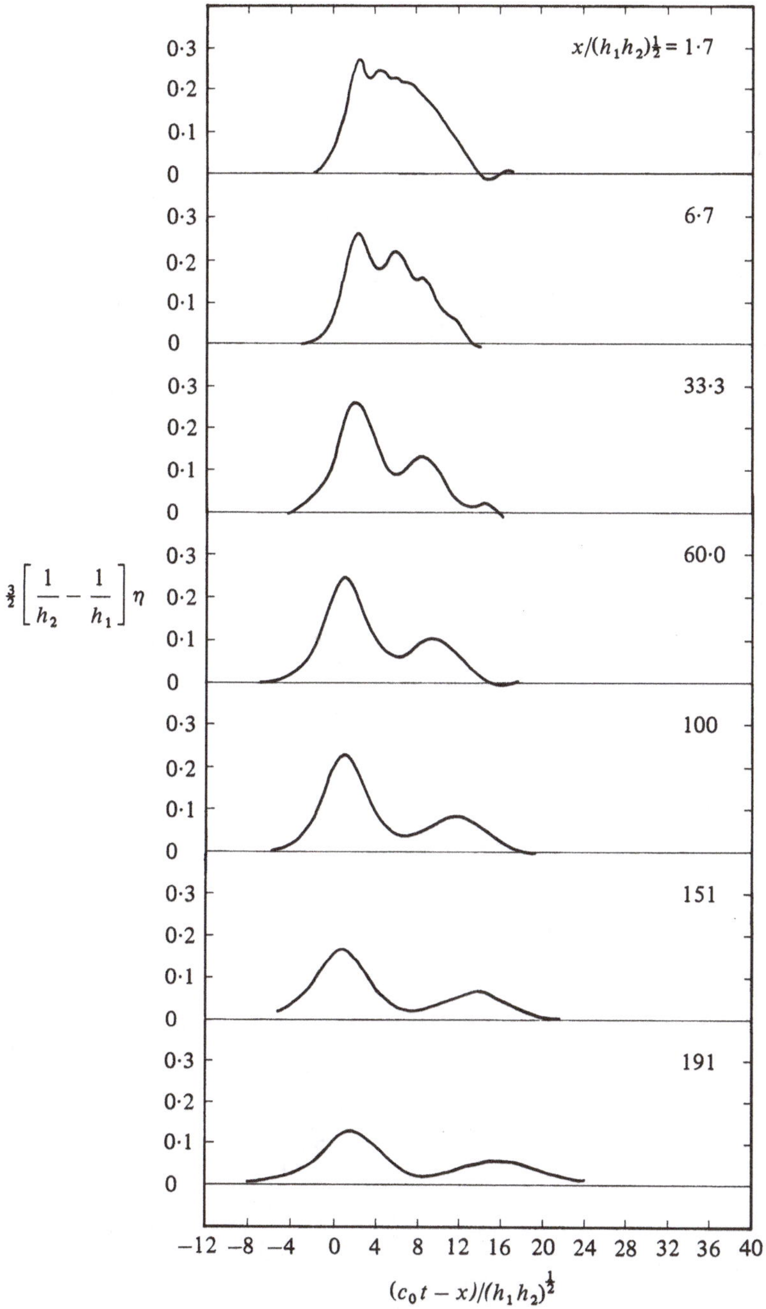

Figure 2.3. **Internal waves, measured in time at seven successive locations. $h_1 = 45$ cm., $h_2 = 5$ cm., $\Delta = 0.048$. A vertical plate inserted during the experiment at $x/\sqrt{h_1h_2} = 125$ refleted the wave train back into the test section for the last two measurements.**

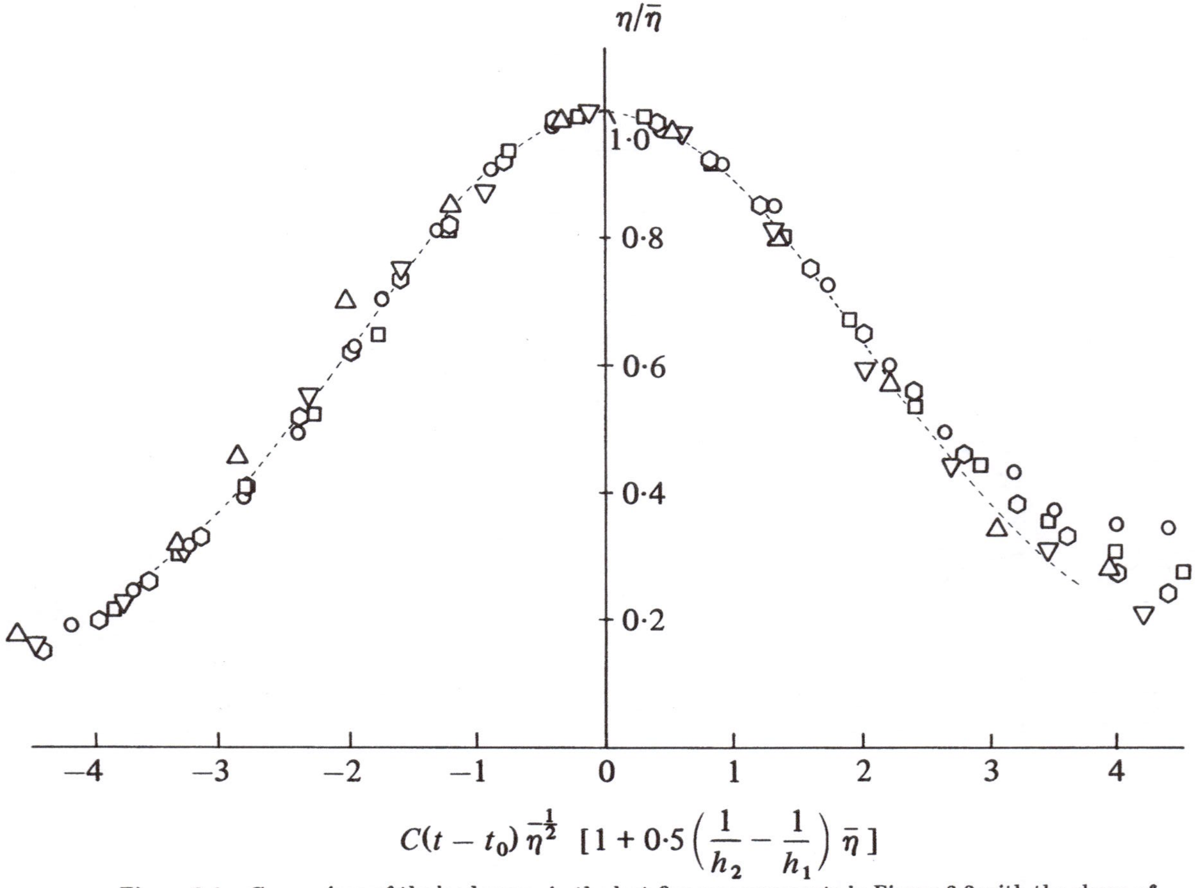

Figure 2.4. Comparison of the lead waves in the last five measurements in Figure 2.3 with the shape of a KdV soliton, according to (14). $C^2 = [3g\,\Delta(h_1-h_2)/4(h_1h_2)^3] \times 8063$. ○, $x/\sqrt{h_1h_2} = 33.3$; □, 60; ⬡, 100; △, 151; ▽, 191; - - - - -, KdV theory.

speeds of the solitons are slower than those predited by KdV theory, and sometimes even slower than c_0. This discrepancy was observed for surface solitons as well (Hammack & Segur 1974). For internal waves, we denote that the predicted wave speeds could be reduced by generalizing the theory to include the influence either of the finite thickness of the pycnocline or of viscosity.

Next, we consider the finite-depth equation, (24). There seems to be no simple way to compare all of the data on one figure for (24), but in Figures 2.5 we show the lead waves measured at two representative points, $x/\sqrt{h_1h_2} = 60$ and 191. Figure 2.5 also shows the wave shapes predicted by KdV theory (14), by the first-order finite-depth theory (27). and by the second-order finite-depth theory (36). In all cases, the free parameter available in the theory [*e.g.* λ in (27)] was chosen to match the peak amplitude of the measure wave. It is apparent that while all of these theorical predictions are reasonably accurate, first-order finite-depth theory is noticeably less accurate than KdV theory. The discrepancy is a apparent even at $x/\sqrt{h_1h_2} = 191$, where $h_2/h_1 = \frac{1}{9}$, $\bar{\eta}/h_2 \sim 0.094$, and one might expected (27) to be quite accurate.

Figures 2.5 support the claim that the range of validity of the asymptotic expansion that leads to the finite-depth equation is small. In Figure 2.5a, where the peak wave amplitude is rather small, the second-order finite-depth theory predicts the measured data about as well as does KdV theory. The peak amplitude in Figure 2.5b is about twice that in Figure 2.5a, and here even the second-order theory is begining to fail. Presumably, third-order corrections now have become important.

As an independent test of the hypotesis that the range of validity of (24) is rather small in a pratical sense, we analyse next the experimental results of Koop & Butler (1981). Figures 2.6 show the relation between soliton amplitude (41) and integral length scale (43), corresponding to figures 10 and 12 of Koop &Butler. The first-order curve is obtained by letting $\epsilon \to 0$ in (41) and (43). Higher-order corrections depend on $\epsilon = h_2/h_1$ and on $\Delta = (1-\rho_1\rho_2)$, and we show in Figures 2.6a and 2.6b two second-order curves, corresponding to the two configurations of Koop & Butler. Note that both second-order curves terminate at finite values of $\bar{\eta}/h_1(1-\Delta)h_2^2$. This termination occurs because $\bar{\eta}/h_1(1-\Delta)h_2^2$ has a maximum in (41) as a function of λ, signalling a breakdown of the asymptotic series. This breakdown occurs earlier that in (34).

A comparison of the first-order theory, second-order theories, and the data in Figures 2.6 reveal the

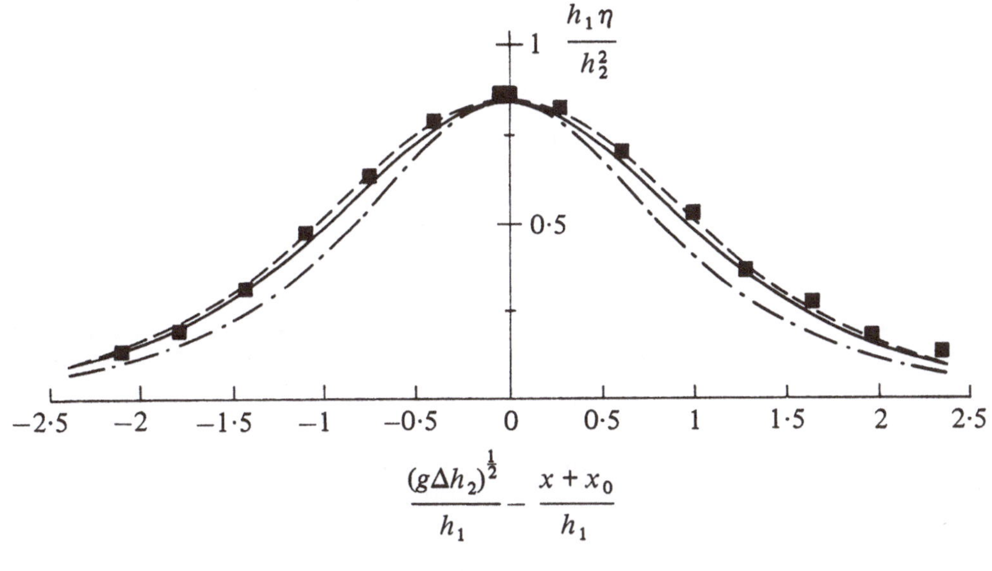

Figure 2.5a. Comparison of two of the measured lead waves in Figure 2.3 with three theoretical shapes, each having the same peak amplitude as that measured. $x/\sqrt{h_1h_2} = 191$; — — —, KdV (14); — - — -, first-order finite-depth (27); ——, second-order finite-depth (36); ■, measured.

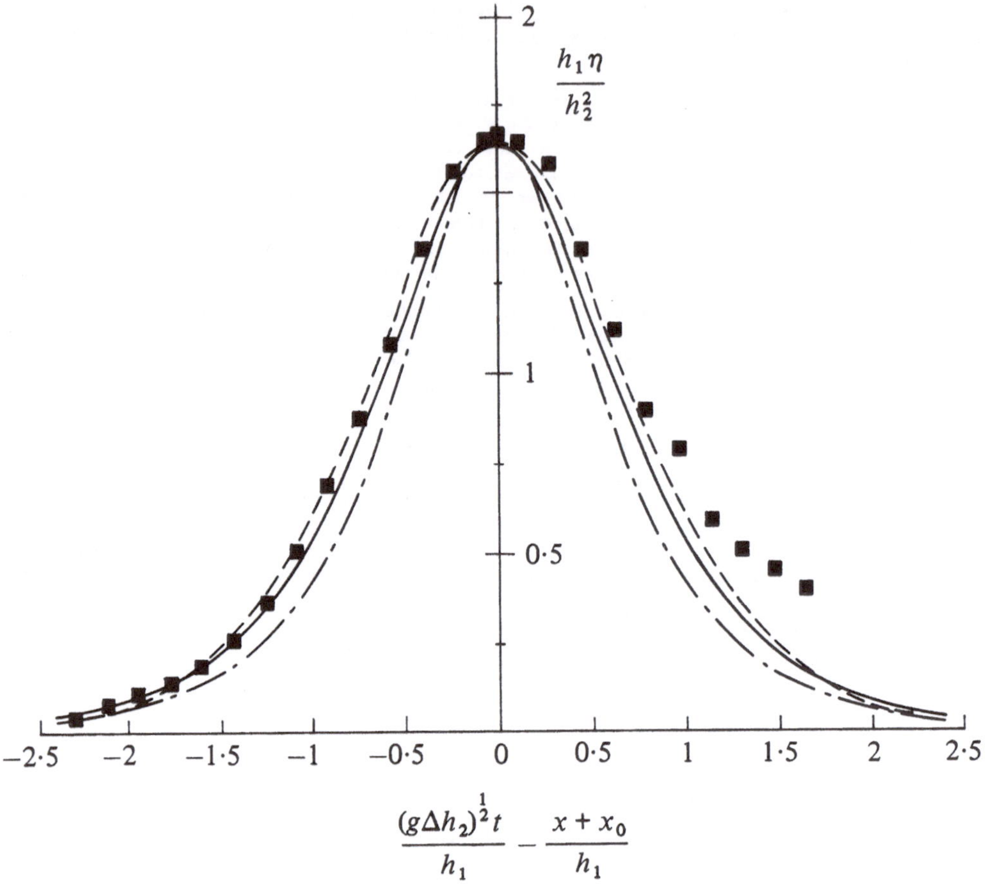

Figure 2.5b. Same symbols are used as in *(a)* for $x/\sqrt{h_1 h_2} = 60$.

following facts.

i. In Figure 2.6a ($\epsilon = 0.197$, corresponding to Figure 10 of Koop & Butler) there are no data with amplitudes sufficiently small that the first- and second-order theories coincide. In this sense, the first-order theory by itself must be considered inadequate to represent these data.

ii. For $\epsilon = 0.197$, the second-order theory predicts the data quite well within its range of validity. However, a significant portion of the data lies outside the range of validity of the theory.

iii. In Figure 2.6b ($\epsilon = 0.029$) the first- and second-order theories are in close agreement over the entire range of validity of the theory. Both predict wave amplitudes somewhat larger than those observed, specially for the longer (and therefore, smaller) waves. This effect is less pronounced but also evident in Figure 2.6a. Koop & Butler noted that the even the KdV equation has this problem for very long waves; they argued that it is a viscous effect.

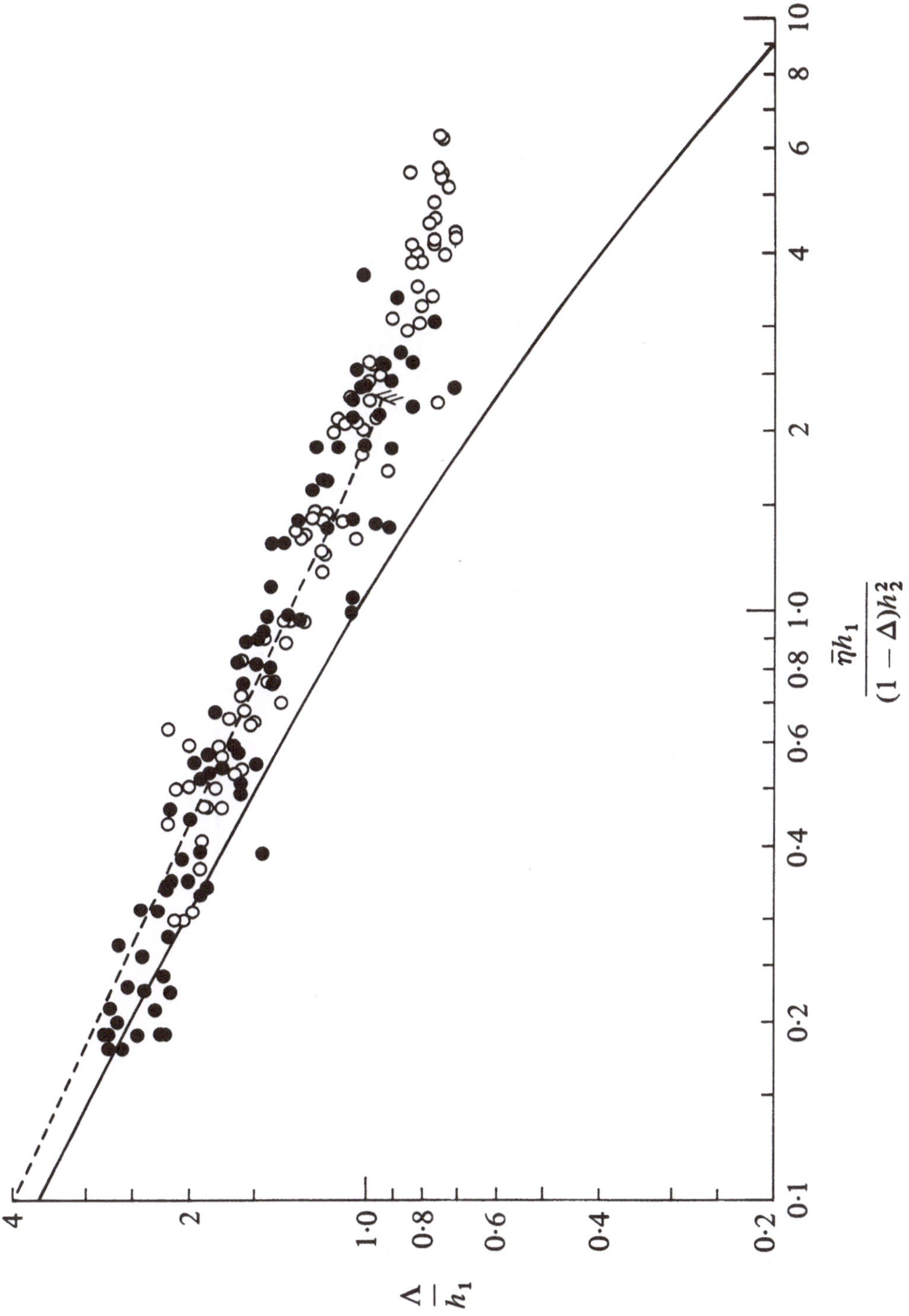

Figure 2.6a. Relation between integral wavelenght (43) and peak amplitude (41) for internal solitary waves in a fluid of finite depth. (a) $\epsilon = h_2/h_1 = 0.197$, $\Delta = 1 - 2 = 0.367$; - - - -, second-order finite-depth theory; ○, incident waves measured by Koop and Butler (1981); •, measured reflected waves.

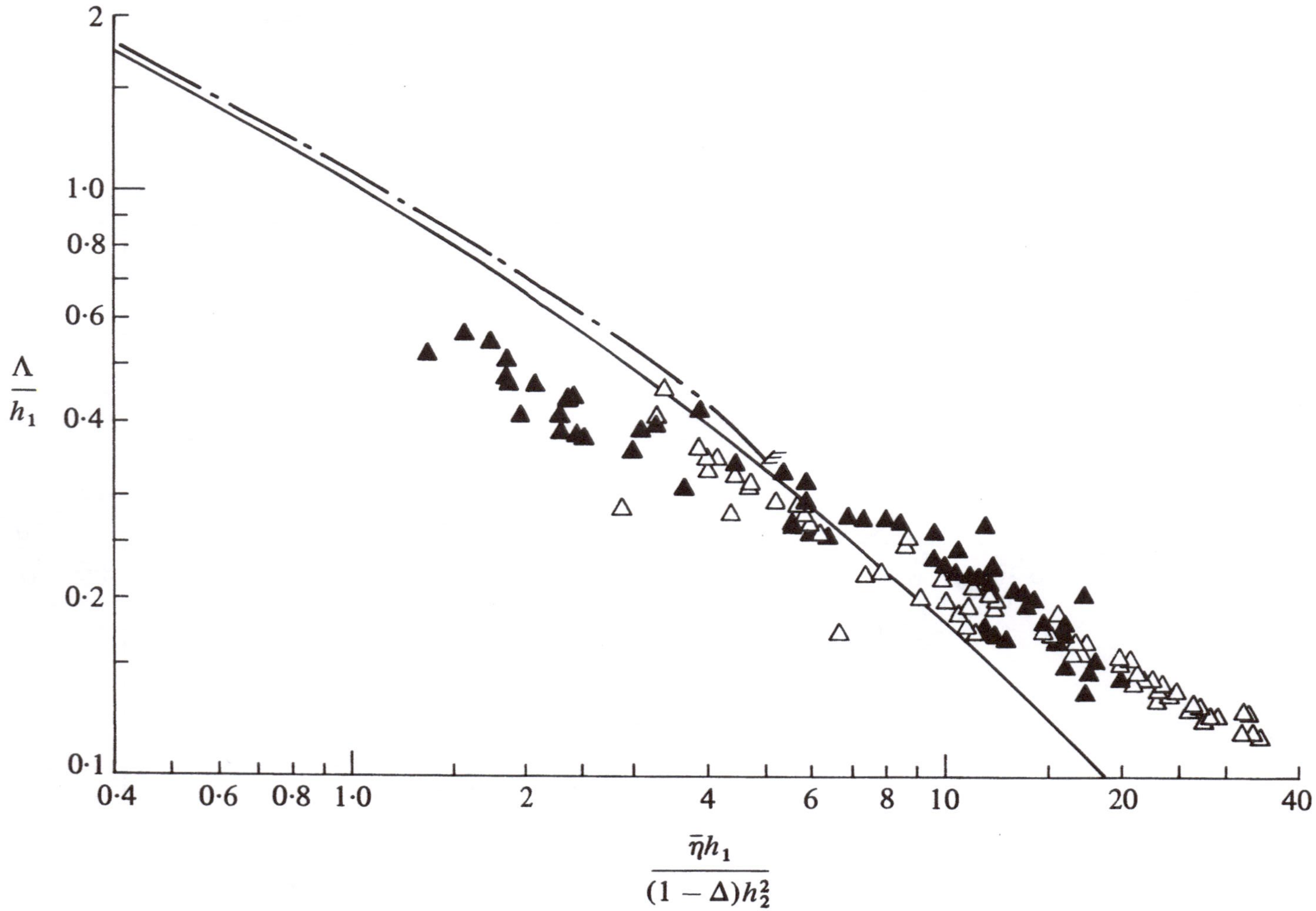

Figure 2.6b. (Cont.) (b) $\epsilon = 0.029$, $\Delta = 0.367$; ——, first-order finite-depth theory; — - — -, second-order finite-depth theory; Δ, measured incident waves; ▲, measured reflected waves.

iv. For $\epsilon = 0.029$, most of the data lies beyond the rage of validity if the theory. There is no obvious physical mechanism, such as wave breaking, that precipitates the breakdown of the theory.

Finally, we summarize our major conclusions.

i. Both (1) and (2) are valid (formally) asymptotic equations that govern the slow evolution of long internal waves of small amplitude that propagate in one direction in an inviscid fluid. However the meaning of *long* is different in the two theories, which describe perturbations to two different wave equations, with two different speeds.

ii. The KdV equation (1) seems to have a relatively large range of validity, and has practical predictive value even when its assumptions are satisfied only marginally.

iii. The finite-depth equation (2) by itself has such a small range of validity that it has been difficult even to find it experimentally.

iv. The asymptotic expansion that generates (2) has a larger of validity, but it is also quite limited. In particular, there is a range of wave amplitude for which the long waves in question are too small to break, but too large to be predicted by the finite-depth theory, of any order.

The experiments on which Figures 2.1–2.5 were based were perfomed at the W. M. Keck Laboratory of Hydraulics and Walter Resources at the California Institute of Technology during 1973–74. We acknowledge our debt to Elton Daly and his staff who assisted in the design, construction, and maintenance of experimental facilities, to Martin Kruskal for several helpful conversations regarding the derivation of (24) and to Gary Koop for providing us with the data in Figures 2.6 prior to its publication. Financial support was provided in part by the Office of Naval Research, Fluid Dynamics Division and the National Science Foundation. The second author would like to acknowledge the first author, whose creative interpretation of the data and persistence made this paper posible.

Appendix

For permanent localized waves, (35) is equivalent to

$$L_1[v] \equiv (3f - 2c_1)v + \frac{\partial}{\partial r}T[v] = f^3 \tag{A1}$$

where $f(r;\lambda)$ denotes a soliton with speed $c_1(\lambda)$. A Galerkin procedure may be used to solve (A1) approximately. Here (A1) is replaced by a finite set of equations fo the form

$$(L_1[v], \phi_n) = (f^3, \phi_n) \qquad (n = 0, \ldots, N), \tag{A2}$$

where $(a,b) = \int ab\,dr$, and $[\phi_n, (r)]$ denotes a set of basis functions of a finite-dimensional space of functions on $-\infty < r < \infty$. We have chosen to use

$$\phi_n(r) = \frac{1}{\sqrt{n}} He_n(r) \exp(-\tfrac{1}{4}r^2), \tag{A3}$$

where $He_n(r)$ are Hermite polynomials (notation as in Abramowitz & Stegun 1964). These basis functions could be generalized by including a scaling factor $(r \to \alpha r)$, but (A3) was adequate for our purposes. These functions are orthogonal, and their Gaussian decay mimics the localized nature of the soliton. Therefore, if the soliton width matches roughly the width of the Gaussian filter then very few basis functions are required to represent a soliton to a high degree of accuracy.

A solition is an even function in its argument. Both $L_1[\ ]$ and the right-hand side of (A1) are also even, and (A1) has a soliton that is even. Therefore, we use only the even basis functions, $\phi_{2n}(r)$, in (A2). This choice excludes the solution of the homogeneous problem in (30a), and seems to make (A2) well-posed. Certainly it is computationally stable if only $[\phi_{2n}]$ are used.

After expanding

$$f(r) = \sum_0^N \hat{f}_{2n}\phi_{2n}(r), \qquad v(r) = \sum_0^N \hat{v}_{2n}\phi_{2n}(r), \tag{A4}$$

(A2) reduces to a set of linear algebraic equations of $[\hat{v}_{2n}]$, the coefficients of which are obtained by evaluating certain integrals. The only delicate question is to decide what accuracy is achieved with a certain N in (A4). We used three main tests of numerical accuracy.

i. For a soliton $f(r,\lambda)$ we have that

$$2c_1(\lambda) = \frac{\frac{3}{2}(f^2, f) + (\partial T[f]/\partial r, f)}{(f,f)} \tag{A5}$$

Using (A4) in (A5) yields a sequence $2c_1(\lambda; N)$, which may be compared to the exact result $2c_1(\lambda) = -\lambda\cot\lambda$. With $N = 7$, the error in $2c_1$ remained below 10^{-2} for

$$\tfrac{1}{10}\pi \le \lambda \le \tfrac{1}{2}\pi,$$

increased to about of 0.2 for $\lambda = \frac{2}{3}\pi$,and continued to increase for larger values of λ. Regardless of how well it approximates $-\lambda\cot\lambda$, the solution of (A5) is the appropriate value of $2c_1$ for truncated problem, and it was used in $L_1[\,]$.

ii. It follows from (30c) that an exact solution of

$$L_1[w] = f^2 \tag{A6}$$

is $w = \frac{2}{3}f$. Let $[\hat{w}_{2n}]$ denote the coefficients obtained by solving (A6) approximately, and measure the error in this approximate solution by

$$e_2 = \max_{0\le n\le N} \frac{|\hat{w}_{2n} - \frac{2}{3}\hat{f}_{2n}|}{|\hat{f}_0|}. \tag{A7}$$

For $N = 7$, we find $e_2 < 10^{-2}$ for $\frac{1}{4}\pi \le \lambda \le \frac{1}{2}\pi$, and $e_2 < 6\times10^{-2}$ for $\frac{1}{10}\pi \le \lambda \le \frac{3}{4}\pi$.

iii. Truncation of the series at N requires that the first few terms dominate the series. Let $\hat{v}_{2n}$ denote the coefficient in the solutions of (A2),and define

$$e_3 = \max \frac{\{|\hat{v}_{12}|,\ |\hat{v}_{14}|\}}{|\hat{v}_0|} \tag{A8}$$

For $N = 7$, $e_3 < 2\times10^{-2}$ if $\frac{1}{4}\pi \le \lambda \le \frac{1}{2}\pi$, and $e_3 < 7\times10^{-2}$ if $\frac{1}{10}\pi \le \lambda \le \frac{3}{4}\pi$.

References

M. J. Ablowitz, A. S. Fokas, J. Satsuma, and H. Segur, On the periodic intermediate long wave equation. *J. Phys. A*, **15**, 781 (1982).

M. Abramowitz and I. A. Stegun *Handbook of Mathematical Functions*. N.B.S., Washington D.C., 1964.

T. B. Benjamin, *J. Fluid Mech.* **25**, 241 (1966).

T. B. Benjamin, *J. Fluid Mech.* **29**, 559 (1967).

D. J. Benney, *J. Math. & Phys.* **45**, 52 (1966).

H. H. Chen and Y. C. Lee, *Phys. Rev. Lett.* **43**, 264 (1979).

V. D. Djordjevic, and L. G. Redekoop, *J. Phys. Oceanog.* **8**, 1016 (1978).

J. L. Hammack, Tsunamis— a model of their generation and propagation. W. M. Keck Lab., Caltech report KH-R-28, 1972.

J. L. Hammack, *J. Phys. Oceanog.* **10**, 1455 (1980).

J. L. Hammack and H. Segur, *J. Fluid Mech.* **65**, 289 (1974).

J. L. Hammack and H. Segur, *J. Fluid Mech.* **84**, 337 (1978).

R. I. Joseph, *J. Phys. A* **10**, L225 (1977).

R. I. Joseph and R. C. Adams, *Phys. Fluids* **24**, 15 (1981).

R. I. Joseph and R. Egri, *J. Phys. A* **11**, L97 (1978).

G. H. Keulegan, *J. Res. Nat. Bur. Stand.* **51**, 133 (1953).

Y. Kodama, J. Satsuma, and M. J. Ablowitz, *Phys. Rev. Lett.* **46**, 687 (1981).

C. G. Koop, and G. Butler, *J. Fluid Mech.* **112** , 225 (1981).

D. J. Korteweg and G. De Vries, *Phil. Mag.* **39** , Ser. 5, 422 (1895).

T. Kubota, D. R. S. Ko, and L. Dobbs, *A. I. A. A. J. Hydronaut.* **12**, 157 (1978).

H. Lamb, *Hydrodynamics. Dover*, 1932.

C. Leone, H. Segur, and J. L. Hammack, *Phys. Fluids* **25**, 942 (1982).

R. R. Long, *Tellus* **8**, 460 (1956).

J. W. Miles, *Tellus* **31**, 456 (1979).

H. Ono, *J. Phys. Soc. Japan* **39**, 1082 (1975).

A. R. Osborne, and T. L. Burch, *Science* **208**, 451 (1980).

A. S. Peters, and J. J. Stoker, *Commun. Pure Appl. Math.* **13**, 115 (1960).

J. Satsuma, M. J. Ablowitz, and Y. Kodama, *Phys. Lett.* **73A**, 283 (1979).

H. Segur, *J. Fluid Mech.* **59**, 721 (1973).

H. Segur, and J. L. Hammack, Long internal waves in layers of equal depth. (Preprint 1982).

Chapter 3

Wobbling Kinks in ϕ^4 and Sine-Gordon Theory***

Abstract

When the ϕ^4 model admits a kink-solution, it also admits a wobbling kink, which satisfies the boundary conditions of a kink, but possesses an internal degree of freedom. In this paper we develop a formal perturbation series for the wobbling kink in ϕ^4 theory, and give the first two terms in the series explicitly. Then we prove that the formal series actually is asymptotic for a rather long time $[O(K \ln(1/\epsilon))$, for a certain $K]$. Finally, we construct an exact 3-soliton solution of the sine-Gordon equation that also has the properties of a wobbling kink. For the sine-Gordon equation, the wobbling kink seems to be mildly unstable.

§3.1 Introduction

The ϕ^4 model of one-dimensional field theory, defined by the Hamiltonian

$$H = \tfrac{1}{2} \int_{-\infty}^{\infty} dX \, [\phi_T^2 + \phi_X^2 + \tfrac{1}{2}(\phi^2 - 1)^2] \,, \tag{1.1}$$

is well know in many branches of physics. In Ref. [1], it was introduced as a simple continuum model of lightly doped polyacetylene $[(CH)_X]$. In that context, a solution of fundamental interest is the kink,

$$\phi(X,T) = \tanh\left\{\tfrac{1}{2}(k + (2k)^{-1})X + \tfrac{1}{2}(k - (2k)^{-1})T + X_0\right\},$$

which separates two adjacent regions of different phase of bond alteration in trans-$(CH)_X$. Further discussion of the physical problem may be found in Ref. [2], in its references, and in Ref. [3].

On the basis of an approximate *ansatz*, Rice and Mele [4] found that (1.1) may admit more complicated kink-like solutions, with internal degrees of freeedom, which we will call *wobbling kinks*. A similar *ansatz*, applied to the sine-Gordon equation,

$$u_{tt} - u_{xx} = \sin u \,, \tag{1.2}$$

indicates that it, too, may admit wobbling kinks [5]. The purpose of this note is to establish rigorously that these two models admits wobbling kink solutions, at least for limited (but large) times.

Two definitions are required to make this discussion meaningful. *(i)* In ϕ^4 theory, an ordinary kink is an exact solution of the equation that satisfies the boundary conditions,

$$\phi \to \pm 1 \quad \text{as} \quad X \to \pm\infty \,,$$

*** H. Segur, *Journal of Mathematical Physics* **24**, 1439–1443 (1983) and is stationary in an appropriate coordinate system. *(ii)* A wobbling kink is a solution of (1.4) whose time-average satisfies the boundary conditions of an ordinary kink, and which is periodic in time in an appropriate coordinate system. [A more general definition might allow almost periodic functions, but this generalization is not needed here.]

In order to state concisely my results for ϕ^4 theory, it is convenient to shift to the rest frame of a given kink. Let

$$x = \tfrac{1}{2}\big(k + (2k)^{-1}\big)X + \tfrac{1}{2}\big(k - (2k)^{-1}\big)T + x_0\,,$$

$$t = \tfrac{1}{2}\big(k - (2k)^{-1}\big)X + \tfrac{1}{2}\big(k + (2k)^{-1}\big)T + t_0\,. \tag{1.3}$$

The governing equation becomes

$$\tfrac{1}{2}\phi_{tt} - \tfrac{1}{2}\phi_{xx} = \phi - \phi^3\,, \tag{1.4}$$

with a stationary kink solution,

$$\phi(x,t) = \phi_0(x) = \tanh(x)\,. \tag{1.5}$$

Let ϵ denote an ordering parameter $(0 < \epsilon \ll 1)$ in a formal series solution of (1.4):

$$\phi(x,t) = \phi_0(x) + \sum_{1}^{\infty} \epsilon^n \phi_n(x,t)\,, \tag{1.6a}$$

and let

$$\Phi_N(x,t) = \phi_0(x) + \sum_{1}^{N} \epsilon^n \phi_n(x,t) \tag{1.6b}$$

denote the Nth truncation of the series. In §3.2 and §3.3, we demonstrate the following results.

i. For sufficiently small ϵ, (1.4) admits a formal series solution corresponding to a wobbling kink. We may choose

$$\epsilon\,\phi_1(x,t) = \epsilon\,\text{sech}\,x \tanh x \sin\sqrt{3}\,t\,, \tag{1.7}$$

so the frequency of oscillation is $\sqrt{3}$ to this order of approximation.

ii. The Green's function of the perturbation problem is given explicitly, so any higher term $(n \geq 2)$ in (1.6a) may be found by quadratures over known functions and/or lower terms in the series. To demonstrate the effectiveness of this procedure, we will find $\phi_2(x,t)$ in closed form.

iii. Both $\phi_1(x,t)$ and $\phi_2(x,t)$ are uniformly bounded in (x,t).

iv. Assume $\Phi_N(x,t)$ is uniformly bounded. Then there exists a constant, K_N, and an exact solution of (1.4), $\phi(x,t)$, such that

$$\phi(x,t) - \Phi_N(x,y) = o(\epsilon^N)\,, \tag{1.8}$$

at least for times up to

$$|t| \leq O\big((2K_N)^{-1}\ln(\epsilon^{-1})\big)\,; \tag{1.9}$$

i.e., the formal series is asymptotic on a relatively long time scale. The constant K_N may be chosen to depend only on N, provided $\epsilon < 1$.

v. The energy of the stationary is $H = 2/3$. The wobbling kinks are more energetic,

$$H = \tfrac{2}{3} + \tfrac{1}{2}\epsilon^2 + O(\epsilon^3)\,. \tag{1.10}$$

The sine-Gordon equation also admits wobbling kinks, but they are rather different from the wobbling kinks of ϕ^4 theory. It is known that (1.2) admits N-soliton solutions [6–8]. In §4, we contruct a particular 3-soliton solution that has the properties of wobbling kink. One may think of it as a nonlinear superposition of kink and a breather, traveling at the same speed.

§3.2 The formal series for ϕ^4

Substituting (1.6a) into (1.4) and equating to zero the coefficients of ϵ^n yields a sequence of linear differential equations to be satisfied. Let

$$Lv \equiv \left[\tfrac{1}{2}\partial_t^2 - \tfrac{1}{2}\partial_x^2 + 2 - 3\,\mathrm{sech}^2 x\right] v\,. \tag{2.1}$$

One finds that

$$L\phi_1 = 0\,, \tag{2.2a}$$

$$L\phi_2 = -3\phi_0(\phi_1)^2\,, \tag{2.2b}$$

and for $n \geq 3$,

$$L\phi_n = -3\phi_0 \sum_{j=1}^{n-1} \phi_j\,\phi_{n-j} - \sum_{j=1}^{n-2}\sum_{k=1}^{n-2} \phi_j\phi_k\phi_{n-j-k} \tag{2.2c}$$

Our objective is to find solutions of these equations that are uniformly bounded in (x, t).

Separating variables in (2.2a), [*i.e.*, $\phi_1(x,t) = p(x)q(t)$], reduces that equation to a classical eigenvalue problem. The solutions that correspond to the discrete eigenvalues and that are bounded in x are

$$\overline{\phi}_1(x,t) = \mathrm{sech}^2 x\,(A + Bt)\,, \tag{2.3}$$

$$\phi_1(x,t) = C\,\mathrm{sech}\,x\tanh x\,\sin\!\left(\sqrt{3}(t+t_0)\right). \tag{1.7'}$$

The ***translation mode***, $\overline{\phi}_1$, represents the fact that (1.4) also admits exact kink solutions with a slightly different position (A) or velocity (B). In the present context, only the ***shape mode***, ϕ_1, is of interest. Without loss of generality, we may set $C = 1$, $t_0 = 0$ in (1.7′).

Once ϕ_1 is chosen, ϕ_2 is found by solving (2.2b). The method of solution of (2.2b) indicates the general procedure for all of the higher order equations. First, we express the forcing terms as a finite sum of simple harmonics in t:

$$L\phi_2 = -\tfrac{3}{2}\left(1 - \cos(2\sqrt{3}t)\right)\mathrm{sech}^2 x\tanh^3 x\,.$$

Next, we require the solution to exhibit the same harmonics as those in the forcing terms:

$$\phi_2(x,t) = 1\,f(x) + \cos(2\sqrt{3}t)\,g(x)\,. \tag{2.4}$$

This assumption precludes unbounded growth in time, but secular tems may still appear if $f(x)$ or $g(x)$ grows without bound as $x \to \infty$. In any case, f and g satisfy ordinary differential equations:

$$\left[-\tfrac{1}{2}\left(\frac{d}{dx}\right)^2 + 2 - 3\,\mathrm{sech}^2 x\right] f = -\tfrac{3}{2}\mathrm{sech}^2 x\tanh^3 x\,. \tag{2.5}$$

Homogeneous solution of (2.5) are

$$f_1 = \mathrm{sech}^2 x\,,$$

$$f_2 = 3x\,\mathrm{sech}^2 x + \tanh x + \sinh 2x\,, \tag{2.6}$$

and (2.5) now may be solved by variation of parameters. A particular solution is

$$f(x) = \tfrac{1}{2}\tanh x\,\mathrm{sech}^2 x - \tfrac{3}{4}x\,\mathrm{sech}^2 x\,. \tag{2.7}$$

Of the homogeneous solutions, $f_2(x)$ is unbounded and therefore must be suppressed; $f_1(x)$ is simply the translation mode reappearing at $O(\epsilon^2)$.

Similarly, $g(x)$ satisfies

$$-\left[\frac{1}{2}\left(\frac{d}{dx}\right)^2 + 4 + 3\,\mathrm{sech}^2 x\right] g = \tfrac{3}{2}\mathrm{sech}^2 x \tanh^3 x\,, \tag{2.8}$$

which can solved by variation of parameters if we know two solutions of the homogeneous problem. More generally, as we go to higher order, we will need the solutions of homogeneous equations of the form

$$\left[\left(\frac{d}{dx}\right)^2 + p^2 + 6\,\mathrm{sech}^2 x\right] G = 0\,. \tag{2.9}$$

But these can all be found, using the *triangular representation* developed by Gel'fand and Levitan [9]. Set

$$G(x;p) = e^{ipx} + \int_{-\infty}^{x} K(x,y) e^{ipy}\, dy\,.$$

then $K(x,y)$ satisfies the Gel'fand–Levitan–Marchenko equation [7], and can be found in closed form for this particular potential:

$$K(x,y) = 3\,\mathrm{sech}\, x\,[e^{y}\tanh x - e^{2y}\mathrm{sech}\, x]\,.$$

It follows that two linearly independent solutions of (2.9) are the real and imaginary parts of

$$\begin{aligned} G(x;p) = e^{ipx}\Bigg[1 + \frac{3(1-ip)}{1+p^2}\tanh x\,(1+\tanh x) \\ - \frac{3(2-ip)}{4+p^2}(1+\tanh x)^2\Bigg], \end{aligned} \tag{2.10}$$

as may be verified by direct substitution. Evaluating (2.10) with $p^2 = 8$ provides the functions required to solve (2.8). The final result at $O(\epsilon^2)$ is that the solution of (2.2b) is

$$\begin{aligned} \phi_2(x,t) = \tfrac{1}{2}\tanh x\,\mathrm{sech}^2 x - \tfrac{3}{4} x\,\mathrm{sech}^2 x + c_1 \mathrm{sech}^2 x + \\ - \tfrac{1}{8}\cos(2\sqrt{3}\,t)\big[(J_c(x)\cos px + J_s(x)\sin px)(3-\tanh^2 x) \\ + p\big(J_s(x)\cos px - J_c(x)\sin px\big)\tanh x \\ + 3\,\mathrm{sech}^2 x\tanh x + c_2 G_c(x,p) + c_3 G_s(x,p)\big]\,, \end{aligned} \tag{2.11}$$

where $p^2 = 8$,

$$J_c(x) + i\,J_s(x) = \int_{-\infty}^{x} e^{ip\xi}\,\mathrm{sech}^2\xi\, d\xi\,, \tag{2.12}$$

$$G_c + iG_s = G(x;p)\,,$$

G is defined by (2.10), and (c_1, c_2, c_3) are arbitrary real contants. It is easy to verify that $\phi_2(x,t)$ is uniformly bounded in (x,t).

Thus $\phi_0(x)$, $\phi_1(x,t)$, and $\phi_2(x,t)$ are each uniformly bounded. Higher terms in the series, (1.6a), may exhibit secular growth. In fact, it seems likely that ϕ_3 may contain secular terms, because (1.7) forces the frequency of the oscillation to be constant ($\omega = \sqrt{3}$), whereas the frequency of a nonlinear oscillator typically depends on amplitude.[1] If $\phi_3(x,y)$ turns out to be secular, it suggest that (1.7) should be replaced with

$$\phi_1(x,t;\epsilon) = \mathrm{sech}\, x\tanh x\sin(\omega t)\,, \tag{2.13a}$$

[1] *N. B.* I have not calculated ϕ_3.

$$\omega \sim \sqrt{3}(1+\epsilon^2\omega_1+\cdots)\,, \tag{2.13b}$$

with a corresponding change in (2.2c). If the constant, ω_1, may be chosen to eliminate the secular terms in ϕ_3, then (2.13) indicates the amplitude-dependence of the frequency. This idea goes back to Stokes [10].

It remains to calculate the energy of the stationary and wobbling kinks. Given ϕ_0, ϕ_1, ϕ_2, it is straightforward to verify (1.10). Note that the $O(\epsilon)$-term in H vanishes identically because of (2.2a).

§3.3 Proof of asymptoticity for ϕ^4

In this section we assume that a formal series representation of a wobbling kink solution of (1.4) has been generated, up to and including $\phi_N(x,t;\epsilon)$, and that $(\phi_0,\phi_1,\ldots,\phi_N)$ all are smooth and uniformly bounded in (x,t). We saw in §2 that this is possible at least for $N=1$ and 2. Denote $\Phi_N(x,t;\epsilon)$ as in (1.6b), and let $\hat{\phi}(x,t;\epsilon)$ denote the solution of (1.4) that satisfies initial conditions:

$$\hat{\phi}=\Phi_N\,,\quad \partial_t\hat{\phi}=\partial_t\Phi_N \qquad \text{at } t=0\,; \tag{3.1a}$$

and boundary conditions:

$$\hat{\phi}-\Phi_N\to 0 \qquad \text{as } |x|\to\infty,\ t>0\,. \tag{3.1b}$$

By the usual theory of hyperbolic differential equations, this solutions exists for some finite time, and it is unique. Our objective here is to show that $\hat{\phi}$ exists for rather long time, and that it remains close to Φ_N during this time. Specifically, we will show that there is a positive constant K_N such that if $|t|\le O\left((2K_N)^{-1}\ln(\epsilon^{-1})\right)$, then

$$\hat{\phi}(x,t;\epsilon)=\Phi_N(x,t;\epsilon)+o(\epsilon^N)\,; \tag{3.2}$$

i.e., that the formal series developed in §2.2 actually is asymptotic for a fairly long time. The method of proof is a simple generalization of a method used by Kruskal [11] for ordinary differential equations.

Because Φ_N is know explicitly, it clearly must satisfy some definite differential equation. One shows that

$$\tfrac{1}{2}(\partial_t^2-\partial_x^2)\Phi_N=\Phi_N-\Phi_N^3+\epsilon^{N+1}F_N(x,t;\epsilon)\,, \tag{3.3}$$

where $F_N=O(1)$ as $\epsilon\to 0$. For $N=1$,

$$\epsilon^2F_1=\epsilon^2[3\phi_0\phi_1^2+\epsilon\phi_1^3]\,.$$

For $N=2$,

$$\epsilon^3F_2=\epsilon[6\phi_0\phi_1\phi_2+\phi_1^3]+3\epsilon^4[\phi_0\phi_2^2+\phi_1^2\phi_2]+3\epsilon^5\phi_1\phi_2^2+\epsilon^6\phi_2^3\,.$$

For $N\ge 3$,

$$\begin{aligned}\epsilon^{N+1}F_N=\epsilon^{N+1}3\phi_0\sum_{j=1}^{N}\phi_j\phi_{N+1-j}&+3\phi_0\sum_{n=N+2}^{2N}\epsilon^n\sum_{j=n-N}^{N}\phi_j\phi_{n-j}\\&+\sum_{n=N+3}^{2N}\epsilon^n\sum_{j=1}^{N}\sum_{k=A}^{B}\phi_j\phi_k\phi_{n-j-k}\\&+\sum_{n=2N+1}^{3N}\epsilon^n\sum_{j=n-2N}^{N}\sum_{k=n-N-j}^{N}\phi_j\phi_k\phi_{n-j-k}\,,\end{aligned}$$

where $A=\max\{n-N-j,1\}$, $B=\min\{n-1-j,N\}$. The only important feature of $F_N(x,t;\epsilon)$ for our purposes is that it be uniformly bounded. Thus, we may pick an $M_N<\infty$ such that if $|\epsilon|<1$, then

$$|F_N(x,t;\epsilon)|\le M_N\,. \tag{3.4}$$

Because F_N is a finite sum of uniformly bounded functions, it is clear that M_N exists. For $N=1$, we may choose $M_1=4$.

Define

$$\delta_N(x,t;\epsilon) = \hat{\phi}(x,t;\epsilon) - \Phi_N(x,t;\epsilon)\,. \tag{3.5}$$

Because Φ_N satisfies (3.3), while $\hat{\phi}$ satisfies (1.4) and (3.1), one shows that for $t \geq 0$, δ_N satisfies

$$\tfrac{1}{2}(\partial_t^2 - \partial_x^2)\delta_N = (1 - 3\Phi_N^2)\delta_N - 3\Phi_N\delta_N^2 - \delta_N^3 - \epsilon^{N+1}F_N\,, \tag{3.6a}$$

$$\delta_N \to 0 \qquad \text{as } x \to \infty\,, \tag{3.6b}$$

$$\text{and at } t = 0, \qquad \delta_N = 0, \quad \partial_t\delta_N = 0\,. \tag{3.6c}$$

We must show that (3.6) has a solution that is $o(\epsilon^N)$ over the time scale in question.

Let $R_N(\delta_N; x, t, \epsilon)$ denote the right hand side of (3.6a). $R_N(\delta)$ is differentiable, so it certainly satisfies a Lipschitz condition. Specifically, if

$$|\delta_N| \leq \epsilon^N K, \qquad |\bar{\delta}_N| \leq \epsilon^N K\,,$$

for some $K < \infty$, then there is a $K_N < \infty$ such that

$$|R_N(\delta_N; x, t, \epsilon) - R_N(\bar{\delta}_N; x, t, \epsilon)| \leq K_N^2\,|\delta_N - \bar{\delta}_N|\,. \tag{3.7}$$

The best possible Lipschitz constant is not of interest here. If $\max(\epsilon)$ is small enough, $K_N^2 = 4$, for all $N \geq 1$.

The problem (3.6) is formally equivalent to

$$\delta_N(x,t,\epsilon) = \int_0^t d\tau \int_{x-t+\tau}^{x+t-\tau} d\xi\, R_N(\delta_N; \xi, \tau, \epsilon)\,, \tag{3.8}$$

as the reader may verify directly. To solve (3.8), we consider an iterated mapping,

$$v_{p+1}(x,t,\epsilon) = \int_0^t d\tau \int_{x-t+\tau}^{x+t-\tau} d\xi\, R_N(v_p; \xi, \tau, \epsilon)\,, \tag{3.9}$$

with $v_0 = 0$. First, we will show that the iterated mapping, (3.9), has a fixed point, which is therefore a solution of (3.8). Second, we will bound the magnitude of this fixed point by controlling the rate of convergence of the mapping (3.9). This leads to the desired result, that the solution of (3.8) remains $o(\epsilon^N)$ over the time scale in question.

Because of (3.4),

$$|v_1 - v_0| \leq 2M_N\,\epsilon^{N+1}\frac{t^2}{2!}\,.$$

Because of (3.7),

$$|v_2 - v_1| \leq \int_0^t d\tau \int_{x-t+\tau}^{x+t-\tau} d\xi\, K_N^2\,|v_1 - v_0| \leq 4K_N^2 M_N\,\epsilon^{N+1}\frac{t^4}{4!}\,.$$

In general,

$$|v_p - v_{p-1}| \leq (2K_N^2)^p\left(\frac{M_N}{K_N^2}\epsilon^{N+1}\right)\frac{t^{2p}}{(2p)!}\,, \tag{3.10}$$

which can be established by induction. As $p \to \infty$, holding $(K_N t)$ fixed, $v_p - v_{p-1} \to 0$, by (3.10). Therefore the sequence v_p converges to a fixed point, and this fixed point is a solution of (3.8). With more effort one may also show that this solution is unique and twice-differentiable, so it solves (3.6) as well.

It remains to estimate the magnitude of δ_N. Because

$$\delta_N = v_0 + \sum_{p=1}^{\infty}(v_p - v_{p-1}) ,$$

it follows that

$$\begin{aligned} |\delta_N| &\le \frac{M_N \epsilon^{N+1}}{K_N^2} \sum_{p=1}^{\infty} \frac{(2K_N^2 t^2)^p}{(2p)!} = \frac{M_N \epsilon^{N+1}}{K_N^2}[\cosh(\sqrt{2}K_N t) - 1] \\ &\le \frac{M_N \epsilon^{N+1}}{2K_N^2} \exp(\sqrt{2}K_N t) . \end{aligned} \tag{3.11}$$

Thus, $\delta_N = o(\epsilon^N)$ provided

$$|t| \le \frac{1}{2K_N} \ln \frac{1}{\epsilon} ; \tag{3.12}$$

this is the desired result.

In summary, we have now established that the formal series representation of the wobbling kink, as found in §3.2, is a proper asymptotic series, at least for times up to that given by (3.12). In other words, a solution of (1.4) exists that behaves approximately like a wobbling kink for a relatively long time.

§3.4 The sine-Gordon equation

Next we construct wobbling kink solutions of the sine-Gordon equation, (1.2). It is know that (1.2) is completely integrable [6–8], and that it possesses N-soliton solutions. A wobbling kink for (1.2) is a special case of an N-soliton solution; one may think of it as a nonlinerar superposition of an ordinary kink and a breather. Like all N-soliton solutions, it is intrinsically nonlinear and cannot be found by the naive sort of perturbation theory described in §3.2.

For the sine-Gordon equation, a stationary kink is given by

$$u(x,t) = -\pi + 4\arctan(me^x) , \tag{4.1}$$

where m is an arbitrary real constant. The corresponding scattering data consist of one discretre eigenvalue at $\varsigma = i/2$, and a *norming constant*, m. The Hamiltonian for (1.2) is

$$H = \int dx \left[\tfrac{1}{2}(u_t)^2 + \tfrac{1}{2}(u_x)^2 + \cos u + 1\right] , \tag{4.2}$$

and $H = 8$ for a stationary kink.

A stationary breather is given by

$$u(x,t) = -\pi + 4\arctan[\beta/\alpha \operatorname{sech}(2\beta x + x_0)\sin(2\alpha t + t_0)] , \tag{4.3a}$$

where

$$\alpha^2 + \beta^2 = \tfrac{1}{4} , \tag{4.3b}$$

and (x_0, t_0) are arbitrary real constants. The corresponding scattering data contain two discrete eigenvalues, $\varsigma_1 = \alpha + i\beta$, $\beta > 0$, and $\varsigma_2 = -\varsigma_1^*$. There are also two complex norming constants, m_1 and $m_2 = m_1^*$, whose real and imaginary parts determine x_0 and t_0, respectively. For a stationary breather, $H = 32\beta$, a fact most easily obtained from a trace formula for an N-soliton solution of (1.2) [12]:

$$H = 2i \sum_{j=1}^{N} \left[\frac{1}{\varsigma_j} - \frac{1}{\varsigma_j^*}\right] . \tag{4.4}$$

The scattering data for a wobbling kink solution are composed of scattering data for a breather plus those for a kink. We need three discrete eigenvalues,

$$\zeta_1 = i/2, \qquad \zeta_2 = \alpha + i\beta, \qquad \zeta_3 = -\zeta_2^*, \tag{4.5a}$$

with $\beta > 0$, and $\alpha^2 + \beta^2 = \frac{1}{4}$. We also need three norming constants:

$$m_1 \text{ real}, \qquad m_2, \qquad m_3 = m_2^*. \tag{4.5b}$$

Use these scattering data to define

$$M_{jk} = \frac{im_k}{\zeta_j + \zeta_k} \exp\left[-\tfrac{1}{2} i(\zeta_j + \zeta_k)(x+t) + i(x-t)/4\zeta_k\right]. \tag{4.6}$$

Then

$$u(x,t) = -\pi + 4 \operatorname{Im}\{\ln[\det(I + iM)]\} \tag{4.7}$$

is an exact solution of (1.2) [8]. If $m_2 = 0$, $m_1 \neq 0$, (4.7) reduces to (4.1); if $m_1 = 0$, $m_2 \neq 0$, it reduces to (4.3). With $m_1 \neq 0$, $m_2 \neq 0$, one shows that

$$\begin{aligned}
\det(I + iM) = |m_2|\, e^{2\beta x}\Bigg[&\frac{2|m_2|\alpha}{\beta}\cosh(2\beta x + x_0) \\
&- \frac{2m_1(1-2\beta)\, e^{x}}{(4\alpha^2 + (1+2\beta)^2)} \cos\left\{2\alpha t + \arg\left(m_2^*(\alpha + i[\beta + \tfrac{1}{2}])\right)\right\}\Bigg] \\
+ i|m_2|e^{2\beta x}\Bigg[&2\sin\left\{2\alpha t + \arg\left(m_2^*(\alpha + i\beta)\right)\right\} \\
&+ \frac{m_1}{|m_2|} e^{(1-2\beta)x} + \frac{4m_1|m_2|\alpha^2(1-2\beta)^2}{\beta^2\left(4\alpha^2 + (1+2\beta)^2\right)^2} e^{(1+2\beta)x}\Bigg],
\end{aligned} \tag{4.8}$$

where $\exp(x_0) = |m_2|\alpha/\beta$.

It is easy to verify from (4.7) and (4.8) that this solution actually is a kink, in sense that $u \to \pm\pi$ as $x \to \pm\infty$, respectively. It is apparent from (4.8) that it oscillates in time, with frequency 2α, and that it is stationary in the sense that it has no mean motion. It therefore qualifies as a stationary wobbling kink, although it is not analogous to the ϕ^4 solution discussed in §3.2.

The energy of this solution, according to (4.4), is simply the sum of the energies of the stationary kink and the stationary breather: $H = 8 + 32\beta$, $0 < \beta < \frac{1}{2}$.

§3.5 Stability

We have now established the existence of wobbling kinks, both for ϕ^4 and the sine-Gordon equations. To be of physical interest they should also be stable to perturbations, and this is a separate issue.

The stability of the wobbling kink for ϕ^4 theory seems to be an open question. An extremely limited form of stability can be established by the following generalization of the proof in §3.3.[2] Replace (1.3a) with the initial conditions:

$$\hat{\phi} = \Phi_N + O(\epsilon^{N+1}), \qquad \partial_t\hat{\phi} = \partial_t\Phi_N + O(\epsilon^{N+1}). \tag{5.1}$$

Then essentially the same proof given in §3.3 shows that the formal series for the wobbling kink is asymptotic to an exact solution of (1.4) over a long time scale, perhaps with an adjusted K_N in (3.13). In this form, the

[2]This observation is due to Martin Kruskal.

small terms in (5.1) constitute an arbitrary small perturbation of the initial conditions [*i.e.*, small $= O(\epsilon^{N+1})$], and the argument in §3.3 establishes both existence and stability for the time scale in question. Unfortunately, the usual definition of stability has no such limitation on time scales, and this argument provides no information about longer times.

For the sine-Gordon equation, the following argument, due to David K. Campbell, indicates that the wobbling kinks are probably unstable. Each wobbling kink is composed of a kink and a breather. In order that these two modes not separate from each other in time, it is necessary that the three discrete eigenvalues in (4.5*a*) satisfy

$$|\zeta_1| = |\zeta_2| = |\zeta_3| , \tag{5.2}$$

where ζ_1 is purely imaginary, and $\zeta_3 = -\zeta_2^*$. Consider initial data for the sine–Gordon equation consisting of a wobbling kink, (4.7), plus a small perturbation. The wobbling kink is stable if and only if the discrete eigenvalues of the perturbed scattering data also satisfy (5.2). If the perturbation is real-valued, then $\zeta_3 = -\zeta_2^*$, both for the original and the perturbed scattering data. However, there seems to be no reason why $|\zeta_1| = |\zeta_2|$ should be preserved by an arbitrary small perturbation. If $|\zeta_1| \neq |\zeta_2|$ in the perturbed scattering data, then the breather and the kink will separate eventually, and the kink will cease to wobble. In this sense, the wobbling nature of the wobbling kink is probably unstable for the sine–Gordon equation. However, this instability is a rather mild one, inasmuch as it grows only algebraically rather exponentially.

Acknowledgments

The author is grateful to David K. Campbell for repeatedly explaining the problem to him, and to both Campbell and Martin Kruskal for several valuable conversations.

This work was supported by the U. S. Army Research Office, under Contract #DAAG29-81-C-0009.

References

[1] M. J. Rice, *Phys. Lett.* **71A**, 152, (1979).

[2] D. K. Campbell and A. R. Bishop, *Nuclear Physics* **B 200**, 297, (1982).

[3] W. P. Su, J. R. Scrieffer, and A. J. Heeger, *Phys. Rev. Lett.* **42**, 1698, (1979).

[4] M. J. Rice and E. J. Mele, *Solid State Comm.* **35**, 487, (1980).

[5] D. K. Campbell, J. F. Schonfeld, and C. A. Wingate, Resonance structure in kink-antikink interaction in ϕ^4 theory. (Preprint.)

[6] M. J. Ablowitz, D. J. Kaup, A. C. Newell, and H. Segur, *Phys. Rev. Lett.* **30**, 1262, (1973).

[7] M. J. Ablowitz and H. Segur, *Solitons and the Inverse Scattering Transform.* SIAM, Philadelphia, 1981.

[8] G. L. Lamb, *Elements of Soliton Theory.* Wiley-Interscience, New York, 1980.

[9] I. M. Gel'fand and B. M. Levitan, American Mathematical Society Translation, Series **2**, **1**, 259, (1955).

[10] G. G. Stokes, *Trans. Cambridge Phil. Soc.* **8**, 81 (1847); (Papers **1**, 197).

[11] M. D. Kruskal, *J. Math. Phys.* **3**, 806, (1962).

[12] D. W. McLaughlin, *J. Math. Phys.* **16**, 96, (1975).

Lie Groups and Solutions of Nonlinear Differential Equations

Pavel Winternitz*

Centre de Recherche de Mathématiques Appliquées
Université de Montréal, Canada

Contents:

---

* Supported in part by the Natural Sciences and Engineering Research Council of Canada and by the Fonds FCAC pour l'Aide et le Soutien à la Recherche du Gouvernement du Québec.

Introduction

The last 15 years or so have seen an explosion of activity in the study of nonlinear phenomena. The motivation has largely come from the physical sciences and engineering, very often from completely practical problems, such as building oil rigs and other constructions in oceans, constructing tokamaks and other devices for containing and heating plasma, etc. The interesting phenomenon of solitons —stable solitary waves [4,13,16,31,32,87,121,139]— is making its appearance in fields ranging from hydrodynamics (where it all started) to elementary particle physics (which by now has joined general relativity in being a fundamentally nonlinear sphere of scientific activity).

Mathematical progress in the treatment of nonlinear differential equations has been swift and broadly based. The most spectacular developments have been in the field of exact analytic solutions of nonlinear equations —the application of linear methods to solve nonlinear problems [1,2,4,56,87,88,139–143].

These lectures will be devoted to the application of group theory to the study of certain types of nonlinear equations. It is useful to recall that the theory of Lie groups [89–92] was originally developed in connection with the theory of differential equations —both linear and nonlinear ones. The emphasis on linear equations is probably due to the fact that they are easier to handle and that hence progress has been faster. By now nonlinar phenomena have caught up with us and new mathematical methods are catching up with nonlinear differential equations.

In more recent times several books have been devoted to group theory and differential equations (see *e.g.*, [13,21,57,74,96,106]) or at least have sections treating the subject in some detail [6–8,44].

Among the applications of group theory in the theory of differential equations, let us mention:

1. The classification of Ordinary Differential Equations (ODE's) and Partial Differential Equations (PDE's) into classes according to the symmetry group they admit. Such a classification in particular helps to identify nonlinear equations that can be linearized by a transformation of variables and, indeed, Lie group theory provides the tools for such a linearization [46–49,85,125].

2. An application often arises in physics when new theories are being developed, *i.e.*, when the equations of motion describing a system are not yet known; the symmetries of the system, however, are at least to some degree understood. The requirement that the equations be invariant under the symmetry group of the system poses strong restrictions on their form. In some important cases, *e.g.* the conformal group acting in Minkowski space, the underlying symmetries are nonlinear ones.

3. The question of symmetry breaking, either explicit or spontaneous, should be mentioned in this context. Given a physical system described by some differential equations invariant under a Lie group G, we ask how the system can be modified, either by including further interaction terms in

the equation or by imposing specific boundary conditions or other supplementary conditions on the solutions. In general such a modification will involve symmetry breaking, *i.e.*, the reduction from G to a subgroup $H \subset G$. A classification of subgroups here provides a classification of possible types of symmetry breaking (or *directions of symmetry breaking*). See [18,30,108–112,135].

4. Once a specific equation or system of equations is agreed upon and does allow for a nontrivial symmetry group, then this group can be used to classify solutions, to generate new solutions from known ones, and also to identify and construct particularly simple solutions [17,20,50,51,54,64–68,83,105,127].

5. For PDE's, the use of group theory and differential geometry has made it possible to provide a systematic theory of the separation of variables, both in the case of linear and nonlinear equations [26–29,77–79,96–98,111,137,138]. This is the subject of Willard Miller's talks at this School and Workshop. These methods are particularly powerful in the physically interesting cases of Hamilton–Jacobi and Schrödinger equations in spaces with nontrivial isometry groups.

6. The problem of *symmetry reduction* for PDE's, *i.e.*, the problem of introducing new independent variables that reduce the considered PDE to a lower-order PDE or to an ordinary differential equation, finds an elegant solution in terms of group theory [60,61].

7. The systematic construction of Bäcklund transformations for certain types of nonlinear PDE's, leads to a realization of finite-dimensional Lie algebras in terms of vector fields in some fixed finite number of variables [45,69,71,80,99–101,113,120,129].

8. Linear equations always allow for linear superposition principles. Thus, the general solution of a finite system of linear ODE's can be expressed in terms of a finite number of particular solutions. Surprisingly, this last property is shared by certain systems of nonlinear ODE's. The theory of Lie groups and Lie algebras makes it possible to identify and construct systems of nonlinear ODE's for which such nonlinear superposition principles exist and also to construct the superposition formulae explicitly [9,11,12,70,136].

In these lectures we shall concentrate on two of the above aspects. Thus Chapter 1 is devoted to superposition principles for systems of nonlinear ordinary ODE's. In Chapter 2 we analyze the problem of symmetry reduction for systems of relativistically invariant PDE's in $n+1$ dimensions. Other problems, in particular the construction and use of Bäcklund transformations are also touched upon, though not treated systematically.

The results presented here which are original, are the result of various scientific collaborations. Thus the work on superposition principles and fundamental sets of solutions was conducted together with my colleagues Robert L. Anderson, John Harnad and Steve Schnider, and with the students Claude Dubois and David Rand. The results on symmetry reduction for nonlinear Klein–Gordon equations were obtained with Michal Grundland and John Harnad. I thank all of them for the collaboration and for helpful discussions in the preparation of these notes.

Chapter 1

Superposition Principles and Fundamental Sets of Solutions for Nonlinear Ordinary Differential Equations

§1.1 Formulation of the problem

For a linear Ordinary Differential Equation (ODE), or a system of such equations, a linear superposition principle allows us to represent the general solution as a linear combination in terms of a finite number of particular solutions.

It makes sense to ask whether this property is restricted to linear ODE's or whether it can be generalized. This question was posed by several mathematicians around 1893: S. Lie [92], M. E. Vessiot [128], and others (see [6,7,126]). The problem can be formulated as follows. Consider a system of first-order ODE's

$$\frac{d\mathbf{u}}{dt} = \boldsymbol{\eta}(\mathbf{u}, t) \quad i.e. \quad \frac{du^\mu}{dt} = \eta^\mu(\mathbf{u}, t), \qquad 1 \leq \mu \leq n.$$

We ask: when can the general solution of this system be expressed as a function of a finite number m of particular solutions $\mathbf{u}_1(t), \mathbf{u}_2(t), \ldots, \mathbf{u}_m(t)$ and a sufficient number n of arbitrary constants $c_1, c_2, \ldots, c_n$:

$$\mathbf{u}(t) = \mathbf{F}(\mathbf{u}_1(t), \mathbf{u}_2(t), \ldots, \mathbf{u}_m(t), c_1, c_2, \ldots, c_n).$$

The answer to this question was given by Sophus Lie in his lecture notes *Vorlesungen über Continuierliche Gruppen* [92].

§1.2 Lie's theorem on fundamental sets of solutions

The system of equations

$$\frac{d\mathbf{u}}{dt} = \boldsymbol{\eta}(\mathbf{u}, t) \qquad \mathbf{u}, \boldsymbol{\eta} \in R^n \text{ or } C^n, \tag{1}$$

admits a *fundamental set of solutions* $\mathbf{u}_1(t), \mathbf{u}_2(t), \ldots, \mathbf{u}_m(t)$, such that its general solution is given by the *nonlinear superposition rule*

$$\mathbf{u}(t) = \mathbf{F}(\mathbf{u}_1(t), \mathbf{u}_2(t), \ldots, \mathbf{u}_m(t), c_1, c_2, \ldots, c_n), \tag{2}$$

where the c_i are arbitrary (significant) constants, *if and only if* the following conditions hold:

i. The right-hand side of (1) can be (possibly after a transformation) factorized to give

$$\frac{d\mathbf{u}}{dt} = Z_1(t)\boldsymbol{\xi}_1(\mathbf{u}) + Z_2(t)\boldsymbol{\xi}_2(\mathbf{u}) + \cdots + Z_r(t)\boldsymbol{\xi}_r(\mathbf{u}). \tag{3}$$

ii. The vector fields

$$\hat{X}_i = \sum_{\mu=1}^{n} \xi_i^{\mu}(\mathbf{u})\frac{\partial}{\partial u^{\mu}}, \tag{4}$$

generate a finite-dimensional Lie algebra

$$[\hat{X}_a, \hat{X}_b] = \sum_{c=1}^{r} C_{ab}^{c}\hat{X}_c, \qquad 1 \leq a, b, c \leq r. \tag{5}$$

The number m of solutions in the fundamental set satisfies

$$n \cdot m \geq r. \tag{6}$$

We shall omit the proof here. See [92].

As a first example consider the case of $n = 1$, *i.e.*, one ordinary first-order equation for a real function $u(t) \in R$. According to Lie's theorem, we must construct a finite-dimensional Lie algebra using operators of the type

$$X_a = \xi_a(u)\frac{d}{du}, \qquad u \in R. \tag{7}$$

It is well known [91] that the only finite-dimensional Lie algebras which can be constructed in terms of vector fields in one real variable (7) are *sl(2,R)* and its subalgebras. The corresponding realization of *sl(2,R)* is unique —up to a change of variables $u = \phi(y)$— and is given by

$$X_1 = \frac{d}{du}, \qquad X_2 = u\frac{d}{du}, \qquad X_3 = u^2\frac{d}{du}. \tag{8}$$

The equation can be written using Lie's theorem and is

$$\frac{du}{dt} = Z_1(t) + Z_2(t)u + Z_3(t)u^2, \tag{9}$$

where $Z_i(t)$, $i = 1, 2, 3$ are arbitrary functions of the independent variable t. This is the well-known and very important Riccati equation. Putting $u = \phi(y)$, where ϕ is some differentiable function, we obtain an equivalent realization of *sl(2,R)* and a "disguised" Riccati equation

$$\frac{dy}{dt} = Z_1(t)\frac{1}{\dot{\phi}(y)} + Z_2(t)\frac{\phi(y)}{\dot{\phi}(y)} + Z_3(t)\frac{\phi^2(y)}{\dot{\phi}(y)}. \tag{10}$$

The superposition principle for the Riccati equation is well known. Indeed, any four solutions of (9), *e.g.*, u, u_1, u_2, and u_3, satisfy the property that their anharmonic ratio is constant:

$$\frac{u-u_1}{u-u_2}\frac{u_3-u_2}{u_3-u_1}=c. \tag{11}$$

Considering $u(t)$ to be the general solution and u_1, u_2, and u_3 to be particular solutions, we can solve (11) for u and obtain the superposition law

$$u(t)=\frac{(u_1-u_3)u_2c+u_1(u_3-u_2)}{(u_1-u_3)c+(u_3-u_2)}, \tag{12}$$

where $c \in R$ is a constant related to the initial value $u(t_0)$ of the solution $u(t)$.

The result seems to be very limited: for $n=1$ only one nonlinear ODE satisfies Lie's theorem, namely the Riccati equation (9). Proper subalgebras of *sl(2,R)* lead to linear inhomogeneous equations [if $Z_3(t)=0$, *i.e.*, $\hat{X}_3$ is absent in the subalgebra], or linear homogeneous equations [if $Z_1=Z_3=0$, *i.e.*, $\hat{X}_1$ and $\hat{X}_3$ are absent].

Let us stress two important points:

1. The Riccati equation and the algebra *sl(2,R)* [or *sl(2,C)*] come up again and again in *soliton physics* in connection with Bäcklund transformations [13,15,19,87] and inverse scattering equations for the Korteweg–de Vries equation, sine–Gordon equation, nonlinear Schrödinger equation, and many others. The algebra *sl(2,R)* underlies the AKNS version [1,2,4] of the inverse scattering method of solving the above PDE's and many others. See the lectures of Mark Ablowitz in these Proceedings.

2. For $n \geq 2$, infinitely many Lie algebras can be written in terms of vector fields in n variables. Indeed, consider $n=2$. In addition to three different semisimple Lie algebras *sl(3,R)*, *o(3,1)*, and *o(2,2)* (corresponding to *sl(2,R)* $\sim$ *o(2,1)* for $n=1$) we can realize families of Lie algebras with arbitrarily large abelian ideals. Indeed, S. Lie has classified all Lie algebras which can be realized in terms of the operators

$$\hat{X}_\alpha=\xi_\alpha(x,y)\frac{\partial}{\partial x}+\eta_\alpha(x,y)\frac{\partial}{\partial y}, \tag{13}$$

where ξ_α and η_α are arbitrary (differentiable) functions [91]. A typical example is given by

$$\begin{aligned}&\hat{X}_1=\partial_x,\quad \hat{X}_2=x\partial_x+\tfrac{1}{2}ny\partial_y,\quad \hat{X}_3=x^2\partial_x+nxy\partial_y,\\ &\hat{Y}_1=\partial_y,\quad \hat{Y}_2=x\partial_y,\ \cdots,\hat{Y}_{n+1}=x^n\partial_y,\quad \hat{Z}=y\partial_y.\end{aligned} \tag{14}$$

The operators $\hat{Y}_i$, $i=1,2,\ldots,n+1$ form an abelian ideal. It turns out that this example is not particularly interesting. Indeed the pair of nonlinear ODE's associated with the algebra (14) by Lie's theorem —see (3) and (4)— is

$$\frac{dx}{dt}=Z_1(t)+Z_2(t)x+Z_3(t)x^2, \tag{15a}$$

$$\frac{dy}{dt}=Z_2(t)\tfrac{1}{2}ny+Z_3(t)nxy+Z_4(t)+\cdots+Z_{4+n}(t)x^n+Z_{5+n}(t)y. \tag{15b}$$

We see that (15) splits into a Riccati equation for $x(t)$ and an equation for $y(t)$ which becomes linear once $x(t)$ is found from (15a) and substituted into (15b). Similar comments hold for the other non-semisimple algebras in Lie's list. On the other hand, the semi-simple Lie algebras *sl(3,R)*, *o(3,1)*, and *o(2,2)* lead to interesting systems of coupled Riccati equations: projective and conformal systems of Riccati equations [9,11,12].

§1.3 Systematic construction of ordinary differential equations with superposition principles

1.3.1 Direct approach

The most direct approach would be to make use of Lie's criterion for a fixed number of real or complex variables $u_1, u_2, \ldots, u_n$ and to classify and construct all Lie algebras which can be realized in terms of the operators

$$\hat{X}_a = \sum_{\mu=1}^{n} \xi_a^\mu(\mathbf{u}) \frac{\partial}{\partial u^\mu}. \tag{16}$$

Notice in particular that if the coefficients of the vector field are linear polynomials

$$\xi_a^\mu(\mathbf{u}) = \sum_{\nu=1}^{n} A_a^{\mu\nu} u_\nu + B_a^\mu, \qquad \mu, \nu = 1, 2, \ldots, n, \quad a == 1, 2, \ldots, r, \tag{17}$$

where $A_a^{\mu\nu}$ and B_a^μ are constants, then Eqs. (3) are linear:

$$\frac{du^\mu}{dt} = \sum_{a=1}^{r} Z_a(t) \left(\sum_{\nu=1}^{n} A_a^{\mu\nu} u_\nu + B_a^\mu \right). \tag{18}$$

In particular, if the B_a^μ vanish, we obtain a system of linear homogeneous equations for which tthe superposition law (2) reduces to an ordinary linear combination. The most general linear homogeneous equation is obtained when $r = n^2$ in (17) and $B_a^\mu = 0$. The underlying algebra is $gl(n,F)$ ($F = R$ or C), and its realization is the defining linear representation on the homogeneous manifold $GL(n,F)/Aff(n\text{-}1,F)$. The trivial case of linear homogeneous or inhomogeneous equations is thus automatically included.

The algebras of interest are those realized by the operators (16) in which $\xi_a^\mu(u)$ are nonlinear functions of u^i and which cannot be linearized by an invertible transformation of variables

$$u^\mu = \phi^\mu(y^1, y^2, \ldots, y^n), \qquad \mu = 1, 2, \ldots, n. \tag{19}$$

To proceed directly by dimension and to classify the obtained Lie algebras turns out to be a very difficult task, which was attempted by Lie himself. For $n = 1$ the problem and result is trivial, for $n = 2$ it is already quite difficult (some 40 pages in Lie's book [91]), for $n \geq 3$ I do not even know whether a classification exists. The direct approach has thus turned out not to be very fruitful.

1.3.2 Structural approach via representations of classical Lie groups

An alternative approach which we shall pursue is a structural one. Thus, restrictions can be impposed either on the form of the coefficients $\xi_a^\mu(u)$, or on the type of Lie algebra to be considered; *e.g.*, we may restrict ourselves to $\xi_a^\mu(u)$ which are polynomials, or even only quadratic polynomials. This leads directly to interesting mathematics, in particular the theory of filtered Lie algebras and transitive differential geometry. See below and in Refs. [58,82,103,122].

A related possibility is to start directly from some Lie group G and some manifold M on which G acts in a nonlinear manner, and to calculate the infinitesimal action of G. This directly yields the vector fields (16) which by construction provide the basis for a Lie algebra, namely, the Lie algebra corresponding to the Lie group G. The nonlinear equations satisfying Lie's criterion can then be read off directly:

$$\begin{aligned}\frac{du^\mu}{dt} &= \sum_{i=1}^{r} Z_i(t)\hat{X}_i u^\mu \\ &= \sum_{i=1}^{r} Z_i(t) \sum_{\nu=1}^{n} \xi_i^\nu(\mathbf{u})\frac{\partial u^\mu}{\partial u^\nu} \\ &= \sum_{i=1}^{r} Z_i(t)\xi_i^\mu(\mathbf{u}).\end{aligned} \tag{20}$$

Questions which arise in this context are:

1. How does one choose the group G and the manifold M on which G acts in a nonlinear manner?
2. How does one obtain the superposition formulas explicitly?
3. Does one obtain interesting equations in this manner?
4. Are the obtained superposition formulae useful?

Our present answer to question **1** is as follows. Consider a Lie group G acting transitively on a homogeneous space M. The space M can always be identified with a quotient space

$$M \sim G/G_0, \tag{21}$$

where G_0 is a subgroup of G, namely the isotropy group of a point $p \in M$, which we choose to be the origin [72]. We shall let G run through the complex and real classical semisimple Lie groups *SL(n,C), O(n,C), Sp(2n,C), SL(n,R), SL(n,H), U(p,q), O(p,,q),O*(2n), Sp(2n,R), SP(p,q)* (see Helgason [72]), and G_0 through their maximal subgroups. It turns out that with these restrictions on G and G_0 the problem becomes manageable and we obtain a multitude of interesting new results.

Whether the restriction to G semisimple means that interesting systmes of equations with superposition principles are left out, is still an open question. (See however [122]). The second question raised above is answered in the following Section.

§1.4 Methods of deriving superposition formulae

Let us consider the system of equations

$$\frac{d\mathbf{u}}{dt} = \boldsymbol{\xi}(\mathbf{u},t) = \sum_{a=1}^{r} Z_i(t)\boldsymbol{\xi}_i(\mathbf{u}) \equiv \mathbf{Y}(y), \tag{22}$$

satisfying the conditions (4) and (5) of Lie's theorem. The right-hand side of (22) determines a curve

$$Y(t) = \sum_{\mu=1}^{n} \xi^\mu(\mathbf{u},t)\frac{\partial}{\partial u^\mu} \tag{23}$$

in the Lie algebra L corresponding to the Lie group G. For t fixed, Y is an element of L; as t changes, $Y(t)$ remains in L following some curve (not, in general, corresponding to any one-parameter subalgebra). Equation (23) is the local coordinate representation of this curve for vector fields on a manifold M induced by the infinitesimal action of the Lie group G, in a suitable neighbourhood of the identity element $e \in G$.

The general solution of (22) can be written as

$$\mathbf{u}(t) = g(t)\,\mathbf{u}(t_0), \tag{24}$$

where $g\mathbf{u}$ denotes the image of $\mathbf{u}$ under $g \in G$, and $g(t)$ is the (unique) curve in the Lie group G satisfying the equation

$$\frac{dg(t)}{dt} g^{-1}(t) = Y(t), \qquad g(t_0) = g_0, \tag{25}$$

for some arbitrary fixed initial g_0.

Our basic formula is (24) and from here we proceed to derive a superposition rule. This rule will have the form (24) in which the path $g(t)$ in the group G is reconstructed in terms of a finite number m of particular solutions.

Indeed, let us assume that we know m solutions

$$\mathbf{u}_1(t), \mathbf{u}_2(t), \ldots, \mathbf{u}_m(t), \tag{26}$$

satisying the initial conditions

$$\mathbf{u}_k(t_0) = \mathbf{u}_{k0}, \qquad k = 1, 2, \ldots, m. \tag{27}$$

Formula (24) tells us how each of the m points $\mathbf{u}_{k0}$ transforms under the action of the group G. To be more precise, we are considering the action of G on the cartesian product $[M]^n$ of n copies of the space M. If the action of G on $[M]^n$ is a free one, *i.e.*, the joint stabilizer $G_0^m \subset G$ of all the initial conditions (27) is only the identity subgroup $G_0^m = e$, then we can reconstruct the element $g(t)$, *i.e.*, the path $g(t)$ in G, from the equations

$$\mathbf{u}_k(t) = g(t)\mathbf{u}_k(t_0), \qquad k = 1, 2, \ldots, m. \tag{28}$$

The (optimal) number m of solutions in a fundamental set of solutions is thus the smallest number of solutions for which we have

$$G_0^m = \{g \mid g\mathbf{u}_k(t_0) = \mathbf{u}_k(t_0), k = 1, 2, \ldots, m\} = \{e\}. \tag{29}$$

Relation (29) provides us with the number m of solutions in a fundamental set and with the *independence* conditions on admissible solutions. Indeed, the m solutions must be such that their joint stabilizer G_0^m is the identity element $e \in G$ and any subset of the solutions $\mathbf{u}_1(t)$, $\mathbf{u}_2(t), \ldots, \mathbf{u}_m(t)$ must have a larger stabilizer.

Our ***prescription*** for obtaining the superposition law is thus the following:

1. Identify the finite group action (24) which gives rise to the local representation (23) of the vector fields $\mathbf{Y}(t)$.

2. Determine how many copies m of the manifold M must be taken in order for the action (24) of $g(t)$ to be a free one, and determine the independence conditions to be imposed on $\mathbf{u}(t_0)$ for (29) to hold. We shall call any set of m solutions satisying the established independence conditions a ***fundamental set of solutions***.

3. Solve equations (28) to express $g(t)$ in terms of the properly chosen particular solutions $\mathbf{u}_1(t)$, $\mathbf{u}_2(t), \ldots, \mathbf{u}_m(t)$:

$$g(t) = f(\mathbf{u}_1(t), \mathbf{u}_2(t), \ldots, \mathbf{u}_m(t)). \tag{30}$$

Comments:

1. Once the group element $g(t)$ in (24) is properly parametrized, solving equations (28) becomes an algebraic problem. In some cases it is advantageous to "prepare" a particularly convenient set of "standardized" fundamental solutions $u_i^s(t)$ and initial conditions $u_i^s(t_0)$ by applying a properly chosen constant transformation $h_0 \in G$ to the given solutions:

$$\mathbf{u}_i^s(t) \equiv h_0 \mathbf{u}_i(t), \qquad \mathbf{u}_i^s(t_0) \equiv h_0 \mathbf{u}_i(t_0), \qquad i = 1, 2, \ldots, m. \tag{31}$$

This amounts to reconstructing $[h_0 g(t) h_0^{-1}] \equiv \tilde{g}(t)$ instead of $g(t)$, *i.e.*, changing the (irrelevant) initial condition g_0 in (25).

2. The fundamental set of solutions (26) is an arbitrary set of solutions satisfying the independence conditions,, established using the requirement (29). The independence conditions exclude sets which satisfy certain algebraic constraints, *i.e.*, the allowed sets form a generic m-tuplet of solutions. If the given solutions do not lie on a generic orbit then more than the minimal number m may be needed in order to reconstruct the transformation $g(t)$. In order to assure stability of the results with respect to small perturbations of the initial data it is important to choose $\mathbf{u}_1(t_0), \mathbf{u}_2(t_0), \ldots, \mathbf{u}_{m_0}(t_0)$ on a generic orbit of solutions.

3. It is sometimes advantageous to use a modification of the procedure described above to obtain a superposition formula. Instead of reconstructing the group element $g(t)$ completely from m solutions, we can use a smaller number of particular solutions $\mathbf{u}_i(t)$, $i = 1, 2, \ldots, m_0 < m$; let these solutions be such that the simultaneous stabilizer $G_0^{m_0}$ of their initial conditions $\mathbf{u}_1(t_0), \mathbf{u}_2(t_0), \ldots, \mathbf{u}_{m_0}(t_0)$, acts linearly on M in suitably defined local coordinates. We may then use $\mathbf{u}_i(t)$, $i = 1, 2, \ldots, m_0$, to reconstruct the coset $g(t)G_0$. This is equivalent to reducing the considered set of equations to a set of linear ordinary differential equations, for which we can then write a linear superposition formula. This *linearization* of the original equations is achieved by a change of the dependent variables; the transformation depends on the m_0 known solutions [11].

4. It is always possible to replace the system of nonlinear differential equations (22) satisfying Lie's criterion by a system of linear equations (the linear system is in general higher-dimensional and has a larger solution space). This can be achieved by considering the action of the group G of (21) either on the group G itself, or on some space $M' \sim G/H$ ($H \subset G_0 \subset G$) on which G acts linearly. Note that the dimension of M' is larger than that of M. We then determine the vector fields corresponding to the infinitesimal action of G on M', and using (20), associate a system of linear ODE's with this action, corresponding to the same curve $\mathbf{Y}(t)$ in the Lie algebra.

An example of such a relationship is provided by the Riccati equation (9):

$$\frac{du}{dt} = a(t) + b(t)u + c(t)u^2, \tag{32}$$

corresponding to $M \sim G/G_0$, $G = SL(2,R)$, $G_0 = \left\{\begin{pmatrix} \alpha & 0 \\ \beta & \alpha^{-1} \end{pmatrix}\right\}$. If we replace M by

$$M' \sim G/H, \qquad G = SL(2,R), \qquad H = \left\{\begin{pmatrix} 1 & 0 \\ \beta & 1 \end{pmatrix}\right\}, \tag{33}$$

we obtain the linear (defining) representation of $SL(2,R)$ and the linear equations

$$\begin{pmatrix} \dot{x} \\ \dot{y} \end{pmatrix} = \begin{pmatrix} \frac{1}{2}b(t) & a(t) \\ -c(t) & -\frac{1}{2}b(t) \end{pmatrix} \begin{pmatrix} x \\ y \end{pmatrix}. \tag{34}$$

Solutions u of the Riccati equation (32) are obtained from solutions (x, y) of the linear equations (34) by putting

$$u = \frac{x}{y}, \qquad y \neq 0. \tag{35}$$

The relation between the linear system [in the example Eq. (34)] and the original nonlinear one should not be confused with the *linearization* of a nonlinear equation by a transformation of the dependent variables that does not change the solution space of the equation [46–49,85]. It is however always possible to obtain solutions of the nonlinear system from those of the linear one. For further details see the original articles [9,11,12,70].

§1.5 The group $SL(N,C)$ and complex matrix Riccati equations

1.5.1 The complex Grassmannian $G_k(C^{n+k})$

Let us now realize the program described in §1.3 and §1.4 for the case when $G = SL(n+k,C)$, where $n+k = N$, n and k are positive integers and G_0 is the group of block-triangular matrices

$$G_0 = \left\{ g \mid g = \begin{pmatrix} G_{11} & 0 \\ G_{21} & G_{22} \end{pmatrix} \right\}, \qquad g_{11} \in GL(n,C), \quad G_{22} \in GL(k,C), \quad G_{21} \in C^{k\times n}, \tag{36}$$

leaving a k-dimensional vector space in C^n invariant; g_0 is a maximal parabolic subgroup of $SL(n+k,C)$. Thus we have

$$g \in G, \quad g = \begin{pmatrix} M & N \\ P & Q \end{pmatrix}, \quad M \in C^{n\times n}, \quad N \in C^{n\times k}, \quad P \in C^{k\times n}, \quad Q \in C^{k\times k}. \tag{37}$$

The homogeneous space $M \sim G/G_0$ has complex dimension $(n+k)^2 - n^2 - k^2 - nk = nk$ and can be identified with the Grassmann manifold $G_k(C^{n+k})$ of k-planes in C^{n+k}.

Let us now introduce two types of coordinates on the Grassmann manifold C^{n+k}. ***Homogeneous coordinates*** of a point $p \in C^{n+k}$ are given by the components of an $(n+k) \times k$-dimensional matrix

$$\begin{pmatrix} X \\ Y \end{pmatrix}, \qquad X \in C^{n\times k}, \quad Y \in C^{k\times k}, \tag{38}$$

whose columns span the k-plane defining p. The point p is identified with the equivalence class $\left[\begin{pmatrix} X \\ Y \end{pmatrix}\right]$ under the relation

$$\begin{pmatrix} X \\ Y \end{pmatrix} \sim \begin{pmatrix} XG \\ YG \end{pmatrix}, \qquad G \in GL(k,C), \tag{39}$$

identifying different bases for the same space. The action of an element $g \in SL(n+k,C)$ on the Grassmannian in homogeneous coordinates is obtained from the linear action

$$g : \begin{pmatrix} X \\ Y \end{pmatrix} \mapsto \begin{pmatrix} X' \\ Y' \end{pmatrix} = \begin{pmatrix} M & N \\ P & Q \end{pmatrix} \begin{pmatrix} X \\ Y \end{pmatrix} \tag{40}$$

by the projection $\pi : \begin{pmatrix} X \\ Y \end{pmatrix} \mapsto \left[\begin{pmatrix} X \\ Y \end{pmatrix}\right]$.

Now consider the affine subspace defined by the conditon $\det Y \neq 0$, and introduce the ***affine coordinates***

$$W = XY^{-1} \in C^{n\times k}. \tag{41}$$

The action of $SL(n+k,C)$ in affine coordinates is given by the matrix fractional linear transformations

$$g : W \mapsto W' = X'Y'^{-1} = (MW+N)(PW+Q)^{-1}. \tag{42}$$

The infinitesimal action is given by the homomorphism

$$\phi : sl(n+k,C) \to \chi(G_k(C^{n+k})) \tag{43}$$

from the Lie algebra $sl(n+k,C)$ to the algebra of smooth vector fields on $G_k(C^{n+k})$ defined by

$$\begin{gathered} \phi(\xi)f(p) = -\frac{d}{dt}f(e^{t\xi}p)\bigg|_{t=0}, \\ \xi \in sl(n+k,C), \qquad f \in C^\infty(G_k(C^{n+k})). \end{gathered} \tag{44}$$

In affine coordinates the image of

$$\xi = \begin{pmatrix} C & A \\ -D & -B \end{pmatrix} \qquad A \in C^{n\times k}, \quad B \in C^{k\times k}, \quad C \in C^{n\times n}, \quad D \in C^{k\times n}, \quad \mathrm{Tr}\, C - \mathrm{Tr}\, B = 0, \tag{45}$$

is the vector field

$$\phi(\xi) = -[A_{\alpha\mu} + W_{\alpha\nu}B_{\nu\mu} + C_{\alpha\beta}W_{\beta\mu} + W_{\alpha\nu}D_{\nu\beta}W_{\beta\mu}]\frac{\partial}{\partial W_{\alpha\mu}}. \tag{46}$$

1.5.2 The matrix Riccati equation

Now consider a curve in $sl(n+k,C)$:

$$\xi(t) = \begin{pmatrix} C(t) & A(t) \\ -D(t) & -B(t) \end{pmatrix}, \tag{47}$$

In homogeneous coordinates we can associate a system of linear equations to this curve

$$\begin{pmatrix} \dot X(t) \\ \dot Y(t) \end{pmatrix} = \begin{pmatrix} C(t) & A(t) \\ -D(t) & -B(t) \end{pmatrix} \begin{pmatrix} X(t) \\ Y(t) \end{pmatrix}. \tag{48}$$

In affine coordinates we can read off the corresponding equations from formula (46) or, equivalently, diffferentiate (41) and use (48) to obtain

$$\dot W = \dot X Y^{-1} - XY^{-1}\dot Y Y^{-1} = A + WB + CW + WDW.$$

Thus, taking $G = SL(n+k,C)$ and G_0 as a maximal parabolic subgroup of $SL(n+k,C)$, we have obtained a system of coupled Riccati equations: the *matrix Riccati equation* (MRE):

$$\begin{gathered} \dot W = A + WB + CW + WDW, \\ W \in C^{n\times k}, \quad A \in C^{n\times k}, \quad B \in C^{k\times k}, \quad C \in C^{n\times n}, \quad D \in C^{k\times n}. \end{gathered} \tag{49}$$

Equation (49) defines the flow of the time-dependent vector field $\phi(\xi(t))$. [See (46) and (47).] A vast literature exists on this equation (see *e.g.* [36,40,84,118,123,133]) and it occurs in many applications —optimal control theory, Bäcklund transformations for a multitude of different nonlinear partial differential equations [33,62,63,104,119,139,140], plasma physics, etc.

According to the method outlined in §1.4, the general solution of the matrix Riccati equation is

$$W(t) = [M(t)U + N(t)][P(t)U + Q(t)]^{-1}, \tag{50}$$

where $U \in C^{n \times k}$ is a constant matrix related to the initial conditions for $W(t)$ and the curve

$$g(t) = \begin{pmatrix} M(t) & N(t) \\ P(t) & Q(t) \end{pmatrix}, \tag{51}$$

in $SL(n+k, C)$, is the solution to

$$\begin{pmatrix} \dot{M} & \dot{N} \\ \dot{P} & \dot{Q} \end{pmatrix} = \begin{pmatrix} C & A \\ -D & -B \end{pmatrix} \begin{pmatrix} M & N \\ P & Q \end{pmatrix}, \tag{52}$$

for some (arbitrarily fixed) initial condition

$$g(t_0) = \begin{pmatrix} M_0 & N_0 \\ P_0 & Q_0 \end{pmatrix}. \tag{53}$$

We must now obtain M, N, P, and Q in terms of a ***fundamental set of solutions*** of the MRE (49).

So far we were considering the general rectangular MRE, *i.e.*, k and n were arbitrary. A superposition principle can be obtained in the general case [124] but it becomes particularly nice in two limiting cases, namely $k = 1$ (projective Riccati equations [9,11,12]) and $k = n$ (square matrix Riccati equations [70]). The superposition rule for n projective Riccati equations ($k = 1$) involves $n + 2$ particular solutions.

Here we shall concentrate on the square matrix Riccati equation $n = k$. Surprisingly, in this case the number of solutions in the fundamental set does not depend on the dimension of the matrices. Indeed, for $n \geq 2$, only *five* particular solutions, generically chosen, are sufficient to reconstruct the group element $g(t)$ in (50) and (51).

To summarize, choosing the group G in (21) to be $SL(N, C)$ and $G_0 \subset G$ to be one of its maximal parabolic subgroups (36), we have arrived at an important system of nonlinear ODE's, the matrix Riccati equation (49). By construction it satisfies the conditions of Lie's theorem and we shall proceed to derive a superposition law

Comment. Returning from affine coordinates W to homogeneous coordinates (X, Y) we associate a system of linear differential equations (48) to the MRE (49). The linear equations have a larger solution set than the MRE, since the general solution of this system of $(n+k) \times k$ equations depends on $(n+k) \times k$ constants, whereas the general solution of the MRE (49) depends on $n \times k$ constants. Thus, each solution (X, Y) of (48) generates a solution

$$W = XY^{-1}$$

of the MRE. Two different solutions of the linear system, related as in (39), generate the same solution of the MRE. This is an example of the relation between a linear and nonlinear system of equations discussed at the end of §1.4.

1.5.3 Fundamental sets of solutions of the square matrix Riccati equation

We are following the prescription given in §1.4 in order to obtain a superposition law for the MRE (49). Step **1** has been completed in that formula (50) identifies the group action of $G = SL(2n, C)$

on the complex Grassmannian $M \sim G_n(C^{2n})$, leading to the vector field (46). Formula (50) corresponds to (24) in the general case.

We now proceed to Step **2**, *i.e.*, to identify the *fundamental set of solutions* of the MRE. This amounts to determining the conditions that these solutions must satisfy.

Let $W_1(t), W_2(t), \ldots, W_m(t)$ be known solutions of the MRE, satisfying the initial conditions:

$$W_i(0) = U_i, \qquad U_i \in C^{n\times n}, \quad i = 1, 2, \ldots, m. \tag{54}$$

The condition on the set $\{W_i\}$ is that the simultaneous stabilizer $G_0^m \subset SL(2n, C)$ of the m initial condition matrices be the identity transformation only, *i.e.*, the equations

$$U_i = (MU_i + N)(PU_i + Q)^{-1}, \qquad i = 1, 2, \ldots, m, \tag{55}$$

should imply

$$\begin{pmatrix} M & N \\ P & Q \end{pmatrix} = \lambda \begin{pmatrix} I & 0 \\ 0 & I \end{pmatrix}, \qquad \lambda \in C, \quad M, N, P, Q \in C^{n\times n}. \tag{56}$$

The result for the square MRE is somewhat surprising, and we formulate it as a theorem.

Theorem 1.1. *A fundamental set of solutions of the matrix Riccati equation (49) for $W(t) \in C^{n\times n}$, $n \geq 2$ consists of five solutions. They can be generically chosen, suject to the following local conditions (in the neighbourhhood of $t = t_0$), defining an open dense set in $[C^{n\times n}]^5$:*

1. $\det(W_k - W_1) \neq 0$, $\det(W_3 - W_2) \neq 0$, $k = 2, \ldots, 5$.
2. *All eigenvalues of the matrix anharmonic ratio*

$$R_4(t) = (W_2 - W_3)^{-1}(W_3 - W_1)(W_1 - W_4)^{-1}(W_4 - W_2) \tag{57}$$

are distinct.

3. *The matrix anharmonic ratios $R_4(t)$ and*

$$R_5(t) = (W_2 - W_3)^{-1}(W_3 - W_1)(W_1 - W_5)^{-1}(W_5 - W_2)$$

regarded as linear maps do not have any common invariant subspaces spanning the entire space C^n.

Proof. a. Let $U_1, \ldots U_5$ be constant matrices $U_1 \in C^{n\times n}$, satisfying the above conditions. Consider them as initial value matrices for the five solutions $W_i(t)$ of the MRE. We have then

$$W_i(t) = [M(t)U_i + N(t)][P(t)U_i + Q(t)]^{-1}, \tag{58}$$

where

$$g(t) = \begin{pmatrix} M & N \\ P & Q \end{pmatrix} \tag{59}$$

satisfies (52). Let us apply a matrix fractional linear transformation with constant coefficients to express the matrices U_i in terms of convenient *standard* initial value matrices U_i^s

$$U_i = (M_c U_i^s + N_c)(P_c U_i^s + Q_c)^{-1} \qquad i = 1, \ldots, 5. \tag{60}$$

We choose

$$U_1^s \to \infty, \qquad U_2^s = 0, \qquad U_3^s = I, \tag{61a}$$

where the limit $U_1 \to \infty$ is to be understood as corresponding to a point on the Grassmannian $G_n(C^{2n})$ with homogeneous coordinates (X_1, Y_1), $\det X_1 \neq 0$, $Y_1 = 0$. further we choose

$$U_4^\bullet \equiv \Lambda, \qquad U_5^\bullet \equiv \Omega, \tag{61b}$$

where Λ is a constant diagonal matrix with all eigenvalues distinct, and Ω is a constant matrix with the property that its nonzero entries define the arcs of a connected graph (we introduce an arc between the points P_i and P_k if $\Omega_{ik} \neq 0$ or $\Omega_{ki} \neq 0$). An example of such a matrix and its graph is

$$\begin{pmatrix} 0 & 0 & * & 0 \\ * & 0 & 0 & * \\ 0 & * & 0 & 0 \\ 0 & 0 & 0 & 0 \end{pmatrix}$$

(62)

The conditions (57) are necessary and sufficient for the existence of the nonsingular constant matrix

$$\begin{pmatrix} M_c & N_c \\ P_c & Q_c \end{pmatrix} = \begin{pmatrix} U_1(U_1-U_3)^{-1}(U_3-U_2)Q_0 & U_2Q_0 \\ (U_1-U_3)^{-1}(U_3-U_2)Q_0 & Q_0 \end{pmatrix}, \tag{63}$$

realizing the simultaneous transition (60) from U_i to $U_i^\bullet$ as in (61). The matrix Q_c is defined by the condition

$$\Lambda = Q_c^{-1}(U_2-U_3)^{-1}(U_3-U_1)(U_1-U_4)^{-1}(U_4-U_2)Q_c, \tag{64}$$

i.e., it diagonalizes the matrix anharmonic ratio $R_4(t_0)$.

The transformation (60) amounts to a redefinition of M, N, P and Q in (58) *i.e.*, we have

$$W_i(t) = (\tilde{M}U_i^\bullet + \tilde{N})(\tilde{P}U_i^\bullet + \tilde{Q})^{-1}, \tag{65}$$

with

$$\begin{pmatrix} \tilde{M} & \tilde{N} \\ \tilde{P} & \tilde{Q} \end{pmatrix} = \begin{pmatrix} M & N \\ P & Q \end{pmatrix}\begin{pmatrix} M_c & N_c \\ P_c & Q_c \end{pmatrix}. \tag{66}$$

We shall drop the tildes.

b. Let us now show that the common stabilizer of the standardized initial value matrices $U_1^\bullet, \ldots, U_5^\bullet$ is simply the identity transformation. Indeed, (55) for $U_1^\bullet \to \infty$ implies $P_0 = 0$ for $U_2^\bullet = 0$ it implies $N_0 = 0$, and for $U_3^\bullet = I$ it implies $M_0 + N_0 = P_0 + Q_0$. The stabilizers of $U_1^\bullet$ alone, of the couple $(U_1^\bullet, U_2^\bullet)$, and of the triplet $(U_1^\bullet, U_2^\bullet, U_3^\bullet)$ are, respectively:

$$G_0^1 = \begin{pmatrix} M & N \\ 0 & Q \end{pmatrix}, \qquad G_0^2 = \begin{pmatrix} M & 0 \\ 0 & Q \end{pmatrix}, \qquad G_0^3 = \begin{pmatrix} Q & 0 \\ 0 & Q \end{pmatrix}, \tag{67}$$

The stabilizer $G_0^3 \sim SL(n,C) \otimes Z^2$ acts linearly and by conjugation. Condition (55) for $g \in G_0^3$ and $U_4^\bullet = \Lambda$ now reduces to

$$\Lambda = Q\Lambda Q^{-1},$$

and hence

$$G_0^4 = \begin{pmatrix} Q_D & \\ & Q_D \end{pmatrix}, \qquad (Q_D)_{ik} = q_i\delta_{i,k}, \quad q_i \in C. \tag{68}$$

Finally,

$$\Omega = Q_D\, \Omega\, Q_D^{-1},$$

implies

$$G_0^5 = \begin{pmatrix} I & 0 \\ 0 & I \end{pmatrix}, \qquad i.e., \quad Q_D = I. \tag{69}$$

Thus, we have proven that for $t = t_0$ the simultaneous stabilizer of five solutions $W_1, \ldots, W_5$ satisfying the conditions (57) (for $t = t_0$) is the identity subgroup $G_0^5 = e \in SL(2n, C)$. *Q. E. D.*

1.5.4 Reduction of $SL(2n,C)$ group action to $SL(n,C)$ conjugacy, and the matrix anharmonic ratio

The knowledge of a given set of solutions permits the reduction of the underlying equations to the type associated with the stabilizer of the initial conditions. In the case of the MRE (49) the knowledge of one, two, or three solutions makes it possible to reduce the equation to a linear inhomogeneous equation, linear homogeneous equation, or commutator-type linear homogeneous equation, respectively.

We shall now use three solutions, $W_1(t)$, $W_2(t)$, and $W_3(t)$, satisfying

$$\det(W_i - W_k) \neq 0, \qquad i,k = 1,2,3, \quad i \neq k, \tag{70}$$

(for $t = t_0$ and hence, by continuity, in a neighbourhood of $t = t_0$), to partially reconstruct the $SL(2n,C)$ group element in (50), and to reduce the MRE to a particularly convenient linear equation.

With no loss of generality we can assume that the initial value matrices

$$W_i(t_0) = U_i, \qquad i = 1,2,3, \tag{71}$$

are chosen in the "standard" form (61a). Substituting these three matrices successively into (50), we obtain

$$W_1 = MP^{-1}, \qquad W_2 = NQ^{-1}, \qquad W_3 = (M+N)(P+Q)^{-1}. \tag{72}$$

Solving for M, N, and P in terms of W_1, W_2, W_3, and Q, and substituting back into (50) we obtain

$$W = [W_1(W_3 - W_1)^{-1}(W_2 - W_3)QU + W_2Q][(W_3 - W_1)^{-1}(W_2 - W_3)QU + Q]^{-1} \tag{73a}$$
$$= [W_1(W_3 - W_1)^{-1}(W_2 - W_3)QUQ^{-1} + W_2][(W_3 - W_1)^{-1}(W_2 - W_3)QUQ^{-1} + I]^{-1} \tag{73b}$$

For $n \geq 2$ this is not yet a superposition principle, since the matrix Q is not determined. For $n = 1$ we have $QUQ^{-1} = U$, where U is a (constant) number and (73b) reduces to the superposition formula (12) for the ordinary Riccati equation, after interchanging solutions **1** and **2**.

We now solve (73b) for QUQ^{-1} and obtain

$$R(t) \equiv (W_2 - W_3)^{-1}(W_3 - W_1)(W_1 - W)^{-1}(W - W_2) = QUQ^{-1}, \tag{74}$$

i.e., the matrix anharmonic ratio $R(t)$ is conjugate to a constant matrix U. Note that the solutions W_1, W_2, W_3, are assumed to be specific known solutions, whereas W is a general solution, corresponding to a general initial value matrix U.

The result (74) is not new [118]. We have put it into a group-theoretical setting and shown that it is a consequence of the fact that the isotropy group G_0^3 [see (67)] acts by conjugation on the initial data. The anharmonic ratio $R(t)$ is defined for solutions $W(t)$ satisfying

$$\det(W - W_1) \neq 0. \tag{75}$$

If (75) is not observed, but we have

$$\det(W - W_2) \neq 0, \tag{76}$$

we can introduce an alternative matrix anharmonic ratio

$$\tilde{R}(t) = (W - W_2)^{-1}(W_1 - W)(W_3 - W_1)^{-1}(W_2 - W_3) = QVQ^{-1}. \tag{77}$$

If both R and $\tilde{R}$ exist in the neighbourhood of $t = t_0$, we have

$$R = \tilde{R}^{-1}, \qquad V = U^{-1}. \tag{78}$$

Let us now rederive (74) using a different method namely linearization of the MRE. In the process we shall establish some further porperties of the matrix anharmonic ratio. We make use again of three solutions of the MRE, satisfying (70).

We perform three successive fractional linear transformations. First, *linearize* the MRE by putting

$$S = (W - W_2)^{-1}. \tag{79}$$

If W and W_2 satisfy the MRE (49), we find that S satisfies a linear inhomogeneous equation

$$\dot{S} = -\tilde{B}S - S\tilde{C} - D, \tag{80}$$

$$\tilde{B} = B + DW_2, \qquad \tilde{C} = C + W_2 D. \tag{81}$$

A second transformation

$$T = S - S_1 = (W_2 - W)^{-1}(W - W_1)(W_1 - W_2)^{-1} \tag{82}$$

introduces the variable T satisfying a linear homogeneous equation:

$$\dot{T} = -\tilde{B}T - T\tilde{C}. \tag{83}$$

[S_1 is a solution of (80), W_1, W_2, and W are solutions of the original MRE.] Finally, use the third particular solution W_3 of the MRE, or equivalently, a solution T_3 of (83), $T_3(t) \neq 0$, to reintroduce the matrix anharmonic ratio of four solutions

$$R = T_3 T^{-1} = (W_2 - W_3)^{-1}(W_3 - W_1)(W_1 - W)^{-1}(W - W_2). \tag{84}$$

Using (83) and (84), we see that R satisfies a linear commutator-type equation corresponding to the action of the stabilizer G_0^3 of three initial values [see (67)]:

$$\dot{R} = [R, \tilde{B}], \tag{85}$$

with $\tilde{B}$ as in (81). The solution of (85) can be written in the form (74):

$$R(t) = Q(t)UQ^{-1}(t), \qquad Q(t) \in SL(n, C), \tag{86}$$

where $U \in C^{n \times n}$ is a constant matrix and (85) implies that Q satisfies a particularly simple linear differential equation

$$\dot{Q} = -\tilde{B}Q, \qquad Q(t_0) = Q_0, \tag{87}$$

for some initial condition Q_0. Equation (87) could of course be solved numerically and in some cases analytically, but our purpose is to derive a superposition law. To do this we shall express $Q(t)$, or alternatively $R(t) = QUQ^{-1}$, in terms of two more solutions of the MRE.

Let us make a comment here on the group-theoretical significance of the matrix anharmonic ratio $R(t)$. For $n = 1$ we have $\dot{R} = 0$ from (85) and (consistently) $R = U =$ constant from (86). Thus, for $n = 1$, R is an $SL(2,C)$ group invariant, formed out of four solutions of the Riccati equation. The existence of this invariant is a consequence of the fact that $SL(2,C)$ acts freely on the cartesian product of three Grassmannians $[G_1(C^2)]^3$. For $n \geq 2$, the $SL(2n,C)$ invariant is not R, but $U = Q^{-1}RQ$, where Q must still be determined in terms of two further solutions, *i.e.*, this n^2 dimensional invariant depends on six solutions, since $SL(2n,C)$ for $n \geq 2$ acts freely on $[G_n(C^{2n})]^5$. From R itself we can form n independent complex elementary trace invariants:

$$I_k = \mathrm{Tr}\, R^k, \qquad k = 1, 2, \ldots, n. \tag{88}$$

For $n \geq 2$, (88) does not suffice to determine $W(t)$ in terms of W_1, W_2, and W_3 alone. Notice that Eq. (85) has the form of a Lax pair and (88) are "integrals of motion" typically derived from a Lax pair [4,32,88].

1.5.5 Derivation of superposition formulae

Equation (74) of the previous Subsection tells us that the matrix anharmonic ratio $R(t)$ of four solutions of the MRE is conjugate to a constant matrix U, where the conjugating matrix $Q(t)$ is independent of the choice of W_1, W_2, W_3 and W. Equation (85) provides a linear commutator-type equation for $R(t)$. Each of these two results, together with the expression (73) for the general solution $W(t)$ can be used to obtain a superposition formula.

1.5.5.1 Reconstruction of the $SL(n,C)$ group element

Consider the matrix Riccati equation (49) for $n \geq 2$, and assume that we know five solutions $W_1, \ldots, W_5$, satisfying the independence conditions (50). The matrix anharmonic ratios R_4 and R_5 of (57) satisfy

$$R_a = QU_aQ^{-1}, \qquad a = 4, 5, \tag{89}$$

for some constant *initial value* matrices U_4 and U_5 which, with no loss of generality we choose as in (61b). We now write Q in the form

$$Q(t) = Q_0(t)\, Q_D(t), \tag{90}$$

where Q_D is an element of the isotropy group G_0^4:

$$Q_D \Lambda Q_D^{-1} = \Lambda = U_4^\circ, \tag{91}$$

i.e., Q_D is a diagonal matrix. The nonsingular matrix $Q_0(t)$ is determined from the condition

$$R_4(t) = Q_0(t) \Lambda Q_0^{-1}(t). \tag{92}$$

Thus, Q_0 is a matrix diagonalizing $R_4(t)$, *i.e.*, the columns of $Q_0(t)$ are the appropriately ordered eigenvectors of $R_4(t)$. Remember that all eigenvalues of $R_4(t)$ in the neighbourhood of $t = t_0$ are distinct. The arbitrariness in the normalization of these eigenvectors is related to the fact that Q_D is so far not determined. This arbitrariness can be removed by setting the first nonzero element of each column in $Q_0(t)$ equal to 1.

Finally, the last solution $W_5(t)$ is used to determine the matrix elements of Q_D:

$$\left(Q_D(t)\right)_{ik} = d_i \delta_{i,k} \tag{93}$$

from the system of linear algebraic equations

$$[Q_0^{-1} R_5(t) Q_0] Q_D = Q_D \Omega, \qquad \Omega = U_5^s, \tag{94}$$

for the nonzero matrix elements d_i of $Q_D(t)$. Substituting $Q = Q_0 Q_D$ back into (73) we obtain the explicit form of the superposition formula

$$W(t) = [W_1(W_3 - W_1)^{-1}(W_2 - W_3)Q_0 Q_D U + W_2 Q_0 Q_D][(W_3 - W_1)^{-1}(W_2 - W_3)Q_0 Q_D U + Q_0 Q_D]^{-1}. \tag{95a}$$

Alternatively, using (77) and solving for W we obtain

$$W(t) = [W_1(W_3 - W_1)^{-1}(W_2 - W_3)Q_0 Q_D + W_2 Q_0 Q_D V][(W_3 - W_1)^{-1}(W_2 - W_3)Q_0 Q_D + Q_0 Q_D V]^{-1}. \tag{95b}$$

These two formulae are equivalent if $\det U \neq 0$ and $\det V \neq 0$; we have then $V = U^{-1}$.

1.5.5.2 Algebra of matrix anharmonic ratios

Let us derive a somewhat different superposition formula, taking equation (85) for the matrix anharmonic ratio as our starting point. Since (85) is a linear differential equation for R, its solutions obviously form a linear space. The commutator form of the right-hand side of this equation assures us that the solutions also form an associative algebra under matrix multiplication. Indeed, if R_1 and R_2 satisfy (85), then so does their product $R_1 \cdot R_2$. The inverse R^{-1} of a solution is a solution, and $R = I$ is a solution.

It follows that we can generate n^2 linearly independent solutions of (85) as polynomials in a smaller number of *basic* solutions. This is best formulated in terms of the initial value matrices U_i. Since $R_a = QU_aQ^{-1}$, $R_b = QU_bQ^{-1}$ implies $R_aR_b = QU_aU_bQ^{-1}$, it is sufficient to construct a basis of constant matrices $U_i \in C^{n\times n}$ from which all matrices $U \in C^{n\times n}$ can be generated as polynomials. To do this we need the following lemma.

Lemma 1.1. *The algebra of all constant matrices $U \in C^{n\times n}$ is generated polynomially by precisely two matrices, say U_4 and U_5, satisfying the following conditions:*

1. *At least one linear combination of U_4 and U_5, say U_4 itself, is nonsingular and has all eigenvalues different.*
2. *The matrices U_4 and U_5 have no common nontrivial invariant subspaces.*

Proof. Since U_4 has n different nonzero eigenvalues, we can, with no loss of generality, take it to be diagonal [94]

$$U_4 = \mathrm{diag}(\lambda_1, \lambda_2, \ldots, \lambda_n), \qquad \lambda_i \neq 0, \quad \lambda_i \neq \lambda_k, \quad \text{for } i \neq k. \tag{96}$$

The powers $U_4, U_4^2, \ldots, U_4^n$ are linearly independent and nonsingular matrices and n appropriately chosen linear combinations of them give us a convenient basis for all diagonal matrices:

$$\{E_a\}, \quad (E_a)_{lm} = \delta_{a,l}\delta_{a,m}, \qquad a, l, m == 1, 2, \ldots, n. \tag{97}$$

The other given matrix U_5 has, by assumption, no common nontrivial invariant eigenspaces with U_4. This can be expressed by the requirement that in the basis where U_4 is diagonal, the nonzero entries of U_5 define the arcs of a strongly connected oriented graph [we introduce an arc from P_i to P_k if $(U_5)_{ik} \neq 0$].

We now construct a basis for the off-diagonal matrices by putting

$$E_{ik} = \frac{1}{(U_5)_{ik}} E_i U_5 E_k, \qquad i,k = 1,2,\ldots,n, \quad (U_5)_{ik} \neq 0. \tag{98}$$

If $(U_5)_{ab} = 0$ for some a and b, we choose a permissible path $ai_1i_2\cdots b$ from a to b on the oriented graph and put:

$$E_{ab} = E_{ai_1} E_{i_1 i_2} \cdots E_{i_k b} \neq 0, \tag{99}$$

thus completing the basis for $C^{n\times n}$ and prooving the lemma. *Q. E. D.*

Example.

$$U_4 = \begin{pmatrix} 1 & 0 & 0 \\ 0 & 2 & 0 \\ 0 & 0 & 3 \end{pmatrix}, \qquad U_5 = \begin{pmatrix} 0 & 0 & 3 \\ 2 & 7 & 0 \\ 0 & 1 & 5 \end{pmatrix}.$$

The graph corresponding to U_5 is

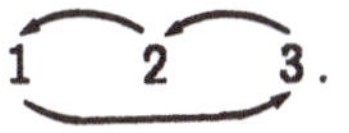

We have

$$E_1 = 3U_4 - \tfrac{5}{2}U_4^2 + \tfrac{1}{2}U_4^3 = \begin{pmatrix} 1 & 0 & 0 \\ 0 & 0 & 0 \\ 0 & 0 & 0 \end{pmatrix},$$

$$E_2 = -\tfrac{3}{2}U_4 + 2U_4^2 - \tfrac{1}{2}U_4^3 = \begin{pmatrix} 0 & 0 & 0 \\ 0 & 1 & 0 \\ 0 & 0 & 0 \end{pmatrix},$$

$$E_3 = \tfrac{1}{3}U_4 - \tfrac{1}{2}U_4^2 + \tfrac{1}{6}U_4^3 = \begin{pmatrix} 0 & 0 & 0 \\ 0 & 0 & 0 \\ 0 & 0 & 1 \end{pmatrix},$$

$$E_{13} = \tfrac{1}{3}E_1U_5E_3, \quad E_{21} = \tfrac{1}{2}E_2U_5E_1, \quad E_{32} = E_3U_5E_2,$$

$$E_{12} = E_{13}E_{32} = \tfrac{1}{3}E_1U_5^2E_2, \quad E_{23} = E_{21}E_{13} = \tfrac{1}{6}E_2U_5^2E_3,$$

$$E_{31} = E_{32}E_{21} = \tfrac{1}{2}E_3U_5^2E_1.$$

Comment. Notice that the conditions on U_4 and U_5 in the above Lemma are somewhat stronger than conditions (61b) for Theorem 1. Indeed, $U_4^s \equiv \Lambda$ had all eigenvalues different, but one of them could be zero. The other matrix $U_5^s \equiv \Omega$ corresponded to a connected graph, not necessarily a strongly connected oriented one.

We now apply the Lemma to derive a superposition formula for the matrix Riccati equation.

Let U_4 and U_5 satisfy the conditions of Lemma 1, and let

$$R_4 = QU_4Q^{-1}, \qquad R_5 = QU_5Q^{-1}, \tag{100}$$

be the corresponding (known) matrix anharmonic ratios. We now generate n^2 linearly independent solutions of (85) as

$$R_{jj} = R_4^j, \quad R_{jk} = R_4^j R_5 R_4^k, \qquad j,k = 1,2,\ldots,n, \quad j \neq k, \tag{101}$$

for $(R_5)_{jk} \neq 0$ and

$$R_{ab} = (R_4^a R_5 R_4^{j_1})(R_4^{j_1} R_5 R_4^{j_2}) \cdots (R_4^{j_l} R_5 R_4^b) \tag{102}$$

for $(R_5)_{ab} = 0$, where $(a\, j_1 \cdots j_l\, b)$ is some admissible path on the graph corresponding to the matrix U_5.

The superposition formula is then obtained by putting

$$R(t) = \sum_{j,k=1}^{n} c_{jk} R_{jk}, \tag{103}$$

and substituting (103) for $R(t)$ into

$$W = [W_1(W_3 - W_1)^{-1}(W_2 - W_3)R + W_2][(W_3 - W_1)^{-1}(W_2 - W_3)R + I]^{-1}. \tag{104}$$

The superposition rule takes a particularly simple form if U_4 satisfies (96) and U_5 has no vanishing matrix elements at all:

$$(U_5)_{ik} \neq 0, \qquad i,k = 1,2,\ldots,n.$$

In this case (103) reduces to

$$R(t) = \sum_{j,k=1}^{n} c_{jk} R_4^j R_5 R_4^k, \qquad c_{jk} \in C, \tag{105}$$

where c_{jk} are arbitrary constants. Finally, we summarize the results of this subsection in the form of two theorems. The arguments presented above constitute their proofs.

Theorem 1.2. *The general solution $W(t)$ of the matrix Riccati equation (49) is expressed in terms of five particular solutions $W_1, \ldots, W_5$, satisfying conditions (57) for some $t = t_0$, by formulae (95), where the matrices Q_0 and Q_D are determined by equations (92) and (94).*

Theorem 1.3. *Let the fundamental set of solutions of the matrix Riccati equation (49) satisfy (57) and in addition let the matrix anharmonic ratio $R_4(t_0)$ be nonsingular and the two ratios $R_4(t_0)$ and $R_5(t_0)$ have no common nontrivial invariant eigenspaces. Then, in addition to the superposition rule of Theorem 2, we can write an alternative superposition formula, namely (104) with $R(t)$ given by (103).*

1.5.6 Special cases of the matrix Riccati equation

So far we have not imposed any particular conditions on the matrices $A, \ldots, D$ in the MRE; the superposition formulae which we have obtained are hence quite general.

In special cases, further results can be obtained. One type of restriction which can be imposed is to require that the curve in the Lie algebra

$$\xi(t) = \begin{pmatrix} C(t) & A(t) \\ -D(t) & -B(t) \end{pmatrix}, \tag{106}$$

be restricted to a different classical Lie algebra than $sl(2n,C)$. Thus, if we require that $A, \ldots, D$ be real, we have $\xi(t) \in sl(2n,R)$, if we impose $\xi K + K\xi^{\top} = 0$ with

$$K = \begin{pmatrix} 0 & I \\ -I & 0 \end{pmatrix},$$

we have $\xi \in sp(2n,C)$, etc. We shall come back to this, below, in §1.6.

Here let us discuss some special cases when less than five solutions are needed in order to obtain the general solution of the MRE.

1.5.6.1 Matrix Riccati equations with constant coefficients

Let us again consider the MRE

$$\dot{W} = A + WB + CW + WDW, \qquad A,B,C,D,W \in C^{n\times n}, \tag{107}$$

however, this time let A, B, C, and D be constant matrices. In this case, Eq. (107) allows constant solutions. Let us choose W_2 as such a constant solution

$$\dot{W}_2 = 0, \qquad A + W_2B + CW_2 + W_2DW_2 = 0, \tag{108}$$

i.e., a solution of the algebraic MRE in (108). Let W_1 be a solution of (107) satisfying

$$\det(W_1 - W_2) \neq 0. \tag{109}$$

If a second constant solution, satisfying (109) exists, then this is a particularly appropriate choice for W_1. Put

$$T = (W_2 - W)^{-1}(W - W_1)(W_1 - W_2)^{-1} \tag{110}$$

as in (82); T will satisfy (83),

$$\dot{T} = -\tilde{B}T - T\tilde{C}, \qquad \tilde{B} = B + DW_2, \qquad \tilde{C} = C + W_2D, \tag{111}$$

where $\tilde{B}$ and $\tilde{C}$ are constant matrices. Let us write a solution of (111) as

$$T(t) = G(t)T_0H(t), \qquad G,H \in GL(n,C), \tag{112}$$

where $T_0 \in C^{n\times n}$ is a constant matrix, and we have

$$\dot{G} = -\tilde{B}G, \qquad \dot{H} = -H\tilde{C}. \tag{113}$$

Equations (113) can be integrated to give

$$G = e^{-\tilde{B}t}, \qquad H = e^{-\tilde{C}t}, \tag{114}$$

where we have imposed

$$G(0) = I, \qquad H(0) = I. \tag{115}$$

Inverting (110) we obtain

$$W = [W_2T(W_1 - W_2) + W_1][T(W_1 - W_2) + I]^{-1}. \tag{116}$$

Substituting (112) and (114) into (116), we obtain a formula expressing the general solution $W(t)$ of the MRE with constant coefficients in terms of two particular solutions [124]:

$$W(t) = [W_2 e^{-\tilde{B}t} T_0 e^{-\tilde{C}t}(W_1 - W_2) + W_1][e^{-\tilde{B}t} T_0 e^{-\tilde{C}t}(W_1 - W_2) + I]^{-1}. \tag{117}$$

Formula (117) is intermediate between a superposition formula and an explicit solution. Two solutions are used, W_2 must be constant (W_1 *may* be constant) and they must satisfy (109). Part of the time dependence of $W(t)$ is displayed explicitly in exponentials (the entire time dependence is explicit if W_1 is also constant).

Formula (117) is particularly convenient. A similar formula has been proposed for operator Riccati equations, when A, B, C, D, and W are linear operators in some infinite-dimensional space, rather than matrices [123]. It is worthwhile mentioning that MRE's with constant coefficients are particularly common in applications, in particular in optimal control theory.

1.5.6.2 Examples of linearizable matrix Riccati equations

We have seen that it is always possible to associate a linear system of ODE's with the MRE (49), namely, the system (48) in $2(n \times n)$ dimensions. It is also always possible to *linearize* the MRE by introducing the variable T of (82), making use of two known solutions, W_1 and W_2, of the MRE.

In special cases it is possible to linearize the MRE by a transformation of dependent variables, without increasing the dimension of the space, or using known solutions. This type of linearization, possible only in special cases, fits into the general framework of linearization, developed on one hand by Moshe Flato and collaborators [46–49], and on the other by Bluman and Kumei [85].

Without studying the problem of linearization systematically, let us consider some examples which are of interest in applications. Consider again the MRE (49) and the associated linear system (48). Let us perform a transformation of variables

$$\begin{pmatrix} \tilde{X} \\ \tilde{Y} \end{pmatrix} = \begin{pmatrix} \alpha I & \beta I \\ \gamma I & \delta I \end{pmatrix} \begin{pmatrix} X \\ Y \end{pmatrix} \equiv S \begin{pmatrix} X \\ Y \end{pmatrix}, \qquad \alpha\delta - \beta\gamma = 1, \tag{118}$$

where α, β, γ, and δ are complex constants, and correspondingly

$$V = \tilde{X}\tilde{Y}^{-1} = (\alpha W + \beta)(\gamma W + \delta)^{-1}. \tag{119}$$

The new variables will satisfy the equation

$$\begin{pmatrix} \dot{\tilde{X}} \\ \dot{\tilde{Y}} \end{pmatrix} = \tilde{\xi} \begin{pmatrix} \tilde{X} \\ \tilde{Y} \end{pmatrix} = S\xi S^{-1} \begin{pmatrix} \tilde{X} \\ \tilde{Y} \end{pmatrix}, \tag{120}$$

with ξ as in (106). The variable V will satisfy a linear homogeneous equation if $\tilde{\xi}$ is block-diagonal, *i.e.*, if

$$\begin{aligned} \alpha^2 A + \beta^2 D &= \alpha\beta(B + C), \\ \gamma^2 A + \delta^2 D &= \gamma\delta(B + C). \end{aligned} \tag{121}$$

Equations (121) can be satisfied in one of two ways:

(i)
$$A=\lambda(B+C), \qquad D+\mu(B+C), \\ B+C\neq 0, \quad \lambda\mu\neq 0, \quad \lambda,\mu\in C \tag{122}$$

Equations (121) in this case imply

$$S=\begin{pmatrix} 1 & \frac{1-\sqrt{1-4\lambda\mu}}{2\mu} \\ \frac{\mu}{\sqrt{1-4\lambda\mu}} & \frac{1+\sqrt{1-4\lambda\mu}}{2\sqrt{1-4\lambda\mu}} \end{pmatrix}\otimes I, \qquad 1-4\lambda\mu\neq 0. \tag{123}$$

The MRE in this case has the special form

$$\dot W=\lambda(B+C)+WB+CW+\mu WDW, \tag{124}$$

where λ and μ are constants, and (119) reduces to

$$V=\left(W+\frac{1-\sqrt{1-4\lambda\mu}}{2\mu}\right)\Big/\left(\frac{\mu}{\sqrt{1-4\lambda\mu}}W+\frac{1+\sqrt{1-4\lambda\mu}}{2\sqrt{1-4\lambda\mu}}\right). \tag{125}$$

The variable $V\in C^{n\times n}$ satisfies the linear homogeneous equation

$$\begin{aligned} \dot V &= V\tilde B+\tilde C V, \\ \tilde B &= \tfrac12\Big((1+\sqrt{1-4\lambda\mu})B+(-1+\sqrt{1-4\lambda\mu})C\Big), \\ \tilde C &= \tfrac12\Big((-1+\sqrt{1-4\lambda\mu})B+(1+\sqrt{1-4\lambda\mu})C\Big). \end{aligned} \tag{126}$$

Thus, by the simple transformation (125) we have reduced the MRE (124) to (126), an equation of the type (83). Following the arguments in §1.5.4 and §1.5.5, above, we can now obtain a superposition law for (126) and hence for the MRE (124) in terms of three solutions of (124) [or (126)].

(ii)
$$C=-B, \quad D=-\lambda^2 A, \qquad \lambda\in C, \quad \lambda\neq 0. \tag{127}$$

The MRE in this case reduces to

$$\dot W=A+WB+BW-\lambda^2 WAW \tag{128}$$

with $\lambda =$ constant, and the transformation (119) to

$$V=(W+\lambda^{-1})(W-\lambda^{-1})^{-1}. \tag{129}$$

The variable V will again satisfy a linear homogeneous equation, namely

$$\dot V=V(\lambda A+B)+(\lambda A-B)V\equiv V\tilde B+\tilde C V. \tag{130}$$

Again, a superposition formula in terms of just three particular solutions can be obtained.

The reason why the MRE's (124) and (128) are simpler than the general case is that the underlying Lie algebra in both cases is the subalgebra $gl(n,C)\otimes gl(n,C)\subset gl(2n,C)$, rather than $gl(2n,C)$ itself.

The explicit superposition formula for (126) and (130) is obtained by defining

$$R\equiv VV_1^{-1}, \tag{131}$$

where

$$V_1 = (\alpha W_1 + \beta)(\gamma W_1 + \delta)^{-1}, \tag{132}$$

with α, β, γ, and δ as in (125) or (129), respectively, and W_1 a solution of the MRE (124) or (128), respectively. The quantity R satisfies

$$\dot{R} = [\tilde{C}, R], \tag{133}$$

and hence

$$R = QR(0)Q^{-1}, \quad Q = \tilde{C}Q, \qquad Q \in GL(n, C). \tag{134}$$

Choosing W_2 and W_3 such that $R_2(0)$ has all eigenvalues different and $R_2(0)$ and $R_3(0)$ do not have any common nontrivial invariant subspaces spanning C^n, we can solve for $Q \equiv Q_0 Q_D$ exactly, as in (90–94). Inverting (131) we obtain the superposition formula

$$\begin{aligned} W = & [-\delta Q_0 Q_D R(0) + \beta(\gamma W_1 + \delta)(\alpha W_1 + \beta)^{-1} Q_0 Q_D] \\ & \times [\gamma Q_0 Q_D R(0) - \alpha(\gamma W_1 + \delta)(\alpha W_1 + \beta)^{-1} Q_0 Q_D]^{-1}. \end{aligned} \tag{135}$$

In the special case of $\tilde{B}$ and $\tilde{C}$ constant, (126) and (130) can be integrated directly without using known particular solutions.

1.5.7 Discussion of superposition formulas for the matrix Riccati equation

The obtained nonlinear superposition rules play a similar role as the linear ones; they reduce the problem of finding infinitely many solutions of the MRE to that of finding a finite number, in general just five solutions. The formulae can be used on one hand to obtain solutions explicitly, on the other hand to study general properties of solutions.

In some cases it may be possible to construct certain specific solutions of the MRE analytically for particularly well-chosen initial conditions. If five such solutions are available, satisfying the conditions (57), then (95) or (104) provides the general solution analytically. Usually this is not the case. Then, a fundamental set of solutions can be constructed numerically and the superposition formulae provide a new numerical method for solving the MRE. This aspect will be discussed further below in §1.8.

One general result which can be extracted directly from the superoposition formula (95) concerns the MRE with constant coefficients (107):

Proposition. *It is not possible for the matrix Riccati equation (107) with constant coefficients, to have five (or more) constant solutions satisfying the independence conditions (57).*

Proof. Assume that five constant solutions $W_1, \ldots, W_5$ satisfying (57) exist. We could use them to calculate Q_0 and Q_D of (90) and these matrices would also be constant. Formulae (95) would then imply that all solutions of the considered MRE are constant. This is an obvious contradiction. *Q. E. D.*

The fact that the general solution $W(t)$ of the MRE (49) can be written in the form of a matrix fractional linear tranformation (50) (with coefficients M, N, P, and Q known in terms of five particular solutions) and the related fact that $W(t)$ is the ratio $W = XY^{-1}$ of two matrices satisfying a system of linear equations (48), has implications for the singularity structure of $W(t)$. It can be used to investigate the Painlevé property [55,73,75,107] of the MRE, *i.e.*, to show that its only moving singularities are poles. Assuming that the coefficientes A, B, C, and D of the MRE are analytical functions of t, we can determine the positions of the poles of $W(t)$ from the superposition formula (95), *e.g.* from the equation

$$\det[(W_3 - W_1)^{-1}(W_2 - W_3)Q_0 Q_D U + Q_0 Q_D] = 0. \tag{136}$$

§1.6 Other Lie groups and other manifolds

1.6.1 General comments

In §1.5 we have treated the complex matrix Riccati equation generated by the action of the Lie group $G = SL(n+k,C)$ on the Grassmann manifold $G_k(C^{n+k}) \sim G/G_0$, where G_0 is a maximal parabolic subgroup of $SL(n+k,C)$, defined in (36). We have concentrated on the case $n = k$, but $k = 1$ and $1 < k < n$ have been treated in the literature [9,11,12,124]. Work on all classical Lie groups and their maximal subgroups is in progress.

One direction which is being pursued is that of restricting from the action of $SL(2n,C)$ on $G_n(C^{2n})$ to the action of certain subgroups of this group. The subgroup will in general not act transitively on the entire Grassmannian. In this case it is necessary to add further constraints characterizing particular orbits under the subgroup. The previously derived superposition rules will not necessarily respect the constraints and therefore a new analysis may be called for. The obtained equations will be particular forms of the square MRE. It should however be stressed that other group actions on other manifolds lead to equations which are not necessarily of the Riccati type, and which involve nonlinearities that are not necessarily quadratic.

Reductions leading to particular MRE's which have so far been considered are given by taking a subgroup $G \subset SL(2n,C)$ with G acting on the Grassmann manifold or some submanifold. So far we have taken $G = SL(2n,R)$, $Sp(2n,C)$, $Sp(2n,R)$, $U(n,n)$, $O(2n,C)$, $O(2n,2n)$, and others [41,70,117].

Let us briefly review the results for $SL(2n,R)$, $Sp(2n,C)$, and $Sp(2n,R)$. Those for $U(n,n)$ and $O(n,n)$ are in the literature.

1.6.2 The real matrix Riccati equation and $SL(2n,R)$

Results similar to those presented in §1.5 have been obtained in [70] for $G = SL(n+k,R)$ and G_0 the group of block-triangular real matrices

$$G_0 = \left\{ g \mid g = \begin{pmatrix} G_{11} & 0 \\ G_{21} & G_{22} \end{pmatrix} \right\}, \qquad G_{11} \in GL(n,R), \quad G_{21} \in R^{k\times n}, \quad G_{22} \in GL(k,R). \tag{137}$$

The homogeneous space $M \sim G/G_0$ in this case is the real Grassmann manifold $G_k(R^{n+k})$ of k-planes in R^{n+k}. The equation obtained by considering the action of G on M in affine coordinates is the real MRE

$$\begin{gathered} \dot{W} = A + WB + CW + WDW, \\ A, W \in R^{n\times k}, \quad B \in R^{k\times k}, \quad C \in R^{n\times n}, \quad D \in R^{k\times n}. \end{gathered} \tag{138}$$

We can again associate the linear system (48) with the MRE (138), again write the superposition formula in the form (50); this time with M, N, P, and Q as real matrices. With the restriction $k = n$ (real square MRE) we can again express the general solution in terms of five particular solutions. Theorem 1 concerning the fundamental set of solutions is still valid, except that the solutions must, of course, be real. The proof of the Theorem in the $SL(2n,R)$ case is slightly more complicated than in the $SL(2n,C)$ case, since $U_4 = \Lambda$ is not necessarily diagonalizable over R (even if all eigenvalues of Λ are different). The possible presence of pairs of complex conjugate eigenvalues reflects itself in the fact that Λ is only guaranteed to be block-diagonalizable with 1×1 *blocks* corresponding to real eigenvalues and 2×2 blocks of the form

$$\begin{pmatrix} a & b \\ -b & a \end{pmatrix}, \qquad a, b \in R, \quad b > 0, \tag{139}$$

corresponding to pairs of complex eigenvalues.

The superposition formulae are still written in the form (95) or (103–105). The reconstruction of the matrix $Q = Q_0 Q_D$ in (95) is slightly more complicated in the real case than in the complex one. The complications are again related to the fact that $R_4(0)$ in (92) is not always diagonalizable over R. They are not essential and they are dealt with in Ref. [70], where the superposition rule for the real square MRE is treated in detail.

1.6.3 The symplectic matrix Riccati equation

A particularly interesting reduction of the MRE (49) is obtained by requiring that the curve $\xi(t)$ of (47) for $n = k$ lie in the symplectic algebra $sp(2n, C)$ defined by

$$sp(2n,C) = \left\{ \xi \in sl(2n,C) \mid \xi K + K \xi^\top = 0 \right\}, \tag{140}$$

where K is an antisymmetric nonsingular matrix which we can choose to be

$$K = \begin{pmatrix} 0 & I \\ -I & 0 \end{pmatrix}. \tag{141}$$

The curve $\xi(t)$ thus has the form

$$\xi(t) = \begin{pmatrix} C(t) & A(t) \\ -D(t) & -C^\top(t) \end{pmatrix}, \tag{142}$$

where

$$A = A^\top, \qquad D = D^\top \tag{143}$$

(the superscript $^\top$ denotes transposition). The corresponding curve in the group $Sp(2n, C)$ is given by (51–53) and satisfies

$$g(t) K g^\top(t) = K, \tag{144}$$

where this condition is imposed on $g(t_0)$. The group $Sp(2n, C)$ preserves the symplectic inner product and hence does not act transitively on $G_n(C^{2n})$:

$$(X^\top \, Y^\top) \begin{pmatrix} 0 & I \\ -I & 0 \end{pmatrix} \begin{pmatrix} X \\ Y \end{pmatrix} = X^\top Y - Y^\top X = \kappa, \tag{145}$$

where κ is a constant matrix.

We restrict ourselves to the Lagrangian subspace $G_n^0(C^{2n})$ for which $\kappa = 0$ in (145): the submanifold of totally isotropic n-panes in C^{2n}. The action of $Sp(2n, C)$ on this $\frac{1}{2}n(n+1)$-dimensional manifold is well defined and transitive. The condition

$$X^\top Y - Y^\top X = 0 \tag{146}$$

in affine coordinates $W = XY^{-1}$ is equivalent to

$$W = W^\top. \tag{147}$$

The complex symplectic MRE is thus

$$\dot{W} = A + WC^\top + CW + WDW, \qquad A = A^\top, \quad D = D^\top, \tag{148}$$

and the symmetry property (147) persists for all t once it is imposed for $t = t_0$.

A fundamental set of solutions for the symplectic MRE (148) consists of only four solutions, rather than five as in the case of the general MRE (49). Let us formulate the result as a theorem.

Theorem 1.4. *The general solution $W(t)$ of the symplectic MRE (148) is given in terms of four particular solutions $W_1, \ldots, W_4$ satisfying (at least in the neighbourhood of some initial point $t = t_0$):*

1.
$$\det(W_i - W_k) \neq 0 \quad i = 1,2,3,$$
$$\det(W_2 - W_4) \neq 0, \tag{149a}$$
$$\det(W_4 - W_1) \neq 0.$$

2. *All eigenvalues of*

$$R \equiv T_4 T_3^{-1} = (W_2 - W_4)^{-1}(W_4 - W_1)(W_3 - W_1)^{-1}(W_2 - W_3) \tag{149b}$$

are distinct, where $T(t)$ is as in (82). The superposition formula can be written as

$$W(t) = [W_2 O_1 D O_2 T_0 + W_1 (W_1 - W_2)^{-1} O_1 (D^\top)^{-1} O_2][O_1 D O_2 T_0 + (W_1 - W_2)^{-1} O_1 (D^\top)^{-1} O_2]^{-1}, \tag{150}$$

where $T_0 = T_0^\top \in \mathbb{C}^{n \times n}$ is a symmetric constant matrix, O_1 is an orthogonal matrix which diagonalizes the nonsingular matrix $T_3(t)$,

$$T_3(t) = (W_2 - W_3)^{-1}(W_3 - W_1)(W_1 - W_2)^{-1}, \tag{151}$$

D is a diagonal matrix

$$D = \operatorname{diag}\left(\sqrt{\lambda_1(t)}, \sqrt{\lambda_2(t)}, \ldots, \sqrt{\lambda_n(t)}\right), \tag{152}$$

where $\lambda_i(t)$ are eigenvalues of $T_3(t)$, and O_2 is an orthogonal matrix which diagonalizes $\tilde{T}_4(t)$:

$$\tilde{T}_4(t) = D^{-1}(t) O_1^\top(t) T_4(t) O_1(t) D^{-1}(t). \tag{153}$$

Proof. Put

$$T = (W_2 - W)^{-1}(W - W_1)(W_1 - W_2)^{-1}, \tag{154}$$

as in (82). It is easy to see that if W_1, W_2, and W satisfy the symplectic MRE (148), then T satisfies the linear homogeneous equation

$$\dot{T} = -\tilde{C}^\top T - T\tilde{C}. \tag{155}$$

The solution of (155) is

$$T(t) = G T_0 G^\top, \tag{156}$$

where G satisfies the linear equation

$$G = -\tilde{C}^\top G, \qquad G(t_0) = G_0. \tag{157}$$

Let us assume that two more solutions, W_3 and W_4, are known and we calculate $T_3(t)$ and $T_4(t)$. Since we are restricting ourselves to symmetric solutions of the symplectic MRE, we also have

$$T(t) = T^\top(t). \tag{158}$$

We can use a constant transformation G_0 of the type (156) to simultaneously transform $T_3(t_0)$ and $T_4(t_0)$ into a convenient form. Putting $G_0 = Q_1 D_0 Q_2$, where Q_1 and Q_2 are constant complex orthogonal matrices and D_0 is constant diagonal, we can simultaneously require

$$G_0 T_3(t_0) G_0^\top = I, \qquad G_0 T_4(t_0) G_0^\top = \Lambda, \tag{159}$$

where Λ is diagonal and has all eigenvalues distinct and nonzero. Use was made of the fact that conditions (149) assure that all eigenvalues of $T_3(t_0)$ are nonzero ($\det T_3(t_0) \neq 0$). The matrix Q_2 diagonalizes $T_3(t_0)$, D_0 is chosen so as to transform the diagonalized $T_3(t_0)$ matrix into I, Q_1 is chosen so as to diagonalize $T_4(t_0)$. The fact that all eigenvalues of Λ are different and nonzero follows from (149b) since we have

$$R(t) = T_4(t)T_3^{-1}(t) = GT_4(t_0)G^\top[GT_3(t_0)G^\top]^{-1} = GT_4(t_0)T_3^{-1}(t_0)G^{-1},$$

i.e.

$$R(t) = GG_0^{-1}\Lambda(G_0^\top)^{-1}[G_0^{-1}(G_0^\top)^{-1}]^{-1}G^{-1} = GG_0^{-1}\Lambda G_0 G^{-1}.$$

Thus the eigenvalues of Λ coincide with those of $R(t)$, all distinct (and nonzero) by assumption. Without loss of generality we can now assume

$$T_3(t_0) = I, \qquad T_4(t_0) = \Lambda, \tag{160}$$

and absorb the constant matrix G_0 into the constant matrix T_0 in (156).

Now put

$$G(t) \equiv O_1(t)D(t)O_2(t), \tag{161}$$

where O_1 and O_2 are orthogonal matrices and $D(t)$ is diagonal. We have

$$\begin{aligned} T_3(t) &= O_1(t)D(t)O_2(t)IO_2^\top(t)D(t)O_1^\top(t) \\ &= O_1(t)D^2(t)O_1^\top(t). \end{aligned} \tag{162}$$

Thus: $O_1(t)$ is the orthogonal matrix that diagonalizes $T_3(t)$; any ambiguity in O_1 can be absorbed into the yet unknown $O_2(t)$. The matrix $D(t)$ is then given in (152). Further, put

$$\tilde{T}_4(t) \equiv D^{-1}O_1^\top T_4(t)O_1D^{-1} = O_2\Lambda O_2^\top. \tag{163}$$

The matrix O_2 is thus determined by the fact that it diagonalizes $\tilde{T}_4(t)$. Since all eigenvalues of the diagonal matrix Λ are distinct this determines O_2 up to a discrete ambiguity, *i.e.*, O_2 can be multiplied from the right by an arbitrary element of the centralizer of Λ in $O(n)$:

$$O_2 \sim O_2 S, \qquad S\Lambda S^\top = \Lambda, \quad S \in O(n),$$

i.e.,

$$S = \mathrm{diag}(\epsilon_1, \epsilon_2, \ldots, \epsilon_n), \qquad \epsilon_i = \pm 1. \tag{164}$$

Again we absorb this ambiguity into T_0 and require that the matrices $O_1(t)$ and $O_2(t)$ be smooth functions of t.

Thus $G(t)$ in (161) and (156) has been completely determined. Inverting (154), we obtain

$$W = [W_2T + W_1(W_1 - W_2)^{-1}][T + (W_1 - W_2)^{-1}]^{-1}. \tag{165}$$

Inserting (156) into (165) and performing some obvious manipulations, we obtain the final superposition formula (150). *Q. E. D.*

The case of the real symplectic MRE, based on the group *Sp(2n,R)* is very similar to the complex one. Some algebraic complications arise, due to the fact that, *e.g.* (159) must be modified and that various real forms of *O(n,C)* may come up, not only the compact form *O(n,R)*. It is however true that the general symmetric solution of the real symplectic MRE can be expressed in terms of four particular generically chosen solutions. For details we refer to the original article [70].

§1.7 Relation to Bäcklund transformations for nonlinear partial differential equations

1.7.1 The method of pseudopotentials and ODE's with superposition principles

The method of *pseudopotentials* provides a certain unity to a variety of different solution techniques for certain classes of nonlinar PDE's [37,38,45,69,71,86,87,99–101,113,120,129]. We have in mind precisely the type of PDE's treated at this School and Workshop in the lectures of Mark Ablowitz, Athanassios Fokas, and others, namely those which can be treated by inverse scattering, Riemann–Hilbert and similar techniques. The pseudopotential approach for such equations makes it possible to derive a finite or infinite number of conservation laws, as well as Bäcklund transformations and the linear equations of the inverse scattering method.

We cannot go into all the details here, but we shall use some examples to establish a relation between this approach to PDE's and the superoposition principles for ODE's discussed in these lectures.

Let us consider a nonlinear ODE, for simplicity, of second order and involving one dependent variable u and two independent ones, ξ and η:

$$H(\xi,\eta,u_{\xi\xi},u_{\eta\eta},u_{\xi\eta},u_\eta,u_\xi,u)=0. \tag{166}$$

We shall call a ***pseudopotential***, a quantity $y^\mu(\xi,\eta)$, $\mu=1,2,\ldots,n$, satisfying an overdetermined system of first order differential equations

$$y^\mu_\xi(\xi,\eta)=F^\mu(\xi,\eta,u,u_\xi,u_\eta,\ldots,u^*,u^*_\xi,u^*_\eta,\ldots,y^\mu,y^{\mu *}), \tag{167a}$$
$$y^\mu_\eta(\xi,\eta)=G^\mu(\xi,\eta,u,u_\xi,u_\eta,\ldots,u^*,u^*_\xi,u^*_\eta,\ldots,y^\mu,y^{\mu *}), \tag{167b}$$

such that the compatibility conditions

$$F^\mu_\eta=G^\mu_\xi,\qquad \mu=1,2,\ldots,n, \tag{168}$$

are equivalent to the original equation (166). Throughout, the subscripts denote derivatives, the superscripts, components of the pseudopotentials. The asterisks in (167) denote complex conjugates (if u and y are allowed to be complex), the dots indicate possible presence of higher derivatives. For a more general and more rigorous fromulation we refer to *e.g.* [113].

Given the equation (166), the problem is to obtain the equations (167), (*i.e.*, the functions F^μ and G^μ), and then to solve them for $y^\mu(\xi,\eta)$.

Several interesting situations can occur:

1. If equations (167) do not involve y^μ on the right-hand side or, more generally, if y^μ can be eliminated from the right-hand side, then y^μ is simply a potential and (168) is a conservation law.
2. The pseudopotential $y^\mu(\xi,\eta)$ may itself satisfy a partial differential equation (or a system of PDE's) obtained by eliminating the original function u from (167) and equations obtained by differentiating

(167) sufficiently many times, and using (166). Let us denote this equation

$$K(\xi,\eta,y^\mu,y^\mu_\xi,y^\mu_\eta,y^\mu_{\xi\xi},y^\mu_{\xi\eta},y^\mu_{\eta\eta},\ldots)=0. \tag{169}$$

The equations (167) now provide a ***Bäcklund transformation***: given a solution of the nonlinear PDE (166) we can put it into (167), solve for y and thus obtain a solution of (169). In the special case when (169) and (166) coincide, we have an *inner Bäcklund transformation* or *auto-Bäcklund transformation* and hence a procedure for generating new solutions of (166) from old ones. If the two equations do not coincide, we have an *outer Bäcklund transformation*. In special cases such transformations are of considerable interest, *e.g.*, they may relate nonlinear equations to linear ones. Such is the case of the relation between the Liouville equation

$$u_{\xi\eta}=e^u, \tag{170}$$

and the wave equation $u_{\xi\eta}=0$ [13].

3. In some cases [35,37,62,63,69,86], pseudopotentials can be used to generate inner Bäcklund transformations even if $y^\mu(\xi,\eta)$ does itself not satisfy the original PDE. Indeed, it may be possible to construct a further transformation

$$z(\xi,\eta)=\phi(\xi,\eta,u,u_\xi,u_\eta,\ldots,y^\mu), \tag{171}$$

such that $z(\xi,\eta)$ satisfies (166) whenever u satisfies this equation and y^μ satisfies the pseudopotential equation (167).

4. If F^μ and G^μ in (167) are linear in the pseudopotential y, then these equations can be interpreted as inverse scattering equations [1,2,4,32,45,56,71,87,88,139] and used to obtain large classes of solutions of (166).

Wahlquist and Estabrook [45,71,113,129] have developed a systematic method of deriving the pseudopotential equations (167) for a given PDE (166) —when they exist. Their method consists of replacing the equation (or equations) (166) by a closed Pfaffian system of differential forms, and requiring that these forms together with

$$\theta^\mu=dy^\mu-F^\mu\,d\xi-G^\mu\,d\eta \tag{172}$$

generate a differential ideal. This requirement leads to a set of commutation relations between vector fields in the variables y^μ. These commutation relations must then be solved to obtain the vector fields explicitly and these in turn determine the functions F^μ and G^μ.

As an example, let (166) be the generalized nonlinear Schrödinger equation [69],

$$iu_t+u_{xx}=f(u,u^*), \tag{173}$$

where f is some function of u and its complex conjugate. Let us look for a pseudopotential $y^\mu(x,t)$ satisfying

$$y^\mu_x(x,t)=F^\mu(u,u^*,u_x,u^*_x,u_t,u^*_t,y^\nu), \tag{174a}$$
$$y^\mu_t(x,t)=G^\mu(u,u^*,u_x,u^*_x,u_t,u^*_t,y^\nu). \tag{174b}$$

Following the Walhquist–Estabrook procedure (for details, see [69]), we quite easily obtain the dependence of F^μ and G^μ on the orginal variable u:

$$\begin{aligned}F^\mu&=i[|u|^2Q^\mu(y)+uP^\mu(y)-u^*R^\mu(y)+U^\mu(y)],\\G^\mu&=(uu^*_x-u^*u_x)Q^\mu(y)-u_xP^\mu(y)-u^*_xR^\mu(y)\\&\quad+i[|u|^2V^\mu(y)-uX^\mu(y)-u^*T^\mu(y)+S^\mu(y)].\end{aligned} \tag{175}$$

The dependence on y^μ is thus isolated in the eight unknown vector-valued functions $Q(y), \ldots, S(y)$ of the vector $y = (y^1, y^2, \ldots, y^n)$. In order to obtain a closed differential ideal, the functions $Q^\mu, \ldots, S^\mu$ must satisfy certain differential equations, best expressed in terms of commutation relations, satisfied by the vector fields [45,129]. Thus we introduce the vector fields

$$\hat{Q} = Q^\mu(y)\frac{\partial}{\partial y^\mu}, \quad \hat{P} = P^\mu(y)\frac{\partial}{\partial y^\mu}, \quad \ldots, \quad \hat{S} = S^\mu(y)\frac{\partial}{\partial y^\mu}, \tag{176}$$

and

$$\hat{F} = y^\mu_x \frac{\partial}{\partial y^\mu}, \qquad \hat{G} = y^\mu_t \frac{\partial}{\partial y^\mu}. \tag{177}$$

Equations (174) and (175) can now be rewritten as operator equations

$$\begin{aligned} \hat{F} &= i[|u|^2\hat{Q} + u\hat{P} - u^*\hat{R} + \hat{U}], \\ \hat{G} &= (uu^*_x - u^*u_x)\hat{Q} - u_x\hat{P} - u^*_x\hat{R} + i[|u|^2\hat{V} - u\hat{X} - u^*\hat{T} + \hat{S}]. \end{aligned} \tag{178}$$

The compatibility condition (168), in this case

$$F^\mu_t = G^\mu_x,$$

implies that the operators in (178) must satisfy the following commutation relations [69]:

$$[\hat{P}, \hat{R}] = \hat{V}, \quad [\hat{P}, \hat{U}] = \hat{X}, \quad [\hat{R}, \hat{U}] = \hat{Y},$$

$$[\hat{Q}, \hat{P}] = 0, \quad [\hat{Q}, \hat{R}] = 0, \quad [\hat{Q}, \hat{U}] = 0, \tag{179}$$

$$[|u|^2\hat{V} - u\hat{X} - u^*\hat{Y} + \hat{S}\,,\, |u|^2\hat{Q} + u\hat{P} - u^*\hat{R} + \hat{U}] = (u^*\hat{Q} + \hat{P})f(u,u^*) + (-u\hat{Q} + \hat{R})f^*(u,u^*),$$

where $f(u,u^*)$ is the function figuring in (173).

A pseudopotential $y^\mu(x,t)$ for the nonlinear Schrödinger equation is obtained in the form (174) from (175) whenever a solution of the commutation relations (179) exists. These equations are compatible only under certain restrictions on the function $f(u,u^*)$ and we shall not go into this here. The usual method for solving equations of the type (179) is to assume that the vector fields involved, in this case given by (176), all lie within some finite-dimensional Lie algebra.

Returning now to Lie's theorem on fundamental sets of solutions, we see that the requirement that the vector fields (176) lie within a finite-dimensional Lie algebra, coincides with condition (5) on the vector fields (4). Equations (174a), considered as ordinary differential equations in the independent variable x (with t as a parameter) and equations (174b) considered as ODE's in the independent variable t (with x as a parameter) will hence by necessity have the form (3) and will satisfy Lie's theorem.

The same is true in the general case of the pseudopotential equation (167) for the PDE (166). We can conclude that a large class of pseudopotential equations, Bäcklund transformations, inverse scattering equations, etc., namely, all those that can be obtained by the Wahlquist–Estabrook method in its usual application, will have the form (3) and will satisfy the conditions of Lie's theorem on fundamental sets of solutions.

In particular, if we are considering a scalar second-order ODE as in (166) and looking for an auto-Bäcklund transformation, then the pseudopotential $y(\xi,\eta)$ must also be a scalar, and we can drop the label μ. The vector fields figuring in the commutation relations of the type (179) will depend on one variable only, *i.e.*, they will have the form

$$\hat{X}_a = X_a(y)\frac{d}{dy}.$$

As we have seen in §1.2, such vector fields must lie in the Lie algebra $sl(2,R)$ (or $sl(2,C)$ if y is complex), if they are contained in a finite-dimensional Lie algebra at all. The pseudopotential equations in this case will by necessity have the form of Riccati equations.

As an illustration, consider a special case of the nonlinear Schrödinger equation (173), namely the *cubic Schrödinger equation*

$$iu_t + u_{xx} = \epsilon u|u|^2, \epsilon = \pm 1. \tag{180}$$

The solution of equations (179) in this case yields an outer Bäcklund transformation [45,69]:

$$\begin{aligned} y_x &= -\tfrac{1}{2}(zy^2 - \epsilon z^* + 2ky), \\ y_t &= -\tfrac{1}{2}ik(zy^2 - \epsilon z^* + 2ky) + \tfrac{1}{2}i(-2xy^2 - \epsilon z^*_x + \epsilon |z|^2 y), \end{aligned} \tag{181}$$

where k is an arbitrary complex parameter. Notice that both equations (181) are Riccati equations for the function y.

Riccati equations also appear as Bäcklund transformations for the sine–Gordon equation, the Korteweg–de Vries equation, and others. As a matter of fact, I do not know of any example in the literature where a Bäcklund transformation of the form (167) does not satisfy the conditions (4) and (5) of Lie's theorem on fundamental sets of solutions.

1.7.2 Bäcklund transformations for the nonlinear σ-model and the matrix Riccati equation

Much attention in elementary particle physics has been devoted over the past few years to a class of nonlinear field theories in two dimensions, known as σ models [42,43,62,63,104,114,119,140]. One version [62,63,104,140] of these models, called the principal chiral fields model, involves fields $g(\xi,\eta)$ taking their values in some Lie group G. We shall consider the $U(n)$ principal chiral model, *i.e.*, the case when G is the unitary group $G = U(n)$.

The field equations for this model are

$$\begin{aligned} (g^\dagger g_\xi)_\eta + (g^\dagger g_\eta)_\xi &= 0, \\ (g_\xi g^\dagger)_\eta + (g_\eta g^\dagger)_\xi &= 0, \end{aligned} \tag{182}$$

where

$$\xi = \tfrac{1}{2}(x+t), \qquad \eta = \tfrac{1}{2}(x-t),$$

are light-cone variables, and

$$gg^\dagger = g^\dagger g = I, \qquad g \in C^{n\times n}, \tag{183}$$

(the superscript † denotes hermitian conjugation). Ogielski *et al.* [104] have presented a Bäcklund transformation for equations (182), namely

$$\begin{aligned} 2(1-\beta)(g'+g)_\xi &= (g'-g)(g^\dagger_\xi g' + g'^\dagger g_\xi), \\ 2(1+\beta)(g'-g)_\eta &= -(g'+g)(g^\dagger_\eta g' + g'^\dagger g_\eta), \end{aligned} \tag{184}$$

where β is a real parameter. By construction, (182) is the compatibility condition for the two equations (184), g' is subject to the constraints

$$g^\dagger g' + g'^\dagger g = 2\beta I, \qquad g'^\dagger g' = I, \quad |\beta| \le 1, \tag{185}$$

thus (184) transforms a solution $g \in U(n)$ into a solution $g' \in U(n)$.

Performing some simple manipulations using the constraint (185), we can transform the Bäcklund transformation (184) into the form

$$\begin{aligned} g'_\xi &= \frac{1}{2(1-\beta)}[-g_\xi + g'g^\dagger g_\xi - gg_\xi^\dagger g' + g'g_\xi^\dagger g'], \\ g'_\eta &= \frac{1}{2(1+\beta)}[g_\eta + g'g^\dagger g_\eta - gg_\eta^\dagger g' - g'g_\eta^\dagger g']. \end{aligned} \tag{186}$$

Equations (186) are now matrix Riccati equations for the matrix g'; the coefficients A, B, C, and D of (47–49), *e.g.* for the first of equations (186), are

$$\xi(t) = \begin{pmatrix} C & A \\ -D & -B \end{pmatrix} = \frac{-1}{2(1-\beta)}\begin{pmatrix} gg_\xi^\dagger & g_\xi \\ g_\xi^\dagger & g^\dagger g_\xi \end{pmatrix}. \tag{187}$$

Notice that $\xi(t)$ satisfies

$$\xi K + K\xi^\dagger = 0, \qquad K = \begin{pmatrix} I & 0 \\ 0 & -I \end{pmatrix}, \tag{188}$$

i.e., we have $D^\dagger = -A$, $B^\dagger + B = 0$, and $C^\dagger + C = 0$, since $gg^\dagger = g^\dagger g = I$. Thus the MRE (186) is based on the action of the group $U(n,n)$ on a submanifold of the complex Grassmannian $G_n(C^{2n})$. As in the general case (49) we can replace (186) by a system of $2n \times 2n$ linear equations (48), namely

$$\begin{pmatrix} X_\xi \\ Y_\xi \end{pmatrix} = \frac{-1}{2(1-\beta)}\begin{pmatrix} gg_\xi^\dagger & g_\xi \\ g_\xi^\dagger & g^\dagger g_\xi \end{pmatrix}\begin{pmatrix} X \\ Y \end{pmatrix}, \tag{189}$$

where $g' = XY^{-1}$ (and similarly for the second equation).

We have thus obtained another example of the type discussed in §1.7.1: the Bäcklund transformation of the nonlinear σ-model can be cast into the form of a MRE, and hence satisfies the condition of Lie's theorem. The general solution g' of Eqs. (186) can be expressed in terms of five particular solutions for each specific g. The system (186) can be replaced by the linear system (189) which in turn can serve as an inverse scattering system for the σ-model field equations (182).

Harnad *et al.* [62,63] have reduced the Bäcklund transformation for the nonlinear σ-model to a MRE in a different manner. Instead of (184) they use an equivalent version of the Bäcklund transformation, namely

$$\begin{aligned} g_\xi g^\dagger - g'_\xi g'^\dagger &= -\lambda(gg'^\dagger)_\xi, \\ g_\eta g^\dagger - g'_\eta g'^\dagger &= \lambda(gg'^\dagger)_\eta, \quad g, g' \in U(n), \end{aligned} \tag{190}$$

with the constraint

$$\lambda gg'^\dagger + \bar\lambda g'g^\dagger = (\lambda + \bar\lambda)I = 2\beta I, \tag{191}$$

where λ is a complex scalar constant. Introducing a new unitary matrix

$$U \equiv gg'^\dagger, \qquad UU^\dagger = I, \tag{192}$$

they rewrite the Bäcklund transformation and constraint (191) as

$$U_\xi = \frac{1}{|1+\lambda|^2}[\bar\lambda A_0 + A_0 U - (1 + \lambda + \bar\lambda)UA_0 + \lambda UA_0U], \tag{193a}$$

$$U_\eta = \frac{1}{|1-\lambda|^2}[-\bar\lambda B_0 + B_0 U - (1 - \lambda - \bar\lambda)UB_0 - \lambda UB_0U], \tag{193b}$$

where

$$A_0 \equiv g'_\xi g'^\dagger = -A_0^\dagger, \qquad B_0 = g'_\eta g'^\dagger = -B_0^\dagger, \tag{194}$$

and the constraint (191) reduces to

$$\lambda U + \overline{\lambda} U^\dagger = \lambda + \overline{\lambda}. \tag{195}$$

Again, the Bäcklund transformation (193) has the form of a pair of matrix Riccati equations, where the matrix (47) is

$$\xi_1(t) = \begin{pmatrix} C & A \\ -D & -B \end{pmatrix} = \frac{1}{|1+\lambda|^2} \begin{pmatrix} A_0 & \overline{\lambda} A_0 \\ -\lambda A_0 & (1+\lambda+\overline{\lambda}) A_0 \end{pmatrix}, \tag{196a}$$

and

$$\xi_2(t) = \frac{1}{|1-\lambda|^2} \begin{pmatrix} B_0 & -\overline{\lambda} B_0 \\ \lambda B_0 & (1-\lambda-\overline{\lambda}) B_0 \end{pmatrix}, \tag{196b}$$

for (193a) and (193b), respectively. The matrices satisfy (188) so that we again have a $U(n,n)$ MRE. In addition to the $U(n,n)$ conditions (188), the matrices $\xi_1(t)$ and $\xi_2(t)$ satisfy further constraints so that the curves $\xi_i(t)$ lie in a subgroup of $U(n,n)$, namely

$$\xi_i(t) \in GL(n,C) \subset U(n,n). \tag{197}$$

This is a situation analogous to the one discussed in § 1.5.6.2, and it makes it possible to diagonalize the two matrices (196) simultaneously, and thus linearize the MRE (193). indeed, put

$$T = \begin{pmatrix} I & \overline{\lambda} I \\ I & \lambda I \end{pmatrix}, \qquad \lambda - \overline{\lambda} \neq 0. \tag{198}$$

We then have

$$T^{-1}\xi_1 T = \begin{pmatrix} 1+\overline{\lambda} & 0 \\ 0 & 1+\lambda \end{pmatrix} \otimes A_0, \qquad T^{-1}\xi_2 T = \begin{pmatrix} 1-\overline{\lambda} & 0 \\ 0 & 1-\lambda \end{pmatrix} \otimes B_0. \tag{199}$$

It follows that if U satisfies (193), then

$$V = -(\lambda U - \overline{\lambda} I)(U - I)^{-1}, \tag{200}$$

satisfies the linear equations

$$\begin{aligned} V_\xi &= (1+\overline{\lambda}) A_0 V - (1+\lambda) V A_0, \\ V_\eta &= (1-\overline{\lambda}) B_0 V - (1-\lambda) V B_0, \end{aligned} \tag{201}$$

Notice that, contrary to Eqs. (189), these equations do not involve a doubling of the dimension of the space of dependent variables: $V \in GL(n,C)$. Again, (200) can serve as the inverse scattering equations for the system (182). The exact relation with the Zakharov–Mikhaĭlov [140] linear system for the σ-model is established in Refs. [62,63], where the authors use the Bäcklund transformation described above, as well as its linearization, to obtain a permutability theorem for the σ-model.

1.7.3 Comments on soliton superposition laws

Unfortunately, the term *nonlinear superposition principles* has been used in the literature on nonlinear equations in at least two different senses.

This entire Chapter, as well as Refs. [9,11,12,41,70,116,117,124,136] are devoted to superposition principles for nonlinear ODE's, as defined in §1.1 and §1.2. We are obtaining the general solutions of a system of ODE's from a finite number of particular solutions, rather than superimposing, say, two solutions to get a third one.

Soliton superposition principles [4,13,32,87,139], on the other hand, refer to solutions of PDE's such as the sine–Gordon equation, the Korteweg–de Vries equation, the σ-model field equations (182), and many others. They make it possible to combine very specific types of solutions into new solutions. No statement about general solutions of the corresponding PDE's is involved. Typically, this type of soliton superposition principle makes it possible to combine two single-soliton solutions into a two-soliton solution, and more generally, to add a further soliton to an n-soliton solution.

The two types of superposition principles are thus quite different. The intriguing fact is that the PDE's which allow for soliton solutions and soliton superposition principles, also allow for Bäcklund transformations. In turn, these Bäcklund transformations invariably have the form of ODE's satisfying Lie's criteria and hence having superposition principles in the sense used throughout these lectures.

The relation between the two types of superposition laws remains to be clarified.

§1.8 Nonlinear superposition principles and numerical solutions of the matrix Riccati equations

One of the applications of the superposition formulas in general, and those for the matrix Riccati equation in particular, is to provide new numerical methods for solving nonlinear ODE's.

For the $SL(2n,R)$ or $SL(2n,C)$ square MRE's with variable coefficients it is necessary to generate five numerical solutions forming a fundamental set of solutions, and then to use the superposition formula to calculate further solutions. Extensive numerical calculations have been performed [116,117], leading to the following conclusions:

1. For relatively large dimensions of the matrices involved, the superposition formula is less computer-time consuming than traditional methods (*e.g.* Runge–Kutta calculations). For $n = 2$ or 3, the Runge–Kutta method is faster, for $n \sim 10$, the superposition formula is preferable.

2. In traditional methods, the equation is solved from point to point. If a pole is encountered for certain initial conditions at some $t = t_0$, the calculation wil stop at some $t_0 - \Delta t$ ($\Delta t > 0$). To continue beyond the pole is a problem that has to be treated separately. The situation is quite different when a superposition formula is used, *e.g.* (95). The solution is then given by a formula of the type
$$W(t) = [C_1(t)U + C_2(t)][C_3(t)U + C_4(t)]^{-1}, \tag{202}$$
where $C_1(t), \ldots, C_4(t)$ are evaluated once and for all using known (well behaved) solutions. Thus a data bank is created and (202) is used to calculate $W(t)$ for any given initial value matrix U and any time t. The analyticity properties of the solution are already built in, so (202) will give the solution for any t, irrespectively of singularities which may have occured at some points $t_0 < t_1 < \cdots < t_k < t$.

3. Formula (202) makes it possible to approach singularities from below and above and thus to pinpoint their positions t_i.

4. The superposition formula (95) based on the reconstruction of the group element is more stable with respect to perturbations of the input solutions than the *polynomial* superposition formula (103), (104).

Let us now look at an example (from an M. Sc. thesis by D. Rand [116]). We restrict ourselves to a low dimension ($n = 3$) simply because we want to present the results on graphs.

First consider a real symplectic MRE (148)

$$\dot{W} = A + WB + B^{\top}W + WDW, \qquad A = A^{\top}, \quad D = D^{\top}, \tag{203}$$

with constant coefficients

$$A = \begin{pmatrix} 0 & 0 & 0 \\ 0 & 0 & 0 \\ 0 & 0 & 1 \end{pmatrix}, \quad B = \begin{pmatrix} -\frac{1}{2} & 1 & 0 \\ -1 & -\frac{1}{2} & 0 \\ 0 & 1 & 0 \end{pmatrix}, \quad D = -\begin{pmatrix} 1 & 2 & 1 \\ 2 & 4 & 2 \\ 1 & 2 & 1 \end{pmatrix}. \tag{204}$$

We shall apply the superposition rule (150), making use of four basic solutions, chosen to satisfy the initial conditions

$$W_1(0) = I, \quad W_2(0) = 0, \quad W_3(0) = \tfrac{1}{2}I, \quad W_4(0) = \begin{pmatrix} 1 & & \\ & \frac{1}{2} & \\ & & \frac{1}{3} \end{pmatrix}. \tag{205}$$

Equations (151) and (149b) then yield

$$T_3(0) = I, \qquad R = \begin{pmatrix} 0 & & \\ & 1 & \\ & & 2 \end{pmatrix}. \tag{206}$$

The conditions of Theorem 1.4 are satisfied, so the matrices O_1, D, and O_2 can be calculated as stated in the Theorem. formula (150) then provides the values of $C_1, \ldots, C_4$ in (202). the basic solutions $W_1(t), \ldots, W_4(t)$ were calculated using the Runge–Kutta integration method. They are presented in Figures 1, ahead. In Figures 2, we present two solutions calculated using the superposition formula, corresponding respectively to

$$W_5(0) = \begin{pmatrix} 0 & 1 & 1 \\ 1 & 0 & 1 \\ 1 & 1 & 0 \end{pmatrix}, \qquad \text{and} \quad W_6(0) = -I. \tag{207}$$

Notice the pole in $W_6(t)$ at $t \sim 0.30$ sec.

For further examples, see [116,117], where matrices with $n = 2$, 3, and 10 were considered, both with constant and variable coefficients. Numerically, the most advantageous approach is to use the exponential rule (117) in the case of constant coefficients and the *diagonalization* method (95) in the case of A, B, C, and D depending on t.

§1.9 General comments and future outlook

In §1.3 we have sen that for every Lie group-subgroup pair (G, G_0) it is possible to construct a Lie algebra of vector fields corresponding to the infinitesimal action of G on the homogeneous space G/G_0. In turn, formula (20) provides us with a system of ODE's corresponding to this Lie algebra and, by Lie's theorem, admitting a fundamental set of solutions and a superposition principle.

We have considered several examples of such pairs, all leading to various types of systems of Riccati equations, *i.e.*, ordinary differential equations with quadratic nonlinearities. Two points should be stressed at this stage:

1. Not all systems of Riccati equations admit superposition principles.
2. Systems of equations which do admit superposition principles are not necessarily Riccati equations, and can have other than second-order nonlinearities.

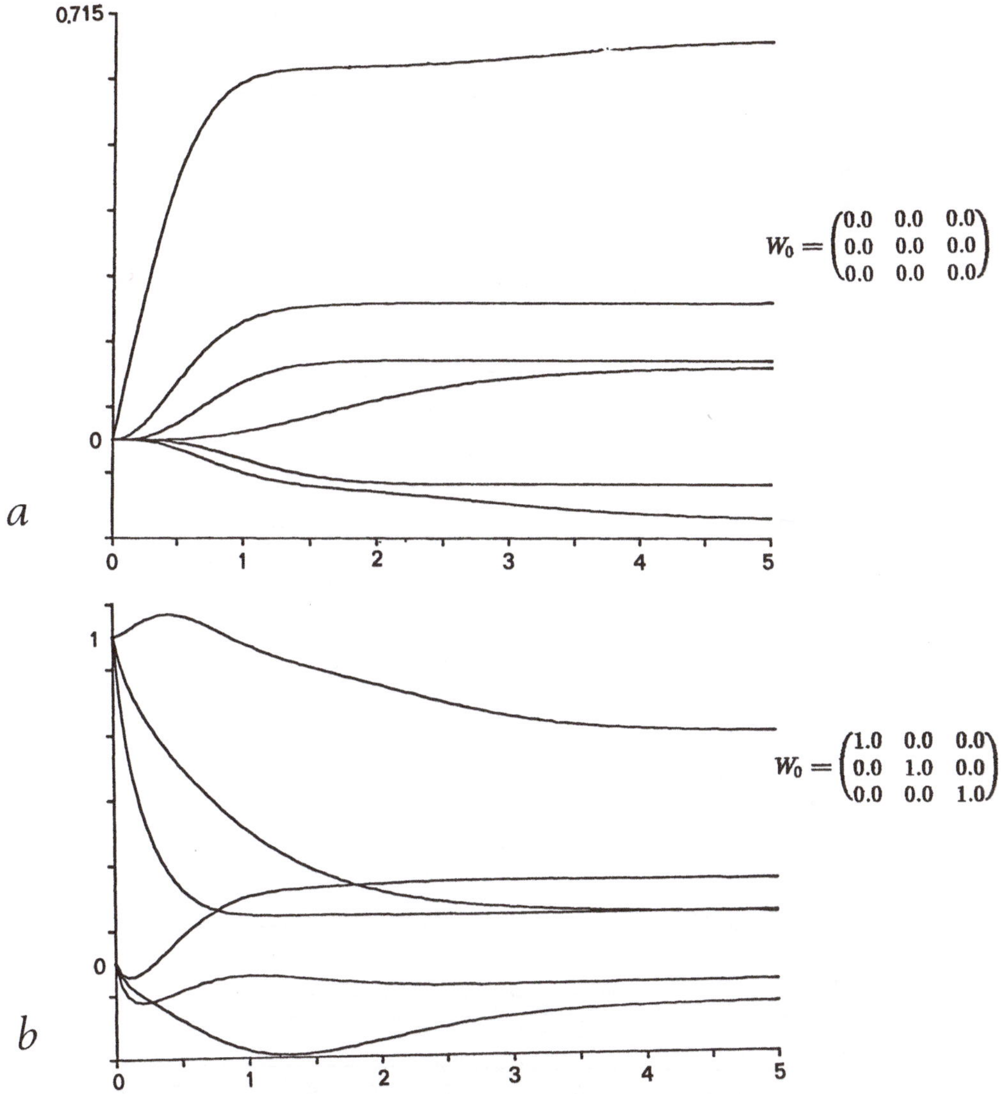

Figures 1. Basic solutions of the symplectic matrix Riccati equation, (203), calculated using the Runge–Kutta method.

To clarify point **1**, consider the most general system of n coupled Riccati equations

$$\dot{y}^{\mu} = a^{\mu}(t) + b^{\mu}_{\nu}(t)y^{\nu} + c^{\mu}_{\nu\sigma}(t)y^{\nu}y^{\sigma}. \tag{208}$$

The vector fields associated with this system are

$$\hat{A}_{\mu} = \frac{\partial}{\partial y^{\mu}}, \qquad \hat{B}^{\nu}_{\mu} = y^{\nu}\frac{\partial}{\partial y^{\mu}}, \qquad \hat{C}^{\nu\sigma}_{\mu} = y^{\nu}y^{\sigma}\frac{\partial}{\partial y^{\mu}}.$$

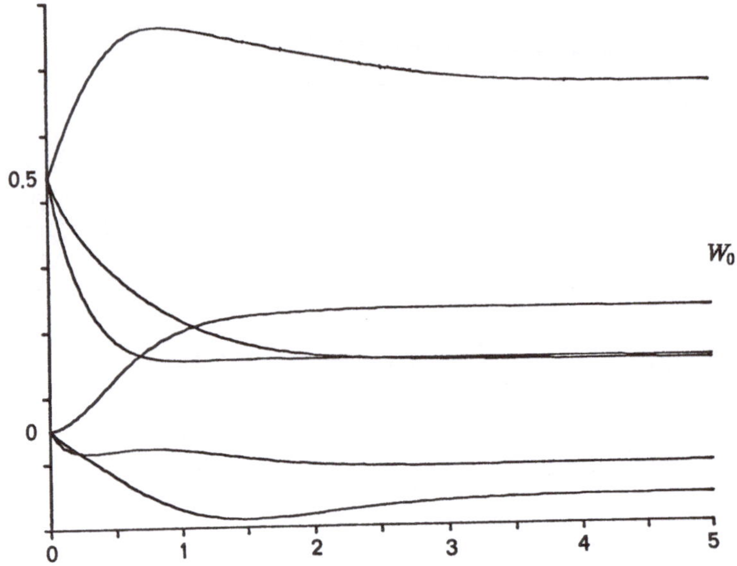

$$W_0 = \begin{pmatrix} 0.5 & 0.0 & 0.0 \\ 0.0 & 0.5 & 0.0 \\ 0.0 & 0.0 & 0.5 \end{pmatrix}$$

c

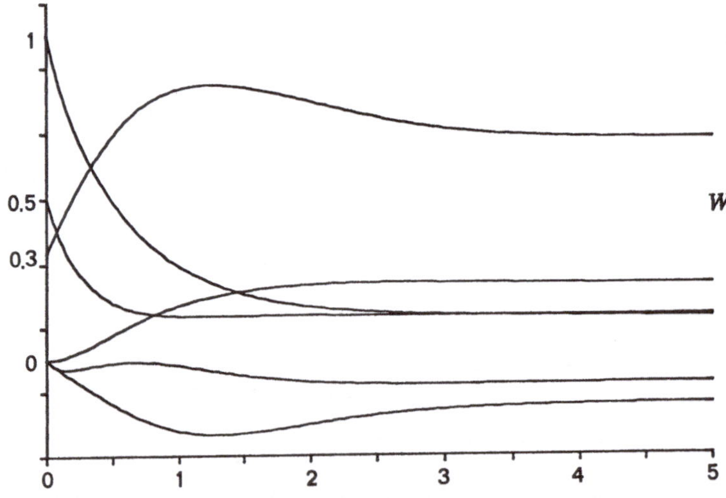

$$W_0 = \begin{pmatrix} 1.0 & 0.0 & 0.0 \\ 0.0 & 0.5 & 0.0 \\ 0.0 & 0.0 & 0.3 \end{pmatrix}$$

d

Figures 1. (Cont.)

Calculating the commutator $[\hat{C}_{\mu}^{\nu\sigma}, \hat{C}_{\mu'}^{\nu'\sigma'}]$ we obtain new vector fields with cubic coefficients. Commuting further we obtain an infinite-dimensional Lie algebra of vector fields with polynomial coefficients of all orders.

We are presently in the process of clarifying the conditions to be imposed on the coefficients a^{μ}, b_{ν}^{μ}, and $c_{\nu\sigma}^{\mu}$ in order to obtain a finite-dimensional Lie algebra. Alternatively, one can ask what are the conditions imposed on the groups G and G_0 for the vector fields representing the Lie algebra of G on $M = G/G_0$ to have polynomial coefficients of second order, in appropriately chosen local coordinates.

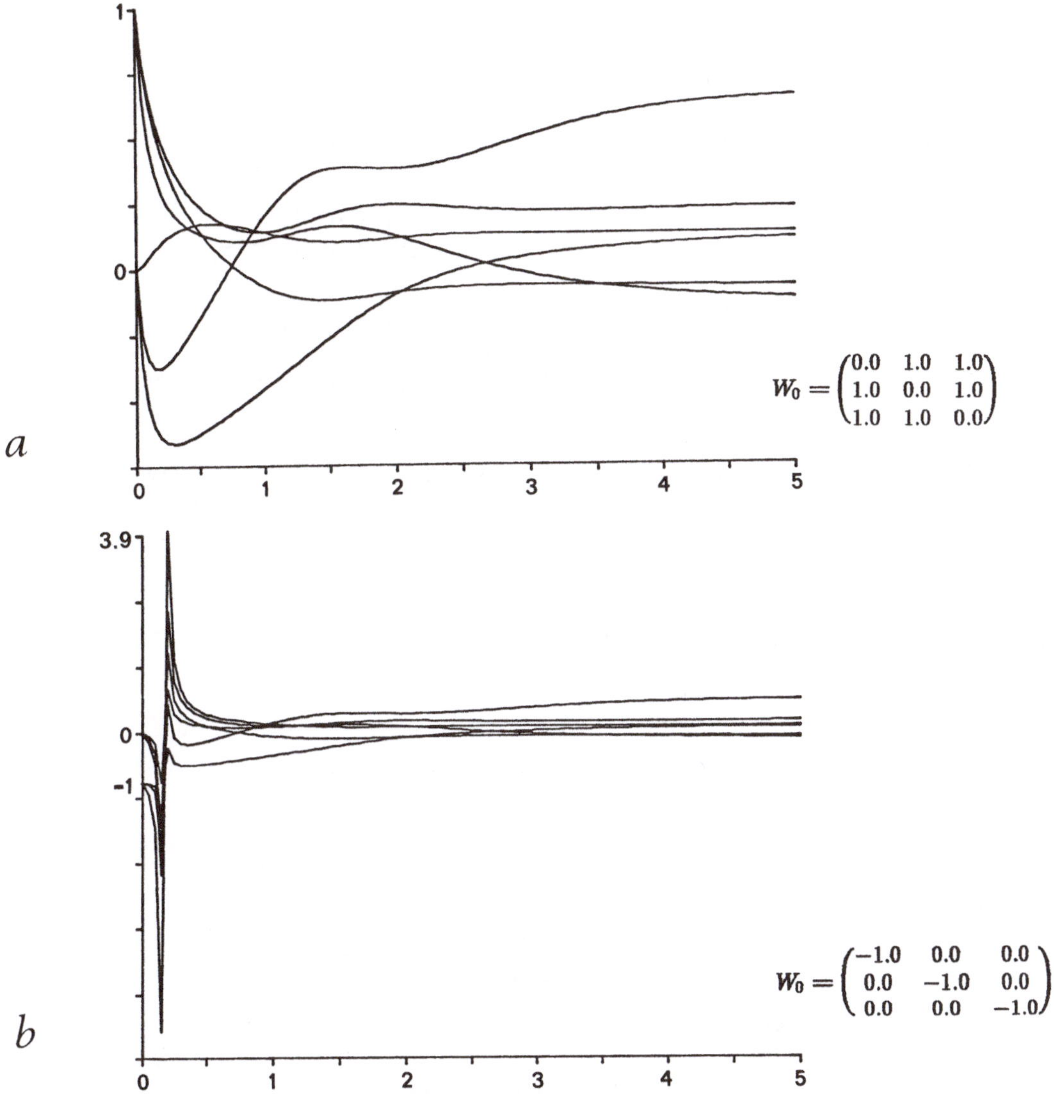

Figures 2. Solutions of the matrix Riccati equation (203), calculated using the superposition formula (150).

An example of a system of nonlinear ODE's admitting a superposition principle and having higher-order nonlinearities is obtained, *e.g.* by taking the pseudo-orthogonal group $G = O(p,q)$ and a subgroup G_0 of $O(p,q)$ leaving a k-dimensional $(1 \leq k \leq q \leq p)$ completely isotropic space invariant. To see this, take a convenient realization of the Lie algebra $o(p,q)$ by real $n \times n$ matrices, with $n = p+q$:

$$\begin{gathered} o(p,q) = \{X \mid X \in R^{(p+q)\times(p+q)}, \qquad XK + KX^\top = 0\}, \\ K = K^\top, \qquad \det K \neq 0, \qquad \operatorname{sign} K = (p,q). \end{gathered} \tag{209}$$

A convenient choice of nonsingular symmetric matrix K with signature (p,q) is

$$K = \begin{pmatrix} & & I_k \\ & \tilde{I} & \\ I_k & & \end{pmatrix}, \qquad \tilde{I} = \begin{pmatrix} I_{p-k} & \\ & -I_{q-k} \end{pmatrix}. \tag{210}$$

The matrix $X \in o(p,q)$ then has the form

$$X = \begin{pmatrix} X_{11} & X_{12} & X_{13} \\ X_{21} & X_{22} & -\tilde{I}X_{12}^{\top} \\ X_{31} & -X_{21}^{\top}\tilde{I} & -X_{11}^{\top} \end{pmatrix}, \qquad \begin{aligned} &X_{11} \in R^{k\times k}, \\ &X_{12} \in R^{k\times(p+q-2k)}, \\ &X_{21} \in R^{(p+q-2k)\times k}, \\ X_{13} = {}&-X_{13}^{\top} \in R^{k\times k}, \\ X_{31} = {}&-X_{31}^{\top} \in R^{k\times k}, \\ X_{22} = {}&\tilde{I}X_{22}^{\top}\tilde{I} \in R^{(p+q-2k)\times(p+q-2k)}. \end{aligned} \tag{211}$$

Putting $X_{12} = 0$ and $X_{13} = 0$ in (211), we obtain a subalgebra $L_0 \subset L = o(p,q)$. The corresponding Lie group $G_0 = \{\exp L_0\}$ will leave a k-dimensional completely isotropic (light-like) space invariant.

To parametrize the space G/G_0, we introduce homogeneous coordinates for a family of k planes in R^{p+q}:

$$U = \begin{pmatrix} U_1 \\ U_2 \\ U_3 \end{pmatrix} \sim \begin{pmatrix} U_1 G \\ U_2 G \\ U_3 G \end{pmatrix}, \qquad \begin{aligned} U_1, U_3 &\in R^{k\times k}, \\ U_2 &\in R^{(p+q-2k)\times k}, \\ G &\in GL(k,R). \end{aligned} \tag{212}$$

The group $O(p,q)$ acts transitively on a submanifold of isotropic vectors:

$$U^{\top}KU = 0, \qquad i.e., \quad U_1^{\top}U_3 + U_3^{\top}U_1 = -U_2^{\top}\tilde{I}U_2. \tag{213}$$

To get rid of the redundancy inherent in (212), we introduce affine coordinates

$$Z = \begin{pmatrix} Z_1 \\ Z_2 \end{pmatrix} = \begin{pmatrix} U_1 U_3^{-1} \\ U_2 U_3^{-1} \end{pmatrix}, \qquad \det U_3 \neq 0. \tag{214}$$

The isotropy condition (213) translates into

$$Z_1^{\top} + Z_1 = -Z_2^{\top}\tilde{I}Z_2. \tag{215}$$

Thus, $Z_2 \in R^{(p+q-2k)\times k}$ is arbitrary, the matrix Z_1 must be decomposed into its symmetric and antisymmetric parts

$$Z_1 = \tfrac{1}{2}[Z_1 + Z_1^{\top}] + \tfrac{1}{2}[Z_1 - Z_1^{\top}] = Z_1^S + Z_1^A; \tag{216}$$

$Z_1^A \in R^{k\times k}$ is an arbitrary antisymmetric matrix, Z_1^S is given by (215). Dropping all further details, let us just write down the final system of equations. In homogeneous coordinates the infinitesimal action of $G = O(p,q)$ on G/G_0 gives rise to the linear system:

$$\begin{pmatrix} \dot{U}_1 \\ \dot{U}_2 \\ \dot{U}_3 \end{pmatrix} = \begin{pmatrix} X_{11} & X_{12} & X_{13} \\ X_{21} & X_{22} & -\tilde{I}X_{12}^{\top} \\ X_{31} & -X_{21}^{\top}\tilde{I} & -X_{11}^{\top} \end{pmatrix} \begin{pmatrix} U_1 \\ U_2 \\ U_3 \end{pmatrix}, \tag{217}$$

with X_{ik} as in (211) and U_i as in (213). Introducing Z_1 and Z_2 as in (214) and using the constraint (215), we obtain in affine coordinates:

$$\begin{aligned}\dot{Z}_1^A &= X_{13} + X_{11} Z_1^A + Z_1^A X_{11}^\top + \tfrac{1}{2}(X_{12} Z_2 - Z_2^\top X_{12}^\top) - Z_1^A X_{31} Z_1^A \\ &\quad + \tfrac{1}{2}(Z_1^A X_{21}^\top \tilde{I} Z_2 + Z_2^\top \tilde{I} X_{21} Z_1^A) - \tfrac{1}{4} Z_2^\top \tilde{I}(Z_2 X_{21}^\top - X_{21} Z_2^\top)\tilde{I} Z_2 \\ &\quad - \tfrac{1}{4} Z_2^\top \tilde{I} Z_2 X_{31} Z_2^\top \tilde{I} Z_2,\end{aligned} \tag{218a}$$

$$\begin{aligned}\dot{Z}_2 &= -\tilde{I} X_{12}^\top + X_{21} Z_1^A + X_{22} Z_2 + Z_2 X_{11}^\top - \tfrac{1}{2} X_{21} Z_2^\top \tilde{I} Z_2 \\ &\quad - Z_2 X_{31} Z_1^A + Z_2 X_{21}^\top \tilde{I} Z_2 + \tfrac{1}{2} Z_2 X_{31} Z_2^\top \tilde{I} Z_2.\end{aligned} \tag{218b}$$

Thus, the constraint (215) has introduced third- and fourth-order terms in (218), in addition to the quadratic ones. Notice also that in (217) and (218) the X_{ik} are, in general, functions of the independent variable t.

In special cases the system (218) reduces to Riccati equations. Thus, if $k = 1$, then $X_{13} = X_{31} = 0$, $Z_1^A = 0$. Equation (218a) reduces to $0 = 0$ and equation (218b) reduces to the *conformal Riccati equations* considered elsewhere [9,11,12]. Similarly, iff $p = q = k$ we have $Z_2 = 0$, $Z_1 + Z_1^\top = 0$, and (218a) reduces to $O(n,n)$ Riccati equations [41]. Work is presently in progress on obtaining the superposition rule for the general case of equations (218).

The research program described in this Chapter is far from completed. On one hand we plan to obtain superposition rules for all equations obtained from pairs (G, G_0), $G_0 \subset G$, where G is a classical Lie group and G_0 a maximal subgroup. On the other hand, we are using the theory of transitive primitive filtered Lie algebras [58,82,103] to identify all finite-dimensional Lie algebras giving rise to vector fields with polynomial coefficients of different orders, and thus to ODE's with polynomial nonlinearities [122]. We are pursuing the relation between nonlinear ODE's with superposition principles and Bäcklund transformations for completely integrable systems of PDE's. Finally, it would be most desirable to extend some aspects of this approach to partial differential equations, or infinite systems of ordinary differential equations.

Chapter 2

Symmetry Reduction for Nonlinear Partial Differential Equations

§2.1 Formulation of the problem and motivation

Consider a partial differential equation, in general nonlinear, for definiteness involving n independent variables $x_1, x_2, \ldots x_n$, and one dependent variable $u = u(x_1, x_2, \ldots, x_n)$:

$$H(x_i, u, u_{x_i}, u_{x_i x_k}, u_{x_i x_k x_l}, \ldots) = 0, \qquad i = 1, 2, \ldots, n, \tag{1}$$

where the subscripts denote partial derivatives, and H is some function.

The problem to be discussed in this Chapter is that of introducing a set of new independent variables

$$\xi_a(x_1, x_2, \ldots, x_n), \qquad a = 1, 2, \ldots, k < n, \tag{2}$$

such that Eq. (1) is reduced to a PDE in the k variables (2), by imposing that u depends on the variables ξ_a only:

$$u = u(\xi_1, \xi_2 \ldots, \xi_k). \tag{3}$$

A particularly important case is the reduction of (1) to an ODE, *i.e.*, the case when the number of variables introduced in (2) is $k = 1$.

We shall consider the case when the equation (1) has a nontrivial symmetry group, *i.e.*, a Lie group G of transformatons acting on the space of independent and dependent variables, and taking solutions of (1) into solutions. We shall present a method of generating the variables ξ_a explicitly as invariants of certain subgroups of the symmetry group G. We shall call this method ***symmetry reduction*** and the variables ξ_a, ***symmetry variables***.

The idea of using the symmetry group of a PDE (or system of PDE's) to reduce the equation to an ODE (or a system of ODE's) is of course a very old one. For a discussion and further references see *e.g.* Ames [8], Bluman and Cole [20,21], Ovsyannikov [106], Ibragimov [74], and others. The symmetry variables and corresponding symmetry solutions of (1) are a generalization of what is often called, *self-similar*, *scaling*, or *similarity* solutions.

The presentation here follows two recent articles (Grundland, Harnad and Winternitz [60,61]) on relativistically invariant nonlinear scalar equations. A similar method was applied earlier to study certain classes of nonlinear time-dependent Schrödinger equations [30], and to determine invariant solutions of Yang–Mills equations in compactified Minkowski space [64–68].

The novelty of the present approach is in that we make full use of systematic classifications of the subgroups of Lie groups which have only become available quite recently [108–112].

Let us say a few words about the motivation for performing symmetry reduction. In general, lower-dimensional equations and specially ODE's are easier to solve than higher-dimensional PDE's. Symmetry reduction to ODE's will provide us with certain classes of solutions to the original PDE. These *symmetry solutions* can be then combined with other methods (*e.g.*, Bäcklund transformations) to obtain more general solutions.

A further reason why it is useful to know all possible non-equivalent reductions of a PDE to different ODE's is the so-called *Painlevé conjecture* [3,4,59,95]. According to this conjecture, a nonlinear PDE is solvable by an inverse scattering transform only if every ODE obtained by a reduction of this PDE is of Painlevé type (possibly after a transformation of variables).

We recall here that by definition an ODE is of the Painlevé type if it has no moveable critical points, *i.e.*, the only singularities of its solutions which depend on the initial conditions are poles. The critical points (branch points, essential singularities), if they exist at all, must be fixed, *i.e.*, their position must be determined by the equation itself, and not depend on the initial conditions. Linear ODE's always have the Painlevé property. Painlevé [107] and Gambier [55] studied equations of the form[1]

$$W'' = F(W', W, Z), \tag{4}$$

where F is rational in W', algebraic in W, and locally analytic in Z. They showed that only 50 canonical classes of equations of type (4) with the Painlevé property exist. Of these, 44 are reducible either to linear equations or first-order nonlinear equations (studied earlier by Fuchs in 1884), or can be integrated directly in terms of known functions. The remaining six classes of equations cannot be reduced to simpler equations and define genuinely new functions, the famous Painlevé transcendants $P_I, \ldots, P_{VI}$.

Symmetry reduction to an ODE can be thus used to test whether a given PDE satisfies the Painlevé conjecture, and could thus be completely integrable. Interestingly, the complementary approach has also proven useful. Bäcklund transformations and other results known for completely integrable PDE's have been used to obtain new properties of the Painlevé transcendants [5,23–25,52,53,115].

Finally, let us remark that symmetry reduction from an n-dimensionsal ($n \geq 3$) PDE to a two-dimensional one can also be of particular interest. Indeed, if this two-dimensional equation turns out to be completely integrable, the techniques of Bäcklund transformations, inverse scattering, etc., can be used to find families of particular solutions of the original PDE.

§2.2 Symmetry reduction for nonlinear relativistically invariant scalar equations

2.2.1 The equations and their invariance group

For the rest of these lectures we restrict ourselves to equations of the form

$$H\big(\Box u, (\nabla u)^2, u\big) = 0, \tag{5}$$

[1] For reviews, see Ince [75] and Hille[73].

where H is some function and

$$\Box u = u_{x_0x_0} - u_{x_1x_1} - \cdots - u_{x_nx_n}, \tag{6}$$

$$(\nabla u)^2 = (u_{x_0})^2 - (u_{x_1})^2 - \cdots - (u_{x_n})^2. \tag{7}$$

The independent variables $x = (x_0, x_1, \ldots, x_n)$ are thus coordinates of a point in the $(n+1)$-dimensional Minkowski space *M(n,1)*, *i.e.*, a real $(n+1)$-dimensional space with the metric

$$ds^2 = g_{\mu\nu}\, dx^\mu\, dx^\nu = (dx^0)^2 - (dx^1)^2 - \cdots - (dx^n)^2. \tag{8}$$

The group of isometries of the space *M(n,1)*, *i.e.*, the group of point transformations leaving the distance between two points in this space invariant, is the Poincaré group *P(n,1)* (the Inhomogeneous Lorentz Group). A general Poincaré transformation $g \equiv (\Lambda, a) \in P(n,1)$ acts in *M(n,1)* according to

$$\begin{gathered} x \mapsto x' = \Lambda x + a, \\ a \in R^n, \qquad \Lambda \in R^{n\times n} \\ \Lambda^\top I_{n,1}\Lambda = I_{n,1}, \qquad I_{n,1} = \begin{pmatrix} -I_n & \\ & 1 \end{pmatrix}. \end{gathered} \tag{9}$$

The superscript $^\top$ on Λ denotes transposition of the matrix. The matrix Λ corresponds to homogeneous Lorentz transformations (including rotations), the vector a to space-time translations.

Equation (5) is the most general second-order equation for a scalar function $u(x_0, x_1, \ldots, x_n)$ invariant under *P(n,1)*. It includes the nonlinear Klein–Gordon equation

$$\Box u = F\big(u, (\nabla u)^2\big), \tag{10}$$

where F is some scalar function, and the relativistic Hamilton–Jacobi equation

$$(\nabla u)^2 + V(u) = E. \tag{11}$$

To verify the Poincaré invariance of (5) it suffices to verify that

$$[T(g)u](x) = u(x') = u(\Lambda x + a), \tag{12}$$

is a solution whenever $u(x)$ is a solution. assuming that $u(x)$ is infinitely differentiable, we can make use of the infinitesimal approach, *i.e.*, represent the operator $T(g)$ as

$$T(g) = e^{\alpha X}, \tag{13}$$

where X is an element of the Lie algebra *p(n,1)* of the Poincaré group, and then expand (13) into a power series in α.

A basis for the Lie algebra *p(n,1)* is given by the infinitesimal Lorentz transformations $M_{\mu\nu}$ and translations P_μ satisfying the commutation relations

$$\begin{aligned} [M_{\mu\nu}, M_{\rho\sigma}] &= g_{\mu\rho}M_{\nu\sigma} - g_{\nu\rho}M_{\mu\sigma} + g_{\mu\sigma}M_{\rho\nu} - g_{\nu\sigma}M_{\rho\mu}, \\ [M_{\mu,\nu}, P_\sigma] &= -g_{\nu\sigma}P_\mu + g_{\mu\sigma}P_\nu, \\ [P_\mu, P_\nu] &= 0, \qquad \mu, \nu = 0, 1, \ldots, n, \\ g_{00} &= 1, \quad g_{aa} = -1, \quad a = 1, 2, \ldots, n, \quad g_{\mu\nu} = 0 \text{ for } \mu \neq \nu. \end{aligned} \tag{14}$$

When acting on scalar functions $u(x)$, $x \in M(n,1)$, the infinitesimal operators (14) of $p(n,1)$ are realized by the differential operators

$$\begin{aligned} M_{ab} &= x_a \partial_b - x_b \partial_a, && \text{(rotations)} \\ M_{0a} &= -x_0 \partial_a - x_a \partial_0, && \text{(Lorentz } boosts\text{)} \\ P_\mu &= \partial_\mu, && \text{(translations)} \\ a,b &= 1,2,\ldots,n, \quad \mu = 0,1,\ldots,n, && \partial_\mu \equiv \frac{\partial}{\partial x_\mu}. \end{aligned} \tag{14}$$

In particular, we shall call the operators

$$X_a = M_{0a} - M_{1a} = -(x_0 - x_1)\partial_a - x_a(\partial_0 - \partial_1), \qquad a = 2,3,\ldots,n, \tag{16}$$

the ***light-cone translations***. They play an important role.

For further information on the Poincaré group, we refer to Wigner's classical article [132], and various reviews [76,102,134].

2.2.2 Subgroups of the Poincaré group and their orbits in Minkowski space

The method of symmetry reduction makes use of subgroups G of the invariance group of the considered PDE, in our case the Poincaré group, and of the invariants $\xi(x_0, x_1, \ldots, x_n)$ of these subgroups in the space of independent variables, in our case $M(n,1)$.

Let us introduce some preliminary concepts. Consider a subgroup $G \subset P(n,1)$, and let it act on some generic point $\{x_p\} \in M(n,1)$. This action will sweep out some manifold in $M(n,1)$: the ***orbit*** of x_p in $M(n,1)$ under the action of G. the ***dimension*** of the orbit will be

$$d = \dim G - \dim G_0, \tag{17}$$

where G_0 is the ***isotropy group*** of x_p, *i.e.*, the subgroup of G leaving x_p invariant. The ***codimension*** k of the orbit is the dimension of the entire space, minus the dimension of the orbit:

$$k = n + 1 - d. \tag{18}$$

A subgroup G with generic orbits of codimension k will have k invariants when acting on Minkowski space

$$\xi_a(x_0, x_1, \ldots, x_n), \qquad a = 1,2,\ldots,k. \tag{19}$$

The problem of finding the invariants of a given subgroup $G \subset P(n,1)$ can be reduced to that of solving a simultaneous set of linear partial differential equations. To do this, consider the Lie algebra L of the Lie group G, and choose a basis $\{X_1, X_2, \ldots, X_m\}$ spanning L. Each of the operators X_i is a linear combination of the operators $\{M_{\mu\nu}, P_\mu\}$ of (15), with constant coefficients, and is hence a simple linear differental operator. A differentiable function $\psi(x_0, x_1, \ldots, x_n)$ will be invariant under G if and only if it satisfies the equations

$$X_i \psi(x_0, x_1, \ldots, x_n) = 0, \qquad i = 1,2,\ldots,m. \tag{20}$$

The general solution of (20) will be an arbitrary function of k *elementary invariants*:

$$\psi = \phi(\xi_1, \xi_2, \ldots, \xi_k), \tag{21}$$

where k is the codimension of the generic orbits in G. The elementary invariants are k arbitrarily chosen, functionally independent solutions of (20).

Example 1. The homogeneous Lorentz group $O(n,1)$. Consider a time-like point x_T in $M(n,1)$, say,

$$x_T = \begin{pmatrix} 0 \\ \vdots \\ 0 \\ 1 \end{pmatrix}.$$

The isotropy group of x_T is $G = O(n,1)$. The orbit of x_T under $G_0 = O(n)$ is the upper sheet of the hyperboloid

$$x_0^2 - x_1^2 - \cdots - x_n^2 = 1, \qquad x_0 \geq 1.$$

the dimension of the orbit is

$$d = \dim O(n,1) - \dim O(n) = \tfrac{1}{2}n(n+1) - \tfrac{1}{2}n(n-1) = n,$$

its codimension is

$$k = (n+1) - d = 1.$$

The group $O(n,1)$ hence has $k = 1$ elementary invariants in $M(n,1)$ satisfying

$$M_{\mu\nu}\psi(x_0, x_1, \ldots, x_n) = 0, \qquad \mu, \nu = 0, 1, \ldots, n,$$

namely,

$$\xi = \sqrt{x_0^2 - x_1^2 - \cdots - x_n^2}. \tag{22}$$

The general invariant is $\phi = \phi(\xi)$.

Example 2. The subgroup G of $P(3,1)$ generated by the Lie algebra $\{M_{01} + aP_2, P_3, a \neq 0\}$.

A general element $\exp\alpha(M_{01} + aP_2) + \beta P_3$ of this Lie group can be represented by the matrix action

$$\begin{pmatrix} \cosh\alpha & 0 & 0 & \sinh\alpha & 0 \\ 0 & 1 & 0 & 0 & a\alpha \\ 0 & 0 & 1 & 0 & \beta \\ \sinh\alpha & 0 & 0 & \cosh\alpha & 0 \\ 0 & 0 & 0 & 0 & 1 \end{pmatrix} \begin{pmatrix} x_1 \\ x_2 \\ x_3 \\ x_0 \\ 1 \end{pmatrix} = \begin{pmatrix} x_1\cosh\alpha + x_0\sinh\alpha \\ x_2 + a\alpha \\ x_3 + \beta \\ x_1\sinh\alpha + x_0\cosh\alpha \\ 1 \end{pmatrix}.$$

The isotropy group of any point x is only the identity $G_0 = e$. The dimension d and codimension k of the orbits is

$$d = \dim G - \dim G_0 = 2, \qquad k = 4 - 2 = 2.$$

There are hence two elementary invariants. To find them, put

$$P_3\psi(x_0, \ldots, x_3) = \frac{\partial\psi}{\partial x_3} = 0,$$
$$(M_{01} + aP_2)\psi = \left[-\left(x_0\frac{\partial}{\partial x_1} + x_1\frac{\partial}{\partial x_0}\right) + a\frac{\partial}{\partial x_2}\right]\psi = 0.$$

The first equation implies $\psi = \psi(x_0, x_1, x_2)$. To solve the second, put

$$dx \equiv -\frac{dx_1}{x_0} = -\frac{dx_0}{x_1} = \frac{dx_2}{a},$$

and solve to obtain

$$\xi_1 = \sqrt{x_0^2 - x_1^2}, \qquad \xi_2 = x_2 + a\ln(x_0 + x_1). \tag{23}$$

The general invariant of G is hence $\phi(\xi_1, \xi_2)$.

2.2.3 Reduction to an ordinary differential equation

In order to reduce (5) to an ODE, let us assume that the solution $u(x_0, x_1, \ldots, x_n)$ depends on one variable only:

$$u = u(\xi), \qquad \xi = \xi(x_0, x_1, \ldots, x_n). \tag{24}$$

We then have

$$\begin{aligned} \Box u &= u_{\xi\xi}(\nabla\xi)^2 + u_\xi \Box\xi, \\ (\nabla u)^2 &= u_\xi^2 (\nabla\xi)^2. \end{aligned} \tag{25}$$

Equation (5) will reduce to an ODE if and only if $\Box\xi$ and $(\nabla\xi)^2$ are functions of ξ alone:

$$\Box\xi = \alpha(\xi), \qquad (\nabla\xi)^2 = \beta(\xi). \tag{26}$$

In order to reduce (5) to an ODE we must thus find functions $\xi(x_0, x_1, \ldots, x_n)$ satisfying (26). With no loss of generality we can simplify our taks by replacing ξ by some function $\phi(\xi)$. We then have

$$(\nabla\phi)^2 = (\phi_\xi)^2(\nabla\xi)^2 = (\phi_\xi)^2\beta(\xi)).$$

Choosing ϕ appropriately, we can transform $\beta(\xi)$ to some canonical form. We can restrict ourselves in (26) to

$$\beta(\xi) = \kappa, \qquad \kappa = \pm 1, \text{ or } 0. \tag{27}$$

A systematic method of providing solutions of (26) is by introducing ***codimension 1 symmetry variables.*** This can be stated in the form of a theorem:

Theorem 2.1. *Let $\xi(x_0, x_1, \ldots, x_n)$ be an invariant of a subgroup $G \subset P(n,1)$ having generic orbits of codimension 1 in Minkowski space $M(n,1)$. The assumption $u = u(\xi)$ will reduce the equation (5),*

$$H(\Box u, (\nabla u)^2, u) = 0,$$

to an ordinary differential equation in ξ.

Proof. It suffices to show that Eqs. (26) are satisfied by ξ. The operators $\Box$ and $(\nabla\cdot, \nabla\cdot)$ are invariant under *P(n,1)*; hence

$$\begin{aligned} X(\Box\xi) &= \Box(X\xi) = 0, \\ X(\nabla\xi)^2 &= 2(\nabla(X\xi), \nabla\xi) = 0, \end{aligned} \tag{28}$$

for any $X \in p(n,1)$, in particular $X \in LG$, where LG is the Lie algebra of G.

The group G has generic orbits of codimension 1, hence any invariant of G is a function of the invariant ξ. Equations (28) hence imply

$$\Box\xi = \alpha(\xi), \qquad (\nabla\xi)^2 = \beta(\xi). \qquad \textit{Q. E. D.}$$

It should be stressed that we do not claim that codimension 1 symmetry variables provide *all* solutions of Eqs. (26). Indeed, in § 2.4, below, we discuss a different type of solution, namely *degenerate codimension 2 symmetry variables.*

2.2.4 Reduction to lower-dimensional partial differential equations

The results of the previous Section, summarized in Theorem 2.1, can easily be generalized to the case of k variables ξ_k. Indeed, let us reduce Eq. (5) to a k-dimensional PDE by postulating that u should depend on $\xi_1, \xi_2, \ldots, \xi_k$ only,

$$u = u(\xi_1, \xi_2, \ldots, \xi_k), \quad \xi_a = \xi_a(x_0, x_1, \ldots, x_n), \quad \begin{array}{l} a = 1, 2, \ldots, k, \\ 2 \le k \le n-1. \end{array} \tag{29}$$

Under this assumption we have

$$\begin{aligned} \Box u &= \sum_{a,b=1}^{k} u_{\xi_a \xi_b} (\nabla \xi_a, \nabla \xi_b) + \sum_{a=1}^{k} u_{\xi_a} \Box_{\xi_a}, \\ (\nabla u)^2 &= \sum_{a,b=1}^{k} u_{\xi_a} u_{\xi_b} (\nabla \xi_a, \nabla \xi_b). \end{aligned} \tag{30}$$

Equation (5) will reduce to a PDE in the k variables ξ_a if and only if $(\nabla \xi_a, \nabla \xi_b)$ and $\Box \xi_a$ are functions of ξ_a only:

$$\Box \xi_a = \alpha_a(\xi_1, \xi_2, \ldots, \xi_k), \quad (\nabla \xi_a, \nabla \xi_b) = \beta_{ab}(\xi_1, \xi_2, \ldots, \xi_k), \qquad a, b = 1, 2, \ldots, k. \tag{31}$$

As in the case of reduction to an ODE, we can generate simultaneous solutions of Eqs. (31) by introducing ***codimension-k symmetry variables***, *i.e.*, variables which are invariants of a subgroup $G \subset P(n,1)$ having generic orbits of codimension k. Let us again formulate the results as a theorem.

Theorem 2.2. *Let $\xi_a(x_0, x_1, \ldots, x_n)$, $a = 1, 2, \ldots, k$, be a set of independent invariants of a subgroup $G \subset P(n,1)$ having generic orbits of codimension k in $M(n,1)$. The assumption $u = u(\xi_1, \xi_2, \ldots, \xi_k)$ will reduce the studied equation (5):*

$$H(\Box u, (\nabla u)^2, u) = 0,$$

to a partial differential equation in k variables ξ_a, $(a = 1, 2, \ldots, k)$.

Proof. It suffices to show that $\xi_1, \xi_2, \ldots, \xi_k$ constructed as invariants of G satisfy (31). By the same argument as used in the proof of Theorem 2.1, it is obvious that they do. *Q. E. D.*

We thus have a systematic method of constructing symmetry variables and performing a symmetry reduction of the relativistically invariant equation (5). We simply let G run through all subgroups of $P(n,1)$ having orbits of the required codimension, calculate the invariants of G and use them to reduce the equation. The relation between groups G and their orbits is not one-to-one. Different groups have the same orbits and hence the same invariants. However, the maximal group G_M leaving a given set of variables ξ_i invariant, is uniquely defined.

2.2.5 Relation to the separation of variables in partial differential equations

The operator approach to the separation of variables in a PDE, originally proposed for linear differential equations of the Laplace–Beltrami (or Schrödinger) type [137,138] and then generalized to other types of equations[2] involves the following procedure. In order to obtain multiplicative separation of variables,

$$\Delta\psi = E\psi, \qquad \psi(x_1, x_2, \dots, x_n) = \prod_{i=1}^{n} \psi_i(x_i, \lambda_1, \lambda_2, \dots, \lambda_n), \tag{32}$$

construct a complete set of n linearly independent symmetric second-order operators $\{L_1, L_2, \dots, L_n\}$ satisfying

$$[L_i, L_k] = 0, \qquad L_1 = \Delta, \tag{33}$$

and possibly some furhter algebraic conditions [27,77,78]. The separable solutions then satisfy the following set of simultaneous equations:

$$L_i\psi(x_1, x_2, \dots, x_n) = \lambda_i\psi(x_1, x_2, \dots, x_n), \qquad i = 1, 2, \dots, n, \tag{34}$$

where the eigenvalues λ_i are the separation constants ($\lambda_1 = E$). For linear equations we can construct complete sets of separated eigenfunctions in this manner. For many spaces (including spaces of constant curvature), the second-order operators L_i are second order operators in the enveloping algebra of the invariance Lie algebra of the equation.

In our approach to symmetry reduction, the symmetry variables satisfy the equations

$$X_i\phi(x_1, x_2, \dots, x_n) = 0, \qquad i = 1, 2, \dots, m,$$

where the X_i are first-order operators forming the basis of an m-dimensional Lie algebra that is not necessarily abelian. Instead of requiring that solutions of the considered PDE be products or sums of functions of one variable, we are looking for specific classes of solutions, depending on one variable only, or more generally, on several variables. We are dealing with nonlinear equations and there is no claim about completeness of the solutions.

§2.3 The special case of the (*2+1*)-dimensional Minkowski space *M(2,1)*

Let us consider in some detail the example of the (*2+1*)-dimensional Minkowski space *M(2,1)*. All subgroups of the Poincaré group *P(2,1)* are known [108], as are those of *P(3,1)* [109,112].

In order to demonstrate how the method of symmetry reduction works, we shall proceed *ab initio* and not make any use of previous subgroup classifications. The space *M(2,1)* has dimension 3, we are hence interested in subgroups with generic orbits of codimension 1 and 2, and actually only in those which are maximal among the different subgroups, having a given invariant or pair of invariants.

2.3.1 Orbits of codimension 1

Let us first consider subgroups of *P(2,1)* with generic orbits of codimension 1. We shall proceed by the number cf free linearly independent translations present in the considered Lie algebra L. We are classifying subgroups and subalgebras up to conjugacy under the Poincaré group and will always choose the algebra representing a certain conjugacy class in some convenient standard form.

---

[2]See [26–30,77-79,96–98] and references therein. See also the lectures of Willard Miller in *these Proceedings*.

(i) 3 free translations. These are P_0, P_1, and P_2. The group of translations generated by these acts transitively on *M(2,1)*; hence there is no invariant. Indeed,

$$P_\mu \phi(x_0, x_1, x_2) = 0, \qquad \mu = 0, 1, 2,$$

implies $\phi =$ constant.

(ii) 2 free translations. The codimension of the orbits of a group generated by two independent translations is $k = 1$, so they determine the invariant completely. Three possibilities occur: both space-like, one space-like and one time-like, one space-like and one light-like. The basis for the translation space consists of two mutually orthogonal translations. The bases and invariants are

•(ii-a) $$\{P_1, P_2\} \qquad \xi = x_0, \tag{35}$$

•(ii-b) $$\{P_0, P_2\} \qquad \xi = x_1, \tag{36}$$

•(ii-c) $$\{P_0 - P_1, P_2\} \qquad \xi = x_0 + x_1. \tag{37}$$

The general subalgebra of *p(2,1)* spanned by two mutually orthogonal translations is obtained from those above by a general Lorentz transformation, as is the general invariant:

$$\xi = (A, x) = A_0 x_0 - A_1 x_1 - A_2 x_2, \tag{38}$$

where the constant vector satisfies $(A, A) = 1, -1$, or 0 for the cases (ii-a), (iib), and (ii-c), respectively. The maximal subalgebra of *p(2,1)* leaving ξ invariant is the euclidean algebra *e(2)* :$\{M_{12}, P_1, P_2\}$ in case (ii-a), the Poincaré algebra *p(1,1)*: $\{M_{02}, P_0, P_2\}$ in case (ii-b), and the Heisenberg algebra $\{M_{02} - M_{12}, P_0 - P_1, P_2\}$ in case (ii-c).

(iii) 1 free translation. We again proceed by signature:

• (iii-a) Space-like P_2. The group $\{\exp P_2\}$ has orbits of codimension 2, the invariant is $\phi(x_0, x_1)$. To reduce the codimension to 1, we must add further elements. In view of the structure of *p(2,1)*, where the translations $\{P_0, P_1, P_2\}$ figure as an ideal, and in view of the assumption that L contains only one free translation operator, P_2 must figure as an ideal in the algebra $L \subset$ *p(2,1)*, that we are constructing. Let us introduce the normalizer of $\{P_2\}$ in *p(2,1)*, *i.e.*, the largest subalgebra of *p(2,1)* containing P_2 as an ideal:

$$\operatorname{nor} P_2 = \{X \mid X \in p(2,1) \text{ and } [X, P_2] = \alpha P_2, \quad \alpha \in R\}.$$

It is a simple problem in linear algebra to find:

$$\operatorname{nor} P_2 = \{M_{01}, P_0, P_1, P_2\}.$$

Since L by assumption contains no further free translations, the only element of $\operatorname{nor} P_2$ we can add to P_2 is

$$M \equiv M_{01} + \sum_{\mu=0}^{2} a_\mu P_\mu, \qquad a_\mu \in R.$$

Since P_2 is independently contained in L, we can put $a_2 = 0$. Since we are classifying subalgebras and their invariants up to conjugacy, we can replace $\{M, P_2\}$ by

$$M' = e^{\alpha P_0 + \beta P_1} M e^{-(\alpha P_0 + \beta P_1)} = M + [\alpha P_0 + \beta P_1, M] + \tfrac{1}{2!}[\alpha P_0 + \beta P_1, [\alpha P_0 + \beta P_1, M]] + \cdots,$$
$$P_2' = e^{\alpha P_0 + \beta P_1} P_2 e^{-(\alpha P_0 + \beta P_1)} = P_2,$$

where we have made use of the Baker–Campbell–Hausdorff formula. Choosing $\alpha = a_0$, $\beta = a_1$, we find that $\{M, P_2\}$ is conjugate to

$$\{M_{01}, P_2\}. \tag{39}$$

The invariant ϕ must satisfy

$$M_{01}\phi = (-x_0\partial_1 - x_1\partial_0)\phi = 0, \qquad P_2\phi = \partial_2\phi = 0. \tag{40}$$

Solving (40) we find $\phi = \phi(\xi)$:

$$\xi = \sqrt{x_0^2 - x_1^2}\,. \tag{41}$$

Performing a general inhomogeneous Lorentz transformation, we transform the *standard* variable (41) into

$$\xi = \sqrt{(x-b, A_0)^2 - (x-b, A_1)^2}\,, \qquad (A_\mu, A_\nu) = g_{\mu\nu}, \tag{42}$$

where b, A_1, and A_2 are vectors in *M(2,1)*.

• (iii-b) Time-like P_0. Skipping the details, we find that the invariant of P_0 is $\phi(x_1, x_2)$, the normalizer of P_0 in *p(2,1)* containing P_0 as the only translation, is

$$\{M_{12}, P_0\}. \tag{43}$$

Its invariant is $\phi(\xi)$ with

$$\xi = \sqrt{x_1^2 + x_2^2}\,. \tag{44}$$

Under a general Lorentz transformation, (44) goes into

$$\xi = \sqrt{(x-b, A_1)^2 + (x-b, A_2)^2}, \qquad (A_j, A_k) = \delta_{j,k}. \tag{45}$$

• (iii-c) Light-like vector $P_0 - P_1$. The normalizer of $P_0 - P_1$ in *p(2,1)* is

$$\operatorname{nor}(P_0 - P_1) = \{M_{01}, M_{02} - M_{12}, P_0 + P_1, P_2, P_0 - P_1\}. \tag{46}$$

The subalgebra L could contain both elements M_{01} and $M_{02} - M_{12}$, or any one of them, in each case possibly extended by translations. If both are present, then up to conjugation under *P(2,1)* the algebra must be of the form

$$\{X_i\} = \{M_{01} + bP_2, M_{02} - M_{12}, P_0 - P_1\}, \qquad b \in R.$$

The corresponding Lie group acts transitively on *M(2,1)*, *i.e.*,

$$X_i\phi(x_0, x_1, x_2) = 0, \qquad i = 1, 2, 3,$$

implies $\phi = \text{constant}$.

If L is two-dimensional and M_{01} is present, then it can, by conjugacy in *P(2,1)* be reduced to

$$\{M_{01} + aP_2, P_0 - P_1\}, \qquad a \in R.$$

The equations

$$\begin{aligned}(M_{01} + aP_2)\phi &= [-(x_0\partial_1 + x_1\partial_0) + a\partial_2]\phi = 0,\\ (P_0 - P_1)\phi &= (\partial_0 - \partial_1)\phi,\end{aligned}$$

imply $\phi = \phi(\xi)$,

$$\xi = x_2 + a\ln(x_0 + x_1). \tag{47}$$

For $a \neq 0$ this is a new symmetry variable. By a Lorentz transformation, (47) can be transformed into:

$$\begin{gathered}\xi = (A, x) + a\ln(B, x),\\ (A,A) = -1, \quad (A,B) = 0, \quad (B,B) = 0,\end{gathered} \tag{48}$$

where A and B are vectors in *M(2,1)*.

If M_{01} is not present in L, then L can be transformed into either $\{M_{02} - M_{12}, P_0 - P_1\}$ or $\{M_{02} - M_{12} + P_0 + P_1, P_0 - P_1\}$. An invariant of the first of these two algebras must satisfy

$$\begin{aligned}(M_{02} - M_{12})\phi = [-(x_0 + x_1)\partial_2 - x_2(\partial_0 - \partial_1)]\phi = 0,\\ (P_0 - P_1)\phi = (\partial_0 - \partial_1)\phi = 0,\end{aligned}$$

yielding $\phi = \phi(x_0 + x_1)$, *i.e.*, an invariant which is not new. The second algebra provides a new result: namely the equations

$$\begin{aligned}[-(x_0 + x_1)\partial_2 - x_2(\partial_0 - \partial_1) + \partial_0 + \partial_1]\phi = 0,\\ (\partial_0 - \partial_1)\phi = 0,\end{aligned}$$

have the solution $\phi = \phi(\xi)$,

$$\xi = x_2 + \tfrac{1}{4}(x_0 + x_1)^2, \tag{49}$$

or, after a Lorentz transformation,

$$\xi = (A, x) + (B, x)^2, \qquad A^2 = -1, \quad B^2 = 0, \quad (A, B) = 0. \tag{50}$$

(iv) No free translations. The algebra L is a subalgebra of *o(2,1)*, possibly extended by translations.

The entire algebra *o(2,1)* is simple and hence does not have any nonsplitting extensions. The invariant of $\{M_{01}, M_{02}, M_{12}\}$ is $\phi(\xi)$,

$$\xi = \sqrt{x_0^2 - x_1^2 - x_2^2}\,. \tag{51}$$

We are looking for subgroups with generic orbits of codimension $k = 1$, hence L must be at least two-dimensional. The only two-dimensional subalgebra of *o(2,1)* is

$$\{M_{01}, M_{02} - M_{12}\}. \tag{52}$$

The commutation relations allow the extension

$$A = M_{01} + a(P_0 - P_1), \qquad B = M_{02} - M_{12} + b(P_0 - P_1), \tag{53}$$

	Standard variable ξ_S	General variable ξ	Algebra	$\kappa = (\nabla\xi)^2$	$\Box\xi = \kappa k/\xi$	Equation
1.	x_0	$(A,x), \quad A^2 = 1$	$M_{12},\ P_1,\ P_2$	1	0	$H(u_{\xi\xi}, u_\xi^2, u) = 0$
2.	x_1	$(A,x), \quad A^2 = -1$	$M_{02},\ P_0,\ P_2$	-1	0	$H(-u_{\xi\xi}, -u_\xi^2, u) = 0$
3.	$x_0 + x_1$	(A_0, x), $A^2 = 0$	$M_{02} - M_{12}$, $P_2,\ P_0 - P_1$	0	0	$H(0,0,u) = 0$
4.	$x_2 + a\ln(x_0 - x_1)$	$(A,x) + a\ln(B,x)$, $A^2 = -1,\ B^2 = 0,\ (A,B) = 0$	$M_{01} + aP_2$, $P_0 - P_1,\ a \neq 0$	-1	0	$H(-u_{\xi\xi}, -u_\xi^2, u) = 0$
5.	$x_2 + \frac{1}{4}(x_0 + x_1)^2$	$(A,x) + (B,x)^2$, $A^2 = -1,\ B^2 = 0,\ (A,B) = 0$	$M_{02} - M_{12} + P_0 +$ $P_1, \quad P_0 - P_1$	-1	0	$H(-u_{\xi\xi}, -u_\xi^2, u) = 0$
6.	$\sqrt{x_1^2 + x_2^2}$	$\sqrt{(x-b,A_1)^2 + (x-b,A_2)^2}$ $(A_i, A_k) = -\delta_{i,k}$	$M_{12},\ P_0$	-1	$\frac{-1}{\xi}$	$H(-u_{\xi\xi} - \xi^{-1}u_\xi,$ $-u_\xi^2, u) = 0$
7.	$\sqrt{x_0^2 - x_1^2}$	$\sqrt{(x-b,A_0)^2 - (x-b,A_1)^2}$ $(A_\mu, A_\nu) = g_{\mu\nu}$	$M_{01},\ P_2$	1	$\frac{1}{\xi}$	$H(u_{\xi\xi} + \xi^{-1}u_\xi,$ $u_\xi^2, 0) = 0$
8.	$\sqrt{x_0^2 - x_1^2 - x_2^2}$	$\sqrt{(x-b,A_0)^2 - (x-b,A_1)^2 - (x-b,A_2)^2}$, $(A_\mu, A_\nu) = g_{\mu\nu}$	$M_{01},\ M_{02},\ M_{12}$	1	$\frac{2}{\xi}$	$H(u_{\xi\xi} + 2\xi^{-1}u_\xi,$ $u_\xi^2, u) = 0$

Table 1. Symmetry variables for *M(2,1)*, their invariance algebras and the reduced ordinary differential equations. The quantities A, A_μ, B, and b in the third column are vectors in *M(2,1)*.

however, conjugacy by an element $g \in P(2,1)$ of the type $\exp[\alpha P_2 + \beta(P_0 + P_1)]$ will transform a and b into zero, so (53) is conjugate to (52). The invariant of (52) coincides with that of $o(2,1)$, *i.e.*, (51).

Let us summarize the results: the codimension 1 symmetry variables provided by the subgroups of $P(2,1)$ are the "obvious" ones:

$$\xi = x_0,\ x_1\ , x_0 + x_1,\ \sqrt{x_1^2 + x_2^2},\ \sqrt{x_0^2 - x_1^2},\ \text{or}\ \sqrt{x_0^2 - x_1^2 - x_2^2}, \tag{54}$$

plus two "nonobvious" ones:

$$\xi = x_2 + a\ln(x_0 + x_1), \quad a \in R, \quad a \neq 0, \qquad \text{or} \quad \xi = x_2 + \tfrac{1}{4}(x_0 + x_1)^2. \tag{55}$$

To obtain the ODE to which the variable ξ reduces the PDE (5), we must calculate $\Box\xi$ and $(\nabla\xi)^2$ for each ξ.

The results are summarized in Table 1, where we give the standard variables ξ_S, the general variable ξ, the Lie algebra of the maximal invariance group for each variable, the values of $(\nabla\xi)^2 = \kappa$ and $\Box\xi = \kappa k/\xi$, and the ODE obtained by the codimension 1 symmetry reduction.

If we restrict to the nonlinear Klein–Gordon equation (10), the similarity of the obtained equations becomes particularly visible. In all cases, we have:

$$\kappa\left(u_{\xi\xi} + \frac{k}{\xi}u_\xi\right) = F(u, \kappa u_\xi^2), \tag{56}$$

with $\kappa = \pm 1, 0$ and $k = 0, 1$, or 2, as in Table 1.

2.3.2 Orbits of codimension 2

The algebra L, in order to correspond to a Lie group with orbits of codimension 2, can contain at most one free translation. If present, this translation can be either time-like, space-like, or light-like, leading to the following invariants ξ and η:

$$\text{(i)} \qquad P_0 \quad : \quad \xi_1 = x_1, \qquad \xi_2 = x_2. \tag{57}$$

$$\text{(ii)} \qquad P_2 \quad : \quad \xi_1 = x_0, \qquad \xi_2 = x_1. \tag{58}$$

$$\text{(iii)} \qquad P_0 - P_1 \quad : \quad \xi_1 = x_2, \quad \xi_2 = x_0 + x_1. \tag{59}$$

If no free translations are present, we have a subalgebra of $o(2,1)$, possibly extended by translations. It is easy to verify that only one-dimensional subalgebras lead to orbits of codimension 2. The algebra $o(2,1)$ has three mutually nonconjugate one-dimensional subalgebras: M_{01}, $M_{02} - M_{12}$, and M_{12}. Let us consider them one by one.

The algebra $\{M_{01} + \alpha_\mu P_\mu\}$ can by translations be simplified to one of the two following algebras which we give with their invariants:

$$\text{(iv)} \qquad M_{01} \quad : \quad \xi = x_2, \qquad \eta = \sqrt{x_0^2 - x_1^2}. \tag{60}$$

$$\text{(v)} \qquad M_{01} + aP_2 \quad : \quad \begin{array}{c} \xi_1 = x_2 + a\ln(x_0 + x_1), \\ a \in R, \quad a \neq 0, \end{array} \quad \xi_2 = \sqrt{x_0^2 - x_1^2}. \tag{61}$$

The algebra $\{M_{12} + \alpha_\mu P_\mu\}$ can be transformed into one of the following:

$$\text{(vi)} \qquad M_{12} \quad : \quad \xi_1 = x_0, \qquad \xi_2 = \sqrt{x_1^2 + x_2^2}. \tag{62}$$

$$\text{(vii)} \qquad M_{12} + aP_0,\ a \neq 0 \quad : \quad \xi_1 = x_0 + a\arcsin\left(x_1/\sqrt{x_1^2 + x_2^2}\right), \quad \xi_2 = \sqrt{x_0^2 - x_1^2}. \tag{63}$$

The algebra $\{M_{02} - M_{12} + a_\mu P_\mu\}$ can be transformed into one of the following:

$$\text{(viii)} \qquad M_{02} - M_{12} \quad : \quad \xi_1 = \sqrt{x_0^2 - x_1^2 - x_2^2}, \quad \xi_2 = x_0 - x_1. \tag{64}$$

$$\text{(ix)} \qquad M_{02} - M_{12} + P_0 + P_1 \quad : \quad \begin{array}{c} \xi_1 = x_0 - x_1 + x_2(x_0 + x_1) + \frac{1}{6}(x_0 + x_1)^3, \\ \xi_2 = x_2 + \frac{1}{4}(x_0 + x_1)^2. \end{array} \tag{65}$$

In order to obtain the two-dimensional PDE, we must in each case calculate $\Box\xi_i$ and $(\nabla\xi_i, \nabla\xi_k)$, $i = 1, 2$, substitute into (30), and then substitute the obtained expressions for $\Box u$ and $(\nabla u)^2$ into (5). The results are presented in Table 2. For greater clarity, we have restricted ourselves to the nonlinear Klein–Gordon equation (10); the results for the more general equation (5) can be written with equal ease. In columns 2 and 3, we only give the *standard* variables ξ_S and their Lie algebras. The general variable ξ can be obtained from each ξ_S by a Lorentz transformation, as in Table 1.

	Variables	Algebra	Equation
1.	$x_1,\ x_2$	P_0	$-(u_{x_1x_1}+u_{x_2x_2})=F(u,-u_{x_1}^2-u_{x_2}^2)$
2.	$x_0,\ x_1$	P_2	$u_{x_0x_0}-u_{x_1x_1}=F(u,u_{x_0}^2-u_{x_1}^2)$
3.	$x_2,\ \eta=x_0+x_1$	P_0-P_1	$-u_{x_2x_2}=F(u,-u_{x_2}^2)$
4.	$x_2,\ \rho=\sqrt{x_0^2-x_1^2}$	M_{01}	$u_{\rho\rho}+\rho^{-1}u-u_{x_2x_2}=F(u,u_\rho^2-u_{x_2}^2)$
5.	$x_2,\ r=\sqrt{x_1^2+x_2^2}$	M_{12}	$-u_{rr}-r^{-1}u_r+u_{x_0x_0}=F(u,u_{x_0}^2-u_r^2)$
6.	$\eta=x_0+x_1,$ $\tau=\sqrt{x_0^2-x_1^2-x_2^2}$	$M_{02}-M_{12}$	$u_{\tau\tau}+2\eta\tau^{-1}u_{\eta\tau}+2\tau^{-1}u_\tau=$ $F(u,u_\tau^2+2\eta\tau^{-1}u_\eta u_\tau)$
7.	$\rho=\sqrt{x_0^2-x_1^2},$ $\xi=x_2+a\ln(x_0+x_1),\ a\neq0$	$M_{01}+aP_2$	$u_{\rho\rho}+2a\rho^{-1}u_{\rho\xi}+\rho^{-1}u_\rho-u_{\xi\xi}=$ $F(u,u_\rho^2+2a\rho^{-1}u_\rho u_\xi)$
8.	$r=\sqrt{x_1^2+x_2^2},$ $\varsigma=x_0+a\arcsin(x_1/\sqrt{x_1^2+x_2^2})$	$M_{12}+aP_0$	$(1-a^2r^{-2})u_{\varsigma\varsigma}-u_{rr}-r^{-1}u_r=$ $F(u,u_r^2+(1-a^2r^{-2})u_\varsigma^2)$
9.	$\xi=x_2+\frac{1}{4}(x_0+x_1)^2$ $\sigma=x_0-x_1+x_2(x_0+x_1)+\frac{1}{6}(x_0+x_1)^3$	$M_{02}-M_{12}+$ P_0+P_1	$4\xi u_{\sigma\sigma}-u_{\xi\xi}=F(u,-u_\xi^2+4\xi u_\sigma^2)$

Table 2. Codimension 2 symmetry variables for *M(2,1)*, their invariance algebras and the reduced nonlinear Klein–Gordon equation.

2.3.3 Degenerate codimension 1 and codimension 2 symmetry variables

Looking at Tables 1 and 2, we see that a degeneracy occurs when a null variable $\eta=x_0+x_1$, or more generally $\eta=(A,x)$, $(A,A)=0$, is present.

Indeed, in the codimension 1 case, the PDE $H(\Box u,(\nabla u)^2,u)=0$ for $u=u(x_0+x_1)$ reduces to a functional equation

$$H(0,0,u)=0, \tag{66}$$

involving no derivatives. The solution $u(x_0+x_1)$ will be constant and is in some sense trivial. On the other hand, such trivial solutions often serve as an input in Bäcklund transformations [13,32] to produce highly nontrivial solutions, such as solitons.

The codimension 2 case is more interesting. The subgroup of translations parallel to a generating line of the light cone, $\exp\alpha(P_0-P_1)$ has two basic invariants, *e.g.*, x_2 and $\eta=x_0+x_1$. This is N° **3** in Table 2.

We have

$$\nabla x_2=(0,0,1), \qquad \nabla\eta=(1,1,0), \tag{67}$$

and hence

$$(\nabla\eta)^2=0, \qquad \Box\eta=0, \qquad (\nabla\eta,\nabla x_2)=0. \tag{68}$$

Hence, although the codimension of the generic orbit is 2, and although we have two invariants, the equation $H(\Box u,(\nabla u)^2,u)=0$ reduces to the ODE

$$H(-u_{x_2x_2},-u_{x_2}^2,u)=0. \tag{69}$$

This in itself is not new, but coincides with one of the codimension 1 cases. However, relations (68) allow us to introduce a new type of variable, which we call a ***degenerate codimension 2 symmetry variable***

$$\xi = x_2 + \phi(x_0 + x_1), \tag{70}$$

where ϕ is an arbitrary twice-differentiable function of its argument. We have

$$(\nabla\xi)^2 = -1, \qquad \Box\xi = 0,$$

so equation (5) reduces to the ODE

$$H(-u_{\xi\xi}, -u_\xi^2, u) = 0, \tag{71}$$

i.e., the same equation as (69), but in a much more general independent variable: one involving the arbitrary function $\phi(x_0 + x_1)$. Performing a Lorentz transformation, we transform the *standard* variable (70) into

$$\xi = (A, x) + \phi(B, x), \quad A^2 = -1, \quad B^2 = 0, \quad (a, B) = 0. \tag{72}$$

Only one such degenerate codimension 2 variable exists in *M(2,1)*; in higher dimensions, as we shall see, new ones appear.

§2.4 Symmetry reduction in *M(n,1)*

The results obtained above for *M(2,1)* can be generalized to *M(n,1)* for n an arbitrary positive integer. The situation is particularly simple for codimension 1 symmetry variables. It can be summed up in the following theorem.

Theorem 2.3. *All codimension 1 symmetry variables ξ for equation (5), $H(\Box u, (\nabla u)^2, u) = 0$, are conjugate under a P(n,1) transformation to one of the following standard variables ξ_S:*

$$\begin{gathered} x_0 + x_1, \quad x_0, \quad x_1, \quad x_2 + a\ln(x_0 + x_1),\ a \neq 0, \quad x_2 + \tfrac{1}{4}(x_0 + x_1)^2, \\ r_k = \sqrt{x_1^2 + x_2^2 + \cdots + x_{k+1}^2}, \qquad k = 1, 2, \ldots, n-1, \\ \rho_k = \sqrt{x_0^2 - x_1^2 - \cdots - x_k^2}, \qquad k = 1, 2, \ldots, n. \end{gathered} \tag{73}$$

The assumption $u = u(\xi)$ reduces (5) to the ODE

$$H\Big(\kappa\big(u_{\xi\xi} + \frac{m}{\xi}u_\xi\big), \kappa u_\xi^2, u\Big) = 0, \qquad \text{with } \kappa = (\nabla\xi)^2 \text{ and } m = \xi\Box\xi. \tag{74}$$

We have

$$\begin{aligned} &(\kappa, m) = (0,0) && \text{for} && \xi = x_0 + x_1, \\ &(\kappa, m) = (1,0) && \text{for} && \xi = x_0, \\ &(\kappa, m) = (-1,0) && \text{for} && \xi = \begin{cases} x_1, \\ x_2 + a\ln(x_0 - x_1),\ a \neq 0, \\ x_2 + \tfrac{1}{4}(x_0 + x_1)^2, \end{cases} \\ &(\kappa, m) = (-1,k) && \text{for} && \xi = r_k, \\ \text{and}\ &(\kappa, m) = (1,k) && \text{for} && \xi = \rho_k. \end{aligned}$$

The standard variables ξ_S, the general variables ξ, their invariance algebras, and the reduced equations are those given in Table 3.

Standard variable ξ_S	General variable ξ	Invariance algebra	Equation
1. x_0	$(A,x), \quad A^2=1$	$M_{ik},\ P_i,\ i,k=1,2,\ldots,n$	$H(u_{\xi\xi},u_\xi^2,u)=0$
2. x_1	$(A,x), \quad A^2=-1$	$M_{\alpha\beta},\ P_\alpha,\ \alpha\beta=2,\ldots,n,0$	$H(-u_{\xi\xi},-u_\xi^2,u)=0$
3. x_0+x_1	$(A_0,x), \quad A^2=0$	$M_{ab},\ M_{0a}-M_{1a},\ P_a,$ $P_0-P_1,\ a,b=2,3,\ldots,n$	$H(0,0,u)=0$
4$_a$. $x_2+a\ln(x_0+x_1)$	$(A,x)+a\ln(B,x),$ $A^2=-1,\ B^2=0,$ $(A,B)=0$	$M_{ab},\ M_{0a}-M_{1a},\ P_a,$ $M_{01}+aP_2,\ P_0-P_1$ $a,b=3,4,\ldots,n$	$H(-u_{\xi\xi},-u_\xi^2,u)=0$
5. $x_2+\frac{1}{4}(x_0+x_1)^2$	$(A,x)+(B,x)^2,$ $A^2=-1,\ B^2=0,$ $(A,B)=0$	$M_{ab},\ M_{0a}-M_{1a},\ P_a$ $M_{02}-M_{12}+P_0+P_1,$ P_0-P_1 $a,b=3,4,\ldots,n$	$H(-u_{\xi\xi},-u_\xi^2,u)=0$
6$_k$. $\sqrt{x_1^2+x_2^2+\cdots+x_{k+1}^2}$ $k=1,\ldots,n-1$	$\sqrt{[\sum_{i=1}^{k+1}(x-B,A_j)^2]}$ $(A_i,A_j)=-\delta_{i,j},$ $i,j=1,2,\ldots,k$	$M_{\alpha\beta},\ P_\alpha,\ M_{ab}$ $\alpha,\beta=k+1,\ldots,n,0$	$H(-u_{\xi\xi}-k\xi^{-1}u_\xi,$ $-u_\xi^2,u)=0$
7$_k$. $\sqrt{x_0^2-x_1^2-\cdots-x_k^2}$ $k=1,2,\ldots,n$	$\sqrt{[\sum_{\alpha=0}^{k}g_{\alpha\alpha}(x-B,A_\alpha)^2]}$ $(A_\alpha,A_\beta)=g_{\alpha\beta}$ $\alpha,\beta=0,1,\ldots,k$	$M_{ab},\ P_a,\ M_{\alpha\beta}$ $a,b=k+1,\ldots,n$ $\alpha,\beta=0,1,\ldots,k$	$H(u_{\xi\xi}+k\xi^{-1}u_\xi,$ $u_\xi^2,0)=0$

Table 3. Codimension 1 symmetry variables for *M(n,1)*, their invariance algebras and reduced equations.

Proof. It is easy to verify that each of the variables ξ_S given in Table 1 and in Eq. (73), does reduce (5) to the indicated equation, and that ξ_S is the invariant of the indicated Lie algebra. The proof that Table 3 is complete, *i.e.*, that no other codimension 1 symmetry variables exist, is quite long. We omit this proof and refer to the original article [61].

All codimension 2 and 3 symmetry variables in *M(3,1)* have been found [61], and the corresponding two- and three-dimensional PDE's written out. We shall not reproduce the results here. The codimension 2, $\ldots, n$ symmetry variables for *M(n,1)* with $n>3$ have so far not been studied, but it is clear that new types of variables appear for each new value of n.

Degenerate codimension 2 symmetry variables are of particular interest, since they reduce the PDE (5) to an ODE, *i.e.*, they perform the same function as the codimension 1 symmetry variables. For $n=2$ only one such variable exists, namely

$$\xi=x_2+\phi(x_0+x_1).$$

See (70). This variable will reduce (5) to the ODE (71) for any $n\geq 2$. For $n=3$, and hence for any $n\geq 3$, we obtain a further degenerate codimension 2 symmetry variable. Indeed, consider the algebra $\{M_{ab},\ M_{03}-M_{13}+P_2,\ P_a,\ P_0-P_1,\ a,b=4,\ldots,n\}$. This algebra has two independent invariants

$$\eta=x_0+x_1, \qquad \varsigma=x_3+(x_0+x_1)x_2, \tag{75}$$

satisfying

$$\nabla\eta = (1,1,0,\ldots,0), \qquad \nabla\varsigma = (x_2, x_2, x_0 + x_1, 1, 0, \ldots, 0),$$

$$(\nabla\eta)^2 = 0, \quad (\nabla\eta, \nabla\varsigma) = 0, \quad (\nabla\varsigma)^2 = -(x_0 + x_1)^2 - 1,$$

$$\Box\eta = 0, \qquad \Box\varsigma = 0.$$

Equation (5) reduces to

$$H\big(-(1+\eta^2)u_{\varsigma\varsigma}, -(1+\eta^2)u_\varsigma^2, u\big) = 0. \tag{76}$$

This allows us to introduce the degenerate codimension 2 variable

$$\xi = \frac{x_3 + (x_0 + x_1)x_2}{\sqrt{1 + (x_0 + x_1)^2}} + \phi(x_0 + x_1), \tag{77}$$

where ϕ is an arbitrary sufficiently smooth function. The *ansatz* $u = u(\xi)$ reduces (5) to the ODE (71).

The question of further degenerate codimension 2 variables for $n \geq 4$ has so far not been investigated.

§2.5 Solutions of the $(n+1)$-dimensional sine–Gordon equation obtained by symmetry reduction

A special case of the Poincaré-invariant equation (5) is the $(n+1)$-dimensional sine–Gordon equation

$$\Box u = \sin u, \qquad u = u(x_0, x_1, \ldots, x_n). \tag{78}$$

This equation is of considerable physical and mathematical interest, particularly in low dimensions.[3]

Equation (78) for $n = 1$ is one of the prime examples of a completely integrable nonlinear partial differential equation, such as the Korteweg–de Vries equation, the nonlinear (cubic) Schrödinger equation, and many others. The entire machinery of Bäcklund transformations, inverse scattering techniques, etc., has been applied to solve this equation [4,88,139]. For $n \geq 2$ the equation is also physically interesting and has been extensively studied [10,34,93,131].

In this Section we shall apply the method of symmetry reduction described above to reduce (78) for arbitrary n to an ODE and discuss the obtained ODE's.

From Theorem 2.3, we see that for all codimension 1 symmetry variables ξ, the sine–Gordon equation reduces to

$$\kappa\left(u_{\xi\xi} + \frac{m}{\xi}u_\xi\right) = \sin u. \tag{79}$$

[3] The 1+1 sine–Gordon equation figures in studies of magnetic flux in Josephson junctions, Bloch wall motion of magnetic crystals, self-induced transparency in optics, the Thirring model in classical and quantum field theory, and many other applications [16,87,121]. The original *Bäcklund transformation* was discovered by Bäcklund and Bianchi in a geometric investigation of surfaces with constant negative curvature, described by the 1+1 sine–Gordon equation [15,19].

Here,

$$\kappa = 0, \qquad \text{for } \xi = x_0 + x_1,$$

and

$$\kappa = 1 \text{ or } -1, \qquad \text{in all other cases;}$$

similarly,

$$m = 0 \qquad \text{for } \xi = \begin{cases} x_0,\ x_1,\ x_2 + a\ln(x_0+x_1),\ x_2 + \frac{1}{4}(x_0+x_1)^2, \\ x_2 + \phi(x_0+x_1), \text{ and } \dfrac{x_3 + (x_0+x_1)x_2}{\sqrt{1+(x_0+x_1)^2}} + \phi(x_0+x_1), \end{cases}$$

and

$$m = k \qquad \text{for } \xi = \begin{cases} r_k = \sqrt{x_1^2 + \cdots x_{k+1}^2}, & k = 1, 2, \ldots, n-1, \\ \rho_k = \sqrt{x_0^2 - x_1^2 - \cdots - x_k^2}, & k = 1, 2, \ldots, n. \end{cases}$$

For $m = 0$, $\kappa = \pm 1$, (79) is the exact rigid pendulum equation. Its solutions are well known [14]. Indeed, three types of solutions exist:

1. Periodic solutions (the pendulum oscillates between two extreme positions $\pm u_m$ with $\cos u_m < 1$):

$$u = 2\arccos[\text{dn}(\xi + \alpha, M)] + \tfrac{1}{2}(1+\kappa)\pi, \qquad 0 < M < 1, \quad \kappa = \pm 1. \tag{80}$$

2. Nonperiodic solutions (the pendulum rotates about a fixed point):

$$u = 2\arccos[\text{cn}(\frac{\xi + \alpha}{M}, M)] + \tfrac{1}{2}(1+\kappa)\pi, \qquad 0 < M < 1, \quad \kappa = \pm 1. \tag{81}$$

3. An intermediate case (the pendulum has precisely the right initial conditions to rotate to its highest point and stop there; the time required for this is $t \to \infty$):

$$u = 4\arctan e^{\pm\xi} - \tfrac{1}{2}(1-\kappa)\pi, \qquad \kappa = \pm 1. \tag{82}$$

The modulus M and the constant α are related to the initial conditions for the pendulum. The range of M is such that for ξ real, the Jacobi elliptic functions $\text{dn}(x, M)$ and $\text{cn}(x, M)$ are real and have one real and one pure-imaginary period.

The above solutions are well known also for the sine–Gordon equation [130], at least for $n = 1$ and

$$\xi = (A, x), \qquad A^2 = \kappa = \pm 1. \tag{83}$$

Equations (80–82) are then travelling wave solutions, in particular (82) is the one-kink solution of soliton theory.

The novelty of our approach (in addition to the fact that n is arbitrary), is that for $n \geq 2$, the variable ξ can be different than (83), namely

$$\xi = (A, x) + \phi((B, x)), \qquad A^2 = -1, \quad B^2 = 0, \quad (A, B) = 0, \quad \kappa = -1, \tag{84}$$

and for $n \geq 3$, also

$$\xi = \frac{(A, x) + (B, x)(C, x)}{\sqrt{1 + (B, x)^2}} + \phi((B, x)), \qquad \begin{matrix} A^2 = -1, \quad B^2 = 0, \quad C^2 = -1, \\ (A, B) = (B, C) = (C, A) = 0, \\ \kappa = -1. \end{matrix} \tag{85}$$

For $m \geq 1$ we can transform (79) into an equation of the type (4) studied by Painlevé [107] and Gambier [55]. Indeed, putting

$$u = 2i \ln y, \tag{86}$$

we reduce (79) to

$$\ddot{y} = \frac{1}{y}\dot{y}^2 - \frac{m}{\xi}\dot{y} + \frac{\kappa}{4}\left(y^3 - \frac{1}{y}\right), \tag{87}$$

which is rational in $\dot{y}$, algebraic in y (actually rational), and analytic in ξ.

For $m = 1$, (87) is a special case of the Painlevé III transcendent, satisfying

$$\ddot{y} = \frac{1}{y}\dot{y}^2 - \frac{1}{\xi}\dot{y} + \frac{a\xi + by + cy^3 + d\xi y^4}{\xi y}, \tag{88}$$

with $a = -\frac{1}{4}\kappa$, $d = \frac{1}{4}\kappa$, $b = c = 0$. A different transformation

$$u = 4 \arctan i\sqrt{z}, \tag{89}$$

transforms (79), for $m = 1$, into a special case of the Painlevé V transcendent.

For $m \geq 2$, Eq. (87) does not have the Painlevé property, *i.e.*, its solutions will have moving critical points. For $m = 2$, the equation has a name, the *Emden equation* [39], and some of its properties are known.

We have thus established that the sine–Gordon equation (78) in *M(n,1)* for $n \geq 2$, does *not* satisfy the conditions of the Painlevé conjecture, since m in (79) can go up to $m = n$. Equation (78) is known to be integrable by the inverse scattering technique for $n = 1$ only.

§2.6 Conclusions and outlook

In this Chapter we have studied the problem of symmetry reduction in detail for Poincaré -invariant equations of the type $H(\Box u, (\nabla u)^2, u) = 0$. We have shown that symmetry variables are systematically generated as invariants of subgroups of the invariance group *P(n,1)*. In particular, if we are interested in reducing (5) to an ODE, we can do this by introducing a ***codimension 1 symmetry variable*** ξ, which is an invariant of a subgroup of *P(n,1)* having generic orbits of codimension 1 in *M(n,1)*. Further reductions to ODE's are provided by ***degenerate codimension 2 symmetry variables***. These are obtained by constructing the invariants ξ and ηdrop out, *i.e.*, when $\nabla\eta)^2 = \Box\eta = \nabla\xi, \nabla\eta) = 0$. The variable η of subgroups with generic orbits of codimension 2, and identifying cases when the derivatives with respect to, say, η will then either not figure at all in the obtained equation, or it will figure as a parameter. If η does not figure at all, or if it can be eliminated by a change of variables, then we obtain an ODE in some new independent variable, involving an arbitrary function of η. See (72) and (77).

The methods discussed in this Chapter are quite general and provide a procedure for reducing PDE's with nontrivial symmetry groups to ODE's or lower-order PDE's. A classification of subgroups of the symmetry group provides a classification of symmetry variables. Work is in progress on a systematic study of symmetry reduction for nonlinear heat equations, and other nonlinear PDE's of interest in applications.

References

[1] M. J. Ablowitz, D. J. Kaup, A. C. Newell, and H. Segur, Nonlinear evolution equations of physical significance. *Phys. Rev. Lett.* **31**, 125–127 (1973).

[2] M. J. Ablowitz, D. J. Kaup, A. C. Newell, and H. Segur, The inverse scattering transform: Fourier analysis for nonlinear problems. *Studies Appl. Math.* **53**, 249–315 (1974).

[3] M. J. Ablowitz, A. Ramani, and H. Segur, A connection between nonlinear evolution equations and ordinary differential equations of *P*-type, I and II. *J. Math. Phys.* **21**, 715–721 and 1006–1015 (1980).

[4] M. J. Ablowitz and H. Segur, *Solitons and the Inverse Scattering Transform.* SIAM, Philadelphia, 1981.

[5] M. J. Ablowitz and H. Segur, Exact linearization of a Painlevé transcendent. *Phys. Rev. Lett.* **38**, 1103–1106 (1977).

[6] W. F. Ames, *Nonlinear Ordinary Differential Equations in Transport Processes.* Academic Press, New York, 1968.

[7] W. F. Ames, in *Nonlinear Equations in Abstract Spaces.* V. Lakshmikantham ed., Academic Press, New York, 1978, pp. 43–66.

[8] W. F. Ames, *Nonlinear Partial Differential Equations in Engineering,* Vols. I and II. Academic Press, New York, 1965 and 1972.

[9] R. L. Anderson, A nonlinear superposition principle admitted by coupled Riccati equations of the projective type. *Lett. Math. Phys.* **4**, 1-7 (1980).

[10] R. L. Anderson, A. O. Barut, and R. Rączka, Bäcklund transformations and new solutions of nonlinear wave equations in four-dimensional space-time. *Lett. Math. Phys.* **3**, 351–358 (1979).

[11] R. L. Anderson, J. Harnad and P. Winternitz, Group-theoretical approach to superposition rules for systems of Riccati equations. *Lett. Math. Phys.* **5**, 143–148 (1981).

[12] R. L. Anderson, J. Harnad and P. Winternitz, Systems of ordinary differential equations with nonlinear superposition principles. *Physica* **4D**, 164–182 (1982).

[13] R. L. Anderson and N. H. Ibragimov, *Lie-Bäcklund Transformations in Applications.* SIAM, Philadelphia (1979).

[14] P. Appel and E. Lacour, *Principes de la Théorie des Fonctions Elliptiques et Applications.* Gauthier–Villars, Paris, 1922.

[15] A. V. Bäcklund, Zur Theorie der Partiellen Differentialgleichungen erster Ordnung. *Math. Ann.* **17**, 285–328 (1880).

[16] A. Barone, F. Esposito, C. J. Magee, and A. C. Scott, Theory and applications of the sine–Gordon equation. *Nuovo Cimento* **1**, 227-267 (1971).

[17] J. Beckers, J. Harnad, M. Perroud, and P. Winternitz, Tensor fields invariant under subgroups of the conformal group. *J. Math. Phys.* **19**, 2126–2153 (1978).

[18] J. Beckers, J. Harnad, M. Perroud, and P. Winternitz, Subgroups of the euclidean group and symmetry breaking in nonrelativistic quantum mechanics. *J. Math. Phys.* **18** 72–83 (1977).

[19] L. Bianchi, Ricerche sulle superficie a curvatura constante e sulle elicoidi. *Ann. Scuola Norm. Sup. Pisa* **2**, 285 (1879).

[20] G. W. Bluman and J. D. Cole, The general similarity solution of the heat equation. *J. Math. Mech.* **18**, 1025–1042 (1969).

[21] G. W. Bluman and J. D. Cole, *Similarity Methods for Differential Equations.* Applied Mathematical Sciences #13, Springer Verlag Lecture Notes in Mathematics, 1974.

[22] G. W. Bluman and S. Kumei, On the remarkable nonlinear diffusion equation $\frac{\partial}{\partial x}[a(u+b)^{-2}\frac{\partial u}{\partial x}]-\frac{\partial u}{\partial t}=0$. *J. Math. Phys.* **21**, 1019–1023 (1980).

[23] M. Boiti and F. Pempinelli, Similarity solutions of the Korteweg–de Vries equation. *Nuovo Cimento* **51B**, 70–78 (1979).

[24] M. Boiti and F. Pempinelli, Nonlinear Schrödinger equation, Bäcklund transformations and Painlevé transcendents. *Nuovo Cimento* **59B**, 40–58 (1980)

[25] M. Boiti and F. Pempinelli, Similarity solutions and Bäcklund transformations of the Boussinesq equation. *Nuovo Cimento* **56B**, 148–156 (1980).

[26] C. P. Boyer, E. G. Kalnins, and W. Miller Jr., Symmetry and the separation of variables for the Hamilton–Jacobi equation $W_t^2-W_x^2-W_y^2=0$. *J. Math. Phys.* **19**, 200–211 (1978).

[27] C. P. Boyer, E. G. Kalnins, and W. Miller Jr., Separable coordinates for four-dimensional Riemannian spaces. *Commun. Math. Phys.* **59**, 285–302 (1978).

[28] C. P. Boyer, E. G. Kalnins, and P. Winternitz, Completely integrable relativistic hamiltonian systems and separation of variables in hermitian hyperbolic spaces. *Preprint* CRMA–1104 (1982), to appear in *J. Math. Phys.* **24** (1983).

[29] C. P. Boyer, E. G. Kalnins, and P. Winternitz, Separation of variables for the Hamilton–Jacobi equation on complex projective spaces. *Preprint* CRMA–1064 (1981), submitted to *SIAM J. Math. Anal.*

[30] C. P. Boyer, R. T. Sharp, and P. Winternitz, Symmetry breaking interactions for the time dependent Schrödinger equation. *J. Math. Phys.* **17**, 1439–1451 (1976).

[31] R. K. Bullough and P. J. Caudrey (eds.), *Solitons.* Springer Verlag, 1980.

[32] F. Calogero and A. Degasperis, *Spectral Transform and Solitons.* North Holland, Amsterdam, 1982.

[33] *(a)*L.–L. Chau, Integrability of self dual Yang–Mills equations and the role of the Kac–Moody algebra. *Lectures in these Proceedings. (b)* L.–L. Chau, Bianchi–Bäcklund transformations, conservation laws, and linearization of various field theories. In *The High Energy Limit*, A. Zichichi ed., Plenum Publ. Corp. 183, pp. 249–279.

[34] P. L. Christiansen and P. S. Lomdahl, Numerical study of *2+1*-dimensional sine–Gordon solitons. *Physica* **2D**, 482–494 (1981).

[35] J. Clairin, Sur les transformations de Bäcklund. *Ann. Sci. Ecole Norm. Sup.* **3**, Supplément, 1–63 (1902).

[36] W. J. Coles, Matrix Riccati differential equations. *SIAM J. Appl. Math.* **13**, 627–634 (1965).

[37] J. Corones, Solitons and simple pseudopotentials. *J. Math. Phys.* **17**, 756–759 (1976).

[38] J. Corones and F. J. Testa, Pseudopotentials and their applications. In *Bäcklund Transformations*, Proceedings, Nashville Tenn., 1974, R. M. Miura, ed. Springer Verlag, 1976.

[39] H. T. Davis, *Introduction to Nonlinear Differential Equations.* Dover, New York, 1962.

[40] M. C. Delfour, E. B. Lee, and A. Manitius, F-reduction of the operator Riccati equation for hereditary differential systems. *Automatica* **14**, 385–395 (1978).

[41] C. Dubois, *Règles de superposition pur les équations de Riccati matricielles générées sous U(n,n) et O(n,n).* M. Sc. Thesis, Université de Montréal, 1982.

[42] H. Eichenherr, *SU(n)*-invariant nonlinear σ-models. *Nucl. Phys.* **B146**, 215–223 (1978).

[43] (a) H. Eichenherr and M. Forger, On the dual symmetry of the nonlinear sigma models. *Nucl. Phys.* **B155**, 381–393 (1979); (b) *ib.*, More about nonlinear sigma models on symmetric spaces. *Nucl. Phys.*

B164, 528–535 (1980); (c) *ib.*, Higher local conservation laws for nonlinear sigma models on symmetric spaces. *Commun. Math. Phys.* **82**, 227-255 (1981).

[44] L. P. Eisenhart, *Continuous Groups of Transformations.* Dover, New York, 1961.

[45] (a) F. B. Estabrook, Some old and new techniques for the practical use of exterior differential forms. In *Bäcklund Transformations*, Proceedings, Nashville Tenn., 1974, R. M. Miura, ed. Springer Verlag, 1976; (b) F. B. Estabrook and H. D. Wahlquist, Prolongation structures of nonlinear evolution equations. *J. Math. Phys.* **17**, 1293–1297 (1976).

[46] M. Flato, G. Pinczon, and J. Simon, Nonlinear representations of Lie groups. *Ann. Scient. Ecole Norm. Sup.* **4** t. 10, 405–418 (1977).

[47] (a) M. Flato and J. Simon, Nonlinear equations and covariance. *Lett. Math. Phys.* **2**, 155–160 (1977); (b) *ib.*, Yang–Mills equations are formally linearizable. *Lett. Math. Phys.* **3**, 279-283 (1979).

[48] M. Flato and J. Simon, Linearization of relativistic nonlinear wave equations. *J. Math. Phys.* **21**, 913–917 (1980).

[49] M. Flato and J. Simon, On a linearization program of nonlinear field equations. *Phys. Lett.* **94B**, 518–522 (1980).

[50] A. S. Fokas,Generalized symmetries and constants of the motion of evolution equations. *Lett. Math. Phys.* **3**, 467–473 (1979).

[51] A. S. Fokas, A symmetry approach to exactly solvable evolution equations. *J. Math. Phys.* **21**, 1318–1325 (1980).

[52] A. S. Fokas and M. J. Ablowitz, On a unified approach to transformations and elementary solutions of Painlevé equations. *J. Math. Phys.* **23**, 2033–2043 (1982).

[53] A. S. Fokas and Y. C. Yortsos, The transformation parameters of the sixth Painlevé equation and one-parameter families of solutions. *Lett. Nuovo Cimento* **30**, 539–544 (1981).

[54] A. S. Fokas and R. L. Anderson, Group theoretical nature of Bäcklund transformations. *Lett. Math. Phys.* **3**, 117–126 (1979).

[55] B. Gambier, Sur les équations différentielles du second ordre et du premier degré dont l'inrégrale générale est á points critiques fixes. *Acta Math.* **33**, 1–55 (1910).

[56] C. S. Gardner, J. M. Greene, M. D. Kruskal, and R. M. Miura, Method for solving the Korteweg–de Vries equation. *Phys. Rev. Lett.* **19**, 1095–1097 (1967).

[57] P. G. Glockner and M. C. Singh (eds.), *Symmetry, Similarity, and Group Theoretic Methods in Mechanics.* University of Calgary, Calgary, 1974.

[58] M. Golubitsky, Primitive actions and maximal subgroups of Lie groups. *J. Diff. Geom.* **7**, 175–191 (1972).

[59] B. Grammaticos, B. Dorizzi, and R. Padjen, Painlevé property and integrals of motion for the Henón–Heiles system. *Phys. Lett.* **89A**, 111–113 (1982).

[60] A. M. Grundland, J. Harnad, and P. Winternitz, Solutions of the multidimensional sine–Gordon equation obtained by symmetry reduction. *Kinam* **4**, 333–344 (1982).

[61] A. M. Grundland, J. Harnad, and P. Winternitz, Symmetry reduction for nonlinear relativistically invariant equations. Preprint CRMA-1162 (1983), submitted to *J. Math. Phys.* .

[62] J. Harnad, Y. Saint-Aubin, and S. Shnider, Superposition of solutions to Bäcklund transformations for the *SU(n)* principal sigma model. *Preprint* CRMA–1074 (1982).

[63] J. Harnad, S. Shnider, and Y. Saint-Aubin, Quadratic pseudopotentials for *GL(n,C)* principal sigma models. *Preprint* CRMA–1075 (1982).

[64] J. Harnad, S. Shnider, and J. Tafel, Group actions on principal bundles and dimensional reduction. *Lett. Math. Phys.* **4**, 107-113 (1980).

[65] J. Harnad, S. Shnider, and L. Vinet, Solutions to Yang–Mills equations on $\overline{M}^4$ invariant under subgroups of *O(4,2)*. In *Complex Manifold Techniques in Theoretical Physics*. Pitman Research Notes in Mathematics **32**, 219–230 (1979).

[66] J. Harnad, S. Shnider, and L. Vinet, The Yang–Mills system in compactified Minkowski space; invariance conditions and *SU(2)* invariant solutions. *J. Math. Phys.* **20**, 931–942 (1979).

[67] J. Harnad, S. Shnider, and L. Vinet, Group actions on principal bundles and invariance conditions for gauge fields. *J. Math. Phys.* **21**, 2719–2724 (1980).

[68] J. Harnad and L. Vinet, On the *U(2)* invariant solutions to Yang–Mills equations in compactified Minkowski space. *Phys. Lett.* **6B**, 589–592 (1978).

[69] J. Harnad and P. Winternitz, Pseudopotentials and Lie symmetries for the generalized nonlinear Schrödinger equation. *J. Math. Phys.* **23**, 517–525 (1982).

[70] J. Harnad, P. Winternitz, and R. L. Anderson, Superposition principles for matrix Riccati equations. *Preprint* CRMA-1024 (1981), to appear in *J. Math. Phys.* **24**, (1983)

[71] B. K. Harrison and F. B. Estabrook, Geometric approach to invariance groups and solutions of partial differential equations. *J. Math. Phys.* **12**, 653–666 (1971).

[72] S. Helgason, *Differential Geometry, Lie groups, and Symmetric Spaces*. Academic Press, New York, 1978.

[73] E. Hille, *Ordinary Differential Equations in the Complex Domain*. John Wiley, 1976.

[74] N. Kh. Ibragimov, *Gruppovyĭe Svoĭstva Nĭekotorykh Differentsĭalnykh Uravnĭeniĭ*, (Group Theoretical Properties of some Differential Equations). Nauka, Novosibirsk, 1967.

[75] E. L. Ince, *Ordinary Differential Equations*. Dover, New York, 1956.

[76] H. Joos, Zur Darstellungstheorie der inhomogenen Lorentzgruppe als Grundlage quantenmechanischer Kinematik. *Fortschr. d. Phys.* **10**, 65–146 (1962).

[77] E. G. Kalnins and W. Miller Jr., Killing tensors and variable separation for Hamilton–Jacobi and Helmholtz equations. *SIAM J. Math. Anal.* **11**, 1011–1026 (1980).

[78] E. G. Kalnins and W. Miller Jr., Killing tensors and nonorthogonal variable separation for Hamilton–Jacobi equations. *SIAM J. Math. Anal.* **12**, 617–629 (1981).

[79] E. G. Kalnins, W. Miller Jr., and P. Winternitz, The group *O(4)*, separation of variables and the hydrogen atom. *SIAM J. Appl. Math.* **30**, 630–664 (1976).

[80] D. J. Kaup, The Estabrook–Wahlquist method with examples of application. *Physica* **1D**, 391–411 (1980).

[81] K. K. Kobayashi and M. Izutsu, Exact solution of the n-dimensional sine–Gordon equation. *J. Phys. Soc. Japan* **41**, 1091–1092 (1976).

[82] K. K. Kobayashi and T. Nagano, On filtered Lie algebras and geometric structures, I and I. *J. Math. Mech.* **13**, 875–908 (1964), and **14**, 513–521 (1965).

[83] (a) Y. Kosmann-Schwarzbach, Sur les transformations de similitude des equations aux dérivées partielles. *C. R. Acad. Sc. Paris* **287A**, 953–956 (1978); (b) *ib.*, Generalized symmetries of nonlinear partial differential equations. *Lett. Math. Phys.* **3**, 395–404 (1979).

[84] V. Kučera, A review of the matrix Riccati equation. *Kybernetika* **9**, 42–61 (1973).

[85] S. Kumei and G. W. Bluman, When nonlinear differential equations are equivalent to linear differential equations. *SIAM J. Appl. Math.* **42**, 1157–1174 (1982).

[86] G. L. Lamb Jr., Bäcklund transformations for certain nonlinear evolution equations. *J. Math. Phys.* **15**, 2157-2165 (1974).

[87] G. L. Lamb Jr., *Elements of Soliton Theory*. John Wiley, 1980.

[88] P. D. Lax, Integrals of nonlinear equations of evolution and solitary waves. *Commun. Pure Appl. Math.* **21**, 467–490 (1968).

[89] S. Lie, Allgemeine Untersuchungen über Differentialgleichungen die eine continuirliche endliche Gruppe gestatten. *Math. Ann.* **25**, 71-151 (1885).

[90] S. Lie, *Vorlesungen über differentialgleichungen mit bekannten infinitesimalen Transformationen*. Teubner, Leipzig, 1891); reprinted by Chelsea Publ. Co., New York, 1967.

[91] S. Lie and F. Engel, *Theorie der Transformationgruppen*. Teubner, Leipzig, 1888 (Vol. 1); 1890 (Vol. 2); 1893 (Vol. 3); reprinted by Chelsea Publ. Co., New York, 1967.

[92] S. Lie and G. Scheffers, *Vorlesungen über continuierliche Gruppen mit geometrischen und anderen Anwend.* Teubner, Leipzig, 1893; reprinted by Chelsea Publ. Co., New York, 1967.

[93] G. Leibrandt, New exact solutions of the classical sine–Gordon equation in *2+1* and *3+1* dimensions. *Phys. Rev. Lett.* **41**, 435–438 (1978).

[94] A. I. Malcev, *Foundations of Linear Algebra.* W. H. Freeman, San Francisco, 1963.

[95] J. B. McLeod and P. J. Olver, The connection between partial differential equations soluble by inverse scattering, and ordinary differential equations of Painlevé type. MRC *Report* 2135, University of Wisconsin, 1980.

[96] W. Miller Jr., *Symmetry and the Separation of Variables.* Addison-Wesley, Reading Mass., 1977.

[97] W. Miller Jr., The Technique of Variable Separation for Partial Differential Equations. *Lectures in these Proceedings.*

[98] W. Miller Jr., J. Patera, and P. Winternitz, Subgroups of Lie groups and the separation of variables. *J. Math. Phys.* **22**, 251–260 (1981).

[99] H. C. Morris, Prolongation structures and generalized inverse scattering problems. *J. Math. Phys.* **17**, 1867-1869 (1976).

[100] H. C. Morris, Prolongation structures and nonlinear evolution equations in two spatial dimensions. *J. Math. Phys.* **17**, 1870–1872 (1976).

[101] H. C. Morris, Prolongation structures and nonlinear evolution equations in two spatial dimensions. I. A generalized nonlinear Schrödinger equation. *J. Math. Phys.* **18**, 285–288 (1977).

[102] T. D. Newton, The inhomogeneous Lorentz group. In *Theory of Groups in Classical and Quantum Physics*, T. Kahan, ed. Am. Elsavier, 1966.

[103] T. Ochiai, Classification of the finite nonlinear primitive Lie algebras. *Trans. AMS* **124**, 313–322 (1966).

[104] A. T. Ogielski, M. K. Prasad, A. Sinha, and L.-L. Chau Wang, Bäcklund transformations and local conservation laws for principal chiral fields. *Phys. Lett.* **91B**, 387–391 (1980).

[105] P. J. Olver, Symmetry groups and group invariant solutions of partial differential equations. *J. Diff. Geometry* **14**, 497–542 (1979).

[106] L. V. Ovsi͡annikov, *Grupovoĭ Analiz Differentsi͡alnikh Uravnteniĭ*, (Group Theoretical Analysis of Differential Equations). Nauka, Moscow, 1978.

[107] P. Painlevé, Sur les équations différentielles du second ordre et d'ordre supérieur dont l'integrale générale est uniforme. *Acta Math.* **25**, 1–85 (1902).

[108] J. Patera, R. T. Sharp, P. Winternitz, and H. Zassenhaus, Subgroups of the similitude group of three-dimensional Minkowski space. *Can. J. Phys.* **54**, 986–994 (1976).

[109] J. Patera, R. T. Sharp, P. Winternitz, and H. Zassenhaus, Subgroups of the Poincaré group and their invariants. *J. Math. Phys.* **17**, 1439–1451 (1976).

[110] J. Patera, R. T. Sharp, P. Winternitz, and H. Zassenhaus, Continuous subgroups of the fundamental groups of physics. III. The de Sitter groups. *J. Math. Phys.* **18**, 2259–2288 (1977).

[111] J. Patera and P. Winternitz, A new basis for the representations of the rotation group. Lamé and Heun polynomials. *J. Math. Phys.* **14**, 1130–1139 (1973).

[112] (a) J. Patera, P. Winternitz, and H. Zassenhaus, Continuous subgroups of the fundamental groups of physics. I. general Method and the Poincaré group. *J. Math. Phys.* **16**, 1597–1614 (1975); *ib.*, II. The similitude group. *J. Math. Phys.* **16**, 1615–1624 (1975).

[113] F. A. E. Pirani, D. C. Robinson, and W. F. Shadwick, *Local Jet Bundle Formulation of Bäcklund Transformations*. D. Reidel, Dordrecht, 1979.

[114] K. Polhmeyer, Integrable hamiltonian systems and interactions through quadratic constraints. *Commun. Math. Phys.* **46**, 207–221 (1976).

[115] Z. Popowicz, Painlevé ping-pong $P^3 - P^5$. *Workshop contribution in these Proceedings.*

[116] D. Rand, *Etude Numérique des Règles de Superposition pour les Equations Matricielles de Riccati.* M. Sc. Thesis, Université de Montréal, 1982.

[117] D. Rand and P. Winternitz, Nonlinear superposition principles: a new numerical method for solving matrix Riccati equations. *Preprint* CRMA–1124 (1982).

[118] W. T. Reid, *Riccati Differential Equations.* Academic Press, New York, 1972.

[119] Y. Saint-Aubin, Bäcklund transformations and soliton type solutions for σ-models with values in real Grassmannian spaces. *Preprint* CRMA–1106 (1982). To appear in *Lett. Math. Phys.*

[120] W. F. Shadwick, The Bäcklund problem for the equation $\frac{\partial^2 z}{\partial x^1 \partial x^2} = f(z)$. *J. Math. Phys.* **19**, 2312–2317 (1978).

[121] A. C. Scott, F. Y. F. Chu, and D. W. McLaughlin, The soliton: a new concept in applied science. *Proc. IEEE* **61**, 1443-1483 (1973).

[122] S. Shnider and P. Winternitz, Classification of systems of nonlinear ordinary differential equations with superposition principles. *Preprint* CRMA–1164, 1983.

[123] M. Sorine, Sur l'équation de Riccati stationnaire associée au probléme de contrôle d'un système parabolique. *C. R. Acad. Sc. Paris* **287A**, 445–448 (1978).

[124] M. Sorine and P. Winternitz, Superposition laws for nonlinear equations arising in optimal control theory. To be published.

[125] E. Taflin, Analytic linearization, hamiltonian formalism, and infinite sequences of constants of motion for the Burgers equation. *Phys. Rev. Lett.* **47**, 1425–1428 (1981).

[126] G. Temple, Lectures on topics in nonlinear differential equations. *Report* 1415, David Taylor Model Basin, Carderock Md., 1960.

[127] Tu Gue-Zhang, The Lie algebra of the invariance group of the KdV, MKdV, or Burgers equation. *Lett. Math. Phys.* **3**, 387–393 (1979).

[128] M. E. Vessiot, Sur les systèmes d'équations différentielles du premier ordre qui ont des systèmes fondamentaux d'intégrales. *Ann. sc. Ecole Norm. sup.* **10**, 53– (1983).

[129] (a) H. De. Wahlquist and F. B. Estabrook, Prolongation structures of nonlinear evolution equations. *J. Math. Phys.* **16** 1–7 (1975); (b) *ib.*, Bäcklund transformation for solutions of the Korteweg–de Vries equation. *Phys. Rev. Lett.* **31**, 1386–1390 (1973).

[130] G. B. Whitham, *Linear and Nonlinear Waves.* John Wiley, 1974.

[131] G. B. Whitham, Comments on some recent multisoliton solutions. *J. Phys.* **A12**, L1–L3 (1979).

[132] E. P. Wigner, On unitary representations of the inhomogeneous Lorentz group. *Ann. Math.* **40**, 149–204 (1939).

[133] J. C. Willems, Least square stationary optimal control and the algebraic Riccati equation. *IEEE Trans. Autom. Control* **AC16**, 621–634 (1971).

[134] P. Winternitz, The Poincaré group, its little groups and their applications in particle physics. *Preprint*, Rutherford Laboratory, RPP–T3 (1969).

[135] P. Winternitz, Subgroups of Lie groups and symmetry breaking. In *Group Theoretical Methods in Physics*, R. T. Sharp and B. Kolman, eds. Academic Press, New York, 1977, pp. 549–572.

[136] P. Winternitz, Nonlinear action of Lie groups and superposition principles for nonlinear differential equations. *Physica* **114A**, 105–113 (1982).

[137] P. Winternitz and I. Friš, Invariant expansions of relativistic amplitudes and the subgroups of the proper Lorentz group. *Yad. Fiz.* **1**, 889–901 (1965); (*Sov. J. Nucl. Phys.* **1**, 636–643 (1965)).

[138] P. Winternitz, I. Lukač, and Ya. Smorodinskiĭ, Quantum numbers in the little groups of the Poincaré group. *Yad. Fiz.* **7**, 192–201 (1968); (*Sov. J. Nucl. Phys.* **7**, 139–145 (1968)).

[139] V. E. Zakharov, S. B. Manakov, S. P. Novikov, and L. P. Pitaevskiĭ, *Teoriâ Solitonov*, (Theory of Solitons). Nauka, Moscow, 1980.

[140] V. E. Zakharov and A. V. Mikhaĭlov, Relativistically invariant two-dimensional models of field theory which are integrable by means of the inverse scattering problem method. *Zh. Eksp. Teor. Fiz.* **74**, 1953–1973 (1978); (*Sov. Phys. JETP* **47**, 1017–1027 (1978)).

[141] V. E. Zakharov and A. B. Shabat, exact theory of two-dimensional self-focusing and one-dimensional self-modulation of waves in nonlinear media. *Zh. Eksp. Teor. Fiz.* **61**, 118–134 (1971); (*Sov. Phys. JETP* **34**, 62–69 (1972)).

[142] V. E. Zakharov and A. B. Shabat, A scheme for integrating nonlinear equations of mathematical physics by the method of the inverse scattering problem. I. *Funkts. Anal. Pril.* **8**, 43–53 (1974); (*Funct. Anal. Appl.* **8**, 226–235 (1974)).

[143] V. E. Zakharov and A. B. Shabat, Integration of nonlinear equations of mathematical physics by the method of the inverse scattering problem. II. *Funkts. Anal. Pril.* **13**, 13–22 (1979); (*Funct. Anal. Appl.* **13**, 116–174 (1979)).

Workshop

Speakers:

Radha Balakrishnan

Francisco Javier Chinea

Daniel David

Manuel de Llano

Bernadette Dorizzi

Daniel Finley, III

Basile Grammaticos

Brosl Hasslacher

Julio Herrera

Darryl D. Holm

Ernesto Lacomba

Eduardo Piña

Ziemowit Popowicz

Pedro Ripa

Rosalia Santoleri

Stanly Steinberg

Kurt Bernardo Wolf

Energy Transport in an Inhomogeneous Heisenberg Ferromagnetic Chain

Radha Balakrishnan

Department of Theoretical Physics
University of Madras, India

Abstract:

The spin evolution equation of a classical inhomogeneous Heisenberg chain is derived and its exact equivalence (in the continuum limit) to a generalized nonlinear Schrödinger equation with x-dependent coefficients is proved. An extension of the AKNS–ZS formalism is given which enables us to solve the latter equation exactly for certain specific inhomogeneities. Energy-momentum transport along the chain is related to the solution of this equation.

§1 Introduction

The classical Heisenberg chain with Hamiltonian

$$H = -J \sum_{i=1}^{N-1} \mathbf{S}_i \cdot \mathbf{S}_{i+1},$$

has in the continuum limit the spin evolution equation:

$$\frac{\partial \mathbf{S}}{\partial t} = \mathbf{S} \times \mathbf{S}_{xx}.$$

It has been shown [1] that this is equivalent to the nonlinear Schrödinger equation (NLS)

$$iq_t + q_{xx} + 2q|q|^2 = 0,$$

whose soliton solutions have been found by Zakharov and Shabat [2], using the method of Inverse Spectral Transforms [3]. Here q is related to $\mathbf{S}$ as follows:

$$q = \tfrac{1}{2}k \,\exp\left(i \int_{-\infty}^{x} \tau(x,t)\, dx\right),$$

where [4]

$$\text{curvature } k = \sqrt{\frac{\partial \mathbf{S}}{\partial x} \cdot \frac{\partial \mathbf{S}}{\partial x}} \quad \text{and torsion } \tau = \frac{1}{k^2} \mathbf{S} \cdot (\mathbf{S}_x \times \mathbf{S}_{xx}).$$

It is obvious from the Hamiltonian that

$$\text{Energy density } E(x,t) = \tfrac{1}{2} k^2 = 2|q|^2, \text{ and}$$
$$\text{Momentum density } P(x,t) = k^2 \tau = 4|q|^2 (\arg q)_x.$$

The equation of continuity $E_t + P_x = 0$ is satisfied.

It is interesting to see what happens to the solitonic behaviour of the energy density when specific inhomogeneities are introduced in the magnetic chain. In other words, we wish to consider the dynamics of the Hamiltonian

$$H = -J \sum_{i=1}^{N-1} f_i \mathbf{S}_i \cdot \mathbf{S}_{i+1}, \tag{1}$$

where the interaction strength between neighbouring spins varies in a specific manner as one moves along the chain. Physical systems which could be described by such a model have been discussed in a recent paper [5].

§2 Spin evolution equation and its equivalence with a generalized nonlinear Schrödinger equation

By calculating $d\mathbf{S}_i/dt = (\mathbf{S}_i, H)$, where (A, B) denotes the Poisson bracket of A with B, we get

$$\frac{d\mathbf{S}_i}{dt} = J f_i (\mathbf{S}_i \times \mathbf{S}_{i+1}) + J f_{i-1} (\mathbf{S}_i \times \mathbf{S}_{i-1}). \tag{2}$$

We mention that this equation is valid in the quantum case as well. Assuming $\mathbf{S}_i$ and f_i to vary slowly over one lattice separation, we may take its continuum limit to give

$$\mathbf{S}_t = f\,(\mathbf{S} \times \mathbf{S}_{xx}) + f_x\,(\mathbf{S} \times \mathbf{S}_x). \tag{3}$$

Lamb [4] has established the connection between the motion of twisted space curves and nonlinear equations with soliton solutions. Equation (3) may be identified with the equation of motion of the tangent to a moving space curve as follows: Let $\mathbf{e}_1(x,t)$, $\mathbf{e}_2(x,t)$, and $\mathbf{e}_3(x,t)$ denote respectively the unit tangent, normal, and binormal vectos of a moving space curve. These satisfy the Serret–Frenet equations

$$\mathbf{e}_{1x} = k\mathbf{e}_2, \quad \mathbf{e}_{2x} = -k\mathbf{e}_1 + \tau\mathbf{e}_3, \quad \text{and } \mathbf{e}_{3x} = -\tau\mathbf{e}_2,$$

where k and τ have been defined in the Introduction. Defining

$$\mathbf{N} = (\mathbf{e}_2 + i\mathbf{e}_3) \exp\left(i \int_{-\infty}^{x} \tau(x,t)\, dx\right), \quad \text{and } q = \tfrac{1}{2} k \exp\left(i \int_{-\infty}^{x} \tau(x,t)\, dx\right),$$

it is easy to prove $\mathbf{N}_x = -2q\,\mathbf{e}_1$, $\mathbf{N}_t = iR\mathbf{N} + \gamma\mathbf{e}_1$, and

$$\mathbf{e}_{1t} = -\tfrac{1}{2}(\gamma^* \mathbf{N} + \gamma \mathbf{N}^*), \tag{4}$$

where R is real and γ is arbitrary thus far and depends on the moving space curve. The compatibility condition $\mathbf{N}_{xt} = \mathbf{N}_{tx}$ leads to the following nonlinear partial differential equation for q:

$$q_t + \tfrac{1}{2}\gamma_x - iRq = 0, \tag{5}$$

with

$$R_x = i(\gamma q^* - \gamma^* q). \tag{6}$$

On renormalizing $\mathbf{S}^2$ to unity, it becomes possible to identify $\mathbf{S}$ with the vector $\mathbf{e}_1$. Equation (3) then becomes

$$\mathbf{e}_{1t} = -k\tau f\,\mathbf{e}_2 + (kf)_x\mathbf{e}_3 = i[(qf)_x\mathbf{N}^* - (q^*f)_x\mathbf{N}]. \tag{7}$$

Comparing Eqs. (7) and (4), γ is *determined* as

$$\gamma = -2i(qf)_x. \tag{8}$$

Substituting Eq. (8) in (6), integrating that equation and substituting the result in (5), we get

$$iq_t + fq_{xx} + 2fq\,|q|^2 + 2q\int_{-\infty}^{x} f_x\,|q|^2\,dx + qf_{xx} + 2f_xq_x = 0. \tag{9}$$

Equation (9) reduces to NLS when $f = 1$, as required. Further, when f is a linear function of x, it reduces to the equation considered by Calogero and Degasperis [6]. They have proved the existence of soliton solutions for q in this case. Lakshmanan and Bullough [7] have written down the equivalent vector equation for linear f. Our analysis has shown that to a generalized NLS such as (9) for any $f(x)$, there exists a corresponding inhomogeneous Heisenberg Hamiltonian (1).

The question arises whether Eq. (9) supports soliton solutions for any other $f(x)$. In what follows, we shall show that the conventional AKNS–ZS formalism [8,2] can be directly applied to Eq. (9) only for f equal to constant or linear function of x. We shall further show that extension of the formalism to other $f(x)$ can be achieved by permitting the *eigenvalue* ς appearing in the theory to be a function of both x and t. This is explained in the next section.

§3 Extension of the AKNS–ZS procedure; evolution of eigenvalue

To solve Eq. (9) by the method of inverse spectral transforms, one first reduces it to the following AKNS–ZS form [8,2]:

$$\begin{pmatrix} v_1 \\ v_2 \end{pmatrix}_x = \begin{pmatrix} -i\varsigma & q \\ -q^* & i\varsigma \end{pmatrix}\begin{pmatrix} v_1 \\ v_2 \end{pmatrix}, \qquad \begin{pmatrix} v_1 \\ v_2 \end{pmatrix}_t = \begin{pmatrix} A & B \\ C & -A \end{pmatrix}\begin{pmatrix} v_1 \\ v_2 \end{pmatrix}. \tag{10}$$

Allow for the possibility $\varsigma = \varsigma(x,t)$! [1] The extended AKNS conditions found from $(v_i)_{xt} = (v_i)_{tx}$ read

$$\begin{aligned} A_x - qC - q^*B &= -i\varsigma_t, \\ B_x + 2i\varsigma B + 2Aq &= q_t, \\ C_x - 2i\varsigma C + 2Aq^* &= -q_t^*. \end{aligned} \tag{11}$$

---

[1] Recall that the conventional AKNS–ZS procedure assumes $\varsigma = $ constant.

Using Eq. (9) for q_t in Eq. (11), we may deduce that for $\varsigma = 0$, a possible solution [9] is

$$\begin{aligned} A &= if|q|^2 + i\int_{-\infty}^{x} f_x|q|^2\,dx, \\ B &= i(qf)_x, \\ C &= i(qf)_x^{*}. \end{aligned} \tag{12}$$

For $\varsigma \neq 0$, let us write

$$A = if|q|^2 + i\int_{-\infty}^{x} f_x|q|^2\,dx + M(f,\varsigma) + W(f,\varsigma,q), \tag{13}$$

$$B = i(fq)_x + 2\varsigma fq + Y(f,\varsigma,q), \tag{14}$$

$$C = i(fq)_x^{*} - 2\varsigma fq^{*} + Z(f,\varsigma,q^{*}). \tag{15}$$

Where M, W, Y, and Z must vanish when $\varsigma = 0$. Substituting Eqs. (13–15) in AKNS conditions (3), it is found that M, W, Y, and Z must satisfy

$$M_x = -i\varsigma_t, \tag{16}$$

$$W_x - qZ - q^{*}Y = 0, \tag{17}$$

$$Y_x + 2i\varsigma Y = -2q[W + M + 2if\varsigma^2 + f\varsigma_x], \tag{18}$$

$$Z_x - 2i\varsigma Z = -2q^{*}[W + M + 2if\varsigma^2 - f\varsigma_x]. \tag{19}$$

To apply the inverse scattering method, we need the asymptotic values of A, B, and C in (13–15). This in turn means that we must know the asymptotic values of M, W, Y, and Z as $|x| \to \infty$.

Two cases are distinguished: $\varsigma_x = 0$, and $\varsigma_x \neq 0$.

i. For $\varsigma_x = 0$, it is obvious from (17–19) that $Y = Z = W = 0$, $M = -2if\varsigma^2$, is a solution for all x. Hence, from (16),

$$\varsigma_t = 2f_x\varsigma^2. \tag{20}$$

Therefore f = constant and f = linear function of x, are the only possibilities.

ii. When $\varsigma_x \neq 0$, we may therefore write

$$M(f,\varsigma) = -2if\varsigma^2 + N_1(\varsigma_x, \varsigma_{xx}, \ldots), \tag{21}$$

where N_1 depends on first and higher derivatives of ς so that when $\varsigma_x = 0$, we get the correct expression for M.

Equations (18) and (19) become

$$Y_x + 2i\varsigma Y = -2q[W + N_1 + f\varsigma_x], \tag{22}$$

$$Z_x - 2i\varsigma Z = -2q^{*}[W + N_1 - f\varsigma_x]. \tag{23}$$

Given that $q, q^{*} \to 0$ sufficiently rapidly as $|x| \to \infty$, Eq. (17) implies that $W \to W_0(t)$ as $|x| \to \infty$, provided Y and Z exist in that limit. If N_1 *and* $f\varsigma_x$ exist asymptotically, then the right-hand sides of

Eqs. (22–23) vanish and $Y \sim C_0 \exp(-2i \int^x \varsigma\, dx)$ as $|x| \to \infty$. Since Y must vanish as $\varsigma \to 0$ for all x, Y vanishes as $|x| \to \infty$. Similarly for Z.

So far N_1 is unspecified except that it depends on derivatives of ς and that it exists asymptotically. choosing $N_1 = \mp f \varsigma_x$ as a constant solution, (16) becomes

$$i\varsigma_t \mp (f\varsigma_x)_x - 2i(f\varsigma^2)_x = 0, \tag{24}$$

a nonlinear evolution equation for the eigenvalue $\varsigma(x,t)$ itself. Also, as $|x| \to \infty$, $A \sim \mp f\varsigma_x - 2i\varsigma^2 f$; $B \sim 0$, and $C \sim 0$.

For the present case $\varsigma_x \neq 0$, and the first AKNS equation in (10) is no longer a straightforward eigenvalue problem. However, the form of this equation immediately suggests that if $\varsigma(x,t)$ is a separable function of x and t, then it can be reduced to an AKNS eigenvalue problem by a suitable change of variables, as we shall see below. Hence we seek solutions of Eq. (24) of the form

$$\varsigma(x,t) = g(x)\, h(t). \tag{26}$$

This immediately yields

$$ih_t \mp (fg_x)_x g^{-1} h - 2i(fg^2)_x g^{-1} h^2 = 0. \tag{27}$$

Therefore

$$fg_x = \lambda \int g\, dx + \lambda_0, \tag{28}$$

$$fg^2 = \mu \int g\, dx + \mu_0, \tag{29}$$

where λ, λ_0, μ, and μ_0 are constants. Hence

$$ih_t \mp \lambda h - 2i\mu h^2 = 0.$$

Consider the first of the two equations in (10) and put $\varsigma(x,t) = g(x)\, h(t)$. On making the transformation $\int^x g(x)\, dx \to y$ and $q(x,t)/g(x) \to Q(y,t)$, this equation becomes

$$\begin{pmatrix} W_1 \\ W_2 \end{pmatrix}_y = \begin{pmatrix} -ih(t) & Q(y,t) \\ -Q^*(y,t) & +ih(t) \end{pmatrix} \begin{pmatrix} W_1 \\ W_2 \end{pmatrix}. \tag{30}$$

Let us impose the requirement that the transformation be such that $y = \int^x g(x)\, dx \to \pm\infty$ as $x \to \pm\infty$. This, along with a mild requirement on $g(x)$ that it not vanish too rapidly as $|x| \to \infty$ ensures that $Q(y,t) \to 0$ as $|y| \to \infty$. These conditions guarantee that the analytic properties of the scattering functions etc. in the complex h-plane are the same as in a conventional AKNS–ZS formalism, and therefore the inverse spectral transform analysis can be carried out with ease, incorporating the time dependence of the eigenvalue [10].

§4 The direct problem

For h real, let ϕ and $\overline{\phi}$ be two independent Jost solutions of (29) satisfying the boundary conditions

$$\phi \sim \begin{pmatrix} 1 \\ 0 \end{pmatrix} e^{-ihy}, \quad \overline{\phi} \sim \begin{pmatrix} 0 \\ -1 \end{pmatrix} e^{ihy} \quad \text{as } y \to -\infty.$$

The t-dependent coefficients of these solutions may be found from the asymptotic behaviour of the second equation in (10). As $x \to -\infty$,

$$A \sim \mp f\varsigma_x - 2if\varsigma^2,$$

or

$$A \sim \mp(fg_x)_x h - 21(fg^2)_x h^2.$$

Using (28) and (29), and setting $\int^x g(x)\,dx = y$, we get

$$A \sim (\mp\lambda_0 h - 2i\mu_0 h^2)_x \mp \lambda h y - 2i\mu_0 h^2 y,$$

or

$$A \sim -H(t) - ih_t y, \quad \text{where } H(t) = \pm\lambda_0 h + 2i\mu_0 h^2.$$

Therefore,

$$W \sim \exp\left(-\int_0^t H(t')\,dt'\right)\phi \qquad \text{as } y \to -\infty.$$

§5 Time evolution of scattering parameters

For $y \to +\infty$, writing the boundary conditions in customary notation as

$$\begin{pmatrix} W_1 \\ W_2 \end{pmatrix} = \begin{pmatrix} a(h,t) & \exp(-ihy) \\ b(h,t) & \exp(+iht) \end{pmatrix} \exp\left(-\int_0^t H(t')\,dt'\right),$$

and substituting in the second equation in (10), we get

$$\begin{aligned} a(h,t) &= a(h_0,t), \\ b(h,t) &= b(h_0,0)\exp\left(+2\int_0^t H(t')\,dt'\right), \end{aligned} \tag{31}$$

where $h_0 = h(0)$. With a knowledge of the spectrum h_0 corresponding to a given initial *potential* $Q(y,0)$, the time evolution of the scattering data is given by (31). The potential $Q(y,t)$ is constructed by using this information in the Gel'fand–Levitan–Marchenko equations. Changing back to the x-variable, we get $q(x,t)$.

§6 An example

As an illustrative example, consider the case $\lambda = 0$, $\mu \neq 0$ in Eqs. (28) and (29), giving $g = \lambda_0 \int^x (1/f(x))\,dx$, along with $f_x g =$ constant. In order to have $y = \int^x g(x)\,dx \to \pm\infty$ as $x \to \pm\infty$, the class of solutions this leads to is

$$f(x) = \frac{f_0}{(x - x_0)^{2n+1}}, \qquad n = 0, 1, 2, \ldots.$$

A complete analysis of Eqs. (28) and (29) should give other classes of $f(x)$ for which (9) can be solved exactly. The recovery of the usual result $a_t = 0$ [see (31)] suggests the presence of an infinite number of conservation laws. From the Hamiltonian (1), it is easy to see that in the continuum limit, the energy density $E(x,t) = 2f|q|^2$. The momentum density derived using the continuity equation $E_t + P_x = 0$ is just $P(x,t) = 4f^2|q|^2(\arg q)_x$. Thus for certain specific inhomogeneities, the energy momentum densities have solitonic behaviour. It should be interesting to study whether the spin density [11], determined from (3), also has such a behaviour for those inhomogeneities. This, and the study of the geometric structure of the problem are under way.

This work was supported by the Council of Scientific and Industrial Research, India.

References

[1] M. Lakshmanan, Continuum spin systems as an exactly solvable dynamical system. *Phys. Lett.* **61A**,, 53–54 (1972).

[2] V. E. Zakharov and A. B. Shabat, Exact theory of two-dimensional self-focusing and one-dimensional self-modulating waves in nonlinear media. *Sov. Phys. JETP* **34**, 62–69 (1972).

[3] G. S. Gardner, M. D. Kruskal, R. M. Miura, and J. M. Greene, Method of solving the Korteweg–de Vries equation. *Phys. Rev. Lett.* **19**, 1095–1097 (1967).

[4] G. L. Lamb, Jr., Solitons on moving space curves. *J. Math. Phys.* **18**, 1654-1661 (1977).

[5] R. Balakrishnan, On the inhomogeneous Heisenberg chain. *J. Phys.* **C15**, L1305–L1308 (1982).

[6] F. Calogero and A. Degasperis, Exact solution via the spectral transform of a generalization with linearly x-dependent coefficients of the nonlinear Schrödinger equation. *Lett. Nuovo Cimento* **22**, 420–424 (1978).

[7] M. Lakshamanan and R. K. Bullough, Geometry of generalized nonlinear Schrödinger and Heisenberg ferromagnetic spin equations with linearly x-dependent coefficients. *Phys. Lett.* **80A**,, 287–292 (1980).

[8] M. J. Ablowitz, D. J. Kaup, A. C. Newell, and H. Segur, The inverse scattering transform analysis for nonlinear problems. *Stud. Appl. Math.* **53**, 249–315 (1974).

[9] R. Balakrishnan, Dynamics of a generalized classical Heisenberg chain. *Phys. Lett.* **92A**,, 243–246 (1982).

[10] M. R. Gupta, Exact inverse scattering solution of a nonlinear evolution equation in a non-uniform medium. *Phys. Lett.* **72A**,, 420–422 (1979).

[11] L. A. Takhtajan, Integration of the continuous Heisenberg spin chain through the inverse scattering method. *Phys. Lett.* **64A**,, 235–237 (1977).

Bäcklund Transformations in General Relativity

Francisco Javier Chinea

Departamento de Métodos Matemáticos de la Física
Facultad de Ciencias Físicas
Universidad de Madrid, Spain

§1 Introduction

The motivation for the present work may be summarized as follows:

$$R_{\alpha\beta} = 0 \quad \Rightarrow \quad g_{\alpha\beta} = ?,$$

where $R_{\alpha\beta}$ is the Ricci tensor of a four-dimensional pseudo-Riemannian space, and $g_{\alpha\beta}$ the corresponding metric. We are thus concerned with finding solutions of the Einstein equations in vacuum. This represents a formidable task in the general case, and as a consequence the situations to be considered are usually special, either in the algebraic or in the isometric sense. In particular, much work has been done in recent years on solutions which possess two commuting Killing fields. Two coordinates will be ignorable, and the problem reduces to the integration of a set of partial differential equations in the two remaining variables. Such a set may be cast in a very elegant form, either as a single equation for a complex field [1] or, alternatively, as an equation for a three-vector of constant length [2]. There is, in addition, another set of equations for a single scalar function; but this second set is compatible and formally integrable, provided the equations in the first set are satisfied, and may thus be considered as subsidiary.

The basic strategy is to construct algorithms for getting new solutions of the equations from known ones (*solution-generating techniques*). The main approaches are:

i. Work with the so-called infinite hierarchy of potentials, such potentials being transformed among themselves by an infinite-dimensional group (*the Geroch group*). See [3–6].

ii. Application of the inverse scattering method [7].

iii. Application of the Riemann–Hilbert problem [8].

iv. Use of Bäcklund transformations [9–14].

There exist close connections among all different techniques [15]. The present work will be concerned with the last method; however, a review of the field is not attempted in what follows, which is mostly based on the author's work.

Bäcklund transformations were introduced for the first time in General Relativity in the inportant references [9] and [10], and applications may be seen in [11] and [12].

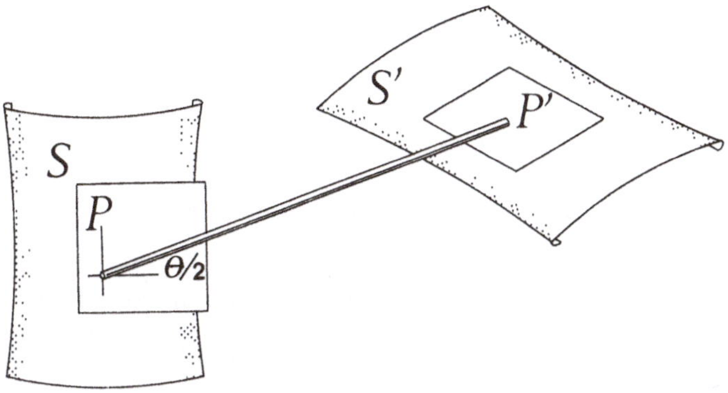

Figure 1. The Bianchi construction.

§2 Bäcklund transformations as gauge transformations

The basic ideas behind what is nowadays known under the generic name of ***Bäcklund transformation*** originated in the researches of several geometers of the last century, who worked on the classical theory of surfaces.[1] Ribaucour had devised a geometric construction relating surfaces of constant curvature (1870), and Bianchi introduced in 1879 the transformation shown in Figure 1. Suppose one has a two-surface S of constant negative curvature,[2] whose metric, when referred to asymptotic coordinates, takes the form:

$$ds^2 = du^2 + 2\cos\varphi(u,v)du\,dv + dv^2. \tag{2.1}$$

It is trivial to check that the constant curvature condition implies:[3]

$$\varphi_{uv} = \sin\varphi, \tag{2.2}$$

where subscripts denote partial derivatives. Suppose now that a line segment of unit length is contained in the tangent plane to the surface S at a point P, at an angle $\frac{1}{2}\theta$ with respect to a line of curvature going through P. Suppose further that P moves about in S in such a way that the motion of the other end point, P', is fixed by the requirement that:

i. PP' be also tangent to the locus S' of P'.

ii. The tangent planes at P and P' be orthogonal.

Then, it is found that the locus S' is again a surface of constant negative curvature. Furthermore, the above construction implies analitically:

$$\begin{aligned}\tfrac{1}{2}(\theta+\varphi)_u &= \sin\tfrac{1}{2}(\theta-\varphi),\\ \tfrac{1}{2}(\theta-\varphi)_v &= \sin\tfrac{1}{2}(\theta+\varphi),\end{aligned} \tag{2.3}$$

[1]The classical theory of Bianchi–Bäcklud transformations may be seen in Ref. [16], where references to the original work will be found.

[2]We take the curvature to be −1 in what follows.

[3]Equation (2.2) was very dear to nineteenth century geometers, being central in the theory of surfaces of constant curvature. Many attempts to integrate it or at least to find methods providing as many new solutions as possible were made. It is in this context that the Bianchi transformation should be considered

and the integrability conditions for (2.3) are:

$$\varphi_{uv} = \sin\varphi, \tag{2.4a}$$
$$\theta_{uv} = \sin\theta. \tag{2.4b}$$

In fact, the metric corresponding to S' is:

$$ds'^2 = du^2 + 2\cos\theta(u,v)du\,dv + dv^2. \tag{2.5}$$

Notice that a new solution θ may be obtained by quadratures from (2.3), because the integrability condition (2.4a) is satisfied by virtue of φ being a solution of that equation. Bäcklund generalized the previous construction by assuming that the tangent planes at P and P' meet at a constant (but not necessarily right) angle; this amounts to introducing a parameter in (2.3). He also addressed much more general situations and equations [17]. For our purposes, the essential feature of Bäcklund transformations is that they provide a way of getting new solutions of a given equation [such as (2.4b)] by means of a known solution [such as (2.4a)] by integrating a (formally integrable) system of partial differential equations [such as (2.3)]. From this point of view, they are nonlinear generalizations of the Cauchy–Riemann equations.

Bäcklund transformations may be considered as a specific type of gauge transformation [18]. In order to see that, the following example may be used: Consider a *O(2,1)*-invariant nonlinear σ-model, characterized by the field equation

$$q_{uv} = (q_u \cdot q_v)q, \qquad q^2 = -1, \tag{2.6}$$

where q is a three-vector, with scalar product $q \cdot p = q^1p^1 + q^2p^2 - q^3p^3$ (For the essential features of nonlinear σ-models, see reference [19]). Let z be defined by

$$q = \frac{1}{z+\bar{z}}(i(z-\bar{z}), 1 - z\bar{z}, 1 + z\bar{z}).$$

It is easy to check that (2.6) is equivalent to:

$$z_{uv} = 2z_uz_v(z+\bar{z})^{-1}. \tag{2.7}$$

Suppose we now want to generate solutions of (2.7). In order to do that, one introduces the matrix system[4]

$$A_u = MA, \qquad A_v = NA, \tag{2.8}$$

where $A \in$ *SU(2)*, with $M, N \in$ *su(2)* given by:

$$M = \frac{1}{z+\bar{z}}\begin{pmatrix} \frac{1}{2}(z-\bar{z})_u & \lambda z_u \\ -\lambda\bar{z}_u & -\frac{1}{2}(z-\bar{z})_u \end{pmatrix}, \qquad N = \frac{1}{z+\bar{z}}\begin{pmatrix} \frac{1}{2}(z-\bar{z})_v & -\lambda^{-1}z_v \\ \lambda^{-1}\bar{z}_v & -\frac{1}{2}(z-\bar{z})_v \end{pmatrix}, \tag{2.9}$$

where λ is a real constant. The integrability condition for (2.8) is

$$M_v - N_u + [M,N] = 0. \tag{2.10}$$

---

[4]This is closely connected with a specific formulation which is very useful for applications of the inverse scattering method. See [20].

But (2.10) is precisely equivalent to (2.7) when M and N are given by (2.9). This scheme may be used for many different equations, the essencial equations being (2.8) and (2.10), with matrices in an appropiate matrix group and its corresponding Lie algebra, depending functionally on a field φ and its derivatives. One may think of $\omega = M\,du + N\,dv$ as a connection on a certain bundle with that group as the structural group, and such that the connection has a vanishing curvature [21],

$$\Omega \equiv d\omega - \omega \wedge \omega = 0. \tag{2.11}$$

Equation (2.11) is, by construction, equivalent to a certain field equation $\mathcal{F}(\varphi) = 0$ for φ.

It is now natural to introduce gauge transformations of the connection:

$$\tilde{\omega} = S\omega S^{-1} + dS\,S^{-1}, \tag{2.12}$$

$$\begin{aligned}\tilde{M} &= SMS^{-1} + S_u S^{-1},\\ \tilde{N} &= SNS^{-1} + S_v S^{-1},\end{aligned} \tag{2.13}$$

where S is a matrix in the group. Such transformations leave (2.8) form-invariant, and guarantee that the transformed $\tilde{\Omega}$ satisfies $\tilde{\Omega} = 0$ whenever $\Omega = 0$. Bäcklund transformations are obtained in the following way:

i. Assume that $\tilde{M}$ and $\tilde{N}$ have the same functional form with respect to the transformed field $\tilde{\varphi}$ as M and N have with respect to the original field.

ii. Assume that $S = S(\varphi, \tilde{\varphi})$.

The classical transformation (2.3) may be obtained in precisely this fashion. We shall continue, however, with the present example: Equations (2.13) imply that the Bäcklund transformations to be found this way are such that $\tilde{z}_u$ has to be expressed linearly in terms of z_u and $\bar{z}_u$, and correspondingly for $\tilde{z}_v$. In particular, one may try

$$\frac{\tilde{z}_u}{\tilde{z}+\bar{\tilde{z}}} = P(z,\bar{z},\tilde{z},\bar{\tilde{z}})\,\frac{z_u}{z+\bar{z}}, \qquad \frac{\tilde{z}_v}{\tilde{z}+\bar{\tilde{z}}} = P(z,\bar{z},\tilde{z},\bar{\tilde{z}})\,\frac{z_v}{z+\bar{z}}, \tag{2.14}$$

and then equations (2.13) imply:

$$[(z+\bar{z})S_z + Sm] + P[(\tilde{z}+\bar{\tilde{z}})S_{\tilde{z}} - mS] = 0, \tag{2.15a}$$

$$[(z+\bar{z})S_z + Sn] + p[(\tilde{z}+\bar{\tilde{z}})S_{\tilde{z}} - nS] = 0, \tag{2.15b}$$

where

$$m = \begin{pmatrix} \frac{1}{2} & \lambda \\ 0 & -\frac{1}{2} \end{pmatrix}, \qquad n = \begin{pmatrix} \frac{1}{2} & -\lambda^{-1} \\ 0 & -\frac{1}{2} \end{pmatrix}.$$

The scalar functions P and p may be obtained from equations (2.15), provided that those equations satisfy the corresponding algebraic compatibility conditions. This results in a certain set of quadratic first-order partial differential equations for S. A particular solution may be found by using the ***Ansatz*** $S_z = S_{\bar{z}}$, $S_{\tilde{z}} = S_{\bar{\tilde{z}}}$. One finds in this manner:

$$P = \frac{\sqrt{r-1} - i}{\sqrt{r-1} + i}, \qquad p = \frac{\sqrt{r-1} + i}{\sqrt{r-1} - i}, \tag{2.16}$$

where $r = c(z+\bar{z})(\tilde{z}+\bar{\tilde{z}})$, and c is a real constant. Bäcklund transformations for (2.7) are thus easily found; they take the form (2.14) with P and p given by (2.16). The same procedure will be used in §4 in order to obtain Bäcklund transformations for the Ernst equation.

§3 Vacuum Einstein field equations with two commuting Killing fields

We shall be concerned from now on with pseudo-Riemannian spaces which admit two independent commuting Killing fields, ξ and χ:

$$\mathcal{L}_\xi g = 0, \qquad \mathcal{L}_\chi g = 0, \qquad [\xi, \chi] = 0,$$

where g is the corresponding metric. Orthogonal transitivity will also be assumed.[5] Two different cases may be considered, according as ξ and χ are both spacelike, *e.g.*, cylindrical symmetry, or as one of them is timelike while the other is spacelike, *e.g.*, stationary and axially symmetric spaces. For the purpose of illustration, the former case will be treated in what follows, in order to derive the relevant equations. The metric may be written in such a case as:

$$ds^2 = g_{ij}(x^0, x^1)\, dx^i\, dx^j + g_{\bar{i}\bar{j}}(x^0, x^1)\, dx^{\bar{i}}\, dx^{\bar{j}}, \qquad i, j = 0, 1; \quad \bar{i}, \bar{j} = 2, 3; \quad x^2, x^3 = \varphi, z.$$

The unbarred block of the metric may always be expressed as

$$g_{ij} = e^{\psi(x^0, x^1)} \eta_{ij},$$

where η_{ij} is the two-dimensional Minkowski metric;[6] on the other hand, $g_{\bar{i}\bar{j}}$ may be written as

$$g_{\bar{i}\bar{j}} = \tau(x^0, x^1)\, \gamma_{\bar{i}\bar{j}}(x^0, x^1),$$

where $\tau = \sqrt{\det(g_{\bar{i}\bar{j}})}$ and $\det \gamma = 1$. The Einstein equations may trivially be seen to reduce to two sets:

$$R_{\bar{i}\bar{j}} = 0 \quad \Leftrightarrow \quad (\tau \gamma^k \gamma^{-1})_k = 0, \qquad \tau^k{}_k = 0, \tag{3.1}$$

where $\gamma^k \equiv \eta^{ki} \gamma_i$ and $\tau^k \equiv \eta^{ki} \tau_i$, and

$$R_{ij} = 0 \quad \Leftrightarrow \quad \begin{cases} \psi_u = \dfrac{\tau_{uu}}{\tau_u} - \dfrac{\tau_u}{2\tau} + \dfrac{\tau}{4\tau_u} \operatorname{Tr}(\gamma^{-1}\gamma_u\gamma^{-1}\gamma_u), \\ \psi_v = \dfrac{\tau_{vv}}{\tau_v} - \dfrac{\tau_v}{2\tau} + \dfrac{\tau}{4\tau_v} \operatorname{Tr}(\gamma^{-1}\gamma_v\gamma^{-1}\gamma_v), \\ \psi_{uv} = \dfrac{\tau_u \tau_v}{2\tau^2} - \frac{1}{4} \operatorname{Tr}(\gamma_{uv}\gamma^{-1}). \end{cases} \tag{3.2}$$

The remaining equations $R_{i\bar{j}} = 0$ are identically satisfied. Notice that equations (3.2) are integrable whenever the equations (3.1) are satisfied, and ψ may be obtained in principle by quadratures; accordingly, we shall concentrate on (3.1) in what follows. Introduce now the $sl(2,\mathbb{R})$ basis:

$$\rho_1 = \begin{pmatrix} 1 & 0 \\ 0 & -1 \end{pmatrix}, \qquad \rho_2 = \begin{pmatrix} 0 & 1 \\ 1 & 0 \end{pmatrix}, \qquad \rho_3 = \begin{pmatrix} 0 & -1 \\ 1 & 0 \end{pmatrix},$$

which satisfies

$$\rho_A \rho_B = \epsilon_{AB}{}^D \rho_D + \eta_{AB}\, I,$$

---

[5]This means that two-surfaces orthogonal to the group orbits exist, and is equivalent to the existence of an extra discrete symmetry. See reference [4].

[6]This can always be done in two dimensions by means of a coordinate change.

where $A, B, D = 1, 2, 3$, and $\eta_{AB} = \mathrm{diag}\,(++-)$. Define $Q = \gamma\rho_3$, so that $\det\gamma = 1 \Leftrightarrow Q^2 = -I$ and $\gamma^\top = \gamma \Leftrightarrow \mathrm{Tr}\, Q = 0$. Hence Q is a traceless matrix and may be expanded as $Q = q^A\rho_A$. Equations (3.1) may then be expressed as

$$(\tau q \times q^k)_k = 0, \qquad q^2 = -1, \qquad \tau_{uv} = 0,$$

or, equivalently, [2]:

$$q_{uv} + \frac{\tau_u}{2\tau} q_v + \frac{\tau_v}{2\tau} q_u = (q_u \cdot q_v) q, \qquad q^2 = -1, \tag{3.3}$$

with $\tau = U(u) + V(v)$. Equation (3.3) may be algebraically reduced by introducing the parametrization:

$$q = \frac{1}{f + \bar{f}} \left(i(f - \bar{f}), 1 - f\bar{f}, 1 + f\bar{f} \right),$$

which solves the constraint $q^2 = -1$ by means of a complex function f; equation (3.3) now reads:

$$f_{uv} + \frac{\tau_u}{2\tau} f_v + \frac{\tau_v}{2\tau} f_u = \frac{2 f_u f_v}{f + \bar{f}}. \tag{3.4}$$

To the global *O(2,1)* invariance of (3.3) there corresponds the following *SL(2,*R*)* invariance of equation (3.4):

$$f \mapsto \frac{af + ib}{-icf + d}, \qquad ad - bc = 1, \quad a, b, c, d \text{ real constants.}$$

Equation (3.4) turns out to look exactly like the Ernst equation for this case [1]:

$$\mathcal{E}_{uv} + \frac{\tau_u}{2\tau} \mathcal{E}_v + \frac{\tau_v}{2\tau} \mathcal{E}_u = \frac{2\mathcal{E}_u \mathcal{E}_v}{\mathcal{E} + \bar{\mathcal{E}}}. \tag{3.5}$$

However, both quantities, f and $\mathcal{E}$, should not be confused. In order to relate one to the other, write the metric in the Lewis–Papapetrou form [22]:

$$ds^2 = h(dz + \omega\, d\varphi)^2 + S^2 h^{-1}\, d\varphi^2 + e^v (d\rho^2 - dt^2),$$

where one may take[7] $S = \rho = u + v$, and then f and $\mathcal{E}$ will be given by:

$$f = h^{-1} S + i\omega, \tag{3.6a}$$
$$\mathcal{E} = h + i\psi, \tag{3.6b}$$

where

$$\psi_u = h^2 S^{-1} \omega_u, \tag{3.7a}$$
$$\psi_v = -h^2 S^{-1} \omega_v. \tag{3.7b}$$

In the case of one timelike and one spacelike Killing field, a derivation entirely similar to the one leading to equation (3.3) yields: [2]

$$q_{\varsigma\bar{\varsigma}} + \frac{\tau_\varsigma}{2\tau} q_{\bar{\varsigma}} + \frac{\tau_{\bar{\varsigma}}}{2\tau} q_\varsigma = -(q_\varsigma \cdot q_{\bar{\varsigma}}) q, \qquad q^2 = 1, \tag{3.8}$$

[7] Due to the conformal invariance of the equations, there is no loss of generality in this assumption.

where $\varsigma = \frac{1}{2}(\rho + iz)$, ρ and z real, and $\tau = \eta(\varsigma) + \overline{\eta(\varsigma)}$, $\eta(\varsigma)$ arbitrary. As previously stressed, there is no loss of generality in taking $\eta(\varsigma) = \varsigma$. The three-vector q may be parametrized as

$$q = \frac{1}{f+g}(-f+g, 1+fg, 1-fg),$$

where f and g are now *real* functions. Equation (3.8) translates into the following ones:

$$f_{\varsigma\bar{\varsigma}} + \frac{\tau_\varsigma}{2\tau} f_{\bar{\varsigma}} + \frac{\tau_{\bar{\varsigma}}}{2\tau} f_\varsigma = \frac{2 f_\varsigma f_{\bar{\varsigma}}}{f+g}, \tag{3.9a}$$

$$g_{\varsigma\bar{\varsigma}} + \frac{\tau_\varsigma}{2\tau} g_{\bar{\varsigma}} + \frac{\tau_{\bar{\varsigma}}}{2\tau} g_\varsigma = \frac{2 g_\varsigma g_{\bar{\varsigma}}}{f+g}. \tag{3.9b}$$

Equations (3.9) are invariant under

$$f \mapsto \frac{af+b}{cf+d}, \qquad g \mapsto \frac{ag-b}{-cg+d}, \qquad ad - bc = 1.$$

Equation (3.8) involves a variable q which is obtained algebraically from the metric. Alternatively, one may derive the following Ernst equation [1]:

$$\mathcal{E}_{\varsigma\bar{\varsigma}} + \frac{\tau_\varsigma}{2\tau} \mathcal{E}_{\bar{\varsigma}} + \frac{\tau_{\bar{\varsigma}}}{2\tau} \mathcal{E}_\varsigma = \frac{2 \mathcal{E}_\varsigma \mathcal{E}_{\bar{\varsigma}}}{\mathcal{E} + \overline{\mathcal{E}}}, \tag{3.10}$$

where $\mathcal{E}$ is related to the components of the metric by equations similar to (3.6b) and (3.7a–b). Equation (3.10) may be cast in a form similar to (3.3) and (3.8), by defining

$$V = \frac{1}{\mathcal{E} + \overline{\mathcal{E}}}(i(\mathcal{E} - \overline{\mathcal{E}}), 1 - \mathcal{E}\overline{\mathcal{E}}, 1 + \mathcal{E}\overline{\mathcal{E}}),$$

which brings (3.10) over into:

$$V_{\varsigma\bar{\varsigma}} + \frac{\tau_\varsigma}{2\tau} V_{\bar{\varsigma}} + \frac{\tau_{\bar{\varsigma}}}{2\tau} V_\varsigma = (V_\varsigma \cdot V_{\bar{\varsigma}}) V, \qquad V^2 = -1. \tag{3.11}$$

The sign differences between (3.8) and (3.11) should be noticed.

§4 Integrability representation and Bäcklund transformations for f and $\mathcal{E}$

Following the scheme described in §2, in the case characterized by the existence of two spacelike Killing vectors, we may introduce *su(2)* matrices[8] M and N as:

$$M = \frac{1}{f+\bar{f}} \begin{pmatrix} \frac{1}{2}(f-\bar{f})_u & \kappa f_u \\ -\bar{\kappa}\bar{f}_u & -\frac{1}{2}(f-\bar{f})_u \end{pmatrix}, \qquad N = \frac{1}{f+\bar{f}} \begin{pmatrix} \frac{1}{2}(f-\bar{f})_v & -\overline{\kappa^{-1}} f_v \\ \kappa^{-1}\bar{f}_v & -\frac{1}{2}(f-\bar{f})_v \end{pmatrix}, \tag{4.1}$$

[8] An *su(1,1)* version also exists. See reference [14].

where $\kappa = \sqrt{(V(v)-\lambda)/(U(u)+\lambda)}$, with λ a real constant, and $U(u)$ and $V(v)$ such that $\tau = U(u) + V(v)$. It is easy to check that

$$M_v - N_u + [M,N] = 0 \quad \Rightarrow \quad f_{uv} + \frac{\tau_u}{2\tau} f_v + \frac{\tau_v}{2\tau} f_u = \frac{2 f_u f_v}{f+\bar{f}}.$$

Using the representation (4.1), Bäcklund transformations may be found in the manner described in §2. Let

$$\frac{\tilde{f}_u}{\tilde{f}+\bar{\tilde{f}}} = P \frac{f_u}{f+\bar{f}} + L, \qquad \frac{\tilde{f}_v}{\tilde{f}+\bar{\tilde{f}}} = p \frac{f_v}{f+\bar{f}} + \ell, \tag{4.2}$$

with P, L, p, and ℓ functions of f, $\bar{f}$, $\tilde{f}$, $\bar{\tilde{f}}$, u, and v. Using the *Ansatz* $S_f = S_{\bar{f}}$, $S_{\tilde{f}} = S_{\bar{\tilde{f}}}$, and parametrizing S as

$$S = \begin{pmatrix} a & b \\ -\bar{b} & \bar{a} \end{pmatrix}, \qquad a\bar{a} + b\bar{b} = 1,$$

the following solution is obtained:

$$\begin{aligned} P &= \frac{-b+\kappa a}{b+\kappa\bar{a}}, & L &= \frac{U_u b}{2(U+\lambda)(b+\kappa\bar{a})}, \\ p &= \frac{b+\kappa^{-1}a}{-b+\kappa^{-1}\bar{a}}, & \ell &= \frac{V_v b}{2(V-\lambda)(b-\kappa^{-1}\bar{a})}, \end{aligned} \tag{4.3}$$

with

$$\begin{aligned} b &= i\sqrt{(U+\lambda)(V-\lambda)/\sigma}, \\ a &= \sqrt{1 - c(\hat{\tau}+2\lambda) - \tfrac{1}{4}\tau^2\sigma^{-1} - c^2\sigma} + i\left(\frac{(\hat{\tau}+2\lambda)}{\sqrt{\sigma}} + c\sqrt{\sigma}\right), \end{aligned}$$

and

$$\hat{\tau} = U - V, \qquad \sigma = (f+\bar{f})(\tilde{f}+\bar{\tilde{f}}), \qquad c \text{ a real constant.}$$

It is remarkable that (with the *Ansatz* used above) f, $\bar{f}$, $\tilde{f}$, and $\bar{\tilde{f}}$ enter in the functions P, L, p, and ℓ through the single combination σ. We have been dealing with equation (3.4), but it is obvious that all this may be used to generate solutions of the formally identical equation (3.5).

Minkowski space is characterized by $f = u + v$. The Bäcklund transformation just described may be integrated with this f as an input, giving $\tilde{f} = u+v$ (we have set $c = 0$ for simplicity); it looks as though the transformation could not move us away from Minkowski. Notice, however, that any $SL(2,\mathbb{R})$ transform of f may be used, as this change may be absorbed by a corresponding $SL(2,\mathbb{R})$ change of the coordinates. Let us take, for instance, $f = (u+v)^{-1}$. One then finds:

$$\tilde{f} = (u+v)^3 + \alpha^2(u+v)^{-1} + 4i\alpha(u-v),$$

(where α is a real constant), which is **not** Minkowski. The $SL(2,\mathbb{R})$ transformations, when applied to the Ernst equation, are usually referred to as the *Ehlers transformations* [23][9] Denoting by E an $SL(2,\mathbb{R})$

[9] In contrast to what happens when applied to f, some $SL(2,R)$ transformations are not trivial when applied to the $\mathcal{E}$ potential.

transformation applied to f, and by B one of the previously found Bäcklund transformations for f, we see that, in general:

$$E \circ B \neq B \circ E.$$

Bäcklund transformations for the stationary axisymmetric case are also found by using the method described in §2 [14], and they are applicable to equation (3.10). We shall not elaborate on this point, and shall rather concentrate on a new type of Bäcklund transformation to be described in the next Section. It is formally more elegant and concise than the one just discussed, and permits an easier approach to the subject of asymptotic conditions and superposition of solutions.

§5 Pohlmeyer-type transformations and the superposition principle

Based on the similarities between equations (3.3) and (3.8) on the one hand, and equation (2.6) on the other, Bäcklund transformations resembling those introduced by Pohlmeyer in reference [19] may be found. Such transformations are relevant for working directly with the vector equations (3.3), (3.8), and (3.11), rather than using the f or $\mathcal{E}$ representations. For equation (3.3) one has[10] [24]:

$$q_u + p_u = \frac{1}{2(U+\lambda)}[(U+V)q_u \cdot p - U_u]q + \frac{1}{2(U+\lambda)}[(U+V)q \cdot p_u - U_u]p, \tag{5.1a}$$

$$q_v - p_v = -\frac{1}{2(V-\lambda)}[(U+V)q_v \cdot p + V_v]q + \frac{1}{2(V-\lambda)}[(U+V)q \cdot p_v + V_v]p, \tag{5.1b}$$

with the compatible constraints

$$q^2 = p^2 = -1, \qquad p \cdot q = \frac{U - V + 2\lambda}{U+V}. \tag{5.1c}$$

For equation (3.8), the following transformation is obtained:

$$q_\varsigma + ip_\varsigma = \frac{1}{2(\eta + i\lambda)}[(\eta+\bar{\eta})p_\varsigma \cdot q - i\eta_\varsigma]p - \frac{i}{2(\eta+i\lambda)}[(\eta+\bar{\eta})q_\varsigma \cdot p - i\eta_\varsigma]q, \tag{5.2a}$$

with

$$q^2 = 1, \qquad p^2 = -1, \qquad q \cdot p = \frac{i(\eta - \bar{\eta}) - 2\lambda}{\eta + \bar{\eta}}. \tag{5.2b}$$

Equations (5.1) and (5.2) are manifestly $O(2,1)$-invariant. Notice that equations (5.2) transform solutions of

$$q_{\varsigma\bar{\varsigma}} + \frac{r_\varsigma}{2r} q_{\bar{\varsigma}} + \frac{r_{\bar{\varsigma}}}{2r} q_\varsigma = -(q_\varsigma \cdot q_{\bar{\varsigma}})q, \qquad q^2 = 1,$$

into solutions of

$$p_{\varsigma\bar{\varsigma}} + \frac{r_\varsigma}{2r} p_{\bar{\varsigma}} + \frac{r_{\bar{\varsigma}}}{2r} p_\varsigma = (p_\varsigma \cdot p_{\bar{\varsigma}})p, \qquad p^2 = -1,$$

---

[10] These transformations may also be derived within the scheme of gauge transformations, but this is not necessary.

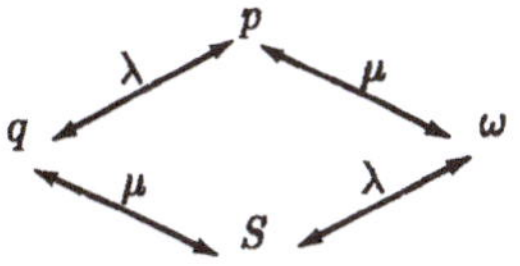

$$q^2 = \omega^2 = 1$$

$$p^2 = s^2 = -1$$

Figure 2.

and viceversa; two steps are thus required in order to go back to the original equation. Iteration of the transformation may be avoided due to the existence of the following permutability property[11] [24]. See Figure 2.

Suppose one starts from a solution q of (3.8) and integrates equations (5.2) with parameter value λ , getting p, and with parameter value μ to get s. Then, it may be shown that there exists a vector ω which is simultaneously a Bäcklund transform of p with parameter μ, and of s with parameter λ, so that ω is again a solution of (3.8). Furthermore, the following **algebraic** relation (*superposition*) among q, p, s, and ω is satisfied:

$$\omega = q + \frac{2(\lambda - \mu)}{(\eta + \bar{\eta})(1 + p \cdot s)} (s - p). \tag{5.3}$$

Alternatively, one may consider (3.11) as the basic equation to be solved, and use equations (5.2) with a known solution p of (3.11). Then, (5.3) may be rewritten as:

$$s = p + \frac{2(\lambda - \mu)}{(\eta + \bar{\eta})(\omega \cdot q - 1)} (\omega - q). \tag{5.4}$$

Similar superposition properties obviously hold for equations (5.1).

It is a remarkable fact that the transformation (5.4) produces new solutions maintaining the correct asymptotic behaviour of the input solution [24]. This may be explicitly checked by starting from Minkowski space [$\mathcal{E} = 1$, $p = (0, 0, 1)$]: Integrating equations (5.2) with $\eta(\varsigma) = \varsigma$, and using (5.4), one gets

$$\frac{1 + \mathcal{E}'}{1 - \mathcal{E}'} = \frac{s^2 - i s^1}{s^3 - 1} = \frac{i}{a - b} (e^{i\beta} r_1 - e^{i\alpha} r_2), \tag{5.5}$$

where

$$r_1 = \sqrt{r^2 + b^2 + 2br\cos\theta}, \qquad r_2 = \sqrt{r^2 + a^2 + 2ar\cos\theta},$$

$$\rho = r\sin\theta, \quad z = r\cos\theta, \qquad a = 2\lambda, \quad b = 2\mu,$$

and α and β being integration constants. Then,

$$\mathcal{E}' = 1 + \frac{2(a - b)}{i(e^{i\beta} - e^{i\alpha})} \frac{1}{r} + O(r^{-2}), \tag{5.6}$$

[11] The permutability property of the classical Bianchi–Bäcklund transformation may be seen in the references quoted under [16]. G. Neugebauer has found a similar property for his transformation in [10].

which shows the right asymptotic behaviour, provided that the coefficient of the $1/r$ term be real. This can always be achieved in one of the following ways:

i. Choose $\beta = -\alpha$.

ii. Perform an Ehlers transformation

$$\mathcal{E}'' = \frac{\cos\delta\,\mathcal{E}' + i\sin\delta}{i\sin\delta\,\mathcal{E}' + \cos\delta},$$

with an appropriate real constant δ. In particular, the Kerr solution may be obtained from (5.5) by taking

$$a = -b = \kappa, \qquad \alpha = -\beta = \nu - \tfrac{1}{2}\pi,$$

and

$$x + y \equiv \kappa^{-1}\sqrt{z^2 + \rho^2 + \kappa^2 + 2\kappa z}, \qquad x - y \equiv \kappa^{-1}\sqrt{z^2 + \rho^2 + \kappa^2 - 2\kappa z},$$

to get

$$\frac{1 + \mathcal{E}'}{1 - \mathcal{E}'} = x\cos\nu + iy\sin\nu.$$

By iterating the process of superposition of solutions just described, new solutions of the Einstein equations with an arbitrary number of parameters may be generated. Work on this subject and on the application of the transformations to non-Abelian gauge fields is currently being pursued.

References

[1] F. J. Ernst, New formulation of the axially symmetric gravitational field problem. *Phys. Rev.* **167**, 1175–1178 (1968).

[2] D. Maison, Are the stationary, axially symmetric Einstein equations completely integrable? *Phys. Rev. Lett.* **41**, 521–522 (1978); *ib.* On the complete integrability of the stationary, axially symmetric Einstein equations. *J. Math. Phys.* **20**, 871–877 (1979).

[3] R. Geroch, A method for generating new solutions of Einstein's equation. II. *J. Math. Phys.* **13**, 394–404 (1972).

[4] W. Kinnersley, Symmetries of the stationary Einstein–Maxwell field equations. I. *J. Math. Phys.* **18**, 1529–1537 (1977).

[5] W. Kinnersley and D. M. Chitre, Group transformation that generates the Kerr and Tomimatsu–Sato metrics. *Phys. Rev. Lett.* **40**, 1608–1610 (1978).

[6] C. Hoenselaers, W. Kinnerley, and B. C. Xanthopoulos, Generation of asymptotically flat, stationary space-times with any number of parameters. *Phys. Rev. Lett.* **42**, 481–482 (1979); *ib.* Symmetries of the stationary Einstein-Maxwell equations. VI. Transformations which generate asymptotically flat spacetimes with arbitrary multipole moments. *J. Math. Phys.* **20**, 2530–2536 (1979).

[7] V. A. Belinskiĭ and V. E. Zakharov, Integration of the Einstein equations by means of the inverse scattering problem technique and construction of exact soliton solutions. *Zh. Eksp. Teor. Fiz.* **75**, 1955–1971 (1978) [*Sov. Phys. JETP* **48**, 985–994 (1978)].

[8] I. Hauser and F. J. Ernst, A homogeneous Hilbert problem for the Kinnersley–Chitre transformations. *J. Math. Phys.* **21**, 1126–1140 (1980); see also Professor Ernst's lectures in *these Proceedings*.

[9] B. K. Harrison, Bäcklund transformation for the Ernst equation of general relativity. *Phys. Rev. Lett.* **41**, 1197–1200 (1978); 1835 (E)(1978).

[10] G. Neugebauer, Bäcklund transformations of axially symmetric stationary gravitational fields. *J. Phys.* **A12**, L67–L70 (1979).

[11] G. Neugebauer and D. Kramer, Generation of the Kerr–NUT solution from flat space-time by Bäcklund transformations. *Exp. Tech. Phys.* **28**, 3-8 (1980); G. Neugebauer, A general integral of the axially symmetric stationary Einstein equations. *J. Phys.* **A13**, L19–L21 (1980); D. Kramer and G. Neugebauer, The superposition of two Kerr solutions. *Phys. Lett.* **A75**, 259–261 (1980).

[12] B. K. Harrison, New large family of vacuum solutions of the equations of general relativity. *Phys. Rev.* **D21**, 1695–1697 (1980).

[13] M. Omote and M. Wadati, Bäcklund transformations for the Ernst equation. *J. Math. Phys.* **22**, 961–964 (1981).

[14] F. J. Chinea, Integrability formulation and Bäcklund transformations for gravitational fields with symmetries. *Phys. Rev.* **D24**, 1053–1055 (1981); **D26**, 2175 (E)(1982); *ib.* Bundle connections and Bäcklund transformations for gravitational fields with isometries. *Physica* **A114**, 151–153 (1982).

[15] C. M. Cosgrove, Relationships between the group-theoretic and soliton-theoretic techniques for generating stationary axisymmetric gravitational solutions. *J. Math. Phys.* **21**, 2417–2447 (1980).

[16] G. Darboux, *Leçons sur la Théorie Générale des Surfaces*, vol. 3, ch. XII–XIII. Gauthier–Villars, Paris, 1894; L. Bianchi, *Lezioni di Geometria Differenziale*, vol. 1, Part 2, ch. XV–XVI. N. Zanichelli, Bologna 1927; L. P. Eisenhart, *A Treatise on the Differential Geometry of Curves and Surfaces*, chapter VIII. Dover, New York 1960.

[17] E. Goursat, *Le Problème de Bäcklund*. Gauthier–Villars, Paris, 1925.

[18] A. Neveu and N. Papanicolaou, Integrability of the classical $[\bar{\psi}_i\psi_i]_2^2$ and $[\bar{\psi}_i\psi_i]_2^2 - [\bar{\psi}_i\gamma_5\psi_i]_2^2$ interactions. *Commun. Math. Phys.* **58**, 31–64 (1978); M. Crampin, Solitons and SL(2,R). *Phys. Lett.* **A66**, 170–172 (1978); R. Sasaki, Soliton equations and pseudospherical surfaces. *Nucl. Phys.* **B154**, 343–357 (1979); F. J. Chinea, On the intrinsic geometry of certain nonlinear equations: The sine–Gordon equation. *J. Math. Phys.* **21**, 1588–1592 (1980).

[19] K. Pohlmeyer, Integrable hamiltonian systems and interactions through quadratic constraints. *Commun. Math. Phys.* **46**, 207–221 (1976).

[20] V. E. Zakharov and A. B. Shabat, Exact theory of two-dimensional self-focusing and one-dimensional self-modulation of waves in nonlinear media. *Zh. Eksp. Teor. Fiz.* **61**, 118–134 (1971) [*Sov. Phys. JETP* **34**, 62–69 (1972)]; M. J. Ablowitz, D. J. Kaup, A. C. Newell, and H. Segur, The inverse scattering transform —Fourier analysis for nonlinear problems. *Stud. Appl. Math.* **53**, 249–315 (1974).

[21] M. Crampin, F. A. E. Pirani, and D. C. Robinson, The soliton connection. *Lett. Math. Phys.* **2**, 15–19 (1977).

[22] T. Lewis, Some special solutions of the equations of axially symmetric gravitational fields. *Proc. Roy. Soc. (London)* **A136**, 176–192 (1932); A. Papapetrou, Eine rotationssymmetrische Lösung in der allgemeinen Relativitätstheorie. *Ann. Phys. (Leipzig)* **12**, 309–315 (1953).

[23] J. Ehlers, in *Les Théories Relativistes de la Gravitation*. CNRS, Paris 1959.

[24] F. J. Chinea, New Bäcklund transformations and superposition principle for gravitational fields with symmetries. *Phys. Rev. Lett.* **50**, 221-224 (1983).

A Prolongation Structure for a Generalization of the Classical Massive Thirring Model

Daniel David

Centre de Recherches en Mathématiques Appliquées
Université de Montréal, Canada

Abstract

We propose a new class of classical field theories, defined in one dimension of space, of the form

$$i\gamma^{\mu}\Psi_{,\mu} - m\Psi - \overline{\Psi}\gamma^{\mu}(g_1 + g_2\gamma^5)\Psi\gamma_{\mu}\Psi - \overline{\Psi}(g_3 + g_4\gamma^5)\Psi\Psi = 0. \tag{1}$$

When we set $g_2 = g_3 = g_4 = 0$, this reduces to the well known classical massive Thirring model. Like the latter, the theories defined by (1) are invariant under the Poincaré group $P(1,1)$ as well as $U(1)$ (*i.e.*, under multiplication by a constant phase factor). However, they do not preserve parity unless $g_2 = g_4 = 0$, and are not invariant under time reversal unless $g_1 = g_3 = 0$. Quantized versions of these theories could be viewed therefore as conceivable models for describing a weak self-interaction of a spinor particle.

An interesting feature of (1) is that it defines nonlinear theories. As such, this property suggests that (1) could possibly have solutions other than those usually associated with linear systems. For instance, it could admit solitons or objects similar to them. We investigated (1) with this perspective in mind; we showed that, under certain restrictions, it has indeed special solutions of that type.

In order to achieve this, we chose to use the *prolongation structure method* (also known as the *Wahlquist–Estabrook method*) in the geometrical setting and with the formalism established in Reference [1]. According to this method, we began by replacing (1) by an equivalent exterior pfaffian system of 2–forms. We then introduced a Bäcklund map

$$\chi : J^1(M, C^2) \times C^n \to J^1(M, C^n)$$

such that $(\psi_i, \overline{\psi}_i, \psi_{i,\nu}, \overline{\psi}_{i,\nu}; y^B) \mapsto y^A_{,\mu}$, $1 \le A, B \le n$. This map has the property of being completely determined by the requirement that its integrability conditions contain (1). As usual, we translated this requirement into an equivalent algebraic construction problem which we solved. We found that exactly six classes of Bäcklund maps are permitted. Of these, one is particularly interesting and is defined when $g_2 = g_4 = 0$, $g_1 \neq 0$. When the pseudopotential y is one-dimensional, it takes the form

$$\begin{aligned} y_{,\xi} &= \psi_1 y^2 - i[m/\sigma - 2(g_1 - g_2)|\psi_1|^2]\, y + 2mg_1/\sigma \cdot \overline{\psi}_1, \\ y_{,\eta} &= \sigma\psi_2 y^2 - i[m\sigma - 2(g_1 + g_2)|\psi_2|^2]\, y + 2mg_1 \cdot \overline{\psi}_2, \end{aligned} \tag{2}$$

where σ is a complex free parameter. When y is two-dimensional, it immediately induces a Lax pair,

$$[\partial_x - M(\Psi,\sigma)]\, y = 0,$$
$$[\partial_t - N(\Psi,\sigma)]\, y = 0,$$

which could be used to solve (1) with the inverse scattering transform formalism. From (2), we have derived many interesting results. We constructed, for instance, an infinite family of conservation laws. This result is important because it implies that the theory is completely integrable. Also, we constructed a nontrivial Bäcklund transformation of the form

$$(\Psi, y, \overline{y}) \mapsto \tilde{\Psi} = F(y,\overline{y})\,\Psi + G(y,\overline{y}).$$

This transformation, in association with (2), can be used to generate new solutions $\tilde{\Psi}$ of (1) from a known one Ψ. From the trivial solution $\Psi = 0$, we have computed a solution $\tilde{\Psi}$ which has several peculiarities. Its components $\tilde{\psi}_i$ are non-periodic oscillating functions which are localized in the sense that their moduli $|\tilde{\psi}_i|$ are hyperbolic lumps of the form $[\alpha_i^2 \cosh^2(Ax+Bt)+\beta_i^2 \sinh^2(Ax+Bt)]^{-1}$. These lumps are solitary objects which propagate with a constant velocity of magnitude $|v| \in (0,1)$, in natural units. Moreover, one finds that they have a finite energy. For more details, one should consult Ref. [2].

References

[1] F. A. E. Pirani, D. C. Robinson, and W. F. Shadwick, *Local Jet Bundle Formulation of Bäcklund Transformations with Applications to Non-Linear Evolution Equations.* D. Reidel, Dordrecht, 1979.

[2] D. David, On an extension of the classical Thirring model. Preprint CRMA-1142, 1983. (To be published.)

Atomic Nuclei as Solitons

Manuel de Llano† and Ernst F. Hefter‡

Instituto de Física
Universidad Nacional Autónoma de México

Abstract:

The N-soliton solution of the Korteweg-de Vries equation, in an extended form, is utilized to represent a reflectionless nuclear mean field potential in which the N-empirical bound state energies are associated with the solitons, and formulate generally the problem of nuclear structure, reactions, and two-nucleon interactions from an inverse scattering theory approach.

We consider a possibly useful formulation [1] of nuclear theory in terms of a nuclear mean field (shell model or optical potential) Schrödinger equation

$$H(x,t)\,\psi_n(x,t) \equiv [-M\partial_{xx} + U(x,t)]\,\psi_n(x,,t) = E_n\psi_n(x,t),$$

$$M \equiv \frac{\hbar^2}{2m}, \quad U(x,t) \equiv U(\vec{r},t) + \frac{\hbar^2 \ell(\ell+1)}{2mr^2} \quad \text{with} \quad r \equiv x \geq 0, \tag{1}$$

$$\text{or} \quad \psi_n(x) = -\psi_n(-x) \Rightarrow \psi_n(0) = 0,$$

where m is the nucleon mass, ℓ the orbital angular momentum, and E_n the (time-independent) energy eigenvalues. The mean field $U(x,t)$, if assumed to evolve according to the Korteweg-de Vries equation with the N-soliton solution representing the N nondegenerate levels of the well, will clearly develop in time into N widely separated "pulses" —in clear contrast to the physics at hand of a spatially compact nucleus.

A time re-scaling [2] remedies this difficulty and produces, by the Lax procedures [3], an extended Korteweg-de Vries equation

$$U_\tau(x,\tau) = \frac{v_0}{L_0}U_x + 6U\,U_x - M\,U_{xxx}, \tag{2}$$

---

†Work supported by the Instituto Nacional de Investigaciones Nucleares and Consejo Nacional de Ciencia y Tecnología.

‡On leave from the Institut für Theoretische Physik, Universität Hannover, Germany.

appropiate for s-waves, in which τ is proportional to the time t; v_0 and L_0 are arbitrary constants. It has been shown in [4] that higher-order Lax procedures yield nonlinear equations more general than (2) which are appropriate for $\ell > 0$ waves. By properly choosing the abovementioned proportionality constant one can mimick a nuclear mean field which is i) a single "droplet", for *structure* studies, or ii) two colliding "droplets" for *reaction* studies, *e.g.*, nucleon-nucleus, α-nucleus, and nucleus-nucleus (including heavy ion) collisions. Inverse scattering theory [5] can then be employed to reconstruct, for all $\tau > 0$, and hence all $t > 0$, the mean field $U(x,t)$.

References

[1] E. F. Hefter and M. de Llano, Nuclei as solitons. (To be published.)

[2] E. F. Hefter, The extended Korteweg-de Vries equation. *Z. Naturforschung* **37a**, 1119–1123 (1982).

[3] P. D. Lax, Integrals of nonlinear equations of evolution and solitary waves. *Commun. Pure Appl. Math.* **21**, 467– (1968).

[4] E. F. Hefter and I. A. Mitropolsky, to be published.

[5] K. Chadan and P. C. Sabatier, *Inverse Problems in Quantum Scattering Theory* (Springer, 199); J. F. Schonfeld *et al*, *Ann. Phys. (N. Y.)* **128**, 1 (1980).

Integrability in Dynamical Systems and the Painlevé Property

Bernadette Dorizzi

Centre National d'Etudes des Télécommunications
Issy les Moulineaux, France

Abstract

The analytic structure of the solution of an ordinary differential equation is intimately related to its integrability. The Painlevé property, *i.e.*, pure poles being the only movable singularities, allows the identification of new integrable dynamical systems. In this paper, we recall briefly the Ablowitz–Ramani–Segur (ARS) algorithm [4], which deals with an ordinary differential equation which possesses the Painlevé property, and applies it to hamiltonian dynamical systems with polynomial potentials of third and fourth order.

§1 Introduction

In this paper we shall focus on the study of integrable dynamical systems, the equations of motion of which are a system of ordinary differential equations. The search of integrability in such systems is difficult, but is rewarded by the nice behaviour of the trajectories. In fact, integrability allows long time predictions on the system. This is not possible when the only information available on the solution is given by numerical integration, or when the system exhibits a chaotic behaviour. In this last case, indeed, two arbitrarily close initial data give rise to rapidly diverging trajectories and, despite some recent progress based on renormalization and scaling, chaotic regimes remain unpredictable.

Unfortunately, it is not easy to detect integrability. The most widely used tools are numerical integration combined with observation of the section surfaces and group theoretical methods based on the study of dynamical symmetries. Substantial progress in this domain was made possible by the *Painlevé conjecture.* It is related to the analytic properties of the solutions of the equations of motion, namely, whenever the solutions possess the Painlevé property, *i.e.*, their only movable singularities on the complex-time plane are poles, the system is integrable. In order to clarify the concept of movable singularities, let us consider a *linear* second order differential equation:

$$\ddot{x} + p(t)\dot{x} + q(t)x = 0.$$

Its general solution is:

$$x(t) = Ax_1(t) + Bx_2(t),$$

where A and B are two arbitrary constants depending on the initial data, and $x_1(t)$ and $x_2(t)$ are two independent solutions. The location of the complex singularities of x does not depend on A and B, but on the singularities of p and q. They are fixed and independent of the initial values of the problem. Now, if the equation is *nonlinear*, a different kind of singularities can appear. If we consider the equation:

$$\dot{x} + x^2 = 0,$$

whose solution reads:

$$x(t) = \frac{1}{t - t_0},$$

we remark that the latter has a singularity at t_0, which is the constant of integration of the equation. This singularity is movable as it depends on the initial data of the problem. This conjecture has allowed the identification of the integrable, onc-dimensional, ordinary differential equations of first and second order. At first order, the only Painlevé-type equations are the linear and the Ricatti ones. At second order, Painlevé and co-workers have identified 50 equations, 44 of which can be reduced to equations which are integrable in terms of elementary transcendental functions. The remaining six are new. Their solutions define the Painlevé trascendents. At higher orders, some partial classification exists (see [1]). In the case of the three-dimensional Lorenz system, Segur [2] has been able to identify the parameters for which the system is partially or fully integrable. Further cases of integrable systems with the Painlevé property have been considered by Bountis *et al.* in [3].

The remaining problem is thus to recognize, from the equation of motion, whether or not the system possesses the Painlevé property. A suitable algorithm has been developed by Ablowitz, Ramani, and Segur in [4]. We will first give a brief sketch of this method and illustrate it on an example. Then we will present some new integrable hamiltonian systems satisfying the Painlevé property.

§2 The Ablowitz–Ramani–Segur algorithm

We present here the method as developed in [4], for a single ordinary differential equation (ODE):

$$\dot{x} = f(x, t). \tag{2.1}$$

At the same time, we will illustrate it on the example of a system of two first order ODE's. At this point, one has to notice that this algorithm does not identify essential singularities and provides only necessary conditions for an equation to be of the Painlevé type. There are three steps in the algorithm:

Step 1. Find the dominant behaviour.
We assume that the solution becomes infinite at the singularity, and look for a solution of the form

$$x = \alpha(t - t_0)^p, \qquad \text{with } \operatorname{Re} p < 0, \tag{2.2}$$

which we substitute into (2.1).
For some values of p, some terms of the equations balance when $t \to t_0$, while the others can be ignored. They are called the ***dominant terms*** of the equations. If any of the possible p's is not an integer, the equation is not of the Painlevé type. Otherwise, one has to go on with the second step.

Example. Let us consider the system

$$\begin{cases} \dot{x} = x(a - x - y) & (2.3a) \\ \dot{y} = (x - 1). & (2.3b) \end{cases}$$

We set

$$\begin{cases} x &= \alpha\tau^p \\ y &= \beta\tau^q, \end{cases} \qquad \tau = (t - t_0). \tag{2.4}$$

Substituting (2.4) into (2.3), we find the following values for p and q:

$$p = -1, \quad q = \alpha, \qquad \text{from (2.3b)},$$

and

$$\alpha = 1, \qquad \beta \text{ undetermined},$$

or

$$\alpha = -1, \qquad \beta = 2.$$

There are thus two distinct leading behaviours:

$$\begin{cases} x &= \tau^{-1} \\ y &= \beta\tau \end{cases} \quad \text{corresponding to the leading terms} \quad \begin{cases} \dot{x} &= -x^2 \\ \dot{y} &= xy, \end{cases} \tag{2.5}$$

$$\begin{cases} x &= -\tau^{-1} \\ y &= 2\tau^{-1} \end{cases} \quad \text{corresponding to the leading terms} \quad \begin{cases} \dot{x} &= -x^2 - xy \\ \dot{y} &= xy. \end{cases} \tag{2.6}$$

Step 2. Find the resonances.
For every negative integer value of p, the solution has an expression —at least formal— in the form of a Laurent series. The resonances are the powers of $(t - t_0)$ at which the different arbitrary constants enter in this expansion. To find them, one substitutes into the equations composed of the leading terms, the following form for x:

$$x = \alpha(t - t_0)^p + \beta(t - t_0)^{p+r}. \tag{2.7}$$

The equation then reduces to:

$$\mathcal{Q}(r)\beta(t - t_0)^q = 0. \tag{2.8}$$

The resonances are the roots of $\mathcal{Q}(r) = 0$. One can note that:

- one root is always -1 (arbitrariness of t_0),
- if α is arbitrary, 0 is also a root,
- a root with $\operatorname{Re} r > 0$, r noninteger, indicates a branch point,
- a root with $\operatorname{Re} r < 0$ is purely formal.

If there are no branch points, one has to go further with Step 3, dealing only with the positive integer resonances.

Example. We have two distinct cases of leading behaviour to consider:

Case 2.5: We set

$$\begin{cases} x &= \tau^{-1}(1 + \gamma\tau^r) \\ y &= \beta\tau(1 + \delta\tau^r). \end{cases} \tag{2.9}$$

We substitute (2.9) in

$$\begin{cases} \dot{x} &= -x^2 \\ \dot{y} &= xy, \end{cases}$$

and obtain the following system for γ and δ :

$$\begin{cases} \gamma(r+1) &= 0 \\ \gamma - \delta r &= 0. \end{cases} \tag{2.10}$$

It has nontrivial solution if and only if $r = -1$ or $r = 0$.

Case 2.6: We set

$$\begin{cases} x &= -\tau^{-1}(1+\gamma\tau^r) \\ y &= 2\tau^{-1}(1+\delta\tau^r). \end{cases} \tag{2.11}$$

We substitute (2.11) in

$$\begin{cases} \dot{x} &= -x^2 - xy \\ \dot{y} &= xy, \end{cases}$$

and obtain the following system for γ and δ :

$$\begin{cases} \gamma(1-r) - 2\delta &= 0 \\ 2\gamma + 2r\delta &= 0. \end{cases} \tag{2.12}$$

It has nontrivial solution if and only if $r = -1$ or $r = 2$.

Step 3. Find the constants of integration.
Let (p, α) given as in Step 1, $r_1 < r_2 < \cdots < r_s$, and the resonances, *i.e.*, the positive integer roots of $\mathcal{Q}(r)$, given from Step 2. We substitute

$$x = \alpha(t-t_0)^p + \sum_{j=0}^{s} \alpha_j (t-t_0)^{p+j}$$

into the full equation and calculate, by recurrence, the coefficients α_j. When $j = r_k$, $(k = 1, 2, \ldots, s)$, α_j is either undetermined or impossible to express. In the last case, one has to introduce *log* terms in the expansion and the equation is not of the Painlevé type.

Example. For the case (2.5) the resonances are -1 and 0, and the resulting Laurent series is thus algebraic. For the case (2.6), we must calculate the coefficients of the series up to the second order, *i.e.*, $r = 2$:

$$\begin{aligned} x &= -\tau^{-1} + a_1 + a_2\tau, & \dot{x} &= \tau^{-2} + a_2, \\ y &= 2\tau^{-1} + b_1 + b_2\tau, & \dot{y} &= -2\tau^{-2} + b_2. \end{aligned}$$

Equations (2.3) read:

$$\begin{aligned} \tau^{-2} + a_2 &= -(\tau^{-2} + a_1^2 - 2a_1\tau^{-1} - 2a_2) - (-2\tau^{-2} - b_1\tau^{-1} - b_2 + 2a_1\tau^{-1} + a_1b_1 + 2a_2), \\ -2\tau^{-2} + b_2 &= -2\tau^{-2} - b_1\tau^{-1} - b_2 + 2a_1\tau^{-1} + a_1b_1 + 2a_2 - 2\tau^{-1} - b_1; \end{aligned}$$

then,

$$\begin{cases} b_1 &= a \\ a_1 &= \frac{1}{2}a + 1, \end{cases} \qquad \begin{cases} a_2 - b_2 &= aa_1 - a_1b_1 \\ 2(a_2 - b_2) &= b_1 - a_1b_1. \end{cases}$$

The last system leads to the compalibility condition for $a_2 - b_2$: $2(aa_1 - a_1b_1) = b_1 - a_1b_1$, *i.e.*, $a_1 = 1 = 1 + a/2$. The system (2.3) is thus Painlevé if and only if $a = 0$.

§3 New cases of hamiltonian integrable systems

We will now consider the case of hamiltonian dynamical systems in two dimensions and show how the Painlevé conjecture allows the identification of new integrable cases. The hamiltonian has the form

$$H = \tfrac{1}{2}(\dot{x}^2 + \dot{y}^2) + V(x,y), \tag{3.1}$$

and is a constant of the motion. Integrability, in such systems, is synonymous with the existence of a second constant of the motion, independent of the hamiltonian.

1. The Hénon–Heiles hamiltonian:

$$H = \tfrac{1}{2}(\dot{x}^2 + \dot{y}^2 + ax^2 + by^2) + dx^2y - \tfrac{1}{3}ey^3. \tag{3.2}$$

This hamiltonian has been studied by Chang *et al.* [5], and by Bountis and Segur [3]. The equations of motion are:

$$\begin{cases} \ddot{x} &= -ax - 2dxy \\ \ddot{y} &= -by - dx^2 + ey^2. \end{cases}$$

The Painlevé-analysis leads to two distinct leading behaviour (Step 1):

$$\begin{cases} x &= \alpha\tau^p \\ y &= \beta\tau^q, \end{cases} \quad \text{with} \quad \begin{matrix} p &= -2 \\ q &= -2, \end{matrix} \qquad \begin{matrix} \alpha &= \pm(3/d)\sqrt{2+e/d} \\ \beta &= -3/d, \end{matrix}$$

and

$$\begin{matrix} p = \tfrac{1}{2} \mp \tfrac{1}{2}\sqrt{1-48d/e}, \\ q = -2, \end{matrix} \qquad \begin{matrix} \alpha \text{ arbitrary}, \\ \beta = 6/e. \end{matrix}$$

The resonances (Step 2) are:

$$n = -1,\ 6,\ \tfrac{5}{2},\ \pm\tfrac{1}{2}\sqrt{1-24(1+\tfrac{e}{d})},$$

or

$$n = -1,\ 0,\ 6,\ \pm\sqrt{1-48d/e}.$$

These values lead to integer values for the following values of the parameters:

$$\text{(a)} \begin{cases} e &= -d \\ a &= b, \end{cases} \qquad \text{(b)}\ e = -2d, \qquad \text{(c)}\ e = -6d, \qquad \text{(d)} \begin{cases} e &= -16d \\ b &= 16a. \end{cases}$$

In fact, case (c) is not a Painlevé case, as can be seen in performing Step 3 (one has to introduce *log* terms). In all other cases, the constants of motion have been calculated:

Case (a) $C = \dot{x}\,\dot{y} + axy + \tfrac{1}{3}dx^3 + dxy^2$. This case is separable in the $x+y$, $x-y$ coordinate system.

Case (b) $C = [(b-4a) + 4dy]\,\dot{x}^2 - 4dx\,\dot{x}\,\dot{y} - d^2x^4 + a(b-4a)x^2 - 4dax^2y - 4d^3x^2y^2$. This integral is due to J. Greene [6].

Case (c) $C = 3\dot{x}^4 + 6(a+2dy)x^2\dot{x}^2 - 4dx^3\dot{x}\,\dot{y} - 4dx^4(ay+dy+3a^2x^4) - \tfrac{2}{3}d^2x^6$. This last constant is of fourth order in the velocities and has been found independently by Hall in [7], and by Grammaticos, Dorizzi, and Padjen in [8].

More generally, we have studied the case of hamiltonians with homogeneous polynomials potentials of third degree in [9].

2. Homogeneous third degree polynomial potentials:

$$V = y^3 + ay^2x + byx^2 + cx^3. \tag{3.3}$$

The coefficient a can be eliminated through a rotation, except for $(x+iy)^3$ and $(x^2+y^2)(x\pm iy)$, which are integrable cases. The Painlevé analysis is rather complicated in that case and the search for resonances leads to the following equation for the N_i's of the form:

$$36(N_1 + N_2 + N_3 - 12) = N_1\, N_2\, N_3\,, \tag{3.4}$$

where $N = m(m+1)$, with m an integer or half-integer. The solution $N_1 = 6$, $N_2 = 6$, N_3 free, is to be rejected. It corresponds to the potential $V = y^3 + \frac{3}{2}\,yx^2 + dx^3$, and leads to logarithmic singularities. The other solutions of (3.4), up to permutations of N_1, N_2, and N_3, which recovers $a = 0$, read

N_1	N_2	N_3	**Potential**	**Order of the constant**
0	0	12	$x^3 + y^3$	$2^{\rm nd}$
2	30	30	$y^3 + \frac{1}{2}\,yx^2$	$2^{\rm nd}$
$\frac{3}{4}$	90	90	$y^3 + \frac{3}{16}\,yx^2$	$4^{\rm th}$
2	20	90	$y^3 + \frac{1}{2}\,yx^2 + i\,\frac{1}{6\sqrt{3}}x^3$	$4^{\rm th}$

The three first cases recover the results of the Hénon–Heiles potential [with $a = b = 0$ in Eq. (3.2)], which we presented in paragraph one. The last case is a new one. The second constant is of degree four in the velocities and reads:

$$\begin{aligned} C = {} & \dot y^{4} + 2\dot y^{2}\dot x^{2} - i\,\tfrac{2}{\sqrt3}\,\dot y\,\dot x^{3} + \left(4y^3 + 2yx^2 - i\,\tfrac{1}{3\sqrt3}\,x^3\right)\dot y^{2} + \left(i\sqrt3\,yx^2 + x^3\right)\dot x\,\dot y \\ & + \left(4y^3 - 2i\sqrt3\,y^2x - i\,\tfrac{1}{3\sqrt3}\,x^3\right)\dot x^{2} + 4y^6 + 4y^4x^2 + i\,\tfrac{1}{3\sqrt3}\,x^3y^3 + \tfrac{5}{4}y^2x^4 + i\,\tfrac{1}{6\sqrt3}\,x^5y + \frac{1}{54}\,x^6 \end{aligned} \tag{3.5}$$

This exhausts *all* the Painlevé cases for potentials of the form (3.3), and the direct approach for the computation of the integrals of motion allows the calculation of the constants in each case.

3. Homogeneous fourth degree polynomial potentials. We begin by introducing a potential V of the form:

$$V = y^4 + ay^3x + by^2x^2 + cyx^3 + dx^4. \tag{3.6}$$

As in the preceding case, the coefficient a can be set to zero by an adequate rotation, except for the potential

$$V = (x \pm iy)^4 + \mu(x^2 + y^2)(x \pm iy)^2,$$

which is indeed integrable. At this point, we do not pretend to be exhaustive and we will limit ourselves to potentials of the form:

$$V = y^4 + ay^2x^2 + bx^4,$$

which are parity-symmetric separately in x and in y. In this form, there are two well-known integrable cases:

Potential	**Order of the constant**
$x^4 + y^4$,	$2^{\rm nd}$: (separable),
$(x^2 + y^2)^2$,	$1^{\rm st}$: (the constant is just the angular momentum).

We have found two other new integrable cases:

$$y^4 + \tfrac{3}{4}x^2y^2 + \tfrac{1}{16}x^4, \qquad 2^{\text{nd}}\colon C = -y\dot{x}^2 + x\dot{x}\dot{y} + 8x^2y^3 + 4x^4y,$$

$$y^4 + \tfrac{3}{4}x^2y^2 + \tfrac{1}{8}x^4, \qquad \begin{aligned} 4^{\text{th}}\colon C = {} & \dot{x}^4 + (24x^2y^2 + 4x^4)\dot{x}^2 - 16x^3y\dot{x}\dot{y} \\ & + 4x^4\dot{y}^2 + 4x^8 + 16x^6y^2 + 16x^4y^4. \end{aligned}$$

§4 Conclusion

In this paper, we have shown how the complex-plane singularity analysis of the equations of motion is interesting for the study of integrability of ODE's. The Painlevé analysis has led us to the discovery of new integrable dynamical systems. They correspond to the motion of a particle in two-dimensional homogeneous polynomial potentials of third and fourth degree. In the case of cubic interaction, we have been able to identify *all* the Painlevé cases and to calculate the second constant of motion for each of them. For the case of quartic interaction, due to the complexity of the problem, our search was not exhaustive but we nevertheless discovered new integrable potentials.

The Painlevé conjecture thus appears to be a powerful tool in the study of dynamical systems, and it is far from having shown the limits of its possibilities.

References.

[1] F. J. Bureau, Integration of some nonlinear systems of ordinary differential equations.*Ann. Matematica N.* **94**, 344-359 (1972).

[2] H. Segur, Solitons and the Inverse Scattering Transform. Lecture at the International School of Physics "Enrico Fermi" (Varenna, Italy, July 1980).

[3] T. Bountis, H. Segur, and F. Vivaldi, Integrable hamiltonian systems and the Painlevé property. *Phys. Rev.* **A25**, 1257–1264 (1982).

[4] M. J. Ablowitz, A. Ramani, and H. Segur, Nonlinear evolution equations and ordinary differential equations of Painlevé type. *Lett. Nuovo Cimento* **23**, 333–338 (1978).

[5] Y. F. Chang, M. Tabor, J. Weiss, and C. Corliss, On the analytic structure of the Hénon–Heiles system. *Phys. Lett.* **85A**, 211-213 (1981).

[6] J. Greene, *quoted in reference 6.*

[7] L. J. Hall, On the existence of a last invariant of conservative motion. *Annals of Physics.* (Submitted.) Preprint (1982).

[8] B. Grammaticos, B. Dorizzi, and R. Padjen, Painlevé property and integrals of motion for the Hénon–Heiles system. *Phys. Lett.* **83A**, 111-113 (1982).

[9] B. Grammaticos, B. Dorizzi, and A. Ramani, Integrability of hamiltonians with third and four degree polynomial potentials. *J. Math. Phys.* (To Appear.) Preprint (1982).

Criteria for the Existence of Bäcklund Transformations

James Daniel Finley, III

Department of Physics and Astronomy
The University of New Mexico
Albuquerque, New Mexico, USA

Abstract

A pedagogical guide to the meaning and purpose of Bäcklund transformations is given. Jet bundles of C^∞ maps of manifolds are used to describe partial differential equations (PDE's) and various kinds of transformations on them. The purposes are twofold. First, to explain to the non-expert, the basic ideas of this approach via intuitive concepts. Second, to describe recently-developed criteria which can be used to determine the nonexistence of a Bäcklund transformation for any particular equation or system of equations. Previous work on Bäcklund transformations is extended to the case of non-quasi-linear PDE's in an arbitrary number of independent variables. The general form (modulo contact transformations) is given that a PDE must have in order that a Bäcklund transform of it may exist. For more than two independent variables, the concept of multiple copies of the original equation is introduced and an explanation for the usefulness of this concept is sketched. As an indication of the utility of such a negative concept, it is also explained how the author came to be interested in such transforms —certain equations which occur in the study of self-dual Einstein spaces— and the conclusions which the above considerations lead to: These equations do not admit Bäcklund transformations.

§1 Introduction

As a general relativist, working in exact solutions to Einstein's field equations, I have had many opportunities to uncover specific nonlinear equations (or systems) which have particular physical interest. In recent years, some of those most interesting to me have been generated as special cases of the Plebański key equation [11], which determines all self-dual Einstein spaces:

$$\Theta_{xx}\Theta_{yy} - (\Theta_{xy})^2 = 0. \tag{1.1}$$

A particular case of this, valid for all such (self-dual Einstein) spaces with one *rotational* symmetry, is given by the *generalization* of the Liouville equation,

$$H_{sq} = \left(e^H\right)_{rr}, \tag{1.2}$$

where $H = H(s, q, r)$ [13].

My interest in Bäcklund transformations developed, therefore, as a tool to help solve these highly nonlinear partial differential equations. The equations given above and, indeed, most of the ones that are likely to be generated in the context of general relativity, are not time-evolution equations, nor are they quasi-linear, both of which are properties shared by the vast majority of equations which have been found to be amenable to treatment by Bäcklund transformations. Therefore, some generalizations were necessary. To recapitulate only briefly the history of these transformations, we note that they originated with Lie [1] and Bäcklund [2] in the late nineteenth century. Other work about the same time was done by Clairin [3] and Goursat [4]. They are non-trivial generalizations of the more familiar methods of change of independent variables, change of dependent variables, and contact transformations, which might also vary the roles of independent and dependent variables. See §2 for more details on this.

Because of the somewhat general audience at this School and Workshop, this talk is intended not only to explain at least the nature of the particular discoveries which we made abount the applicability of Bäcklund transformations to equations of this type, but also to give a more general background about the nature of Bäcklund transformations for those who have not previously known of their great need for them and who, like ourselves, view them mostrly as a tool for solving physically relevant PDE's rather than as a subject for research in and of themselves. Therefore, before going into the background-level discussion about these transformations, I wish to note several other people whose work has been instrumental in our approach to the subject. This list is not intended to be exhaustive nor categorized by technique since, as already mentioned, my approach will be more pedagogical than technical or historical. For a much more mathematical approach with more detailed references, see the article by this author and P. Denes, *Bäcklund transformations for general PDE's*, [12]. We follow some combination of approaches culled from the work of Forsyth [5] and Pirani [6], with significant input from Harrison [7], Estabrook and Wahlquist [9], Hermann [9] and Robert Bryant [10].

§2 The basic idea in pictures

Given a PDE to solve, I may describe the situation in terms of some set of independent variables, some set of dependent variables, and some derivatives of the dependent variables with respect to the independent ones:

derivatives of old dependent variables	$p = \Theta_x, \quad q = \Theta_y,$ $r = \Theta_{xx}, \quad s = \Theta_{xy}.$
old dependent variables	Θ
old independent variables	$x,\ y,\ z,\ t.$

The entire structure (considered as a space or ***manifold***) —-the independent variables, the dependent variables, and the derivatives of the dependent variables, all considered as independent quantities— is called a *jet bundle* (in this case, the $2^{\rm nd}$ jet bundle because ***derivatives*** up through second order are considered). A PDE, in this context, is most easily considered as a surface, Σ, in this whole space; *i.e.*, a relation between some of these variables. An example might be $r^3 + xs = 0$.

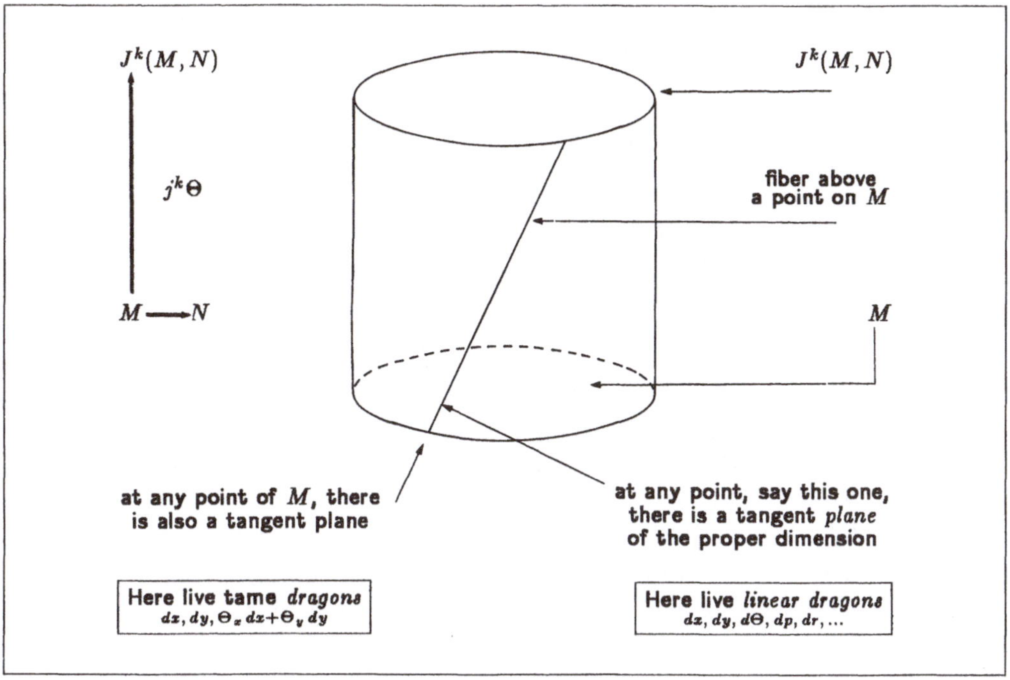

Figure 1.

Figure 1, above, is a rather more formal view of the same kind of thing. We use $\Theta : M \longrightarrow N$ to denote an arbitrary C^∞ function, and denote the kth order jet bundle by $J^k(M,N)$. Two such maps are said to agree to order k at x in M (to be ***equivalent***) if there are coordinate charts such that their Taylor expansions are the same up through the kth order terms. The equivalence class of maps which agree with f (to order k, at x) is denoted by $j^k{}_x f$ —the k-jet of f at x. The k-jet bundle, $J^k(M,N)$ is the set of all such k-jets, for all $x \in M$, for all $f : U \subset M \longrightarrow N$, and for all $U \subset M$. (This set has a natural structure as a C^∞ manifold.) Since this is a manifold, at any point of M, we have a tangent plane (and a cotangent plane). On the cotangent plane of the manifold (at any point) live tame ***dragons***: dx, dy, $\Theta_x\, dx + \Theta_y\, dy$, $\Theta_{xx}\, dx + \Theta_{xy}\, dy$, $\ldots$, etc. (The ***dragons*** referred to are the one-forms which live in the cotangent space; they will be the ***beasts*** that do all of our work for us, and also cause us trouble.) On the other hand, at any point on the larger manifold, the jet bundle, we also have a tanglent ***plane*** of the proper dimension. In a cotangent plane at any point of the jet bundle, live *linear* ***dragons*** such as dx, dy, $d\Theta$, dp, dr, dx, $\ldots$. Here the symbols dp, dr, etc., are meant to be independent variables which, in the projection down to the base manifold from the jet bundle itself, would become derivatives of Θ. Relative to this picture, a solution to a PDE is a function $\Theta : M \longrightarrow N$, such that $j^k\Theta \subset \Sigma$.

However, in order to really understand the need for this complicated superstructure, it is necessary to go back over the figure with some historical feeling. We use the picture below, showing various jet bundles over two base manifolds, to describe the various kinds of changes which have been tried, historically, to put a PDE into a form more amenable to solution (the kinds of changes already mentioned somewhat earlier).

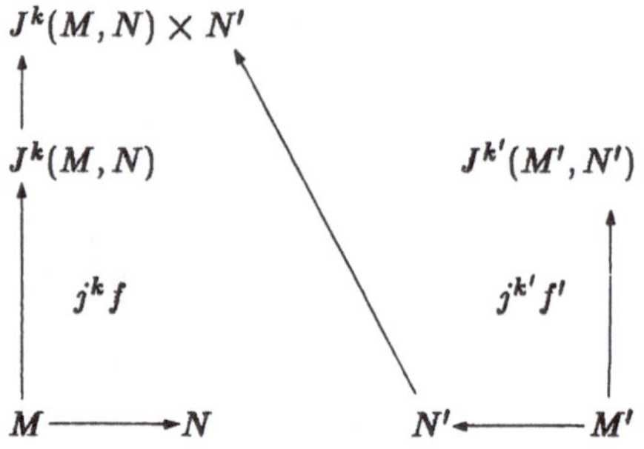

1. $\Phi : M \longrightarrow M$ ....changes of independent variables.
2. $\Phi : M \longrightarrow M, \quad \Psi : N \longrightarrow N$ ...... separate changes of independent and dependent variables.
3. $\Phi : M \times N \longrightarrow M \times N$ .... mixed changes of independent and dependent variables.
4. $\Psi : J^k(M,N) \longrightarrow J^k(M,N)$ ....*contact transformations.*
5. $\Upsilon : J^k(M,N) \times N' \longrightarrow J^1(M',N')$ .... a Bäcklund transformation.

In order to better see what a Bäcklund transformation actually entails, we set up some coordinates, in appropriate local charts, on these various manifolds. we let

$$
\begin{aligned}
x^a &: \text{coordinates in } M, \\
z^\beta &: \text{coordinates in } N, \\
z^\beta{}_a, z^\beta{}_{ab}, z^\beta{}_{abc}, \ldots &: \text{coordinates in } J^k(M,N), \\
x'^r &: \text{coordinates in } M', \\
z'^A &: \text{coordinates in } N', \\
z'^A{}_r &: \text{coordinates in } J^1(M',N').
\end{aligned}
\tag{2.1}
$$

Then we may write the Bäcklund transformation itself as a mapping

$$
\Upsilon : J^{k-1}(M,N) \times N' \longrightarrow J^1(M',N'), \tag{2.2}
$$

where

$$
\Upsilon : \begin{cases} x'^r = X^r(x^a, z^\beta, z^\gamma{}_a, z^\gamma{}_{ab}, \ldots, z'^A), \\ z'^A = z'^A, \\ z'^A{}_r = Z^A{}_r(x^a, z^\beta, z^\gamma{}_a, z^\gamma{}_{ab}, \ldots, z'^A). \end{cases} \tag{2.3}
$$

In order to describe how such a map may be put to the purposes I have in mind, we must have a few technical tools. the first of these is the ***contact module over*** $J^k(M,N)$: We define a particular set of one-forms in the cotangent bundle over $J^k(M,N)$ as follows:

$$
\begin{aligned}
\theta^\alpha &= dz^\alpha - z^\alpha{}_a\, dx^a, \\
\theta^\alpha{}_a &= dz^\alpha{}_a - z^\alpha{}_{ab}\, dx^b, \\
&\cdots \\
\theta^\alpha{}_{a_1 a_2 \cdots a_{k-1}} &= dz^\alpha{}_{a_1 a_2 \cdots a_{k-1}} - z^\alpha{}_{a_1 a_2 \cdots a_{k-1}}\, dx^a.
\end{aligned}
\tag{2.4}
$$

The subspace of the cotangent bundle spanned by the set of all such one-forms will be referred to as $\Omega^{(k)}$. We may see immediately that for any such function $f : M \longrightarrow N$, we have

$$(j^k f)^*[\Omega^{(k)}] = 0\,! \tag{2.5}$$

Therefore, these objects allow us to test whether some particular function defined on the jet bundles could actually have come from a map on the base manifolds.

The next concept we can build up, with the use of the contact module, is the ***total derivative***. Let u be some arbitrary function defined on at least some subset of $J^k(M,N)$. Then, on the appropriate cotangent bundle, we have

$$\begin{aligned} du &= \frac{\partial u}{\partial x^a}\,dx^a + \frac{\partial u}{\partial z^\alpha}\,dz^\alpha + \frac{\partial u}{\partial z^\alpha{}_a}\,dz^\alpha{}_a + \cdots + \frac{\partial u}{\partial z^\alpha{}_{a_1a_2\cdots a_k}}\,dz^\alpha{}_{a_1a_2\cdots a_k}, \\ &= \frac{\partial u}{\partial x^a}\,dx^a + \frac{\partial u}{\partial z^\alpha}\,[z^\alpha{}_a\,dx^a + \theta^\alpha] + \frac{\partial u}{\partial z^\alpha{}_a}\,[z^\alpha{}_{ab}\,dx^b + \theta^\alpha{}_a] + \cdots, \end{aligned} \tag{2.6}$$

or

$$du \equiv (D^{(k)}{}_a u)\,dx^a \quad \bmod \Omega^{(k)}. \tag{2.7}$$

Recall that the notation mod $\Omega^{(k)}$ means that we ignore any terms in $\Omega^{(k)}$! This means that this quantity $D^{(k)}{}_a$ is just the usual ***physicist's notion*** of a ***total derivative***, made legitimate on the jet bundle by this device. This of course means that

$$(j^k f)^*(du) = D^{(k)}u = (D^{(k)}{}_a u)\,dx^a. \tag{2.8}$$

Also, of course, in the cotangent bundle over $J^1(M', N')$, we have

$$\theta'^A = dz'^A - Z'^A{}_r\,dx'^r, \tag{2.9}$$

whose function is the same, but over the space over M'.

Using the picture in Figure 2, we may envision, finally, what it is that a Bäcklund transformation does. We note, first, that if $H : M \longrightarrow N$, then $H^* : N^* \longrightarrow M^*$ maps (inversely) the cotangent spaces (spaces where the contact forms live). The important object for us to consider is the mapping of contact forms generated by

$$[(j^{k-1}f)^* \times (j^1 f')^*](\Upsilon^*(\theta'^A)). \tag{2.10}$$

Clearly, one must have the relation

$$\begin{aligned} \Upsilon^*(\theta'^A) &= dz'^A - Z^A{}_r\,dX^r \\ &= dz'^A - Z^A{}_r[(D^{(k)}{}_b X^r)\,dx^b + X^r{}_{,z'^B}\,dz'^B] \quad \bmod \Omega^{(k)}, \end{aligned} \tag{2.11}$$

where the notation $A_{,z}$ means $\partial A/\partial z$. When pulled back by two solutions —f and f'— this constitutes a set of PDE's to be solved, of $(k-1)^{st}$ order in f and first order in f'. This set is a Bäcklund transformation! If one is actually given a solution f, then he may find an f', assuming that the above set of equations is integrable. Therefore a very important criterion here is the set of integrability conditions for these equations.

§3 Integrability conditions

In order to most easily describe the desired results, it is necessary to extend the notion of ***total derivative*** defined earlier. We extend it so that it includes the z'-dependence of the functions involved:

$$D^{(k)}{}_a \;\rightarrow\; \tilde{D}^{(k)}{}_a = D^{(k)}{}_a + U^A{}_a \frac{\partial}{\partial z'^A}, \tag{3.1}$$

where

$$U^A{}_a = G^A{}_B\,Z^B{}_c\,[D^{(k)}{}_a X^c], \tag{3.2}$$

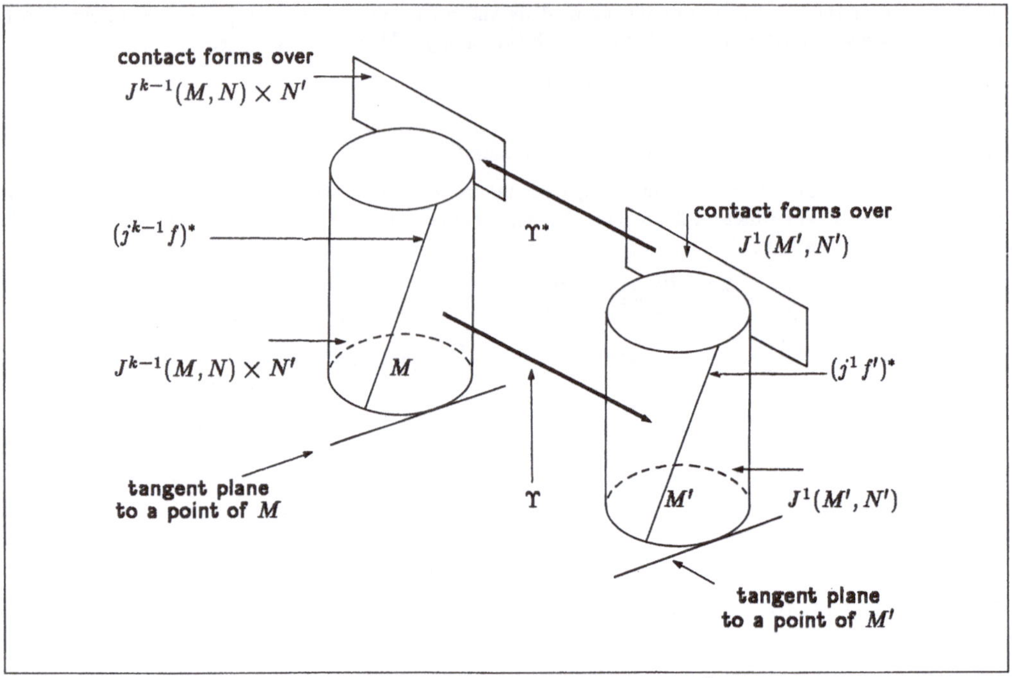

Figure 2.

and the matrix $G^A{}_B$ is that inverse to $\delta^B{}_A - Z^B{}_a\,(\partial X^a/\partial z'^A)$. This object is rather like a covariant derivative in that it no longer commutes with itself; it describes some sort of *curvature* that the jet bundle has obtained by being constrained to have this extra dependence on the *new* unknown functions, z'^A. Defining

$$\Omega^k{}_\Upsilon = \Omega^k + \Upsilon^*\Omega'^1, \tag{3.3}$$

we may write, for $u : J^k(M,N) \times N'$,

$$du = (\tilde{D}^{(k)}{}_a u)\, dx^a \quad \mathrm{mod}\ \Omega^k{}_\Upsilon. \tag{3.4}$$

Poincaré's Lemma ($dd = 0$) then gives us the integrability conditions (over M) for the system, as

$$\tilde{D}^{(k)}{}_{[a} U^A{}_{b]}\, dx^a \wedge dx^b = 0 \quad \mathrm{mod}\ \Omega^k{}_\Upsilon. \tag{3.5}$$

We notice immediately that this condition is very much as if it were the *vanishing of the curvature* of the bundle. There is also an integrability condition over M' [over $J^{k'}(M',N')$ to be more precise, where k' is in principle unknown, but finite]. In order to acquire this other condition, one must extend the process *up the ladder* on the primed side until the equations *behave* themselves sufficiently well —until they can be expressed in terms of only primed variables. For example, extension to $J^2(M',N')$ gives the following:

$$Z^a{}_{rs} = U^b{}_{(r}\,\tilde{D}^{(k)}{}_b\, Z^A{}_{s)}, \tag{3.6}$$

where $U^b{}_r$ is the matrix inverse to $\tilde{D}^{(k)}{}_r X^b$.

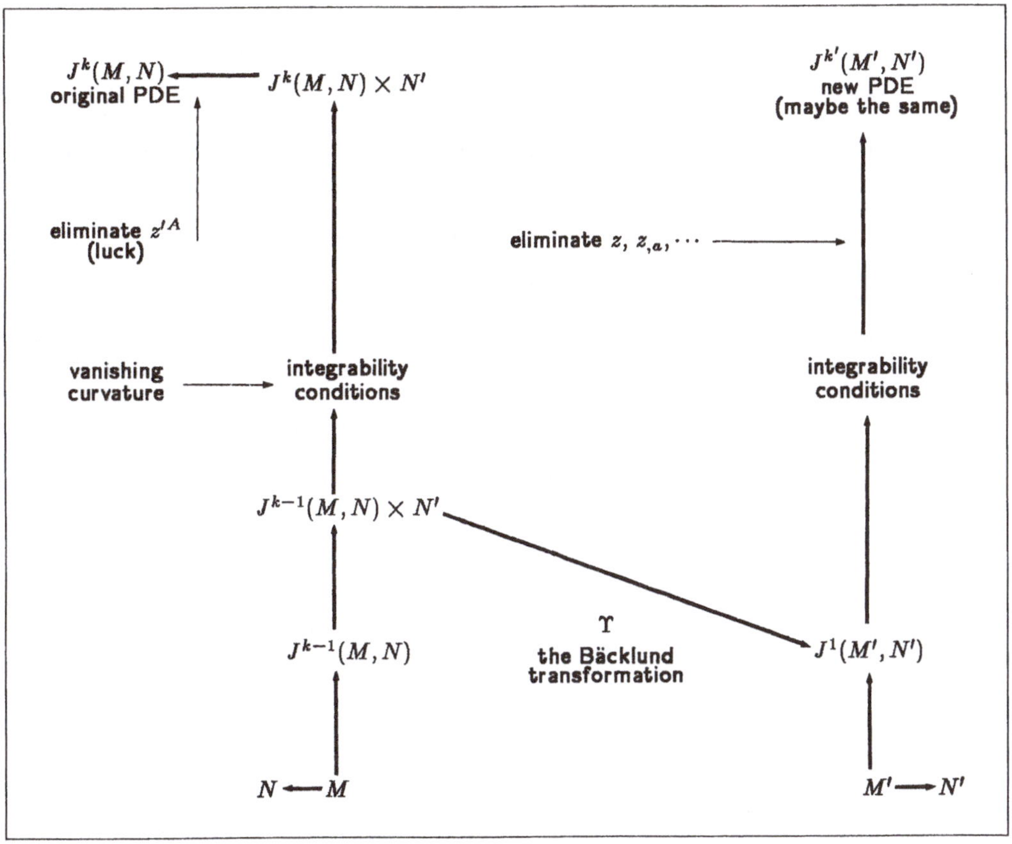

Figure 3.

To understand this somewhat better, we refer first to Figure 3 and then to the explanatory phraseology given below (a summary of much of the preceding): Given an original PDE, the problem is to find Υ, a map which will relate solutions of that PDE to those for some new PDE. The integrability conditions for Υ should be

1. the original PDE when pulled back over M,
2. the new PDE when pushed forward over M'.

In order for this to work, it is necessary that all dependence on z'^A disappear when those integrability conditions over M are actually evaluated; likewise it is necessary that only dependence on the primed quantities remains when the conditions are evaluated over M'. It is this last condition which determines k': k' is the minimal integer for which this dependency condition occurs.

As a pedagogical attempt to clarify this fairly complicated situation, we now present an example of about the simplest possible kind: dimension of N = dimension of $N' = 1$, $k = 2$, dimension of M = 2. In that case, some of the sums involved in the more general expressions have only one term:

$$\Upsilon^*\theta' = A\,dz' - B_b\,dx^b, \tag{3.7}$$

with

$$A = 1 - Z_a X^a{}_{,x'}, \qquad B_b = Z_a \left(X^a{}_{,x^b} + z_{bc} X^a{}_{,z_c}\right). \tag{3.8}$$

Dividing by A and applying the exterior derivative to the system, one acquires

$$\tilde{D}^{(2)}{}_{[a}y_{b]} = \cdots = \left(z_{11}z_{22} - (z_{12})^2\right)U + R^{ab}z_{ab} + L^a z_a + W, \tag{3.9}$$

where U, R^{ab}, L^a, and W are made of Z_a, X^a, and their derivatives. for example, denoting the expression $Z_a X^a{}_{,z_c}$ by E^c, we may write explicitly the expression

$$U = E^{[1}E^{2]}{}_{,x'} + A\,E^{[1}{}_{,z_2]} - E^{[1}A_{,z_2]}.$$

The others are similar, but will not be written out here. The idea, then, at this level, is to pick U, R^{ab}, L^a, and W so that the integrability condition above is in fact the equation whose transform is desired. That is, we generate the desired equation by proper choice of these coefficients; however, choosing of these coefficients generates partial differential equations for the quantities which make up Υ. An explicit example is given by the following choice:

$$U = 0, \quad R^{12} = 0, \quad L^a = 0, \quad W = 0, \quad R^{11} = R^{22} = 0.$$

This particular choice generates the two-dimensional Laplace equation, *i.e.*, it causes the combination $D^{(2)}{}_{[1}\Upsilon_{2]}$ to be proportional to the Laplace operator. the set of equations above has many solutions. A particular one is given by $x' = X'$, $B_b = Z_b$, and the set

$$B_{1,z_2} + B_{2,z_1} = 0, \qquad B_{2,z_2} - B_{1,z_1} = 0. \tag{3.10}$$

This gives the familiar Cauchy–Riemann equations as a method of generating new solutions of the Laplace equation from old solutions.

§4 Negative results

We summarise some negative results which are already moderately well known about Bäcklund transformations and then discuss the problem in more detail.

1. In the case of two independent variables and only one dependent variable, the non-linearity in the equation may not be worse than that found in the Monge–Ampère equation. In fact, it must be slightly better behaved than the most general Monge–Ampère equation [5].
2. When there are more than two independent variables, the ***curvature*** gets more than one component, which causes considerable trouble in trying to cause this ***curvature*** to ***be*** the desired PDE.

To see better the difficulties involved in more than two independent variables, we consider explicitly the case of dimension N = dimension $N' = 1$, dimension of $M = 4$, $k = 2$, applicable, for example, to Eq. (1.1). We use, for convenience, a spinorial notation, where the four coordinates are given by two spinors, p^A, q_B, where each of A and B is free to range over the values 1, 2. There are then three independent set of curvature forms:

$$\begin{aligned}[\tilde{D}_{p^A}, \tilde{D}_{p^B}] = \epsilon_{AB}\big(&U\,\theta_{p_C p_D}\,\theta_{p^C p^D} + W^{EF}\,\theta_{p_C p^E}\,\theta_{p^C p^F} \\ &+ X\,\theta_{p_C q_D}\,\theta_{p^C q^D} + R^{EF}\,\theta_{p_E p^F} + S^{EF}\theta_{p_E q^F} + V\big),\end{aligned} \tag{4.1}$$

$$\begin{aligned}[\tilde{D}_{q_A}, \tilde{D}_{q_B}] = \epsilon^{AB}\big(&U\,\theta_{q^C q^D}\,\theta_{q_C q_D} + W^{EF}\,\theta_{q^C p^E}\,\theta_{q_C q^F} \\ &+ X\,\theta_{q_C q_D}\,\theta_{q^C q^D} + R'^{EF}\,\theta_{q_E q_F} + S'^{EF}\theta_{q_E q^F} + V'\big),\end{aligned} \tag{4.2}$$

$$\begin{aligned}[\tilde{D}_{q_A}, \tilde{D}_{p^B}] = &U\,\theta_{p_A p^C}\,\theta_{p_C q^B} + W^{EF}\left(\theta_{p_A q^F}\,\theta_{q^B p^E} - \theta_{p_A p^E}\,\theta_{q^B q^F}\right) \\ &+ X\,\theta_{p_A q_C}\,\theta_{q^B q^C} + R^{DA}\,\theta_{p_D q^B} + R'{}_{DB}\,\theta_{p^D q_A} \\ &+ S^{DA}\theta_{q^B q^D} + S'{}_{DB}\theta_{p_A p_D} + V''^{A}{}_{B}.\end{aligned} \tag{4.3}$$

In general, it would of course be very difficult to make all of these equations do what one wants. However, certain special cases may present no particular problems. In order to make this into the four-dimensional Laplace equations, one may take

$$U = 0 = X, \quad W^{EF} = 0, \quad V = 0 = V', \quad V''^{A}{}_{B} = 0,$$

$$R^{EF},\ S^{EF},\ R'^{EF}, \text{ and } S'^{EF} \text{ proportional to } \epsilon^{EF}.$$

This generates the Bäcklund transformation

$$\theta'_{q^A} = \theta_{p^A}, \qquad \theta'_{p^A} = \theta_{q^A}.$$

In order for a Bäcklund transformation to exist for a given PDE, it is necessary that one be able to write the equation so as to be the ***curvature***. Therefore, in particular, in more than two independent variables, one must be able to satisfy all the ***curvature*** equations at once. However, there is a very important caveat. The general form (given above for the special case quoted) for a Bäcklund transformation is form-invariant under all possible contact transformations, but the particular desired final form is not. Therefore, it is necessary, before deciding that a given particular PDE cannot be so transformed, to show that it cannot be put into the form the ***curvature*** has, even after having done all possible contact transformations. Again, an example will illustrate, hopefully, the kinds of problems that might be encountered: We consider the particular contact transformation:

$$\Phi^1 : J^1(M,N) \longrightarrow J^1(M,N)$$
$$\text{by } \hat{x}^1 = x^1, \quad \hat{x}^2 = z_2, \quad \hat{z} = x^2 z_2 - z, \quad \hat{z}_1 = -z_1, \quad \hat{z}_2 = x^2. \tag{4.4}$$

This particular contact transformation maps

$$\begin{array}{ccc} z_{11} + z_{22} = 0 & & z_{11}\, z_{22} - (z_{12})^2 + 1 = 0 \\ \text{Laplace equation} & \longrightarrow & \text{simple Monge–Ampère} \\ \Big\downarrow \Upsilon \ \ \text{auto-Bäcklund transformation} & & \Big\downarrow \Upsilon^{\Phi} \equiv \tilde{\Phi}^1 \circ \Upsilon \circ (\tilde{\Upsilon}^{k-1})^{-1} \\ z'_{1'1'} + z'_{2'2'} = 0 & & z'_{1'1'} + z'_{2'2'} = 0 \end{array}$$

Here the symbol $\tilde{\Phi}^1$ refers to the extension of Φ^1 to the appropriate jet bundle. Therefore, we see that the rather more complicated form of this simple form of the Monge–Ampère equation may be contact transformed, or Bäcklund transformed, into the two-dimensional Laplace equation. Similarly, much more complicated-appearing equations may be contact transformed into ones which have the form necessary for a Bäcklund transformation to exist. This process —doing all posible contact transformations— must be done in order to assure oneself that no Bäcklund transformation exists for the PDE in question.

§5 The multiple-copy concept

In the case of more than two independent variables, a plausible method for avoiding, at least some of the time, the problems generated by having to satisfy several different equations simultaneously, is to consider a set of several copies of the original equation, instead of the original PDE. The idea, of course, is that the several components of the ***curvature*** are to generate all these copies [10]. the difference, here, between this concept and the more usual idea of Bäcklund transformations is the following: Given two PDE's related by a Bäcklund transformation, a solution of the one PDE can be inserted into a set of lower-order PDE's to generate a solution of the other. In the case of multiple copies, one must insert several solutions of the original (multiply-copied) PDE into the Bäcklund transformation equations, which then generates several solutions of the other PDE. Again, we illustrate this process by an example. We look at the three-dimensional Laplace equation. By careful consideration of the allowed forms for three-dimensional Bäcklund transformations, it can be checked that, in fact, there is no single-copy Bäcklund transformation for this equation. However, we now let

$$\Upsilon : J^1(M, N \times N) \times N' \times N' \longrightarrow J^1(M', N' \times N'). \tag{5.1}$$

This allows sufficient extra freedom for the Bäcklund transformation to exist; additionally, one gets mixing of the kinds of terms in the *curvature*. To be more precise, we note that the one-copy *curvature* equations (forced to be quasi-linear) are simply

$$Z_{[c,z_a} z_{d]a} = 0, \tag{5.2}$$

which cannot be satisfied. However, for two copies we have, analogously, the equations

$$Z^1{}_{[c,z^1{}_a} z^1{}_{d]a} + Z^1{}_{[c,z^2{}_a} z^2{}_{d]a} = 0, \tag{5.3}$$
$$Z^2{}_{[c,z^1{}_a} z^1{}_{d]a} + Z^2{}_{[c,z^2{}_a} z^2{}_{d]a} = 0, \tag{5.4}$$

Notice that the first term of the first equation and the second term of the second equation are simply (two) duplicates of the equation which one obtains for the single-copy approach. However, the other terms in each of the two equations are mixing terms, where derivatives of one set of dependent variables with respect to the other set of independent variables occur; these terms allow considerable more freedom in finding ways to match up the PDE's to the allowed Bäcklund form. Indeed, for the three-dimensional Laplace equation, a simple example solution is given by the following:

$$\begin{aligned} Z^1{}_2 + Z^2{}_3 &= -z^2{}_1, & Z^1{}_3 - Z^2{}_2 &= +z^1{}_1, \\ Z^2{}_1 &= z^1{}_2 + z^2{}_3, & -Z^1{}_1 &= z^1{}_3 - z^2{}_2. \end{aligned} \tag{5.5}$$

We note that these particular equations [10] can be written in a much neater form if quaternions are introduced. We let the four basis quaternions be denoted by the symbols $\mathbf{1}$, $\hat{\imath}$, $\hat{\jmath}$, and $\hat{k}$. Then, we may define, just for this particular section, the symbols $\hat{D}$ and $\hat{Z}$ by

$$\hat{D} \equiv \hat{\imath}\partial_1 + \hat{\jmath}\partial_2 + \hat{k}\partial_3, \qquad \hat{Z} \equiv \mathbf{1}z^1 + \hat{\imath}z^2 + \hat{\jmath}Z^1 + \hat{k}Z^2. \tag{5.6}$$

The set of four equations given above reduces to the very simple

$$\hat{D}\,\hat{Z} = 0, \qquad \hat{D}^2 = \begin{array}{l}\text{three-dimensional} \\ \text{Laplace operator.}\end{array} \tag{5.7}$$

Following the techniques given in the example, one could hope to determine Bäcklund transformations for considerably more complicated equations than the three-dimensional Laplace equation. The various *matrix sine–Gordon* equations in the literature [14], for example, are various interesting special cases of this approach. The Dirac equation (in four independent variables) can also be considered as an example of this kind of situation, for the four-dimensional Laplace equation. However, the basic idea of the multi-copy concept is obscured by the (not-highly-motivated) introduction of matrices into the equations.

Unfortunately, it turns out that this approach does not seem to help with the original set of equations (generated from the Plebański equation) which started this work. the general forms generated by all allowed contact transformations can be obtained on the first two equations given in §1. It can then be seen that, in fact, they are ***too nonlinear*** to have Bäcklund transformations. Therefore, no ***solution-generators*** for these equations can be found in this way. this talk must therefore end on a negative note. Nonetheless, this fact does indeed exhibit the usefulness of the present approach to Bäcklund transformations. This approach —determining the most general form of PDE which is allowed to have a Bäcklund transformation, modulo contact transformations— allows one to actually answer for certain, but only in the negative, the question as to the existence of a Bäcklund transformation for any given PDE, or system of PDE's. In the event that there is a positive answer to the question of existence, this method does not really help to find any particular solution to that existence problem. the methods of Estabrook and Wahlquist are then probably the most helpful.

For an ending, then, I will write down the most general form (obtained by the ***curvature*** method given above) which will permit a Bäcklund transformation for the situation of two independent variables, but a third-order PDE, which therefore includes the Korteweg–de Vries equation:

$$G^{rscd}\, z_{rs[b}\, z_{a]cd} + K_{[a}{}^{cd}\, z_{b]cd} + L^{cd}\, z_{[a}\, z_{b]cd} + M^{cdr}\, z_{[ar}\, z_{b]cd} + \text{lower-order terms!}$$

References

[1] S. Lie, Theorie der Transformationsgruppen. V. *Archiv für Math.* **4**, 232–261 (1879).

[2] A. Bäcklund, Zur Theorie der Flächentransformationen. *Math. Ann.* **19**, 387–422 (1882).

[3] J. Clairin, *Ann. Scient. École Norm. Supplément* **19**, (1902).

[4] E. Goursat, Leçons sur l'integration des equations aux derivées partielles du second ordre. *Mem. Sci. Math.* **6** (1925).

[5] A. Forsyth, *Theory of Differential Equations.* Vol. 6. Cambridge University Press, 1906.

[6] F. Pirani, D. Robinson, and W. Shadwick, *Local Jet Bundle Formulation of Bäcklund Transformations.* Mathematical Physics Studies, Vol. 1, Reidel, Dordrecht, 1979.

[7] B. Harrison and F. Estabrook, Geometric approach to invariance groups and solutions of partial differential systems. *J. Math. Phys.* **12**, 653–665 (1971).

[8] F. Estabrook and H. Wahlquist, Prolongation structures of nonlinear evolution equations. *J. Math. Phys.* **16**, 1–7 (1975).

[9] R. Hermann, *The Geometry of Non-Linear Differential Equations, Bäcklund Transformation, and Solitons. Part A.* Interdisciplinary Mathematics, Vol. 12, Math-Sci Press, Brookline, 1976.

[10] R. Bryant, Exterior differential systems. NSF Research Conference, 1982. (Unpublished.)

[11] J. F. Plebański, Some solutions of complex Einstein equations. *J. Math. Phys.* **16**, 2395–2402 (1975).

[12] P. Denes and J. D. Finley, III, Bäcklund Transformations for General PDE's. (To be published in *Physica D.*)

[13] J. D. Finley, III and J. F. Plebański, The classification of all $\mathcal{H}$ spaces admitting a Killing vector. *J. Math. Phys.* **20**, 1938–1945 (1979).

[14] G. Liebbrandt, R. Morf, and S. Wang, Solutions of the sine–Gordon equation in higher dimensions. *J. Math. Phys.* **21**, 1613–1624 (1980); H. Morris, Prolongation structures and nonlinear evolution equations in two spatial dimensions. *J. Math. Phys.* **17**, 1870–1872 (1976).

The "Weak Painlevé" Property and Integrability of Two-Dimensional Hamiltonian Systems

Basile Grammaticos

Centre National d'Etudes des Tèlècommunications
Issy les Moulineaux, France

Abstract

The integrability of dynamical systems, described by nonlinear differential equations, is associated to the singularity structure of the solutions in the complex-time plane. In this work, the usual Painlevé property, *i.e.*, existence of poles as the only movable singularities, is somewhat extended for the case of two-dimensional hamiltonian systems. Integrability in this case is compatible with the presence of algebraic branch points of a specific nature.

§1 Introduction

The discovery of completely integrable dynamical systems described through partial differential equations (PDE's) [1], together with the observation that their reductions to ordinary differential equations (ODE's) possessed specific analytic properties in the complex-time plane, have led to the formulation of the Painlevé conjecture. Ablowitz, Ramani, and Segur in [2] have conjectured, and abundantly verified, that whenever a PDE is integrable, its reduction possesses the Painlevé property, *i.e.*, their only singularities in complex t are poles. Segur and Bountis in [3] have extended the Painlevé criterion to systems described by ODE's. The work of various group (see [4]) has demonstrated the usefulness of the Painlevé conjecture as integrability detector.

Integrability in the case of a system of N coupled nonlinear first-order differential equations is synonymous with the existence of N analytic single-valued integrals which can be explicitly time-dependent. In the case of a hamiltonian system with N degrees of freedom, complete integrability requires the existence of N single-valued analytic integrals of motion, which are time independent and in involution. The philosophy in the latter case is that once the N constants of motion are given, there exists a transformation to action-angle variables which reduces the problem to free motion in an N-dimensional space. This reduction, however, cannot be carried through in closed form in most cases. There is thus a difference between integrability for ODE's, where the solution to the system is obtained through the solution of an algebraic problem once the integrals are known, and hamiltonian systems.

In the case of one-dimensional hamiltonian systems, integrability is automatically assured by the existence of the energy integral. The Painlevé property plays no role in this case. On the other hand, it could be satisfied through a suitable global transformation of the dependent variable. The important question we address in this work is the following: is it possible to find a dynamical system which, although integrable, does *not* possess the Painlevé property, and where the latter *cannot* be recovered through some reasonably simple global transformation of the dependent variables? We will show the answer to this question to be affirmative for two-dimensional hamiltonian systems.

§2 The "weak Painlevé" property

Let us consider the equations of motion for the system described by the hamiltonian

$$H = \tfrac{1}{2}(\dot{x}^2 + \dot{y}^2) - (y^5 + x^2y^3 + \tfrac{3}{16}x^4y). \tag{1}$$

They read:

$$\begin{aligned} \ddot{x} &= 2y^3x + \tfrac{3}{4}yx^3, \\ \ddot{y} &= 5y^4 + 3x^2y^2 + \tfrac{3}{16}x^4. \end{aligned} \tag{2}$$

In order to perform a leading-order analysis near a singularity, we put $x = \alpha\tau^\nu$, $y = \beta\tau^\mu$ ($\tau = t - t_0$), and we find

$$\begin{aligned} \nu(\nu-1)\alpha\tau^{\nu-2} &= 2\beta^3\alpha\tau^{3\mu+\nu} + \tfrac{3}{4}\beta\alpha^3\tau^{\mu+3\nu}, \\ \mu(\mu-1)\beta\tau^{\mu-2} &= 5\beta^4\tau^{4\mu} + 3\alpha^2\beta^2\tau^{2\mu+2\nu} + \tfrac{3}{16}\alpha^4\tau^{4\nu}. \end{aligned} \tag{3}$$

Two possibilities exist. Either $\mu = \nu$, in which case we find readily $\mu = \nu = -\frac{2}{3}$ and α, β are determined through the nonlinear system (3), or $\mu < \nu$, in which case the first term in the right-hand side is dominant. We find thus: $\mu = -\frac{2}{3}$, $\beta^3 = \frac{2}{9}$, and $\nu(\nu-1) = \frac{4}{9}$, *i.e.*, $\nu = -\frac{1}{3}$ or $\nu = \frac{4}{3}$, with α being undetermined. Thus, the solution has in general singularities of the type $(t-t_0)^{1/3}$, *i.e.*, algebraic branch points. This in itself is not incompatible with the Painlevé property, since a transformation to x^3 and y^3 could give pure poles. This turns out not to be the case, however, as can be shown through the study of the resonances. A search for the undetermined coefficients of the expansion in the form

$$\begin{aligned} x &= \alpha\tau^\nu(1 + \gamma\tau^n), \\ y &= \beta\tau^\mu(1 + \delta\tau^n), \end{aligned}$$

leads to a resonance $n = \frac{10}{3}$. This means that the cubes of x an y are not pure poles, and so the classical Painlevé property is not satisfied. The expansion, however, does not contain anything *worse* than $(t-t_0)^{1/3}$. This has led to the following generalization of the Painlevé conjecture for two-dimensional hamiltonian systems. We claim that in this case the "weak Painlevé" property is compatible with integrability. We will say that a system has the *"weak Painlevé"* property whenever the solution in the neighourhood of a singularity at t_0 can be expressed as an expansion in powers of $(t-t_0)^{1/r}$, where r is an integer determined solely by the leading behaviour of the singularity.

Once the conjecture has been reformulated, one must question its predictive character. It turns out that the "weak Painlevé" concept is indeed very rich and leads to new families of integrable systems. One of the most important results is the following (see [5]): Let us consider a homogeneous potential V of degree $p+2$ in two dimensions. The condition for the system to possess the "weak Painlevé" property, *i.e.*, for the solutions to have an expansion in terms of $(t-t_0)^{1/p}$, combined with a simplifying assumption $V_{xx}+V_{yy}=\lambda V_x/x$, leads to the following equation for the potential:

$$x(V_{yy}-V_{xx})-2yV_{xy}-3V_x=0. \tag{4}$$

As we shall see in the next paragraph, this PDE is exactly the integrability condition, *i.e.*, the condition for this sytem to posess a second integral of motion quadratic in velocities, and it is valid for a class of potentials far larger than homogeneous polynomials. Thus, for two-dimensional systems, the "weak Painlevé" condition is synonymous with integrability.

§3 The direct approach to integrability

For the complete integrability of a two-dimensional hamiltonian system, one needs a second constant of motion beyond the energy. The general form of an integral quadratic in velocities is

$$c=g^{(0)}\dot{x}^2+g^{(1)}\dot{x}\,\dot{y}+g^{(2)}\dot{y}^2+h, \tag{5}$$

where $g^{(i)}$ and h are functions of x and y. The condition of constancy of c, $\frac{dc}{dt}=0$, leads to:

$$\begin{aligned} g^{(0)}&=\alpha y^2+\beta y+\gamma,\\ g^{(1)}&=-2\alpha xy-\beta x-\delta y-\epsilon,\\ g^{(2)}&=\alpha x^2+\delta x+\varsigma, \end{aligned} \tag{6}$$

and

$$2(g^{(0)}-g^{(2)})V_{xy}+(2g_y{}^{(0)}-g_x{}^{(1)})V_x-(2g_x{}^{(2)}-g_y{}^{(1)})V_y-g^{(1)}(V_{xx}-V_{yy})=0. \tag{7}$$

The solution of Eq. (7) leads to eight classes of integrable systems. When the quadratic terms in the $g^{(i)}$'s are present, one has four possibilities.

The first was initially obtained by Darboux, and reads

$$V=\frac{F(u)+G(v)}{u^2-v^2}, \tag{8}$$

where

$$\begin{aligned} 2u^2&=\rho^2+\gamma+\sqrt{\rho^2+\gamma^2-4\gamma x^2},\\ 2v^2&=\rho^2+\gamma-\sqrt{\rho^2+\gamma^2-4\gamma x^2},\\ \rho^2&=x^2+y^2, \end{aligned}$$

and F, G are arbitrary functions.

In the degenerate case $\gamma=0$, one has

$$V=F(\rho)+\frac{1}{\rho^2}G\left(\frac{x}{y}\right). \tag{9}$$

Two complex solutions exist also at the particular combination $\epsilon^2+\gamma^2=0$ of the parameters in (6):

$$V=\frac{F(\rho^2+\sqrt{\rho^4-2\gamma z^2})+G(\rho^2-\sqrt{\rho^4-2\gamma z^2})}{\sqrt{\rho^4-2\gamma z^2}}, \tag{10}$$

with $z = x + iy$. In the limit $\gamma \to 0$, the solution is

$$V = F(\rho) + \frac{1}{\rho^2} G(z). \tag{11}$$

When only the constant terms are present in the $g^{(i)}$'s, one obtains a separable potential

$$V = F(x) + G(y), \tag{12}$$

or a quasiseparable one (for $\epsilon = \pm i/2$):

$$V = F(z) + \rho^2 G(z). \tag{13}$$

Finally, in the case of $g^{(i)}$ linear in x and y, one obtains for V, after an adequate translation and rotation, the equation

$$x(V_{yy} - V_{xx}) - 2yV_{xy} - 3V_x = 0,$$

i.e., the same as equation (3), which was derived from the "weak Painlevé" condition. The solution to this equation reads:

$$V = \frac{F(y+\rho) + G(y-\rho)}{\rho}. \tag{14}$$

In the singular case $\delta = \pm i\beta$, one finds

$$V = F(z) + \frac{1}{\rho} G(z). \tag{15}$$

It is quite instructive to examine some particular families of solutions for the potential (14). A most interesting case is the potential $V = 1/\rho$. The latter, besides being axially symmetric, belongs also to the class (14). It possesses thus three integrals of motion: energy, square of angular momentum, and the y-component of the Laplace-Runge-Lenz vector $\dot{\vec{r}} \times \vec{L} + \vec{r}/r$, which is the well-known third integral of the Kepler problem. A family of polynomial potentials can also be computed:

$$V_n = \sum_{k=0}^{[n/2]} 2^{n-2k} C_{n-k}^{k} x^{2k} y^{n-2k}. \tag{16}$$

The quintic potential (1) which was at the origin of the "weak Painlevé" concept is a member of this family. The integrable case of the Hénon-Heiles potential

$$V = ax^2 + by^2 + x^2y + 2y^3, \tag{17}$$

can be written as a superposition of V_1, V_2, and V_3.

Another interesting limit of the integrable potentials described above, are integrable *billiards*. In the case of the potential (8), the coordinates u and v define confocal ellipses and hyperbolas. By choosing $F(u) = 0$ for $u < u_0$ and $F(u) = \infty$ for $u > u_0$, we obtain an elliptical billiard which is a well-known integrable case. Circular and rectangular billiards are limits of the potentials (9) and (12). Potential (14) furnishes a new integrable billiard: it corresponds to the region contained between two confocal parabolas.

§4 Conclusion

In this work we have introduced the "weak Painlevé" concept, which is a most useful criterion of integrability for two-dimensional hamiltonian systems. This has led to the discovery of classes of integrable hamiltonians for which the second integral of motion is quadratic in the velocities. These systems possess a particularly simple structure which would allow, in principle, the reduction of the trajectory to quadratures. A question which arises naturally is whether the "weak Painlevé" type singularities cannot be reduced to true Painlevé ones through some simple change of variables. At the present stage it seems that this is not possible.

Throughout this work, the dual approach, Painlevé analysis and direct search of integrals, has been adopted. It has been proven particularly powerful since for each case of predicted integrability (through the singularity analysis), the integrals of motion have been directly computed. This of course need not be always the case. Although the only example we can produce are the Painlevé equations themselves, it might turn out that a system possesses the Painlevé property and no integral of motion can be made explicit. Such a situation would of course hint at the existence of new transcendents.

Acknowledgements

The results presented in this paper were obtained in association with B. Dorizzi and A. Ramani. The author is grateful to them for this fruitful collaboration. He wishes also to acknowledge most stimulating discussions with M. Kruskal and P. Winternitz.

References

▶ For a thorough review of integrable partial differential equations and inverse scattering transform techniques, see the courses of Mark Ablowitz and Athanasios Fokas, in *these Proceedings*.

▶ For an introduction to the Painlevé method and its application to the study of integrability of dynamical systems, see the contribution of Bernadette Dorizzi in *these Proceedings*.

[1] M. Ablowitz and H. Segur, *Solitons and the Inverse Scattering Transform*. SIAM Studies in Applied Mathematics, 1981.

[2] M. Ablowitz, A. Ramani, and H. Segur, A connection between nonlinear evolution equations and ordinary differential equations of *P*-type. *J. Math. Phys.* **21**, 715–721 (1980).

[3] T. Bountis and H. Segur, Logarithmic singularities and chaotic behavior in hamiltonian systems *AIP Conference Proceedings*, # 88, 279–292 (1982).

[4] M. Tabor and J. Weiss, Analytic structure of the Lorentz system. *Phys. Rev.* **A24**, 2157–2167 (1981); C. R. Menyuk, H. H. Chen, and Y. C. Lee, Restricted multiple three-wave interactions: Painlevé analysis, Plasma Preprint PL82–052, University of Maryland (1982).

[5] B. Dorizzi, B. Grammaticos, and A. Ramani, A new class of integrable systems. *J. Math. Phys.* (To appear.)

On the Zakharov Equations in One Dimension

Julio E. Herrera

Centro de Estudios Nucleares
Universidad Nacional Autónoma de México

Abstract

In the context of Plasma Physics some equations arise which are not completely integrable, but have solitary wave solutions. The interactions between these waves differ from those of ***aristocratic solitons***. A short review is made of the case of Langmuir solitons, as described by the Zakharov equations in one dimension.

§1 Introduction

One of the most fundamental problems in Plasma Physics, and currently an area of active research, is the one of strongly turbulent plasmas. In an unmagnetized plasma, Langmuir waves predominate if the temperature of ions is negligible in comparison with the one of electrons. These are essentially high frequency electrostatic oscillations which produce a ponderomotive force which depletes the ion density. Such depression may play the role of a well which traps self-consistently the high-frequency oscillations. Under certain circumstances, these structures may collapse, as shown by V. E. Zakharov in 1972, in a seminal work where he derived the equations that describe this phenomenon. They are thus known as *Zakharov equations* [1]. Excellent reviews on this subject have beeen made by Rudakov and Tsytovich [2], and by Thornhill and ter Haar [3]. Further advancement in the case of magnetized plasmas has been made recently [4–6].

In this contribution we shall restrict ourselves to the case of Zakharov equations for an unmagnetized plasma in one dimension. Generally speaking, they describe the evolution of an almost monochromatic pump wave in a homogeneous medium which is strongly dispersive, weakly nonlinear, and whose response time is finite. In dimensionless units they may be written as

$$iE_t + E_{xx} - nE = 0, \tag{1.1}$$

$$n_{tt} - n_{xx} = |E|^2_{xx}, \tag{1.2}$$

where E is the electric field, and n is the ion density in units of the uniform unperturbed density n_0 [3]. In the static approximation, where n_t may be neglected, $n = -|E|^2$, provided we ask for E, $n \to 0$ as

$|x| \to \infty$. Then, Eqs. (1.1) and (1.2) reduce to

$$iE_t + E_{xx} + |E|^2 E = 0, \tag{1.3}$$

the Schrödinger equation with a cubic nonlinearity. This equation is known to be completely integrable, to have multisoliton solutions, and its initial value problem may be solved by the inverse scattering method [7]. It is also known that Eqs. (1.1) and (1.2) have a one- solitary wave solution

$$E = E_0 \operatorname{sech}\left[\frac{|E_0|(x - x_0 - vt)}{\sqrt{2(1-v^2)}}\right] \exp\left[\tfrac{1}{2}ivx - i\left(\tfrac{1}{4}v^2 - \frac{|E_0|^2}{2(1-v^2)}\right)t + i\phi\right], \tag{1.4}$$

$$n = -\frac{|E|^2}{1-v^2}, \tag{1.5}$$

where E_0, v, x_0, and ϕ are constants. The amplitude of the wave is E_0, v is its speed, and x_0 and ϕ are parameters which allow us to fix its position and phase at a given time. It is clear that Eqs. (1.4) and (1.5) tend to the one-soliton solution for (1.3) as the speed v tends to zero. Some experimental evidence for the existence of these solitary waves has been produced by Antipov *et al.* in Refs. [8] and [9].

In contrast with (1.3), the Zakharov equations are not completely integrable. Although there is work where it has been claimed that there exist multisoliton solutions for them [10], it has turned out to be incorrect. The fact that the Zakharov equations in one dimension have only three conserved quantities, in contrast with (1.3) which has an infinite number, may be regarded as the reason why they are not completely integrable. Such conserved quantities may be derived from a lagrangian, as has been shown by Gibbons *et al.* in [11]. They may be written as

$$N = \int_{-\infty}^{\infty} |E|^2\, dx, \tag{1.6}$$

$$P = \tfrac{1}{2}\int_{-\infty}^{\infty} [i(E\,E_x^* - E^*\,E_x) + 2nV]\, dx, \tag{1.7}$$

$$H = \int_{-\infty}^{\infty} [|E_x|^2 + n|E|^2 + \tfrac{1}{2}n^2 + \tfrac{1}{2}V^2]\, dx, \tag{1.8}$$

where V is a hydrodynamic flux that satisfies the continuity equation

$$n_{tt} + V_x = 0, \tag{1.9}$$

and N, P, and H may be interpreted as the plasmon number, the momentum, and the kinetic energy.

Thus, the interactions between two solitary waves from the Zakharov equations differ in general from those of ***real*** or ***aristocratic*** solitons, such as the solutions to Eq. (1.3). While ***aristocratic solitons*** pass through each other without merging or affecting their shapes or speeds, these solitary waves interact in a way we shall call ***nontrivial*** [11–13]. In some cases two of them may fuse into a single one, radiating the excess energy in the form of ion sound [11,12].

In §2 we shall review some work which describes, and allows us to undestand, such ***nontrivial*** interactions. In §3, some comments are made regarding the integrable limits of the Zakharov equations, in the context of the Estabrook–Wahlquist method. Since Reference [10] is misleading, it is worthwhhile to show why it is incorrect. This will be done in §4. Concluding remarks are offered in §5.

§2 The nontrivial interactions

2.1 Numerical work

Equations (1.1) and (1.2) have been integrated numerically for several initial conditions by Degtyarev *et al.* [12], and more recently by Payne *et al.* [13]. A case which has been studied in both references is the one of two colliding solitons with equal amplitude, and opposite velocities of equal magnitude. The main results may be summarized as follows:

i. When $|E| < 1$, the solitary waves pass through each other and are slowed down slightly. The momentum and energy which are lost in the collision transform into ion-sound waves that satisfy the homogeneous wave equation (1.2).

ii. If $|E| > 1$, corresponding to the case of strong turbulence, the two waves merge into a single solitary wave, provided their initial speed is below a critical value. Again, the lost energy and momentum are radiated away in the form of ion-sound waves. This phenomenon is independent of the value of $|E_0|$.

iii. When the initial speed of the two solitary waves is above a certain critical value, they pass again through each other, slightly slowed down, and emmiting ion-sound radiation.

Reference [12] describes other interesting phenomena such as the interaction of ion-sound pulses with standing solitons, and the breakup of a soliton by an ion-sound wavetrain.

2.1.1 The mechanism of radiation

The radiation of the solitary waves is a consequence of their acceleration, as may be understood through the following argument [14]:

Consider an inhomogeneity $s(x)$ in the ion density, so that instead of (1.1), we have

$$iE_t + E_{xx} - \big(n + x(x)\big)\, E = 0, \tag{2.1}$$

and let us study the behaviour of a single soliton. It has been shown by Chen and Liu [15], in the context of Eq. (1.3), that if $s(x)$ is of the form

$$s(x) = e(t) + x\, f(t), \tag{2.2}$$

then (2.1) may be reduced to the Schrödinger equation in a homogeneous medium by means of the transformation

$$\begin{aligned} E(x,t) &= u(y,\tau)\,\exp[iy\,p(\tau) + ir(\tau)], \\ x &= y - m(\tau), \\ t &= \tau, \end{aligned} \tag{2.3}$$

where

$$p(\tau) = \frac{dm}{d\tau} = 2\int_0^\tau f(t')\,dt', \tag{2.4}$$

$$r(\tau) = \int_0^\tau [\tfrac{1}{4}p^2(t') - e(t') + m(t')\,f(t')]\,dt'. \tag{2.5}$$

If we call $\overline{x}(t)$ the position of the centre of the soliton at a given time, it may be seen that $s(x)$ may be written in the form (2.2), provided the width of the soliton is much smaller than the scale of the inhomogeneity:

$$\begin{aligned} s(x) &= s[(x-\bar{x}(t))+\bar{x}(t)] \\ &\approx s[\bar{x}(t)] + (x-\bar{x}(t))\, s'[\bar{x}(t)]. \end{aligned} \tag{2.6}$$

Then, $e(t) = s[\bar{x}(t)] - \bar{x}(t)\, s'[\bar{x}(t)]$ and $f(t) = s'[\bar{x}(t)]$. From (2.3) and (2.4) it may be see that the acceleration of the soliton at time t is given by

$$\ddot{\bar{x}} = -\frac{d^2 m}{d\tau^2} = -2f(t) = -2s'[\bar{x}(t)]. \tag{2.7}$$

In other words, the solitary wave is accelerated by the slope of the inhomogeneity, very much as a classical particle is accelerated in a potential well.

Let us now propose for (2.1) and (1.2) the following one-soliton solution:

$$E = \mathcal{E}(x-\bar{x}(t)) \exp[i\phi(x,t)], \tag{2.8}$$

$$n = \frac{\mathcal{E}^2(x-\bar{x}(t))}{\dot{\bar{x}}^2-1} + N, \tag{2.9}$$

where N is the ion-sound out of the soliton. From Eq. (1.2) we get

$$N_{tt} - N_{xx} = -\frac{\ddot{\bar{x}}(3\dot{\bar{x}}^2+1)}{(\dot{\bar{x}}^2-1)^2}\frac{d}{dx}\mathcal{E}^2 + \frac{[2\dot{\bar{x}}\,\dddot{\bar{x}}(\dot{\bar{x}}^2-1) - 2\ddot{\bar{x}}^2(3\dot{\bar{x}}^2+1)]\mathcal{E}}{(\dot{\bar{x}}^2-1)^3}, \tag{2.10}$$

where $\dot{\bar{x}}$, $\ddot{\bar{x}}$, and $\dddot{\bar{x}}$ may be computed from (2.7). Assuming $\dot{\bar{x}} \ll 1$ and neglecting higher-order terms, Eq. (2.10) takes the simpler form

$$N_{tt} - N_{xx} = -\ddot{\bar{x}}\frac{d\mathcal{E}^2}{dx}, \tag{2.11}$$

which essentially means that ion-sounnd is generated due to the acceleration of the soliton. Kaw *et al.* have solved (2.11) for examples where $s(x)$ represents linear and quadratic inhomogeneities.

It is clear that this argument is valid under the approximation that the deceleration of the soliton produced by the loss of energy due to radiation is negligible in comparison with the acceleration due to the inhomogeneity.

Although this method cannot be applied directly to the problem of the two colliding solitary waves because of the approximations involved, it sheds light on the mechanism that produces the ion-sound waves. When the two packets interact, each is affected by the ion density depression of the other one and, as a consequence, of the acceleration, ion-sound is generated. It is worthwhile to note in relation with this, that for the case of fusion Degtÿarev *et al.* [12] mention that the fusion process speeds up sharply as soon as the solitons start to overlap.

2.2 The conserved quantities

The question of whether the colliding solitons will merge or not, will depend on how much ion-sound is generated during the interaction. Gibbons *et al.* [11] used the conserved quantities (1.6)–(1.8) in order to study which interaction processes were possible, and found the amount of energy which should be generated for fusion to occur. This was used in order to make a rough model of the interaction. Let us outline some of their results.

They note that the conserved quantities for the one-soliton solution may be rewritten as

$$N = 2m, \tag{2.12}$$
$$P = mv + \frac{2m^3 v}{3(1-v^2)^3}, \tag{2.13}$$
$$H = \tfrac{1}{2}mv^2 + \frac{m^3(5v^2-1)}{6(1-v^2)^3}, \tag{2.14}$$

where $m = 2E_0(1-v^2)$. This results from substituting (1.1) and (1.2) into (1.6)–(1.8). They also note that *pure ion-sound, i.e.*, solutions to $n_{tt} - n_{xx} = 0$ contribute to the conserved quantities as

$$N = 0, \tag{2.15}$$
$$P = p_+ - p_-, \tag{2.16}$$
$$H = p_+ + p_-, \tag{2.17}$$

where

$$p_\pm = \int_{-\infty}^{\infty} n_\pm^2 \, dx, \tag{2.18}$$

and n_+ (n_-) represents ion-sound waves traveling to the right (left).

In the case when ion-sound and more than one-soliton are present, assuming that in the asymptotic limit $|t| \to \infty$, the initial and final states may be described by isolated solitons and ion-sound, the conserved quantities may be written as

$$N = 2\sum_i m_i = 2\sum_f m_f, \tag{2.19}$$
$$\begin{aligned} P &= \sum_i \left\{ m_i v_i + \frac{2m_i^3 v_i}{3(1-v_i^2)^3} \right\} + p_+^i - p_-^i \\ &= \sum_f \left\{ m_f v_f + \frac{2m_f^3 v_f}{3(1-v_f^2)^3} \right\} + p_+^f - p_-^f, \end{aligned} \tag{2.20}$$
$$\begin{aligned} H &= \sum_i \left\{ \tfrac{1}{2}m_i v_i^2 + \frac{m_i^3(5v_i^2-1)}{6(1-v_i^2)^3} \right\} + p_+^i + p_-^i \\ &= \sum_f \left\{ \tfrac{1}{2}m_f v_f^2 + \frac{m_f^3(5v_f^2-1)}{6(1-v_f^2)^3} \right\} + p_+^f + p_-^f, \end{aligned} \tag{2.21}$$

where the indices i and f stand for the initial and final states, respectively. The assumption of asymptotic independence of solitons and ion-sound is well supported on theoretical considerations and numerical work [3].

This leads to the following consequences

i. Solitary waves cannot decay into pure sound, or viceversa, since this would violate the conservation of N.

ii. It is proven in Ref. [11] that for a single soliton with given N and P, H is minimum. Thus, it cannot break up into smaller solitons and ion sound. Likewise, a number of solitons and ion-sound cannot merge to form a single one, without ion-sound.

iii. Scattering of solitons and ion-sound is always possible.

iv. A number of solitons can merge into a single soliton, provided ion-sound is radiated away. The inverse process can also occur, *i.e.*, a soliton may be broken up into smaller ones, if enough ion-sound is radiated upon it.

Consequence **iv** is particularly relevant to the phenomena described in **2.1**. In that case, H would be

$$H = mv^2 + \tfrac{1}{3}\frac{m^3(5v^2-1)}{(1-v^2)^3} = mv'^2 + \tfrac{1}{3}\frac{m^3(5v'^2-1)}{(1-v^2)^3} + 2p, \tag{2.22}$$

where $p = 2p_+ = p_+ + p_-$ is the ion sound generated, and in the case of fusion,

$$H_f = mv^2 + \tfrac{1}{3}\frac{m^3(5v^2-1)}{(1-v^2)^3} = -\tfrac{4}{3}m^3 + p. \tag{2.23}$$

Then, it may be seen that fusion will be achieved only if

$$p_g \gtrsim p = mv^2 + \tfrac{1}{3}\frac{m^3(5v^2-1)}{(1-v^2)^3} + \tfrac{4}{3}m^3, \tag{2.24}$$

where p_g is the ion-sound generated during the interaction.

Gibbons *et al.* [11] made a rough estimate for p_g using a continuity equation related to H, assuming the time during which the solitons interact to be the one it would take the two undisturbed packets to pass through each other. This leads to the following condition on the speed [3]:

$$F(v) \equiv \tfrac{4}{3}(1-v^2)^2 + \frac{5v^2-1}{3(1-v^2)} \lesssim 1. \tag{2.25}$$

The function $F(v)$ is 1 for $v = 0$, and is smaller than one up to $v \sim 0.6$, with a minimum value of 0.85 at $v \sim 0.45$. After $v \sim 0.6$ it grows up rapidly. This model is very successful since it shows that there is a critical speed, above which no fusions can occur; this result is independent of E_0.

The piece of information that has been included in an artificial way is the time of the interaction. This could be improved if the acceleration of each soliton could be computed together with the ion-sound generation, as in the case shown in the previous subsection. However, this has not been done due to the obstacles which appear when one leaves the adiabatic approximation.

§3 The integrable limits

Although the system of equations (1.1) and (1.2) is not completely integrable in general, there are certain limits in which they may become completely integrable. One of them is the Schrödinger equation with a cubic nonlinearity (1.3). Yajima and Oikawa [16] have found that a system for unidirectional sonic waves may be also solved by means of the inverse scattering method. In this case $n(x,t) = n(x-t)$ if they move to the right, and then

$$n_{tt} - n_{xx} = \left(\frac{\partial}{\partial t} - \frac{\partial}{\partial x}\right)\left(\frac{\partial n}{\partial t} + \frac{\partial n}{\partial x}\right) \approx -2(n_t + n_x)_x, \tag{3.1}$$

so instead of Eq. (1.2) we get

$$n_t + n_x = -\tfrac{1}{2}|E|^2, \tag{3.2}$$

after integrating in space. Yajima and Oikawa solved the system formed by (1.1) and (3.2). See [16].

The question of whether a given nonlinear evolution equation is exactly solvable can be answered up to a certain extent by the Estabrook–Wahlquist method. If it is, the method also helps to find the way to solve it [17]. If it is expected that the equation may be solved by the inverse scattering method, then the Estabrook–Wahlquist method may be formulated in a simplified way. In this Section we present, just for didactical purposes, how the method may be applied to Eq. (1.3) and to the Yajima-Oikawa system.

The first problem that appears when one wishes to use the inverse scattering method in order to solve a nonlinear evolution equation, is to find the linear transformation associated with it. In other words, given a nonlinear evolution equation

$$u_t = K(u), \tag{3.3}$$

where $K(u)$ is a nonlinear operator, then one wishes to find matrices P and Q that define a system, when applied on a vector V,

$$\begin{aligned} V_x &= PV, \\ V_t &= QV, \end{aligned} \tag{3.4}$$

such that the condition

$$(V_x)_t - (V_t)_x = (P_t - Q_x + [P,Q])V = 0, \tag{3.5}$$

where $[P,Q] \equiv PQ - QP$, is equivalent to Eq. (3.3).

One way to approach this problem is to propose an *ansatz* for P in terms of u, substitute it in (3.5), and, using (3.3), find a form for Q such that Eqs. (3.3) and (3.5) are equivalent. This will yield a set of commutation relations. If it is possible to find an algebra whose elements satisfy these relations, then one can proceed to solve the scattering problem. Otherwise, one can try a different *ansatz*. If it is not possible to find P and Q with the properties required, the system is not exactly solvable.

Take the case of Eq. (1.3) for example. Let us rename $E \mapsto q$ and $E^* \mapsto r$. Then, this equation is equivalent to the system

$$q_t = i(q_{xx} + q^2 r), \tag{3.6a}$$

$$r_t = -i(r_{xx} + qr^2). \tag{3.6b}$$

As an *ansatz*, we propose $P = X_1 + qX_2 + rX_3$, where the X_i are constant matrices. Then, from condition (3.5) and the system (3.6), we get

$$i(q_{xx} + q^2 r)X_2 - i(r_{xx} + qr^2)X_3 - Q_x + [P,Q] = 0. \tag{3.7}$$

Now, Q is chosen in this case in such a way that all derivatives are cancelled,

$$Q = iq_x X_2 - ir_x X_3 + iq[X_1, X_2] - ir[X_1, X_3] - iqr[X_2, X_3] + X_0. \tag{3.8}$$

Then, the following commutation relations arise if (3.7) is to be satisfied:

$$X_2 = [X_2, [X_2, X_3]], \tag{3.9a}$$

$$X_3 = -[X_3, [X_2, X_3]], \tag{3.9b}$$

$$[X_1, [X_2, X_3]] + [X_2, [X_1, X_3]] - [X_3, [X_1, X_2]] = 0, \tag{3.9c}$$

$$[X_2, [X_1, X_2]] = 0, \tag{3.9d}$$

$$[X_3, [X_1, X_3]] = 0, \tag{3.9e}$$

$$[X_2, X_0] = -i[X_1, [X_1, X_2]], \tag{3.9f}$$

$$[X_3, X_0] = i[X_1, [X_1, X_3]], \tag{3.9g}$$

$$[X_1, X_0] = 0. \tag{3.9h}$$

The purpose is to find an algebra that satisfyies Eqs. (3.9). Using the Jacobi identity, (3.9c) is transformed into

$$[X_1, [X_2, X_3]] = 0, \tag{3.9c'}$$

from where we may propose

$$[X_2, X_3] = \alpha X_1, \tag{3.10}$$

where α is a scalar. This allows us immediately to find, from (3.9a) and (3.9b),

$$[X_1, X_2] = -\alpha^{-1} X_2, \tag{3.11}$$

$$[X_1, X_3] = \alpha^{-1} X_3. \tag{3.12}$$

On the other hand, from (3.9h), let us propose $X_1 = \beta X_0$, with β a scalar. Comparing $[X_1, [X_1, X_2]]$ from (3.11) (or $[X_1, [X_1, X_2]]$ from (3.12)) with the same commutator from (3.9f) [(3.9g)], one finds $\beta = i\alpha$. Equations (3.10), (3.11), and (3.12) form an algebra we were searching for.

If we take the representation for *SL(2,R)*

$$Y_1 = \begin{pmatrix} 1 & 0 \\ 0 & -1 \end{pmatrix}, \qquad Y_2 = \begin{pmatrix} 0 & 1 \\ 0 & 0 \end{pmatrix}, \qquad Y_3 = \begin{pmatrix} 0 & 0 \\ 1 & 0 \end{pmatrix}, \tag{3.13}$$

we may write

$$X_1 = \alpha^{-1} Y_1, \qquad X_2 = -2\alpha Y_2, \qquad X_3 = -2\alpha Y_3. \tag{3.14}$$

For the Yajima–Oikawa system, the same procedure follows, although it is somewhat more complex. We now take the system

$$q_t = i(q_{xx} - nq), \tag{3.15a}$$

$$r_t = i(-r_{xx} + nr), \tag{3.15b}$$

$$n_t = -(n + qr)_x, \tag{3.15c}$$

where the factor 1/2 in (3.2) has been neglected. This will make no essential difference for our purpose. Here the *ansatz* $P = x_1 + qx_2 + rx_3 + nx_4$ yields

$$Q = iq_x X_2 - ir_x X_3 - (n + qr) X_4 + iq[X_1, X_2] - ir[X_1, X_3] - iqr[X_2, X_3] + X_0, \tag{3.16}$$

and the algebra that satisfies the commutator relations is shown in Table 1. Their representation for it is given (with some modifications) in Ref. [16].

For the case of Zakharov equations, something similar may be attempted. They are equivalent to the system

$$q_t = i(q_{xx} - nq), \tag{3.17a}$$

$$r_t = i(-r_{xx} + nr), \tag{3.17b}$$

$$n_t = v_x, \tag{3.17c}$$

$$v_t = (n + qr)_x, \tag{3.17d}$$

and the *ansatz* $P = X_1 + qX_2 + rX_3 + nX_4 + vX_5$ yields

$$Q = iq_x X_2 - ir_x X_3 + vX_4 + (n + qr) X_5 + iq[X_1, X_2] - ir[X_1, X_3] - iqr[X_2, X_3] + X_0. \tag{3.18}$$

	X_1	X_2	X_3	X_4	X_5	X_6	X_7	X_0
X_1	0	X_5	X_6	X_7	$\alpha X_2 + iX_5$	$\alpha X_3 - iX_6$	$\frac{2}{\alpha}(1-\alpha)X_1 + (4\alpha - 1)X_4 + \frac{2}{\alpha}X_0$	0
X_2	·	0	$\frac{1}{2}iX_4$	0	0	$-\frac{3}{4\alpha}X_1 + \frac{1}{4}X_4 + \frac{1}{2}iX_7 - \frac{3}{4\alpha}X_0$	X_2	$-\alpha iX_2 + X_5$
X_3	·	·	0	0	$-\frac{3}{4\alpha}X_1 + \frac{1}{4}X_4 - \frac{3}{4\alpha}X_0$	0	X_3	$\alpha iX_3 + X_6$
X_4	·	·	·	0	X_2	X_3	$2X_4$	X_7
X_5	·	·	·	·	0	$-\frac{i}{2\alpha}(\frac{1}{2} - \alpha)X_1 + \frac{1}{2}i\alpha X_4 - \frac{1}{4}X_7 - \frac{i}{4\alpha}X_0$	$iX_2 - X_5$	$\alpha X_2 + i(1-\alpha)X_5$
X_6	·	·	·	·	·	0	$-iX_3 - X_6$	$\alpha X_3 - i(1-\alpha)X_6$
X_7	·	·	·	·	·	·	0	$\frac{2}{\alpha}(1-\alpha)X_1 + (4\alpha - 1)X_4 + \frac{2}{\alpha}X_0$
X_0	·	·	·	·	·	·	·	0

Table 1. Algebra for the P, Q pair in the Yajima–Oikawa system.

However, it may be shown in this case, that there is no nontrivial algebra that satisfies the commutator relations that arise.

One might suppose that a different *ansatz* could probably be more lucky, however we know that the Zakharov equations have only a finite number of conserved quantities. Thus they are not completely integrable, and there exist no P, Q pair for them.

§4 On bilinear forms for the Zakharov equations

One method that is relatively simple and allows to find exact solutions for nonlinear differential equations is Hirota's technique [18,19]. In [10] it is used in order to find multisoliton solutions for the Zakharov equations, following closely the steps of [18] for the case of (1.3). The former result turns out to be incorrect however.

The procedure may be outlined in the following way: Let us take

$$E(x,t) = \frac{G(x,t)}{F(x,t)}, \tag{4.1}$$

where G is complex and F is real. Then the Zakharov equations may be rewritten as the following system of bilinear differential equations:

$$iD_t G \cdot F + D_x^2 G \cdot F = 0, \tag{4.2}$$
$$-D_t^2 F \cdot F + D_x^2 F \cdot F = G \cdot G^*, \tag{4.3}$$

where we have defined

$$D_h^n(k\cdot k') \equiv \left(\frac{\partial}{\partial h}-\frac{\partial}{\partial h'}\right)^n k(h)k'(h')\bigg|_{h=h'}, \tag{4.4}$$

with

$$n=-2[\ln F(x,t)]_{xx}. \tag{4.5}$$

It is important to note that the only difference between this case and the one of (1.3) is that the first term in (4.3) does not appear in the latter one.

Solutions for equations (4.2) and (4.3) may be attempted by expanding F and G in terms of a parameter ϵ:

$$F=1+\epsilon^2 f_2+\epsilon^4 f_4+\cdots, \tag{4.6}$$

$$G=\epsilon g_1+\epsilon^3 g_3+\cdots. \tag{4.7}$$

This yields the following hierarchy:

$$i\frac{\partial g_1}{\partial t}+\frac{\partial^2 g_1}{\partial x^2}=0, \tag{4.8a}$$

$$2\left(-\frac{\partial^2 f_2}{\partial t^2}+\frac{\partial^2 f_2}{\partial x^2}\right)=|g_1|^2, \tag{4.8b}$$

$$i\frac{\partial g_3}{\partial t}+\frac{\partial^2 g_3}{\partial x^2}=-\left[iD_t+D_x^2\right](g_1\cdot f_2), \tag{4.8c}$$

$$2\left(-\frac{\partial^2 f_4}{\partial t^2}+\frac{\partial^2 f_4}{\partial x^2}\right)=g_1g_3^*+g_3g_1^*-\left[-D_t^2+D_x^2\right](f_2\cdot f_2), \tag{4.8d}$$

and so on. Thus the idea is to solve for g_1 in (4.8a), then for f_2 in (4.8b), etc. If it were possible to terminate the series, one would get an exact solution.

Equation (4.8a) may be solved as a sum of exponentials

$$g_1=\sum_{k=1}^{N}\exp\left[\hat{\eta}_k(x,t)\right], \tag{4.9}$$

with $\hat{\eta}_k(x,t)\equiv P_kx-\Omega_kt-\eta_k^0$, where P_k and Ω_k are complex values that satisfy

$$-i\Omega_k+P_k^2=0, \tag{4.10}$$

and η_k^0 are arbitrary complex constants.

If we take $N=1$, then $f_2=a\exp(\hat{\eta}_1+\hat{\eta}_1^*)$, where $a=\{2[(P_1+P_1^*)^2-(\Omega_1+\Omega_1^*)^2]\}^{-1}$, and the right hand side of (4.8c) is

$$-\left[iD_t+D_x^2\right](g_1\cdot f_2)=-(i\Omega^*+P_k^{*2})a\exp(\hat{\eta}_1+\hat{\eta}_1^*)=0, \tag{4.11}$$

because of (4.10). Thus the series terminates, and it may be seen that if we take $P_1=\eta_1+i\xi_1$, where η_1 and ξ_1 are real, then

$$F=1+a\exp(\hat{\eta}_1+\hat{\eta}_1^*),\qquad G=\exp\hat{\eta}_1, \tag{4.12}$$

yield the one-soliton solution (1.4) and (1.5). The parameters η_1 and ξ_1 are proportional to the amplitude and the velocity respectively.

When taking $N = 2$, one would find a two-soliton solution if the series could be terminated in a similar way. Now,

$$f_2 = a(1,1^*)\exp(\hat{\eta}_1 + \hat{\eta}_1^*) + a(1,2^*)\exp(\hat{\eta}_1 + \hat{\eta}_2^*) + a(2,1^*)\exp(\hat{\eta}_2 + \hat{\eta}_1^*) + a(2,2^*)\exp(\hat{\eta}_2 + \hat{\eta}_2^*), \quad (4.13)$$

where

$$a(k,l^*) = \frac{1}{2[(P_k + P_l^*)^2 - (\Omega_k + \Omega_l^*)^2]}. \quad (4.14)$$

The right hand side of (4.8c) is

$$\begin{aligned} -[iD_t + D_x^2](g_1 \cdot f_2) = \Big\{ & a(1,1^*)\Big[-i(-\Omega_2 + \Omega_1 + \Omega_1^*) - (P_2 - P_1 - P_1^*)^2\Big] \\ & + a(2,1^*)\Big[-i(-\Omega_1 + \Omega_2 + \Omega_1^*) - (P_1 - P_2 - P_1^*)^2\Big]\Big\}\exp(\hat{\eta}_1 + \hat{\eta}_2 + \hat{\eta}_1^*) \\ + \Big\{ & a(2,2^*)\Big[-i(-\Omega_1 + \Omega_2 + \Omega_2^*) - (P_1 - P_2 - P_2^*)^2\Big] \\ & + a(1,2^*)\Big[-i(-\Omega_2 + \Omega_1 + \Omega_2^*) - (P_2 - P_1 - P_2^*)^2\Big]\Big\}\exp(\hat{\eta}_1 + \hat{\eta}_2 + \hat{\eta}_2^*). \end{aligned} \quad (4.15)$$

According to reference [10]

$$g_3 = a(1,2)\,a(1,1^*)\,a(2,1^*)\exp(\hat{\eta}_1 + \hat{\eta}_2 + \hat{\eta}_1^*) + a(1,2)\,a(1,2^*)\,a(2,2^*)\exp(\hat{\eta}_1 + \hat{\eta}_2 + \hat{\eta}_2^*), \quad (4.16)$$

but when substituting this in (4.8c), one finds that there is a difference between the right hand side and the left hand side which, using (4.10) may be written as

$$\Delta = \frac{(P_1 + p_1^*)(P_2 + P_1^*)(P_1 - P_2)\left[P_2(P_1^{*2} - P_1^2) + P_1(P_2^2 - P_1^{*2}) + P_1^*(P_2^2 - P_1^2)\right]}{[(P_1 + P_1^*)^2 + (P_1^2 - P_1^{*2})^2][(P_2 + P_1^*)^2 + (P_2^2 - P_1^{*2})^2]}. \quad (4.17)$$

In fact, if the procedure is followed properly, the series F and G do not terminate in this case. It will be a different matter for the case of (1.3), where the first terms in (4.8b), (4.8d), and so on, do not exist, and the term in square brackets in Δ is zero. In other words, Hirota's technique allows to find a one-soliton solution for the Zakharov equations in a rather simple way, but an N soliton solution cannot be found by induction, as it claimed in [10].

Asymptotic solutions when $|t| \to \infty$ might seem valid in that work, since the error enters only as a change of the position of each soliton, and does not introduce a major change in the form of the functions. However, when they come closer, it is found that the solutions of [10] do not conserve the quantities (1.6–8).

Conclusion

The Zakharov equations are not completely integrable, but contain in certain limits equations such as the nonlinear Schrödinger equation (1.3), and system (1.1), (3.2), that have been solved using the inverse scattering method. This makes them particularly interesting because their numerical solutions can be partially understood in terms of the properties of the integrable limits. A special feature is that, in the general case, when two solitary waves collide, energy and momentum can be transfered from electrostatic waves to ion-sound waves. The mechanism through which this is done is roughly understood, as it was seen in §2. However, we lack a detailed description of the bifurcation process in which two solitons merge. This could be achived if it were possible to compute more accurately the ion-sound that is generated.

Since this system of equations is not completely integrable, it will not be possible to find a P, Q pair. Still, we may surmise that if it were written as a perturbed nonlinear Schrödinger equation, such as it has been done by Gibbons [20], then the collision process may probably produce a motion of the poles that represent the solitons. If this were the case, it would be interesting to find their trajectories.

Acknowledgement

I wish to thank Dr. Antonmaría Minzoni, for useful discussions.

References

[1] V. E. Zakharov, Collapse of Langmuir waves. *Sov. Phys. JETP* **35**, 908–914 (1972).

[2] L. I. Rudakov and V. N. Tsytovich, Strong Langmuir turbulence. *Phys. Rep.* **40**, 1–73 (1978).

[3] S. G. Thornhill and D. ter Haar, Langmuir turbulence and modulational instability. *Phys. Rep.* **43**, 43–99 (1978).

[4] D. ter Haar, Physics of hot plasmas. *Physica Scripta* **T2/1**, 5–9 (1982).

[5] C. R. Ovenden, G. Statham, and D. ter Haar, Strong Turbulence of a Magnetized Plasma. I. The Generalized Zakharov Equations. University of Oxford, Department of Theoretical Physics (preprint 31/82).

[6] G. Statham and D. ter Haar, Strong Turbulence of a Magnetized Plasma. II. The Ponderomotive Force. University of Oxford, Department of Theoretical Physics (preprint 17/82).

[7] V. E. Zakharov and A. B. Shabat, Exact theory of two-dimensional self-focusing and one-dimensional self-modulation of waves in nonlinear media. *Sov. Phys. JETP* **34**, 62–69 (1972).

[8] S. V. Antipov, M. V. Nezlin, E. N. Snezhkin, and A. S. Trubnikov, Langmuir Solitons. *Sov. Phys. JETP* **47**, 506–516 (1978).

[9] S. V. Antipov, M. V. Nezlin, E. N. Snezhkin, and A. S. Trubnikov, Excitation of Langmuir solitons by monoenergetic electron beams. *Sov. Phys. JETP* **49**, 797–804 (1979).

[10] Y.-C. Ma, On the multi-soliton solutions of some nonlinear evolution equations. *Stud. App. Math.* **60**, 73–79 (1979).

[11] J. Gibbons, S. G. Thornhill, M. J. Wardrop, and D. ter Haar, On the theory of Langmuir solitons. *J. Plasma Phys.* **17**, 153–170 (1977).

[12] L. M. Degtyarev, V. G. Nakhan'kov, and L. I. Rudakov, Dynamics of the formation and interaction of Langmuir solitons and strong turbulence. *Sov. Phys. JETP* **40**, 264–268 (1975).

[13] G. L. Payne, D. R. Nicholson, and R. M. Downie, Numerical Solution of the Zakharov Equations. University of Iowa, Department of Physics and Astronomy, preprint 2/82 Rev.

[14] P. K. Kaw, N. L. Tsintsadze, and D. D. Tskhakaya, Radiation of ionacoustic waves by Langmuir soliton due to its acceleration. Proceedings of the 1982 International Conference on Plasma Physics, pp. 225, Göteborg, Sweden (1982). pp. 225.

[15] H. H. Chen and C. S. Liu, Nonlinear wave and soliton propagation in media with arbitrary inhomogeneities. *Phys. Fluids* **21**, 377–380 (1978).

[16] N. Yajima and M. Oikawa, Formation and interaction of sonic-Langmuir solitons. *Prog. Theor. Phys.* **56**, 1719–1739 (1976).

[17] D. J. Kaup, The Estabrook–Wahlquist method with examples of application. *Physica D* **1D**, 391–411 (1980), and references therein.

[18] R. Hirota, Exact envelope-soliton solutions of a nonlinear wave equation. *J. Math. Phys.* **14**, 805–809 (1973).

[19] R. Hirota, Direct Methods in Soliton Theory. In *Solitons*. R. K. Bullough and P. J. Caudrey eds., Topics in Current Physics # 17, Springer Verlag, 1980; pp. 157–176, and references therein.

[20] J. Gibbons, Behaviour of slow Langmuir solitons. *Phys. Lett.* **67A**, 22–24 (1978).

Infinity Manifolds on Energy Levels for Celestial Mechanics

Ernesto A. Lacomba

Departamento de Matemáticas
Universidad Autónoma Metropolitana, México

Abstract

We consider the n-body problem of celestial mechanics. Our goal will be to describe its energy surfaces with some added asymptotic boundaries, as well as to picture the kind of orbits arising in some simple examples. Among the asymptotic boundaries, we will mainly focus on the infinity manifolds [5], which describe escape orbits.

§1 Statement of the problem —examples

The n-body problem in $\Re^d$ (for $d = 1$, 2 or 3 dimensions), with particles of masses m_k located at $\mathbf{x}_k \in \Re^d$ for $k = 1, 2, \ldots, n$ is defined by the following system of second-order ordinary differential equations in R^d:

$$m_k \frac{d^2\mathbf{x}_k}{dt^2} = \sum_{\substack{j=1 \\ j \neq k}}^{n} G \frac{m_j m_k}{r_{jk}^2} \frac{\mathbf{x}_j - \mathbf{x}_k}{r_{jk}}, \qquad k = 1, 2, \ldots, n, \tag{1}$$

where $r_{jk} = |\mathbf{x}_j - \mathbf{x}_k|$. The right-hand sides can be written as $\dfrac{\partial U}{\partial \mathbf{x}_k}$, if we define the potential energy of the system by

$$U(\mathbf{x}_1, \ldots, \mathbf{x}_n) = \sum_{j<k} G \frac{m_j m_k}{r_{jk}}. \tag{2}$$

We remark that (1) has *no* equilibrium points, characterized as those where

$$\frac{\partial U}{\partial \mathbf{x}_k} = 0, \qquad \frac{d\mathbf{x}_k}{dt} = 0, \qquad k = 1, 2, \ldots, n.$$

The reason being that the forces acting on the system are internal and attracting. We assume the center of mass is fixed at the origin.

Calogero studied in his course [1] the one-dimensional case $d = 1$, with potentials U such that the system is integrable. However, in celestial mechanics (1) is *non*integrable for any $n > 2$ bodies. Only the case of $n = 2$ bodies renders an integrable system equivalent to a central force problem with the same law. The solutions are always plane conic curves in configuration space.

Notice that the Newtonian potential (2) is *homogeneous* of degree -1, an important fact for the sequel. We will comment at the end of this article about the case when U is a more general homogeneous function.

Regarding bad behaviour of the orbits of (1), we have two special features: appearance of *singularities* in finite time (from U), and *escape* of one or more particles. To escape motions we will associate exactly the infinity manifolds. (See § 2, below.) Singularities are divided in turn into *collisions* of some or all of the particles, or a possible *wild oscillatory motion* if $n > 4$ and $d > 1$. See [8] and [9].

Regarding collisions: if they involve two bodies, we can regularize the solution and prolongate it beyond the singularity by the methods of Levi-Civita or Sundman. If they involve more than two bodies, there exists in general no regularization.

Denoting $\mathbf{q} = (\mathbf{x}_1, \ldots, \mathbf{x}_n)$, $\mathbf{p} = (m_1\dot{\mathbf{x}}_1, \ldots, m_n\dot{\mathbf{x}}_n) \in (\Re^d)^n$, system (1) can be written as

$$\dot{\mathbf{q}} = \mathbf{A}^{-1}\mathbf{p}, \qquad \dot{\mathbf{p}} = \nabla U(\mathbf{q}), \tag{3}$$

where $\mathbf{A}$ is the block-diagonal matrix

$$\mathbf{A} = \begin{pmatrix} m_1\mathbf{1} & \mathbf{0} & \cdots & \mathbf{0} \\ \mathbf{0} & m_2\mathbf{1} & \cdots & \mathbf{0} \\ \vdots & \vdots & & \vdots \\ \mathbf{0} & \mathbf{0} & \cdots & m_n\mathbf{1} \end{pmatrix}.$$

This is a hamiltonian system

$$\dot{\mathbf{q}} = \frac{\partial H}{\partial \mathbf{p}}, \qquad \dot{\mathbf{p}} = -\frac{\partial H}{\partial \mathbf{q}}, \tag{4}$$

in terms of the total energy function

$$H(\mathbf{q},\mathbf{p}) = \tfrac{1}{2}\mathbf{p}^{\top}\mathbf{A}^{-1}\mathbf{p} - U(\mathbf{q}),$$

Then H is a first integral, *i.e.*,

$$\frac{dH\big(\mathbf{q}(t), \mathbf{p}(t)\big)}{dt} = 0,$$

for any solution $\gamma(t) = \big(\mathbf{q}(t), \mathbf{p}(t)\big)$, so $\gamma(t)$ is kept in a fixed *energy surface*

$$E_h = \{(\mathbf{q},\mathbf{p}) \mid H(\mathbf{q},\mathbf{p}) = h\}.$$

In the sequel we will often refer to the *energy equation* $H = h$, written in the following form:

$$\tfrac{1}{2}\mathbf{p}^{\top}\mathbf{A}^{-1}\mathbf{p} = U(\mathbf{q}) + h. \tag{5}$$

The projection of E_h to configuration $\mathbf{q}$-space is clearly the region $U + h \geq 0$. In general, E_h is a sphere $S^{dn-1} \subset \Re^{dn}$ bundle over $\{\mathbf{q} : U + h \geq 0\}$, pinched to points over the boundary (if $h < 0$) $\{\mathbf{q} : U + h = 0\}$.

Another first integral for the system is the angular momentum defined for $d \geq 2$ by the formula

$$J(\mathbf{q},\mathbf{p}) = \sum_i m_i \mathbf{x}_i \times \dot{\mathbf{x}}_i.$$

(For $d = 2$, this is a scalar.) A result by Sundman states that $J = 0$ if total collapse is to occur. In the three-body example we consider below, $J \equiv 0$ holds.

Examples:

1. Central force in one dimension. The system (1) is written as

$$\dot{x} = y, \qquad \dot{y} = -x^{-2},$$

with energy equation

$$\tfrac{1}{2}y^2 = x^{-1} + h. \tag{6}$$

See Figure 1, right.

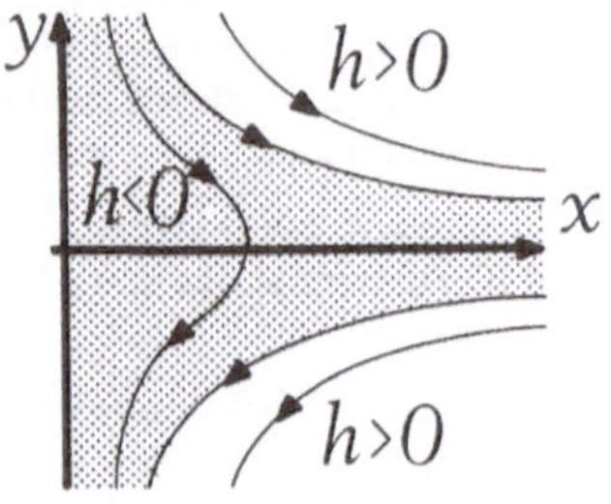

Figure 1.

2. Central force in two dimensions. Here $U(\mathbf{q}) = |\mathbf{q}|^{-1}$, and (1) reads as

$$\dot{\mathbf{q}} = \mathbf{p}, \qquad \dot{\mathbf{p}} = \frac{\partial U}{\partial \mathbf{q}} = -\frac{\mathbf{q}}{|\mathbf{q}|^3},$$

with energy equation

$$\tfrac{1}{2}|\mathbf{p}|^2 = |\mathbf{q}|^{-1} + h$$

Let $h < 0$, then the projection of E_h to the $\mathbf{q}$-plane is the punctured disk defined by $0 < |\mathbf{q}| \leq -h^{-1}$ (equality corresponds to the zero velocity circle). By considering each ray of the disk, we check that E_h is just an open solid torus [3]. If we further fix the angular momentum J,, we get a torus $S^1 \times S^1$ (as expected because of integrability). Collision orbits in E_h (in this case the only ones for $J = 0$) project over the rays in Figure 2, below. We show there neighboring orbits for $J > 0$ and $J < 0$.

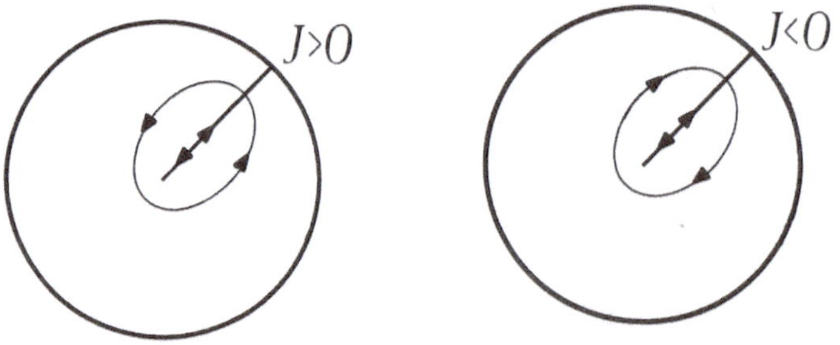

Figure 2.

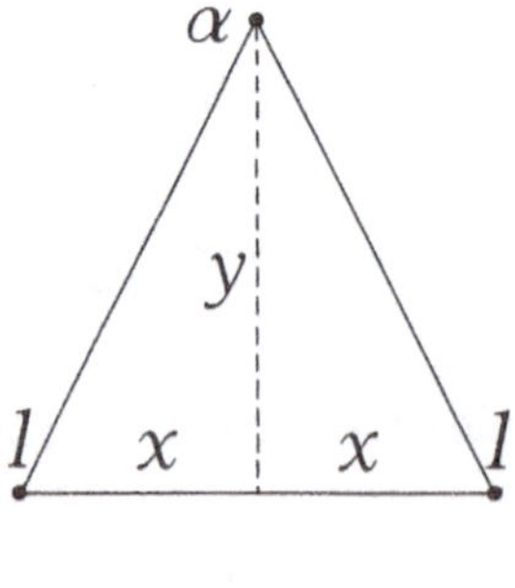

Figure 3.

3. The isosceles (planar) three-body problem ($J \equiv 0$). We are given a particle of mass α moving along an axis and two particles of unit mass symmetrically situated with respect to the axis (see Figure 3, left). The energy equation is

$$\tfrac{1}{2}\mathbf{p}^\top \mathbf{A}^{-1}\mathbf{p} = \frac{1}{2x} + \frac{2\alpha}{\sqrt{x^2 + y^2}} + h,$$

where

$$\mathbf{A} = \operatorname{diag}(2, \frac{2\alpha}{2+\alpha})$$

The simplest collision orbits are the so-called *homothetic* solutions: homotheties in time of the two possible equilateral triangles or the collinear configurations. They are presented in the configuration plane of Figure 4, below:

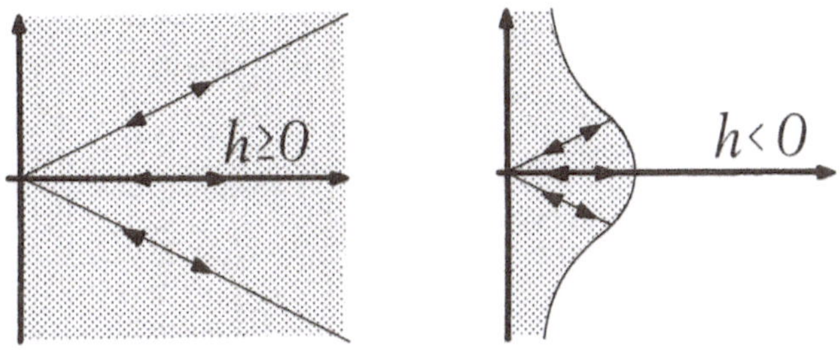

Figure 4.

Such orbits are regularizable but without a continuous dependence on initial conditions to nearby orbits [2]. This is clear from the possible behaviour of near-collision orbits in configuration space shown in Figure 5, below:

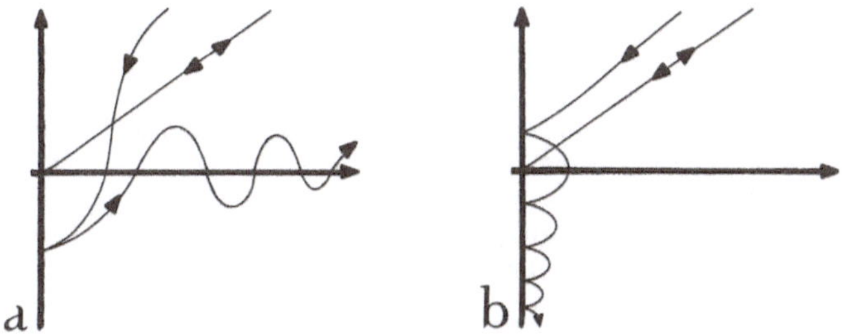

Figure 5.

§2 Treatments of singularity and escape

For simplicity, we use for illustration the otherwise trivial one-dimensional central force problem. In all cases it is enough to look at the transformed energy equation.

1. For regularization, due to Levi-Civita and Sundman, we write (6) as $x^2y^2 = 2x + 2hx^2$. If we let $u = xy$, it becomes

$$u^2 = 2x + 2hx^2, \tag{7}$$

with an additional time change

$$\frac{dt}{d\tau} = x$$

in the differential equations. Collision is described as an elastic bouncing. See Figure 6, right.

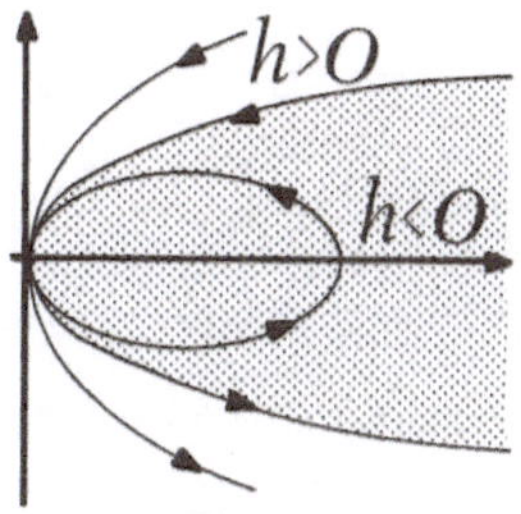

Figure 6.

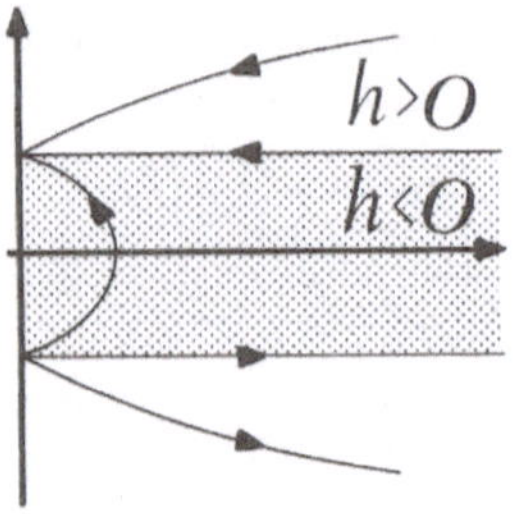

Figure 7.

2. Blow-up of origin in configuration space was treated by Mc Gehee in [6]. Writing (6) as $xy^2 = 2 + 2hx$ and letting $v = \sqrt{x}y$, we get

$$v^2 = 2 + 2hx, \tag{8}$$

with a time change

$$\frac{dt}{d\tau} = x^{3/2}.$$

See Figure 7, left.

3. For the escape behaviour we apply the blow-up at the infinity point in configuration space, as described by Lacomba and Simó in [5]. Setting $z = x^{-1}$ in (6) gives

$$y^2 = 2z + 2h \tag{9}$$

with a time change

$$\frac{dt}{d\tau} = \frac{1}{z}.$$

This transformation is good if $h > 0$.

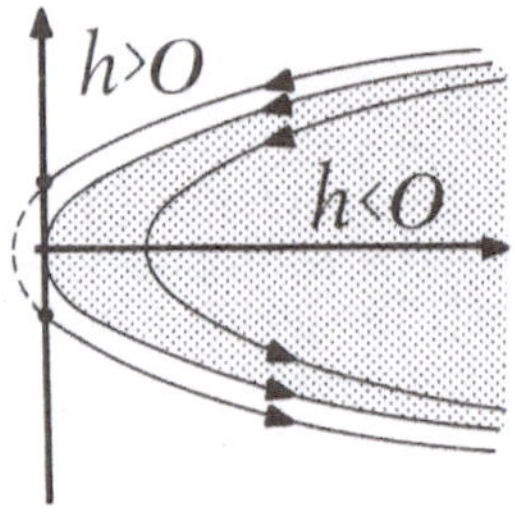

Figure 8.

Of the above methods, only regularization is very particular, since it works in general only for binary collisions. We now consider **2.** and **3.**.

A. Blow-up of total collision $\mathbf{q} = 0$ in (2), ([6,2]). Let $r = \sqrt{\mathbf{q}^\top \mathbf{A}\mathbf{q}}$ be the radius of inertia and $\mathbf{Q} = \mathbf{q}/r$, to get generalized polar coordinates. In (5) we get

$$\tfrac{1}{2}(\sqrt{r}\mathbf{p})^\top \mathbf{A}^{-1}(\sqrt{r}\mathbf{p}) = U(\mathbf{Q}) + rh,$$

suggesting the complementary momentum change $\mathbf{P} = \sqrt{r}\mathbf{p}$ to obtain the new energy equation

$$\tfrac{1}{2}\mathbf{P}^\top \mathbf{A}^{-1}\mathbf{P} = U(\mathbf{Q}) + rh. \tag{10}$$

After a time change $\dfrac{dt}{d\tau} = r^{3/2}$ in (3) we obtain

$$\begin{aligned} r' &= rv, \\ \mathbf{Q}' &= \mathbf{A}^{-1}\mathbf{P} - v\mathbf{Q}, \\ \mathbf{P}' &= \nabla U(\mathbf{Q}) + \tfrac{1}{2}v\mathbf{P}, \end{aligned} \tag{11}$$

where

$$v = \mathbf{P}\cdot\mathbf{Q}, \quad \mathbf{Q}^\top \mathbf{A}\mathbf{Q} = 1.$$

We remark that $r = 0$ has already sense in (11), defining a fictitious flow on the (fictitious) ***total collision manifold***:

$$C = \{(r, \mathbf{Q}, \mathbf{P}) : r = 0,\ \mathbf{Q}^\top \mathbf{A}\mathbf{Q} = 1,\ \mathbf{P}^\top \mathbf{A}^{-1}\mathbf{P} = 2U(\mathbf{Q})\}.$$

It appears as a boundary to any fixed E_h, which in the new coordinates is defined by

$$E_h = \{(r, \mathbf{Q}, \mathbf{P}) : r > 0,\ \mathbf{Q}^\top \mathbf{A}\mathbf{Q} = 1,\ \text{and (10) holds}\}.$$

The new system on $E_h \cup C$ does have equilibrium points (in C) defined by

$$r = 0, \qquad \nabla_{\mathbf{Q}}\, U = 0, \qquad v = \pm\sqrt{2U(\mathbf{Q})}.$$

The equilibrium points are usually (generically) hyperbolic with stable and unstable invariant submanifolds (see [2]), giving local information about orbits neighboring total collision.

B. Infinity manifolds [4,5] We have to consider here three cases, according to the sign of the energy. In any event, the idea is to blow up the infinity of configuration space with appropriate momentum transformations. We set $\rho = r^{-1}$, $\mathbf{Q} = \rho\mathbf{q}$.

B.1. If $h = 0$, let $\mathbf{P} = \rho^{-1/2}\mathbf{p}$. This amounts to a radial inversion in Mc Gehee's transformation, with energy equation

$$\mathbf{P}^\top \mathbf{A}^{-1}\mathbf{P} = 2U(\mathbf{Q}), \tag{12}$$

and a time change $\dfrac{dt}{d\tau} = \rho^{-3/2}$. In these new coordinates there appears also a (fictitious) manifold as a boundary to E_0, the ***infinity manifold***:

$$N_0 = \{(\rho, \mathbf{Q}, \mathbf{P}) : \rho = 0,\ \mathbf{Q}^\top \mathbf{A}\mathbf{Q} = 1,\ \text{and (12) holds}\},$$

carrying a fictitious flow, and equilibrium points defined as on C.

B.2. If $h > 0$, we let $\mathbf{P} = \mathbf{p}$ and $\dfrac{dt}{d\tau} = \rho^{-1} = r$. The energy equation, generalizing (9), becomes

$$\tfrac{1}{2}\mathbf{P}^\top \mathbf{A}^{-1}\mathbf{P} = \rho\, U(\mathbf{Q}) + h. \tag{13}$$

The added ***infinity manifold*** is now

$$N_h = \{(\rho, \mathbf{Q}, \mathbf{P}) : \rho = 0,\ \mathbf{P}^\top \mathbf{A}^{-1}\mathbf{P} = 2h\},$$

with the two submanifolds of equilibrium points

$$S_h^\pm = \{(\rho, \mathbf{Q}, \mathbf{P}) \in N_h : \mathbf{A}^{-1}\mathbf{P} = v\mathbf{Q},\ v = \pm\sqrt{2h}\}.$$

B.3. If $h < 0$, we have in general several components: none for $n = 2$, two for the isosceles three-body, and three for the general plane three-body systems.

§3 Interpretation of the examples

We now interpret the various boundary manifolds in the examples of § 1. In the one-dimensional central force problem we have the two cases shown in Figure 9, next page:

Figure 9.

In the isosceles problem with regularization of binary collisions, C is topologically $S^2 - 4$ points. The same is true for N_h if $h \geq 0$. However, N_h is two copies of $S^2 - 2$ points for $h < 0$. In Figure 10, below, $E_h \cup C \cup N_h$ is shown for the two cases. Notice the forbidden dark line segments and the homothetic solutions connecting equilibrium points:

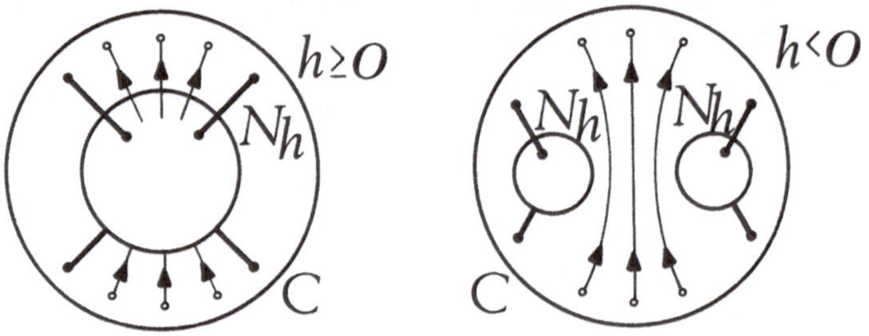

Figure 10.

For $h = 0$ one shows the existence of an orbit ejected from equilateral triple collision and going to collinear escape, thus connecting equilibrium points from C to N_0 without being homothetical (see [5]). This is shown in Figure 11a, below. For $h < 0$ we can even introduce some symbolic dynamics (see [2,7]), to show the existence of, say, ejection collision orbits as in Figure 11b, or periodic orbits as in Figure 11c:

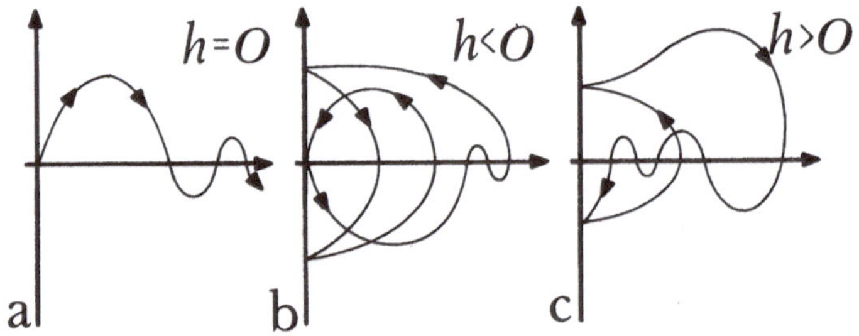

Figure 11.

§4 Comments

The extension of this material to the study of mechanical systems with more general homogeneous potentials U, with or without singularities, is possible. The case of negative degree is very much like in celestial mechanics, but when the degree is positive, the voriginal Mc Gehee transformation works at total collision only for $h = 0$, and at infinity for any h. If U takes positive and negative values, there is a restriction in the values of the normalized coordinate $\mathbf{Q}$.

References

[1] F. Calogero, Integrable dinamical systems and related mathematical results. *These Proceedings*.

[2] R. Devaney, Singularities in classical mechanical systems. In *Ergodic Theory and Dynamical Systems*. Vol. I. (A. Katok, ed.) pp. 211–333, Birkhauser, Basel, 1981.

[3] W. Kaplan, Topology of the two-body problem. *Amer. Math. Monthly* **49**, 316–323 (1942).

[4] E. Lacomba, Variétés de l'infini por une énergie non nulle en mécanique célèste. *C.R.A.S., Paris* **I295**, 503–506 (1982).

[5] E. Lacomba and C. Simó, Boundary manifolds for energy surfaces in celestial mechanics. *Celestial Mechanics* **28**, 37–48 (1982).

[6] R. Mc Gehee, Triple collision in the collinear three-body problem. *Inv. Math.* **27**, 191–227 (1974).

[7] R. Moeckel, Orbits of the three-body problem which pass infinitely close to triple collision. *Amer. J. Math.* **103**, 1323–1341 (1981).

[8] D. Saari, Singularities and collisions of Newtonian graviational systems. *Arch. Rational Mech. Anal.* **49**, 311–320 (1973).

[9] H. Sperling, On the real singularities of the n-body problem. *J. Reine Angew. Math.* **145**, 14–50 (1970).

Order in the Chaotic Region

Eduardo Piña[†]

Departamento de Física
Universidad Autónoma Metropolitana
Iztapalapa D. F., Mexico

Abstract:

A simple algorithm is constructed for the ***quadratic*** one-parameter family of maps. It predicts the number of periodic orbits of arbitrary period ocurring for parameter values lower than any other corresponding to a given stable periodic orbit. This algorithm produces the number of periodic unstable points coexisting with the stable periodic orbit. This method associates an important polynomial in a one-to-one correspondence with the symbol of Metropolis ***et al.*** for ordering the periodic orbits, with the permutation matrix of the dynamics of points in the stable cycle, and with the matrix of regions separated by these points.

These polynomials coincide with those polynomials used by many authors in connection with the triangular map and with the topological entropy.

§1 Introduction

Many different physical systems have been recently observed to have a periodic behavior represented by a simple mapping of a nonlinear iteration function. These physical systems include hydrodynamics [1], chemical reactions [2], electronic circuits [3,4], optical devices [5], and acoustic phenomena [6], among others. The theoretical analysis of these physical systems presents mathematical difficulties because of intrinsic non-linearities. At the same time, the reported experiments follow a strange parallelism with very simple mathematical nonlinear models studied some years ago with the aid of computers [7–9].

When iterating a one-parameter family of functions for a fixed value of the parameter and a large number of iterations, one obtains stable points which repeat their value every n iterations. By changing the value of the fixed parameter, the same n-period —or any other different period— may be obtained. The order of the periods occurring is reproduced with different functions [8], and in various different experiments [2,3]. The order of the points in each cycle is also reproduced in the experiment with the Belusov–Žhabotinskiĭ reaction [2].

The purpose of this paper is to study the mathematical properties of this order discovered by Metropolis *et al.* [8], through extending and unifying the work of several authors. Our main interest

---

[†]Also at the Escuela Superior de Física y Matemáticas, Instituto Politécnico Nacional, Zacatenco, D.F., México.

centers on the mathematical conditions for a mapping to reproduce theoretically the experimental findings, including the order of the periods, and the order of the points in a period.

The subharmonic bifurcation [7–9], following any stable cycle when the value of the parameter is increased, will be included as a particular case of this study. This bifurcation has been found in all the experiments referred to above [1–6]. The Feigenbaum scaling properties [9] will not be considered in this paper.

§2 The Metropolis–Stein-Stein Order

In this section we present the order of periodic cycles of points in the so-called chaotic region of a quadratic one-parameter family of iterated maps. We denote by $\psi_\mu(x)$ the iteration function corresponding to a value μ of the parameter, while $\psi_\mu{}^{(n)}(x)$ denotes the nth iterate of $\psi_\mu(x)$.

A real solution $x(\mu)$ of the equation

$$x = \psi_\mu^{(n)}(x)\,, \tag{2.1}$$

is an n-periodic point of a cycle with prime period n, provided x is not a solution for a smaller value of n. Such a point x is called a ***fixed point*** of ψ_μ for $n = 1$. The cycle is formed by the n iterates obtained from a particular solution $x(\mu)$.

The solution to (2.1) is either an element of a stable cycle or a stable fixed point ($n = 1$) whenever the derivative of $\psi_\mu{}^{(n)}(x)$ at $x(\mu)$ has an absolute value less than one:

$$\left|\frac{d}{dx}\psi_\mu{}^{(n)}(x)\right| < 1. \tag{2.2}$$

The iteration function $\psi_\mu(x)$ has the following properties:

i. $\psi_\mu(x) = 0$ for $x = 0$ and $x = 1$.

ii. $\psi_\mu(x)$ has a unique maximum in the interval (0,1).

iii. There exists not more than one stable fixed point, or one stable periodic cycle, for each value of the parameter μ.

iv. The first derivative with respect to x of the nth iterate $\psi_\mu{}^{(n)}(x)$, at an n-periodic point, is a monotonic function of the parameter μ. It is monotonically decreasing when the first derivative or slope is smaller than 1, and monotonically increasing for values of the slope greater than 1.

A very rich bifurcation structure has been discovered for these maps, which is discussed by many authors [7–11]. First we shall describe some features of this structure [7,8,11].

In the one-parameter family of maps we found arbitrary stable periods for appropriate values of μ. This period appears for a particular μ-value giving a slope equal to 1. As μ grows, the cycle remains stable until this derivative decreases to -1 at the μ-bifurcation value μ_1, then it remains unstable for all the μ values larger than μ_1. At this μ there appears a stable cycle with double period $2n$, which also bifurcates for a larger value of μ (say μ_2), and the $2n$ solution (2.1) become again unstable when the decreasing slope of the $\psi_\mu{}^{(2n)}(x)$ function attains the value -1. This bifurcation process repeats itself, thus generating a sequence $\{\mu_k\}$ of bifurcation values which accumulate at one limit value of the parameter μ.

The first bifurcation cascade begins at a value μ when the non-zero fixed point becomes unstable. After this first cascade is accumulated at one characteristic μ, pairs of cycles of *any* period k, one stable and one unstable, appear by tangent bifurcation. The unstable cycle remains unstable. The stable one bifurcates as discussed above and starts to be unstable when it bifurcates.

For a fixed μ, the stable cycle is the limit of a large number of iterations, for most of the initial-value points of the interval which form the basin of the stable cycle. A set of points of zero measure is iterated into the unstable cycles. These unstable cycles are detected during small time intervals through precise computation and they coexist with the stable one.

The order on μ of the occurring period k is a universal property of these maps disvovered by Metropolis, Stein, and Stein in [8] (hereafter referred as MSS) for stable cycles in terms of symbols formed with the letters R and L, according to the position of the points of the cycle with respect to the critical point characterized by the property of having zero slope, and assuming the value of the parameter is selected in such a way that this maximum belongs to the cycle.

Instead of using MSS symbols for one cycle, we introduce an equivalent polynomial formed according to the rules:

i. The order of the polynomial is $n-1$, where n is the period of the cycle.

ii. The n coefficients of this polynomial are 1 or -1.

iii. Beginning with the largest power with a coefficient equal to one, we associate a change of sign with an R in the MSS symbols, and no change of sign with an L.

For example, to the symbol RLR there corresponds the polynomial $x^3 - x^2 - x + 1$.

These polynomials play an important role in the theory. In particular, the MSS order may be described directly in terms of these polynomials as follows.

1. The MSS symbols RL^{n-2} are ordered on μ by n. These cycles are the maximum cycles corresponding to the largest value of μ for a given n, and have as representing polynomials

$$x^{n-1} - x^{n-2} - x^{n-3} - \cdots - x - 1 = \frac{x^n - 2x^{n-1} + 1}{x - 1}. \tag{2.3}$$

2. Given an admisible polynomial $P(x)$ representing a cycle, the harmonic of this cycle is the product of the previous admissible polynomial $P(x)$, times a cyclic polynomial, *i.e*

$$P(x)\,(x^k - 1)\,, \tag{2.4}$$

where k is the period of $P(x)$.

The coefficients of the harmonic polynomial repeat their sign reversed. For example, the harmonic of the polynomial $x^2 - x - 1$ is

$$(x^2 - x - 1)(x^3 - 1) = x^5 - x^4 - x^3 - x^2 + x + 1. \tag{2.5}$$

From this follows a cascade of period-doubling harmonics represented by the family of polynomials

$$P(x) \prod_{m=0}^{\ell} (x^{2^m k} - 1), \qquad \ell = 0, 1, 2, \ldots\,. \tag{2.6}$$

Any harmonic cycle is stable in a contiguous region to the stable region of the original cycle.

3. For each admisible cycle of period k represented by the polynomial $P(x)$, we define the ***antiharmonic*** polynomial through

$$P(x)\,(x^k+1)\,, \tag{2.7}$$

and obtain in a similar way the cascade of antiharmonics

$$P(x)\prod_{m=0}^{\ell}(x^{2^m k}+1), \qquad \ell=0,1,2,\ldots. \tag{2.8}$$

We note that the coefficient of the antiharmonic polynomials are repeated in the same order. For example, if x^3-x^2-x-1 is the polynomial corresponding to a cycle of period four, then the antiharmonic polynomial is $x^7-x^6-x^5-x^4+x^3-x^2-x-1$. These antiharmonics are merely auxiliar polynomials and never represent real periods for the quadratic-like mappings.

4. New admisible polynomials are now constructed by recurrence. Starting from any two contiguous admissible polynomials ordered on μ, we construct the harmonic series of the first and the antiharmonic series of the second. A new admissible polynomial is constructed between them with the maximal common coefficients in both the harmonics and antiharmonics, starting with the largest power of each polynomial and increasing the antiharmonic family until there is a difference with the new polynomial.

For example, given the polynomials x^2-x-1 and x^3-x^2-x-1, the harmonic and antiharmonic polynomials are, respectively, $x^5-x^4-x^3-x^2+x+1$, and $x^7-x^6-x^5-x^4+x^3-x^2-x-1$; the polynomial ordered on μ lying between x^2-x-1 and x^3-x^2-x-1 according to the order given by μ, is the polynomial $x^4-x^3-x^2-x+1$, having the same first five coefficients as the harmonic, and the antiharmonic.

This process may be continued step by step, until all the allowed cycles represented by polynomials are thus obtained.

These polynomials are related to the polynomials previously introduced by several authors; for example, the maximal polynomials are also found by Guckenheimer [11]. In particular, our polynomials coincide with the polynomials introduced by Derrida, Gervois, and Pomeau in [10] (hereafter referred to as DGP). These authors defined the polynomials in connection with the triangular map in a slightly different way. They write

$$P(x)=x^{n-1}-\sum_{m=0}^{n-2}\alpha_m x^{n-2+m}\,, \tag{2.9}$$

where

$$\alpha_0=1, \qquad \alpha_j=\prod_{m=1}^{j}\beta_m\,, \tag{2.10}$$

and

$$\beta_m=\begin{cases}1, & \text{when the } (m+1)^{\text{th}} \text{ term in the MSS sequence is } \mathsf{L},\\ -1, & \text{when it is } \mathsf{R}.\end{cases} \tag{2.11}$$

Notwithstanding the differing definitions, the polynomials obtained are the same, and many of the properties derived by DGP for these polynomials are used in this paper. These polynomials will be very useful in the next section, where they are the generators of the periodic unstable points coexisting with the corresponding stable cycle.

§3 The periodic points

For a given μ or a particular map, Sharkovskiĭ [12] found an order relation between the occurring periods of the iteration function. The periods are ordered in the sequence

$$3,\ 5,\ 7,\ \ldots, 2\cdot 3, 2\cdot 5, 2\cdot 7, \ldots, 2^2\cdot 3, 2^2\cdot 5, 2^2\cdot 7, \ldots, 2^3, 2^2, 2, 1. \tag{3.1}$$

The existence of any period implies the coexistence with any of the periods placed to the right of that period in this ordered sequence. Sharkovskiĭ's theorem predicts the order in which these periods appear on μ for the first time, in stable form with its unstable partner. When at a given μ a period appears for the first time no period to its left in the Sharkovskiĭ sequence coexists with it in unstable form, since no one has yet appeared on μ. The Sharkovskiĭ theorem may be verified from the well-know MSS table for stable cycles up to period eleven. However, the same table shows a structure more complex than the Sharkovskiĭ order, because many of the stable periods occur several times and, after the minimal cycle appears, the Sharkovskĭ order is no longer respected; the MSS prescription for ordering the periods becomes then most important.

A more general result follows from the paper of Smale and Williams [13], which is relatively unknown. The same idea was rediscovered shortly afterwards by Guckenheimer *et al.* [14], and independently by Stefan [10]. It is now well known.

Smale and Williams consider the case of stable period three, and explain an algorithm which generates all the unstable periodic points coexisting with it. The Smale-Williams formalism is given by the following outline. First, they construct a non-negative matrix $\mathbf{M}$, with entries 0 and 1, generating the dynamics of the four regions separated by the points of the three-cycle, as shown in the following figure:

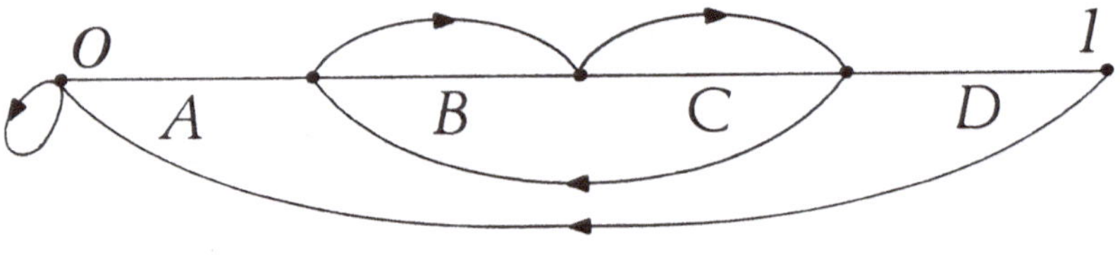

i.e.,

$$(A\,B\,C\,D)\begin{pmatrix}1&0&0&1\\1&0&1&0\\0&1&1&0\\0&0&0&0\end{pmatrix}=(A\cup B,\ C,\ B\cup C,\ A). \tag{3.2}$$

Then they use the theory of dynamic systems, based on an old paper by Fatou, to deduce the following results:

1. The number N_j of periodic points asociated with unstable periods of order j is given by

$$N_j = \operatorname{trace}\mathbf{M}^j\ , \tag{3.3}$$

where $\mathbf{M}$ is the Smale–Williams matrix of regions in (3.2), and N_j includes the periodic points of periods dividing j.

2. The above equation may be replaced by the equivalent expression

$$N_j = \sum_{m=1}^{n+1} (\lambda_m)^j \,, \tag{3.4}$$

where λ_m are the eigenvalues of the Smale–Williams matrix $\mathbf{M}$. Zero and one are two eigenvalues corresponding to wandering points in D and the source 0, respectively. The equation may be therefore simplified to

$$N_j = 1 + \operatorname{trace} \tilde{\mathbf{M}}^j = 1 + \sum_{m=1}^{n-1} (\tilde{\lambda}_m)^j \,, \tag{3.5}$$

where $\tilde{\mathbf{M}}$ is the Smale–Williams matrix which excludes both the non-wandering unstable points and the zero source at the left end point of the interval, *i.e.*

$$\tilde{\mathbf{M}} = \begin{pmatrix} 0 & 1 \\ 1 & 1 \end{pmatrix} . \tag{3.6}$$

The eigenvalues of $\tilde{\mathbf{M}}$ are

$$\tilde{\lambda}_{1,2} = \tfrac{1}{2}(1 \pm \sqrt{5}) \,, \tag{3.7}$$

and they express the solution is given in terms of the Fibonacci-Lucas numbers L_j

$$N_j - 1 = \left(\frac{1+\sqrt{5}}{2}\right)^j + \left(\frac{1-\sqrt{5}}{2}\right)^j = L_j \,. \tag{3.8}$$

3. Smale and Williams note also that the generating function of the numbers N_j's is the zeta function ς related to the characteristic polynomial of the matrix $\mathbf{M}$,

$$\varsigma(t) = \exp\left(\sum_{m=0}^{\infty} \frac{N_m}{m} t^m\right) = \frac{1}{(1-t)(1-t-t^2)} \,. \tag{3.9}$$

The Smale–Williams treatment of the stable three-period can be of course repeated for any other cycle. Some of these ideas were rediscovered by Guckenheimer *et al* [14] and Stefan [10].

Our next step is to generalize this theory, to complete and connect it with the use of our polynomials. The starting point is the study of the periodic cycles ordered on μ by our polynomials. For, each cycle can be represented by non-negative permutation matrix formed by 0's and 1's. This is the matrix representing the cycle; it respects both the order R, L of the MSS symbol and the convexity of the iteration function, as explained below.

It is possible to associate with the convexity of the quadratic-like iteration function a corresponding concavity of the cyclic matrix which has a U-shape of the 1's in the matrix. This matrix has a 1 on the last row dividing it in so many colums to the left as letters L and so many columns to the right as letters R in the MSS symbol. The 1's are placed to the right and to the left of the lowest 1 in a monotonic fashion, forming always a U-shape comming from the convex character of the iteration function.

We present a second example. Corresponding to the symbol $\mathsf{RL^2R}$, there is a 5-period cycle represented by the permutation matrix

$$\begin{pmatrix} 0 & 0 & 0 & 0 & 1 \\ 1 & 0 & 0 & 0 & 0 \\ 0 & 0 & 0 & 1 & 0 \\ 0 & 1 & 0 & 0 & 0 \\ 0 & 0 & 1 & 0 & 0 \end{pmatrix}, \tag{3.10}$$

having two columns to the right of the lowest 1 and two columns to the left. The matrix (3.10) cyclically transforms the points from the order (1 2 4 3 5) to the order (2 3 5 4 1). This is the order of the five points of the cycle in the interval (0,1). This is the pattern reproduced in the dynamics of a very complicated chemical reaction for different cycles [2].

Notice the U-shape of this example is a one-to-one correspondence between this matrix and the symbol $\mathsf{RL^2R}$. Given the symbol, we associate the first R to 1, the two succesive L's to the numbers 2 and 3, the last R to 4, and 5 to the center. Shift the numbers cyclically. Demand the U-shape monotonicity, and you will find the cyclic matrix as shown. This construction can be repeated for any MSS symbol.

The Smale–Williams matrix for regions corresponding to any cycle follows from the matrix of cyclic points. This matrix is formed by column blocks of 1's separated by verteces in positions occupied by the non-zero elements of the cyclic matrix. This property comes from continuity as the boundary of the regions are the cyclic points. The dynamics of the regions is determined by the dynamics of the boundaries.

The generalized Smale–Williams matrix has two non-zero elements in each row, for every cycle, except for the last row of zero values –as in the three-cycle example. It has also the U-shape. The matrix is again separated into columns to the right and to the left of two columns determined by the lower non-zero portion matrix. These columns correspond to the number of R 's and L's in the MSS symbol.

For example, the matrix corresponding to the previously illustrated $\mathsf{RL^2R}$ period is

$$\begin{pmatrix} 1 & & 0 & & 0 & & 0 & & 0 & & 1 \\ & & & & & & & & & \bullet & \\ 1 & & 0 & & 0 & & 0 & & 1 & & 0 \\ & \bullet & & & & & & & & & \\ 0 & & 1 & & 0 & & 0 & & 1 & & 0 \\ & & & & & & & \bullet & & & \\ 0 & & 1 & & 0 & & 1 & & 0 & & 0 \\ & & & \bullet & & & & & & & \\ 0 & & 0 & & 1 & & 1 & & 0 & & 0 \\ & & & & & \bullet & & & & & \\ 0 & & 0 & & 0 & & 0 & & 0 & & 0 \end{pmatrix}. \tag{3.11}$$

In this matrix the verteces separating columns of 1's are shown by dots • in the same place occuped by 1's in the matrix (3.10).

For any period, the Smale–Williams matrix has two eigenvalues, 0 and 1, associated with the extreme regions of the interval and with the first and last rows and columns. Sustracting this ***frame***, the remaining submatrix corresponds to the Stefan's matrices referred to by DGP in [10].

The periodic points N_j of period j coexisting with any stable period can be obtained from the Smale–Williams equations, which remain valid for any period, as traces of the jth power of the Smale–Williams matrix [see Eq. (3.3)]. Also, in terms of the sum over eigenvalues of the matrix to the power j^{th} power, as in Eq. (3.4). Thirdly, they can be given in terms of the characteristic polynomial of the matrix by means of the ς function

$$\varsigma(t) = \exp\left(\sum_{m=0}^{\infty} \frac{N_m}{m} t^m \right) = \frac{1}{(1-t)\, P(1/t)\, t^n} \,. \tag{3.12}$$

Here $P(x)$ is the characteristic polynomial of the Stefan matrix (the Smale–Williams matrix without frame), and n is the period of the stable cycle represented by this matrix. An important fact is that $P(x)$, the characteristic polynomial of the Stefan matrix, is also our polynomial previously used to order the stable periods.

In this context and when referring to Stefan's unpublished work on these matrices, DGP note that these characteristic polynomials are the same polynomials (2.9) which they found in connection with their piecewise linear mapping.

Through direct computation, the characteristic polynomial of the Stefan matrix of a given region may be verified in many cases to be simply related to the MSS symbol. We will prove in the next section that, on the other hand, the characteristic polynomials of the matrix of regions are ordered on μ by the values of the coefficients of these polynomials, showing that these polynomials determine the order of periodc points in the interval and the order on μ of each period.

From the practical point of view, the connections between the MSS symbol and the characteristic polynomial, and between this polynomial and the coexisting periodic unstable points [via equation (3.12) or any equivalent method], provides a powerful algorithm to obtain the periodic points N_j from the MSS symbol. A version of this algorithm will be presented in the next section.

§4 The method of periodic points and the topological entropy

In this section we present a new way of calculating the number of periodic points N_j through a recurrence relation determined by the characteristic polynomial of the matrix of regions.

Instead of using any of the proposed Smale–Williams expressions for the numbers N_j, it is advantageous to use Newton's equations [15] for the traces of the powers of the matrix of terms in the coefficients of its characteristic polynomial. For the Stefan matrix, the coefficients of the characteristic polynomial are determined by the MSS symbol; the trace of the jth power of the matrix is the number $N_j - 1$ which we denote by L_j, as in the three-period case considered by Smale and Williams.

If

$$P(x) = x^{k-1} + \sum_{m=1}^{k-1} \alpha_m x^{k-m-1} \tag{4.1}$$

is the equation of the characteristic polynomial, with coefficients α_j , then Newton's equations for the first $L_j = N_j - 1$ numbers ($j = 1, 2, \ldots$) are given in terms of the α_j's in the form [15]

$$\begin{aligned} L_1 &= -\alpha_1 \,, \\ L_2 &= -\alpha_1 L_1 - 2\alpha_2 \,, \\ L_3 &= -\alpha_1 L_2 - \alpha_2 L_1 - 3\alpha_3 \,, \\ L_4 &= -\alpha_1 L_3 - \alpha_2 L_2 - \alpha_3 L_1 - 4\alpha_4 \,, \\ &\vdots \\ L_{n-1} &= -\alpha_1 L_{n-2} - \alpha_2 L_{n-3} - \cdots - (n-1)\alpha_{n-1} \,. \end{aligned} \tag{4.2}$$

For values of j greater than $n-1$, the numbers L_j obey the recurrence relation [15]

$$L_q = -\sum_{j=1}^{k-1} \alpha_j L_{q-j} \,. \tag{4.3}$$

These numbers N_j have the important property of being, for fixed j, monotonic increasing, discontinuous functions of the parameter μ. They are ordered by μ because, after they appear in stable or unstable form —and eventually the stable becomes unstable— they never disappear as μ increases, but remain unstable instead. The monotonicity of the N_j with the parameter μ is determined by the coefficients of the characteristic polynomial. The L_j are the same numbers for two polynomials having the same coefficients α_k for $k \leq j$.

According to Newton's equations, when the first α_k is different for two polynomials or is zero for one of them, the N_k will be different; N_k is largest if α_k is negative, smallest when α_k is positive, and intermediate if there is no α_k because of the polynomial ending at the previous non-zero coefficient. This agrees with the MSS ordering procedure. The intermediate polynomial between a harmonic and an antiharmonic polynomial is ordered by the fact the harmonic has a $+1$ coefficient, and the antiharmonic a -1 coefficient in the first differing coefficient. The order for any two polynomials is determined in this way by the first differing coefficient. The increasing order on μ and on any N_j is then 1, 0, -1 for that differing coefficient.

The Smale–Williams region matrix is a non-negative matrix. A classical result by Gerschgorin [16], predicts that the maximal root is real, and that it lies between by 1 and 2 (the sum of elements in a column or in a row). The maximal root is unique and real. This maximal root may have a larger absolute value than any other root when the ***graph*** of the matrix is irreducible. The graph of a matrix, whose elements are either 0 or 1 [16], is formed by drawing an oriented line joining the verteces from p to q whenever the matrix entry a_{pq} is non-zero (1 in this case). This matrix has the same number of verteces, rows, and columns. The graph is irreducible when any pair of verteces can be joined by an oriented cycle in the graph.

The full Smale–Williams matrix has always a reducible graph. Supressing the border, the Stefan matrix may be irreducible or not. If it is irreducible, the largest root dominates the other roots in the calculation (3.4) for N_j and one has, for large j,

$$N_j \doteq (\lambda_{\max})^j \,. \tag{4.4}$$

The largest root $\lambda_{\max}$ is also a monotonically increasing function of the parameter μ. This last fact was discovered by DGP in a different way.

As an example, consider the maximal polynomial (2.3) used to construct the MSS order:

$$P(x) = \frac{x^n - 2x^{n-1} + 1}{x-1} \,. \tag{4.5}$$

For large n, this polinonial has its largest root very close to the value 2. This is also the maximal value allowed by the Perron-Frobenius theorem [16]. The extreme value 2 for the maximal root corresponds to the largest value of the μ parameter.

In this example all the coefficients α_j have a value equal to -1 while the numbers L_j, obtained by induction from (4.2) and (4.3), become

$$L_j = 2^j - 1\,, \qquad j \leq n \,. \tag{4.6}$$

This produces $N_k = 2^k$, the largest number of real roots in eq. (2.1) for the quadratic map, and confirms the maximal properties of this family of polynomials.

A factorization of a particular set of MSS symbols was found by DGP which may be represented by a factorization of our polynomials. Given two admisible polynomials, $P(x)$ of period n and $Q(x)$ of any period, the polynomial

$$P(x)\,Q(x^n) \tag{4.7}$$

represents also a MSS symbols, and corresponds to a periodic cycle.

The graph of a factorizable polynomial is reducible, and for that case it may have multiple roots with maximal magnitude, as in the case of $(x-1)(x^4-x^2-1)$, representing the cycle RLR^3. The DGP product (4.7) defines a mapping of all polynomial $Q(x)$ into a subset of them: $P(x)Q(x^n)$. This map, defined by the first factor $P(x)$, transforms any polynomial into a factorizable polynomial. DGP observed that the largest root of a factorizable polynomial is equal to the largest root of the first factor, provided it is not 1. It follows that the family of polynomials defined by this mapping is present in contiguous order on μ determined by this largest root. It defines an infinite family of cyclic periods called the ***window*** of the first factor. In particular, this window begins with a harmonic cascade of polynomials which is a particular portion of the complete window. The graph of the Stefan's matrix of the first factor is irreducible. The order of polynomials in the family is the order of polynomials $Q(x)$ on μ.

The window of the polynomial $P(x) = x-1$ is specially interesting. The factor polynomials are

$$(x-1)\,Q(x^2)\,, \tag{4.8}$$

and the maximal root, instead of beeing the largest root of the first factor, is the square root of the largest root of $Q(x)$. To analyze how the set of polynomials is mapped by this factor $(x-1)$, we studied the family of maximal polynomials (4.5), because they are ordered on μ. We obtained the factorizable polynomials

$$(x-1)\frac{x^{2n}-2x^{2n-2}+1}{x^2-1} = \frac{x^{2n}-2x^{2n-2}+1}{x+1}\,. \tag{4.9}$$

These polynomials are ordered on μ by order of increasing n. For large n, this polynomial has a root very near to the value $\sqrt{2}$:

$$\lambda_{\max} \doteq \sqrt{2}\left(1-\frac{1}{(n+1)2^{n+1}}\right). \tag{4.10}$$

A second example of the polynomials is the family of minimal odd periods considered by Stefan [12] and other authors [17,18]:

$$P(x) = \frac{x^n-2x^{n-2}-1}{x+1}\,, \tag{4.11}$$

They are ordered on μ by decreasing order of n. For a large n, the maximal root of this polynomial has a value near $\sqrt{2}$:

$$\lambda_{\max} \doteq \sqrt{2}\left(1+\frac{1}{2^{1+n/2}}\right). \tag{4.12}$$

The two families of polynomials, (4.9) and (4.11), have, furthermore, the same limit value for the largest root. They are in a contiguous region on μ, and the limit μ value corresponds to the Ruelle ergodic point [19], the Hoppenstead and Hyman m_2^* [20], the Grossman and Thomae $\tilde{a}_1$ [21], and the Chang and Wright transition parameter [22].

For the minimal polynomials (4.11) it is possible to obtain the value of their numbers L_j by induction:

$$L_{2k+1} = 1, \qquad L_{2k} = 2^{k+1} - 1, \qquad k < j\,. \tag{4.13}$$

This result represents a proof of Sharkovskiĭ's theorem [12] for odd periods. This is implicit in the first equation of (4.13). The rest of the proof of this theorem for even periods may be given by considering the factorization maps (4.7) of the polynomial $x - 1$ and its harmonics as first factor, and the minimal polynomials (4.11) as second factor.

Other families of interesting polynomials could be discovered and ordered by studying the MSS table, but this analysis will not be pursued any further in the present paper.

To end this section, we will mention the topological entropy or larger Lyapunov number, providing a numerical for the sensitive dependence of iterations on initial conditions [23].

The topological entropy is bounded by the inequality [18]

$$\text{Topological entropy} > \lim_{j\to\infty} \frac{\ln L_j}{j} = \ln \lambda_{\max} \tag{4.14}$$

In published work [18,24], some families of our polynomials were used differently in order to bound topological entropy as in (4.14). It is important to note that the MSS symbol provides the characteristic polynomial in the way presented here, thus providing the bound (4.14) on the topological entropy through its largest root.

§5 The number of stable orbits before a stable cycle

In this section we analyze the numbers $N_j = L_j + 1$ of periodic points of period j. The objetive is to find out how many orbits of an arbitrary period are present in stable form for lower values of the parameter μ corresponding to these numbers N_j. These data will be compared with the MSS table for periods with $j \leq 11$.

The number N_j of periodic points associated with period j includes all the points in periodic orbits of periods dividing j. If M_j is the number of cycles of period j, excluding the divisors of j, there holds

$$N_k = \sum_{j\backslash k} j M_j\,, \tag{5.1}$$

where the symbol $j\backslash k$ represents all the divisors of the number k.

The summation (5.1) may be inverted by using the Möbius function [25]

$$\mu(j) = \begin{cases} 1, & \text{if } j = 1, \\ (-1)^k, & \text{if } j \text{ is the product of } k \text{ different primes,} \\ 0, & \text{if } j \text{ is divisible by one square larger than one.} \end{cases}$$

One has [25]

$$M_n = \frac{1}{n} \sum_{j\backslash n} \mu(n/j) N_j\,, \tag{5.2}$$

and this number of cycles includes all the cycles of prime period n.

Not all of these cycles are formed by points which were stable for a lower value of the parameter μ. This number includes all the cycles formed by unstable points.

The Number M_j of cycles of period j is formed by the cycles which appear on μ as harmonics of another cycle of half period, and the number of cycles which appear on μ as a pair of stable and unstable cycles of period j

Calling E_n the number of harmonic cycles of period n, and F_n the number of cycles which were stable for the same period, we obtain the splitting of the total number of cycles of period n,

$$M_n = 2F_n - E_n \,. \tag{5.3}$$

This number will be reduced by 1 when the period n is the period of the coexisting stable cycle. This cycle is not included in M_n until it becomes unstable.

All the stable orbits become unstable by bifucation, and the number of harmonic cycles of period $2n$ is equal to the number of stable orbits of half-period n,

$$F_n = E_{2n} \,. \tag{5.4}$$

The total number of non-harmonic cycles of period j is given by

$$F_j - E_j = \tfrac{1}{2}(M_j - E_j) \,. \tag{5.5}$$

For $j = 2n$ the total number of cycles which were stable is

$$F_{2n} = \tfrac{1}{2}(M_{2n} + F_n) \,, \tag{5.6}$$

where we use property (5.4). For $j = 2n + 1$ the total number of cycles is

$$F_{2n+1} = \tfrac{1}{2} M_{2n+1} \,. \tag{5.7}$$

The total number of stable cycles H_k associated with period k (including cycles whose period divides k) was found by the Gilbert–Riordan method [26] to be

$$H_k = \sum_{d\backslash k} F_d = \frac{1}{4k} \sum_{d\backslash k} \varphi(d)[3 + (-1)^d] N_{k/d} \,, \tag{5.8}$$

where $\varphi(d)$ is the Euler Totient function [25], equal to the number of integers not exceeding and relatively prime to d,

$$\varphi(n) = \sum_{d\backslash n} \mu(n/d) d \,. \tag{5.9}$$

The numbers F_d are then expressed in terms of H_k through [25]

$$F_n = \sum_{d\backslash n} H_d \mu(n/d) \,. \tag{5.10}$$

We have equations to calculate these numbers in terms of periodic numbers N_j.

All points become unstable for the largest value of the parameter and then N_k attains its largest value 2^k. In this extreme case all these numbers have been tabulated by several authors [7,8,26] for $k \leq 12$. We calculated them for values of small periods, finding complete agreement with the computed MSS table. Generalization of these ideas for the cubic map are in progress and show a broad field of validity [27].

Aknowledgements

The research reported in this paper was started while the author was a visitor at the Center of Statistical Mechanics, University of Texas at Austin. It is a pleasure to thank Profs. I. Prigogine and L. Reichl for their kind hospitality. I also want to express my appreciation to Prof. W. Schieve for suggesting this subject of research and for many interesting conversations. I am grateful to Profs. M. Feigenbaum, R. Balescu, and T. de la Selva for some remarks, and to D. B. Hernández for useful bibliography.

References

[1] A. Libchaber and J. Maurer, Local probe in a Rayleigh-Benard experiment in liquid helium. *J. Phys. (Paris) Lett.* **39** L, 369–372; Yu N. Belyaev, A. A. Monakhov, S. A. Sherbakov, and I. M. Yavorskaya, Onset of turbulance in rotating fluids. *JETP Lett.* **29**, 295–298 (1979); J. P. Gollub, S. V. Benson, and J. Steinman, A subharmonic route to turbulent convection. *Ann. N.Y. Acad. Sci.* **357**, 22–27 (1981); M. Giglio, S. Musazzi, and U. Perini, Transition to chaotic behavior via a reproducible sequence of period-doubling bifurcations, *Phys. Rev. Lett.* **47**, 243–246 (1981).

[2] R. H. Simoyi, A. Wolf and H. L. Swinney, One-dimensional dynamics in a multicomponent chemical reaction, *Phys. Rev. Lett.* **49**, 245–248 (1982).

[3] J. Testa, J. Perez, and C. Jeffries, Evidences for universal chaotic behavior of a driven nonlinear oscillator, *Phys. Rev. Lett.* **48**, 714–717 (1982).

[4] P. S. Linsay, Period doubling and chaotic behavior in a driven anharmonic oscillator, *Phys. Rev. Lett.* **47**, 1349–1352 (1981); F. T. Arecchi and F. Lisi, Hopping mechanism generating $1/f$ noise in nonlinear systems. *Phys. Rev. Lett.* **49**, 94–98 (1982); R. W. Rollins and E. R. Hunt, Exactly solvable model of a physical system exhibiting universal chaotic behavior. *Phys. Rev. Lett.* **49**, 1295–1298 (1982).

[5] H. M. Gibbs, F. A. Hopf, D. L. Kaplan, and R. L. Shoemaker, Observation of chaos in optical bistability. *Phys. Rev. Lett.* **46**, 474–477 (1981); F. T. Arecci, R. Meucci, G. Puccioni, and J. Tradicce, Experimental evidence of subharmonic bifurcations, multistability and turbulence in a Q-switched gas laser. *Phys. Rev. Lett.* **49**, 1217–1220 (1982).

[6] R. Keolian, L. A. Turkevich, S. J. Putterman, I. Rudnick, and J. A. Rudnick, Subharmonic sequences in the Faraday experiment: departures from period doubling. *Phys. Rev. Lett.* **47**, 1133–1136 (1981); W. Lauterborn and E. Cramer, Subharmonic route to chaos observed in acoustics. *Phys. Rev. Lett.* **47**, 1445–1448 (1981); C. W. Smith, M. J. Tejwani and D. A. Farris, Bifurcation universality for first-sound subharmonic generation in superfluid helium 4. *Phys. Rev. Lett.* **48**, 492–494 (1982).

[7] R. M. May, Simple mathematical models with very complicated dynamics. *Nature* **261**, 459–467 (1976); P. J. Myrberg, Iteration von Quadratwurzeloperationen, Iteration der Reelen Polynome Zweiten Grades. *Ann. Akad. Sci. Fennicæ* **A,** I Nos. 259, (1958), 336/3, (1963); E. N. Lorenz, The problem of deducing the climate from the governing equations. *Tellus* **16**, 1–11 (1964); P. Collet and J. P. Eckman, *Iterated maps on the interval as dynamical systems.* Birkhäuser, Basel, 1980.

[8] N. Metropolis, M. L. Stein, and P. R. Stein, On finite limit sets for transformations on the unit interval. *J. Combinatorial Theory* **15A**, 25–44 (1973).

[9] M. J. Feigenbaum, Quantitative universality for a class of nonlinear transformations. *J. of Stat. Phys.* **19**, 25–52 (1978); *ib.*, The universal metric properties of nonlinaer transformations. *J. of Stat. Phys.* **21**, 669–706 (1979); *ib.*, The transition to aperiodic behavior in turbulent systems. *Commun. Mat. Phys.* **77**, 65–68 (1980).

[10] B. Derrida, A. Gervois, and Y. Pomeau, Iteration of endomorphisms on the real axis and representation of numbers. *Ann. Inst. Henri Poincaré* **A29**, 305–356 (1978).

[11] J. Guckenheimer, On the bifurcation of maps of the interval. *Invent. Math.* **39**, 165–178 (1977).

[12] O. Stefan, A theorem of Sarkovskiĭ on the existence of periodic orbits of continuous endomorphisms of the real line. *Commun. Math. Phys.* **54**, 237–248 (1977).

[13] S. Smale and R. Williams, The quantitative analysis of a difference equation of population growth. *J. Math. Biol.* **3**, 1–4 (1976).

[14] J. Guckenheimer, G. Oster, and A. Ipaktchi, The dynamics of density dependent population models. *J. Math. Biol.* **4**, 101–147 (1977).

[15] C. Jordan, *Calculus of Finite Differences*, Chelsea Publishing Co., 1979. p. 593.

[16] R. S. Varga, *Matrix Iterative Analysis*. Prentice-Hall, 1962.

[17] J. Guckenheimer, Sensitive dependence to initial conditions for one dimensional maps. *Commun. Math. Phys.* **70**, 133-160 (1979).

[18] L. Block, J. Guckenheimer, M. Misiurewicz, and L. S. Young, *Periodic Points and Topological Entropy of One Dimensional Maps*. Lecture Notes in Mathematics, **#819**, Spriger Verlag, 1980. pp. 18–34.

[19] D. Ruelle, Applications consevant une measure absolument continue por rapport a *dx* sur [0,1]. *Commun. Math. Phys.* **55**, 47 (1977).

[20] F. C. Hoppenstead and J. M. Hyman, Periodic solutions of a logistic difference equation. *SIAM J. Appl. Math.* **32**, 73–81 (1977).

[21] S. Grossman and S. Thomae, Invariant distributions and stationary correlation functions of one-dimensional discrete processes. *Z. Naturforsch.* **32a**, 1353–1363 (1977).

[22] S. J. Chang and J. Wright, Transitions and distribution functions for chaotic systems. *Phys. Rev.* **A23**, 1419–1433 (1981).

[23] R. Shaw, Stange attractors, chaotic behavior, and information flow. *Z. Naturforsch.* **36a**, 80–112 (1981).

[24] J. Dias de Deus, R. Dilão, and J. Taborda Duarte, Topological entropy and approaches to chaos in dynamics of the interval. *Phys. Lett.* **90A**, 1–4 (1982).

[25] K. Goldberg, M. Newman, and E. Haynsworth, in *Handbook of Mathematical Functions*. (M. Abramowitz and I. A. Stegun eds.) Dover Publ. 1965. p. 826.

[26] E. N. Gilbert and J. Riordan, Symmetry types of periodic sequences. *Illinois J. Math.* **5**, 657–665 (1961).

[27] O. Chavoya, F. Angulo, and E. Piña. (To be published.)

Painlevé Ping-Pong P^3–P^5

Ziemowit Popowicz

Institute of Theoretical Physics
University of Wrocław, Poland

Abstract

In this lecture I should like to discuss the concept of the Ping-Pong which appeared in the Painlevé equations. First let me explain that the name "Painlevé Ping-Pong", in this context, is not mine but belongs to Prof. F. J. Bureau [1]. In the first part of my talk I shall give the explicit construction of this Ping-Pong for the third and fifth Painlevé equations. The Ping-Pong, fairly sketchily, is a special kind of the transformation and exists also for the other Painlevé equations [1,2]. It can be considered as the Bäcklund transformation for the third and fifth Painlevé equations [3]. In the second part I shall describe the possible physical applications of this Ping-Pong.

§1 Construction of the Ping-Pong

Let us consider the special case of the third Painlevé equation

$$\omega'' = \frac{1}{\omega}(\omega')^2 - \frac{1}{r}\omega' + \frac{1}{r}(\alpha\omega^2 + \beta) - \frac{1}{\omega} + \omega^3, \tag{1}$$

where α and β are arbitrary constants. We denote this equation by $P^3_{(\alpha,\beta,-1,1)}$. It can be written as the system of the first order differential equations [4,5]

$$r\frac{d\omega}{dr} = (\alpha\epsilon - 1)\omega + r\vartheta + \epsilon r\omega^2, \tag{2}$$

$$r\omega\frac{d\vartheta}{dr} = \beta\omega - r + (\alpha\epsilon - 2)\omega\vartheta + r\vartheta^2, \tag{3}$$

where $\epsilon^2 = 1$. If we eliminate ϑ from (2) we obtain (1), while eliminating ω from (3) we obtain ξ defined by

$$\vartheta = -\frac{\xi + 1}{\xi - 1}, \tag{4}$$

which satisfies

$$\xi'' = \frac{3\xi - 1}{2\xi(\xi-1)}\xi'^2 - \frac{1}{\tau}\xi' + \frac{\xi - 1}{\tau^2}\left[a\xi + \frac{b}{\xi}\right] - \frac{\epsilon}{\tau}\xi, \tag{5}$$

where

$$r = \sqrt{2\tau}, \qquad a = \tfrac{1}{32}(\beta - \alpha\epsilon + 2)^2, \qquad b = -\tfrac{1}{32}(\beta + \alpha\epsilon - 2)^2. \tag{6}$$

Equation (5) is a special case of the fifth Painlevé equation; we denote it by $P^5_{(a,b,-\epsilon,0)}$. Using (2) one can easily find that

$$\vartheta = \frac{d\omega}{dr} - \epsilon\omega^2 - \frac{\alpha\epsilon - 1}{r}\omega. \tag{7}$$

In this manner we have started playing Ping-Pong. Equation (7) defines the transformation between $P^3_{(\alpha,\beta,-1,1)}$ and $P^5_{(a,b,-\epsilon,0)}$. Let us denote this transformation by

$$\vartheta = P^5_{(a,b,-\epsilon,0)}\omega. \tag{8}$$

On the other hand, a given solution of the fifth Painlevé equation generates four different solutions of the Painlevé equation of the third kind with different values of the parameters. This conclusion follows immediately from (6) if we describe α and β in terms of a and b. In other words, if ξ is the solution of (5), then

$$\omega_1 = -\frac{r(1-\vartheta^2)}{r\,d\vartheta/dr - \beta' - (\alpha'\epsilon - 2)\vartheta} \tag{9}$$

is the solution of $P^3_{(\alpha',\beta',-1,1)}$, ϑ is defined by (4), and

$$\alpha' = -2\epsilon(\eta\sqrt{2a} - \sigma\sqrt{-2b} - 1), \tag{10}$$
$$\beta' = 2(\eta\sqrt{2a} + \sigma\sqrt{-2b}), \tag{11}$$
$$\sigma^2 = \eta^2 = 1. \tag{12}$$

In this manner we define the transformation between $P^5_{(a,b,-\epsilon,0)}$ and $P^3_{(\alpha,\beta,-1,1)}$, which we denote by

$$\omega = P^3_{(\alpha,\beta,-1,1)}\vartheta. \tag{13}$$

Let us consider now the superposition of these two transformations which we denote by

$$\omega_1 = P^3_{(\alpha',\beta',-1,1)}P^5_{(a,b,-\epsilon,0)}\omega. \tag{14}$$

To obtain the explicit form of ω_1 as a funtion of ω, we introduce (7) into (9). We then obtain

$$\omega_1 = \frac{r\mathcal{R}(\mathcal{R}-2)}{r\,d\mathcal{R}/dr - (\alpha'\epsilon - 2)\mathcal{R} - (\beta' - \alpha'\epsilon + 2)}, \tag{15}$$

where

$$\mathcal{R} = \frac{d\omega}{dr} - \epsilon\omega^2 - \frac{1}{r}(\alpha\epsilon - 1)\omega + 1. \tag{16}$$

This formula allows us to generate the eight solutions of the Painlevé equation of the third kind starting from the third Painlevé transcendent. Two of them are equal to ω, and can be found by choosing

$$\epsilon^2 = 1, \qquad \eta = \sigma = 1. \tag{17}$$

The other six solutions are the solutions of the third Painlevé equation with different values of the parameters. In order to construct the transformation which can generate the new solutions from the same Painlevé equation, we introduce one discrete transformation to our Ping-Pong. First, let us notice that if ω is a solution of $P^3_{(\alpha,\beta,-1,1)}$, then $1/\omega$ is a solution of $P^3_{(-\beta,-\alpha,-1,1)}$. Let us denote this discrete transformation by $\mathcal{T}$, *e.g.* $\mathcal{T}_\omega = 1/\omega$ or symbolically $\mathcal{T} P^3_{(\alpha,\beta,-1,1)} = P^3_{(-\beta,-\alpha,-1,1)}$. We now combine this transformation $\mathcal{T}$ with the previous ones into the following transformation:

$$\mathcal{T} P^3_{(\alpha',\beta',-1,1)} P^5_{(a,b,-\epsilon,0)} \omega = f_{(-\beta',-\alpha',-1,1)}, \tag{18}$$

where $f_{(-\beta',-\alpha',-1,1)}$ is a solution of $P^3_{(-\beta',-\alpha',-1,1)}$ such that $\alpha' \neq \alpha$, and $\beta' \neq \beta$, ω is a solution of $P^3_{(\alpha,\beta,-1,1)}$, and a, b, α', β' are defined by (6) and (10–12). Moreover we now assume that $f_{(-\beta',-\alpha',-1,1)}$ is also a solution of $P^3_{(\alpha',\beta',-1,1)}$, and

$$P^3_{(\alpha',\beta',-1,1)} P^5_{(a,b,-\epsilon,0)} \omega \neq 1. \tag{19}$$

Such assumptions give us:

for $\epsilon = 1$,

$$\alpha + \beta = 0, \qquad \alpha' + \alpha = 2, \qquad \alpha' = -\beta', \tag{20}$$

or

$$\alpha + \beta = 4, \qquad \alpha' + \alpha = -2, \qquad \alpha' = -\beta'; \tag{21}$$

and for $\epsilon = -1$

$$\alpha + \beta = 0, \qquad \alpha' + \alpha = -2, \qquad \alpha' = -\beta', \tag{22}$$

or

$$\alpha + \beta = -4, \qquad \alpha' + \alpha = -3, \qquad \alpha' = -\beta'. \tag{23}$$

In this way we restrict our consideration to the three classes of the third kind Painlevé equations $P^3_{(\alpha,-\alpha,-1,1)}$ and $P^3_{(\alpha,\beta,-1,1)}$, where $\alpha + \beta = \pm 4$, and α is an arbitrary parameter.

From the previous considerations we may deduce that

$$P^3_{(\alpha,\beta,-1,1)} P^5_{(a,b,-\epsilon,0)} f_{(-\beta',-\alpha',-1,1)} = \omega_2 \tag{24}$$

can generate a solution of $P^3_{(\alpha,\beta,-1,1)}$. This is our Ping-Pong, see Figure 1 below.

$$\begin{array}{ccccccccccc} P^3_{(\alpha,\beta,-1,1)} & & & & & & & & P^3_{(-\beta',-\alpha',-1,1)} & & \\ \uparrow & & & & & & & & \uparrow & & \\ P^3_{(\alpha,\beta,-1,1)} \longrightarrow & P^5_{(a,b,-\epsilon,0)} & \longrightarrow & P^3_{(\alpha',\beta',-1,1)} & \xrightarrow{\mathcal{T}} & P^3_{(-\beta',-\alpha',-1,1)} & \longrightarrow & P^5_{(a,b,-\epsilon,0)} & \longrightarrow & P^3_{(\alpha,\beta,-1,1)} \end{array}$$

$$-\beta' = \alpha' \qquad \alpha + \beta = 0 \quad \text{or} \quad \alpha + \beta = \pm 4, \qquad \epsilon = \pm 1$$

Figure 1. Painlevé Ping-Pong P^3–P^3

To be more precise, let us define this Ping-Pong for $P^3_{(\alpha,-\alpha,-1,1)}$; it can be symbolically denoted by

$$P^3_{(\alpha,-\alpha,-1,1)} P^5_{(a,b,-\epsilon,0)} \mathcal{T} P^3_{(\alpha',\beta',-1,1)} P^5_{(a,b,-\epsilon,0)} \omega = \omega, \tag{25}$$

where

$$a = \tfrac{1}{8}(1-\alpha)^2, \quad b = -\tfrac{1}{8}, \quad \epsilon = 1, \quad \text{or} \quad a = \tfrac{1}{8}, \quad b = -\tfrac{1}{2}(\alpha+1)^2, \quad \epsilon = -1, \tag{26}$$

and $\alpha' + \beta' = \pm 1$ for $\epsilon = \pm 1$, α being an arbitrary parameter, ω a (seed) solution of $P^3_{(\alpha,-\alpha,-1,1)}$, and ω_2 the new solution of the same equation. To be more precise we prove the following

Theorem. *The transformation (25) generates a new solution of the $P^3_{(\alpha,-\alpha,-1,1)}$ equation such that $\omega_2 \neq \omega$ and $\omega_2 \neq 1/\omega$.*

Proof. The transformation (25) can be written as a system of first-order differential equations. Indeed, using (2–3) and the definition of $\mathcal{T}$ we obtain, for $\epsilon = 1$,

$$r\frac{d\omega}{dr} = (\alpha-1)\omega + r\vartheta + r\omega^2, \tag{27}$$

$$r\omega\frac{d\vartheta}{dr} = -\alpha\omega - r + (\alpha-2)\omega\vartheta + r\vartheta^2, \tag{28}$$

$$r\frac{d\omega_1}{dr} = (1-\alpha)\omega_1 + r\vartheta + r{\omega_1}^2, \tag{29}$$

$$r\omega_1\frac{d\vartheta}{dr} = (2-\alpha)\omega_1 - r - \alpha\omega_1\vartheta + r\vartheta^2, \tag{30}$$

$$-r\frac{d\omega_1}{dr} = (1-\alpha)\omega_1 + r\vartheta{\omega_1}^2 + r, \tag{31}$$

$$r\omega_1\frac{d\vartheta_1}{dr} = (2-\alpha)\omega_1 - r{\omega_1}^2 - \alpha\omega_1\vartheta_1 + r\vartheta^2{\omega_1}^2, \tag{32}$$

$$r\frac{d\omega_2}{dr} = (\alpha-1)\omega_2 + r\vartheta_1 + r{\omega_2}^2, \tag{33}$$

$$r\omega_2\frac{d\vartheta_1}{dr} = -\alpha\omega_2 - r + (\alpha-2)\omega_2\vartheta_1 + r{\vartheta_1}^2. \tag{34}$$

Let us suppose that our theorem is not true, and that $\omega_2 = \omega$ or $\omega_2 = 1/\omega$. In the first case, subtracting (27) from (33), we obtain $\vartheta = \vartheta_1$. Next, subtracting (30) from (31), we obtain $\vartheta^2 = 1$. Introducing this into (27), we obtain

$$r\frac{d\omega}{dr} = (\alpha-1)\omega \pm r + r\omega^2. \tag{35}$$

Equation (35) contradicts the assumption that ω is a solution of $P^3_{(\alpha,-\alpha,-1,1)}$. In the second case, if we compute ϑ_1 as a funtion of ω through (27–31), and compare it with ϑ_1 computed from (33), we obtain a contradiction. These considerations prove our theorem.

Analogously, one can verify that a similar theorem holds for $P^3_{(\alpha,\beta,-1,1)}$ where $\alpha + \beta = \pm 4$.

In this manner, transformations (27–34) are Bäcklund transformations. These Bäcklund transformations allow us to generate a whole hierarchy of new solutions of $P^3_{(\alpha,-\alpha,-1,1)}$ and $P^3_{(\alpha,\beta,-1,1)}$ where $\alpha + \beta = \pm 4$. As an example, let us write the recurrence formula for the $P^3_{(\alpha,-\alpha,-1,1)}$ equation. Using

(27–34) we obtain

$$\omega_2 = -\frac{r(1-\vartheta_1{}^2)}{r\,d\vartheta_1/dr + \alpha - (\alpha-2)\vartheta_1}, \tag{36}$$

where

$$\vartheta_1 = -\frac{r\,d\omega_1/dr + (1-\alpha)\omega_1 + r}{r\omega_1{}^2}, \tag{37}$$

$$\omega_1 = -\frac{r(1-\vartheta^2)}{r\,d\vartheta/dr - (2-\alpha) + \alpha\vartheta}, \tag{38}$$

$$\vartheta = \frac{d\omega}{dr} + (1-\alpha)\frac{\omega}{r} - \omega^2. \tag{39}$$

The knowledge of the Ping-Pong for $P^3_{(\alpha,-\alpha,-1,1)}$, or for $P^3_{(\alpha,\beta,-1,1)}$, where $\alpha+\beta = \pm 4$, allows us immediately to construct the Ping-Pong for the special cases of the fifth Painlevé equation, see Figure 2 below.

$$P^5_{(a,b,-\epsilon,0)} \longrightarrow P^3_{(\alpha',\beta',-1,1)} \xrightarrow{\ \mathcal{T}\ } P^3_{(-\beta',-\alpha',-1,1)} \longrightarrow P^5_{(a,b,-\epsilon,0)}$$

$$\alpha' = -\beta'$$

$$\begin{array}{llcll} b=\frac{1}{8}, & a=\frac{1}{8}(5-\alpha)^2 & \text{or} & a=\frac{1}{8}(1-\alpha)^2 & \epsilon=1 \\ a=\frac{1}{8}, & b=-\frac{1}{8}(\alpha+3)^2 & \text{or} & b=-\frac{1}{8}(\alpha+1)^2 & \epsilon=-1 \end{array}$$

Figure 2. Painlevé Ping-Pong P^5–P^5.

Indeed, if we apply the transformation $P^5_{(a,b,-\epsilon,0)}$ to (18), we obtain a solution of the $P^5_{(a,b,-\epsilon,0)}$ equation. Therfore let us denote these transformations by

$$P^5_{(a,b,-\epsilon,0)}\,\mathcal{T}\,P^3_{(\alpha',\beta',-1,1)}\mathcal{K} = \vartheta, \tag{40}$$

where:

for $\epsilon = 1$,

$$a = \tfrac{1}{8}(1-\alpha)^2, \qquad b = -\tfrac{1}{8}, \tag{41}$$

$$\alpha' = -\alpha + 2, \qquad \beta' = \alpha - 2, \tag{42}$$

or

$$a = \tfrac{1}{8}(5-\alpha)^2, \qquad b = -\tfrac{1}{8}, \tag{43}$$

$$\alpha' = -\beta' = \alpha - 2; \tag{44}$$

and for $\epsilon = -1$,

$$a = \tfrac{1}{8}, \qquad b = -\tfrac{1}{8}(\alpha + 3)^2, \tag{45}$$

$$\alpha' = -\beta' = (\alpha + 1), \tag{46}$$

or

$$a = \tfrac{1}{8}, \qquad b = -\tfrac{1}{8}(\alpha + 1)^2, \tag{47}$$

$$\alpha' = -(\alpha + 2) = -\beta'; \tag{48}$$

and

$$K = -\tfrac{\xi+1}{\xi-1}, \qquad \vartheta = -\tfrac{\xi_1+1}{\xi_1-1}. \tag{49}$$

Here α is an arbitrary parameter, and ξ, ξ_1 are solutions of $P^5_{(a,b,-\epsilon,0)}$. Now one can prove in a similar manner a theorem analogous to that of the third Painlevé equation, namely, that these transformations create new solutions of the fifth Painlevé equations. One can similarly deduce the recurrence formula for these new solutions.

§2 Applications

The Ping-Pong system constructed in the previous section allows us to create a new family of solutions for the third and fifth Painlevé equation. From the physical point of view, this new solutions are interesting and the Ping-Pong can be applied to them. The present author knows two such kinds of systems. One of them is the well known two-dimensional Ising model. In [6] Wu, McCoy, Tracy and Barouch have shown that the spin-spin correlation funtions for the two-dimensional Ising model can be expressed, in closed form, in terms of the solutions of the third Painlevé equation $P^3_{(0,0,1,-1)}$. The application of our Ping-Pong technology to this system allows us to obtain more complicated correlation funtions for the Ising model. The second example which I would like to disscus in more detail is the less known two dimensional self-dual $SU(2)$ Yang–Mills equation. This equation describes string-like solutions in the pure non-abelian theory in a manner similar to Nielson–Oleson vortices [7] in the abelian theory. It has been a recent idea [8,9] to construct such solutions using the Bogomolny–Prasad–Sommerfield limit [10,11]. Indeed, since the solutions are naturally independent of the two coordinates, the third and the fourth components of the potential can be taken as the two sets of Higgs fields. Then, the self-dual conditions, as we show, solve this problem. This knowledge suffices to show that the flux is -2π, just as in the Nielson–Oleson work, and to compute the 'tension', *i.e.*, the action per unit distance along the vortex. In order to solve the two-dimensional self-dual $SU(2)$ Yang–Mills equation we used a different method than in [9]. We follow Yang [12] in considering an analytic continuation of $\mathbf{A}_\mu$ into the complex space; the self-duality equation is also valid in this complex space.

Now consider four new complex variables defined by

$$y\sqrt{2} = x_1 + ix_2, \qquad \bar{y}\sqrt{2} = x_1 - ix_2, \tag{50}$$
$$z\sqrt{2} = x_3 - ix_4, \qquad \bar{z}\sqrt{2} = x_3 + ix_4. \tag{51}$$

It is simple to check that the self-duality equatioons reduce to

$$F_{yz} = F_{\bar{y}\bar{z}} = 0, \tag{52}$$
$$F_{y\bar{y}} + F_{z\bar{z}} = 0. \tag{53}$$

Equation (52) implies that the potentials $\mathbf{A}_y, \mathbf{A}_z, \mathbf{A}_{\overline{y}}$, and $\mathbf{A}_{\overline{z}}$ are pure gauges for fixed $y, z, \overline{y}$, and $\overline{z}$, so we can find two 2×2 complex matrices $\mathbf{D}$ and $\overline{\mathbf{D}}$ such that

$$\mathbf{A}_y = \mathbf{D}^{-1}\mathbf{D}_{,y}, \qquad \mathbf{A}_z = \mathbf{D}^{-1}\mathbf{D}_{,z}, \tag{54}$$

$$\mathbf{A}_{\overline{y}} = \overline{\mathbf{D}}^{-1}\overline{\mathbf{D}}_{,\overline{y}}, \qquad \mathbf{A}_{\overline{z}} = \overline{\mathbf{D}}^{-1}\overline{\mathbf{D}}_{,\overline{z}}, \tag{55}$$

where $\mathbf{D}_{,y} = \partial_y \mathbf{D}$, etc. We now define the matrix $\mathbf{J}$ through

$$\mathbf{J} = \mathbf{D}\overline{\mathbf{D}}^{-1}. \tag{56}$$

Clearly $\det \mathbf{J} = \mathbf{1}$. The remaining self-duality equation (53) can be written as

$$(\mathbf{J}^{-1}\mathbf{J}_{,y})_{,\overline{y}} + (\mathbf{J}^{-1}\mathbf{J}_{,z})_{,\overline{z}} = 0. \tag{57}$$

Thus far we have not chosen any particular gauge. In the rest of this paper we work in Yang's R-gauge [11], which is defined by choosing the matrix $\mathbf{J}$ for the $SU(2)$ gauge theory, in the form

$$\mathbf{J} = \frac{1}{\phi}\begin{pmatrix} 1 & e \\ f & \phi^2 + ef \end{pmatrix}, \tag{58}$$

where ϕ, e, and f are independent complex funtions of $y, z, \overline{y}$, and $\overline{z}$. Substitution of (58) in (57) gives

$$(\partial_y \partial_{\overline{y}} + \partial_z \partial_{\overline{z}}) \ln \phi + \frac{f_y e_{\overline{y}} + f_z e_{\overline{z}}}{\phi^2} = 0, \tag{59}$$

$$\phi(\partial_y \partial_{\overline{y}} + \partial_z \partial_{\overline{z}}) f - 2(f_y \phi_{\overline{y}} + f_z \phi_{\overline{z}}) = 0, \tag{60}$$

$$\phi(\partial_y \partial_{\overline{y}} + \partial_z \partial_{\overline{z}}) e - 2(e_{\overline{y}} \phi_y + f_{\overline{z}} \phi_z) = 0. \tag{61}$$

The R-gauge potentials are given by

$$\mathbf{A}_u = -\frac{1}{2\phi}\begin{pmatrix} \phi_{,u} & 0 \\ 2f_{,u} & -\phi_{,u} \end{pmatrix}, \tag{62}$$

$$\mathbf{A}_{\overline{u}} = \frac{1}{2\phi}\begin{pmatrix} \phi_{,\overline{u}} & 2e_{,\overline{u}} \\ 0 & -\phi_{,\overline{u}} \end{pmatrix}, \tag{63}$$

where $u = y, z$, and $\overline{u} = \overline{y}, \overline{z}$. According to our previous considerations, we now assume that

$$\partial_z \mathbf{A}_u = \partial_{\overline{z}} \mathbf{A}_u = 0, \tag{64}$$

$$\partial_z \mathbf{A}_{\overline{u}} = \partial_{\overline{z}} \mathbf{A}_{\overline{u}} = 0. \tag{65}$$

This can be achieved choosing

$$\phi = \phi(y, \overline{y}) = \exp \alpha, \qquad \alpha = \alpha(y, \overline{y}), \tag{66}$$

$$e_{\overline{y}} = \frac{1}{\sqrt{2}} i\phi^2, \qquad e_{\overline{z}} = \frac{1}{\sqrt{2}}, \tag{67}$$

$$f_y = \frac{1}{\sqrt{2}} i\phi^2, \qquad f_z = \frac{1}{\sqrt{2}}, \tag{68}$$

Then our equations (59–61) are reduced to

$$\partial_y \partial_{\overline{y}} \alpha = \sinh \alpha. \tag{69}$$

The well known soliton-like solutions of this equation are not good in our context because they do not produce the finite flux, and the third and the fourth components do not behave as the Higgs fields. On the other hand, assuming that α depends only on r, where $r^2 = 2y\overline{y}$, we obtain that our Sinh–Gordon equation is reduced to

$$\phi'' + \frac{1}{r}\phi' = \frac{\phi'^2}{\phi} + \phi^3 - \frac{1}{\phi}, \tag{70}$$

i.e., the third Painlevé equation $P^3_{(0,0,-1,1)}$.Let me remark that the same equation in the same context has been obtained in [9] using quite different gauges and techniques. With the results of this paper, one may easily compute the flux

$$F = \int \vec{F}_{12} \vec{A}_3 \times \vec{A}_4 \, dx_1 dx_2 = -2\pi \tag{71}$$

for the solution found in [6]. Moreover, for this solution, the third and the fourth components of the potential behave as the Higgs fields. However, for this solution, 'tension' is divergent; for this reason the solution is called string-like. On the other hand, if one can found the solution of the third Painlevé equation $P^3_{(0,0,-1,1)}$ which gives the finite flux and finite tension, then our Ping-Pong can help to find multi-string solutions in this model.

References

[1] F. J. Bureau, *private communication.*

[2] M. J. Ablowitz and A. S. Fokas, On a unified approach to transformations and elementary solutions of Painlevé equations. *J. Math. Phys.* **23**, 2033–2042 (1982).

[3] Z. Popowicz, *The Bäcklund transformations for the third and fifth Painlevé equations*, IFT University of Wrocław preprint (1982).

[4] N. Lukashevich, *Differential Equations* **3**, 11 (1967); in Russian: 1913, (1967).

[5] V. Gromak, *Differential Equations* **9**, 2–373 (1975).

[6] T. T. Wu, B. M. McCoy, C. A. Tracy, and E. Barouch, Spin-spin correlation functions for the two-dimensional Ising model: exact theory in the scaling region. *Phys. Rev.* **B13**, 316–374 (1976).

[7] H. B. Nielsen and P. Oleson, Vortex-line models for dual strings. *Nucl. Phys.* **B61**, 45–61 (1973).

[8] M. A. Lohe, Two- and three-dimensional instantons. *Phys. Lett.* **70B**, 325–328 (1977).

[9] C. Sacliğlou, A string-like self-dual solution of Yang–Mills theory. *Nucl. Phys.* **B178**, 361–372 (1981).

[10] E. B. Bogomol'nyǐ, The stability of classical solutions. *Soviet J. Nucl. Phys.* **24**, 449–454 (1976).

[11] M. K. Prasad and C. M. Sommerfeld, Exact classical solution for the 't Hooft monopole and the Julia–Zee dyon. *Phys. Rev. Lett.* **35**, 760–762 (1975).

[12] C. N. Yang, Conditions of self-duality for SU(2) gauge fields on Euclidean four-dimensional space. *Phys. Rev. Lett.* **38**, 1377–1379 (1977).

Chaotic Evolution of a System of Weakly Interacting Ocean Waves

Pedro Ripa

Centro de Investigación Científica
y Estudios Superiores de Ensenada
Ensenada, B. C. N., Mexico

Ocean variabiliy at large scales (greater, say, than a hundred kilometers and longer than a few days) is caused by several types of agents:

1. The ocean is *forced* by the wind, the differential solar heating, etc.
2. Most of the energy is ultimately *dissipated*, presumably through a cascade toward smaller scales, but in any case by processes poorly understood and parametrized.
3. *Restoring* agents, acting in the interior of the fluid, tend to bring it to an equilibrium state.

At the scales of interest here, these are mainly due to the *buoyancy*, produced by raising heavier fluid or sinking lighter one, and the earth's *rotation* which deflects horizontal motion to the right (left) in the northern (southern) hemisphere. Rotation effects vary with latitude due to the earth's *curvature*, and, in particular, go through a zero at the equator: this is why the low latitudes have peculiar dynamics (*e.g.*, the equatorial region acts as a waveguide). Finally, the equations of motion are intrinsically *nonlinear*, even though truly nonlinear effects may sometimes be neglected for small amplitude perturbations

In summary, the physical evolution of the ocean is controlled by equations of the form

Rate of change of the dynamical fields = Linear effects
+ Nonlinear effects + Dissipation + external Forcing,

or, in short,

$$R = L + N + D + F. \tag{1}$$

This represents a set of coupled partial differential equations.

Classical *turbulence* models correspond to energies large enough to approximate (1) as $R \sim N + (D + F)$; linear *waves*, on the other hand, are solutions of $R \sim L + (D + F)$, presumably valid for small aplitudes. In both cases, the term $D + F$ is enclosed in parenthesis to indicate that it may be absent.

A *free* wave results from the balance $R = L$. As an example, the meridional velocity field v due to an equatorial wave has the form (see [1])

$$v \approx F_\lambda(z)\, H_n(\lambda y)\, \exp(-\tfrac{1}{2}\lambda^2 y^2)\, \cos(kx - \omega t), \tag{2}$$

where F depends upon the density structure, H_n is a Hermite polynomial, and (x, y, z, t) denote the coordinates *(eastward, northward, upward, time)*. A very important fact is that, because of the balance

$R = L$, the frequency ω is not arbitrary, but given by a certain *dispersion relation*

$$\omega = \Omega(k, n, \lambda). \tag{3}$$

A *forced wave*, on the other hand, results from the balance $R = (L) + F$. An equatorial forced wave might be represented by fields as in (2), but such that ω is *not* given by (3); instead, ω and the other parameters are determied by the forcing.

Let us consider now the nonlinear effects. One of the nonlinear terms in the equation (1) corresponding to v, is the advection $v\partial_y v$, *i.e.*, the meridional momentum equation is of the form

$$\partial_t v = L(v) - v\partial_y v + \cdots, \tag{4}$$

where the dots represent the other nonlinear terms and the case $F = D = 0$ is considered. Now, let us assume that the energy is sufficiently small so that we may take the v field as a *first approximation*, as a superposition of free waves as in (2) and (3), and replace it, as a *second approximation* in the nonlinear term in (4). Consider the contribution of two waves, say, with parameters $(k_j, \lambda_j, n_j, \omega_j)$, $j = 1, 2$. The (y, z)-dependent part may be *projected* into the standard functions $F_\lambda(z)H_n(\lambda y)\exp(-\frac{1}{2}\lambda^2 y^2)$ of the linear waves, using their orthogonality. The (x, t)-dependent part is of the form $\cos(k_1 x - \omega_1 t)\times \cos(k_2 x - \omega_2 t) = \frac{1}{2}\cos[(k_1 + k_2)x - (\omega_1 + \omega_2)t] + \cos[(k_1 - k_2)x - (\omega_1 - \omega_2)t]$; therefore the nonlinear term in (4) may be seen as *forcing* a new wave with wavenumber and frequency given by $k_f = k_1 + k_2$ and $\omega_f = \omega_1 + \omega_2$. Th amplitude of the forced wave will be negligible (*i.e.*, bounded by a constant times the product of the amplitude of the original waves) unless the forcing frequency happens to coincide with the free frequency of the wave, namely if

$$\omega_1 + \omega_2 \equiv \Omega(k_1 + k_2, n, \lambda), \tag{5}$$

for some n and λ such that the projection of the (y, z) structure function does not vanish.

The reasoning above is only meant as a heuristic argument to introduce the concept of wave *resonance*, first predicted by Phillips in 1960 [2]. A systematic derivation, using multiple time-scales, was done by Brotherton [3]. Under the approximation of only taking resonant triads, the weak-interactions limit, we introduce time-dependent amplitudes replacing $\cos(\cdots)$ by $\mathrm{Re}[X(t)\exp(i\cdots)]$ in (2) and similar equations for other fields. The evolution equation for a *discrete* set of amplitudes is found to be of the form

$$\dot{X}_a(t) = \tfrac{1}{2}\omega_a \sum_{bc} \gamma_{abc} X_b^*(t) X_c^*(t), \tag{6}$$

where the summation is restricted to *resonant triads*, namely those which satisfy[1]

$$k_a + k_b + k_c = 0, \tag{7a}$$
$$\omega_a + \omega_b + \omega_c = 0. \tag{7b}$$

The set (6) may be derived from Euler's equations $(\partial L/\partial \dot{X}_a)^{\cdot} = \partial L/\partial X_a$ for the Lagrangian

$$L = -\tfrac{1}{2}i \sum_a{}' \omega_a^{-1} X_a^* \dot{X}_a + \tfrac{1}{6}i \sum_{a,b,c} \gamma_{abc} X_a X_b X_c, \tag{8}$$

[1]The interaction coefficient γ is real and completely symmetric in the subscripts for a normalization of the X_a exemplified in Eq. (9b) below. For each mode a in the summation there is anoher one $-a$ such that $(k_{-a}, \omega_{-a}, X_{-a}) = -(k_a, \omega_a, X_a^*)$.

where the prime indicates sum over physically independent modes[2] The integrals of motion can be obtained using Noether's theorem [4], from the symmetries of L. It follows from (7) that L is invariant under the transformations $X_a \to \exp(ik_a\delta x)X_a$ and $X_a \to \exp(-i\omega_a\delta t)X_a$, which results in the conservation of total *pseudomomentum* and *energy*, *viz.*

$$P = \sum_a{}' \frac{k_a}{\omega_a}|X_a|^2 = \text{constant}, \tag{9a}$$

$$E = \sum_a{}' |X_a|^2 = \text{constant}. \tag{9b}$$

Moreover, $\partial_t L = 0$ implies the conservation of the value of the Hamiltonian

$$H = \frac{1}{3!}\sum_{a,b,c} \gamma_{abc}\,\mathrm{Im}(X_a X_b X_c) = \text{const}. \tag{10}$$

Generalized coordinates q_a and momenta p_a, may be defined through $X_a = (\omega_a q_a + ip_a)/\sqrt{2}$ so that (6) is obtained using Hamilton–Jacobi equations. Moreover, it is easy to show that E, P, and H are ***in involution***[3] and therefore the three-wave problem is integrable (an N-wave system has N degrees of freedom).

The results of Eqs. (6–10) are quite general: they are found in fluid systems with (at least) one homogeneous coordinate. We finish by briefly describing some results of the application of (6) and (7) to the equatorial waveguide.

One of the simplest problems in wave-wave interactions which we may ask, is what happens if initially there is only a single wave, say, with parameters (k_1, ω_1). From the analysis of the complete evolution equation (1) it follows that all harmonics and x-independent modes will be excited; this energy transfer is enhanced if a second harmonic is a resonance, *e.g.*, Eq. (5) with $\omega_2 = \omega_1$, $k_2 = k_1$. We have investigated [5] a somewhat more general problem, namely, the existence of chains of resonant harmonics

$$\Omega(jk_1, n_j, \lambda) = j\Omega(k_1, n_1, \lambda), \tag{11}$$

for some integer j, and the solutions of the corresponding systems (6). Notice that the meridional structure of the different modes may not be the same ($n_j \neq n_1$), but the vertical structure has been chosen the same for all components ($\lambda_j = \lambda$) for simplicity.

Kinematics. For equatorial waves of the second kind —Rossby or rotational modes— there are solutions of (11) for a finite number of harmonics, *e.g.* ($j = 1, 2$), ($j = 1, 2, 4$), etc. For equatorial waves of the second kind —inertia-gravity or potential modes— the solutions of (11) are on infinite chains, *e.g.* ($j = 1, 2, 4, 5, 7, 8, 10, \ldots$) of $j \not\equiv 0 \bmod(3)$; the value of the interaction coefficient γ varies considerably from triad to triad. Last but not least, equatorial Kelvin waves constitute a rather trivial solution of (11): because they are non-dispersive (hyperbolic) for these modes, (7a) implies (7b); therefore (11) is valid for all integer j and γ is the same for all triads.

Dynamics. For a system restricted to only resonant harmonics, energy and pseudomomentum are not independent integrals of motion, *viz.* $P = E(k_1/\omega_1)$. The two-wave problem is then integrable; for instance, the solution of (6) for ($j = 1, 2$), and an initial condition corresponding to all the energy in

---

[2] Either a or $-a$ for each pair $(a - a)$

[3] See the contribution by Francesco Calogero in *these Proceedings*.

the fundamental wave, is

$$X_1(t) = X_1(0)\,\mathrm{sech}(\mu t), \qquad X_2(t) = X_1(0)\tanh(\mu t), \tag{12}$$

where

$$\mu = \gamma_{1,1,-2}\,\omega_1\sqrt{E}. \tag{13}$$

If the fourth harmonic is also in resonance ($j = 1, 2, 4$) in (11), the energy ultimately goes to that mode (for the same initial condition) but much more slowly than in (12), *viz.* $X_1 \sim 1/\sqrt{t}$ and $X_2 \sim 1/t$.

The system of Kelvin modes[4] is in some sense the other extreme case of chain resonant harmonics[5] Starting as before with all the energy in the fundamental wave, it is found that all harmonics gain energy in a coherent way, to the point that the solution is not valid for $\mu t > \frac{1}{2}$: the approximation of neglecting (off-resonant) interactions with other modes, as in (6), breaks down.

The last system we considered is the infinite chain of inertia-gravity modes [$j \neq 0$ mod(3) in (11)]. Since the interaction coefficient γ has a different value in each triad, (6) has to be solved numerically with the system truncated to a finite number of components. Two kinds of initial conditions were used: (i) all energy in the fundamental wave, $X_j(0) = \sqrt{E}\,\delta_{j,1}$; this implies $X_j(t)$ real, and (ii) $X_j(0)$ chosen *at random*, but constrained to be real, which is presumably the class of solutions with maximum energy exchange, because $H \equiv 0$.

The evolution of the system truncated to the first *five* harmonics ($j = 1, 2, 4, 5, 7$) goes rapidly [in a time $\mu t = O(1)$] to a certain solution of the three-wave problem $X_2 \simeq 0$, $X_7 \simeq 0$, and there is *catalytic* energy exchange between X_1 and X_5 in a time scale determined by the energy of X_4, which is quite unaffected by the process.

Adding one more component ($j = 1, 2, 4, 5, 7$), another unexpected result was found: the orbit looks ergodic, unstable, and sensitive to initial conditions, in sum, *chaotic*[6] It is interesting to find this type of solution in the weak-interactions approximation, which is the limit opposite to that of classical turbulence, namely $E \to 0$ instead of $E \to \infty$.

References

[1] B. W. Moore and S. G. H. Philander, Modelling of the tropical oceanic circulation. *The Sea* **6**, 319–361 (1977).

[2] O. M. Phillips, On the dynamics of unsteady gravity waves of finite amplitude. *J. Fluid Mech.* **9**, 193–217 (1960).

[3] F. P. Bretherton, Resonant interactions between waves. The case of discrete oscillations. *J. Fluid Mech.* **20**, 457–479 (1964).

[4] E. Nœther, Invariante Variationsprobleme. *Nachr. Kgl. Ges. Wiss. Göttingen, Math. Phys. Kl.*, 235–257 (1918).

[5] P. Ripa, Weak interactions of equatorial waves in a one-layer model. Parts I and II. *J. Phys. Ocean.*, (submitted for publication).

[4] All integer j in (11) and uniform interaction coefficient γ in (6).

[5] The system (6) is the Fourier transform of the one-dimensional advection equation, $\partial_t U + U\,\partial_x U$.

[6] Work is in progress to put this characterization of the behaviour as that of *conservative chaos* on a more firm foundation.

Preliminary Observations of Large Amplitude Tidal Internal Waves near the Strait of Messina

Rosalia Santoleri

Istituto di Fisica
Università di Roma "La Sapienza", Italy

§1 Introduction

In this note we study large amplitude internal waves generated by the tide near the sill of the Strait of Messina, a long and narrow channel between the Italian peninsula and Sicily. See Figure 1.

The existence of strong currents of tidal origin inside the Strait of Messina is fairly well established [1]. Messina is a tidal enphidrome point [2,3]. This means that the phase of the tide changes approximately five hours from North to South in the Strait. Therefore, when a high tide occurs in tthe Ionian Sea, a low tide is present in the Tyrrhenian Sea, determining a strong northward current. The situation is reversed about six hours later and the current flows southward. The reversal time is called *dead water*.

It is important to note that either in the Tyrrhenian or the Ionian Sea, two water layers are present. The superficial Atlantic water forms the upper layer and the Levantine inttermediate water forms the lower layer [4]. These layers move in opposite directions in the Strait of Messina: the first southward ($v = 10$ cm/s), the second, northward ($v \sim 15$cm/s). The interface between the two water masses is 150m deep at the ends of the Strait. During the tidal flow from the Thyrrenian to the Ionian Sea, only Atlantic water can be found in the sill and, during the opposite flow, only Levantine water can be observed. Therefore, the interface between the two layers reaches the surface and the bottom during the two opposite tidal phases [5,6]. In this way the flow over the sill can be considered a borotrope.

A recent SEASAT–SAR satellite image shows long semidiurnal striations with focus about five miles East of the Northern mouth of the Strait of Messina [7]. These are surface manifestations of presumably large-amplitude internal waves [8] propagating away from the Strait into the Tyrrhenian Sea. The correlation between tidal flow in a strait or over the continental shelves and the generation of nonlinear internal waves is well known. Refs. [8–12] show a clear recurrence of nonlinear internal wave trains with tidal periodicity.

A reasonable hypothesis is that the internal waves of the SEASAT–SAR image are generated on the sill at the moment of dead water.

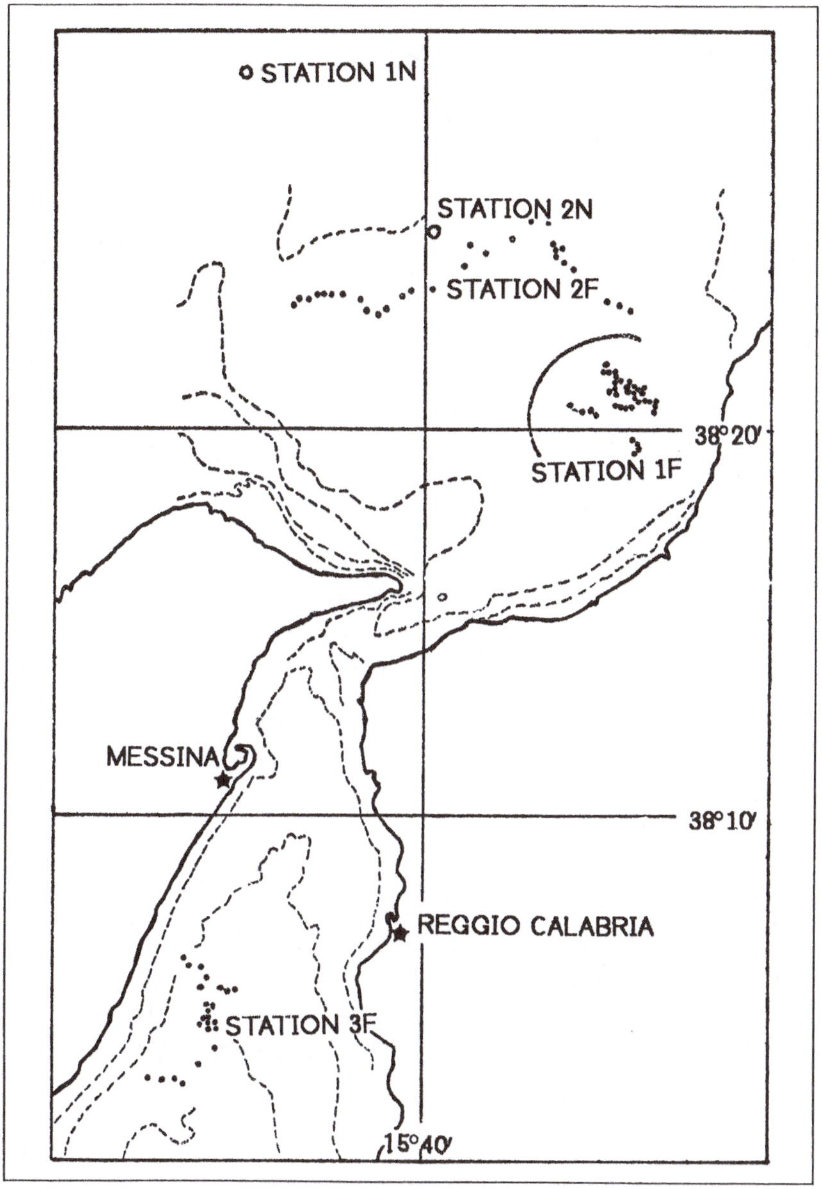

Figure 1. Bathimetric map of the Strait of Messina and positions of the stations.

We have analyzed the data of the *Judith 1980* and *Mark 1* cruises organized by the Istituto di Fisica (Università di Roma), in which large amplitude internal waves had been measured at 10 and 15 miles North of the sill of the Strait of Messina, and at 11 miles South of the sill. The internal waves measured in the Tyrrhenian Sea are similar to the ones observed by satellite. The data analysis suggests their interpretation in the framework of the Korteweg–de Vries (KdV) equation.

§2 Field observations

The existence of large amplitude internal waves was confirmed by the observations made during the *Judith 1980* (November 17–26, 1980) and *Mark 1* (February 10–21, 1981) cruises.

Judith 1980. During the *Judith 1980* cruise, temperature T, salinity S, and sigma-T σ vertical profiles were found by CTD NEILBROWN 2 at two stations (Figure 1): Station 1N (14 miles North of the sill, November 24–25), and station 2N (10 miles North of the sill, November 25). A representative profile of vertical structure of T, S, and σ is shown in Figure 2. In station 1N, the time series of temperature show the passage of packets of large amplitude internal waves occurring approximately every 12 hours (Figures 3a and 3b). The waves resemble the one observed in the SEASAT–SAR image [7]. These wave packets consist in $\sim$ 15m thermocline depression at 45m depth, lasting for two hours and modulating in several shorter waves. Station 2N data show the passage of a wave train similar to the one of station 1N. Its amplitude is $\sim$ 12m, but the temperature jump is $\sim$ 1°C, while in station 1N this jump is $\sim$ 3°C. The mean physical quantities describing hydrological variables are shown in Table 1.

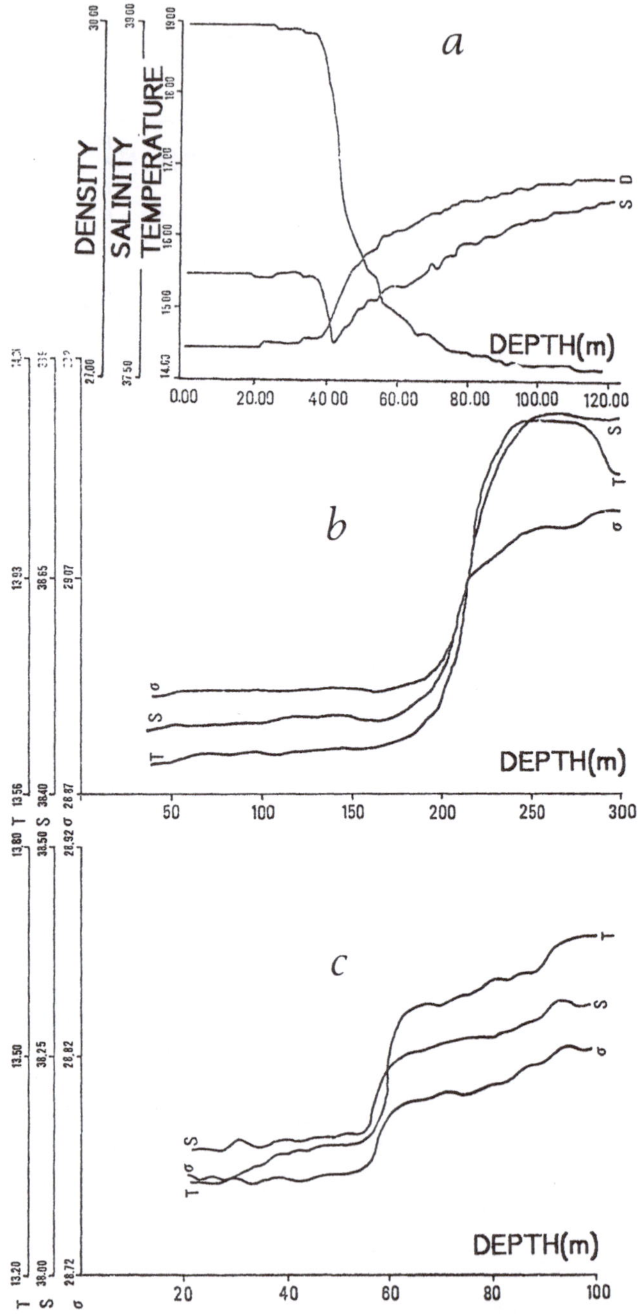

Figure 2. Typical vertical profile of T, S, and σ during *(a)* the *Judith 1980* cruise, *(b)* the *Mark 1* cruise in the Tyrrhenian Sea, and *(c)* the *Mark 1* cruise in the Ionian Sea.

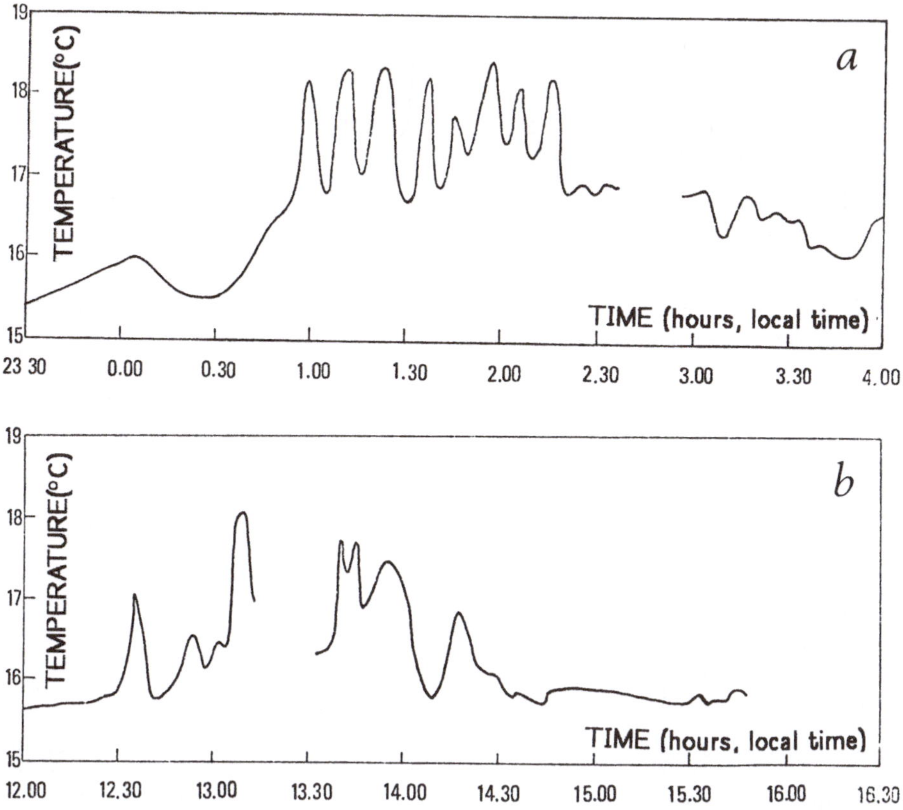

Figure 3. Temperature versus time at 45m relative to station 1N (November 24–25, 1980).

Station	h_1	h_2	g'	N	T
1N	60	430	1.4×10^{-2}	1.4×10^{-2}	$7\div5$
2N	60	410	1.5×10^{-2}	1.4×10^{-2}	$7\div5$
1F	120	280	1.4×10^{-3}	4.7×10^{-3}	22
2F	130	350	2.8×10^{-3}	5.2×10^{-3}	20
3F	200	650	1.5×10^{-3}	3.7×10^{-3}	28

h_1 = thickness of the upper layer (in m), h_2 = thickness of the lower layer (in m), $g' = g\,\Delta\rho/\rho$ = reduced gravity (in ms^{-2}), N = Brunt–Väisële frequency (in s^{-1}), $T = 2\pi/N$ = period (in s).

Table 1.

Mark 1. During the *Mark 1* cruise, T, S, and σ were measured using a CTP GUIDELINE at three different stations (Figure 1): Station 1F (10 miles North of the sill, February 15), Station 2F (10 miles North of the sill, February 15), and station 3F (11 miles south of the sill, February 18, 1981). The weather and sea conditions were rather unpleasant, with strong winds of ~ 18 Knots.

The vertical temperature profiles were inverted (Figure 4) probably due to the extremely cold 1980–81 winter in Southern Italy. The vertical temperature gradient of the upper 100m was relatively weak and the density fairly constant. See Table 1.

The temperature time series North of the sill (Figures 5e and 5b) show a semidiurnal periodicity thermocline due to oscillations, as in stations 1N and 2N. However, the internal wave shape is quite different; instead of a modulated wave train, we can observe a depression of the interface with an amplitude of ~ 60m with respect to the mean thermocline depth (120–130m). It lasts for five to six hours showing a rather irregular pattern; a slow decrease followed by a rather sharp increase in the interface. It is important to note that the measurements in station 2F were made for 12 hours, drifting along a line of approximately constant distance from the Northern mouth of the Strait of Messina, *i.e.*, a line of constant phase for the internal waves. The data show a rather remarkable similarity with those of station 1F. This confirms that the internal wave signal issues from the Northern mouth of the Strait of Messina with circular symmetry.

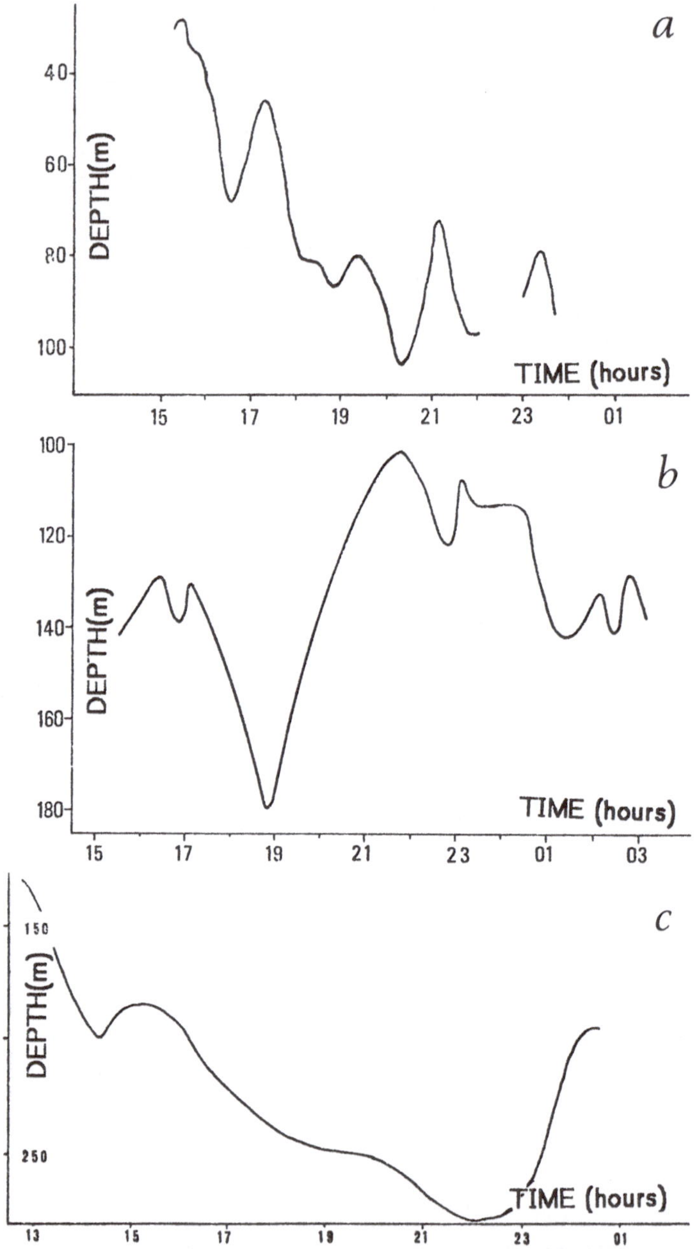

Figure 4. *(a)* Station 1F, February 14: Isothermal relative to $T = 15.5°C$. *(b)* Station 2F, February 14; Isothermal relative to $T = 13.60°C$. *(c)* Station 3F, February 17: Isothermal relative to $T = 13.63°C$.

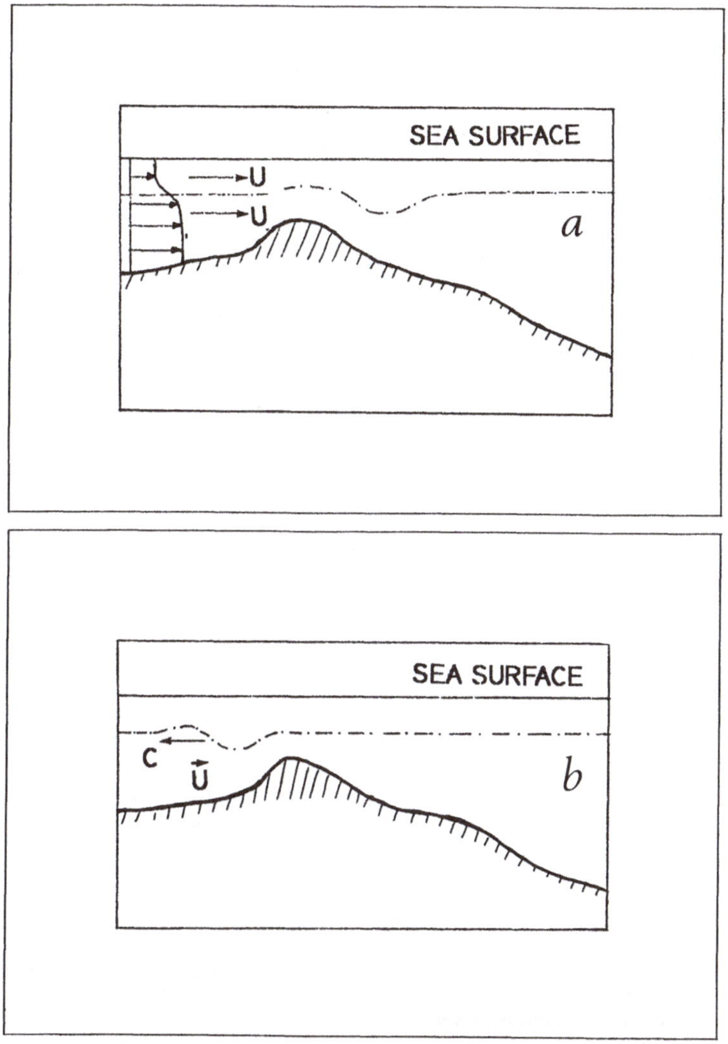

Figure 5. **Representation of Maxworthy's mechanism of the formation of tidal internal waves over the sill of the Strait of Messina. *(a)* Formation of quasi-steady lee waves due to the tidal flow ($u =$ current velocity). *(b)* Upstream propagation ($c =$ wave velocity) of the lee wave at the tide reversal time.**

Station 3F is the only station located in the Ionian Sea, South of the sill. The analysis refers to the data taken over a period of 12 hours, but some general features of the situation may be observed. The internal waves (Figure 5c) have 100m in amplitude, two times larger than the amplitudes in stations 1F and 2F. This difference can be easily explained considering the energy spreading, which is greater in the Tyrrhenian Sea, where the waves propagate radially, than in the Ionian Sea, where they are constricted by the parallel banks of the narrow Strait. The wave pattern is quite different from the

North one. It does not seem to show the passage of any definite wave train (as in 1N and 2N), and of any impulsive and localized disturbance (as in 1F and 2F). This aspect will be discussed in more detail in the following.

§3 Theoretical problems and internal wave packets

The data analysis presented in the previous paragraph indicates two interesting discussion points:

1. The theoretical interpretation of the wave pattern, as measured to the North of the sill.

2. A closer look at the generation mechanism on the sill.

A theoretical description of long nonlinear internal waves is given by the KdV equation. This is a third-order approximation to the Navier–Stokes equations. In a two-layer system with flat bottom, the one-dimensional equation is [12]

$$\eta_t + c_0 \eta_x + \alpha \eta \eta_x x + \gamma \eta_{xxx} = 0, \tag{1}$$

where $\eta(x,t)$ is the height of the interface from the mean level, c_0 is the phase speed of the associate linear internal waves,

$$\begin{aligned} c_0 &= \sqrt{g \frac{\Delta\rho}{\rho} \frac{h_1 h_2}{h_1 + h_2}}, \\ \alpha &= -\tfrac{3}{2}\, c_0 \frac{h_2 - h_1}{h_1 h_2}, \\ \gamma &= \tfrac{1}{6}\, c_0\, h_1 h_2, \end{aligned} \tag{2}$$

where $\Delta\rho = \rho_2 - \rho_1$, $\rho = \frac{1}{2}(\rho_1 + \rho_2)$, with ρ_1, h_1, ρ_2, and h_2 respectively the density and the thickness of the first and second layer. Equation (1) is valid under the following conditions:

i. Long wave approximation

$$\delta = (D/L)^2 \ll 1, \tag{3}$$

where $D = \sqrt{\frac{4}{3} \frac{h_1 h_2}{1 - h_1/h_2}}$ is the characteristic depth and $L = \sqrt{\frac{-12\gamma}{\eta_0 \alpha}}$ is the lenght scale, with η_0 the mean amplitude of the nonlinear waves.

ii. Small waves, but finite amplitude

$$\epsilon = \eta_0 / h_1 \ll 1. \tag{4}$$

iii. Balance of nonlinearity and dispersion

$$\epsilon \sim \delta. \tag{5}$$

The KdV equation gives the time evolution for a sufficiently localized initial wave $\eta(x,0)$. This equation predicts two completely different kinds of behaviour for initial waves with positive and with negative aplitude; if $\eta(x,0) > 0$, $\eta(x,t)$ will disperse into an oscillatory wave train. If $\eta(x,0) < 0$, after an initial interactive time, $\eta(x,t)$ will evolve into one or more solitons described by the following equation:

$$\eta(x,t) = -\eta_0 \operatorname{sech}^2 [(x - ct)/L], \tag{6}$$

Station	c_0	L	D	η_0	ϵ	δ	c	τ
1N	0.85	400	200	15	0.25	0.25	0.94	7÷8
2N	0.88	440	195	12	0.20	0.20	0.96	7÷8
1F	0.40	400	280	60	0.50	0.50	0.50	13–14
2F	0.50	320	425	70	0.50	0.50	0.65	13

c_0 = phase velocity of the linear internal waves associated to the nonlinear one, L = characteristic lenght scale, D = characteristic depth, η_0 = mean amplitude of the observed internal waves, ϵ = nonlinear term of the KdV equation, δ = dispersive term of the KdV equation, c = phase velocity of the nonlinear wave, τ = period of the nnonlinear waves ($\tau = L/c$).

Table 2.

and into a dispersive tail. These waves have two peciuliar characteristics:

1. Stability, *i.e.*, they to not vanish at $x \to \infty$.

2. Quasi-linearity, *i.e.*, they are not modified by interaction with other solitons.

From the amplitude of the initial wave, the number and amplitude of the solitons existing in the asymptotic solution could be evaluated. Qualitatively, the amplitude decreases progressively from the first solution to the others.

In order to check if the KdV equation can be applied to the waves observed North of the sill of the Strait of Messina, the parameters ϵ and δ are presented in Table 2. In November, the values of ϵ and δ satisfy the condition of KdV applicability. The value of the wavelenght and the time scale (Table 2) fit the observed ones very well ($L_{\text{exp}} = 500\text{m}$, $T_{\text{exp}} \simeq 10'$). The shape of the recorded waves, as shown in Figure 3, is quite similar to the characteristic pattern predicted by the KdV equation —with some differences. In our case the amplitude of the waves in the train does not decrease from the first soliton to the others, and the distance between the peaks is not exactly like the one predicted. These features could be explained considering that the theoretical results deal with asymptotic behaviour, while our data were taken close to the source, where the interaction effects are still present.

To evaluate the relationship between the observed waves and the inversion of the tide in the Strait of Messina, we have computed the time necessary for a wave comming from the sill to reach the station. This time is given by $\Delta t = d/cgr$, where $cgr = \frac{1}{2}(1 + Kh_1)c$ is the group velocity of the nonlinear internal waves, and d is the distance of the station from the sill. The generation time resulting from this computation is in agreement with the time of dead water on the sill. See Table 3.

The profile of the isothermal time evolution of the February data set appears to be quite different from those observed in November. The 1F and 2F station data show tidal signals of amplitude 60–70m, without any wave-train structure (Figures 4a and 4b). This is probably the initial depression of the thermocline due to the tide. It seems reasonable to think that the compact form of the waves is correlated to the greater nonlinear effect —ϵ is double the November one. For want of wave structure, it is quite impossible to correctly evaluate the group velocity of our waves; it is therefore impossible to check also qualitatively our generation hypothesis.

Station	c	K	cgr	d	t	Δt	t_0	d.w.t.
1N	0.94	8.6×10^{-3}	0.72	14.0	0 – 1	10^h	14^h – 15^h	15.38
1st observation					11–25–1980		11–24–1980	11–24–1980
1N	0.94	5.1×10^{-3}	0.65	14.0	12 –13	11^h	1^h – 2^h	3.01
2nd observation					11–25–1980		11–25–1980	11–25–1980
2N	0.94	6.3×10^{-3}	0.73	10.8	$8{:}00^h$ – $8{:}30^h$	7^h	1^h – 1.30^h	3.01
					11–25–1980		11–25–1980	11–25–1980

Comparison between dead water time (d.w.t.), as given by the tidal tables of the Istituto Idrografico delle Marine (Genoa, Italy) and the estimated generation time of the initial thermocline depression at the sill $t_0 = t^* - \Delta t$. c = phase speed of the nonlinear waves, cgr = group velocty of the nonlinear waves, K = wave number, d = distance between the examined station and the sill (in nautical miles), t^* = time at which the first nonlinear wave arrived at the station, $\Delta t = d/cgr$ = time required for waves starting from the sill to reach the station.

Table 3.

§4 Generation mechanism and comparison of the interface time evolution in the Ionian and Tyrrhenian Seas

The generation mechanism of the nonlinear internal waves over the sill can be explained following the Maxworthy idea [13]. The tidal current from the Tyrrhenian towards the Ionian Sea is assumed to generate a steady lee wave immediately South of the sill (Figures 5). This is associated to a depression of the thermocline and is characterized by a phase speed opposite to the direction of the current. For this reason the depression cannot be propagated during the flood tide. When the tidal flow changes, the current and the phase speed of the lee waves are in the same direction —Northward. Therefore the depression can be propagated and evolves as a solitary wave train into the Tyrrhenian Sea. Maxworthy showed [13] that this mechanics can be also applied to a barotropic situation over the sill.

The current from the Ionian towards the Tyrrhenian Sea does not seem to determine the parameter of a strong lee wave over the sill, because of the asymmetry of the bottom topography (Figure 5). This could explain the difference between the two sets of measurements at the North and South ends of the Strait. With respect to this phenomenon it is interesting to note that a similar feature was observed by Lacombe in the Strait of Gibraltar. There is a close analogy between Lacombe's data [14] East of this Strait, and the waves observed during the *Mark 1* cruise North of the Strait of Messina. Both display a strong impulsive semidiurnal signal. Lacombe's data West of the sill of the Strait of Gibraltar and Figure 5c in our data at the South end of the Strait of Messina show a more modulated quasilinear periodic tidal behaviour.

§5 Conclusions

During two cruises (*Judith 1980* and *Mark 1*), we have investigated the existence of high frequency packets of large amplitude internal waves at both sides of the Strait of Messina. These gave a more clear-cut high frequency character in the Northern basin; at the South of the Strait they display a rather smooth behaviour. The nonlinear internal wave packers occur in the Tyrrhenian Sea with a semidiurnal periodicity (12 hours); their generation mechanism seems to be strongly correlated to dead water periods over the sill, followed by the flow from the Ionian towards the Tyrrhenian Sea. These internal waves can be described, at least in the presence of strong stratification, by the KdV equation as the time evolution of an initial strong and localized thermocline depression. The Maxworthy mechanism of generation of nonlinear waves gives a simple and satisfactory interpretation of the two different kinds of wave behaviour observed to the North and to the South of the sill.

References

[1] A. Defont, *Physical Oceanography. Vol. I.* Pergamon Press, 1961.

[2] F. Vercelli, Le regime delle correnti e delle maree nello Stretto di Messina. Commissione Internazionale del Meditterraneo, Venice, 1925.

[3] F. Vercelli and M. Picotti, Le regime chimico fisico delle acque nello Stretto di Messina. Commisione Internazionale del Mediterraneo, Venice, 1925.

[4] J. N. Nielsen Hydrography of the Mediterranean and adyacent waters. *Rep. Danish Ocean. Exped. Med.* **1**, 1908–1910 (1912).

[5] R. del Ricco, A numerical model of the vertical circulation of tidal straits and its application to the Messina Strait. *Nuovo Cimento* **5C**, No. 1 (1982).

[6] T. Hopkins, E. Salusti, and D. Settimi, Tidal currents and internal waves in the Strait of Messina. Work in progress.

[7] W. Alpers and E. Salusti, Scylle and Charybdis observed from space. *J. Geophys. Res.* **88**, 1800–1808 (1983).

[8] S. Apel, private communication, 1980.

[9] D. Farmer and J. D. Smith, In *Hydrodynamics of Estmerine and Fjords*, p. 465. J. Nihoul, ed., Edsevier, New York, 1978.

[10] A. E. Gargett, Generation of internal waves in the Strait of Georgia. *British Columbia Deep Sea Res.* **23**, 17–32 (1976).

[11] D. Halpern, Observation of short period internal waves in Masachusetts Bay. *J. Marine Res.* **29** (1971).

[12] A. Osborne and T. Burch, Internal solitons in the Andaman Sea. *Science* **208**, 451–460 (1980).

[13] T. Maxworthy, A note on internal solitary waves produced by tidal flow over a three-dimensional ridge. *J. Geophys. Res.* **84**, 338 (1979).

[14] H. Lacombe, Le détroit de Gibraltar. Notes et Memoires du Service Géologique du Mer, N^{0} 222. (1980).

Lie Series and Riccati Equations*

Ladislav Hlavatý,† Stanly Steinberg,¶‡
and Kurt Bernardo Wolf

Instituto de Investigaciones en Matemáticas Aplicadas y en Sistemas
Universidad Nacional Autónoma de México

AbstractØ

Lie series and a special matrix notation for first-order differential operators are used to show that the Lie group properties of matrix Ricatti equations arise in a natural way. The Lie series notation makes it evident that the solutions of the matrix Riccati equations are curves in a group of nonlinear transformations which is a generalization of the linear fractional transformations familiar from classical complex analysis. It is easy to obtain a linear representation of the Lie algebra of the group of nonlinear transformations, and this linearization leads directly to the standard linearization of the matrix Riccati equations. We note that the matrix Riccati equations considered here are of the general rectangular type.

---

*This work has been partially supported by the Consejo Nacional de Ciencia y Tecnología (CONACyT, México) Project ICCBCHE 790373 (covering the stay of L. Hlavatý at IIMAS), and partially by the National Science Foundation (NSF, USA) grant # MCS-8102683 (covering the sabbatical leave of S. Steinberg).

†On leave from the Institute of Physics, Czechoslovak Academy of Sciences, Prague, ČSSR.

¶Presentor.

‡On leave from the Department of Mathematics and Statistics, The University of New Mexico, Albuquerque N. M., USA.

ØThis work appeared as a *preprint* of the University of New Mexico, and will be published in *Journal of Mathematical Analysis and Applications*.

Linear and Nonlinear Differential Equations as Invariants on Coset Bundles*

Ladislav Hlavatý,† **Stanly Steinberg,**‡
and **Kurt Bernardo Wolf**¶

Instituto de Investigaciones en Matemáticas Aplicadas y en Sistemas
Universidad Nacional Autónoma de México

Abstract

Among the ways in which group theory adapts its methods to study nonlinear systems, we have come upon a rather natural construction —natural for a group theorist— whereby certain families of differential equations become identified with invariants built within a group. The equations embedded in this way thus far are tensor generalizations of Burgers equation, third- or higher-order-derivative Korteweg–de Vries equations, and similar generalizations of the diffusion and Hirota equations. The groups which harbour these are subgroups of inhomogeneous linear groups and some of their normal extensions. The construction is natural for a group theorist because *any* space $\mathcal{C}$, *homogeneous* under the action of a group $\mathcal{G}$ may be identified with a *coset space* $\mathcal{C} = \mathcal{H}\backslash\mathcal{G}$ by some subgroup $\mathcal{H} \subset \mathcal{G}$. We are able to find the appropriate coordinates of this space so that they become the set of *independent* and *dependent* variables satisfying the above differential equations.

§1 Introduction

Lie's methods for the analysis of a differential equation $\Phi(\mathbf{u}, \mathbf{x}) = 0$ with solutions $\mathbf{u}(\mathbf{x})$ lead us to find its *similarity algebra* [1]; subsequent exponentiation yields the *similarity* or *symmetry* group $\mathcal{G}$,

$$\mathbf{u} \xrightarrow{g} \overline{\mathbf{u}}(\mathbf{u}, \mathbf{x}), \quad \mathbf{x} \xrightarrow{g} \overline{\mathbf{x}}(\mathbf{u}, \mathbf{x}), \qquad g \in \mathcal{G}, \text{ such that } \quad \Phi(\mathbf{u}, \mathbf{x}) = 0 \Rightarrow \Phi(\overline{\mathbf{u}}, \overline{\mathbf{x}}) = 0. \tag{1.1}$$

These operations may be cast in a convenient ***matrix-cum-vector*** realization. We may now proceed and, with some ingenuity, find a subgroup $\mathcal{H}$ and coordinates of the coset space [2,Chapter 4] $\mathcal{C} = \mathcal{H}\backslash\mathcal{G}$ such that they be $\{\mathbf{u}, \mathbf{x}\}$. This is always possible: see [3, Ch. 2, Thm. 3.2].

---

* This work has been partially supported by the Consejo Nacional de Ciencia y Tecnología (CONACyT, México) Project ICCBCHE 790373 (covering the stay of L. Hlavatý at IIMAS), and partially by the National Science Foundation grant # MCS-8102683 (covering the sabbatical leave of S. Steinberg).

† On leave from the Institute of Physics, Czechoslovak Academy of Sciences, Prague, ČSSR.

‡ On leave from the Department of Mathematics and Statistics, The University of New Mexico, Albuquerque N. M., USA.

¶ Presentor.

Where is now the differential equation we started with? To answer this propely, we must first define functions $\mathbf{u}(\mathbf{x})$ as *sections* in C, and then the *derivatives* of these with respect to $\mathbf{x}$. We need thus an appropriate *fiber bundle of sections*, Z. Finally, we *prolong* the action of $\mathcal{G}$ from C to Z [4]. This is done in more detail in Reference [5,§2], of which this contribution is an overview written with the intention that formalities be kept to a minimum. Basically, the convenience of this construction is to have $z = \{\mathbf{x}, \mathbf{u}(\mathbf{x}), \mathbf{u}_{\mathbf{x}}(\mathbf{x}), \mathbf{u}_{\mathbf{xx}}(\mathbf{x}), \ldots\}$ as coordinates for Z [$\mathbf{u}_{\mathbf{x}}(\mathbf{x})$ is the set of all partial derivatives of $\mathbf{u} = \{u_i\}$ with respect to $\mathbf{x} = \{x_i\}$]. The *prolongation* of the action of $\mathcal{G}$ to Z supplements (1.1) with information on the transformation of partial derivatives through chainruling:

$$\frac{\partial u_i}{\partial x_j} \xrightarrow{g} \frac{\partial \overline{u}_i}{\partial \overline{x}_j} = \left(\frac{\partial u_k}{\partial \overline{x}_j(\mathbf{u},\mathbf{x})} \frac{\partial}{\partial u_k} + \frac{\partial x_l}{\partial \overline{x}_j(\mathbf{u},\mathbf{x})} \frac{\partial}{\partial x_l} \right) \overline{u}_i(\mathbf{u},\mathbf{x}), \tag{1.2}$$

and correspondingly for higher derivatives.

In Z, thus, a differential equation $\Phi(\mathbf{u},\mathbf{x}) = 0$ is associated to a function on the bundle coordinates, $\Phi(z)$, which may be acted upon by the elements of $\mathcal{G}$. Since $\mathcal{G}$ is here the similarity group of $\Phi(\mathbf{u},\mathbf{x}) = 0$, its prolongued action on $\Phi(z)$ should map the latter onto another function $\overline{\Phi}(z)$ of Z in the following way:

$$\Phi(z) \xrightarrow{g} \overline{\Phi}(z) = \mu(z; g)\, \Phi(\overline{z}). \tag{1.3}$$

This formula characterizes a ***multiplier realization*** of $\mathcal{G}$ on Z, where $\mu(z;g)$ is a *multiplier function* over $Z \times \mathcal{G}$. The surface $\Phi(z) = 0$ will thus map onto $\Phi(\overline{z}) = 0$, providing the group-theoretic characterization of the original differential equation as an *invariant surface* in the bundle. This hinges on: *(i)* the structure of $\mathcal{G}$, *(ii)* the placing of $\mathcal{H}$ in $\mathcal{G}$, and *(iii)* the coordinates chosen for C, *i.e.*, the identification of dependent and independent variables with coordinate sets.

In order to keep this presentation as simple and straightforward as possible, in §2 we shall give a collection of concrete examples of groups, subgroups, and coset coordinates which will yield the Burgers [6;7,Chapter 4], diffusion [1,§2.7;8,§10.1], Korteweg–de Vries (see *e.g.*, [7]), and Hirota [9] equations. We shall not insist at all on the formal fiber bundle structure except as remarks ***post factum***. The groups we shall be using are **not** semisimple.[1] Being non-semisimple, they are prone to the process of *extension* (discussed in **2.1**), and this gives a group-theoretic meaning to the Hopf–Cole [10] and Hirota [9] transformations, presented in §3 under the generic name of the former. In this last Section we add some comments prompted by discussions held during the Workshop. Some conjectures are freely made, within the more informal spirit of these proceedings.

[1] In his contribution to *these Proceedings*, Paul Winternitz discusses Lie's approach to the classification of groups according to the dimension of their coset spaces. For dimension one, the only semisimple group is *SL(2,R)*, the rest being solvable, nilpotent or abelian. This semisimple group leads to Riccati's equation $du/dt = a(t) + b(t)\,u + c(t)\,u^2$ (also given in Stanly Steinberg's presentation), and its treatment follows purely from group theory. This embedding of an ordinary differential equation in the group is *different*, however, from our approach. Paul *does* identify the dependent variable u with a coset coordinate —by the solvable subgroup— *but not* t, the independent coordinate. Rather, t is the parameter of a line of acting evolution group elements $g(t)$, connected to the identity, following a Lie field $\{a(t), b(t), c(t)\}$ specified by the Riccati equation. The group is *not* the similarity group of the equation, since in general t is not involved if the coefficients depend on time in an arbitrary way.

§2 Groups, cosets, and equations

2.1 Some structural preliminaries

We shall work with groups $\mathcal{G}$ whose structure is that of a *semidirect product* [2],

$$\mathcal{G} = \mathcal{V} \stackrel{\bullet}{\otimes} \mathcal{S}, \tag{2.1}$$

with $\mathcal{V}$ normal in $\mathcal{G}$ and $\mathcal{S}$ semisimple, solvable, or otherwise decomposable in turn. The elements $g \in \mathcal{G}$ may be written as pairs

$$g := \{\sigma, v\} = \{\sigma, e_{\mathcal{V}}\}\{e_{\mathcal{S}}, v\}, \qquad \sigma \in \mathcal{S}, \quad v \in \mathcal{V}, \tag{2.2}$$

with an identity $e_{\mathcal{G}} = \{e_{\mathcal{S}}, e_{\mathcal{V}}\}$. The product rule for $g = g_1 g_2$ is then of the form

$$\{\sigma, v\} = \{\sigma_1, v_1\}\{\sigma_2, v_2\}, \tag{2.3a}$$

$$\sigma = \sigma_1 \cdot \sigma_2, \qquad v = v_1^{\sigma_2} \circ v_2, \tag{2.3b}$$

where "$\cdot$" and "$\circ$" are the products in $\mathcal{S}$ and $\mathcal{V}$, respectively, and $v_1 \stackrel{\sigma_2}{\longrightarrow} v_1^{\sigma_2}$ is a mapping $\mathcal{V} \times \mathcal{S}$ onto $\mathcal{V}$, associated to $\sigma_2 \in \mathcal{S}$, isomorphic in $\mathcal{V}$ and homomorphic in $\mathcal{S}$. All the groups to be presented below have this structure as its maximal common characterization. In particular, one example is relevant to us:

The Inhomogeneous Linear Group $I_N SL(N,R) = I_N \stackrel{\bullet}{\otimes} SL(N,R)$. Here $\mathcal{S}$ is $SL(N,R)$, the group of $N \times N$ real unimodular matrices $\mathbf{S}$ with $\cdot$ = matrix multiplication, and $\mathcal{V}$ is the abelian N-parameter translation group I_N of N-element row vectors $\mathbf{v}$ with $\circ$ = vector sum. The semidirect product mapping (2.3b), is $\mathbf{v} \stackrel{\mathbf{S}}{\longrightarrow} \mathbf{vS}$ —matrix multiplication from the right. In *matrix-cum-vector* notation, the product of two elements of $I_N SL(N,R)$ reads

$$\{\mathbf{S}_1, \mathbf{v}_1\}\{\mathbf{S}_2, \mathbf{v}_2\} = \{\mathbf{S}_1\mathbf{S}_2, \mathbf{v}_1\mathbf{S}_2 + \mathbf{v}_2\}. \tag{2.4}$$

We can easily produce subgroups of $I_N SL(N,R)$ by choosing subgroups of $SL(N,R)$ —upper-triangular matrices, for example— which close under the product law (2.4).

Semidirect product groups $\tilde{\mathcal{G}}$ containing a group $\mathcal{G}$ of the form (2.4) will be also of interest in the sequel. These will be *normal extensions* of $\mathcal{G}$, the extension being made on $\mathcal{V}$ to a group $\mathcal{E}$. The idea is that $\mathcal{V}$ be the factor group of some $\mathcal{E}$ by a normal subgroup $\mathcal{N}$, *i.e.*,

$$\mathcal{V} = \mathcal{N}\backslash\mathcal{E}, \quad \text{so that} \quad \tilde{\mathcal{G}} = \mathcal{E} \stackrel{\bullet}{\otimes} \mathcal{S} \subseteq \mathcal{G} = \mathcal{V} \stackrel{\bullet}{\otimes} \mathcal{S}, \text{ and thus } \mathcal{G} = \mathcal{N}\backslash\tilde{\mathcal{G}}. \tag{2.5}$$

Probably the most famous normal extension is the one-parameter central extension of the $2N$-dimensional translation group I_{2N} to the Heisenberg–Weyl group W_N of quantum mechanics. This we shall now exhibit explicitly:

The Heisenberg–Weyl group W_N. We consider the case where $\mathcal{V}$ is I_{2N}, $\mathcal{N}$ is I_1 —a one-parameter (z) group, and $\mathcal{E} = W_N$ whose product law we write as

$$\{\mathbf{v}_1, z_1\}\{\mathbf{v}_2, z_2\} = \{\mathbf{v}_1 + \mathbf{v}_2, z_1 + z_2 + A\mathbf{v}_1\mathbf{K}\mathbf{v}_2^\top\}, \tag{2.6}$$

where $\mathbf{v}_i$ are $2N$-dimensional row-vectors, $\mathbf{v}_i^\top$ their transpose, A is any real parameter (which we shall choose as $A = \frac{1}{2}$), and $\mathbf{K}$ is the $2N \times 2N$ symplectic metric matrix

$$\mathbf{K} := \begin{pmatrix} \mathbf{0} & -\mathbf{1} \\ +\mathbf{1} & \mathbf{0} \end{pmatrix}. \tag{2.7}$$

The product law in (2.6) shows that although I_1 is normal in W_N, the latter **does not** *split* (*i.e.*, it is **not** a semidirect product $I_1 \overset{\bullet}{\otimes} I_{2N}$); I_1 *extends* I_{2N} to W_N and $I_{2N} = I_1 \backslash W_N$. Here it also happens that I_1 is *central* to (*i.e.*, lies in the centre of) W_N.

Associated to the Hirota equation we shall also encounter a generalized extension of I_{2N} to a $(4N+1)$-dimensional group generalizing the W_N structure. We shall call this group

The Hirota group H_N. Let $\mathbf{v}$ and $\boldsymbol{\varsigma}$ be $2N$-dimensional row vectors, and z a scalar. We construct triplets $\{\mathbf{v}, z, \boldsymbol{\varsigma}\}$, with the product law

$$\begin{aligned} &\{\mathbf{v}_1, z_1, \boldsymbol{\varsigma}_1\}\{\mathbf{v}_2, z_2, \boldsymbol{\varsigma}_2\} \\ &\quad = \{\mathbf{v}_1 + \mathbf{v}_2, z_1 + z_2 + A\mathbf{v}_1\mathbf{K}\mathbf{v}_2^\top, \boldsymbol{\varsigma}_1 + \boldsymbol{\varsigma}_2 + Bz_1\mathbf{v}_2 + \tfrac{1}{3}AB(\mathbf{v}_1 + 2\mathbf{v}_2)\mathbf{v}_1\mathbf{K}\mathbf{v}_2^\top\}, \end{aligned} \tag{2.8}$$

where A and B are fixed parameters (we shall set $A = \frac{1}{2}$, $B = -1$). Here, the extending group $\mathcal{N}$ is I_{2N+1} with elements $\{\mathbf{0}, z, \boldsymbol{\varsigma}\}$, normal but not central to $\tilde{\mathcal{G}}$, thus generalizing the Heisenberg–Weyl extension above. This group $\mathcal{E} = H_N$ is also a $2N$-parameter normal extension of W_N, which takes the role of $\mathcal{V}$ in (2.5a), by a group $\mathcal{N} = I_{2N}$ with elements $\{\mathbf{0}, 0, \boldsymbol{\varsigma}\}$.

The Heisenberg–Weyl extension (2.6) of I_{2N} will correspond to the classical Hopf–Cole map between the Burgers and diffusion equations —in their tensor versions, while the Hirota extension (2.8) will correspond to the Hirota map between the KdV and Hirota equations. Higher-derivative KdV-type equations require generalizations of (2.8) to tensor $\boldsymbol{\varsigma}$ which are yet to be written in explicit form.

2.2 The scalar Burgers equation [11]

We let our basic group $\mathcal{G}$ be $I_2\,SL(2,R)$ and the subgroup $\mathcal{H}$ be $S_L\,T(2,R)$, the group of unimodular lower-triangular 2×2 real matrices. Typical elements are

$$g := \left\{\begin{pmatrix} a & b \\ c & d \end{pmatrix}, (x, y)\right\} \qquad \in \mathcal{G}, \qquad ad - bc = 1, \tag{2.9a}$$

$$h := \left\{\begin{pmatrix} a & 0 \\ c & a^{-1} \end{pmatrix}, (0, 0)\right\} \qquad \in \mathcal{H}. \tag{2.9b}$$

Any element $g \in \mathcal{G}$ with $a \neq 0$ may be written as $g = h\,c$, with

$$c = c(u,q,t) = \left\{\begin{pmatrix} 1 & -t \\ 0 & 1 \end{pmatrix}, (u, q - ut)\right\} \qquad \in C = \mathcal{H}\backslash\mathcal{G}, \tag{2.9c}$$

this being a specifically coordinatized representative of the coset space $C = \mathcal{H}\backslash\mathcal{G}$. This space is homogeneous under the action of $\mathcal{G}$ by right multiplication, *i.e.*,

$$c(u,q,t) \xrightarrow{\;g\;} c(u,q,t)\,g = h_g\,c(\bar{u},\bar{q},\bar{t}). \tag{2.10}$$

Replacing (2.9) in (2.10) and using (2.6) for the products, we obtain by elementary *matrix-cum-vector* algebra the group action on the coset coordinates

$$t \xrightarrow{g} \bar{t} = \frac{td-b}{a-ct}, \tag{2.11a}$$

$$q \xrightarrow{g} \bar{q} = (qd+cq+x)\frac{td-b}{a-ct} + y, \tag{2.11b}$$

$$u \xrightarrow{g} \bar{u} = u(a-ct) + cq + x. \tag{2.11c}$$

Chain-ruling prolongs the group action to the rest of the bundle coordinates u_q, u_t, u_{qq}, u_{qt}, u_{tt}, ... We may display the corresponding transformations in matrix form, denoting $M := a - ct = (d+c\bar{t})^{-1}$ and $S := cq + x$, as

$$\begin{pmatrix} \bar{u}_{tt} \\ \bar{u}_{tq} \\ \bar{u}_{qq} \\ \bar{u}_t \\ \bar{u}_q \\ \bar{u} \\ 1 \end{pmatrix} = \begin{pmatrix} M^5 & -2M^4S & M^3S^2 & -4M^4c & 4M^3Sc & 2M^3c^2 & 2M^2Sc^2 \\ 0 & M^4 & -M^3S & 0 & -2M^3c & 0 & -M^2c^2 \\ 0 & 0 & M^3 & 0 & 0 & 0 & 0 \\ 0 & 0 & 0 & M^3 & -M^2S & -Mc & -Sc \\ 0 & 0 & 0 & 0 & M^2 & 0 & Mc \\ 0 & 0 & 0 & 0 & 0 & M & S \\ 0 & 0 & 0 & 0 & 0 & 0 & 1 \end{pmatrix} \begin{pmatrix} u_{tt} \\ u_{tq} \\ u_{qq} \\ u_t \\ u_q \\ u \\ 1 \end{pmatrix}. \tag{2.12}$$

This matrix form displays neatly two features: *(i)* the way to search for invariant functions on Z and *(ii)* the fact that the prolongued action of $\mathcal{G}$ nests according to the order of the bundle, *i.e.*, according to derivative order. The latter will be examined at the end of this subsection; as to the former, we need a function Φ of the coordinates z of Z such that it transforms under $\mathcal{G}$ as (1.3). This need not be a linear function, so it is not a matter of diagonalizing some submatrix of (2.12). On can use the diagonal, nevertheless, to conduct a search for nonlinear functions Φ according to the degree of its elements M. For instance, $\Phi_2(z) := u_{qq}$ alone, transforms according to (1.3) with a multiplier $\mu(z;g) = M^3 = (a-ct)^3$. Indeed, $u_{qq} = 0$ is a linear differential equation whose symmetry group *contains* $I_2SL(2,R)$, and it does so *properly*. Nonlinear differential equations should be more interesting. Thus we note that u_t and uu_q both transform into a common subbundle of order one, with the same power of M in the leading summand: M^3. To have invariant functions, we must get rid of the rest of the summands, so it is a matter of a short search for all possibilities to see that $\Phi_1(z) := u_t + uu_q$ is such a function, and that any linear combination of them,

$$\Phi(z) := -Cu_{qq} + u_t + uu_q, \tag{2.13}$$

with C fixed, is an invariant function satisfying (1.3) with $\mu(z,g) = (a-ct)^3 = M^3$. Of course, $\Phi(z) = 0$ is Burgers equation [6], which is now exhibited in a purely group-theoretic context. Are there any other invariants? Φ_1 and Φ_2 are, but there may be others. Trivially, we may propose powers $(\Phi_1)^n$ and $(\Phi_2)^m$, and not so trivially, Φ^n, $-C(\Phi_2)^n + (\Phi_1)^n$, or $\partial_q^m\Phi$. No invariants seem to involve u_{tt}, for example, but the search in bundles of order higher than two, transcendental functions in z, for example, is open.

Two compelling arguments which associate $I_2SL(2,R)$ strongly with the Burgers equation through (2.13) are: *(i)* the former is the Lie symmetry group of the latter, and *(ii)* the latter is the seemingly unique nontrivial invariant built on the second-order coset bundle of the former. The first argument is the one hereforeto followed by group theorists working with differential equations. Engaging in the second (reverse) argument, was here motivated by the better-known first one, *i.e.*, recognizing[2] (2.11) as a group action (2.12) on a coset space. This allows us to make informed guesses and find the

[2] One of us (K.B.W.) is indebted with G. W. Bluman for a discussion back in 1976 which set this problem on the right track.

equations we are after. The two other *trivial* equations $u_{qq} = 0$ and $u_t + uu_q = 0$;[3] however, show that this method may yield other equations with symmetry groups larger than the one we started with.

2.3 The scalar diffusion equation

Let our basic group $\mathcal{G}$ be now $W_2SL(2,R)$, the central extension of $I_2SL(2,R)$ through I_1, the *vector part* of (2.4) being replaced by (2.6), with elements [8,Chapter 10]

$$\tilde{g} := \left\{ \begin{pmatrix} a & b \\ c & d \end{pmatrix},\ (x,\, y),\ z \right\} = \{\mathbf{M},\, \mathbf{v},\, z\}, \quad \det \mathbf{M} = 1, \tag{2.14}$$

and product law

$$\{\mathbf{M}_1,\, \mathbf{v}_1,\, z_1\}\, \{\mathbf{M}_2,\, \mathbf{v}_2,\, z_2\} = \{\mathbf{M}_1\mathbf{M}_2,\, \mathbf{v}_1\mathbf{M}_2 + \mathbf{v}_2,\, z_1 + z_2 + \tfrac{1}{2}\mathbf{v}_1\mathbf{M}_2\mathbf{K}\mathbf{v}_2^{\top}\}. \tag{2.15}$$

Hence $\tilde{g} = \{\mathbf{M},\, \mathbf{v},\, z\} = \{\mathbf{M},\, \mathbf{0},\, 0\}\, \{\mathbf{1},\, \mathbf{v},\, z\} = \{\mathbf{M},\, \mathbf{0},\, 0\}\, \{\mathbf{1},\, \mathbf{v},\, 0\}\, \{\mathbf{1},\, \mathbf{0},\, 0\}\, \{\mathbf{1},\, \mathbf{0},\, z\}$, and the identity and inverse are easily found. We now choose the subgroup $\tilde{\mathcal{H}}$ to be

$$\tilde{h} := \left\{ \begin{pmatrix} a & 0 \\ c & a^{-1} \end{pmatrix},\ (x,\, 0),\ k \ln a \right\} \in \mathcal{H}, \tag{2.16}$$

with $k \neq 0$ fixed. Again, $a \neq 0$ elements may be written as $\tilde{g} = \tilde{h}\,\tilde{c}$, with coset representatives

$$\tilde{c} = \tilde{c}(v,q,t) := \left\{ \begin{pmatrix} 1 & -t \\ 0 & 1 \end{pmatrix},\ (0,\, q),\ K \ln v \right\} \in \tilde{\mathcal{C}} = \tilde{\mathcal{H}} \backslash \tilde{\mathcal{G}}, \tag{2.17}$$

with K nonzero and fixed. The coordinates we have imposed on $\tilde{\mathcal{C}}$ are such that when the group action on them is investigated, it yields (2.11*a*) and (2.11*b*) for q and t; for v, however, one finds the *multiplier* transformation

$$v \xrightarrow{\ g\ } \bar{v} = v\,(a - tc)^{-k/K} \exp\left[\frac{1}{K} \left(\frac{\tfrac{1}{2}cq^2 + xq}{a - tc} + \frac{1}{2}\frac{td - b}{a - tc}x^2 \right) + \tfrac{1}{2}xy + z \right]. \tag{2.18}$$

Prolonging the group action to the bundle we find an analogue of (2.12), and within, the invariant

$$\Phi(z) = v\,(-kv_{qq} + v_t) + (k + \tfrac{1}{2}K)\,(v_q)^2. \tag{2.19}$$

This leads in general to a nonlinear differential equation[4] but, if the coset coordinates (2.17) are chosen properly —here meaning $k = -\frac{1}{2}K$— we find the diffusion equation

$$-k\, v_{qq} + v_t = 0, \tag{2.20}$$

to be an invariant. This is true since also $\Phi_0(z) := v$ is a group invariant according to (1.3) and we can divide by nonzero v; the last feature is the one which allows —but does not imply— that the invariant may be *linear* in the coordinates of Z, and the associated $\Phi(z) = 0$ to be a *linear* differential equation.

---

[3]The invariance group of the three-dimensional version of this equation was studied by Rosen and Ullrich in Ref. [12].

[4]In all honesty, it should be stated that the differential equation obtained from (2.19) was gleaned out of (2.14)–(2.17) *post factum*, after Burgers' equation was submitted to the Hopf–Cole transformation. The choice of $W_2SL(2,R)$, however, is natural: it is well known to be the symmetry group of the diffusion equation.

2.4 The tensor and vector Burgers equation [11]

Consider now $\mathcal{G} = I_N SL(N,R)$ as detailed in **2.1**, dividing the $N \times N$ matrix part into blocks of size m and n, $m+n = N$, and letting the subgroup $\mathcal{H}$ be $S_L T(m,n,R)$, lower block-triangular matrices. Typical elements are, as in (2.9),

$$g := \left\{ \begin{pmatrix} \mathbf{a} & \mathbf{b} \\ \mathbf{c} & \mathbf{d} \end{pmatrix}, (\mathbf{x}, \mathbf{y}) \right\} \qquad \in \mathcal{G}, \quad \det \begin{pmatrix} \mathbf{a} & \mathbf{b} \\ \mathbf{c} & \mathbf{d} \end{pmatrix} = 1, \tag{2.21a}$$

$$h := \left\{ \begin{pmatrix} \mathbf{a} & \mathbf{0} \\ \mathbf{c} & \mathbf{d} \end{pmatrix}, (\mathbf{0}, \mathbf{0}) \right\} \qquad \in \mathcal{H}. \tag{2.21b}$$

The coset space is parametrized as

$$c = c(\mathbf{u}, \mathbf{q}, \mathbf{t}) := \left\{ \begin{pmatrix} \mathbf{1} & -\mathbf{t} \\ \mathbf{0} & \mathbf{1} \end{pmatrix}, (\mathbf{u}, \mathbf{q} - \mathbf{u}\mathbf{t}) \right\} \qquad \in \mathcal{C} = \mathcal{H} \backslash \mathcal{G}, \tag{2.21c}$$

where $\mathbf{u} := \{u_\alpha\}$ is a row m-vector, $\mathbf{q} := \{q_\mu\}$ a row n-vector, and $\mathbf{t} = \{t_{\alpha\mu}\}$ an $n \times m$ matrix.[5] From there, the group action corresponding to (2.11) may be calculated. We write below the relevant Z-coordinate transformations, abbreviating

$$\mathbf{M} = \mathbf{a} - \mathbf{t}\mathbf{c}, \quad \mathbf{L} = (\mathbf{d} + \mathbf{c}\bar{\mathbf{t}})^{-1}, \quad \mathbf{S} = \mathbf{q}\mathbf{c} + \mathbf{x}, \tag{2.22}$$

for easy comparison with (2.11)–(2.12). In the $N = 2$-dimensional case, $M = L = a - tc$; in general, however,

$$\mathbf{t} \xrightarrow{g} \bar{\mathbf{t}} = (\mathbf{a} - \mathbf{t}\mathbf{c})^{-1}(\mathbf{t}\mathbf{d} - \mathbf{b}), \tag{2.23a}$$

$$\mathbf{q} \xrightarrow{g} \bar{\mathbf{q}} = \mathbf{q}\mathbf{d} + (\mathbf{q}\mathbf{c} + \mathbf{x})(\mathbf{a} - \mathbf{t}\mathbf{c})^{-1}(\mathbf{t}\mathbf{d} - \mathbf{b}) + y, \tag{2.23b}$$

$$\mathbf{u} \xrightarrow{g} \bar{\mathbf{u}} = \mathbf{u}(\mathbf{a} - \mathbf{t}\mathbf{c}) + \mathbf{q}\mathbf{c} + \mathbf{x} = \mathbf{u}\mathbf{M} + \mathbf{S}, \tag{2.23c}$$

$$\frac{\partial u_\alpha}{\partial q_\mu} \xrightarrow{g} \frac{\partial \bar{u}_\alpha}{\partial \bar{q}_\mu} = L_{\mu\nu} \left[\frac{\partial u_\delta}{\partial q_\nu} M_{\delta\alpha} + c_{\nu\alpha} \right], \tag{2.23d}$$

$$\frac{\partial u_\alpha}{\partial t_{\beta\mu}} \xrightarrow{g} \frac{\partial \bar{u}_\alpha}{\partial \bar{t}_{\beta\mu}} = L_{\mu\nu} \left[\left(M_{\gamma\beta} \frac{\partial u_\delta}{\partial t_{\gamma\nu}} - S_\beta \frac{\partial u_\delta}{\partial q_\nu} \right) M_{\delta\alpha} - c_{\nu\alpha}(u_\delta M_{\delta\beta} + S_\beta) \right], \tag{2.23e}$$

$$\frac{\partial^2 u_\alpha}{\partial q_\mu \partial q_\rho} \xrightarrow{g} \frac{\partial^2 \bar{u}_\alpha}{\partial \bar{q}_\mu \partial \bar{q}_\rho} = L_{\mu\nu} L_{\rho\sigma} \frac{\partial^2 u_\delta}{\partial q_\nu \partial q_\sigma} M_{\delta\alpha}. \tag{2.23f}$$

The transformation properties for $\partial^2 u_\alpha / \partial q_\mu \partial t_{\beta\nu}$ and $\partial^2 u_\alpha / \partial t_{\beta\mu} t_{\gamma\nu}$ may be obtained from the Jacobian elements

$$\frac{\partial q_\mu}{\partial \bar{q}_\nu} = L_{\nu\mu}, \qquad \frac{\partial t_{\alpha\mu}}{\partial \bar{q}_\nu} = 0, \tag{2.24a, b}$$

$$\frac{\partial q_\mu}{\partial \bar{t}_{\alpha\nu}} = -S_\alpha L_{\nu\mu}, \qquad \frac{\partial t_{\alpha\mu}}{\partial \bar{t}_{\beta\nu}} = M_{\alpha\beta} L_{\nu\mu}. \tag{2.24c, d}$$

As in the two-dimensional case, two (tensor) invariant functions may be built:

$$\Phi^1_{\alpha\beta\mu}(z) := \frac{\partial u_\alpha}{\partial t_{\beta\mu}} + \frac{\partial u_\alpha}{\partial q_\mu} u_\beta \xrightarrow{g} L_{\mu\rho} M_{\delta\beta} M_{\gamma\alpha} \Phi^1_{\gamma\delta\rho}(z), \tag{2.25a}$$

$$\Phi^2_{\alpha\nu\mu}(z) := \frac{\partial^2 u_\alpha}{\partial q_\nu \partial q_\mu} \xrightarrow{g} L_{\mu\rho} L_{\nu\sigma} M_{\gamma\alpha} \Phi^2_{\gamma\sigma\rho}(z). \tag{2.25b}$$

---

[5] We reserve the first Greek letters for the range $1, 2, \ldots, m$ and the middle ones for $1, 2, \ldots, n$.

These do **not** give rise to a Burgers-type linear combination invariant (2.13) since the multipliers are different for the general $\mathcal{G}$ group element. The class of differential equations *contained* in $I_N SL(N,R)$ generalizing the two-dimensional case is thus too small.

2.4.1 The symplectic Burgers equation

A subgroup of $I_N SL(N,R)$ may be found for which the two multipliers in (2.25) are equal. This happens when in (2.22)

$$\mathbf{L} = \mathbf{M}^\top, \tag{2.26a}$$

which implies [5]

$$\mathbf{ab}^\top = \mathbf{ba}^\top, \quad \mathbf{ac}^\top = \mathbf{ca}^\top, \quad \mathbf{bd}^\top = \mathbf{db}^\top, \quad \mathbf{cd}^\top = \mathbf{dc}^\top, \quad \mathbf{ad}^\top - \mathbf{bc}^\top = \mathbf{1}, \tag{2.26b}$$

and defines the *symplectic* subgroup $I_N Sp(N,R)$ of $I_N SL(N,R)$, where $m = n = N/2$, and which will be taken as a new group $\mathcal{G}$. A corresponding lower-triangular subgroup $\mathcal{H}$ is thus defined for (2.21b). Here $\mathbf{t}$ is symmetric and $t_{\mu\nu} = t_{\nu\mu}$ for the *time* variables, while $\mathbf{u}$ and $\mathbf{q}$ remain $N/2$-vectors. Then, for arbitrary C, the tensor function

$$\Phi^{Sp}_{\alpha\beta\gamma}(z) = -C\frac{\partial^2 u_\alpha}{\partial q_\beta \partial q_\gamma} + \frac{\partial u_\alpha}{\partial t_{\beta\gamma}} + \tfrac{1}{2}\left(u_\beta \frac{\partial u_\alpha}{\partial q_\gamma} + u_\gamma \frac{\partial u_\alpha}{\partial q_\beta}\right) \tag{2.27}$$

is an invariant giving rise to a *symplectic tensor Burgers equation*, which may be written down immediately.

2.4.2 The vector Burgers equation

A tensor set of time variables is not too atractive. A further *Schrödinger* subgroup $I_N \circledast [SL(2,R) \otimes SO(N/2)]$ of $I_N SL(N,R)$, where the submatrices in (2.14) are multiples of the same orthogonal $N \times N$ matrix (*i.e.* $\mathbf{a} = a\mathbf{O}$, $\mathbf{b} = b\mathbf{O}$, $\mathbf{c} = c\mathbf{O}$, $\mathbf{d} = d\mathbf{O}$, with $ad - bc = 1$, $\mathbf{O}^\top\mathbf{O} = \mathbf{1}$), may be another group $\mathcal{G}$ worth considering; the corresponding $\mathcal{H}$ is defined with $c = 0$, $\mathbf{x} = \mathbf{y} = \mathbf{0}$. In this case, $\mathbf{t}$ becomes a scalar variable t, while $\mathbf{u}$ and $\mathbf{q}$ remain as $N/2$-vectors. One thus finds

$$\Phi^{Sch}_\alpha(z) := -C\frac{\partial^2 u_\alpha}{\partial q_\beta \partial q_\beta} + \frac{\partial u_\alpha}{\partial t} + u_\beta \frac{\partial u_\alpha}{\partial q_\beta} \tag{2.28}$$

to be an invariant in the sense (1.3).

2.5 The tensor and vector diffusion equations

We may introduce the group $\tilde{\mathcal{G}} = W_N SL(N,R)$, written in *matrix-cum-vector* notation to have elements $\{\mathbf{M}, \mathbf{v}, z\}$ joining (2.4) and (2.6). The subgroup and coset space are now chosen as

$$\tilde{h} = \left\{\begin{pmatrix} \mathbf{a} & \mathbf{0} \\ \mathbf{c} & \mathbf{d}\end{pmatrix}, (\mathbf{x}, \mathbf{0}), k \ln \det \mathbf{a}\right\} \quad \in \tilde{\mathcal{H}}, \tag{2.29a}$$

$$\tilde{c} = \tilde{c}(v, \mathbf{q}, \mathbf{t}) = \left\{\begin{pmatrix} \mathbf{1} & -\mathbf{t} \\ \mathbf{0} & \mathbf{1}\end{pmatrix}, (\mathbf{0}, \mathbf{q}), K \ln v\right\} \quad \in \tilde{\mathcal{C}} = \tilde{\mathcal{H}}\backslash\tilde{\mathcal{G}}. \tag{2.29b}$$

The mechanics of the computation is quite cumbersome, but one may see that the N-dimensional analogue of (2.12) is **not** obtained for the same reason as in (2.18), so we must search for subgroups.

2.5.1 The symplectic diffusion equation

We restrict the *SL(N,R)* part of W_N *SL(N,R)* to *Sp(N,R)* as in (2.26) and find

$$\Phi^{Sp}_{\beta\gamma}(z) = v\left(-k\frac{\partial^2 v}{\partial q_\beta \partial q_\gamma} + \frac{\partial v}{\partial t_{\beta\gamma}}\right) + (k + \tfrac{1}{2}K)\frac{\partial v}{\partial q_\beta}\frac{\partial v}{\partial q_\gamma}. \tag{2.30}$$

This, for $K = -2k$ and $v \neq 0$, yields a linear diffusion-type set of equations, where the dependent variable v is a scalar by construction.

2.5.2 The vector diffusion equation

We restrict W_N *Sp(N,R)* to the Schrödinger subgroup as in **2.4.2**, finding a scalar-time $N/2$-vector space invariant

$$\Phi^{Sch}(z) = v\,(-k\nabla^2 v + v_t) + (k + \tfrac{1}{2}K)\nabla v \cdot \nabla v, \tag{2.31}$$

which for $K = -2k$ is the ordinary $N/2$-dimensional diffusion equation.

2.6 The Korteweg–de Vries equation

The symmetry group of the KdV equation has the structure of a semidirect product $\mathcal{G}^{KdV(n)} = I_2 \overset{\bullet}{\otimes} S_L T(2,R)$, a generalization of the usual inhomogeneous matrix groups. Its elements may be written as

$$g := \{\mathbf{M}, \mathbf{v}\} \in \mathcal{G}^{KdV(n)}, \qquad \mathbf{M} = \begin{pmatrix} a & b \\ 0 & a^{-1} \end{pmatrix}, \quad \mathbf{v} = (x, y), \tag{2.32a}$$

with product law

$$\{\mathbf{M}_1, \mathbf{v}_1\}\{\mathbf{M}_2, \mathbf{v}_2\} = \{\mathbf{M}_1\mathbf{M}_2,\, a_1^{1-2/n}\mathbf{v}_1\mathbf{M}_2 + \mathbf{v}_2\}. \tag{2.32a$'$}$$

The case $n = 2$ reproduces a subgroup of the Burgers equation case (2.4); the $n = 3$ case applies for the usual KdV equation, and this is the one we shall concentrate upon here. In principle, though, n could be any constant.

The subgroup

$$h := \left\{\begin{pmatrix} a & 0 \\ 0 & a^{-1} \end{pmatrix},\ (x, 0)\right\} \in \mathcal{H}, \tag{2.32b}$$

yields the coset space whose representatives we choose to be

$$c = c(u, q, t) := \left\{\begin{pmatrix} 1 & -t \\ 0 & 1 \end{pmatrix},\ (u,\, q - ut)\right\} \in C = \mathcal{H}\backslash\mathcal{G}. \tag{2.32c}$$

These representatives appear to be similar to (2.9c), but the q- and u-coordinates now transform differently under (2.32a$'$) than in the Burgers case, namely:

$$t \xrightarrow{g} \bar{t} = ta^{-2} - ba^{-1}, \tag{2.33a}$$

$$q \xrightarrow{g} \bar{q} = qa^{-2/n} + txa^{-2} - xba^{-1} + y, \tag{2.33b}$$

$$u \xrightarrow{g} \bar{u} = ua^{2-2/n} + x. \tag{2.33c}$$

The prolongation of this action to the bundle yields the family of invariant functions —for general n—

$$\Phi(z) := -k\, u_{(n)q} + u_t + u\, u_q, \tag{2.34}$$

where $u_{(n)q} = \partial^n u/\partial q^n$, and k is any constant. For $n = 3$, $\Phi(z) = 0$ is the KdV equation.

2.7 Hirota's equation

Consider a group containing the Hirota group H_2 in (2.8) as a normal factor,

$$\tilde{g} := \left\{ \begin{pmatrix} a & b \\ 0 & a^{-1} \end{pmatrix},\ (x,\, y),\ [\alpha, \beta, \gamma] \right\} \qquad \in \tilde{\mathcal{G}}, \tag{2.35a}$$

with subgroup and coset space representatives

$$\tilde{h} := \left\{ \begin{pmatrix} a & 0 \\ 0 & a^{-1} \end{pmatrix},\ (x,\, 0),\ [\alpha, \beta, -12k \ln a] \right\} \qquad \in \tilde{\mathcal{H}}, \tag{2.35b}$$

$$\tilde{c} = \tilde{c}(v, q, t) := \left\{ \begin{pmatrix} 1 & -t \\ 0 & 1 \end{pmatrix},\ (0,\, q),\ [0, 0, K \ln v] \right\} \qquad \in \tilde{\mathcal{C}} = \tilde{\mathcal{H}} \backslash \tilde{\mathcal{G}}, \tag{2.35c}$$

reminiscent of (2.17), where k and K are fixed parameters. The transformations of t and q are given by (2.32a) and (2.32b), while that of v may be computed to be

$$\begin{aligned} v \xrightarrow{g} \overline{v} =& v\, a^{-12k/K} \\ & \times \exp\{\tfrac{1}{6}x(Q^2 + 6xQ\bar{t} + 2x^2\bar{t}^2) + Q(\tfrac{1}{3}xy - \alpha) + \bar{t}(x[\tfrac{2}{3}xy - \alpha] - \beta) + (\tfrac{1}{3}xy - \alpha)y + \gamma\}, \end{aligned} \tag{2.36}$$

where $Q := a^{-2/3}q$ and $\bar{t} = a^{-2}(t - ab)$.

A bundle invariant is found to be

$$\begin{aligned} \Phi(z) =& \partial_q\, v^{-2}\, \{v(-kv_{qqq} + v_t)_q - v_q(-kv_{qqq} + v_t) \\ & - 3k(v_{qq}^2 - v_q v_{qqq}) \\ & + [6k + \tfrac{1}{2}K][v_{qq}^2 + v^{-2} v_q^2\, (v_q^2 - 2v\, v_q q)]\}. \end{aligned} \tag{2.37}$$

Hirota's equation [7,§17.2;9] is obtained for $v \neq 0$, $K = -12k$. Higher Hirota groups H_N, $n > 2$ have been described in principle [5], but not explicitly explored.

The collection of examples we have presented may be enlarged through including oscillator or free-fall Schrödinger equations which also share $\tilde{\mathcal{G}} = W_2 SL(2,R)$ as their dynamical group with the diffusion equation; they require only different subgroups $\tilde{\mathcal{H}}$ which are easily guesed [8,§10.2]. In fact, any equation whose symmetry group properly contains the group of independent variable invariance transformations is amenable to this construction. Triviality descends when the dependent variable is excluded from transformations and is thus not a coset-space coordinate (this is the case of the sine–Gordon equation, for example). It also looms for linear equations when the symmetry group is infinite-parametric —say the wave equation, and perhaps even the Liouville equation— but the essential part may be rescued through factoring by the normal subgroup of additive solutions. The Hamilton–Jacobi equation may be a particularly relevant case [13] to consider further.

2.8 Nesting

The intent of the above examples has been to provide concrete realizations of the very natural idea given in the Introduction, and they point to a second feature in the coset bundle construction which may be important: the group action here ***nests*** the independent and dependent variables: q and t transform among themselves, while u (or v) transforms into $\overline{u}(u,q,t)$. [In the KdV case $\overline{u}(u)$ only.] A similar nesting in the independent variables yields $\overline{t}(t)$ and $\overline{q}(q,t)$. For definiteness, we refer the reader to the scalar Burgers equation in **2.2**, where this feature is most evident. To make it explicit, instead of a single subgroup $\mathcal{H}\subset\mathcal{G}$, we consider a subgroup chain $\mathcal{H}=\mathcal{H}_0\subset\mathcal{H}_1\subset\mathcal{H}_2\subset\mathcal{G}$, and their coset representatives in that chain,

$$\left\{\begin{pmatrix} a & 0\\ c & a^{-1}\end{pmatrix},\ (0,0)\right\}\in\mathcal{H}_0,\qquad c_0(u,q,t)=\left\{\begin{pmatrix} 1 & -t\\ 0 & 1\end{pmatrix},\ (u,\ q-ut)\right\}\quad\in C_0=\mathcal{H}_0\backslash\mathcal{G},\tag{2.38a}$$

$$\left\{\begin{pmatrix} a & 0\\ c & a^{-1}\end{pmatrix},\ (x,0)\right\}\in\mathcal{H}_1,\qquad c_1(q,t)=\left\{\begin{pmatrix} 1 & -t\\ 0 & 1\end{pmatrix},\ (0,q)\right\}\quad\in C_1=\mathcal{H}_1\backslash\mathcal{G},\tag{2.38b}$$

$$\left\{\begin{pmatrix} a & 0\\ c & a^{-1}\end{pmatrix},\ (x,y)\right\}\in\mathcal{H}_2,\qquad c_2(t)=\left\{\begin{pmatrix} 1 & -t\\ 0 & 1\end{pmatrix},\ (0,0)\right\}\quad\in C_2=\mathcal{H}_2\backslash\mathcal{G}.\tag{2.38c}$$

The dependent variable u is thus displayed as the coordinate of a fiber over the *space-time* coset C_1 (and u and q as fiber coordinates over C_2), *i.e.*,

$$c_d(u)=\left\{\begin{pmatrix} 1 & 0\\ 0 & 1\end{pmatrix},\ (u,0)\right\}\in C_d=\mathcal{H}_0\backslash\mathcal{H}_1,\qquad c_0(u,q,t)=c_d(u)\,c_1(q,t),\tag{2.38d}$$

here globally. The bundle actually being considered —and this was left deliberately vague in the Introduction— is the bundle built in the following way: we set C_1 as the base space, and construct fibers where the coordinate u transforms *as it does in* C_0. This allows us to build the *sections* $u(q,t)$ in this original bundle, and ***prolong*** the group action to the partial derivatives $\{u_q(q,t),\ u_t(q,t),\ \ldots\}$, which constitute the coordinates of a higher bundle (together with q and t), which we have called ***coset bundle*** for short. This has been the bundle of our interest. The fact that $\{u,q\}$ are similarly fiber coordinates over C_2 implies that further nested action occurs in the coset bundle, yielding the triangular matrix shape in (2.12). This is probably essential only for evolution equations, while in others, where q and t appear on the same footing, it should be absent. We have not worked with cases where no nesting at all occurs, as in the Hamilton–Jacobi equation.

§3 Group-theoretical Hopf–Cole transformations

Our small "butterfly collection" of groups, cosets, and equations, repeats the following pattern:

1. The independent-variable coset spaces C_1 —the coset bundle base space— are the same for the Burgers and diffusion equation cases [Eqs. (2.11*a*, *b*)], and for the KdV and Hirota equation cases [Eqs. (2.33*a*, *b*)]. On the fiber coordinate u, the actions are different [(2.11*c*) *vs.* (2.18) for Burgers/diffusion and (2.33*c*) *vs.* (2.36) for KdV/Hirota]. In every case, though, the action of the group on the fiber coordinates, u or v, is ***effective***, *i.e.*, there are no superfluous group parameters and the latter may thus be considered to be maximal.

2. Considered as homogeneous spaces for group action, the section $v(q,t)$ fiber coordinate for the diffusion and Hirota cases provide a ***multiplier representation*** for the group:[6]

$$v(c)\xrightarrow{\ g\ }v_g(c)=v(c_g)\,\mu(c;g),\tag{3.1a}$$

[6]Here we denote by v_g and $c_g(q,t)=c(q_g,t_g)$ what has been called $\overline{v}$ and $\overline{c}(q,t)=c(\overline{q},\overline{t})$ up to now, so as to keep track of composition properties. We also note that only the base space coordinates $c(q,t)$ enter into the multiplier.

$$\mu(c; g_1 g_2) = \mu(c; g_1)\, \mu(c_{g_1}; g_2), \qquad \mu(c, e) = 1 \tag{3.1b}$$

The Burgers and KdV equation cases on the other hand, do not; they provide what may be called *summator realizations*, since a summand is *added* to the variable:

$$u(c) \xrightarrow{g} u_g(c) = u(c_g)\, \nu(c; g) + \sigma(c; g), \tag{3.2a}$$

$$\nu(c; g) \text{ a multiplier, and } \sigma(c; g_1 g_2) = \sigma(c, g_1)\, \nu(c_{g_1}; g_2), \quad \sigma(c, e) = 0. \tag{3.2b}$$

3. The multiplier factor $\nu(c; g)$ is a power of the Jacobian of the space transformation: $\partial q/\partial \bar{q} = (a - tc)$ for the scalar Burgers case [as can be seen from the one-dimensional version of (2.24)], and $(\partial q/\partial \bar{q})^2 = a^{4/3}$ for the KdV case [as can be seen from (2.33b)–(2.33c) for $n = 3$]. This holds also for the N-dimensional symplectic Burgers case, but **not** for the general linear group embodied in (2.23c)–(2.24a).

4. The two equations in v (diffusion and Hirota) stem from groups which are *normal extensions* of the groups yielding the two equations in u (Burgers and KdV).

5. It may or not be relevant, but it is certainly interesting, to note that it was possible to choose the coset representatives in all cases so as to lie on a subgroup manifold. The diffusion and Hirota cases have $\tilde{c}(v, q, t)$ lying on maximal abelian subgroup manifolds [*c.f.* (2.17) and (2.35c)]; the one-dimensional Burgers and KdV have $c(u, q, t)$ lying on Heisenberg–Weyl subgroup manifolds [*c.f.* (2.9c) in correspondence with (2.6) through $\{x, y, z\} \leftrightarrow \{u, t, q - \frac{1}{2}ut\}$ and similarly in (2.32c)]. In the N-dimensional Burgers case the subgroup is also nilpotent. The similarity group transformations [(2.11), (2.18), (2.33), and (2.36)] are **not** *automorphisms* of these groups, however, only continuous mappings of the manifolds.

For the purpose of setting up a Hopf–Cole map, the fourth remark assures us that a homomorphic map should exist between coset bundle points of the unextended groups (Burgers and KdV), and *orbits* in the extended groups (diffusion and Hirota) by the extending group (I_1 and I_3, respectively). The third remark tells us that the mapping is achieved in the fiber coordinates only, and by a *derivative* (of first order between Burgers and diffusion cases, of second order between the KdV and Hirota cases), preceded —due to the second remark— by a *logarithm*, to turn multipliers into summators. These mappings thus read:

$$\text{scalar diffusion to Burgers:} \qquad u(q, t) = K \frac{\partial}{\partial q} \ln v(q, t), \tag{3.3a}$$

$$\text{symplectic diffusion to Burgers:} \qquad u_\alpha(\mathbf{q}, t) = K \frac{\partial}{\partial q_\alpha} \ln v(\mathbf{q}, t), \tag{3.3b}$$

$$\text{Hirota to KdV:} \qquad u(q, t) = K \frac{\partial^2}{\partial q^2} \ln v(q, t). \tag{3.3c}$$

Some comments are in order:

i. Equation (3.3a) is the well-known classic Hopf–Cole transformation mapping (one-parameter scaling orbits of) solutions of the diffusion equation to solutions of Burgers equation, and nothing need be added about its virtues.

ii. The symplectic and vector diffusion-to-Burgers transformation in (3.3b) maps solutions of $\Phi(z) = 0$ associated to (2.30) or (2.31), which for $K = -2k$ are linear and interesting, into solutions of the Burgers-type equations, associated to (2.27) or (2.28). As was pointed out to the presentor during the Workshop, not *all* solutions of the Burgers-type equations may be obtained in this way, but only *curl-free* solutions (*i.e.*, $\partial u_\beta/\partial q_\alpha - \partial u_\alpha/\partial q_\beta = 0$). To obtain solutions of the vector or symplectic Burgers equations with curl, one should search in the coset bundle coordinates (2.23) or operators (2.24) for objects transforming as (3.1a) with multiplier $\mathbf{M} = \mathbf{A} - \mathbf{T}\mathbf{C}$, to be applied to the logarithm of the (scalar) solution of the vector or tensor diffusion equation. Such a search —if possible at all— has not yet been conducted.

iii. Equation (3.3*c*) is the mapping introduced by Hirota [9] in his work on multisoliton solutions for the KdV equation. They correspond to determinant functions of exponentials which are solutions to his equation, associated to (2.37) with $K = -12k$. Similar $(n-1)^{\text{th}}$ order derivative mappings apply to the nth derivative KdV equations (2.34), but no new information has been provided yet by our construction.

Much of the work reported in this contribution is tentative, and should be seen as an exploration within a potentially interesting structure. Experience with classical applications of group theory leads one to expect that sooner or later representation theory —harmonic analysis in subgroup bases, coupling coefficients, completeness relations, and other well-polished tools— should become relevant. The subgroups introduced here are the very same subgroups which yield a group-theoretic intepretation of a class of integral transforms termed *canonical* [8,Part IV]. Many results are available which could be of use, provided they are required by the system under study. This does not seem to be the case yet with nonlinear systems in their *sui generis* need of group theory. Since alphabetical coincidence makes this the last paragraph within the present volume, we should emphasize that although the extent of this and similar open questions is not explicitly intended to apply beyond the problem at hand, perhaps it *does* have some bearing on the status of group-theoretical methods in a wider context.

References

[1] G. W. Bluman and J. D. Cole, *Similarity Methods for Differential Equations.* Applied Mathematical Sciences #13, Springer Verlag, 1974.

[2] A. O. Barut and R. Rączka, *Theory of Group Representations and Applications.* Polish Scientific Publishers, Warszaw, 1977.

[3] S. Helgason, *Differential Geometry and Symmetric Spaces.* Academic Press, New York, 1962.

[4] R. Hermann, *The Geometry of Non-Linear Differential Equations, Bäcklund Transformations and Solitons. Part A.* Math-Sci Press, Interdisciplinary Mathematics, Vol. XII, Brookline Ma., 1976.

[5] K. B. Wolf, L. Hlavatý, and S. Steinberg, Nonlinear differential equations as invariants under group action on coset bundles. I. Burgers and Korteweg–de Vries equation families. *preprint* Comunicaciones Técnicas IIMAS #319 (1982).

[6] J. M. Burgers, A mathematical model illustrating the theory of turbulence. *Adv. Appl. Mech.* **1**, 171–199 (1948).

[7] G. B. Whittham, *Linear and Nonlinear Waves* John Wiley & Sons, New York, 1974.

[8] K. B. Wolf, *Integral Transforms in Science and Engineering.* Plenum Press, New York, 1979.

[9] R. Hirota, Exact solutions of the Korteweg–de Vries equation for multiple collisions of solitons. *Phys. Rev. Lett.* **27**, 1192–1194 (1971).

[10] E. Hopf, The partial differential equation $u_t + uu_x = \mu u_{xx}$. *Commun. Pure Appl. Math.* **3**, 201–230 (1950); J. D. Cole, On a quasilinear parabolic equation occurring in thermodynamics. *Q. Appl. Math.* **9**, 225–236 (1951).

[11] K. B. Wolf, Nonlinear group action and differential equations. *Talk delivered at the topical AMS meeting, Albuquerque NM, November 1976.*

[12] G. Rosen and G. W. Ullrich, Invariance group of the equation $\partial \mathbf{u}/\partial t = -\mathbf{u}\cdot\nabla\mathbf{u}$. *SIAM J. Appl. Math.* **24**, 286–288 (1973).

[13] C. P. Boyer and M. Peñafiel, Conformal symmetry of the Hamilton–Jacobi equation. *Nuovo Cimento* **31B**, 195–210 (1976).

Author Index

THIS
VOLUME
WAS FINISHED
ON JULY 11TH, 1983
AT IIMAS–UNAM
MEXICO
CITY

Lecture Notes in Physics

Vol. 152: Physics of Narrow Gap Semiconductors. Proceedings, 1981. Edited by E. Gornik, H. Heinrich and L. Palmetshofer. XIII, 485 pages. 1982.

Vol. 153: Mathematical Problems in Theoretical Physics. Proceedings, 1981. Edited by R. Schrader, R. Seiler, and D.A. Uhlenbrock. XII, 429 pages. 1982.

Vol. 154: Macroscopic Properties of Disordered Media. Proceedings, 1981. Edited by R. Burridge, S. Childress, and G. Papanicolaou. VII, 307 pages. 1982.

Vol. 155: Quantum Optics. Proceedings, 1981. Edited by C.A. Engelbrecht. VIII, 329 pages. 1982.

Vol. 156: Resonances in Heavy Ion Reactions. Proceedings, 1981. Edited by K.A. Eberhard. XII, 448 pages. 1982.

Vol. 157: P. Niyogi, Integral Equation Method in Transonic Flow. XI, 189 pages. 1982.

Vol. 158: Dynamics of Nuclear Fission and Related Collective Phenomena. Proceedings, 1981. Edited by P. David, T. Mayer-Kuckuk, and A. van der Woude. X, 462 pages. 1982.

Vol. 159: E. Seiler, Gauge Theories as a Problem of Constructive Quantum Field Theory and Statistical Mechanics. V, 192 pages. 1982.

Vol. 160: Unified Theories of Elementary Particles. Critical Assessment and Prospects. Proceedings, 1981. Edited by P. Breitenlohner and H.P. Dürr. VI, 217 pages. 1982.

Vol. 161: Interacting Bosons in Nuclei. Proceedings, 1981. Edited by J.S. Dehesa, J.M.G. Gomez, and J. Ros. V, 209 pages. 1982.

Vol. 162: Relativistic Action at a Distance: Classical and Quantum Aspects. Proceedings, 1981. Edited by J. Llosa. X, 263 pages. 1982.

Vol. 163: J. S. Darrozes, C. Francois, Mécanique des Fluides Incompressibles. XIX, 459 pages. 1982.

Vol. 164: Stability of Thermodynamic Systems. Proceedings, 1981. Edited by J. Casas-Vázquez and G. Lebon. VII, 321 pages. 1982.

Vol. 165: N. Mukunda, H. van Dam, L.C. Biedenharn, Relativistic Models of Extended Hadrons Obeying a Mass-Spin Trajectory Constraint. Edited by A. Böhm and J.D. Dollard. VI, 163 pages. 1982.

Vol. 166: Computer Simulation of Solids. Edited by C.R.A. Catlow and W.C. Mackrodt. XII, 320 pages. 1982.

Vol. 167: G. Fieck, Symmetry of Polycentric Systems. VI, 137 pages, 1982.

Vol. 168: Heavy-Ion Collisions. Proceedings, 1982. Edited by G. Madurga and M. Lozano. VI, 429 pages. 1982.

Vol. 169: K. Sundermeyer, Constrained Dynamics. IV, 318 pages. 1982.

Vol. 170: Eighth International Conference on Numerical Methods in Fluid Dynamics. Proceedings, 1982. Edited by E. Krause. X, 569 pages. 1982.

Vol. 171: Time-Dependent Hartree-Fock and Beyond. Proceedings, 1982. Edited by K. Goeke and P.-G. Reinhard. VIII, 426 pages. 1982.

Vol. 172: Ionic Liquids, Molten Salts and Polyelectrolytes. Proceedings, 1982. Edited by K.-H. Bennemann, F. Brouers, and D. Quitmann. VII, 253 pages. 1982.

Vol. 173: Stochastic Processes in Quantum Theory and Statistical Physics. Proceedings, 1981. Edited by S. Albeverio, Ph. Combe, and M. Sirugue-Collin. VIII, 337 pages. 1982.

Vol. 174: A. Kadić, D.G.B. Edelen, A Gauge Theory of Dislocations and Disclinations. VII, 290 pages. 1983.

Vol. 175: Defect Complexes in Semiconductor Structures. Proceedings, 1982. Edited by J. Giber, F. Beleznay, J. C. Szép, and J. László. VI, 308 pages. 1983.

Vol. 176: Gauge Theory and Gravitation. Proceedings, 1982. Edited by K. Kikkawa, N. Nakanishi, and H. Nariai. X, 316 pages. 1983.

Vol. 177: Application of High Magnetic Fields in Semiconductor Physics. Proceedings, 1982. Edited by G. Landwehr. XII, 552 pages. 1983.

Vol. 178: Detectors in Heavy-Ion Reactions. Proceedings, 1982. Edited by W. von Oertzen. VIII, 258 pages. 1983.

Vol. 179: Dynamical Systems and Chaos. Proceedings, 1982. Edited by L. Garrido. XIV, 298 pages. 1983.

Vol. 180: Group Theoretical Methods in Physics. Proceedings, 1982. Edited by M. Serdaroğlu and E. İnönü. XI, 569 pages. 1983.

Vol. 181: Gauge Theories of the Eighties. Proceedings, 1982. Edited by R. Raitio and J. Lindfors. V, 644 pages. 1983.

Vol. 182: Laser Physics. Proceedings, 1983. Edited by J. D. Harvey and D. F. Walls. V, 263 pages. 1983.

Vol. 183: J. D. Gunton, M. Droz, Introduction to the Theory of Metastable and Unstable States. VI, 140 pages. 1983.

Vol. 184: Stochastic Processes – Formalism and Applications. Proceedings, 1982. Edited by G. S. Agarwal and S. Dattagupta. VI, 324 pages. 1983.

Vol. 185: H. N. Shirer, R. Wells, Mathematical Structure of the Singularities at the Transitions between Steady States in Hydrodynamic Systems. XI, 276 pages. 1983.

Vol. 186: Critical Phenomena. Proceedings, 1982. Edited by F.J.W. Hahne. VII, 353 pages. 1983.

Vol. 187: Density Functional Theory. Edited by J. Keller and J.L. Gázquez. V, 301 pages. 1983.

Vol. 188: A. P. Balachandran, G. Marmo, B.-S. Skagerstam, A. Stern, Gauge Symmetries and Fibre Bundles. IV, 140 pages. 1983.

Vol. 189: Nonlinear Phenomena. Proceedings, 1982. Edited by K. B. Wolf. XII, 453 pages. 1983.

Selected Issues from

Lecture Notes in Mathematics

Vol. 836: Differential Geometrical Methods in Mathematical Physics. Proceedings, 1979. Edited by P. L. García, A. Pérez-Rendón, and J. M. Souriau. XII, 538 pages. 1980.

Vol 837: J. Meixner, F. W. Schäfke and G. Wolf, Mathieu Functions and Spheroidal Functions and their Mathematical Foundations Further Studies. VII, 126 pages. 1980.

Vol. 838: Global Differential Geometry and Global Analysis. Proceedings 1979. Edited by D. Ferus et al. XI, 299 pages. 1981.

Vol. 840: D. Henry, Geometric Theory of Semilinear Parabolic Equations. IV, 348 pages. 1981.

Vol. 841: A. Haraux, Nonlinear Evolution Equations- Global Behaviour of Solutions. XII, 313 pages. 1981.

Vol. 842: Séminaire Bourbaki vol. 1979/80. Exposés 543-560. IV, 317 pages. 1981.

Vol. 843: Functional Analysis, Holomorphy, and Approximation Theory. Proceedings. Edited by S. Machado. VI, 636 pages. 1981.

Vol. 845: A. Tannenbaum, Invariance and System Theory: Algebraic and Geometric Aspects. X, 161 pages. 1981.

Vol. 849: P. Major, Multiple Wiener-Itô Integrals. VII, 127 pages. 1981.

Vol. 851: Stochastic Integrals. Proceedings, 1980. Edited by D. Williams. IX, 540 pages. 1981.

Vol. 852: L. Schwartz, Geometry and Probability in Banach Spaces. X, 101 pages. 1981.

Vol. 856: R. Lascar, Propagation des Singularités des Solutions d'Equations Pseudo-Différentielles à Caractéristiques de Multiplicités Variables. VIII, 237 pages. 1981.

Vol. 858: E. A. Coddington, H. S. V. de Snoo: Regular Boundary Value Problems Associated with Pairs of Ordinary Differential Expressions. V, 225 pages. 1981.

Vol. 861: Analytical Methods in Probability Theory. Proceedings 1980. Edited by D. Dugué, E. Lukacs, V. K. Rohatgi. X, 183 pages. 1981.

Vol. 866: J.-M. Bismut, Mécanique Aléatoire. XVI, 563 pages. 1981.

Vol. 878: Numerical Solution of Nonlinear Equations. Proceedings, 1980. Edited by E. L. Allgower, K. Glashoff, and H.-O. Peitgen. XIV, 440 pages. 1981.

Vol. 881: R. Lutz, M. Goze, Nonstandard Analysis. XIV, 261 pages. 1981.

Vol. 888: Padé Approximation and its Applications. Proceedings, 1980. Edited by M. G. de Bruin and H. van Rossum. VI, 383 pages. 1981.

Vol. 898: Dynamical Systems and Turbulence, Warwick, 1980. Proceedings. Edited by D. Rand and L.-S. Young. VI, 390 pages. 1981.

Vol. 901: Séminaire Bourbaki vol. 1980/81 Exposés 561-578 III, 299 pages. 1981.

Vol. 904: K. Donner, Extension of Positive Operators and Korovkin Theorems. XII, 182 pages. 1982.

Vol. 905: Differential Geometric Methods in Mathematical Physics. Proceedings, 1980. Edited by S.J. Andersson, H.-D. Doebner, and. H.R. Petry. VI, 309 pages. 1982.

Vol. 909: Numerical Analysis. Proceedings, 1981. Edited by J.P. Hennart. VII, 247 pages. 1982.

Vol. 912: Numerical Analysis. Proceedings, 1981. Edited by G. A. Watson. XIII, 245 pages. 1982.

Vol. 920: Séminaire de Probabilités XVI, 1980/81. Proceedings. Edité par J. Azéma et M. Yor. V, 622 pages. 1982.

Vol. 921: Séminaire de Probabilités XVI, 1980-81 Supplément: Géométrie Différentielle Stochastique. Proceedings. Edité par J. Azéma et M. Yor. III, 285 pages. 1982.

Vol. 922: B. Dacorogna, Weak Continuity and Weak Lower Semicontinuity of Non-Linear Functionals. V, 120 pages. 1982.

Vol. 923: Functional Analysis in Markov Processes. Proceedings, 1981. Edited by M. Fukushima. V, 307 pages. 1982.

Vol. 926: Geometric Techniques in Gauge Theories. Proceedings, 1981. Edited by R. Martini and E.M.de Jager. IX 219 pages. 1982.

Vol. 927: Y. Z. Flicker, The Trace Formula and Base Change for GL (3). XII, 204 pages. 1982.

Vol. 928: Probability Measures on Groups. Proceedings 1981. Edited by H. Heyer. X, 477 pages. 1982.

Vol. 929: Ecole d'Eté de Probabilités de Saint-Flour X - 1980. Proceedings, 1980. Edited by P.L. Hennequin. X, 313 pages. 1982.

Vol. 930: P. Berthelot, L. Breen, et W. Messing, Théorie de Dieudonné Cristalline II. XI, 261 pages. 1982.

Vol. 931: D.M. Arnold, Finite Rank Torsion Free Abelian Groups and Rings. VII, 191 pages. 1982.

Vol. 932: Analytic Theory of Continued Fractions. Proceedings, 1981. Edited by W.B. Jones, W.J. Thron, and H. Waadeland. VI, 240 pages. 1982.

Vol. 934: M. Sakai, Quadrature Domains. IV, 133 pages. 1982.

Vol. 935: R. Sot, Simple Morphisms in Algebraic Geometry. IV, 146 pages. 1982.

Vol. 936: S.M. Khaleelulla, Counterexamples in Topological Vector Spaces. XXI, 179 pages. 1982.

Vol. 937: E. Combet, Intégrales Exponentielles. VIII, 114 pages. 1982.

Vol. 938: Number Theory. Proceedings, 1981. Edited by K. Alladi. IX, 177 pages. 1982.

Vol. 942: Theory and Applications of Singular Perturbations. Proceedings, 1981. Edited by W. Eckhaus and E.M. de Jager. V, 363 pages. 1982.

Vol. 953: Iterative Solution of Nonlinear Systems of Equations. Proceedings, 1982. Edited by R. Ansorge, Th. Meis, and W. Törnig. VII, 202 pages. 1982.

Vol. 956: Group Actions and Vector Fields. Proceedings, 1981. Edited by J.B. Carrell. V, 144 pages. 1982.

Vol. 957: Differential Equations. Proceedings, 1981. Edited by D.G. de Figueiredo. VIII, 301 pages. 1982.

Vol. 963: R. Nottrot, Optimal Processes on Manifolds. VI, 124 pages. 1982.

Vol. 964: Ordinary and Partial Differential Equations. Proceedings, 1982. Edited by W.N. Everitt and B.D. Sleeman. XVIII, 726 pages. 1982.

Vol. 968: Numerical Integration of Differential Equations and Large Linear Systems. Proceedings, 1980. Edited by J. Hinze. VI, 412 pages. 1982.

Vol. 970: Twistor Geometry and Non-Linear Systems. Proceedings, 1980. Edited by H.-D. Doebner and T.D. Palev. V, 216 pages. 1982.

Vol. 972: Nonlinear Filtering and Stochastic Control. Proceedings, 1981. Edited by S.K. Mitter and A. Moro. VIII, 297 pages. 1983.

Vol. 978: J. Ławrynowicz, J. Krzyż, Quasiconformal Mappings in the Plane. VI, 177 pages. 1983.

Vol. 979: Mathematical Theories of Optimization. Proceedings, 1981. Edited by J.P. Cecconi and T. Zolezzi. V, 268 pages. 1983.